Sixth Edition

Prealgebra

Margaret L. Lial
American River College

Diana L. Hestwood
Minneapolis Community and Technical College

With the assistance of
Linda C. Russell, Minneapolis Community and Technical College
Sara E. Van Asten, North Hennepin Community College

Pearson

VP, Courseware Portfolio Management:	Chris Hoag
Director, Courseware Portfolio Management:	Michael Hirsch
Courseware Portfolio Manager:	Matthew Summers
Courseware Portfolio Assistant:	Shannon Bushee
Content Producer:	Sherry Berg
Managing Producer:	Karen Wernholm
Producer:	Shana Siegmund
Manager, Courseware Quality Assurance:	Mary Durnwald
Manager, Content Development:	Rebecca Williams
Product Marketing Manager:	Alicia Frankel
Product Marketing Assistant:	Hanna Lafferty
Field Marketing Managers:	Jennifer Crum, Lauren Schur
Senior Author Support/Technology Specialist:	Joe Vetere
Manager, Rights and Permissions:	Gina Cheselka
Manufacturing Buyer:	Carol Melville, LSC Communications
Associate Director of Design:	Blair Brown
Program Design Lead:	Barbara Atkinson
Text Design, Production Coordination, Composition, and Illustrations:	Cenveo® Publisher Services
Cover Design:	Studio Montage
Cover Image:	lonumirus/Fotolia

Library of Congress Cataloging-in-Publication Data

Names: Lial, Margaret L. | Hestwood, Diana L.
Title: Prealgebra / Margaret Lial, Diana Hestwood.
Description: 6th edition. | Boston : Pearson, [2018]
Identifiers: LCCN 2016003625 | ISBN 9780134539805 (pbk. : alk. paper)
Subjects: LCSH: Mathematics.
Classification: LCC QA39.3 .L53 2018 | DDC 513/.12—dc23
LC record available at http://lccn.loc.gov/2016003625

2 17

 Pearson

ISBN 13: 978-0-13-453980-5
ISBN 10: 0-13-453980-X

This book is dedicated to Margaret L. Lial

Always passionate about mathematics and teaching,

Always a valued colleague, a mentor, and a friend,

Always in our memory.

This book is also dedicated to my husband, Earl Orf, who is the wind beneath my wings, and to my students at Minneapolis Community and Technical College, from whom I have learned so much.

Diana Hestwood

Contents

Preface

It is with great pleasure that we offer the sixth edition of *Prealgebra*. We have remained true to the original goal that has guided us over the years—to provide the best possible text and supplements package to help students succeed and instructors teach. Extensive classroom testing has helped us mold what we believe is the most student-friendly and student-focused book on the market—written clearly and accessibly for developmental students, always delivering extra help precisely when needed, and continually reinforcing the ideas that need it most. This edition faithfully continues that process through enhanced explanations of concepts, new and updated examples and exercises, student-oriented features like Vocabulary Tips, Concept Checks, Cautions, Pointers, Relating Concepts, and Guided Solutions, as well as an extensive package of helpful supplements and study aids.

This text is part of a series that also includes

- *Basic College Mathematics,* Tenth Edition, by Lial, Salzman, and Hestwood
- *Introductory Algebra*, Eleventh Edition, by Lial, Hornsby, and McGinnis
- *Intermediate Algebra,* Eleventh Edition, by Lial, Hornsby, and McGinnis
- *Introductory and Intermediate Algebra,* Sixth Edition, by Lial, Hornsby, and McGinnis
- *Developmental Mathematics: Basic Mathematics and Algebra,* Fourth Edition, by Lial, Hornsby, McGinnis, Salzman, and Hestwood

WHAT'S NEW IN THIS EDITION

The scope and sequence of topics in *Prealgebra* has stood the test of time and rates highly with our reviewers. Therefore, you will find the table of contents intact, making the transition to the new edition easier.

▶ *Examples and Exercises*　Throughout the text, examples and exercises have been adjusted or replaced to reflect current data and practices. Applications have been updated and cover a wider variety of topics, such as the fields of technology, ecology, and health sciences.

▶ *Relating Concepts Exercises*　Conceptual exercise sets have been expanded to help students tie concepts together and develop higher-level problem-solving skills as they compare and contrast ideas, identify and describe patterns, and extend concepts to new situations. These exercises make great collaborative activities for pairs or small groups of students. Additionally, each of these exercise sets is now covered and assignable in MyMathLab and tagged for easy location and assignment.

▶ *Study Skills Reminders*　Contextualized activities called Study Skills are integrated into the text itself. To help students get off to a successful start, these activities are located primarily in the first four chapters. Also, in this edition, special Study Skills Reminders have been strategically placed throughout the text (especially in the second half) to encourage students to *revisit* the Study Skills activity most appropriate at that point.

▶ *Learning Catalytics*　Learning Catalytics is an interactive student response tool that uses students' own mobile devices to engage them in the learning process. Learning Catalytics is accessible through MyMathLab and is designed to be customized to each instructor's specific needs. Instructors can use Learning Catalytics to generate class discussion and promote peer-to-peer learning, and they can employ the real-time data generated to adjust their instructional strategy. As an introduction to this exciting new tool, we have provided questions drawing on prerequisite skills at the start of each section to check students' preparedness for the new material. Learn more about Learning Catalytics in the Instructor Resources tab in MyMathLab.

▶ *Data Analytics* We analyzed aggregated student usage and performance data from MyMathLab for the previous edition of this text. The results of this analysis helped improve the quality and quantity of exercises that matter the most to instructors and students.

▶ *What Went Wrong* Earlier editions of the text included exercises designed to help students find and fix errors, but in this edition these exercises have been updated and expanded with explicit instructions to emphasize the importance of this aspect of the learning process. When students can find and correct errors, they are demonstrating a higher level of understanding and conceptual knowledge.

▶ *Concept Teaching Tips* These tips point out the underlying mathematical concepts presented to students through the worked examples and margin problems. They highlight the importance of covering certain topics and suggest ways to help students deepen their understanding of key concepts. The Concept Teaching Tips are printed in the margins of the Annotated Instructor's Edition but are enclosed in a box to set them apart from regular Teaching Tips.

▶ *Enhanced MyMathLab Features* and *Lial Video Workbook* Videos have been updated and expanded throughout the course, including many new worked-through Solution Clips for exercises in every section. The corresponding workbook guides students through the videos for conceptual reinforcement — providing extra practice and Guided Examples linked to the videos. In addition, MyMathLab homework exercises have been refined using analyzed aggregated student usage and performance data.

HALLMARK FEATURES

We believe students and instructors will welcome these familiar hallmark features.

▶ *Chapter Openers* The new and engaging Chapter Openers portray real-life situations that make math relevant for students.

▶ *Real-Life Applications* We are always on the lookout for interesting data to use in real-life applications. As a result, we have included many new or updated examples and exercises throughout the text that focus on real-life applications of mathematics. Students are often asked to find data in a table, chart, graph, or advertisement. These applied problems provide an up-to-date flavor that will appeal to and motivate students.

▶ *Figures and Photos* Today's students are more visually oriented than ever. Thus, we have made a concerted effort to include mathematical figures, diagrams, tables, and graphs whenever possible. Many of the graphs appear in a style similar to that seen by students in today's print and electronic media. Photos have been incorporated to enhance applications in examples and exercises.

▶ *Emphasis on Problem Solving* Chapter 3 introduces students to our six-step process for solving application problems algebraically: *Read, Assign a Variable, Write an Equation, Solve, State the Answer,* and *Check.* Devoting an entire chapter to this process allows students to build a strong foundation for problem solving, which is then reinforced through specific problem-solving examples.

▶ *Learning Objectives* Each section begins with clearly stated, numbered objectives, and the material within sections is keyed to these objectives so that students know exactly what concepts are covered.

▶ *Guided Solutions* Selected exercises in the margins and in the exercise sets, marked with a ⓖⓢ icon, show the first few solution steps. Many of these exercises can be found in the MyMathLab homework, providing guidance to students as they start learning a new concept or procedure.

▶ *Pointers* More pointers have been added to examples to provide students with important on-the-spot reminders and warnings about common pitfalls.

▶ *Cautions and Notes*   These color-coded and boxed comments, one of the most popular features of previous editions, warn students about common errors and emphasize important ideas throughout the exposition. Cautions are highlighted in yellow and Notes are highlighted with blue tabs.

▦ ▶ *Calculator Tips*   These optional tips, marked with a red calculator icon, offer helpful information and instruction for students using calculators in the course.

▶ *Margin Problems*   Margin problems, with answers immediately available at the bottom of the page, are found in every section of the text. This key feature allows students to immediately practice the material covered in the examples in preparation for the exercise sets.

▶ *Ample and Varied Exercise Sets*   The text contains a wealth of exercises to provide students with opportunities to practice, apply, connect, and extend the skills they are learning. Numerous illustrations, tables, graphs, and photos help students visualize the problems they are solving. Problem types include skill building, writing, estimation, and calculator exercises, as well as applications and correct-the-error problems. In the Annotated Instructor's Edition of the text, the writing exercises are marked with an icon 🖉 so that instructors may assign these problems at their discretion. Exercises suitable for calculator work are marked in both the student and instructor editions with a calculator icon ▦ . Students can watch an instructor work through the complete solution for all exercises marked with a Play Button icon ▶ in MyMathLab.

▶ *Teaching Tips*   Although the mathematical content in this text is familiar to instructors, they may not all have experience in teaching the material to adult students. The Teaching Tips provide helpful comments from colleagues with successful experience at this level and offer cautions about common trouble spots. The Teaching Tips are printed in the margins of the Annotated Instructor's Edition.

▶ *Solutions*   Solutions to selected section exercises are included in MyMathLab. This provides students with easily accessible step-by-step help in solving the exercises that are most commonly missed. Solutions are provided for the exercises marked with a square of blue color around the exercise number, for example, **15.**

▶ *Summary Exercises*   All chapters now include this helpful mid-chapter review. These exercises provide students with the all-important *mixed* practice they need at these critical points in their skill development.

▶ *Ample Opportunity for Review*   Each chapter ends with a Chapter Summary featuring Key Terms with definitions and helpful graphics, New Formulas, New Symbols, Test Your Word Power, and a Quick Review of each section's content with additional examples. Also included is a comprehensive set of Chapter Review Exercises keyed to individual sections, a set of Mixed Review Exercises, and a Chapter Test. Students can watch an instructor work out the full solutions to the Chapter Test problems in the Chapter Test Prep Videos.

▶ *Test Your Word Power*   This feature, incorporated into each Chapter Summary, helps students understand and master mathematical vocabulary. Key terms from the chapter are presented, along with three possible definitions in a multiple-choice format. Answers and examples illustrating each term are provided.

▶ *Written with developmental readers in mind*   A significant proportion of developmental math students are also developmental reading students. With this in mind, we are thrilled to be working with a developmental reading instructor on this text, making sure that the material is as accessible as possible to our developmental students. This not only helps those students with weak reading skills, but helps *all* students by ensuring that the mathematics is accurate, but not obscured by unnecessarily complex writing.

Resources for Success

MyMathLab Online Course for Lial/Hestwood
Prealgebra, 6th edition

The corresponding MyMathLab course tightly integrates the authors' approach, giving students a learning environment that encourages conceptual understanding and engagement.

NEW! Learning Catalytics

Integrated into MyMathLab, Learning Catalytics use students' mobile devices for an engagement, assessment, and classroom intelligence system that gives instructors real-time feedback on student learning. LC annotations for instructors in the text provide corresponding questions that they can use to engage their classrooms.

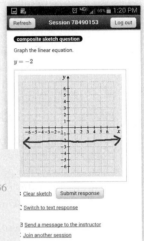

LC LEARNING CATALYTICS

1. In 50 • 36, what are the numbers 50 and 36 called?

   (a) products (b) places (c) factors

2. Simplify: 35 • 120

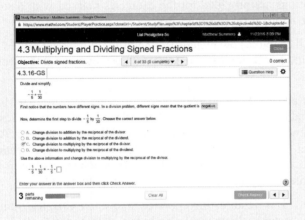

Expanded! Conceptual Exercises

In addition to MyMathLab's hallmark interactive exercises, the Lial team provides students with exercises that tie concepts together and help students problem-solve. Guided Solutions exercises, marked with a "GS" in the Assignment Manager, test student understanding of the problem-solving steps while guiding them through the solution process. Relating Concepts exercises in the text help students make connections and problem-solve at a higher level. These sets are assignable in MyMathLab, with expanded coverage.

NEW! Workspace Assignments

These new assignments allow students to naturally write out their work by hand, step-by-step, showing their mathematical reasoning as they receive instant feedback at each step. Each student's work is captured in the MyMathLab gradebook so instructors can easily pinpoint exactly where in the solution process students struggled.

www.mymathlab.com

Resources for Success

NEW! Adaptive Skill Builder

When students struggle on an exercise, Skill Builder assignments provide just-in-time, targeted support to help them build on the requisite skills needed to complete their assignment. As students progress, the Skill Builder assignments adapt to provide support exercises that are personalized to each student's activity and performance throughout the course.

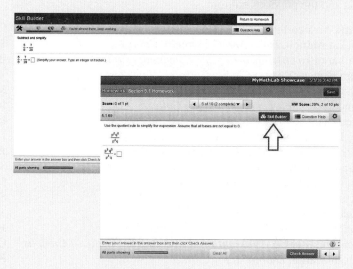

Instructor Resources

Annotated Instructor's Edition

ISBN 10: 0-13-454115-4 **ISBN 13:** 978-0-13-454115-0
The AIE provides annotations for instructors, including answers, Learning Catalytics suggestions, and vocabulary and teaching tips.

The following resources can be downloaded from www.pearsonhighered.com or in MyMathLab:

Instructor's Solutions Manual

This manual provides solutions to all exercises in the text.

Instructor's Resource Manual

This manual includes Mini-Lectures to provide new instructors with objectives, key examples, and teaching tips for every section of the text.

PowerPoints

These editable slides present key concepts and definitions from the text.

TestGen

TestGen® (www.pearsoned.com/testgen) enables instructors to build, edit, print, and administer tests using a computerized bank of questions developed to cover all the objectives of the text.

Student Resources

Student Solutions Manual

ISBN 10: 0-13-454103-0 **ISBN 13:** 978-0-13-454103-7
This manual contains completely worked-out solutions for all the odd-numbered exercises in the text.

Lial Video Workbook

ISBN 10: 0-13-454098-0 **ISBN 13:** 978-0-13-454098-6
This workbook/note-taking guide helps students develop organized notes as they work along with the videos. The notebook includes

- Guided Examples to be used in conjunction with the Lial Section Lecture Videos and/or Objective-Level Video clips, plus corresponding Now Try This Exercises for each text objective.

- Extra practice exercises for every section of the text, with ample space for students to show their work.

- Learning objectives and key vocabulary terms for every text section, along with vocabulary practice problems.

www.mymathlab.com

ACKNOWLEDGMENTS

The comments, criticisms, and suggestions of users, nonusers, instructors, and students have positively shaped this text over the years, and we are most grateful for the many responses we have received. The feedback gathered for this revision of the text was particularly helpful, and we especially wish to thank the following individuals who provided invaluable suggestions for this and the previous edition:

Carla Ainsworth, *Salt Lake Community College*
Vickie Aldrich, *Dona Ana Community College*
Barbara Brown, *Anoka-Ramsey Community College*
Kim Brown, *Tarrant County College—Northeast Campus*
Hien Bui, *Hillsborough Community College*
John Close, *Salt Lake Community College*
Jane Cuellar, *Taft College*
Ky Davis, *Zane State College*
Patricia Donovan, *San Joaquin Delta College*
Randy Gallaher, *Lewis and Clark Community College*
Nancy Graham, *Rose State College*
Lynn Hargrove, *Sierra College*
Susan Kautz, *Lone Star College Cy-Fair*
Nancy Ketchum, *Moberly Area Community College*
Lou Ann Mahaney, *Tarrant County College—Northeast Campus*
Susan Martin, *Diablo Valley College*
Pam Miller, *Phoenix College*
Elizabeth Morrison, *Valencia College—West Campus*

Faith Peters, *Broward College*
Sara Pries, *Sierra College*
Brooke Quinlan, *Hillsborough Community College—Dale Mabry Campus*
Manoj Raghunandanan, *Temple University*
Janalyn Richards, *Idaho State University*
Heather Roth, *Nova Southeastern University*
Julia Simms, *Southern Illinois University—Edwardsville*
Janis Cimperman, *St. Cloud State University*
Sounny Slitine, *Palo Alto College*
Marie E. St. James, *St. Clair County Community College*
Sharon Testone, *Onondaga Community College*
Shae Thompson, *Montana State University*
Joe Timpe, *Tarrant County College*
Cheryl Wilcox, *Diablo Valley College*
Kevin Yokoyama, *College of the Redwoods*
Karl Zilm, *Lewis and Clark Community College*

Our sincere thanks go to the dedicated individuals at Pearson who have worked hard to make this revision a success: Matt Summers, Michael Hirsch, Sherry Berg, Barbara Atkinson, Michelle Renda, Shana Siegmund, Beth Kaufman, Alicia Frankel, Rebecca Williams, and Ruth Berry. We are also grateful to Marilyn Dwyer and Carol Merrigan of Cenveo, Inc. for their excellent production work; to Connie Day for her copyediting; to Lucie Haskins for producing a useful index; and to Alicia Gordon for her accuracy checking.

Linda Russell developed and wrote the Study Skills activities that appear throughout this text. She also worked to make the text more readable for developmental-level students, and provided much-needed help throughout the production phase. Her 25+ years of experience teaching reading and study skills to students at this level was invaluable.

We are pleased to welcome Sara Van Asten to our team. She is the math department chair at North Hennepin Community College, and has a special interest and talent for teaching mathematics to developmental-level students. She contributed a great deal to revising and updating application problems, and she wrote new material for Relating Concepts, Learning Catalytics, and What Went Wrong exercises. The insights she has developed from years of teaching at this level were most helpful.

The ultimate measure of this text's success is whether it helps students master algebra skills, develop problem-solving techniques, and increase their confidence in learning and using mathematics. In order for us, as authors, to know what to keep and what to improve for the next edition, we need to hear from you, the instructor, and you, the student. Please tell us what you like and where you need additional help by sending an e-mail to *math@pearson.com*. We appreciate your feedback!

Diana L. Hestwood

Pretest: Whole Numbers Computation

> **⊘ CAUTION**
>
> This test will check your skills in doing whole numbers computation using paper and pencil. Each part of the test is keyed to a section in the **Whole Numbers Review** chapter, which follows Chapter 10. Based on your test results, work the appropriate section or sections in the Whole Numbers Review chapter *before you begin this course*. You need these skills to be successful!

Adding Whole Numbers   (*Do not use a calculator.*)

1.
$$368 \\ + \ 22$$

2.
$$7093 \\ + 6073$$

3.
$$85 \\ + 2968$$

4. $57{,}208 + 915 + 59{,}387$

5. $714 + 3728 + 9 + 683{,}775$

Subtracting Whole Numbers   (*Do not use a calculator.*)

1.
$$426 \\ - \ 76$$

2.
$$3358 \\ - 2729$$

3.
$$30{,}602 \\ - \ 5\ 708$$

4. $4006 - 97$

5. $679{,}420 - 88{,}033$

Multiplying Whole Numbers   (*Do not use a calculator.*)

1. $3 \times 3 \times 0 \times 6$

2.
$$3841 \\ \times \ \ \ 7$$

3. $(520)(3000)$

(Continued)

Multiplying Whole Numbers *(Continued)*
Do not use a calculator; show your work.

4.   71
    × 26

5. Multiply 359 and 48.

6. 853 × 609

Dividing Whole Numbers   *(Do not use a calculator; show your work.)*

1. 3)69

2. 12 ÷ 0

3. $\dfrac{25{,}036}{4}$

4. 7)5655

5. 52)1768

6. 45,000 ÷ 900

7. 38)2300

8. 83)44,799

Now check your answers on page A–7 in the Answers section. Record the number of problems you worked correctly in each part of the test.

Adding Whole Numbers: _____ correct out of 5.

   If you got 0, 1, or 2 correct, work **Adding Whole Numbers** in the Review chapter.

Subtracting Whole Numbers: _____ correct out of 5.

   If you got 0, 1, or 2 correct, work **Subtracting Whole Numbers** in the Review chapter.

Multiplying Whole Numbers: _____ correct out of 6.

   If you got 0, 1, 2, or 3 correct, work **Multiplying Whole Numbers** in the Review chapter.

Dividing Whole Numbers: _____ correct out of 8.

   If you got 0, 1, or 2 correct, work **Dividing Whole Numbers** and **Long Division** in the Review chapter.

   If you got 3 or 4 correct, work **Long Division** in the Review chapter.

Introduction to Algebra: Integers

In this chapter, you'll see how we use both *positive* and *negative* numbers. For example, we use them to report temperatures, check our bank account balance, and interpret medical test results.

Study Skills

YOUR BRAIN *CAN* LEARN MATHEMATICS

OBJECTIVES

1 Describe how practice fosters dendrite growth.

2 Explain the effect of anxiety on the brain.

Your brain knows how to learn, just as your lungs know how to breathe; however, there are important things you can do to maximize your brain's ability to do its work. This short introduction will help you choose effective strategies for learning mathematics. This is a simplified explanation of a complex process.

Your brain's outer layer is called the **neocortex**, which is where higher-level thinking, language, reasoning, and purposeful behavior occur. The neocortex has about 100 billion (100,000,000,000) brain cells called **neurons.**

Learning Something New

▶ As you learn something new, threadlike branches grow out of each neuron. These branches are called **dendrites**.

▶ When the dendrite from one neuron grows close enough to the dendrite from another neuron, a connection is made. There is a small gap at the connection point called a **synapse.** One dendrite sends an electrical signal across the gap to another dendrite.

▶ *Learning = growth and connecting of dendrites.*

Remembering New Skills

▶ When you practice a skill just once or twice, the connections between neurons are very weak. If you do not practice the skill again, the dendrites at the connection points wither and die back. You have forgotten the new skill!

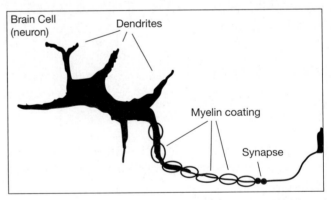

A neuron with several dendrites: one dendrite has developed a myelin coating through repeated practice.

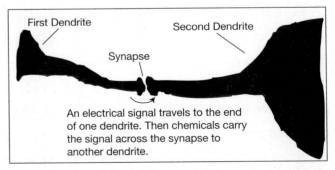

A close-up view of the gap (synapse) between two dendrites.

- If you practice a new skill many times, the dendrites for that skill become coated with a fatty protein called **myelin.** Each time one dendrite sends a signal to another dendrite, the myelin coating becomes thicker and smoother, allowing the signals to move faster and with less interference. Thinking can now occur more quickly and easily, and *you will remember the skill for a long time* because the dendrite connections are very strong.

Become an Effective Student

- You grow dendrites specifically for the thing you are studying. If you practice dividing fractions, you will grow specialized dendrites just for dividing fractions. If you *watch other people* solve fraction problems, *you will grow dendrites for watching, not for solving.* So be sure you are actively learning and practicing.

- If you practice something the *wrong* way, you will develop strong dendrite connections for doing it the wrong way! So as you study, check frequently that you are getting correct answers.

- As you study a new topic that is related to things you already know, you will grow new dendrites, but your brain will also send signals throughout the network of dendrites for the related topics. In this way, you build a complex **neural network** that helps you to apply concepts, see differences and similarities between ideas, and understand relationships between concepts.

In the first few chapters of this text you will find "brain friendly" activities that are designed to help you grow and develop your own reliable neural networks for mathematics. Since you must grow your own dendrites (no one can grow them for you), these activities show you how to

- develop new dendrites,

- strengthen existing ones, and

- encourage the myelin coating to become thicker so signals are sent with less effort.

When you incorporate the activities into your regular study routine, you will discover that you understand better, remember longer, and forget less.

Also remember that *it does take time for dendrites to grow.* Trying to cram in several new concepts and skills at the last minute is not possible. Your dendrites simply can't grow that quickly. You can't expect to develop huge muscles by lifting weights for just one evening before a body-building competition! In the same way, practice the study techniques *throughout the course* to facilitate strong growth of dendrites.

When Anxiety Strikes

If you are under stress or feeling anxious, such as during a test, your body secretes **adrenaline** into your system. Adrenaline in the brain blocks connections between neurons. In other words, you can't think! If you've ever experienced "blanking out" on a test, you know what adrenaline does. You'll learn several solutions to that problem in later activities.

Start Your Course Right!

- Attend all class sessions (especially the first one).

- Gather the necessary supplies.

- Carefully read the syllabus for the course, and ask questions if you don't understand.

1.1 Place Value

OBJECTIVES

1 Identify whole numbers.

2 Identify the place value of a digit through hundred-trillions.

3 Write a whole number in words or digits.

Make a Strong Start!

Before you begin this course, take the Whole Numbers Computation Pretest. Based on your test results, work the appropriate sections in the Whole Numbers Review chapter, which follows Chapter 10. You need these skills to succeed in this course.

VOCABULARY TIP

Value The meaning of the word "value" in mathematics is slightly different than in everyday use. In math, "value" refers to "how much" or "a quantity." You will see the word "value" in many places in this text.

It would be nice to earn millions of dollars like some of our favorite entertainers or sports stars. But how much is a million? If you received $1 every second, 24 hours a day, day after day, how many days would it take for you to receive a million dollars? How long to receive a billion dollars? Or a trillion dollars? Make some guesses and write them here.

It would take _____ to receive a million dollars.

It would take _____ to receive a billion dollars.

It would take _____ to receive a trillion dollars.

The answers are at the bottom left of the page. Later in this chapter, you'll find out how to calculate the answers.

OBJECTIVE **1** **Identify whole numbers.** First we have to be able to write the number that represents *one million*. We can write *one* as 1. How do we make it 1 *million?* Our number system is a **place value system.** That means that the location, or place, in which a number is written gives it a different value. Using money as an example, you can see that

$1 is one dollar.

$10 is ten dollars.

$100 is one hundred dollars.

$1000 is one thousand dollars.

Each time the 1 moved to the left one place, it was worth *ten times* as much. Can you keep moving it to the left? Yes, as many times as you like.

The chart below shows the *value* of each *place*. In other words, you write the 1 in the correct place to represent the number you want to express. It is important to memorize the place value names shown on the chart.

Whole Number Place Value Chart

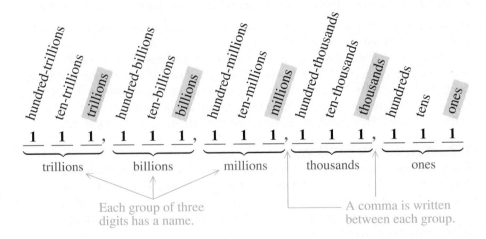

If there had been more room on this page, we could have continued to the left with quadrillions, quintillions, sextillions, septillions, octillions, and more.

Of course, we can use other *digits* besides 1. In our decimal system of writing numbers, we can use these ten **digits:** 0, 1, 2, 3, 4, 5, 6, 7, 8, and 9. In this section we will use the digits to write **whole numbers.**

Answers

About $11\frac{1}{2}$ days to receive a million dollars; nearly 32 *years* to receive a billion dollars; about 31,700 *years* to receive a trillion dollars.

These are whole numbers.

0   8   37   100   24,014

These are *not* whole numbers.

-6 $\frac{3}{4}$ 7.528   0.3 $5\frac{2}{3}$

EXAMPLE 1 **Identifying Whole Numbers**

Circle the whole numbers in this list.

> Zero is a whole number.

75   -4 300   1.5 $\frac{5}{8}$ 0   0.666 $7\frac{1}{2}$ 2

Whole numbers *do* include zero. If we started a list of *all* the whole numbers, it would look like this: 0, 1, 2, 3, 4, 5, . . . with the three dots indicating that the list goes on and on. So the whole numbers in this example are: **75**, **300**, **0**, and **2**.

────── **Work Problem ❶ at the Side.** ▶

OBJECTIVE ❷ Identify the place value of a digit through hundred-trillions. The estimated world population in 2016 was 7,394,582,817 people. There are two 7s in this number, but the value of each 7 is very different. The 7 on the right is in the ones place, so its value is simply 7. But the 7 on the left is worth a great deal more because of the *place* where it is written. Looking back at the place value chart, we see that this 7 is in the *billions* place, so its value is *7 billion*.

7 , 3 9 4 , 5 8 2 , 8 1 7 people in the world
↑   ↑
Value of   Value of
7 billion   *7 ones*

EXAMPLE 2 **Identifying Place Value**

Identify the place value of each 8 in the number of people in the world.

7 , 3 9 4 , 5 8 2 , 8 1 7
              ↑   ↑
Ten-thousands place ┘   └ Hundreds place

────── **Work Problem ❷ at the Side.** ▶

OBJECTIVE ❸ Write a whole number in words or digits. To write a whole number in words, or to say it aloud, begin at the left. Write or say the number in each group of three, followed by the name for that group. When you get to the ones group, do *not* include the group name. **Hyphens** (dashes) are used whenever you write a number from 21 to 99, like twenty–one thousand, or thirty–seven million, or ninety–four billion.

EXAMPLE 3 **Writing Numbers in Words**

(a) Write 6,058,120 in words.

Start at the left.

6 , 0 5 8 , 1 2 0

> Do **not** say "one hundred *and* twenty" here.

six **million**, fifty-eight **thousand**, one hundred twenty
Group   Group   └ Do not use the
name   name   group name "ones."

────── **Continued on Next Page**

---

❶ Circle the whole numbers. (*Hint:* There are five whole numbers.)

0.8 -14 502

$\frac{7}{9}$ 3 $\frac{3}{2}$

14   0 $6\frac{4}{5}$

9.082 $-\frac{8}{3}$ 60,005

VOCABULARY TIP

Place value Think of the number 3333. Each "3" has a different value, even though it is the same digit. The **place** where the 3 is located determines its value: three thousands, three hundreds, three tens, and three ones.

❷ Identify the place value of the digit 8 in each number. Use the place value chart on the previous page if you need help.

(a) 45,628,665

(b) 800,503,622

(c) 428,000,000,000

(d) 2,835,071

Answers

1. 502; 3; 14; 0; 60,005

2. **(a)** thousands **(b)** hundred-millions
(c) billions **(d)** hundred-thousands

3 Write these numbers in words.

(a) 23,605

twenty-three _____,

six hundred _____

(b) 400,033,007

(c) 193,080,102,000,000

4 Write each number using digits.

(a) Eighteen **million**, two **thousand**, three hundred five

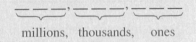

millions, thousands, ones

(b) Two hundred billion, fifty million, six hundred sixteen

(c) Five trillion, forty-two billion, nine million

(d) Three hundred six million, seven hundred thousand, nine hundred fifty-nine

(b) Write 50,588,000,040,000 in words.

Start at the left.

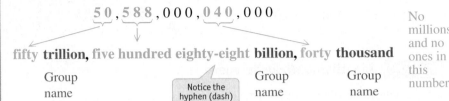

fifty trillion, five hundred eighty-eight billion, forty thousand

Group name · Group name · Group name

Notice the hyphen (dash) in eighty-eight.

No millions and no ones in this number

> **⊘ CAUTION**
>
> You often hear people say "and" when reading a group of three digits. For example, you may hear 120 as "one hundred **and** twenty," but this is **not** correct. The word **and** is used when reading a **decimal point,** which we do not have here. The correct wording for 120 is "one hundred twenty." See **Example 3(a).**

◀ **Work Problem ❸ at the Side.**

When you read or hear a number and want to write it in digits, look for the group names: **trillion, billion, million,** and **thousand.** Write the number in each group, followed by a comma. Do *not* put a comma at the end of the ones group.

EXAMPLE 4 **Writing Numbers in Digits**

Write each number using digits.

(a) Five hundred sixteen **thousand,** nine

The first group name is *thousand,* so you need to fill *two groups* of three digits: thousands and ones.

$$5 \quad 1 \quad 6 \,,\, 0 \quad 0 \quad 9$$

thousands | ones

comma between groups

Write 009 in the ones group. If you wrote 900 it would be 9 *hundred.*

The number is **516,009**.

(b) Seventy-seven **billion,** thirty **thousand,** five hundred

The first group name is *billion,* so you need to fill *four groups* of three digits: billions, millions, thousands, and ones.

$$0 \quad 7 \quad 7 \,,\, 0 \quad 0 \quad 0 \,,\, 0 \quad 3 \quad 0 \,,\, 5 \quad 0 \quad 0$$

billions   millions   thousands   ones

There are no millions, so fill the millions group with zeros.

When writing the number, you can omit the leading 0 in the billions group.

The number is **77,000,030,500**.

◀ **Work Problem ❹ at the Side.**

Answers

3. **(a)** twenty-three *thousand,* six hundred <u>five</u>
   (b) four hundred *million,* thirty-three *thousand,* seven
   (c) one hundred ninety-three *trillion,* eighty *billion,* one hundred two *million*

4. **(a)** 18,002,305 **(b)** 200,050,000,616
   (c) 5,042,009,000,000 **(d)** 306,700,959

1.1 Exercises

FOR EXTRA HELP

Go to MyMathLab *for worked-out, step-by-step solutions to exercises enclosed in a square* ▢ *and video solutions to* ▶ *exercises.*

CONCEPT CHECK *In Exercises 1–4, decide whether each statement is* true *or* false. *If a statement is* false, *fix it.*

1. The digits we can use to write numbers are

  1, 2, 3, 4, 5, 6, 7, 8, and 9.

2. All of these numbers are whole numbers.

  16,565   2   0   400

3. None of these numbers are whole numbers.

  $\dfrac{7}{10}$ -21 0.03 $1\dfrac{1}{2}$

4. Each 7 in this number has the same value.

  1,372,075

Circle the whole numbers. **See Example 1.**

5.   15 $8\dfrac{3}{4}$ 0   3.781

  83,001 -8 $\dfrac{7}{16}$ $\dfrac{9}{5}$

6. 33.7 -5 457 $\dfrac{8}{5}$

  0   6 $1\dfrac{3}{4}$ -14.1

7. 5.8 -6 7 $\dfrac{5}{4}$

  $\dfrac{1}{10}$ 362,049   0.1 $7\dfrac{7}{8}$

8. 75,039 $\dfrac{1}{3}$ -87 6.49

  -0.5 $2\dfrac{7}{10}$ $\dfrac{15}{8}$ 4

Identify the place value of the digit 2 in each number. **See Example 2.**

9. 61,284

10. 82,110

11. 284,100

12. 823,415

13. 725,837,166

14. 442,653,199

15. 253,045,701,000

16. 823,000,419,567

CONCEPT CHECK *In Exercises 17–18, name the place value for each 0 in the number, going from left to right.*

17. 302,016,450,098,570

18. 810,704,069,809,035

In Exercises 19–30, write each number in words. **See Example 3.**

19. 8421

(GS) eight _____ , four _____ twenty-_____

20. 1936

(GS) one _____ , nine _____ thirty-_____

21. 46,205

22. 75,089

23. 3,064,801

24. 7,900,408

25. 840,111,003

26. 304,008,401

27. 51,006,888,321

28. 99,046,733,214

29. 3,000,712,000,000

30. 50,918,000,000,600

In Exercises 31–40, write each number using digits. **See Example 4.**

31. Forty-six thousand, eight hundred five

___ ___ ___ , ___ ___ ___
  thousands   ones

32. Seventy-nine thousand, forty-six

___ ___ ___ , ___ ___ ___
  thousands   ones

33. Five million, six hundred thousand, eighty-two

34. One million, thirty thousand, five

35. Two hundred seventy-one million, nine hundred thousand

36. Three hundred eleven million, four hundred

37. Twelve billion, four hundred seventeen million, six hundred twenty-five thousand, three hundred ten

38. Seventy-five billion, eight hundred sixty-nine million, four hundred eighty-eight thousand, five hundred six

39. Six hundred trillion, seventy-one million, four hundred

40. Four hundred forty trillion, thirty-six thousand, one hundred two

In Exercises 41–52, if the number is given in digits, write it in words. If the number is given in words, write it in digits. ***See Examples 3 and 4.***

41. Every second, 9641 tweets are sent on Twitter. (Data from www.internetlivestats.com)

42. Every second, 2367 photos are posted to Instagram. (Data from www.internetlivestats.com)

43. During a bad winter it cost the state of Minnesota $130,100,000 to remove snow and ice from the roads. (Data from *Star Tribune*.)

44. Since it started, Nintendo has sold 669,360,000 gaming systems. (Data from www.ign.com)

45. Bill Gates, the founder of Microsoft, is the richest person in the world. He is worth $79,200,000,000. (Data from www.forbes.com)

46. The founder of Facebook, Mark Zuckerberg, is worth $33,400,000,000. (Data from www.forbes.com)

47. There are seventy-four million, fifty-nine thousand pet cats in the United States. (Data from www.avma.org)

48. The YouTube video with the most views was watched four million, one hundred sixty-seven thousand, thirty-four times. (Data from www.youtube.com)

49. There are 3,005,000 Google searches performed every minute. (Data from www.internetlivestats.com)

50. There are 2,401,333 emails sent every second. (Data from www.internetlivestats.com)

51. The Mars company makes over fifteen million Snickers candy bars each day. That is more than five billion, four hundred seventy-five million Snickers candy bars in one year! (Data from www.nbcnews.com)

52. The Mars company makes four hundred million M&M's candies each day. That adds up to one hundred forty-six billion each year. (Data from www.mymms.com)

Relating Concepts (Exercises 53–56) For Individual or Group Work

Use your knowledge of place value to **work Exercises 53–56 in order.** *Answers to both odd and even exercises are in the Answers section.*

53. Here is a group of digits: 6, 0, 9, 1, 5, 0, 7, 1. Using each of these digits exactly once, arrange them to make the largest possible whole number and the smallest possible whole number. Then write each number in words.

54. Write these numbers in digits and in words.

    (a) Your house number or apartment building number

    (b) Your phone number, including the area code

    (c) The approximate cost of your tuition and books for this quarter or semester

    (d) Your zip code

    Now tell how you *usually* say each of the numbers in parts (a)–(d) above. Why do you think we ignore the rules when saying these numbers in everyday situations?

55. Look again at the Whole Number Place Value Chart at the beginning of this section. As you move to the left, each place is worth *ten* times as much as the previous place. Computers work on a *binary* system, where each place is worth *two* times as much as the previous place. Complete this place value chart based on 2s.

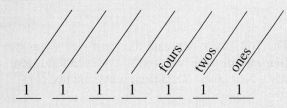

$$\underline{1} \quad \underline{1} \quad \underline{1} \quad \underline{1} \quad \underline{1} \quad \underline{1} \quad \underline{1}$$

The *only* digits you may use in the binary system are 0 and 1. Here is an example.

$$6 = 110 \text{ in binary}$$

one 4 + one 2 + zero 1s = 4 + 2 + 0 = 6

Now try writing some numbers using the 2s (binary) place value chart.

    (a) $5 = \underline{\quad\quad}$

    (b) $10 = \underline{\quad\quad}$

    (c) $15 = \underline{\quad\quad}$

56. (a) Explain in your own words why our number system is called a *place value system*. Include an example as part of your explanation.

    (b) Find information on the Roman numeral system. Write these numbers using Roman numerals.

        $8 = \underline{\quad\quad\quad}$ $38 = \underline{\quad\quad\quad}$

        $275 = \underline{\quad\quad\quad}$ $3322 = \underline{\quad\quad\quad}$

    (c) Explain why the Roman system is *not* a place value system. What are the disadvantages of the Roman system?

Study Skills

USING YOUR TEXT

Be sure to read *Your Brain **Can** Learn Mathematics* before starting this activity. You'll find out how your brain learns and remembers.

OBJECTIVES

1 Explain the meaning of text features such as section numbering, objectives, margin exercises.

2 Locate the Answers, Solutions, and Index sections.

Your text can be very helpful. Find out what it has to offer. First, let's look at some general features that will help you in all chapters.

Table of Contents

Find the Table of Contents. Before you start this course, you'll want to take the Whole Numbers Computation Pretest at the front of the text.

Look at the Preface, which includes a list of supplementary resources for students. If you are interested in any of these, ask your instructor if they are available.

Section Numbering

Each chapter is divided into sections, and each section has a number, such as 1.3 or 3.5. Your instructor will use these numbers to assign homework.

Chapter 3 $\longrightarrow$ **3.5** $\longleftarrow$ Section 5 within Chapter 3

Chapter Features

There are four features to pay special attention to as you work in your text.

▶ **Objectives.** Each section lists the objectives in the upper corner of the first page. The objectives are listed again as each one is introduced. An objective tells you *what you will be able to do after you complete the section*. An excellent way to check your learning is to go back to the list of objectives when you are finished with a section and ask yourself if you can do them all.

▶ **Margin exercises.** The exercises in the shaded margins of the pages in your text give you immediate practice. **This is a perfect way to get your dendrites growing right away!** Often you will be given the first few steps to help you start solving margin exercises. The answers are given at the bottom of the page, so you can check yourself easily.

▶ **Cautions, Pointers, Notes, and Calculator Tips.**

- **Caution!** A yellow box is a comment about a common error that students make, or a common trouble spot you will want to avoid.

- **Pointers** are little "clouds" next to worked examples. They point to specific places where common mistakes are made and give on-the-spot reminders.

- Look for the specially marked **Note** boxes with blue tabs. They contain hints, explanations, or interesting side comments about a topic.

- A small picture of a red calculator 🖩 appears in several places. In the main part of the chapter, the icon means that there is a **Calculator Tip**, which helps you learn more about using your calculator. A calculator in an Exercise section is a recommendation to use your calculator to work that exercise.

▶ **Vocabulary Tips** appear in the margins to help you learn the language of mathematics. Use them to create study cards for terminology that is new to you.

List a page number from this chapter for each of these features:

A *Caution* appears on page _____.

A *Pointer* appears on page _____.

A *Note* appears on page _____.

A *Calculator Tip* appears on page _____.

Study Skills

How will you make good use of the features at the end of each chapter?

_____

_____

_____

Flag the Answers section with a sticky note or other device, so that you can turn to it quickly.

Why Are These Features Brain Friendly?

The text authors included text features that make it easier for you to understand the mathematics. **Your brain naturally seeks organization and predictability.** When you pay attention to the regular features of your text, you are allowing your brain to get familiar with all of the helpful tips, suggestions, and explanations that your book has to offer. You will make the best possible use of your text.

End-of-Chapter Features

Go back to the Table of Contents again. What is listed at the end of each chapter?

▶ **Summary** Find the Summary for this chapter. It lists the chapter's **Key Terms** (arranged in the order that they appear in the chapter) and **New Symbols** and/or **New Formulas.** Then, **Test Your Word Power** checks your understanding of the math vocabulary. Next is a **Quick Review** section. It lists each topic in the chapter and shows a worked example, with tips.

▶ **Review Exercises** Use these exercises as a way to check your understanding of all the concepts in the chapter. You can practice every type of problem. If you get stuck, the numbers in the dark green rectangles tell you which section of the chapter to go back to for more explanations.

▶ **Mixed Review Exercises** help you practice for tests. Make sure you do them.

▶ **Test** Plan to take the chapter test as a practice exam. That way you can be sure you really know how to work all types of problems without looking back at the chapter.

▶ **Cumulative Review** These exercises help you maintain the skills you've learned in all previous chapters. Practicing previous skills throughout the course will be a big help on the final exam.

Answers

How do you find out if you've worked the exercises correctly? Your text provides many of the answers. Throughout each chapter you should work the sample problems in the **margins.** The answers for those are at the **bottom of each page** in the margin area.

For homework, you can find the answers to all of the **odd-numbered section exercises** in the **Answers to Selected Exercises** section near the end of your text. Also, *all* of the answers are given for the Chapter Review Exercises and Chapter Tests. Check your textbook now, and find the page on which the Answers section begins.

Solutions

Look in MyMathLab for worked out, step-by-step solutions to some of the harder odd-numbered exercises. These exercises have a square of blue shading around the problem numbers.

Index

The **Index** is the last thing in your text. All of the topics, vocabulary, and concepts are listed in alphabetical order in the Index. For example, look up the words below. Go to the page or pages listed and find each word. Write down the page that introduces or defines each one. There may be several subheadings listed under the main word, or several page numbers may be listed. Usually, the *first* place that a word appears in the text is where it is introduced and defined. So the earliest page number is a good place to start.

Commutative property of multiplication is defined on page _____ .

Factors of numbers are defined on page _____ .

Rounding of mixed numbers is explained on page _____ .

1.2 | Introduction to Integers

OBJECTIVE ▶ **1** **Write positive and negative numbers used in everyday situations** The whole numbers in the previous section were either 0 or greater than 0. Numbers *greater* than 0 are called *positive numbers*. But many everyday situations involve numbers that are *less* than 0, called *negative numbers*. Here are a few examples.

OBJECTIVES

1 Write positive and negative numbers used in everyday situations.

2 Graph numbers on a number line.

3 Use the $<$ and $>$ symbols to compare integers.

4 Find the absolute value of integers.

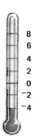

At midnight the temperature dropped to 4 degrees *below* zero.

-4 degrees

Jean had $30 in her bank account. She paid a bill for $40.75. We say she is now $10.75 "in the hole," or $10.75 "in the red," or *overdrawn* by $10.75.

$-\$10.75$

The Packers football team *gained* 6 yards on the first play. On the second play they *lost* 9 yards. We can write these results using a positive number and a negative number.

$+6$ **yards** and -9 **yards**

A plane took off from the airport and climbed to 20,000 feet *above* sea level. We can write this using a positive number.

$+20,000$ **feet**

A scuba diver swam down to $25\frac{1}{2}$ feet *below* the surface. We can write this using a negative number.

$-25\frac{1}{2}$ **feet**

Note

To write a negative number, put a negative sign (a dash) in front of it: -10. Notice that the negative sign looks exactly like the subtraction sign, as in $5 - 3 = 2$. The negative sign and subtraction sign do *not* mean the same thing. See the example below.

-10 means **negative 10** $14 - 10$ means 14 minus 10

Positive numbers can be written two ways:

1. Write a positive sign in front of the number: $+2$ is *positive 2*.

2. Do not write any sign. For example, 16 is understood to be *positive 16*.

1 Write each negative number with a negative sign. Write each positive number in two ways.

(a) The temperature is $5\frac{1}{2}$ degrees below zero.

(b) Cameron lost 12 pounds on a diet.

(c) I deposited $210.35 in my bank account.

(d) I made too many charges, so my account is overdrawn by $65.

(e) The submarine dives to 100 feet below the surface of the sea.

(f) In this round of the card game, I won 50 points.

VOCABULARY TIPS

Graphs show information in the form of a drawing. A number line is one type of graph.

Graph integers by putting a dot at the correct place on a number line. Negative numbers are to the **left** of zero on a number line. Positive numbers are to the right of zero.

2 Graph each set of numbers.

(a) −2 **(b)** 2 **(c)** 0

(d) −4 **(e)** 4

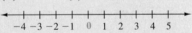

(f) $-3\frac{1}{2}$ **(g)** 1 **(h)** −1 **(i)** 3

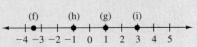

Answers

1. (a) $-5\frac{1}{2}$ degrees **(b)** −12 pounds
   (c) $210.35 or +$210.35 **(d)** −$65
   (e) −100 feet **(f)** 50 points or + 50 points

2.

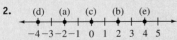

EXAMPLE 1 **Writing Positive and Negative Numbers**

Write each negative number with a negative sign. Write each positive number in two ways.

(a) The river rose to 8 feet above flood stage.

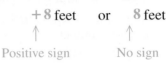

$+8$ feet   or   8 feet

↑ Positive sign   ↑ No sign

(b) Michael lost $500 in the stock market.

$-\$500$

↑ Negative sign

◀ **Work Problem 1 at the Side.**

OBJECTIVE 2 **Graph numbers on a number line.** Mathematicians often use a **number line** to show how numbers relate to each other. A number line is like a thermometer turned sideways. *Zero* is the dividing point between the positive and negative numbers.

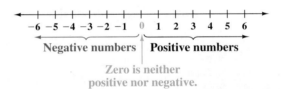

Negative numbers │ Positive numbers

Zero is neither positive nor negative.

The number line could be shown with positive numbers on the left side of 0 instead of the right side. But it helps if everyone draws it the same way, as shown above. This method will also match what you will do later when graphing points and lines.

EXAMPLE 2 **Graphing Numbers on a Number Line**

Graph each number on the number line.

(a) −5 **(b)** 3 **(c)** $1\frac{1}{2}$ **(d)** 0 **(e)** −1

Draw a dot at the correct location for each number.

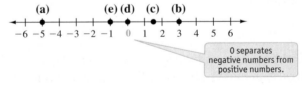

0 separates negative numbers from positive numbers.

◀ **Work Problem 2 at the Side.**

OBJECTIVE 3 **Use the < and > symbols to compare integers.** Later, you will work with fractions and decimals. For now, you will work only with *integers*. A list of **integers** can be written like this:

$$\ldots, -6, -5, -4, -3, -2, -1, 0, 1, 2, 3, 4, 5, 6, \ldots$$

The dots show that the list goes on forever in both directions.

We can use the number line to compare two integers.

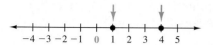   

When comparing 1 and 4,	When comparing 0 and −3,
1 is to the *left* of 4.	0 is to the *right* of −3.
1 is *less than* 4.	0 is *greater than* −3.
Use < to mean "is less than."	Use > to mean "is greater than."

1 < 4

1 **is less than** 4

0 > −3

0 **is greater than** −3

Note

One way to remember which symbol to use is that the "smaller end of the symbol" points to the "smaller number" (the number that is less).

1 < 4   0 > −3

Smaller number ⌐ ⌐ Smaller end   Smaller end ⌐ ⌐ Smaller number
             of symbol   of symbol

EXAMPLE 3   Comparing Integers, Using the < and > Symbols

Write < or > between each pair of integers to make a true statement.

(a) 0 ____ 2

  0 is to the *left* of 2 on the number line,
  so 0 is *less than* 2. Write 0 < 2.

> The smaller end of the symbol
> points to the smaller number.

(b) −1 ____ −4

  1 is to the *right* of −4, so 1 is *greater than* −4. Write 1 > −4.

(c) −4 ____ −2

  −4 is to the *left* of −2, so −4 is *less than* −2. Write −4 < −2.

——————— **Work Problem ❸ at the Side.** ▶

OBJECTIVE ❹ Find the absolute value of integers. In order to graph a number on the number line, you need to ask two things:

1. Which *direction* is it from 0? It can be in a positive direction or a negative direction. You can tell the direction by looking for a positive sign (or no sign, which means positive) or a negative sign.

2. How *far* is it from 0? The distance from 0 is the *absolute value* of a number.

❸ Write < or > between each pair of integers to make a true statement.

(a) 5 ____ 4

  5 is to the *right* of 4
  on the number line, so
  5 *is greater than* 4.

(b) 0 ____ 2

(c) −3 ____ −2

(d) −1 ____ −4

(e) 2 ____ −2

(f) −5 ____ 1

Answers

3. (a) >   **(b)** <   **(c)** <   **(d)** >
     (e) >   **(f)** <

④ Find each absolute value.

GS **(a)** $|13|$

How far away is 13 from 0 on the number line?

GS **(b)** $|-7|$

How far away is -7 from 0 on the number line?

(c) $|0|$

(d) $|-350|$

(e) $|6000|$

Absolute Value

The **absolute value** of a number is its distance from 0 on the number line. *Absolute value* is indicated by two vertical bars. For example,

$$|6| \text{ is read "the absolute value of 6."}$$

The absolute value of a number will *always* be positive (or 0), because it is the *distance* from 0. A distance is never negative. (You wouldn't say that your living room is -16 feet long.) So absolute value concerns only *how far away* the number is from 0; we don't care which direction it is from 0.

EXAMPLE 4 **Finding Absolute Values**

Find each absolute value.

(a) $|4|$ The distance from 0 to 4 on the number line is 4 spaces. So, $|4| = \mathbf{4}$.

(b) $|-4|$ The distance from 0 to -4 on the number line is also 4 spaces. So, $|-4| = \mathbf{4}$.

> Absolute value is always positive (or 0), never negative.

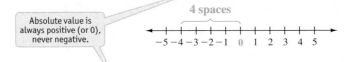

(c) $|0|$ $|0| = \mathbf{0}$ because the distance from 0 to 0 on the number line is 0 spaces.

◀ **Work Problem ④ at the Side.**

1.2 Exercises

FOR EXTRA HELP Go to MyMathLab for worked-out, step-by-step solutions to exercises enclosed in a square ▢ and video solutions to ▶ exercises.

CONCEPT CHECK *Write a positive or negative number for each situation.* **See Example 1.**

1. Mount Everest, the tallest mountain in the world, rises 29,029 feet **above** sea level.

The word "above" indicates a *positive* number.

(Data from www.nationalgeographic.com)

2. The bottom of Lake Baikal in Siberia, Russia, is 5353 feet **below** the surface of the water.

The word "below" indicates a *negative* number.

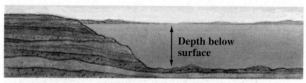

(Data from www.lakebaikal.org)

3. The coldest temperature ever recorded on Earth is 135.8 degrees below zero in Antarctica. (Data from www.theguardian.com)

4. Normal body temperature is 98.6 degrees Fahrenheit, although it varies slightly for some people. (Data from Mayo Clinic *Health Letter*.)

5. During the first three plays of the football game, the Trojans lost a total of 18 yards.

6. The Jets gained 25 yards on a pass play.

7. Angelique won $100 in a prize drawing at the shopping mall.

8. Derice overdrew his bank account by $37.

9. Keith lost $6\frac{1}{2}$ pounds while he was sick with the flu.

10. The mice in an experiment gained $2\frac{1}{2}$ ounces.

Graph each set of numbers. **See Example 2.**

11. $-3, 3, 0, -5$

12. $-2, 2, 0, 5$

13. $-1, 4, -2, 5$

14. $3, -4, 1, -5$

CONCEPT CHECK *Write each statement in words.*

15. (a) $0 < 5$

16. (a) $3 > -7$

(b) $-10 > -17$

(b) $12 < 22$

Write < or > between each pair of numbers to make a true statement.
See Example 3.

17. 10 ___ 2

18. 6 ___ 0

19. −1 ___ 0

20. −3 ___ −1

21. −10 ___ 2

22. −9 ___ 7

23. −3 ___ −6

24. 0 ___ −1

25. −10 ___ −2

26. −1 ___ −5

27. 0 ___ −8

28. 6 ___ −4

29. 10 ___ −2

30. −2 ___ 1

31. −4 ___ 4

32. 9 ___ −9

Find each absolute value. **See Example 4.**

33. $|15|$

34. $|10|$

35. $|-3|$

36. $|-8|$

37. $|0|$

38. $|100|$

39. $|200|$

40. $|-99|$

41. $|-75|$

42. $|-6320|$

43. $|-8042|$

44. $|0|$

Relating Concepts (Exercises 45–48) For Individual or Group Work

A special type of X-ray, called a DEXA test, is a quick way to screen patients for osteoporosis (which causes bones to become weak and brittle). Use the information on the Patient Report Form to **work Exercises 45–48 in order.**

Patient Report Form

Your T score measures your bone density compared to that of a young, healthy woman when peak bone mass is achieved.

T Score	Interpretation
Above −1	Normal bone density
−1 to −2.5	Greater risk of developing osteoporosis
Below −2.5	Osteoporosis

Data from National Osteoporosis Foundation.

45. Here are the T scores for four patients. Draw a number line and graph the four scores.

Patient A: −3   Patient C: −2

Patient B: 0   Patient D: −1

46. List the patients' scores in order from lowest to highest.

47. What is the interpretation of each patient's score?

48. (a) What could happen if patient A did not understand the importance of a negative sign?

(b) For which patient does the sign of the score make no difference? Explain your answer.

Taking a DEXA test of a patient's spine.

HOMEWORK: HOW, WHY, AND WHEN

I t is best for your brain if you keep up with the reading and homework in your math class. Remember that the more times you work with the information, the more dendrites you grow! So give yourself every opportunity to read, work problems, and review your mathematics.

You have two choices for reading your math text. Read the short descriptions below and decide which will be best for you.

Preview before Class; Read Carefully after Class

Maddy learns best by listening to her teacher explain things. She "gets it" when she sees the instructor work problems on the board. She likes to ask questions in class and put the information in her notes. She has learned that it helps if she has *previewed* the section and read the definitions of new vocabulary before the lecture, so she knows generally what to expect in class. *But after the class instruction,* Maddy finds that she can understand the math text easily. She remembers what her teacher said, and she can double-check her notes if she gets confused. So Maddy does her **careful** reading of the section in her text **after** hearing the classroom lecture on the topic.

Read Carefully before Class

De'Lore, on the other hand, feels he learns well by reading on his own. He prefers to read the section, learn the new vocabulary, and try working the example problems before the lecture. That way, he already knows what the teacher is going to talk about. Then, he can follow the teacher's examples more easily. It is also easier for him to take notes in class. De'Lore likes to have his questions answered right away, which he can do if he has already read the chapter section. So De'Lore **carefully** reads the section in his text **before** he hears the classroom lecture on the topic.

Notice that there is **no one right way** to work with your text. You must figure out what works best for you. Note also that both Maddy and De'Lore work with one section at a time. **The key is that you read the text regularly!** The rest of this activity will give you some ideas of how to make the most of your reading.

Try the following steps as you **read** your math text.

▶ Pay attention to every word, even small ones.
"15 divided **by** 3" is **not** the same as "15 divided **into** 3."

$$\begin{array}{r} 5 \\ 3\overline{)15} \end{array} \qquad\qquad \begin{array}{r} 0.2 \\ 15\overline{)3.0} \end{array}$$

One little word makes a difference in how you solve a problem!

▶ Read slowly. Read only one section—or even part of a section—at a time. You may even need to read some sentences two or three times. This is what **all** math students must do to succeed. Mathematics texts require a different type of reading than most other subjects. You must read every word carefully and accurately so you don't miss any important details.

▶ Do the sample problems in the margins **as you go.** Check them right away. The answers are at the bottom of the page.

▶ If your mind wanders, work problems on separate paper and write explanations in your own words.

Why Are These Reading Techniques Brain Friendly?

The steps at the left encourage you to be **actively working with the material** in your text. Your brain **grows dendrites when it is doing something.**

These methods require you to **try several different techniques,** not just the same thing over and over. Your brain loves variety!

Also, the techniques help you to take small breaks in your learning. Those **rest periods are crucial for good dendrite growth.**

Study Skills

Why Are These Homework Suggestions Brain Friendly?

Your brain will grow dendrites as you study the worked examples in the text and try doing them yourself on separate paper. So when you see similar problems in the homework, you will already have dendrites to work from.

Giving yourself a practice test by trying to remember the steps (without looking at your study card) is an excellent way to reinforce what you are learning.

Correcting errors right away is how you learn and reinforce the correct procedures. It is hard to unlearn a mistake, so always check to see that you are on the right track!

▶ Make study cards as you read each section. Make cards for new vocabulary, rules, procedures, formulas, and sample problems.

▶ **NOW,** you are ready to do your homework assignment!

Now Try This
Which steps for reading this text will be most helpful for you?

1 _____

2 _____

3 _____

Homework

Teachers assign homework so you can grow your own dendrites (learn the material) and then coat the dendrites with myelin through practice (remember the material). In learning, you get good at what you practice. So completing homework every day will strengthen your neural network and prepare you for exams.

If you have read each section in your text according to the steps above, you will probably encounter few difficulties with the exercises in the homework. Here are some additional suggestions that will help you succeed with the homework.

▶ If you **have trouble with a problem,** find a similar worked example in the section. Pay attention to *every line* of the worked example to see how to get from step to step. Work it yourself too, on separate paper; don't just look at it.

▶ If it is **hard to remember the steps** to follow for certain procedures, write the steps on a separate card. Then write a short explanation of each step. Keep the card nearby while you do the exercises, but try *not* to look at it.

▶ If you **aren't sure you are working the assigned exercises correctly,** choose two or three odd-numbered problems that are a similar type and work them. Then check the answers in the Answers section at the back of your text to see if you are doing them correctly. If you aren't, go back to the section in the text, review the examples, and find out how to correct your errors. When you are sure you understand, try the assigned problems again.

▶ **Make sure you do some homework every day,** even if the math class does not meet each day!

Now Try This
What are your biggest homework concerns?
List your two main concerns and a **brain friendly solution** for each one.

1 Concern: _____

   Solution: _____

2 Concern: _____

   Solution: _____

1.3 | Adding Integers

OBJECTIVE ▸ 1 **Add integers.** Numbers that you are adding are called **addends,** and the result is called the **sum.** You can add integers while watching a football game. On each play, you can use a *positive* integer to stand for the yards *gained* by your team, and a *negative* integer for yards *lost.* Zero indicates no gain or loss. For example,

Our team gained 6 yards on the first play	and	gained 4 yards on the next play.		Total gain of 10 yards
6	+	4	=	10

Our team lost 3 yards on the first play	and	lost 5 yards on the next play.		Total loss of 8 yards
−3	+	−5	=	−8

A drawing of a football field can help you add integers. Notice how similar the drawing is to a number line. Zero marks your team's starting point.

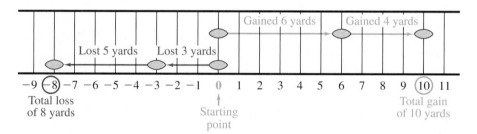

EXAMPLE 1   Using a Number Line to Add Integers

Use a number line to find −5 + (−4).

Think of the number line as a football field. Your team started at 0. On the first play it lost 5 yards. On the next play it lost 4 yards. The total loss is 9 yards.

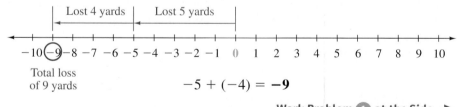

$$-5 + (-4) = -9$$

—— **Work Problem 1 at the Side.** ▶

Do you see a pattern in the margin problems you just did? The first two sums are −4 and 4. They are the same, *except for the sign.* The same is true for the next two sums, −11 and 11, and for the last two sums, −10 and 10. This pattern leads to a rule for adding two integers when the signs are the same.

Adding Two Integers with the Same Sign

Step 1   **Add** the absolute values of the numbers.

Step 2   Use the *common sign* as the sign of the sum. If both numbers are positive, the sum is positive. If both numbers are negative, the sum is negative.

1 Find each sum. Use the number line to help you.

(a) −2 + (−2)

(b) 2 + 2

(c) −10 + (−1)

(d) 10 + 1

(e) −3 + (−7)

(f) 3 + 7

2 Find each sum.

GS (a) $-6 + (-6)$

Both negative,
so the sum
is negative.

GS (b) $9 + 7$

Both positive,
so the sum
is positive.

(c) $-5 + (-10)$

(d) $-12 + (-4)$

(e) $13 + 2$

EXAMPLE 2 **Adding Two Integers with the Same Sign**

Add.

(a) $-8 + (-7)$

Step 1 *Add* the absolute values.

$$|-8| = 8 \quad \text{and} \quad |-7| = 7$$

Add $8 + 7$ to get 15.

Step 2 Use the *common sign* as the sign of the sum. Both numbers are negative, so the sum is negative.

$$-8 + (-7) = -15$$

Both negative   Sum is negative.

(b) $3 + 6 = 9$

Both positive   Sum is positive.

◀ **Work Problem 2 at the Side.**

You can also use a number line (or drawing of a football field) to add integers with *different* signs. For example, suppose that your team gained 2 yards on the first play and then lost 7 yards on the next play. We can represent this as $2 + (-7)$.

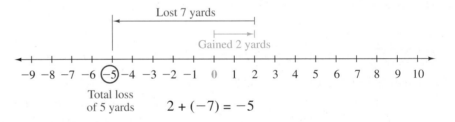

$$2 + (-7) = -5$$

Or, try this one. On the first play your team gained 10 yards, but then it lost 4 yards on the next play. We can represent this as $10 + (-4)$.

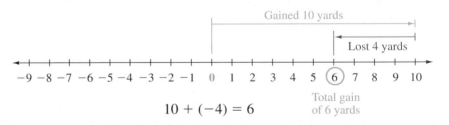

$$10 + (-4) = 6$$

These examples illustrate the rule for adding two integers with unlike, or different, signs.

Adding Two Integers with Unlike Signs

Step 1 *Subtract* the lesser absolute value from the greater absolute value.

Step 2 Use the sign of the number with the *greater absolute value* as the sign of the sum.

Answers

2. **(a)** -12 **(b)** 16 **(c)** -15
   (d) -16 **(e)** 15

| EXAMPLE 3 | Adding Two Integers with Unlike Signs |

Add.

(a) $-8 + 3$

Step 1 $|-8| = 8$ and $|3| = 3$

Subtract $8 - 3$ to get 5.

Step 2 -8 has the *greater absolute value* and is negative, so the sum is also negative.

$$-8 + 3 = \mathbf{-5}$$

(b) $-5 + 11$

Step 1 $|-5| = 5$ and $|11| = 11$

Subtract $11 - 5$ to get 6.

Step 2 11 has the *greater absolute value* and is positive, so the sum is also positive.

$$-5 + 11 = \mathbf{+6} \quad \text{or} \quad \mathbf{6}$$

— Work Problem ❸ at the Side. ▶

| EXAMPLE 4 | Adding Several Integers |

A football team has to gain at least 10 yards during four plays in order to keep the ball. Suppose that your college team lost 6 yards on the first play, gained 8 yards on the second play, lost 2 yards on the third play, and gained 7 yards on the fourth play. Did the team gain enough to keep the ball?

When you're adding several integers, work from left to right.

The team **gained 7 yards,** which is *not* **enough yards to keep the ball.**

— Work Problem ❹ at the Side. ▶

OBJECTIVE ▶ ❷ Identify properties of addition. In our football model for adding integers, we said that 0 indicated no gain or loss, that is, no change in the position of the ball. This example illustrates one of the *properties of addition*. The properties of addition apply to all addition problems, regardless of the specific numbers you use.

Addition Property of 0

Adding 0 to any number leaves the number unchanged.

Some examples are shown below.

$$0 + 6 = 6 \qquad -25 + 0 = -25 \qquad 72{,}399 + 0 = 72{,}399$$

$$0 + (-100) = -100$$

---

❸ Find each sum.

⒢ **(a)** $-3 + 7$

$|-3| = $ ____ and $|7| = $ ____.

Subtract $7 - 3$ to get ____.

The sum is positive because 7 has the greater absolute value and is positive.

$-3 + 7 = $ ____

(b) $6 + (-12)$

(c) $12 + (-7)$

(d) $-10 + 2$

(e) $5 + (-9)$

❹ Write an addition problem and solve it for each situation.

(a) The temperature was -15 degrees this morning. It rose 21 degrees during the day, then dropped 10 degrees. What is the new temperature?

(b) Andrew had $60 in his bank account. First he deposited an $85 tax refund in his account. Then he used his debit card to pay $20 for gas and $75 for groceries. What is the balance in his account?

Answers

3. (a) $|-3| = 3$ and $|7| = 7$; $7 - 3$ is 4; $-3 + 7 = 4$
(b) -6 **(c)** 5 **(d)** -8 **(e)** -4

4. (a) $-15 + 21 + (-10) = -4$ degrees
(b) $60 + 85 + (-20) + (-75) = 50

5 Rewrite each sum using the commutative property of addition. Check that the sum is unchanged.

(a) $175 + 25 = \underline{25} + \underline{\hspace{1cm}}$

Both sums are ___.

(b) $7 + (-37) = \underline{\hspace{1cm}} + \underline{\hspace{1cm}}$

Both sums are ___.

(c) $-16 + 16 = \underline{\hspace{1cm}} + (\underline{\hspace{0.5cm}})$

Both sums are ___.

(d) $-9 + (-41) = \underline{\hspace{1cm}} + (\underline{\hspace{0.5cm}})$

Both sums are ___.

Another property of addition is that you can change the *order* of the addends and still get the same sum. For example,

Gaining 2 yards, then losing 7 yards, gives a result of -5 yards.

Losing 7 yards, then gaining 2 yards, also gives a result of -5 yards.

Commutative Property of Addition

Changing the **order** of two addends does *not* change the sum.

Here are some examples.

$$84 + 2 = 2 + 84 \qquad\qquad -10 + 6 = 6 + (-10)$$

Both sums are 86. ┃ Both sums are -4.

EXAMPLE 5   Using the Commutative Property of Addition

Rewrite each sum, using the **commutative property of addition.** Check that the sum is unchanged.

(a) $65 + 35$

$$65 + 35 = 35 + 65$$
$$100 \quad = \quad 100$$

Both sums are **100**, so the **sum is unchanged.**

> Adding two numbers in a different order does *not* change the sum.

(b) $-20 + (-30)$

$$-20 + (-30) = -30 + (-20)$$
$$-50 \quad = \quad -50$$

Both sums are **-50**, so the **sum is unchanged.**

◀ **Work Problem ⑤ at the Side.**

When there are three addends, parentheses may be used to tell you which pair of numbers to add first, as shown below.

$(3 + 4) + 2$ First add $3 + 4$	┃	$3 + (4 + 2)$ First add $4 + 2$
$7 \quad + 2$ Then add $7 + 2$	┃	$3 + \quad 6$ Then add $3 + 6$
9	┃	9

Both sums are 9. This example illustrates another property of addition.

Associative Property of Addition

Changing the **grouping** of addends does *not* change the sum.

Some examples are shown below.

$$(-5 + 5) + 8 = -5 + (5 + 8) \qquad\qquad 3 + (4 + 6) = (3 + 4) + 6$$
$$0 \quad + 8 = -5 + \quad 13 \qquad\qquad\qquad 3 + \quad 10 \quad = \quad 7 \quad + 6$$
$$8 \quad = \quad 8 \qquad\qquad\qquad\qquad 13 \quad = \quad 13$$

Both sums are 8. ┃ Both sums are 13.

Answers

5. **(a)** $175 + 25 = 25 + 175;\ 200$
(b) $7 + (-37) = -37 + 7;\ -30$
(c) $-16 + 16 = 16 + (-16);\ 0$
(d) $-9 + (-41) = -41 + (-9);\ -50$

We can use the associative property to make addition problems easier. Notice in the first example in the green box at the bottom of the previous page that it is easier to group $-5 + 5$ (which is 0) and then add 8. In the second example in the box, it is helpful to group $4 + 6$ because the sum is 10, and it is easy to work with multiples of 10.

EXAMPLE 6 **Using the Associative Property of Addition**

In each addition problem, pick out the two addends that would be easiest to add. Write parentheses around those addends. Then find the sum.

(a) $-9 + 9 + 6$

Group $-9 + 9$ because the sum is 0.

> Use the associative property to add the easiest numbers first.

$$(-9 + 9) + 6$$
$$0 \quad + 6$$
$$\mathbf{6}$$

(b) $-25 + 17 + 3$

Group $17 + 3$ because the sum is 20, which is a multiple of 10.

$$-25 + (17 + 3)$$
$$-25 + \quad 20$$
$$\mathbf{-5}$$

Note

When using the associative property to make the addition of a group of numbers easier:

1. Look for two numbers whose sum is 0.

2. Look for two numbers whose sum is a multiple of 10 (the sum ends in 0, such as 10, 20, 30, or -100, -200, etc.).

If neither of these occurs, look for two numbers that are easier for you to add. For example, in $98 + 43 + 5$, you may find that adding $43 + 5$ is easier than adding $98 + 43$.

— Work Problem ⑥ at the Side. ▶

VOCABULARY TIP

Commutative vs. associative property of addition To tell the difference between these properties, think of the words "commute" and "associate." When you *commute* to and from work, it doesn't matter which way you are going; you still drive the same number of miles. The commutative property says you may change the order (or direction) of the addends and still get the same sum.

For the associative property, think of *associating* with friends. You get together in groups. The associative property says you may group (or associate) addends in any way and still get the same sum.

---

⑥ In each problem, find the two addends that would be easiest to add. Then find the sum.

(a) $-12 + 12 + (-19)$

(b) $31 + (-75) + 75$

(c) $1 + 9 + (-16)$

(d) $-38 + 5 + 25$

Answers

6. **(a)** $(-12 + 12) + (-19)$
$$0 \quad + (-19)$$
$$-19$$

(b) $31 + (-75 + 75)$
$$31 + \quad 0$$
$$31$$

(c) $(1 + 9) + (-16)$
$$10 \quad + (-16)$$
$$-6$$

(d) $-38 + (5 + 25)$
$$-38 + \quad 30$$
$$-8$$

1.3 Exercises

FOR EXTRA HELP

Go to MyMathLab *for worked-out, step-by-step solutions to exercises enclosed in a square* [ ] *and video solutions to* ▶ *exercises.*

Add by using the number line. **See Example 1.**

1. $-2 + 5$

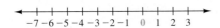

2. $-3 + 4$

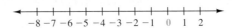

3. $-5 + (-2)$

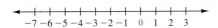

4. $-2 + (-2)$

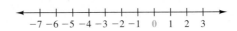

5. $3 + (-4)$
▶

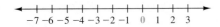

6. $5 + (-1)$

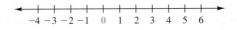

Add. **See Example 2.**

7. (a) $-5 + (-5)$
▶

8. (a) $-9 + (-9)$

9. (a) $7 + 5$

10. (a) $3 + 6$

(b) $5 + 5$

(b) $9 + 9$

(b) $-7 + (-5)$

(b) $-3 + (-6)$

11. (a) $-25 + (-25)$

12. (a) $-30 + (-30)$

13. (a) $48 + 110$

14. (a) $235 + 21$

(b) $25 + 25$

(b) $30 + 30$

(b) $-48 + (-110)$

(b) $-235 + (-21)$

15. CONCEPT CHECK What pattern do you see in your answers to **Exercises 7–14?** Explain why this pattern occurs.

16. CONCEPT CHECK In your own words, explain how to add two integers that have the same sign.

Add. See Example 3.

17. (a) $-6 + 8$

(b) $6 + (-8)$

18. (a) $-3 + 7$

(b) $3 + (-7)$

19. (a) $-9 + 2$

(b) $9 + (-2)$

20. (a) $-8 + 7$

(b) $8 + (-7)$

21. (a) $20 + (-25)$

(b) $-20 + 25$

22. (a) $30 + (-40)$

(b) $-30 + 40$

23. (a) $200 + (-50)$

(b) $-200 + 50$

24. (a) $150 + (-100)$

(b) $-150 + 100$

25. CONCEPT CHECK   What pattern do you see in your answers to **Exercises 17–24?** Explain why this pattern occurs.

26. CONCEPT CHECK   In your own words, explain how to add two integers that have different signs.

Add. See Examples 1–4.

27. $-8 + 5$

28. $-3 + 2$

29. $-1 + 8$

30. $-4 + 10$

31. $-2 + (-5)$

32. $-7 + (-3)$

33. $6 + (-5)$

34. $11 + (-3)$

35. $4 + (-12)$

36. $9 + (-10)$

37. $-10 + (-10)$

38. $-5 + (-20)$

39. $-17 + 0$

40. $0 + (-11)$

41. $1 + (-23)$

42. $13 + (-1)$

43. $\underbrace{-2 + (-12)} + (-5)$
$\underbrace{\quad -14 \quad + (-5)}$
_____

44. $\underbrace{-16 + (-1)} + (-3)$
$\underbrace{\quad -17 \quad + (-3)}$
_____

45. $8 + 6 + (-8)$

46. $-5 + 2 + 5$

47. $-7 + 6 + (-4)$

48. $-9 + 8 + (-2)$

49. $-3 + (-11) + 14$

50. $15 + (-7) + (-8)$

51. $10 + (-6) + (-3) + 4$

52. $2 + (-1) + (-9) + 12$

53. $-7 + 28 + (-56) + 3$

54. $4 + (-37) + 29 + (-5)$

Write an addition problem for each situation and find the sum.

55. The football team gained 13 yards on the first play and lost 17 yards on the second play. How many yards did the team gain or lose in all?

56. At penguin breeding grounds on Antarctic islands, temperatures routinely drop to $-15\,°C$. Temperatures in the interior of the continent may drop another $60\,°C$ below that. What is the temperature in the interior?

57. Nick's bank account was overdrawn by $62. He deposited $50 in his account. What is the balance in his account?

58. Cynthia had $100 in her checking account. She wrote a check for $73 and was charged $27 for overdrawing her account last month. What is her account balance?

59. $88 was stolen from Jay's car. He got $35 of it back. What was his net loss? (*Hint:* net means Jay's final loss)

60. Marion lost 4 pounds in April, gained 2 pounds in May, and gained 3 pounds in June. What was her net gain or loss in weight?

61. Use the score sheet to find each player's point total after three rounds in a card game.

	Jeff	Terry
Round 1	Lost 20 pts	Won 42 pts
Round 2	Won 75 pts	Lost 15 pts
Round 3	Lost 55 pts	Won 20 pts

62. Use the information in the table on flood water depths to find the new flood level for each river.

	Red River	Mississippi
Monday	Rose 8 ft	Rose 4 ft
Tuesday	Fell 3 ft	Rose 7 ft
Wednesday	Fell 5 ft	Fell 13 ft

The table below shows the number of pounds gained or lost by several health club members during a four-month period. A negative number indicates a loss of pounds. A positive number indicates a gain. Find the net gain or loss for each member at the end of the four months.

	Member	Month 1	Month 2	Month 3	Month 4	Net gain or loss
63.	Angela	−2	0	5	−5	
64.	Syshe	−1	2	−6	0	
65.	Brittany	3	−2	−2	3	
66.	Nicole	1	1	−4	2	

Rewrite each sum, using the commutative property of addition. Show that the sum is unchanged. **See Example 5.**

67. $-18 + (-5) = $ _____ + _____

   Both sums are _____ .

68. $-12 + 20 = $ _____ + _____

   Both sums are _____ .

69. $-4 + 15 = $ _____ + _____

   Both sums are _____ .

70. $17 + 1 = $ _____ + _____

   Both sums are _____ .

In each addition problem, find the two addends that would be easiest to add. Then find the sum. **See Example 6.**

71. $6 + (-14) + 14$ *Hint:* Find two addends whose sum is 0.

72. $-9 + 9 + (-8)$

73. $14 + 6 + (-7)$

74. $-18 + 3 + 7$

75. Make up three of your own examples that illustrate the addition property of 0.

76. Make up three of your own examples that illustrate the associative property of addition. Show that the sum is unchanged.

Find each sum.

77. $-7081 + 2965$

78. $-1398 + 3802$

79. $-179 + (-61) + 8926$

80. $36 + (-6215) + 428$

81. $86 + (-99{,}000) + 0 + 2837$

82. $-16{,}719 + 0 + 8878 + (-14)$

1.4 | Subtracting Integers

OBJECTIVES

1. Find the opposite of an integer.
2. Subtract integers.
3. Combine adding and subtracting of integers.

OBJECTIVE ▶ 1 Find the opposite of an integer. Look at how the integers match up on this number line.

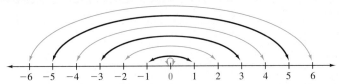

Each integer is matched with its *opposite*. **Opposites** are the same *distance* from 0 on the number line but are on *opposite sides* of 0.

$$+2 \text{ is the opposite of } -2 \quad \text{and} \quad -2 \text{ is the opposite of } +2$$

When you add opposites, the sum is always 0. The opposite of a number is also called its *additive inverse*.

$$2 + (-2) = 0 \quad \text{and} \quad -2 + 2 = 0$$

EXAMPLE 1 Finding the Opposites of Integers

First find the opposite (additive inverse) of each number.
Then show that the sum of the number and its opposite is 0.

> The sum of opposites is 0.

(a) 6 The opposite of 6 is **−6** and $6 + (-6) = 0$

(b) −10 The opposite of −10 is **10** and $-10 + 10 = 0$

(c) 0 The opposite of 0 is **0** and $0 + 0 = 0$

◀ **Work Problem 1 at the Side.**

OBJECTIVE ▶ 2 Subtract integers. Now that you know how to add integers and how to find opposites, you can subtract integers. Every subtraction problem has the same answer as a related addition problem. The examples below illustrate how to change subtraction problems into addition problems.

$$6 - 2 = 4 \qquad 8 - 3 = 5$$
$$\downarrow \quad \downarrow \qquad \qquad \downarrow \quad \downarrow$$
$$6 + (-2) = 4 \qquad 8 + (-3) = 5$$

Same answer (for both)

Subtracting Two Integers

To subtract two numbers, *add* the first number to the *opposite* of the second number. Remember to change *two* things:

Step 1 Make one pencil stroke to change the subtraction symbol to an addition symbol.

Step 2 Make a second pencil stroke to change the *second* number to its *opposite*. If the second number is positive, change it to negative. If the second number is negative, change it to positive.

VOCABULARY TIP

Opposite Just as your left hand is the **opposite** of your right hand, if two numbers are **opposite**, they are on **opposite sides of zero** and have **opposite signs**. So, if one number is positive (like +6), then the opposite will be negative (−6).

1 Find the opposite (additive inverse) of each number. Show that the sum of the number and its opposite is 0.

GS (a) 5

The opposite of 5 is ____.

$5 + ($____$) = 0$

(b) 48

(c) 0

(d) −1

(e) −24

⚠ CAUTION

When changing a subtraction problem to an addition problem, do **not** make any change in the *first* number. The pattern is

1st number − 2nd number = 1st number + opposite of 2nd number.

Answers

1. (a) −5; 5 + (−5) = 0
   (b) −48; 48 + (−48) = 0 (c) 0; 0 + 0 = 0
   (d) 1; −1 + 1 = 0 (e) 24; −24 + 24 = 0

| EXAMPLE 2 | Subtracting Two Integers |

Make *two* pencil strokes to change each subtraction problem into an addition problem. Then find the sum.

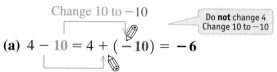

Change 10 to −10

Do not change 4
Change 10 to −10

(a) $4 - 10 = 4 + (-10) = -6$

Change subtraction to addition.

Change −6 to +6

Do not change −9
Change −6 to +6

(b) $-9 - (-6) = -9 + (+6) = -3$

Change subtraction to addition.

Change −5 to +5

(c) $3 - (-5) = 3 + (+5) = 8$ Make *two* pencil strokes.

Change subtraction to addition.

Do not change the first number when subtracting.

Change 9 to −9

(d) $-2 - 9 = -2 + (-9) = -11$ Make *two* pencil strokes.

Change subtraction to addition.

——— **Work Problem ② at the Side.** ▶

OBJECTIVE ▶ ③ Combine adding and subtracting of integers. When adding and subtracting more than two signed numbers, first change all subtractions to adding the opposite. Then add from left to right.

| EXAMPLE 3 | Combining Addition and Subtraction |

Simplify by completing all the calculations.

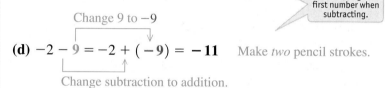

$-5 - \quad 10 - \quad 12 + 1$ Change all subtractions to adding the opposite.
Change 10 to −10 and change 12 to −12

$-5 + (-10) + (-12) + 1$ Add from left to right. First add −5 + (−10)

$-15 \quad + (-12) + 1$ Then add −15 + (−12)

$-27 \quad\quad + 1$ Finally, add −27 + 1

-26

——— **Work Problem ③ at the Side.** ▶

🖩 **Calculator Tip**

Use the *change of sign* key ⊕/⊖, or the negative sign key ⊖, on a *scientific* calculator to enter negative numbers. To enter −5, press ⑤ ⊕/⊖, or ⊖ 5 (check the instructions). To enter +5, just press ⑤. For **Example 3** above, press these keys.

⊖ 5 ⊖ 10 ⊖ 12 ⊕ 1 ⊜   The answer is −26

−5   Subtract.

---

② Subtract by changing subtraction to adding the opposite. (Make *two* pencil strokes.)

GS **(a)** $-6 - 5$

$-6 + (-5) = \underline{\quad\quad}$

GS **(b)** $3 - (-10)$

$3 + \underline{\quad\quad} = \underline{\quad\quad}$

(c) $-8 - (-2)$

(d) $0 - 10$

(e) $-4 - (-12)$

(f) $9 - 7$

③ Simplify.

GS **(a)** $6 - 7 + (-3)$

$6 + (\underline{\quad\quad}) + (-3)$

$-1 \quad + (-3)$

$\underline{\quad\quad}$

(b) $-2 + (-3) - (-5)$

(c) $7 - 7 - 7$

(d) $-3 - 9 + 4 - (-20)$

Answers

2. (a) −11 **(b)** 3 + (+10) = 13 **(c)** −6
(d) −10 **(e)** 8 **(f)** 2

3. (a) 6 + (−7) + (−3); −1 + (−3); −4
(b) 0 **(c)** −7 **(d)** 12

1.4 Exercises

FOR EXTRA HELP

Go to MyMathLab *for worked-out, step-by-step solutions to exercises enclosed in a square* ▢ *and video solutions to* ▶ *exercises.*

First find the opposite (additive inverse) of each number. Then show that the sum of the number and its opposite is 0. **See Example 1.**

1. 6 ▶

2. -3

3. -13 ▶

4. 0

5. CONCEPT CHECK The work shown below has a mistake in it.

What Went Wrong? First write a sentence explaining what the mistake is. Then fix the mistake and find the correct solution.

Valerie calculated $-6 - 6$ this way:

$$-6 - 6$$
$$\downarrow \downarrow$$
$$-6 + 6 = 0$$

6. CONCEPT CHECK The work shown below has a mistake in it.

What Went Wrong? First write a sentence explaining what the mistake is. Then fix the mistake and find the correct solution.

Victor calculated $-9 - 5$ this way:

$$-9 - \quad 5$$
$$\downarrow \quad \downarrow$$
$$9 + (-5) = 4$$

Subtract by changing subtraction to addition. **See Example 2.**

7. $19 - 5$
GS
$$\downarrow \downarrow$$
$$19 + (-5) = \underline{\quad}$$

8. $24 - 11$
GS
$$\downarrow \downarrow$$
$$24 + (-11) = \underline{\quad}$$

9. $10 - 12$

10. $1 - 8$

11. $7 - 19$

12. $2 - 17$

13. $-15 - 10$

14. $-10 - 4$

15. $-9 - 14$ ▶

16. $-3 - 11$

17. $-3 - (-8)$
GS
$$\downarrow \quad \downarrow$$
▶ $$-3 + (\underline{\quad}) = \underline{\quad}$$

18. $-1 - (-4)$
GS
$$\downarrow \quad \downarrow$$
$$-1 + (\underline{\quad}) = \underline{\quad}$$

19. $6 - (-14)$ ▶

20. $8 - (-1)$

21. $1 - (-10)$ ▶

22. $6 - (-1)$

23. $-30 - 30$

24. $-25 - 25$

25. $-16 - (-16)$

26. $-20 - (-20)$

27. $13 - 13$

28. $19 - 19$

29. $0 - 6$

30. $0 - 12$

31. (a) $3 - (-5)$
  (b) $3 - 5$
  (c) $-3 - (-5)$
  (d) $-3 - 5$

32. (a) $9 - 6$
  (b) $-9 - 6$
  (c) $9 - (-6)$
  (d) $-9 - (-6)$

33. (a) $4 - 7$
  (b) $4 - (-7)$
  (c) $-4 - 7$
  (d) $-4 - (-7)$

34. (a) $8 - (-2)$
  (b) $-8 - (-2)$
  (c) $8 - 2$
  (d) $-8 - 2$

Simplify. **See Example 3.**

35. $-2 - 2 - 2$
GS
$$\downarrow \downarrow \downarrow$$
$$-2 + (-2) + (-2)$$
$$\underbrace{\qquad}$$
$$-4 \quad + (-2) = \underline{\quad}$$

36. $-8 - 4 - 8$
GS
$$\downarrow \downarrow \downarrow$$
$$-8 + (-4) + (-8)$$
$$\underbrace{\qquad}$$
$$-12 \quad + (-8) = \underline{\quad}$$

37. $9 - 6 - 3 - 5$

38. $12 - 7 - 5 - 4$

39. $3 - (-3) - 10 - (-7)$

40. $1 - 9 - (-2) - (-6)$

41. $-2 + (-11) - (-3)$

42. $-5 - (-2) + (-6)$

43. $4 - (-13) + (-5)$

44. $6 - (-1) + (-10)$

45. $6 + 0 - 12 + 1$

46. $-10 - 4 + 0 + 18$

This windchill table shows how wind increases a person's heat loss. For example, find the temperature of 15 °F along the top of the table. Then find a wind speed of 20 mph along the left side of the table. The column and row intersect at −2 °F, the "windchill temperature." The actual temperature is 15 °F but the wind makes it feel like −2 °F. The difference between the actual temperature and the windchill temperature is 15 − (−2) = 15 + (+2) = 17 degrees difference.

Use the table to find the windchill temperatures in Exercises 47–48. Then write and solve a subtraction problem to calculate the difference between the actual temperature and the windchill temperature.

WINDCHILL
Temperature (degrees Fahrenheit)

Calm	40	35	30	25	20	15	10	5	0	−5	−10	−15	−20	−25	−30
5	36	31	25	19	13	7	1	−6	−11	−16	−22	−28	−34	−40	−46
10	34	27	21	15	9	3	−4	−10	−16	−22	−28	−35	−41	−47	−53
15	32	25	19	13	6	0	−7	−13	−19	−26	−32	−39	−45	−51	−58
20	30	24	17	11	4	−2	−9	−15	−22	−29	−35	−42	−48	−55	−61
25	29	23	16	9	3	−4	−11	−17	−24	−31	−37	−44	−51	−58	−64
30	28	22	15	8	1	−5	−12	−19	−26	−33	−39	−46	−53	−60	−67
35	28	21	14	7	0	−7	−14	−21	−27	−34	−41	−48	−55	−62	−69
40	27	20	13	6	−1	−8	−15	−22	−29	−36	−43	−50	−57	−64	−71

Wind Speed (miles per hour)

Shaded area: Frostbite occurs in 30 minutes or less.

Data from National Weather Service.

47. (a) 30 °F; 10 mph wind

(b) 15 °F; 15 mph wind

(c) 5 °F; 25 mph wind

(d) −10 °F; 35 mph wind

48. (a) 40 °F; 20 mph wind

(b) 20 °F; 35 mph wind

(c) 10 °F; 15 mph wind

(d) −5 °F; 30 mph wind

Simplify. Begin each exercise by working inside the absolute value bars or the parentheses.

49. $-2 + (-11) + |-2|$

50. $5 - |-3| + 3$

51. $0 - |-7 + 2|$

52. $|1 - 8| - |0|$

53. $-3 - (-2 + 4) + (-5)$

54. $5 - 8 - (6 - 7) + 1$

Relating Concepts (Exercises 55–56) For Individual or Group Work

*Use your knowledge of the properties of addition to **work Exercises 55 and 56 in order**.*

55. Look for a pattern in these pairs of subtractions.

$-3 - 5 \ = $ ___ | $-4 - (-3) = $ ___
$5 - (-3) = $ ___ | $-3 - (-4) = $ ___

Explain what happens when you try to apply the commutative property to subtraction.

56. Recall the addition property of 0. Can 0 be used in a subtraction problem without changing the other number? Explain what happens and give several examples. (*Hint:* Think about *order* in a subtraction problem.)

1.5 | Problem Solving: Rounding and Estimating

One way to get a rough check on an answer is to *round* the numbers in the problem. **Rounding** a number means finding a number that is close to the original number, but easier to work with.

For example, a superintendent of schools in a large city might be discussing the need to build new schools. In making her point, it probably would not be necessary to say that the school district has 152,807 students. It probably would be sufficient to say that there are about 153,000 students, or even 150,000 students.

OBJECTIVE ❶ **Locate the place to which a number is to be rounded.** The first step in rounding a number is to locate the *place to which the number is to be rounded.*

EXAMPLE 1 Finding the Place to Which a Number Is to Be Rounded

Locate and draw a line under the place to which each number is to be rounded. Then answer the question.

(a) Round −23 to the nearest ten. Is −23 closer to −2̲0 or −3̲0?

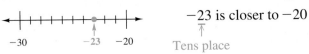

−23 is closer to −20
Tens place

(b) Round $381 to the nearest hundred. Is it closer to $3̲00 or $4̲00?

$381 is closer to $400
Hundreds place

(c) Round −54,702 to the nearest thousand. Is it closer to −5̲4,000 or −5̲5,000?

−54,702 is closer to −55,000

Thousands place

◀ **Work Problem** ❶ **at the Side.**

---

❶ Locate and draw a line under the place to which the number is to be rounded. Then answer the question.

GS **(a)** −746 (nearest ten)

⤒
└── Tens place

Is it closer to −740 or −750? _____

(b) 2412 (nearest thousand)

Is it closer to 2000 or 3000? _____

(c) −89,512 (nearest hundred)

Is it closer to −89,500 or −89,600? _____

(d) 546,325 (nearest ten-thousand)

Is it closer to 540,000 or 550,000? _____

OBJECTIVE ❷ **Round integers.** Use these steps to round integers.

Rounding an Integer
Step 1 Locate the *place* to which the number is to be rounded. Draw a line under that place.
Step 2 Look only at the next digit to the right of the one you underlined. If the next digit is *5 or more, increase* the underlined digit by 1. If the next digit is *4 or less,* do *not* change the digit in the underlined place.
Step 3 Change all digits to the right of the underlined place to zeros.

❗ CAUTION

If you are rounding a negative number, be careful to write the negative sign in front of the rounded number. For example, −79 rounds to −80.

| EXAMPLE 2 | **Using the Rounding Rule for 4 or Less** |

Round 349 to the nearest hundred.

Step 1 Locate the place to which the number is being rounded. Draw a line under that place.

349
└──── Hundreds place

Step 2 Because the next digit to the right of the underlined place is 4, which is *4 or less,* do *not* change the digit in the underlined place.

┌──── Next digit is *4 or less.*
349
└──── 3 remains 3

Step 3 Change all digits to the right of the underlined place to zeros.

┌──── Change to 0 ────┐
349 rounded to the nearest hundred is 300
└──── Leave 3 as 3 ────┘

> Think of money. $349 is closer to $300 than to $400

In other words, 349 is closer to **300** than to 400

──────── **Work Problem 2 at the Side.** ▶

| EXAMPLE 3 | **Using the Rounding Rule for 5 or More** |

Round 36,833 to the nearest thousand.

Step 1 Find the place to which the number is to be rounded. Draw a line under that place.

36,833
└──── Thousands place

Step 2 Because the next digit to the right of the underlined place is 8, which is *5 or more,* add 1 to the underlined place.

┌──── Next digit is *5 or more.*
36,833
└──── Change 6 to 7

Step 3 Change all digits to the right of the underlined place to zeros.

┌──── Change to 0 ────┐
36,833 rounded to the nearest thousand is 37,000
└──── Change 6 to 7 ────┘

In other words, 36,833 is closer to **37,000** than to 36,000

──────── **Work Problem 3 at the Side.** ▶

---

2 Round to the nearest ten.

(a) ┌ Next digit is *4 or less.*
34
└ Tens place

34 rounds to ____
└ Leave 3 as 3

(b) −61

(c) −683

(d) 1792

3 Round to the nearest thousand.

(a) ┌ Next digit is *5 or more.*
1725
└ Thousands place

┌ Change to 0
1725 rounds to _____
└ Change 1 to 2

(b) −6511

(c) 58,829

(d) −83,904

Answers

2. (a) 30 **(b)** −60 **(c)** −680 **(d)** 1790
3. (a) 2000 **(b)** −7000 **(c)** 59,000
 (d) −84,000

④ Round as indicated.

(a) −6036 to the nearest ten

GS (b) 34,968 to the nearest hundred

Next digit is *5 or more.*
34,968
— Change 9 to 10 and regroup 1 into thousands place.
— 4 + regrouped 1 is 5

So 34,968 rounds to _____

(c) −73,077 to the nearest thousand

GS (d) 9852 to the nearest thousand

Next digit is *5 or more.*
9852
— Change 9 to 10; write 0 and regroup 1 into ten-thousands place.

So 9852 rounds to _____

(e) 85,949 to the nearest hundred

(f) 40,387 to the nearest thousand

EXAMPLE 4   Using the Rules for Rounding

(a) Round −2382 to the nearest ten.

Step 1 −2382
          ⊥— Tens place

Step 2 The next digit to the right is 2, which is *4 or less.*

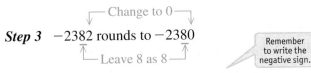

— Next digit is *4 or less.*
−2382
⊥— Leave 8 as 8

Step 3 −2382 rounds to −2380
          — Change to 0 ⟶
          ⊥— Leave 8 as 8 ⟶

> Remember to write the negative sign.

−2382 rounded to the nearest ten is **−2380**

(b) Round 13,961 to the nearest hundred.

Step 1 13,961
          ⊥— Hundreds place

Step 2 The next digit to the right is 6, which is *5 or more.*

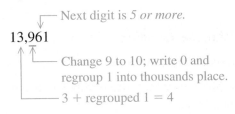

— Next digit is *5 or more.*
13,961
— Change 9 to 10; write 0 and regroup 1 into thousands place.
— 3 + regrouped 1 = 4

Step 3 13,961 rounds to 14,000

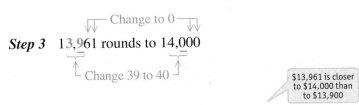

— Change to 0 ⟶
— Change 39 to 40

> $13,961 is closer to $14,000 than to $13,900

13,961 rounded to the nearest hundred is **14,000**

Note

In *Step 2* above, when you added 1 to the hundreds place, notice that the first three digits increased from 139 to 140.

13,961 rounded to 14,000

◀ **Work Problem ④ at the Side.**

Answers

4. **(a)** −6040 **(b)** 35,000 **(c)** −73,000
   (d) 10,000 **(e)** 85,900 **(f)** 40,000

EXAMPLE 5   Rounding Large Numbers

(a) Round −37,892 to the nearest ten-thousand.

Step 1   −37,892

⎯⎯⎯ Ten-thousands place

Step 2   The next digit to the right is 7, which is *5 or more.*

⎯ Next digit is *5 or more.*

−37,892

⎯⎯⎯ Change 3 to 4

⎯ Change to 0 ⎯

Step 3   −37,892 rounds to −40,000

⎯ Change 3 to 4 ⎯

> Remember to write the negative sign.

−37,892 rounded to the nearest ten-thousand is **−40,000**

(b) Round 528,498,675 to the nearest million.

Step 1   528,498,675

⎯⎯⎯ Millions place

⎯ Next digit is *4 or less.*

Step 2   528,498,675

⎯⎯⎯ Leave 8 as 8

⎯ Change to 0 ⎯

Step 3   528,498,675 rounds to 528,000,000

⎯ Leave 8 as 8 ⎯

528,498,675 rounded to the nearest million is **528,000,000**

───── **Work Problem 5 at the Side.** ▶

OBJECTIVE ▶ 3 Use front end rounding to estimate answers in addition and subtraction. In many everyday situations, we can round numbers and **estimate** the answer to a problem. For example, suppose that you're thinking about buying a sofa for $988 and a chair for $209. You can round the prices and estimate the total cost as $1000 + $200 ≈ $1200. The ≈ symbol means "approximately equal to." The estimated total of $1200 is close enough to help you decide whether you can afford both items. Of course, when it comes time to pay the bill, you'll want the *exact* total of $988 + $209 = $1197.

Sofa		**Chair**			
$988	+	$209	=	$1197	← Exact cost
|		|			
Rounds to		Rounds to			
↓		↓			
$1000	+	$200	=	$1200	← Estimated cost

5 Round as indicated.

(a) −14,679 to the nearest ten-thousand

(b) 724,518,715 to the nearest million

GS (c) −49,900,700 to the nearest million

⎯Next digit is *5 or more.*

−49,900,700

⎯ Millions place

−49,900,700

⎯ Change 9 to 10; write 0 and regroup 1 into ten-millions place.

⎯ 4 + regrouped 1 is 5

−49,900,700 rounds

to _____

(d) 306,779,000 to the nearest hundred-million

Answers

5. **(a)** −10,000 **(b)** 725,000,000
   (c) −50,000,000 **(d)** 300,000,000

6 Use front end rounding to round each number.

(a) −94

(b) 508

(c) −2522

(d) 9700

(e) 61,888

(f) −963,369

In **front end rounding,** each number is rounded to the highest possible place, so all the digits become 0 except the first digit. Once the numbers have lots of zeros, working with them is easy. Front end rounding is often used to estimate answers.

EXAMPLE 6 Using Front End Rounding

Use front end rounding to round each number.

(a) −216

Round to the highest possible place, that is, the leftmost digit. In this case, the leftmost digit, 2, is in the hundreds place, so round to the nearest hundred.

The rounded number is **−200**. Notice that all the digits in the rounded number are 0, except the first digit. Also, remember to write the negative sign.

(b) 97,203

The leftmost digit, 9, is in the ten-thousands place, so round to the nearest ten-thousand.

┌─── Next digit is *5 or more.*   ┌─── Change to 0 ───┐
97,203   97,203 rounds to 100,000
└─── Change 9 to 10   └─ Change 9 to 10 ─┘
                                         Regroup 1 into hundred-thousands place.

The rounded number is **100,000**. Notice that all the digits in the rounded number are 0, except the first digit.

◀ **Work Problem 6 at the Side.**

EXAMPLE 7 Using Front End Rounding to Estimate an Answer

Use front end rounding to estimate an answer. Then find the exact answer.

Meisha's paycheck showed gross pay of $823. It also listed deductions of $291. What is her net pay after deductions?

Estimate: Use front end rounding to round $823 and $291.

┌─ Next digit is *4 or less.*   ┌─ Next digit is *5 or more.*
$823 rounds to $800 $291 rounds to $300
└─Leave 8 as 8─┘   └─Change 2 to 3─┘

Use the rounded numbers and subtract to *estimate* Meisha's net pay.

$$\$800 - \$300 = \textbf{\$500} \leftarrow \text{Estimate}$$

Exact: Use the original numbers and subtract to find the *exact* amount.

$$\$823 - \$291 = \textbf{\$532} \leftarrow \text{Exact}$$

Meisha's paycheck will show the *exact* amount of $532. Because $532 is fairly close to the *estimate* of $500, Meisha can quickly see that the amount shown on her paycheck probably is correct. She might also use the estimate when talking to a friend, saying, "My net pay is about $500."

Answers

6. **(a)** −90 **(b)** 500 **(c)** −3000
(d) 10,000 **(e)** 60,000 **(f)** −1,000,000

─── Continued on Next Page

❶ CAUTION

Always *estimate* the answer first. Then, when you find the *exact* answer, check that it is close to the estimate. If your exact answer is very far off, rework the problem because you probably made an error.

🔲 Calculator Tip

It's easy to press the wrong key when using a calculator. If you use front end rounding and estimate the answer *before* entering the numbers, you can catch many such mistakes. For example, a student thought that he entered **Example 7** from the previous page correctly.

823 ⊖ 291 ⊜ ▮ **1114**

Front end rounding gives an estimated answer of 800 − 300 = 500, which is very different from 1114. Can you figure out which key the student pressed incorrectly?

Answer: The student pressed ⊕ instead of ⊖.

——————— **Work Problem ❼ at the Side.** ▶

❼ Use front end rounding to
GS estimate an answer. Then find the exact answer.

Pao Xiong is a bookkeeper for a small business. The company bank account is overdrawn by $3881. He deposits $2090. What is the balance in the account?

Estimate: Use front end rounding.

−$4000 + _____ = _____

Exact:

Answers

7. *Estimate:* −$4000 + $2000 = −$2000
   Exact: −$3881 + $2090 = −$1791
   The account is overdrawn by $1791, which is fairly close to the estimate of −$2000.

1.5 Exercises

FOR EXTRA HELP

Go to MyMathLab for worked-out, step-by-step solutions to exercises enclosed in a square ▢ and video solutions to ▶ exercises.

CONCEPT CHECK *For each number in Exercises 1 and 2, underline the digit to which you are rounding. Then explain the next step in rounding the number.*

1. (a) Round 3702 to the nearest ten.

2. (a) Round 65,081 to the nearest hundred.

(b) Round 908,546 to the nearest thousand.

(b) Round 723,900 to the nearest ten-thousand.

Round each number to the indicated place. **See Examples 1–5.**

3. 625 to the nearest ten

4. 206 to the nearest ten

5. −1083 to the nearest ten
▶

6. −2439 to the nearest ten

7. 7862 to the nearest hundred

8. 6746 to the nearest hundred

9. −86,813 to the nearest hundred

10. −17,211 to the nearest hundred

11. 42,495 to the nearest hundred

12. 18,273 to the nearest hundred

13. −5996 to the nearest hundred
▶

14. −8451 to the nearest hundred

15. −78,499 to the nearest thousand

16. −14,314 to the nearest thousand

17. 5847 to the nearest thousand

18. 49,706 to the nearest thousand

19. 595,008 to the nearest ten-thousand
GS
┌─ Next digit is *5 or more.*
↓
595,008
↑
└─ Change 9 to 10; write 0 and regroup 1 into hundred-thousands place.

20. 725,182 to the nearest ten-thousand
GS
┌─ Next digit is *5 or more.*
↓
725,182
↑
└─ Change 2 to 3; digits to the right become zeros.

21. −8,906,422 to the nearest million

22. −13,713,409 to the nearest million

23. 139,610,000 to the nearest million

24. 609,845,500 to the nearest million

25. 19,951,880,500 to the nearest hundred-million

26. 5,993,505,000 to the nearest hundred-million

27. 8,608,200,000 to the nearest billion

28. 703,750,678,005 to the nearest billion

29. CONCEPT CHECK   The work shown below has a mistake in it.

What Went Wrong?   First write a sentence explaining what the mistake is. Then fix the mistake and find the correct solution.

Mateo rounded 897,005 to the nearest ten-thousand this way:

$$897,005 \approx 800,000$$

30. CONCEPT CHECK   The work shown below has a mistake in it.

What Went Wrong?   First write a sentence explaining what the mistake is. Then fix the mistake and find the correct solution.

Gabriela rounded 99,533,187 to the nearest million this way:

$$99,533,187 \approx 100,533,187$$

Use front end rounding to round each number. ***See Example 6.***

31. Tyrone's truck shows this number on the odometer.

32. Ezra bought a used car with this odometer reading.

33. From summer to winter the average temperature drops 56 degrees.

34. The flood waters fell 42 inches yesterday.

35. Jan earned $9942 working part time.

36. Carol deposited $285 in her bank account.

37. There are 2,485,000,000 global Internet users. (Data from www.wearesocial.org)

38. 40,030,000 U.S. adults have active Twitter accounts. (Data from www.pewinternet.org)

39. The population of Alaska is 736,732 people, and the population of California is 38,802,500 people. (Data from U.S. Census Bureau.)

40. Within forty years, it is estimated that the U.S. population will be 398,328,349 people and Canada's population will be 41,135,648 people. (Data from U.S. Census Bureau.)

Step 1: *First, use front end rounding to estimate each answer.*

Step 2: *Then find the exact answer. In Exercises 49–54, change subtraction to adding the opposite.* **See Example 7.**

41. $-42 + 89$

Estimate: $-40 + 90 = 50$

Exact:

42. $-66 + 25$

Estimate: $-70 + 30 = -40$

Exact:

43. $16 + (-97)$

Estimate:

Exact:

44. $58 + (-19)$

Estimate:

Exact:

45. $-273 + (-399)$

Estimate:

Exact:

46. $-311 + (-582)$

Estimate:

Exact:

47. $3081 + 6826$

Estimate:

Exact:

48. $4904 + 1181$

Estimate:

Exact:

49. $23 - 81$

Estimate:

Exact:

50. $72 - 84$

Estimate:

Exact:

51. $-39 - 39$

Estimate:

Exact:

52. $-91 - 91$

Estimate:

Exact:

53. $-106 + 34 - (-72)$

Estimate:

Exact:

54. $52 - (-87) - 139$

Estimate:

Exact:

Step 1: First, use front end rounding to estimate the answer to each application problem.

Step 2: Then find the exact answer. **See Example 7.**

55. The community has raised $52,882 for the homeless shelter. If the goal is $78,650, how much more needs to be collected?

Estimate:

Exact:

56. A truck weighs 9250 pounds when empty. After being loaded with firewood, it weighs 21,375 pounds. What is the weight of the firewood?

Estimate:

Exact:

57. Dorene Cox is establishing a monthly budget. She will spend $845 for rent, $425 for food, $365 for child care, $182 for transportation, and $240 for other expenses. She will put the remainder in savings. If her monthly take-home pay is $2120, find her monthly savings.

Estimate:

Exact:

58. Jared Ueda had $2874 in his bank account. He used his debit card to pay $308 for auto repairs, $580 for a laptop computer, and $778 for tuition. Find the amount remaining in his account.

Estimate:

Exact:

59. In a laboratory experiment, a mixture started at a temperature of −102 degrees. First the temperature was raised 37 degrees and then raised 52 degrees. What was the final temperature?

Estimate:

Exact:

60. A scuba diver was photographing fish at 65 feet below the surface of the ocean. She swam up 24 feet and then swam down 49 feet. What was her final depth?

Estimate:

Exact:

61. The White House in Washington, D.C., has 132 rooms, 412 doors, and 147 windows. What is the total number of doors and windows? (Data from www.whitehouse.gov)

Estimate:

Exact:

62. There are 36,258 McDonald's restaurants throughout the world in 118 different countries. Burger King has 12,997 restaurants in 87 countries. How many restaurants do the two companies have together? (Data from mcdonalds.com and bk.com)

Estimate:

Exact:

1.6 Multiplying Integers

OBJECTIVES

1. Use a raised dot or parentheses to express multiplication.
2. Multiply integers.
3. Identify properties of multiplication.
4. Estimate answers to application problems involving multiplication.

OBJECTIVE ▸ 1 Use a raised dot or parentheses to express multiplication. In arithmetic we usually use "×" when writing multiplication problems. But in algebra, we use a raised dot or parentheses to show multiplication. The numbers being multiplied are called **factors** and the answer is called the **product**.

Arithmetic   **Algebra**

$3 \times 5 = 15$ $3 \cdot 5 = 15$ or $3(5) = 15$ or $(3)(5) = 15$

Factors Product   Factors Product   Factors Product   Factors Product

---

EXAMPLE 1 **Expressing Multiplication in Algebra**

Rewrite each multiplication in three different ways, using a dot or parentheses. Also identify the factors and the product.

(a) 10×7

Raised dot
↓

Rewrite it as $10 \cdot 7$ or $10(7)$ or $(10)(7)$

The factors are **10** and **7**. The product is **70**.

(b) 4×80

Rewrite it as $4 \cdot 80$ or $4(80)$ or $(4)(80)$

The factors are **4** and **80**. The product is **320**.

◀ **Work Problem ① at the Side.**

VOCABULARY TIPS

Factor In everyday language, a factor influences the result, or outcome of something. In math, factors are the numbers being multiplied in a multiplication problem, **resulting** in the product.

Product Think of a product as the **result** that you get when you multiply numbers together.

① Rewrite each multiplication in three different ways using a dot or parentheses. Also identify the factors and the product.

GS (a) 100×6

$100(6)$

Write it two more ways.

_____ _____

The factors are 100 and ____

The product is ____

(b) 7×12

Note

Parentheses are used to show several different things in algebra. When we discussed the associative property of addition earlier in this chapter, we used parentheses as shown below.

Parentheses show which numbers to add first. → $\underbrace{(-9 + 9)}_{0} + 6$

$\underbrace{ + 6}_{6}$

Now we are using parentheses to indicate multiplication, as in $3(5)$ or $(3)(5)$.

OBJECTIVE ▸ 2 Multiply integers. Suppose that our football team gained 5 yards on the first play, gained 5 yards again on the second play, and gained 5 yards again on the third play. We can add to find the result.

$$5 \text{ yards} + 5 \text{ yards} + 5 \text{ yards} = 15 \text{ yards}$$

A quick way to add the same number several times is to multiply.

Our team made 3 plays	and	gained 5 yards each time.		Our team gained a total of 15 yards.
3	•	5	=	15

Answers

1. **(a)** $100 \cdot 6$ or $(100)(6)$;
   The factors are 100 and 6;
   the product is 600.
   (b) $7 \cdot 12$ or $7(12)$ or $(7)(12)$
   The factors are 7 and 12;
   the product is 84.

When multiplying two integers, first multiply the absolute values. Then attach a positive sign or negative sign to the product according to the rules below.

Multiplying Two Integers

If two factors have *different signs,* the product is *negative.*
For example,

$$-2 \cdot 6 = -12 \quad \text{and} \quad 4 \cdot (-5) = -20$$

If two factors have the *same sign,* the product is *positive.* For example,

$$7 \cdot 3 = 21 \quad \text{and} \quad -3 \cdot (-10) = 30$$

There are several ways to illustrate these rules. First we'll continue with football. Remember, you are interested in the results for *our* team. We will designate our team with a positive sign and **their team** with a **negative sign.**

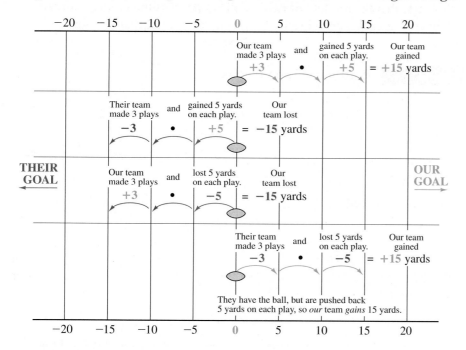

Here is a summary of the football examples.

When two factors have the *same* sign, the product is *positive.*

Both positive

$$3 \cdot 5 = 15$$

Product is positive.

Two factors with *matching* signs give a *positive* product.

$$-3 \cdot (-5) = 15$$

Both negative

When two factors have *different* signs, the product is *negative.*

$$-3 \cdot 5 = -15$$

Product is negative.

Two factors with *different* signs give a *negative* product.

$$3 \cdot (-5) = -15$$

2 Multiply.

(a) $7(-2)$

The factors have **different** signs, so the product is **negative.**

$7(-2) =$ _____

(b) $-5 \cdot (-5)$

(c) $-1(14)$

(d) $10 \cdot 6$

(e) $(-4)(-9)$

There is another way to look at these multiplication rules. In mathematics, the rules or patterns must always be consistent.

Look for a pattern in this list of products.

$$4 \cdot 2 = 8$$
$$3 \cdot 2 = 6$$
Blue numbers decrease by 1 · $\quad 2 \cdot 2 = 4 \quad$ · Red numbers decrease by 2
$$1 \cdot 2 = 2$$
$$0 \cdot 2 = 0$$
$$-1 \cdot 2 = ?$$

To keep the red pattern going, replace the **?** with a number that is 2 *less than* 0, which is -2.

So, $-1 \cdot 2 = -2$. This pattern illustrates that the product of two numbers with *different* signs is *negative*.

Look for a pattern in this list of products.

$$4 \cdot (-2) = -8$$
$$3 \cdot (-2) = -6$$
Blue numbers decrease by 1 · $\quad 2 \cdot (-2) = -4 \quad$ · Red numbers increase by 2
$$1 \cdot (-2) = -2$$
$$0 \cdot (-2) = \quad 0$$
$$-1 \cdot (-2) = \quad ?$$

To keep the red pattern going, replace the **?** with a number that is 2 *more than* 0, which is $+2$.

So, $-1 \cdot (-2) = +2$. This pattern illustrates that the product of two numbers with the *same* sign is *positive*.

EXAMPLE 2 **Multiplying Two Integers**

(a) $-2 \cdot 8 = -16$ The factors have *different signs,* so the product is *negative.*
  Negative · Positive

(b) $-10(-6) = 60$ The factors have the *same sign,* so the product is positive.
  Both negative

(c) $(9)(-11) = -99$ The factors have *different signs,* so the product is *negative.*
  Positive · Negative

◀ **Work Problem 2 at the Side.**

Sometimes there are more than two factors in a multiplication problem. If there are parentheses around two of the factors, multiply them first. If there aren't any parentheses, start at the left and work with two factors at a time.

EXAMPLE 3 **Multiplying Several Factors**

(a) $-3 \cdot (4 \cdot 5)$ Parentheses tell you to multiply $4 \cdot 5$ first. The factors have the *same* sign, so the product is *positive.*
  $-3 \cdot \quad 20$ Then multiply $-3 \cdot 20$. The factors have *different* signs, so the product is *negative.*
  -60

(b) $-2 \cdot (-2) \cdot (-2)$ There is no work to do inside parentheses, so multiply $-2 \cdot (-2)$ first. The factors have the *same* sign, so the product is *positive.*
  $4 \quad \cdot (-2)$ Then multiply $4 \cdot (-2)$. The factors have *different* signs, so the product is *negative.*
  -8

Answers

2. (a) -14 **(b)** 25 **(c)** -14
  (d) 60 **(e)** 36

--- **Continued on Next Page**

❗ CAUTION

In **Example 3(b)** on the previous page, you may be tempted to think that the final product will be *positive* because all the factors have the *same* sign. Be careful to **work with two factors at a time** and keep track of the sign at each step.

——————————— Work Problem ❸ at the Side. ▶

OBJECTIVE ❸ Identify properties of multiplication. Addition involving 0 is unusual because adding 0 does *not* change the number. For example, $7 + 0$ is still 7. (See the section on adding integers.) But what happens in multiplication? Let's use our football team as an example.

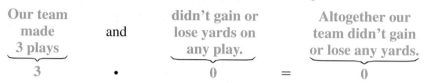

Our team made 3 plays	and	didn't gain or lose yards on any play.	Altogether our team didn't gain or lose any yards.
3	•	0	= 0

This example illustrates one of the properties of multiplication.

Multiplication Property of 0

Multiplying any number by 0 gives a product of 0.
Some examples are shown below.

$$-16 \cdot 0 = 0 \qquad (0)(5) = 0 \qquad 32{,}977(0) = 0$$

So, can you multiply a number by something that will *not* change the number?

$$6 \cdot ? = 6 \qquad -12(?) = -12 \qquad (?)(5876) = 5876$$

The number 1 can replace the ? in each example. This illustrates another property of multiplication.

Multiplication Property of 1

Multiplying a number by 1 leaves the number unchanged.
Some examples are shown below.

$$6 \cdot 1 = 6 \qquad -12(1) = -12 \qquad (1)(5876) = 5876$$

EXAMPLE 4 Using Properties of Multiplication

Multiply. Then name the property illustrated by each example.

(a) $(0)(-48) = 0$ Illustrates the **multiplication property of 0.**

(b) $615(1) = 615$ Illustrates the **multiplication property of 1.**

——————————— Work Problem ❹ at the Side. ▶

---

❸ Multiply.

(a) $-5 \cdot \underbrace{(10 \cdot 2)}$

$$-5 \cdot (\underline{\quad})$$

$$(\underline{\quad})$$

(b) $-1 \cdot 8 \cdot (-5)$

(c) $-3\underbrace{(-2)}(-4)$

$$(\underline{\quad})\,(-4)$$

$$(\underline{\quad})$$

(d) $-2(7)(-3)$

(e) $(-1)(-1)(-1)$

❹ Multiply. Then name the property illustrated by each example.

(a) $819 \cdot 0$

(b) $1(-90)$

(c) $25 \cdot 1$

(d) $(0)(-75)$

Answers

3. (a) $-5 \cdot (20); -100$ **(b)** 40
  (c) $(6)(-4); -24$ **(d)** 42 **(e)** -1

4. (a) 0; multiplication property of 0
  (b) -90; multiplication property of 1
  (c) 25; multiplication property of 1
  (d) 0; multiplication property of 0

5 Show that the product is unchanged and name the property that is illustrated in each case.

GS (a) $(3 \cdot 3) \cdot 2 = 3 \cdot (3 \cdot 2)$

$\underline{\quad} \cdot 2 = 3 \cdot \underline{\quad}$

$\underline{\quad} = \underline{\quad}$

This illustrates the

_____ property

of multiplication.

(b) $11 \cdot 8 = 8 \cdot 11$

(c) $2 \cdot (-15) = -15 \cdot 2$

(d) $-4 \cdot (2 \cdot 5) = (-4 \cdot 2) \cdot 5$

When adding, we said that changing the order of the addends did not change the sum (commutative property of addition). We also found that changing the *grouping* of addends did not change the sum (associative property of addition). These same ideas apply to multiplication.

Commutative Property of Multiplication

Changing the **order** of two factors does not change the product.
For example,

$$2 \cdot 5 = 5 \cdot 2 \qquad \text{and} \qquad -4 \cdot 6 = 6 \cdot (-4)$$

Associative Property of Multiplication

Changing the **grouping** of factors does not change the product.
For example,

$$9 \cdot (3 \cdot 2) = (9 \cdot 3) \cdot 2$$

EXAMPLE 5 Using the Commutative and Associative Properties

Show that the product is unchanged and name the property that is illustrated in each case.

Multiplying two numbers in a different order does not change the product.

(a) $-7 \cdot (-4) = -4 \cdot (-7)$

$28 = 28 \longleftarrow$ Both products are 28

This example illustrates the **commutative property of multiplication.**

Use the associative property to multiply the easiest numbers first.

(b) $5 \cdot (10 \cdot 2) = (5 \cdot 10) \cdot 2$

$5 \cdot 20 = 50 \cdot 2$

$100 = 100 \longleftarrow$ Both products are 100

This example illustrates the **associative property of multiplication.**

◀ **Work Problem 5** at the Side.

Now we look at a property that involves multiplication and addition.

Distributive Property

Multiplication *distributes* over addition. An example is shown below.

$$3(6 + 2) = 3 \cdot 6 + 3 \cdot 2$$

What is the **distributive property** really saying? Notice that there is an understood multiplication symbol between the 3 and the parentheses. To "distribute" the 3 means to multiply 3 times each number inside the parentheses.

This is understood to be multiplying by 3.

$$3(6 + 2)$$

$$3 \cdot (6 + 2)$$

Using the distributive property,

$$3 \cdot (6 + 2) \qquad \text{can be rewritten as} \qquad 3 \cdot 6 + 3 \cdot 2$$

| EXAMPLE 6 | Using the Distributive Property |

Rewrite each product, using the distributive property. Show that the result is unchanged.

(a) $4(3 + 7)$

> *Careful! Multiply both numbers by 4*

$$4(3 + 7) = 4 \cdot 3 + 4 \cdot 7$$
$$4(10) = 12 + 28$$
$$40 = 40 \longleftarrow \text{Both results are 40}$$

(b) $-2(-5 + 1)$

> *Multiply both numbers by −2*

$$-2(-5 + 1) = (-2) \cdot (-5) + (-2) \cdot (1)$$
$$-2(-4) = (10) + (-2)$$
$$8 = 8 \longleftarrow \text{Both results are 8}$$

Work Problem **6** *at the Side.* ▶

OBJECTIVE ▶ 4 Estimate answers to application problems involving multiplication. Front end rounding can be used to estimate answers in multiplication, just as we did when adding and subtracting (see the previous section). Once the numbers have been rounded so that there are lots of zeros, we can use a multiplication shortcut. Look at the pattern in these examples.

$-3 \cdot 2$ is -6

$$-30 \cdot 200 = -6000$$

Total of three zeros · Write three zeros after the −6

$-2 \cdot (-5)$ is 10

$$-2000 \cdot (-5000) = 10{,}000{,}000$$

Total of six zeros · Write six zeros after the 10

| EXAMPLE 7 | Using Front End Rounding to Estimate an Answer |

Use front end rounding to estimate an answer. Then find the exact answer.

Last year Video Land had to replace 392 defective DVDs at a cost of $19 each. How much money was lost on defective DVDs? (*Hint:* It's a loss, so use a negative number for the cost.)

Estimate: Use front end rounding: 392 rounds to 400 and −$19 rounds to −$20. Use the rounded numbers and multiply to estimate the total amount of money lost.

$4 \cdot (-2)$ is -8

$$400 \cdot (-\$20) = -\$8000 \quad \text{Estimate}$$

Total of three zeros · Write three zeros after the −8

Exact: $392 \cdot (-\$19) = -\$7448$

Because the exact answer of −$7448 is fairly close to the estimate of −$8000, you can see that −$7448 probably is correct. Video Land could also use the estimate to report, "We lost about $8000 on defective DVDs last year."

Work Problem **7** *at the Side.* ▶

---

6 Rewrite each product, using the distributive property. Show that the result is unchanged.

(a) $3\underbrace{(8 + 7)} = \underbrace{3 \cdot 8} + \underline{\quad} \cdot 7$

$3(\underline{\quad}) = \underline{\quad} + \underline{\quad}$

$\underline{\quad} = \underline{\quad}$

(b) $10(-6 + 9)$

(c) $-6(4 + 4)$

7 Use front end rounding to estimate an answer. Then find the exact answer.

An average of 27,095 baseball fans attended each of the 81 home games during the season. What was the total home game attendance for the season?

Estimate:

Exact:

Answers

6. **(a)** $3(8 + 7) = 3 \cdot 8 + 3 \cdot 7$;
   $3(15) = 24 + 21; 45 = 45$;
   both results are 45.
   (b) $10(-6) + 10(9)$; both results are 30.
   (c) $-6 \cdot 4 + (-6) \cdot 4$; both results are −48.

7. *Estimate:* $30{,}000 \cdot 80 = 2{,}400{,}000$ fans
   Exact: $27{,}095 \cdot 81 = 2{,}194{,}695$ fans

1.6 Exercises

FOR EXTRA HELP

Go to MyMathLab for worked-out, step-by-step solutions to exercises enclosed in a square ▢ and video solutions to ▶ exercises.

CONCEPT CHECK *Watch for patterns as you work Exercises 1–4.*

1. (a) $9 \cdot 7$
  (b) $-9 \cdot (-7)$
  (c) $-9 \cdot 7$
  (d) $9 \cdot (-7)$

2. (a) $-6 \cdot 9$
  (b) $6 \cdot (-9)$
  (c) $-6 \cdot (-9)$
  (d) $6 \cdot 9$

3. (a) $7(-8)$
  (b) $-7(8)$
  (c) $7(8)$
  (d) $-7(-8)$

4. (a) $8(6)$
  (b) $-8(-6)$
  (c) $-8(6)$
  (d) $8(-6)$

Multiply. ***See Examples 1–4.***

5. $-5 \cdot 7$ ▶

6. $-10 \cdot 2$

7. $(-5)(9)$

8. $(-9)(4)$

9. $3(-6)$

10. $8(-9)$

11. $10(-5)$ ▶

12. $5(-11)$

13. $(-1)(40)$

14. $(75)(-1)$

15. $-56 \cdot 1$

16. $1 \cdot (-87)$

17. $-8(-4)$ ▶

18. $-3(-9)$

19. $11 \cdot 7$

20. $4 \cdot 25$

21. $25 \cdot 0$

22. $0 \cdot 30$

23. $-19(-7)$

24. $-21(-3)$

25. $-13(-1)$

26. $-1(-31)$

27. $(0)(-25)$

28. $(-50)(0)$

29. $\underbrace{-4 \cdot (-6)}_{24} \cdot 2$ GS ▶
  ___

30. $\underbrace{-9 \cdot 3}_{-27} \cdot (-3)$ GS
  ___

31. $(-4)(-2)(-7)$ ▶

32. $(-6)(-2)(-3)$

33. $5(-8)(4)$

34. $5(4)(-6)$

Write an integer in each blank to make a true statement.

35. $(-3)(\underline{}) = -15$

36. $6 \cdot (\underline{}) = -24$

37. $\underline{} \cdot 10 = -30$

38. $(\underline{})(-4) = 16$

39. $-17 = 17(\underline{})$

40. $29 = -29(\underline{})$

41. $(\underline{})(-350) = 0$ ▶

42. $\underline{} \cdot 99 = 99$

43. $5 \cdot (-4) \cdot \underline{} = -100$

44. $\underline{} \cdot 2 \cdot (-2) = -24$

45. $-40 = (\underline{})(-5)(-2)$

46. $-27 = (-3)(\underline{})(-3)$

47. In your own words, explain the difference between the commutative and associative properties of multiplication. Show an example of each.

48. CONCEPT CHECK The work shown below has a mistake in it.

What Went Wrong? First write a sentence explaining what the mistake is. Then fix the mistake and find the correct solution.

Angelo did this multiplication.

$$(-3) \cdot (-3) \cdot (-3) = 27$$

Angelo knew that $3 \cdot 3 \cdot 3$ is 27, and because all the factors have the same sign he made the product positive.

Rewrite each multiplication, using the stated property. Then simplify each expression to show that the result is unchanged. **See Examples 5 and 6.**

49. Distributive property
▶ $9(-3 + 5)$

50. Distributive property
$-6(4 + 5)$

51. Commutative property
▶ $25 \cdot 8$

52. Commutative property
$-7 \cdot (-11)$

53. Associative property
▶ $-3 \cdot (2 \cdot 5)$

54. Associative property
$(5 \cdot 5) \cdot 10$

First use front end rounding to estimate the answer to each application problem. Then find the exact answer. **See Example 7.**

55. Alliette receives \$324 per week for doing child care in her home. How much income will she have for an entire year? There are 52 weeks in a year.

Estimate:

Exact:

56. Enrollment at our community college has increased by 875 students each of the last four semesters. What is the total increase?

Estimate:

Exact:

57. A new computer software store had losses of \$9950 during each month of its first year. What was the total loss for the year?

Estimate:

Exact:

58. A cable company estimates that it is losing 9500 customers each week to other providers and to services like Hulu and Netflix. What is the change in the number of customers during one year?

Estimate:

Exact:

59. Tuition at the state university is $182 per credit for undergraduates. How much tuition will Carissa pay for 13 credits?

Estimate:

Exact:

60. Pat ate a dozen crackers as a snack. Each cracker had 17 calories. How many calories did Pat eat?

Estimate:

Exact:

61. There are 24 hours in one day. How many hours are in one year (365 days)?

Estimate:

Exact:

62. There are 5280 feet in one mile. How many feet are in 17 miles?

Estimate:

Exact:

Simplify.

63. $-8 \cdot |-8 \cdot 8|$

$-8 \cdot \ |-64|$

$-8 \cdot \ \ \ 64$

_____

64. $-7 \cdot |7| \cdot |-7|$

$-7 \cdot 7 \cdot \ \ 7$

$-49 \ \ \cdot \ \ 7$

_____

65. $(-37)\,(-1)\,(85)\,(0)$

66. $-1\,(9732)\,(-1)\,(-1)$

67. $|6-7| \cdot (-355{,}299)$

68. $987 \cdot (-65{,}432) \cdot |9-9|$

Each of these application problems requires several steps and may involve addition and subtraction as well as multiplication.

69. Each of Maurice's four cats needed a $24 rabies shot and a $29 shot to prevent respiratory infections. There was also one $35 office visit charge. What was the total amount of Maurice's bill?

70. Chantele has three children. Her older daughter had a throat culture taken at the clinic today. Her baby needed two prescription medicines and her son received two immunization shots. The co-pay amounts were a $22 office charge for each child, $20 for the throat culture, and $12 for each prescription medicine. There was no co-pay for the shots. How much did Chantele pay?

71. There is a 3-degree drop in temperature for every thousand feet that an airplane climbs into the sky. If the temperature on the ground is 50 degrees, what will the temperature be when the plane reaches an altitude of 24,000 feet? (Data from Lands' End.)

72. An unmanned research submarine descends to 175 feet below the surface of the ocean and takes a water sample. Then it continues to go deeper, taking a water sample every 25 feet. What is its depth when it takes the 15th sample?

73. In Ms. Zubero's algebra class, there are six tests of 100 points each, eight quizzes of six points each, and 20 homework assignments of five points each. There are also four "bonus points" on each test. What is the total number of possible points?

74. In Mr. Jackson's prealgebra class, there are three group projects worth 25 points each, five 100-point tests, a 150-point final exam, and seven quizzes worth 12 points each. Find the total number of possible points.

Relating Concepts (Exercises 75–76) For Individual or Group Work

*Look for patterns as you **work Exercises 75 and 76 in order.***

75. Write three numerical examples for each of these situations:

(a) A positive number multiplied by -1

(b) A negative number multiplied by -1

Now write a rule that explains what happens when you multiply a signed number by -1.

76. Do these multiplications.

$$-2 \cdot (-2) = \underline{\hspace{1cm}}$$
$$-2 \cdot (-2) \cdot (-2) = \underline{\hspace{1cm}}$$
$$-2 \cdot (-2) \cdot (-2) \cdot (-2) = \underline{\hspace{1cm}}$$
$$-2 \cdot (-2) \cdot (-2) \cdot (-2) \cdot (-2) = \underline{\hspace{1cm}}$$

Describe the pattern in the products. Then find the next three products without multiplying all the -2s.

1.7 Dividing Integers

OBJECTIVE ▶ 1 Divide integers. In arithmetic, we usually use $\overline{)}$ to write division problems so that we can do them by hand. Calculator keys use the $\div$ symbol for division. In algebra, we usually show division by using a fraction bar, a slash mark, or the $\div$ symbol. The answer to a division problem is called the **quotient.**

Arithmetic   **Calculator and Algebra**

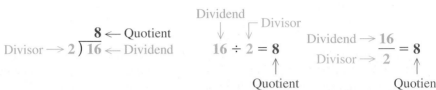

For every division problem, we can write a related multiplication problem. Because of this relationship, the sign rules for dividing integers are the same as the rules for multiplying integers. Examples:

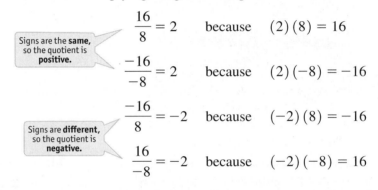

Signs are the **same,** so the quotient is **positive.**

$$\frac{16}{8} = 2 \quad \text{because} \quad (2)(8) = 16$$

$$\frac{-16}{-8} = 2 \quad \text{because} \quad (2)(-8) = -16$$

$$\frac{-16}{8} = -2 \quad \text{because} \quad (-2)(8) = -16$$

Signs are **different,** so the quotient is **negative.**

$$\frac{16}{-8} = -2 \quad \text{because} \quad (-2)(-8) = 16$$

---

1 Divide.

(a) $\dfrac{40}{-8}$ The integers have **different signs,** so the quotient is **negative.**

(b) $\dfrac{49}{7}$

(c) $\dfrac{-32}{4}$

(d) $\dfrac{-10}{-10}$ The integers have the **same sign,** so the quotient is **positive.**

(e) $-81 \div 9$

(f) $-100 \div (-50)$

Dividing Two Integers

If two numbers have **different signs,** the quotient is **negative.** Some examples are shown below.

$$\frac{-18}{3} = -6 \qquad \frac{40}{-5} = -8$$

If two numbers have the **same sign,** the quotient is **positive.** Some examples are shown below.

$$\frac{-30}{-6} = 5 \qquad \frac{48}{8} = 6$$

EXAMPLE 1 Dividing Two Integers

(a) $\dfrac{-20}{5}$ Numbers have *different* signs, so the quotient is *negative.* $\dfrac{-20}{5} = -4$

The sign rules for dividing and multiplying are the same.

(b) $\dfrac{-24}{-4}$ Numbers have the *same* sign, so the quotient is *positive.* $\dfrac{-24}{-4} = 6$

(c) $60 \div (-2)$ Numbers have *different* signs, so the quotient is *negative.* $60 \div (-2) = -30$

◀ **Work Problem** 1 **at the Side.**

Answers

1. (a) −5 (b) 7 (c) −8 (d) 1
   (e) −9 (f) 2

OBJECTIVE ▶ 2 Identify properties of division. You have seen that 0 and 1 are used in special ways in addition and multiplication. This is also true in division.

Examples	Pattern (Division Property)
$\dfrac{5}{5} = 1 \qquad \dfrac{-18}{-18} = 1 \qquad \dfrac{-793}{-793} = 1$	When a nonzero number is divided by itself, the quotient is 1.
$\dfrac{5}{1} = 5 \qquad \dfrac{-18}{1} = -18 \qquad \dfrac{-793}{1} = -793$	When a number is divided by 1, the quotient is the number.
$\dfrac{0}{5} = 0 \qquad \dfrac{0}{-18} = 0 \qquad \dfrac{0}{-793} = 0$	When 0 is divided by any other number (except 0), the quotient is 0.
$\dfrac{5}{0}$ is undefined. $\qquad \dfrac{-18}{0}$ is undefined.	Division by 0 is *undefined*. There is no answer.

The most surprising property is that division by 0 *cannot be done*. Let's review the reason for that by rewriting this division problem as a related multiplication problem.

$$\frac{-18}{0} = ? \quad \text{can be written as the multiplication} \quad ? \cdot 0 = -18$$

If you thought the answer to $\frac{-18}{0}$ should be 0, try replacing ? with 0. It doesn't work in the related multiplication problem! Try replacing ? with any number you like. The result in the related multiplication problem is always 0 instead of -18. That is how we know that *dividing by 0 cannot be done*. Mathematicians say that it is **undefined** and have agreed never to divide by 0.

EXAMPLE 2   Using the Properties of Division

Divide. Then state the property illustrated by each example.

(a) $\dfrac{-312}{-312} = \mathbf{1}$ Any nonzero number divided by itself is 1

(b) $\dfrac{75}{1} = \mathbf{75}$ Any number divided by 1 is the number.

(c) $\dfrac{0}{19} = \mathbf{0}$ Zero divided by any nonzero number is 0

You *cannot* divide by 0. Write "undefined."

(d) $\dfrac{48}{0}$ is **undefined.**   Division by 0 is *undefined*.

🖩 **Calculator Tip**

Try **Examples 2(c)** and **2(d)** above on your calculator.

0 ÷ 19 =   Answer is 0

48 ÷ 0 =   Calculator shows "Error" or "ERR" or "E" for error because it cannot divide by 0

—————— **Work Problem ② at the Side.** ▶

2 Divide. Then state the property illustrated by each division.

GS (a) $\dfrac{-12}{0}$ is _____ .

  Hint: Division by 0 is _____ .

(b) $\dfrac{0}{39}$

(c) $\dfrac{-9}{1}$

(d) $\dfrac{21}{21}$

Answers

2. **(a)** undefined; Division by 0 is undefined.
   (b) 0; 0 divided by any nonzero number is 0.
   (c) -9; any number divided by 1 is the number.
   (d) 1; any nonzero number divided by itself is 1.

③ Simplify.

(a) $\underbrace{60 \div (-3)}(-5)$

$\underbrace{(\underline{\quad}) (-5)}$

$\underline{\quad}$

(b) $-6(-16 \div 8) \cdot 2$

(c) $-8(10) \div 4(-3) \div (-6)$

(d) $56 \div (-8) \div (-1)$

OBJECTIVE **③** **Combine multiplying and dividing of integers.** When a problem involves both multiplying and dividing, first check to see if there are any parentheses. Do what is inside parentheses first. Then start at the left and work toward the right, using two numbers at a time.

EXAMPLE 3 **Combining Multiplication and Division of Integers**

Simplify.

(a) $6(-10) \div \underbrace{(-3 \cdot 2)}$ Do operations inside parentheses first: $-3 \cdot 2$ is -6. The signs are *different*, so the product is *negative*.

$\underbrace{6(-10)} \div (-6)$ Start at the left: $6(-10)$ is -60. The signs are *different*, so the product is *negative*.

$\underbrace{-60 \div (-6)}$ Finally, $-60 \div (-6)$ is 10. The signs are the *same*, so the quotient is *positive*.

$\mathbf{10}$

(b) $\underbrace{-24 \div (-2)}(4) \div (-6)$ No operations inside parentheses, so start at the left: $-24 \div (-2)$ is 12 (*same* sign, *positive* quotient).

$\underbrace{12(4)} \div (-6)$ Next, $12(4)$ is 48 (*same* sign, *positive* product).

$\underbrace{48 \div (-6)}$ Finally, $48 \div (-6)$ is -8 (*different* signs, *negative* quotient).

$\mathbf{-8}$

(c) $\underbrace{-50 \div (-5)} \div (-2)$ No operations inside parentheses, so start at the left: $-50 \div (-5)$ is 10 (*same* sign, *positive* quotient).

$\underbrace{10 \div (-2)}$ Now, $10 \div (-2)$ is -5 (*different* signs, *negative* quotient).

$\mathbf{-5}$

◀ **Work Problem ③ at the Side.**

OBJECTIVE **④** **Estimate answers to application problems involving division.** Front end rounding can be used to estimate answers in division just as you did when multiplying in the previous section. Once the numbers have been rounded so that there are lots of zeros, you can use a division shortcut. Look at the pattern in these examples.

$$4000 \div (-50) = 400 \div (-5) = -80$$

Drop one 0 from both
dividend and divisor.

$$-6000 \div (-3000) = -6 \div (-3) = 2$$

Drop three zeros from both
dividend and divisor.

Answers

3. **(a)** $(-20)(-5)$; 100   **(b)** 24   **(c)** -10
   (d) 7

EXAMPLE 4 Estimating an Answer in Division

First use front end rounding to estimate an answer. Then find the exact answer.

During a 24-hour laboratory experiment, the temperature of a solution dropped 96 degrees. What was the average drop in temperature each hour?

Estimate: Use front end rounding: -96 degrees rounds to -100 degrees and 24 hours rounds to 20 hours. To estimate the average, divide the rounded number of degrees by the rounded number of hours.

$$-100 \text{ degrees} \div 20 \text{ hours} = \textbf{ }-\textbf{5 degrees each hour} \leftarrow \text{Estimate}$$

Exact: $\qquad -96 \text{ degrees} \div 24 \text{ hours} = \textbf{ }-\textbf{4 degrees each hour} \leftarrow \text{Exact}$

Because the exact answer of -4 degrees is close to the estimate of -5 degrees, you can see that -4 degrees probably is correct.

> ▦ **Calculator Tip**
>
> The answer in **Example 4** above "came out even." In other words, the quotient was an integer. Suppose that the drop in temperature had been 97 degrees. Do the division on your calculator.
>
> $\boxed{(-)}$ 97 $\boxed{\div}$ 24 $\boxed{=}$ Calculator shows -4.041666667
>
> -97
>
> The quotient is *not* an integer. We will work with positive and negative decimals later.

——— Work Problem ④ at the Side. ▶

OBJECTIVE ➎ Interpret remainders in division application problems.

In arithmetic, division problems often have a remainder, as shown below.

$$\begin{array}{r} 14 \ \ \textbf{R10} \\ 25\overline{)360} \\ 25 \\ \hline 110 \\ 100 \\ \hline 10 \end{array}$$ ← Remainder

But what does **R10** really mean? Let's look at this same problem by using money amounts.

EXAMPLE 5 Interpreting Remainders in Division Applications

Divide; then interpret the remainder in each application.

(a) The math department at Lake Community College has \$360 to buy scientific calculators for the math lab. If the calculators cost \$25 each, how many can be purchased? How much money will be left over?

We can solve this problem by using the same division as shown above. But this time we can decide what the remainder really means.

$$\begin{array}{r} \textbf{14} \leftarrow \text{Number of calculators purchased} \\ \text{Cost of one calculator} \rightarrow \$25\overline{)\$360} \leftarrow \text{Budget} \\ 25 \\ \hline 110 \\ 100 \\ \hline \$10 \leftarrow \text{Money left over} \end{array}$$

The remainder is the money that is left over.

The department can buy **14 calculators**. There will be **\$10 left over**.

——— Continued on Next Page

④
GS First use front end rounding to estimate an answer. Then find the exact answer.

Laurie and Chuck Struthers lost \$2724 on their stock investments last year. What was their average loss each month?

Estimate:

$$-\$3000 \div \underline{\qquad} = \underline{\qquad}$$

Exact:

Answers

4. *Estimate:* $-\$3000 \div 10 = -\$300$ each month
Exact: $-\$2724 \div 12 = -\$227$ each month

5 Divide; then interpret the remainder in each of these applications.

(a) Chad and Martha are baking cookies for a fund-raiser. They baked 116 cookies and are putting them into packages of a dozen each. How many packages will they have for the fund-raiser? How many cookies will be left over for them to eat?

$$\left.\begin{array}{c}\text{Cookies}\\\text{in each}\\\text{package}\end{array}\right\}\overline{)116}\leftarrow\left\{\begin{array}{c}\text{Packages}\\\\\text{Total}\\\text{number}\\\text{of cookies}\end{array}\right.$$

(b) Coreen is a dispatcher for a bus company. A group of 249 senior citizens is going to a baseball game. If the buses each hold 44 people, how many buses should she send to pick up the seniors?

▦ Calculator Tip

You can use your calculator to solve **Example 5(a)** on the previous page. Recall that digits on the *right* side of the decimal point show *part* of one whole. You cannot order *part* of one calculator, so ignore those digits and use only the *whole number part* of the quotient.

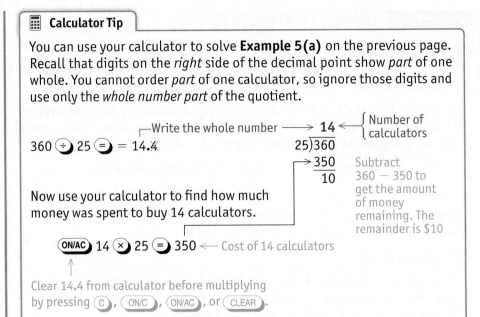

Now use your calculator to find how much money was spent to buy 14 calculators.

ON/AC 14 ⊗ 25 ⊜ 350 ← Cost of 14 calculators

Clear 14.4 from calculator before multiplying by pressing Ⓒ, ON/C, ON/AC, or CLEAR.

(b) Luke's son is going on a Scout camping trip. There are 135 Scouts. Luke is renting tents that sleep 6 people each. How many tents should he rent?

We again use division to solve the problem. There is a remainder, but this time it must be interpreted differently than in the calculator example.

$$\begin{array}{r}22\\6\overline{)135}\\\underline{12}\\15\\\underline{12}\\3\end{array}$$

22 ← Number of tents with 6 Scouts each
Each tent holds → 6)135 ← Total number of Scouts
3 ← Scouts left over

22 tents is not enough for all the Scouts.

If Luke rents 22 tents, 3 Scouts will have to sleep out in the rain. He must rent **23 tents** to accommodate all the Scouts. (One tent will have only 3 Scouts in it.)

◀ **Work Problem 5 at the Side.**

Answers

5. **(a)**
$$\begin{array}{r}9\\12\overline{)116}\\\underline{-108}\\8\end{array}$$
9 ← 9 packages of cookies
8 ← 8 cookies left over to eat

(b) 6 buses, because 5 buses would leave 29 seniors standing on the curb

1.7 Exercises

FOR EXTRA HELP Go to MyMathLab for worked-out, step-by-step solutions to exercises enclosed in a square ▢ and video solutions to ▶ exercises.

CONCEPT CHECK *Watch for patterns as you work Exercises 1–4.*

1. (a) $14 \div 2$

   (b) $-14 \div (-2)$

   (c) $14 \div (-2)$

   (d) $-14 \div 2$

2. (a) $-18 \div (-3)$

   (b) $18 \div 3$

   (c) $-18 \div 3$

   (d) $18 \div (-3)$

3. (a) $-42 \div 6$

   (b) $-42 \div (-6)$

   (c) $42 \div (-6)$

   (d) $42 \div 6$

4. (a) $45 \div 5$

   (b) $45 \div (-5)$

   (c) $-45 \div (-5)$

   (d) $-45 \div 5$

Divide. See Examples 1 and 2.

5. (a) $\dfrac{35}{35}$

   (b) $\dfrac{35}{1}$

   (c) $\dfrac{-13}{1}$

   (d) $\dfrac{-13}{-13}$

6. (a) $\dfrac{-23}{1}$

   (b) $\dfrac{-23}{-23}$

   (c) $\dfrac{17}{1}$

   (d) $\dfrac{17}{17}$

7. (a) $\dfrac{0}{50}$

   (b) $\dfrac{50}{0}$

   (c) $\dfrac{-11}{0}$

   (d) $\dfrac{0}{-11}$

8. (a) $\dfrac{-85}{0}$

   (b) $\dfrac{0}{-85}$

   (c) $\dfrac{6}{0}$

   (d) $\dfrac{0}{6}$

9. $\dfrac{-8}{2}$

10. $\dfrac{-14}{7}$

11. $\dfrac{21}{-7}$

12. $\dfrac{30}{-6}$

13. $\dfrac{-54}{-9}$

14. $\dfrac{-48}{-6}$

15. $\dfrac{55}{-5}$

16. $\dfrac{70}{-7}$

17. $\dfrac{-28}{0}$

18. $\dfrac{-40}{0}$

19. $\dfrac{14}{-1}$

20. $\dfrac{25}{-1}$

21. $\dfrac{-20}{-2}$

22. $\dfrac{-80}{-4}$

23. $\dfrac{-48}{-12}$

24. $\dfrac{-30}{-15}$

25. $\dfrac{-18}{18}$

26. $\dfrac{50}{-50}$

27. $\dfrac{0}{-9}$

28. $\dfrac{0}{-4}$

29. $\dfrac{-573}{-3}$

30. $\dfrac{-580}{-5}$

31. $\dfrac{163,672}{-328}$

32. $\dfrac{-69,496}{1022}$

Simplify. ***See Example 3.***

33. $-60 \div 10 \div (-3)$
$\underbrace{\qquad\qquad}$
$\underbrace{-6 \div (-3)}$
_____

34. $36 \div (-4) \div 3$
$\underbrace{\qquad\quad}$
$\underbrace{-9 \div 3}$
_____

35. $-64 \div (-8) \div (-2)$

36. $-72 \div (-9) \div (-4)$

37. $100 \div (-5)(-2)$

38. $-80 \div 4(-5)$

39. $48 \div 3 \cdot (-12 \div 4)$

40. $-2 \cdot (-3) \cdot (-7) \div 7$

41. $-5 \div (-5)(-10) \div (-2)$

42. $-9(4) \div (-36)(50)$

43. $64 \cdot 0 \div (-8)(10)$

44. $-88 \div (-8) \div (-11)(0)$

CONCEPT CHECK *Use your knowledge of the properties of multiplication and division as you work Exercises 45–48.*

45. Explain whether or not division is commutative like multiplication. Start by doing these two divisions on your calculator: $2 \div 1$ and $1 \div 2$.

46. Explain whether or not division is associative like multiplication. Start by doing these two divisions: $(12 \div 6) \div 2$ and $12 \div (6 \div 2)$.

47. Explain what is different and what is similar about multiplying and dividing two signed numbers.

48. In your own words, describe at least three division properties. Include examples to illustrate each property.

49. CONCEPT CHECK The work shown below has a mistake in it.

What Went Wrong? First write a sentence explaining what the mistake is. Then fix the mistake and find the correct solution.

Thong said $\frac{0}{12}$ was undefined.

50. CONCEPT CHECK The work shown below has a mistake in it.

What Went Wrong? First write a sentence explaining what the mistake is. Then fix the mistake and find the correct solution.

Lien said $(-25) \div 25 = 0$

Solve these application problems by using addition, subtraction, multiplication, or division. First use front end rounding to estimate the answer. Then find the exact answer. **See Example 4.**

51. The greatest ocean depth is 35,836 feet below sea level. If an unmanned research sub dives to that depth in 17 equal steps, how far does it dive in each step? (Data from *The Top 10 of Everything*.)

Estimate: −40,000 ÷ ____ = _____

Exact:

52. Our college enrollment dropped by 3245 students over the last 11 years. What was the average drop in enrollment each year?

Estimate: −3000 ÷ ____ = _____

Exact:

53. When Ashwini discovered that her checking account was overdrawn by $238, she quickly transferred $450 from her savings to her checking account. What is the new balance in her checking account?

Estimate:

Exact:

54. Sarah lost 48 points during the first round of a card game. During the second round she won 191 points. How many points does she have now?

Estimate:

Exact:

55. The foggiest place in the United States is Cape Disappointment, Washington. It is foggy there an average of 165 days each year. How many days is it not foggy each year? (Data from National Weather Service.)

Estimate:

Exact:

56. The number of cell phone users in the United States in 1994 was 24 million. The number of users reached 328 million in 2014. What was the increase in the number of users during this 20-year period? (Data from CTIA - The Wireless Association.)

Estimate:

Exact:

57. A plane descended an average of 730 feet each minute during a 37-minute landing. How far did the plane descend during the landing?

Estimate:

Exact:

58. A discount store found that 174 items were lost to shoplifting last month. The average value of each item was $24. What was the total loss due to shoplifting?

Estimate:

Exact:

59. Mr. and Mrs. Martinez drove on the Interstate for five hours and traveled 315 miles. What was the average number of miles they drove each hour?

Estimate:

Exact:

60. Rochelle has a 48-month car loan for $15,072. How much is her monthly payment?

Estimate:

Exact:

Find the exact answer in Exercises 61–66. Solving these problems requires more than one step.

61. Clarence bowled four games and had scores of 143, 190, 162, and 177. What was his average score? (*Hint:* To find the average, add all the scores and divide by the number of scores.)

62. Sheila kept track of her grocery expenses for six weeks. The amounts she spent were $184, $111, $136, $110, $98, and $153. What was the average weekly cost of her groceries?

63. On the back of an oatmeal box, it says that one serving weighs 40 grams and that there are 13 servings in the box. On the front of the box, it says that the weight of the contents is 510 grams. What is the difference in the total weight on the front and the back of the box? (Data from Quaker Oats.)

64. Emmanuel's doctor recommends that he eat a 2000-calorie-per-day diet with no more than 62 grams of fat. If each gram of fat is 9 calories, how many calories can he consume in other types of food?

65. Stephanie had $302 in her bank account. She made payments of $116 for day care and $718 for rent. She also deposited her $517 paycheck. What is the balance in her account?

66. Gary started a new checking account with a $500 deposit. The bank charged him $18 to print his checks. He also wrote a $193 check for car repairs and a $289 check to his credit card company. What is the balance in his account?

Divide; then interpret the remainder in each application. ***See Example 5.***

67. Stephanie needs to do 1000 minutes of community service at a soup kitchen. How many hours of community service does she need to do?

68. Nikki is catering a large party. If one pie will serve eight guests, how many pies should she make for 100 guests?

69. Hurricane victims are being given temporary shelter in a hotel. Each room can hold five people. How many rooms are needed for 163 people?

70. A college received a $250,000 donation for scholarships. How many $3500 scholarships can be given to students?

Simplify:

71. $|-8| \div (-4) \cdot |-5| \cdot |1|$

72. $-6 \cdot |-3| \div |9| \cdot (-2)$

73. $-6(-8) \div (-5 + 5)$

74. $-9 \div (-9)(-9 \div 9) \div (12 - 13)$

75. Can you guess how many days it would take to receive a million dollars if you got $1 each second? Here's how to use your calculator to get the answer. If you got $1 per second, it would take 1,000,000 seconds to receive $1,000,000. Press these keys.

1000000	÷ 60	= 16666.66667	÷ 60	= 277.7777778	÷ 24	= 11.57407407
	There are 60 seconds in one minute.	About 16,667 minutes	There are 60 minutes in one hour.	About 278 hours	There are 24 hours in one day.	About $11\frac{1}{2}$ days (11.5 is equivalent to $11\frac{1}{2}$)

Notice that you do *not* have to re-enter the intermediate answers. When the answer 16666.66667 appears on your calculator display, just go ahead and enter ÷ 60.

Now use a *scientific* calculator to find how long it takes to receive a *billion* dollars. Start by entering 1000000000. Then follow the pattern shown above. You will need to do one more division step to get the number of years. (Assume that there are 365 days in one year.)

Summary Exercises *Operations with Integers*

Simplify each expression.

1. $2 - 8$

2. $(-16)(0)$

3. $-14 - (-7)$

4. $\dfrac{-42}{6}$

5. $-9(-7)$

6. $\dfrac{-12}{12}$

7. $(1)(-56)$

8. $1 + (-23)$

9. $5 - (-7)$

10. $-88 \div (-11)$

11. $-18 + 5$

12. $\dfrac{0}{-10}$

13. $-40 - (-40)$

14. $-17 + 0$

15. $8(-6)$

16. $-1 - 9$

17. $-5(10)$

18. $\dfrac{30}{0}$

19. $0 - 14$

20. $\dfrac{18}{-3}$

21. $-13 + 13$

22. $\dfrac{-16}{-1}$

23. $20 - 50$

24. $\dfrac{-7}{0}$

25. $(-4)(-6)(2)$

26. $-2 + (-12) + (-5)$

27. $-60 \div 10 \div (-3)$

28. $-8 - 4 - 8$

29. $64(0) \div (-8)$

30. $2 - (-5) + 9$

31. $-9 + 8 + (-2)$

32. $(-6)(-2)(-3)$

33. $8 + 6 + (-8)$

34. $9 - 0 - 16$

35. $-25 \div (-1) \div (-5)$

36. $1 - 32 + 0$

37. $-72 \div (-9) \div (-4)$

38. $-7 + 28 + (-56) + 3$

39. $9 - 6 - 3 - 5$

40. $-6(-8) \div (-5 - 7)$

41. $-1(9732)(-1)(-1)$

42. $-80 \div 4(-5)$

43. $-10 - 4 + 0 + 18$

44. $-7 \cdot |7| \cdot |-7|$

45. $5 - |-3| + 3$

46. $-2(-3)(7) \div (-7)$

47. $-3 - (-2 + 4) - 5$

48. $0 - |-7 + 2|$

49. **CONCEPT CHECK** Describe what happens in each situation.

(a) Zero is divided by a nonzero number.

(b) Any number is multiplied by 0.

(c) A nonzero number is divided by itself.

50. **CONCEPT CHECK** The work shown below has **two** mistakes in it. *What Went Wrong?* First write a sentence explaining what each mistake is. Then fix the mistakes and find the correct solution.
Asho simplified two expressions.

She simplified the first expression this way: $\dfrac{8}{0} = 8$

She simplified the second expression this way:
$-10 \div (-2) \div (-5) = 1$

Study the set of sample math notes below from the next section on exponents, and read the comments about them. Then try to incorporate the techniques into your own math note taking during lectures.

▶ The **date and title** of the day's lecture topic are always at the top of every page. **Always begin a new day with a new page.**

▶ Note the **definitions** of base and exponent are written in parentheses—don't trust your memory!

▶ **Skipping lines** makes the notes easier to read.

▶ See how the **direction word** (*simplify*) is emphasized and explained.

▶ A **star marks an important concept.** This warns you to avoid future mistakes. **Note the underlining,** too, which highlights the importance.

▶ Notice the *Example* and *Explanation* columns, which allow for the example and its explanation to be close together. **Whenever you know you'll be given a series of steps to follow, try the two-column method.**

▶ Note the **braces and arrows,** which clearly show how the problem is set up to be simplified.

January 2   *Exponents*

Exponents used to show repeated multiplication.

$3 \cdot 3 \cdot 3 \cdot 3$ can be written 3^4 — exponent (how many times it's multiplied)
base (the number being multiplied)

Read 3^2 as 3 to the 2nd power or 3 squared

3^3 as 3 to the 3rd power or 3 cubed

3^4 as 3 to the 4th power

etc.

Simplifying an expression with exponents
→ actually do the repeated multiplication

2^3 means $2 \cdot 2 \cdot 2$ and $2 \cdot 2 \cdot 2 = 8$

★ Careful! [5^2 means $5 \cdot 5$ $\underline{NOT}$ $5 \cdot 2$
so $5^2 = 5 \cdot 5 = 25$ BUT $5^2 \neq 10$]

Example	Explanation
simplify $(2^4) \cdot (3^2)$	Exponents means $\underline{multiplication}$.
$2 \cdot 2 \cdot 2 \cdot 2 \cdot 3 \cdot 3$	Use 2 as a factor 4 times. Use 3 as a factor 2 times.
$16 \cdot 9$	$2 \cdot 2 \cdot 2 \cdot 2$ is 16, $3 \cdot 3$ is 9, $16 \cdot 9$ is 144
144	Simplified result is 144 (no exponents left)

Study Skills

Why Are These Notes Brain Friendly?

The notes are **easy to look at,** and you know that the brain responds to things that are visually pleasing. Other techniques that are visually memorable are the use of spacing (the two columns), stars, underlining, and circling. All of these methods **allow your brain to take note of important concepts and steps.**

The notes are also **systematic,** which means that they use certain techniques regularly. This way, your brain easily recognizes the topic of the day, the signals that show an important point, and the steps to follow for procedures. When you develop a system that you always use in your notes, your notes are easy to understand later when you are reviewing for a test.

Now Try This

Find one or two people in your math class to work with. Compare each other's lecture notes over a period of a week or so. Ask yourself the following questions as you examine the notes.

1 What are you doing in your notes to show the **main points** or larger concepts? (Such as underlining, boxing, using stars, capital letters, etc.)

2 In what ways do you **set off the explanations** for worked problems, examples, or smaller ideas (subpoints)? (Such as indenting, using arrows, circling or boxing)

3 What does **your instructor do** to show that he or she is moving from one idea to the next? (Such as saying "Next" or "Any questions," or "Now" or erasing the board.)

4 **How do you mark** that in your notes? (Such as skipping lines, using dashes or numbers, etc.)

5 What **explanations (in words) do you give yourself** in your notes, so that when those new dendrites you grew in lecture are fading, you can read your notes and still remember the new concepts later when you try to do your homework?

6 What **did you learn** by examining your classmates' notes?

▶ _____

▶ _____

▶ _____

7 What **will you try** in your own note taking? List **four** techniques that you will use next time you take notes in math class.

▶ _____

▶ _____

▶ _____

▶ _____

1.8 Exponents and Order of Operations

OBJECTIVE ▶ 1 Use exponents to write repeated factors. An **exponent** is a quick way to write repeated multiplication. Here is an example.

$2 \cdot 2 \cdot 2 \cdot 2 \cdot 2$ can be written $2^5 \leftarrow$ Exponent
$\uparrow$
Base

The *base* is the number being multiplied over and over, and the exponent tells how many times to use the number as a factor. This is called *exponential notation* or *exponential form*.

To simplify 2^5, actually do the multiplication.

$$2^5 = \underbrace{2 \cdot 2 \cdot 2 \cdot 2 \cdot 2}_{} = 32$$

Exponential form   Factored form   Simplified form

Here are some more examples, using 2 as the base.

$2 = 2^1$	is read	"2 to the **first power**."
$2 \cdot 2 = 2^2$	is read	"2 to the **second power**" or, more commonly, "2 **squared**."
$2 \cdot 2 \cdot 2 = 2^3$	is read	"2 to the **third power**" or, more commonly, "2 **cubed**."
$2 \cdot 2 \cdot 2 \cdot 2 = 2^4$	is read	"2 to the **fourth power**."
$2 \cdot 2 \cdot 2 \cdot 2 \cdot 2 = 2^5$	is read	"2 to the **fifth power**."

and so on.

We usually don't write an exponent of 1, so if no exponent is shown, it is understood to be 1. For example, 6 is actually 6^1, and 4 is actually 4^1.

EXAMPLE 1 Using Exponents

Complete this table.

	Exponential Form	Factored Form	Simplified	Read as
(a)	5^3	$5 \cdot 5 \cdot 5$	125	5 cubed, or 5 to the third power
(b)	4^2	$(4)(4)$	16	4 squared, or 4 to the second power
(c)	7^1	7	7	7 to the first power

Work Problem **1** at the Side. ▶

OBJECTIVE ▶ 2 Simplify expressions containing exponents. Exponents are also used with signed numbers, as shown below.

$(-3)^2 = (-3)(-3) = 9$ The factors have the same sign, so the product is positive.

$(-4)^3 = \underbrace{(-4)(-4)}(-4)$ Multiply two numbers at a time.
$\underbrace{16}(-4)$ First, $(-4)(-4)$ is positive 16
-64 Then, $16(-4)$ is -64

VOCABULARY TIP

Exponent "Ex" means "out." The exponent is the small, raised number sitting "outside" of another number or outside parentheses, such as 2^3 or $(-3)^2$ or $(1-5)^3$.

---

1 Write each multiplication using exponents. Indicate how to read the exponential form.

(a) $3 \cdot 3 \cdot 3 \cdot 3$

$3 - \leftarrow$ exponent
$\uparrow$
$\llcorner$ base

Read it as 3 to the _____ power.

(b) $6 \cdot 6$

(c) 9

(d) $(2)(2)(2)(2)(2)(2)$

Answers

1. **(a)** 3^4; Read it as "3 to the fourth power."
   (b) 6^2 is read "6 squared" or "6 to the second power."
   (c) 9^1 is read "9 to the first power."
   (d) 2^6 is read "2 to the sixth power."

2 Simplify.

(a) $(-2)^3$

$$\underbrace{(-2)(-2)}(-2)$$
$$\underbrace{(\_\_)(-2)}$$
$$\overline{}$$

(b) $(-6)^2$

(c) $2^4(-3)^2$

(d) $3^3(-4)^2$

Simplify exponents before you do other multiplications, as shown below. Notice that the exponent applies only to the *first* thing to its *left*.

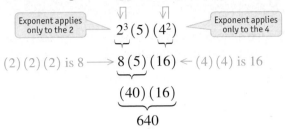

Exponent applies only to the 2 → $2^3(5)(4^2)$ ← Exponent applies only to the 4

$(2)(2)(2)$ is 8 ⟶ $8(5)(16)$ ← $(4)(4)$ is 16

$$\underbrace{(40)(16)}$$
$$640$$

EXAMPLE 2 **Using Exponents with Negative Numbers**

Simplify.

(a) $(-5)^2 = (-5)(-5) = \mathbf{25}$

(b) $(-5)^3 = \underbrace{(-5)(-5)}(-5)$

Be careful! Work with two factors at a time and watch the signs.

$$\underbrace{(25)(-5)}$$
$$\mathbf{-125}$$

(c) $(-2)^4 = (-2)(-2)(-2)(-2) = \mathbf{16}$

(d) $2^3(-3)^2 = \underbrace{(2)(2)(2)}\underbrace{(-3)(-3)}$

$$\underbrace{(8)(9)}$$
$$\mathbf{72}$$

▦ Calculator Tip

On a *scientific* calculator, use the exponent key ⓨˣ (or ⌃) to enter exponents. To enter 5^8, press the following keys.

5   ⓨˣ   8   ⊜   Answer is 390,625

↑   ↑

Base   Exponent

Be careful when using your calculator's exponent key with a negative number, such as $(-5)^3$. Different calculators use different keystrokes, so check the instruction manual, or experiment to see how your calculator works.

◀ **Work Problem 2 at the Side.**

OBJECTIVE ▶ 3 Use the order of operations. In earlier sections, you worked examples that mixed addition and subtraction or mixed multiplication and division. In those situations, you worked from left to right. **Example 3** below is a review.

EXAMPLE 3 **Working from Left to Right**

Simplify.

(a) $-8 - (-6) + (-11)$ Change all subtractions to adding the opposite.

$-8 + 6 + (-11)$ Add from left to right. First add $-8 + 6$

$\underbrace{-2 + (-11)}$ Finally, add $-2 + (-11)$

$\mathbf{-13}$

Continued on Next Page

(b) $-15 \div (-3)(6)$ Do multiplications and divisions from left to right.

$(5)(6)$

30

— Work Problem ❸ at the Side. ▶

Now we're ready to do problems that use a mix of the four operations, parentheses, and exponents. Let's start with a simple example: $4 + 2 \cdot 3$.

If we work from left to right

$4 + 2 \cdot 3$

$6 \quad \cdot 3$

18

If we multiply first

$4 + 2 \cdot 3$

$4 + \quad 6$

10

Which answer is correct?

To be sure that everyone gets the same answer to a problem like this, mathematicians have agreed to do things in a certain order. The following order of operations shows that multiplying is done ahead of adding, so *the correct answer is 10*.

Order of Operations

Step 1   Work inside **parentheses** or **other grouping symbols**.

Step 2   Simplify expressions with **exponents**.

Step 3   Do the remaining **multiplications and divisions** as they occur from left to right.

Step 4   Do the remaining **additions and subtractions** as they occur from left to right.

▦ **Calculator Tip**

Enter the example above in your calculator.

4 ⊕ 2 ⊗ 3 ⊜

Which answer do you get? If you have a scientific calculator, it automatically uses the order of operations and multiplies first to get the correct answer of 10. Some standard, four-function calculators may *not* have the order of operations built into them and will give the **incorrect** answer of 18.

---

❸ Simplify.

(a) $-9 + (-15) - 3$

$(\underline{}) - 3$

$\underline{}$

(b) $-4 - 2 + (-6)$

(c) $3(-4) \div (-6)$

(d) $-18 \div 9(-4)$

$(\underline{})(-4)$

Answers

3. **(a)** $(-24) - 3; -27$ **(b)** -12 **(c)** 2
    (d) $(-2)(-4); 8$

4 Simplify.

(GS) (a) $8 + 6(14 \div 2)$

$8 + \quad 6\,(7)$

$8 + \quad \underline{\quad}$

$\underline{\quad}$

(b) $4(1) + 8(9 - 2)$

(c) $3(5 + 1) + 20 \div 4$

5 Simplify.

(GS) (a) $2 + 40 \div (-5 + 3)$

$2 + 40 \div \quad (-2)$

$2 + \quad (\underline{\quad})$

$\underline{\quad}$

(b) $-5(5) - (15 + 5)$

(c) $(-24 \div 2) + (15 - 3)$

(d) $-3(2 - 8) - 5(4 - 3)$

(e) $3(3) - (10 \cdot 3) \div 5$

(f) $6 - (2 + 7) \div (-4 + 1)$

Answers

4. (a) $8 + 42; 50$ **(b)** 60 **(c)** 23
5. (a) $2 + (-20); -18$ **(b)** -45
   (c) 0 **(d)** 13 **(e)** 3 **(f)** 9

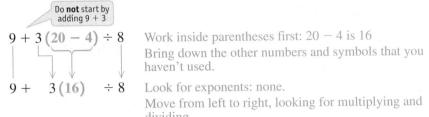

EXAMPLE 4 **Using the Order of Operations with Whole Numbers**

Simplify.

*Do **not** start by adding $9 + 3$*

$9 + 3\,(20 - 4) \div 8$ — Work inside parentheses first: $20 - 4$ is 16
Bring down the other numbers and symbols that you haven't used.

$9 + \quad 3\,(16) \quad \div 8$ — Look for exponents: none.
Move from left to right, looking for multiplying and dividing.

$9 + \quad 3\,(16) \quad \div 8$ — Yes, here is multiplying: $3(16)$ is 48

$9 + \quad 48 \quad \div 8$ — Here is dividing: $48 \div 8$ is 6. There is no other multiplying or dividing, so look for adding and subtracting.

$9 + \quad 6$ — Add last: $9 + 6$ is 15

$\mathbf{15}$

◀ **Work Problem 4 at the Side.**

EXAMPLE 5 **Using the Order of Operations with Integers**

Simplify.

(a) $-8 \div (7 - 5) - 9$ — Work inside parentheses first: $7 - 5$ is 2
Bring down the other numbers and symbols that you haven't used.
Look for exponents: none.

$-8 \div \quad (2) \quad - 9$ — Move from left to right, looking for multiplying and dividing.

$-8 \div \quad (2) - \quad 9$ — Here is dividing: $-8 \div 2$ is -4
No other multiplying or dividing, so look for adding and subtracting.

$-4 \quad - \quad 9$ — Change subtracting to adding. Change 9 to its opposite.

$-4 \quad + (-9)$ — Add $-4 + (-9)$

$\mathbf{-13}$

(b) $3 + 2\,(6 - 8) \cdot (15 \div 3)$ — *Do **not** start by adding $3 + 2$*
Work inside first set of parentheses.
Change $6 - 8$ to $6 + (-8)$ to get -2

$3 + 2\,(-2) \cdot (15 \div 3)$ — Work inside second set of parentheses: $15 \div 3$ is 5

$3 + 2\,(-2) \cdot \quad 5$ — Multiply and divide from left to right.
First multiply $2(-2)$ to get -4

$3 + \quad (-4) \quad \cdot \quad 5$ — Then multiply $-4 \cdot 5$ to get -20

$3 + \quad (-20)$ — Add last: $3 + (-20)$ is -17

$\mathbf{-17}$

◀ **Work Problem 5 at the Side.**

EXAMPLE 6   **Using the Order of Operations with Exponents**

Simplify.

(a) $4^2 - (-3)^2$ The only parentheses are around -3, but there is no work to do inside these parentheses.

$4^2 - (-3)^2$ Simplify the exponents. $4^2 = (4)(4) = 16$, and $(-3)^2 = (-3)(-3) = 9$

$16 - 9$ There is no multiplying or dividing, so add and subtract. $16 - 9$ is 7

7

(b) $(-4)^3 - (4 - 6)^2(-3)$ Work inside parentheses: $4 - 6$ becomes $4 + (-6)$, which is -2

$(-4)^3 - (-2)^2(-3)$ Simplify the exponents next. $(-4)^3$ is $(-4)(-4)(-4) = -64$, and $(-2)^2$ is $(-2)(-2) = 4$

$-64 - 4(-3)$ Look for multiplying and dividing. Multiply $4(-3)$ to get -12

$-64 - (-12)$ Change subtraction to addition. Change -12 to its opposite.

$-64 + (12)$ Add: $-64 + 12$ is -52

-52

⚠ CAUTION

To help in remembering the order of operations, you may have memorized the letters **PEMDAS,** or the phrase "Please Excuse My Dear Aunt Sally."

Please   **E**xcuse   **M**y   **D**ear   **A**unt   **S**ally

Parentheses; **E**xponents; **M**ultiply **and D**ivide; **A**dd **and S**ubtract

Be careful! Do **not** automatically do all multiplication before division. Multiplying and dividing are done **from left to right** (after parentheses and exponents).

——— Work Problem ⑥ at the Side. ▶

⑥ Simplify.

GS (a) $2^3 - 3^2$

[2 • 2 • 2]   [3 • 3]

(b) $6^2 \div (-4)(-3)$

(c) $(-4)^2 - 3^2(5 - 2)$

(d) $(-3)^3 + (3 - 9)^2$

Answers

6. **(a)** $8 - 9$ is -1 **(b)** 27   **(c)** -11 **(d)** 9

7 Simplify.

GS **(a)** $\dfrac{-3\,(2^3)}{-10 - 6 + 8}$

Numerator: $-3\,\underbrace{(2^3)}$

$$-3\,(\underline{}) = \underline{}$$

Denominator: $-10 - 6 + 8$

$$\underbrace{-10 + (-6)} + 8$$

$$\underline{} + 8 = \underline{}$$

Numerator $\longrightarrow$
$\dfrac{}{} =$
Denominator $\longrightarrow$

(b) $\dfrac{(-10)\,(-5)}{-6 \div 3\,(5)}$

(c) $\dfrac{6 + 18 \div (-2)}{(1 - 10) \div 3}$

(d) $\dfrac{6^2 - 3^2\,(4)}{5 + (3 - 7)^2}$

OBJECTIVE ▸ 4 Simplify expressions with fraction bars. A fraction bar indicates division, as in $\dfrac{-6}{2}$, which means $-6 \div 2$. In the expression below, the fraction bar also acts as a grouping symbol, like parentheses.

$$\frac{-5 + 3^2}{16 - 7\,(2)}$$

The fraction bar tells us to do the work in the numerator (above the bar) first. Then do the work in the denominator (below the bar). The last step is to divide the results.

$$\frac{-5 + 3^2}{16 - 7\,(2)} \longrightarrow \frac{-5 + 9}{16 - 14} \longrightarrow \frac{4}{2} \longrightarrow 4 \div 2 = 2$$

The final result is **2**.

EXAMPLE 7 **Using the Order of Operations with Fraction Bars**

Simplify $\dfrac{-8 + 5\,(4 - 6)}{4 - 4^2 \div 8}$

First do the work in the numerator.

Do **not** start by adding $-8 + 5$ $-8 + 5\,\underbrace{(4 - 6)}$ Work inside the parentheses.

$-8 + \underbrace{5\,(-2)}$ Multiply.

$\underbrace{-8 + (-10)}$ Add.

Numerator $\longrightarrow -18$

Now do the work in the denominator.

$4 - \underbrace{4^2} \div 8$ There are no parentheses; simplify the exponent.

$4 - \underbrace{16 \div 8}$ Divide.

$\underbrace{4 - 2}$ Subtract.

Denominator $\longrightarrow 2$

The last step is the division.

Numerator $\longrightarrow$
$\dfrac{-18}{2} = -9$ Signs are **different**, so the quotient is **negative**.
Denominator $\longrightarrow$

◂ **Work Problem 7 at the Side.**

Answers

7. (a) Numerator: $-3\,(8) = -24$;
Denominator: $-16 + 8 = -8$;
$\dfrac{-24}{-8} = 3$

(b) $\dfrac{50}{-10} = -5$ **(c)** $\dfrac{-3}{-3} = 1$

(d) $\dfrac{0}{21} = 0$

1.8 Exercises

FOR EXTRA HELP

Go to MyMathLab *for worked-out, step-by-step solutions to exercises enclosed in a square* ⬜ *and video solutions to* ▶ *exercises.*

CONCEPT CHECK   *Complete this table.*

	Exponential Form	Factored Form	Simplified	Read as
▶ 1.	4^3		64	
2.	10^2		100	
3.		$2 \cdot 2 \cdot 2 \cdot 2 \cdot 2 \cdot 2 \cdot 2$		
4.		$3 \cdot 3 \cdot 3 \cdot 3 \cdot 3$		
▶ 5.		$5 \cdot 5 \cdot 5 \cdot 5$		
6.		$2 \cdot 2 \cdot 2 \cdot 2 \cdot 2 \cdot 2$		
▶ 7.				7 squared
8.				6 cubed
9.				10 to the first power
10.				4 to the fourth power

*Simplify. **See Examples 1 and 2.***

11. (a) 10^1

(b) 10^2

(c) 10^3

(d) 10^4

12. (a) 5^1

(b) 5^2

(c) 5^3

(d) 5^4

13. (a) 4^1

(b) 4^2

(c) 4^3

(d) 4^4

14. (a) 3^1

(b) 3^2

(c) 3^3

(d) 3^4

15. 5^{10} 🖩

16. 4^9 🖩

17. 2^{12} 🖩

18. 3^{10} 🖩

19. $(-2)^2$ ▶

20. $(-4)^2$

21. $(-5)^2$

22. $(-10)^2$

23. $(-4)^3$

24. $(-2)^3$

25. $(-3)^4$ ▶

26. $(-2)^4$

27. $(-10)^3$ ▶

28. $(-5)^3$

29. 1^4

30. 1^5

31. $3^3 \cdot 2^2$ **32.** $4^2 \cdot 5^2$ **33.** $2^3(-5)^2$ **34.** $3^2(-2)^2$

35. $6^1(-5)^3$ **36.** $7^1(-4)^3$ **37.** $(-2)(-2)^4$ **38.** $-6(-6)^2$

39. CONCEPT CHECK Simplify.

$(-2)^2 =$ _____ $(-2)^6 =$ _____

$(-2)^3 =$ _____ $(-2)^7 =$ _____

$(-2)^4 =$ _____ $(-2)^8 =$ _____

$(-2)^5 =$ _____ $(-2)^9 =$ _____

(a) Describe the pattern you see in the signs of the answers.

(b) What would be the sign of $(-2)^{15}$ and the sign of $(-2)^{24}$?

40. CONCEPT CHECK Explain why it is important to have rules for the order of operations. Why do you think our "natural instinct" is to just work from left to right?

Simplify. See Examples 3–7.

41. $12 \div 6(-3)$ Multiply and divide from left to right.

     ___ (-3)

     ___

42. $10 - 30 \div 2$ Divide before subtracting.

     $10 -$ ___

     ___

43. $-1 + 15 - 7 - 7$

44. $9 + (-5) + 2(-2)$ **45.** $10 - 7^2$ **46.** $5 - 5^2$

47. $2 - (-5) + 3^2$ **48.** $6 - (-9) + 2^3$ **49.** $3 + 5(6 - 2)$

50. $4 + 3(8 - 3)$ **51.** $-7 + 6(8 - 14)$ **52.** $-3 + 5(9 - 12)$

53. $2(-3 + 5) - (9 - 12)$ **54.** $3(2 - 7) - (-5 + 1)$ **55.** $-5(7 - 13) \div (-10)$

56. $-4(9 - 17) \div (-8)$

57. $9 \div (-3)^2 + (-1)$

58. $-48 \div (-4)^2 + 3$

59. $2 - (-5)(-2)^3$

60. $1 - (-10)(-3)^3$

61. $-2(-7) + 3(9)$

62. $4(-2) + (-3)(-5)$

63. $40 \div (-2) - 3(-6)$

64. $36 \div (-3) - 8(-2)$

65. $2(5) - 3(4) + 5(3)$

66. $9(3) - 6(4) + 3(7)$

67. $4(3^2) + 7(3 + 9) - (-6)$

68. $5(4^2) - 6(1 + 4) - (-3)$

69. $(-4)^2 \cdot (7 - 9)^2 \div 2^3$

70. $(-5)^2 \cdot (9 - 17)^2 \div (-10)^2$

71. $\dfrac{-1 + 5^2 - (-3)}{-6 - 9 + 12}$

72. $\dfrac{-6 + 3^2 - (-7)}{7 - 9 - 3}$

73. $\dfrac{-2\,(4^2) - 4\,(6-2)}{-4\,(8-13) \div (-5)}$

74. $\dfrac{3\,(3^2) - 5\,(9-2)}{8\,(6-9) \div (-3)}$

75. $\dfrac{2^3 \cdot (-2-5) + 4\,(-1)}{4 + 5\,(-6 \cdot 2) + (5 \cdot 11)}$

76. $\dfrac{3^3 + 4\,(-1-2) - 25}{-4 + 4\,(3 \cdot 5) + (-6 \cdot 9)}$

77. $5^2\,(9-11)\,(-3)\,(-3)^3$

78. $4^2\,(13-17)\,(-2)\,(-2)^3$

79. $|-12| \div 4 + 2 \cdot |(-2)^3| \div 4$

80. $6 - |2 - 3 \cdot 4| + (-5)^2 \div 5^2$

81. $\dfrac{-9 + 18 \div (-3)\,(-6)}{32 - 4\,(12) \div 3\,(2)}$

82. $\dfrac{-20 - 15\,(-4) - (-40)}{14 + 27 \div 3\,(-2) - (-4)}$

Study Skills
REVIEWING A CHAPTER

This activity is really about **preparing for tests.** Some of the suggestions are ideas that you will learn to use a little later in the term, but get started trying them out now. Often, the first chapter in your math text will be review, so it is good to practice some of the study techniques on material that is not too challenging.

OBJECTIVES

1. Use the Chapter Summary to practice every type of problem.
2. Create study cards for vocabulary.
3. Practice by doing review and mixed review exercises.
4. Take the Chapter Test as a practice test.

Reviewing a Chapter

Use these techniques for reviewing a chapter.

▶ **Make a study card for each vocabulary word and concept.** Include a definition, an example, a sketch, and a page reference. Include the symbol or formula if there is one. See the *Using Study Cards* activity for a quick look at some sample study cards.

▶ **Go back to the section** to find more explanations or information about any new vocabulary, formulas, or symbols.

▶ **Use the Chapter Summary** to practice each type of problem. Do not expect the Summary to substitute for reading and working through the whole chapter! First, take the "Test Your Word Power" quiz to check your understanding of new vocabulary. The answers are at the end of the quiz. Then read the Quick Review. **Pay special attention to the green headings.** Check the explanations for the solutions to problems given. Try to think about the **whole chapter.**

▶ **Study your lecture notes** to see what your instructor has emphasized in class. Then review that material in your text.

▶ **Do the Review Exercises.**
  ✓ Check your answers **after** you're done with each **section of exercises.**
  ✓ If you get stuck on a problem, **first** check the Chapter Summary. If that doesn't clear up your confusion, then check the section and your lecture notes.
  ✓ Pay attention to **direction words** for the problems, such as *simplify, round, solve,* and *estimate.*
  ✓ Make **study cards for especially difficult problems.**

▶ **Do the Mixed Review exercises.** This is a good check to see whether you can still do the problems when they are in mixed-up order. **Check your answers carefully** in the Answers section in the back of your text. Are your answers **exact** and **complete?** Make sure you are **labeling** answers correctly, using the right **units.** For example, does your answer need to include *$, cm², ft,* and so on?

Study Skills

Why Are These Review Activities Brain Friendly?

You have already become familiar with the features of your text. This activity requires you to make good use of them. Your **brain needs repetition** to strengthen dendrites and the connections between them. By following the steps outlined here, you will be reinforcing the concepts, procedures, and skills you need to use for tests (and for the next chapters).

The combination of techniques provides repetition in different ways. That **promotes good branching of dendrites** instead of just relying on one branch or route to connect to the other dendrites. A thorough review of each chapter will **solidify your dendrite connections**. It will help you be sure that you understand the concepts **completely and accurately**. Also, taking the chapter Test will **simulate the testing situation,** which gives you practice in test-taking conditions.

▶ **Take the Chapter Test as if it is a real test.** If your instructor has skipped sections in the chapter, figure out which problems to skip on the test before you start.

✓ **Time yourself** just as you would for a real test.

✓ **Use a calculator or notes** just as you would be permitted to (or not) on a real test.

✓ **Take the test in one sitting,** just like a real test is given in one sitting.

✓ **Show all your work.** Practice showing your work just the way your instructor has asked you to show it.

✓ **Practice neatness.** Can someone else follow your steps?

✓ **Check your answers** in the back of the book.

Notice that reviewing a chapter will take some time. Remember that it takes time for dendrites to grow! You cannot grow a good network of dendrites by rushing through a review in one night. But if you use the suggestions over a few days or evenings, you will notice that you understand the material more thoroughly and remember it longer.

Now Try This

Follow the reviewing techniques listed above for your next test. For each technique, write a comment *about how it worked for you in the spaces below.*

1 Make a study card for each vocabulary word and concept.

2 Go back to the section to find more explanations or information.

3 Take the Test Your Word Power quiz and use the Quick Review to review each concept in the chapter.

4 Study your lecture notes to see what your instructor has emphasized in class.

5 Do the Review Exercises, following the specific suggestions on the previous page.

6 Do the Mixed Review exercises.

7 Take the Chapter Test as if it is a real test.

Chapter 1 Summary

Key Terms

1.1

place value system A place value system is a number system in which the location, or place, where a digit is written gives it a different value.

digits The 10 digits in our number system are 0, 1, 2, 3, 4, 5, 6, 7, 8, and 9.

whole numbers The whole numbers are 0, 1, 2, 3, 4, 5, . . .

1.2

number line A number line is like a thermometer turned sideways. It is used to show how numbers relate to each other.

integers The integers are . . . −3, −2, −1, 0, 1, 2, 3, . . .

absolute value The absolute value of a number is its distance from 0 on the number line. Absolute value is indicated by two vertical bars and is always positive (or 0) but never negative.

1.3

addends In an addition problem, the numbers being added are called addends.

sum The answer to an addition problem is called the sum.

addition property of 0 Adding 0 to any number leaves the number unchanged.

commutative property of addition Changing the *order* of two addends does not change the sum.

associative property of addition Changing the *grouping* of addends does not change the sum.

1.4

opposite The opposite of a number is the same distance from 0 on the number line but on the opposite side of 0. It is also called the *additive inverse* because a number plus its opposite equals 0.

1.5

rounding Rounding a number means finding a number that is close to the original number but easier to work with.

estimate Use rounded numbers to get an approximate answer, or estimate.

front end rounding Front end rounding is rounding numbers to the highest possible place, so all the digits become 0 except the first digit.

1.6

factors In a multiplication problem, the numbers being multiplied are called factors.

product The answer to a multiplication problem is called the product.

multiplication property of 0 Multiplying any number by 0 gives a product of 0.

multiplication property of 1 Multiplying a number by 1 leaves the number unchanged.

commutative property of multiplication Changing the *order* of two factors does not change the product.

associative property of multiplication Changing the *grouping* of factors does not change the product.

distributive property Multiplication distributes over addition. For example, $3(6 + 2) = 3 \cdot 6 + 3 \cdot 2$.

1.7

quotient The answer to a division problem is called the quotient.

1.8

exponent An exponent tells how many times a number is used as a factor in repeated multiplication.

New Symbols

$<$	is less than
$>$	is greater than
$\lvert 6 \rvert$ and $\lvert -2 \rvert$	absolute value of 6; absolute value of −2
$\approx$	approximately equal to

2^5 and 10^3 exponential form

Test Your Word Power

See how well you have learned the vocabulary in this chapter.

1 In $(4)(-6) = -24$, the 4 and -6 are called
A. products
B. factors
C. addends.

2 A list of the **whole numbers** is
A. $1, 2, 3, 4, \ldots$
B. $\ldots, -4, -3, -2, -1, 0, 1, 2, 3, \ldots$
C. $0, 1, 2, 3, 4, \ldots$

3 The **absolute value** of a number is
A. its distance from 0 on the number line
B. always negative
C. equal to 0.

4 An **exponent**
A. tells how many times a number is added
B. is the number being multiplied
C. tells how many times a number is used as a factor.

5 The **opposite** of a number is
A. never negative
B. called the additive inverse
C. called the absolute value.

6 **Front end rounding** is rounding numbers
A. to the nearest thousand
B. so there are no zeros
C. to the highest possible place.

7 A list of the **integers** is
A. $1, 2, 3, 4, \ldots$
B. $\ldots, -4, -3, -2, -1, 0, 1, 2, 3, \ldots$
C. $0, 1, 2, 3, 4, \ldots$

8 The **associative property of multiplication** says that
A. changing the order of two factors does not change the product
B. multiplication distributes over addition
C. changing the grouping of factors does not change the product.

Answers to Test Your Word Power

1. B; *Example:* In $7(10) = 70$, the factors are 7 and 10.

2. C; *Example:* $12, 0, 710$, and $89{,}475$ are all whole numbers.

3. A; *Example:* $|-3| = 3$ and $|3| = 3$ because both -3 and 3 are 3 steps away from 0 on the number line.

4. C; *Example:* In 2^5, the exponent is 5, which indicates that 2 is used as a factor 5 times, so $2^5 = 2 \cdot 2 \cdot 2 \cdot 2 \cdot 2 = 32$.

5. B; *Example:* The opposite of 5 is -5; it is the additive inverse because $5 + (-5) = 0$.

6. C; *Example:* Using front end rounding, round $48{,}299$ to the highest possible place, which is ten-thousands. So, $48{,}299$ rounds to $50{,}000$. All digits are 0 except the first digit.

7. B; *Example:* $-10, 0$, and 6 are all integers.

8. C; *Example:* $-4 \cdot (2 \cdot 3)$ can be rewritten as $(-4 \cdot 2) \cdot 3$; both products are -24.

Quick Review

Concepts

1.1 Reading and Writing Whole Numbers

Do not use the word *and* when reading whole numbers. Commas separate groups of three digits. The first few group names are ones, thousands, millions, billions, trillions.

Examples

Write $3, 008, 160$ in words.

three **million**, eight **thousand**, one hundred sixty

Write this number using digits:
twenty **billion**, sixty-five **thousand**, eighteen.

$$\underline{2\ 0}, \underline{0\ 0\ 0}, \underline{0\ 6\ 5}, \underline{0\ 1\ 8}$$
billions **millions** **thousands** **ones**

1.2 Graphing Numbers

Place a dot at the correct location on the number line.

Graph: **(a)** -4 **(b)** 0 **(c)** -1 **(d)** $\dfrac{1}{2}$ **(e)** 2

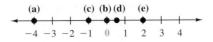

Concepts	Examples

1.2 Comparing Integers

When comparing two integers, the one that is farther to the left on the number line is less than the other.
Use the $<$ symbol for "is less than" and the $>$ symbol for "is greater than."

Write $<$ or $>$ between each pair of numbers to make a true statement.

$$-3 < -2 \qquad\qquad 0 > -4$$

-3 is less than -2 because -3 is to the *left* of -2 on the number line.

0 is greater than -4 because 0 is to the *right* of -4 on the number line.

1.2 Finding the Absolute Value of a Number

Find the distance on the number line from 0 to the number. The absolute value is always positive (or 0) but never negative.

Find each absolute value.

$|-5| = 5$ because -5 is 5 steps away from 0 on the number line.

$|3| = 3$ because 3 is 3 steps away from 0 on the number line.

1.3 Adding Two Integers

When both integers have the *same sign,* add the absolute values and use the common sign as the sign of the sum.

When the integers have *different signs,* subtract the lesser absolute value from the greater absolute value. Use the sign of the number with the greater absolute value as the sign of the sum.

Add.

(a) $-6 + (-7)$ Integers have the *same sign.*

Add the absolute values.

$$|-6| = 6 \quad \text{and} \quad |-7| = 7$$

Add $6 + 7 = 13$ and use the common sign as the sign of the sum: $-6 + (-7) = -13$

(b) $-10 + 4$ Integers have *different signs.*

Subtract the lesser absolute value from the greater.

$$|-10| = 10 \quad \text{and} \quad |4| = 4$$

Subtract $10 - 4 = 6$; the number with the greater absolute value is negative, so the sum is negative: $-10 + 4 = -6$

1.3 Using Properties of Addition

Addition Property of 0: Adding 0 to any number leaves the number unchanged.

Commutative Property of Addition: Changing the *order* of two addends does not change the sum.

Associative Property of Addition: Changing the *grouping* of addends does not change the sum.

Name the property illustrated by each case.

(a) $-16 + 0 = -16$

(b) $4 + 10 = 10 + 4$ Both sums are 14

(c) $2 + (6 + 1) = (2 + 6) + 1$ Both sums are 9

(a) Addition property of 0

(b) Commutative property of addition

(c) Associative property of addition

Concepts	Examples

1.4 Subtracting Two Integers

To subtract two numbers, add the first number to the opposite of the second number.

Step 1 Make one pencil stroke to change the subtraction symbol to an addition symbol.

Step 2 Make a second pencil stroke to change the *second* number to its *opposite*.

Subtract.

(a) $7 - (-2)$ Change subtraction to addition.
$7 + (+2)$ Change −2 to its opposite, +2
$\underbrace{}$
9

(b) $-9 - 12$ Change subtraction to addition.
$-9 + (-12)$ Change 12 to its opposite, −12
$\underbrace{}$
−21

1.5 Rounding Integers

Step 1 Draw a line under the place to which the number is to be rounded.

Step 2 Look only at the next digit to the right of the underlined place. If the next digit is *5 or more*, increase the underlined digit by 1. If the next digit is *4 or less*, do not change the digit in the underlined place.

Step 3 Change all digits to the right of the underlined place to zeros.

Round 36,833 to the nearest thousand.

Next digit
is *5 or more.*   Changed to 0

36,833   rounds to   **37,000**

Thousands place   Change 6 to 7

Round −3582 to the nearest ten.

Next digit
is *4 or less.*   Changed to 0

−3582   rounds to   **−3580**

Tens place   Leave 8 as 8

1.5 Front End Rounding

Round to the highest possible place so that all the digits become 0 except the first digit.

Use front end rounding.

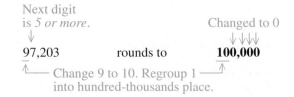

Next digit
is *5 or more.*   Changed to 0

97,203   rounds to   **100,000**

Change 9 to 10. Regroup 1
into hundred-thousands place.

1.5 Estimating Answers in Addition and Subtraction

Use front end rounding to round the numbers in a problem. Then add or subtract the rounded numbers to estimate the answer.

First use front end rounding to estimate the answer. Then find the exact answer.

The temperature was 48 degrees below zero. During the morning it rose 21 degrees. What was the new temperature?

Estimate: $-50 + 20 = $ **−30 degrees**

Exact: $-48 + 21 = $ **−27 degrees**

Concepts	Examples

1.6 Multiplying Two Integers

If two factors have *different signs*, the product is *negative*.

If two factors have the *same sign*, the product is *positive*.

Multiply.

(a) $-5(6) = -30$ — The factors have *different* signs, so the product is *negative*.

(b) $(-10)(-2) = 20$ — The factors have the *same* sign, so the product is *positive*.

1.6 Using Properties of Multiplication

Multiplication property of 0: Multiplying any number by 0 gives a product of 0.

Multiplication property of 1: Multiplying any number by 1 leaves the number unchanged.

Commutative property of multiplication: Changing the *order* of two factors does not change the product.

Associative property of multiplication: Changing the *grouping* of factors does not change the product.

Distributive property: Multiplication distributes over addition.

Name the property illustrated by each case.

(a) $-49 \cdot 0 = 0$

(b) $1(675) = 675$

(c) $(-8)(2) = (2)(-8)$ — Both products are -16

(d) $(-3 \cdot 2) \cdot 4 = -3 \cdot (2 \cdot 4)$ — Both products are -24

(e) $5(2 + 4) = 5 \cdot 2 + 5 \cdot 4$ — Both results are 30

(a) Multiplication property of 0

(b) Multiplication property of 1

(c) Commutative property of multiplication

(d) Associative property of multiplication

(e) Distributive property

1.6 Estimating Answers in Multiplication

First use front end rounding. Then multiply the rounded numbers using a shortcut: multiply the nonzero digits in each factor; count the total number of zeros in the two factors and write that number of zeros in the product.

First use front end rounding to estimate the answer. Then find the exact answer.

At a PTA fund-raiser, Lionel sold 96 photo albums at $22 each. How much money did he take in?

Estimate: $100 \cdot \$20 = $ **$2000**

Exact: $96 \cdot \$22 = $ **$2112**

1.7 Dividing Two Integers

Use the same sign rules for dividing two integers as for multiplying two integers.

If two numbers have *different signs*, the quotient is *negative*.

If two numbers have the *same sign*, the quotient is *positive*.

Divide.

$\dfrac{-24}{6} = -4$ — Numbers have *different* signs, so the quotient is *negative*.

$-72 \div (-8) = 9$ — Numbers have the *same* sign, so the quotient is *positive*.

$\dfrac{50}{-5} = -10$ — Numbers have *different* signs, so the quotient is *negative*.

1.7 Using Properties of Division

(a) When a nonzero number is divided by itself, the quotient is 1.

(b) When a number is divided by 1, the quotient is the number.

(c) When 0 is divided by any other number (except 0), the quotient is 0.

(d) Division by 0 is *undefined*. There is no answer.

State the property illustrated by each case.

(a) $\dfrac{-4}{-4} = 1$ **(b)** $\dfrac{65}{1} = 65$

(c) $\dfrac{0}{9} = 0$ **(d)** $\dfrac{-10}{0}$ is undefined.

The examples are in the same order as the properties listed at the left. (Division is *not* commutative and *not* associative.)

Concepts	**Examples**

1.7 Estimating Answers in Division

First use front end rounding. Then divide the rounded numbers, using a shortcut: Drop the same number of zeros in both the divisor and the dividend.

First use front end rounding to estimate the answer. Then find the exact answer.

Joan has one year to pay off a $1020 loan. What is her monthly payment?

Estimate: $1000 \div 10 = \$100$

Exact: $1020 \div 12 = \$85$

1.7 Interpreting Remainders in Division

In some situations the remainder tells you how much is left over. In other situations, you must increase the quotient by 1 in order to accommodate the "left over."

Divide; then interpret the remainder.

Each chemistry student needs 35 milliliters of acid for an experiment. How many students can be served from a bottle holding 500 milliliters of acid?

$$
\begin{array}{r}
14 \\
35\overline{)500} \\
35 \\
\hline
150 \\
140 \\
\hline
10
\end{array}
$$

$14 \longrightarrow$ 14 students served

$10 \longrightarrow$ 10 milliliters of acid left over

1.8 Using Exponents

An exponent tells how many times a number is used as a factor in repeated multiplication. An exponent applies only to its base (the first thing to the left of the exponent).

Simplify.

Exponent

(a) $2^5 = 2 \cdot 2 \cdot 2 \cdot 2 \cdot 2 = \textbf{32}$

(b) $(-3)^2 = (-3)(-3) = \textbf{9}$

1.8 Order of Operations

Mathematicians have agreed to follow this order.

Step 1 Work inside **parentheses** or other **grouping symbols.**

Step 2 Simplify expressions with **exponents.**

Step 3 Do the remaining **multiplications and divisions** as they occur **from left to right.**

Step 4 Do the remaining **additions and subtractions** as they occur **from left to right.**

Simplify.

$(-2)^4 + 3(-4 + 2)$ Work inside parentheses.

$(-2)^4 + \quad 3(-2)$ Simplify exponents.

$16 \quad + \quad 3(-2)$ Multiply.

$16 \quad + \quad (-6)$ Add.

$\textbf{10}$

1.8 Using the Order of Operations with Fraction Bars

When there is a fraction bar, do all the work in the numerator. Then do all the work in the denominator. Finally, divide numerator by denominator.

Simplify.

$$
\frac{-10 + 4^2 - 6}{2 + 3(1 - 4)} = \frac{-10 + 16 - 6}{2 + 3(-3)} = \frac{0}{-7} = \textbf{0}
$$

Chapter 1   *Review Exercises*

If you need help with any of these Review Exercises, look in the section indicated in the green rectangles.

1.1 **1.** Circle the whole numbers: 86   2.831 -4 0 $\frac{2}{3}$ 35,600

Write these numbers in words.

2. 806

3. 319,012

4. 60,003,200

5. 15,749,000,000,006

Write these numbers using digits.

6. Five hundred four thousand, one hundred

7. Six hundred twenty million, eighty thousand

8. Ninety-nine billion, seven million, three hundred fifty-six

1.2 **9.** Graph these numbers: $-3, 2, -5, 0$.

Write $<$ or $>$ between each pair of numbers to make a true statement.

10. 0 _____ -4

11. -3 _____ -1

12. 2 _____ -2

13. -2 _____ 1

Find each absolute value.

14. $|-5|$

15. $|9|$

16. $|0|$

17. $|-125|$

1.3 *Add.*

18. $-9 + 8$

19. $-8 + (-5)$

20. $16 + (-19)$

21. $-4 + 4$

22. $6 + (-5)$

23. $-12 + (-12)$

24. $0 + (-7)$

25. $-16 + 19$

26. $9 + (-4) + (-8) + 3$

27. $-11 + (-7) + 5 + (-4)$

1.4 *Find the opposite (additive inverse) of each number. Show that the sum of the number and its opposite is 0.*

28. −5

29. 18

Subtract by changing subtraction to addition.

30. $5 - 12$

31. $24 - 7$

32. $-12 - 4$

33. $4 - (-9)$

34. $-12 - (-30)$

35. $-8 - 14$

36. $-6 - (-6)$

37. $-10 - 10$

38. $-8 - (-7)$

39. $0 - 3$

40. $1 - (-13)$

41. $15 - 0$

Simplify.

42. $3 - 12 - 7$

43. $-7 - (-3) + 7$

44. $4 + (-2) - 0 - 10$

45. $-12 - 12 + 20 - (-4)$

1.5 *Round each number as indicated.*

46. 205 to the nearest ten

47. 59,499 to the nearest thousand

48. 85,066,000 to the nearest million

49. −2963 to the nearest hundred

50. −7,063,885 to the nearest ten-thousand

51. 399,712 to the nearest thousand

Use front end rounding to round each number.

52. The combined weight loss of ten dieters was 197 pounds.

53. The land on the shore of the Dead Sea in the Middle East is 1360 feet below sea level. (Data from www.nps.gov)

54. There are 4,099,137,000 Internet users in the world. (Data from www.internetlivestats.com)

55. Within forty years, the Census Bureau predicts a world population of 9,502,000,000 people. (Data from U.S. Census Bureau.)

1.6 *Multiply.*

56. $-6(9)$

57. $(-7)(-8)$

58. $10(-10)$

59. $-45 \cdot 0$

60. $-1(-24)$

61. $17 \cdot 1$

62. $4(-12)$

63. $(-5)(-25)$

64. $-3(-4)(-3)$

65. $-5(2)(-5)$

66. $(-8)(-1)(-9)$

1.7 *Simplify.*

67. $\dfrac{-63}{-7}$

68. $\dfrac{70}{-10}$

69. $\dfrac{-15}{0}$

70. $-100 \div (-20)$

71. $18 \div (-1)$

72. $\dfrac{0}{12}$

73. $\dfrac{-30}{-2}$

74. $\dfrac{-35}{35}$

75. $-40 \div (-4) \div (-2)$

76. $-18 \div 3(-3)$

77. $0 \div (-10)(5) \div 5$

78. Divide; then interpret the remainder. It took 1250 hours to build the set for a new play. How many working days of eight hours each did it take to build the set?

1.8 *Simplify.*

79. 10^4

80. 2^5

81. 3^3

82. $(-4)^2$

83. $(-5)^3$

84. 8^1

85. $6^2 \cdot 3^2$

86. $5^2(-2)^3$

87. $-30 \div 6 - 4(5)$

88. $6 + 8(2 - 3)$

89. $16 \div 4^2 + (-6 + 9)^2$

90. $-3(4) - 2(5) + 3(-2)$

91. $\dfrac{-10 + 3^2 - (-9)}{3 - 10 - 1}$

92. $\dfrac{-1(1-3)^3 + 12 \div 4}{-5 + 24 \div 8 \cdot 2(6-6) + 5}$

Chapter 1 Mixed Review Exercises

Practicing exercises in mixed-up order helps you prepare for tests.
Name the property illustrated by each case.

1. $-3 + (5 + 1) = (-3 + 5) + 1$

2. $-7(2) = 2(-7)$

3. $0 + 19 = 19$

4. $-42 \cdot 0 = 0$

5. $2(-6 + 4) = 2 \cdot (-6) + 2 \cdot 4$

6. $(-6 \cdot 3) \cdot 1 = -6 \cdot (3 \cdot 1)$

First use front end rounding to estimate each answer. Then find the exact answer.

7. Apple sold 192 limited-edition gold Apple watches for $11,900 each. What was the total value of the limited-edition gold Apple watches?

Estimate:

Exact:

8. Chad had $185 in his bank account. He deposited his $428 paycheck and then paid $706 for car repairs. What is the balance in his account?

Estimate:

Exact:

9. Georgia's hybrid car used 22 gallons of gas on her 880-mile trip. What was the average number of miles she drove on each gallon of gas?

Estimate:

Exact:

10. When inventory was taken at Mathtronic Company, 19 calculators and 12 computer modems were missing. Each calculator is worth $39 and each modem is worth $85. What is the total value of the missing items?

Estimate:

Exact:

Elena Sanchez opened a shop that does alterations and designs custom clothing. Use the table of her income and expenses to answer Exercises 11–14.

Month	Income	Expenses	Profit or Loss
Jan.	$2400	−$3100	
Feb.	$1900	−$2000	
Mar.	$2500	−$1800	
Apr.	$2300	−$1400	
May	$1600	−$1600	
June	$1900	−$1200	

11. Complete the table by finding Elena's profit or loss for each month. To show a loss, write a negative sign in front of the number.

12. Which month had the greatest loss? Which month had the greatest profit?

13. What was Elena's average monthly income?

14. What was the average monthly amount of expenses?

1. Write this number in words: 20,008,307

2. Write this number using digits: thirty billion, seven hundred thousand, five

3. Graph the numbers $3, -2, 0, -\dfrac{1}{2}$ on the number line below.

$$\begin{array}{c|c|c|c|c|c|c} \hline -3 & -2 & -1 & 0 & 1 & 2 & 3 \end{array}$$

4. Write $<$ or $>$ between each pair of numbers to make a true statement.

0 _____ -3 $\qquad$ -2 _____ -1

5. Find each absolute value: $|10|$ and $|-14|$.

Add, subtract, multiply, or divide.

6. $3 - 9$

7. $-12 + 7$

8. $\dfrac{-28}{-4}$

9. $-1\,(40)$

10. $-5 - (-15)$

11. $(-8)(-8)$

12. $-25 + (-25)$

13. $\dfrac{17}{0}$

14. $-30 - 30$

15. $\dfrac{50}{-10}$

16. $5\,(-9)$

17. $0 - (-6)$

Simplify.

18. $-35 \div 7\,(-5)$

19. $-15 - (-8) + 7$

20. $3 - 7\,(-2) - 8$

21. $(-4)^2 \cdot 2^3$

22. $\dfrac{5^2 - 3^2}{(4)(-2)}$

23. $-2\,(-4 + 10) + 5\,(4)$

24. $-3 + (-7 - 10) + 4\,(6 - 10)$

25. Explain how an exponent is used. Include two examples, one using a positive number as the base and one using a negative number as the base.

26. Explain the commutative property of addition and the associative property of addition. Also give an example to illustrate each property.

Round each number as indicated.

27. 851 to the nearest hundred

28. 36,420,498,725
to the nearest million

29. 349,812 to the
nearest thousand

First use front end rounding to estimate each answer. Then find the exact answer.

30. Lorene had $184 in her checking account. She deposited her $293 paycheck and then wrote a $506 check for tuition. What is the balance in her account?

Estimate:

Exact:

31. Twenty-two people were in a weight loss program for a year. The total loss for the entire group of people was 1144 pounds. What was the average weight loss for each person?

Estimate:

Exact:

32. One kind of cereal has 220 calories in each serving. Another kind has 110 calories in each serving. During a month with 31 days, how many calories would you save by eating the second kind of cereal each morning for breakfast?

Estimate:

Exact:

Write and solve a subtraction problem to answer this question.

33. On the planet Mars the average high temperature is −6 °F and the average low is −79 °F. What is the difference between the high and low temperatures? (Data from NASA.)

Divide; then interpret the remainder.

34. Anthony has a part-time job as a shipping clerk. He is sending 1276 mathematics books to a college bookstore. Each shipping carton can safely hold 48 books. What is the minimum number of cartons he will need?

Study Skills
MANAGING YOUR TIME

Many college students find themselves juggling a difficult schedule and multiple responsibilities. Perhaps you are going to school, working part time, and managing family demands. Here are some tips to help you develop good time management skills and habits.

▶ **Read the syllabus for each class.** Check on class policies, such as attendance, late homework, and make-up tests. Find out how you are graded. Keep the syllabus in your notebook.

▶ **Make a semester or quarter calendar.** Put test dates and major due dates for *all* your classes on the same calendar. That way you will see which weeks are the really busy ones. Try using a different color pen for each class. Your brain responds well to the use of color. A semester calendar is on the next page.

▶ **Make a weekly schedule.** After you fill in your classes and other regular responsibilities (such as work, picking up kids from school, etc.), block off some study periods during the day that you can guarantee you will use for studying. Aim for 2 hours of study for each 1 hour you are in class.

▶ **Make "To Do" lists.** Then use them by crossing off the tasks as you complete them. You might even number them in the order they need to be done (most important ones first).

▶ **Break big assignments into smaller chunks.** They won't seem so big that way. Make deadlines for each small part so you stay on schedule.

▶ **Give yourself small breaks in your studying.** Do not try to study for hours at a time! Your brain needs rest between periods of learning. Try to give yourself a 10 minute break each hour or so. You will learn more and remember it longer.

▶ **If you get off schedule, just try to get back on schedule tomorrow.** We all slip from time to time. All is not lost! Make a new "to do" list and start doing the most important things first.

▶ **Get help when you need it.** Talk with your instructor during office hours. Also, most colleges have some kind of learning center, tutoring center, or counseling office. If you feel lost and overwhelmed, ask for help. Someone can help you decide what to do first and what to spend your time on right away.

> What two or three of the suggestions above will you try this week? How do you think they will help you?
>
> 1. _____
>
> 2. _____
>
> 3. _____

OBJECTIVES

1 **Create a semester schedule.**

2 **Create a "to do" list.**

Why Are These Techniques Brain Friendly?

Your brain appreciates some order. It enjoys a little routine; for example, choosing the same study time and place each day. You will find that you quickly settle in to your reading or homework.

Also, your brain **functions better when you are calm.** Too much rushing around at the last minute to get your homework and studying done sends hostile chemicals to your brain and makes it more difficult for you to learn and remember. So a little planning can really pay off.

Building rest into your schedule is good for your brain. Remember, it takes time for dendrites to grow.

We've suggested using color on your calendars. This too, is brain friendly. Remember, your brain **likes pleasant colors and visual material** that are nice to look at. Messy and hard-to-read calendars will not be helpful, and you probably won't look at them often.

Study Skills

SEMESTER CALENDAR

WEEK	MON	TUES	WED	THUR	FRI	SAT	SUN
1							
2							
3							
4							
5							
6							
7							
8							
9							
10							
11							
12							
13							
14							
15							
16							

2

Understanding Variables and Solving Equations

Each year, lightning strikes somewhere in the United States 25,000,000 times, and an average of 300 people are struck by lightning. In this chapter, you'll see how you can use algebraic expressions to estimate your distance from a thunderstorm, the weight of a fish you caught, or the temperature on a summer day.

2.1 Introduction to Variables

OBJECTIVE ▶ 1 Identify variables, constants, and expressions. You probably know that algebra uses letters, especially the letter x. But why use letters when numbers are easier to understand? Here is an example.

Suppose you manage your college bookstore. When deciding how many books to order for a certain class, you first find out the class limit, that is, the maximum number of students allowed in the class. You will need at least that many books. But you decide to order 5 extra copies for emergencies.

Rule for ordering books: Order the class limit + 5 extra

How many books would you order for a prealgebra class with a limit of 25 students?

Class limit ⌐↓   ↓⌐ Extra

$$25 + 5$$ You would order 30 prealgebra books.

How many books would you order for a geometry class that allows 40 students to register?

Class limit ⌐↓   ↓⌐ Extra

$$40 + 5$$ You would order 45 geometry books.

You could set up a table to keep track of the number of books to order for various classes.

Class	Rule for Ordering Books: Class Limit + 5 Extra	Number of Books to Order
Prealgebra	25 + 5	30
Geometry	40 + 5	45
College algebra	35 + 5	40
Calculus 1	50 + 5	55

A shorthand way to write your rule is shown below.

$$c + 5$$
↑
The c stands for class limit.

You can't write your rule by using just numbers because the class limit varies, or changes, depending on which class you're talking about. So you use a letter, called a **variable,** to represent the part of the rule that varies. Notice the similarity in the words *varies* and *variable*. When part of a rule does *not* change, it is called a **constant.**

The variable, or the part of
the rule that varies or changes
↓
$$c + 5$$
↑
The constant, or the part of
the rule that does *not* change

$c + 5$ is called an **expression.** It expresses (tells) the rule for ordering books. You could use any letter you like for the variable part of the expression, such as $x + 5$, or $n + 5$, and so on. But one suggestion is to use a letter that reminds you of what it stands for. In this situation, the letter c reminds us of "class limit."

EXAMPLE 1 Writing an Expression and Identifying the Variable and Constant

Write an expression for this rule. Identify the variable and the constant.

Order the class limit minus 10 books because some students will buy used books.

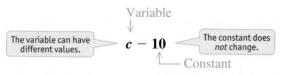

Variable

The variable can have different values. → $c - 10$ ← The constant does *not* change.

↑ Constant

── **Work Problem ❶ at the Side.** ▶

OBJECTIVE ❷ Evaluate variable expressions for given replacement values. When you need to figure out how many books to order for a particular class, you use a specific value for the class limit, like 25 students in prealgebra. Then you **evaluate the expression;** that is, you follow the rule.

Ordering Books for a Prealgebra Class

$c + 5$ Expression (rule for ordering books) is $c + 5$
↓   Replace c with 25, the class limit for prealgebra.

$25 + 5$ Follow the rule. Add $25 + 5$

30   Order 30 prealgebra books.

EXAMPLE 2 Evaluating an Expression

Use this rule for ordering books: Order the class limit minus 10.
The expression is $c - 10$.

(a) Evaluate the expression when the class limit is 32.

$c - 10$ Replace c with 32
↓
$32 - 10$ Follow the rule. Subtract $32 - 10$

22   Order 22 books.

(b) Evaluate the expression when the class limit is 48.

$c - 10$ Replace c with 48
↓
$48 - 10$ Follow the rule. Subtract $48 - 10$

38   Order 38 books.

── **Work Problem ❷ at the Side.** ▶

In any career you choose, there will be many useful "rules" that need to be written using variables (letters) because part of the rule changes depending on the situation. This is one reason algebra is such a powerful tool. Here is another example.

Suppose that you work in a landscaping business. You are putting a fence around a square-shaped garden. Each side of the garden is 6 feet long. How much fencing material should you bring for the job? You could add the lengths of the four sides.

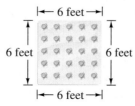

6 feet + 6 feet + 6 feet + 6 feet = 24 feet of fencing

---

❶ Write an expression for this rule.
Identify the variable and the constant.

Order the class limit plus 15 extra books because it is a very large class.

Variable
↳ $c +$ _____

The expression is _____ .

The variable is _____ .

The constant is _____ .

❷ Use this expression for ordering books: $c + 3$

 (a) Evaluate the expression when the class limit is 25.

$c + 3$
↓
___ $+ 3$
───
Order _____ books.

(b) Evaluate the expression when the class limit is 60.

Answers

1. $c + 15$. The expression is $c + 15$.
   The variable is c. The constant is 15.
2. **(a)** $25 + 3$; 28; order 28 books
   (b) $60 + 3$ is 63; order 63 books

3 **(a)** Evaluate the expression $4s$ when the length of one side of a square table is 3 feet.

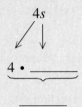

(b) Evaluate the expression $4s$ when the length of one side of a square park is 7 miles.

Or, recall that multiplication is a quick way to do repeated addition. The square garden has 4 sides, so multiply by 4.

$$4 \cdot 6 \text{ feet} = 24 \text{ feet of fencing}$$

So the rule for calculating the amount of fencing for a square garden is:

$$4 \cdot \text{length of one side}$$

Other jobs may require fencing for larger or smaller square shapes. The following table shows how much fencing you will need.

Length of One Side of Square Shape	Expression (Rule) to Find Total Amount of Fencing: 4 • Length of One Side	Total Amount of Fencing Needed
6 feet	4 • 6 feet	24 feet
9 feet	4 • 9 feet	36 feet
10 feet	4 • 10 feet	40 feet
3 feet	4 • 3 feet	12 feet

The expression (rule) can be written in shorthand form as shown below.

Length of one side
↓
$$4 \cdot s$$
Coefficient ↗ ↖ Variable

The number part in a *multiplication* expression is called the numerical coefficient, or just the **coefficient.** We usually don't write multiplication dots in expressions, so we do the following.

$$4 \cdot s \quad \text{is written as} \quad 4s$$

You can use the expression $4s$ any time you need to know the *perimeter* of a square shape, that is, the total distance around all four sides of the square.

> **!** **CAUTION**
>
> If an expression involves adding, subtracting, or dividing, then you **do** have to write $+, -,$ or $\div$. It is *only* multiplication that is understood without writing an operation symbol.
>
> | $4 + s$ | $4 - s$ | $4 \div s$ | $4s$ |
> |---|---|---|---|
> | ↑ | ↑ | ↑ | |
> | Add s | Subtract s | Divide by s | Multiply by s |

> **EXAMPLE 3** **Evaluating an Expression with Multiplication**
>
> The expression (rule) for finding the perimeter of a square shape is $4s$. Evaluate the expression when the length of one side of a square parking lot is 30 yards.

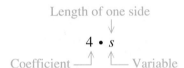

$4s$ Replace s with 30 yards.

$4 \cdot 30 \text{ yards}$ There is no operation symbol between the 4 and the s, so it is understood to be multiplication. 4 times 30 is 120

120 yards ⟵ Total distance around the lot (perimeter) is 120 yards.

◀ **Work Problem** **3** **at the Side.**

Answers

3. **(a)** 4 • 3 feet is 12 feet
   (b) 4 • 7 miles is 28 miles

Some expressions (rules) involve several different steps. An expression used to estimate the amount of time needed for a long car trip is shown below.

Hours needed to get gas and food $\Big\}$ → $2 + \dfrac{d}{60}$ ← $\begin{cases} \text{Distance traveled in} \\ \text{one day (the variable)} \end{cases}$
$\begin{cases} \text{Average driving speed} \\ \text{is 60 miles per hour} \end{cases}$

Remember that a fraction bar means division, so $\dfrac{d}{60}$ is the distance traveled in one day divided by 60. You also need to remember the order of operations, which means doing division before addition.

EXAMPLE 4 **Evaluating an Expression with Several Steps**

Evaluate the expression $2 + \dfrac{d}{60}$ when the distance traveled is 300 miles.

$2 + \dfrac{d}{60}$ — Replace d with 300, the distance traveled that day.

$2 + \dfrac{300}{60}$ ← Follow the rule using the order of operations. First divide: $300 \div 60$ is 5

$\underbrace{2 + \ \ 5}$ Now add: $2 + 5$ is 7

7 ← It will take 7 hours to travel 300 miles.

Work Problem ④ at the Side. ▶

Note

If you like to fish, you can use an expression (rule) like the one below to find the approximate weight (in pounds) of a fish you catch. Measure the length of the fish (in inches) and then use the correct expression for that type of fish. For a northern pike, the weight expression is shown below.

Variable (length of fish) → $\dfrac{l^3}{3600}$

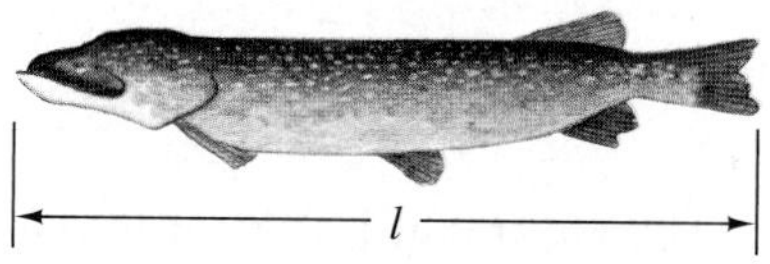

where l is the length of the fish in inches. (Data from *InFisherman*.)
To evaluate this expression for a fish that is 43 inches long, follow the rule by calculating as follows.

$\dfrac{43^3}{3600}$ Replace l with 43, the length of the fish in inches.

In the numerator, multiply $43 \cdot 43 \cdot 43$. Or, use the $\boxed{y^x}$ or $\boxed{\wedge}$ key on a calculator. Then divide by 3600.

Enter 43 $\boxed{y^x}$ 3 $\boxed{\div}$ 3600 $\boxed{=}$ Calculator shows 22.08527778
   ↑   ↑
  Base Exponent

The fish weighs about 22 pounds.

Now use the expression to find the approximate weight of a northern pike that is 37 inches long. (Answer: about 14 pounds.)

(Continued)

---

④ Evaluate the expression $2 + \dfrac{d}{60}$ when the distance traveled is 480 miles.

Answer

4. $2 + \dfrac{480}{60}$ ← Replace d with 480.

$\underbrace{2 + 8}$

10 ← It will take 10 hours to travel 480 miles.

⑤ (a) Use the expression for finding your average bowling score. Evaluate the expression if your total score for 4 games is 532.

$$\frac{t \rightarrow \boxed{}}{g \rightarrow \boxed{}}$$

Your average score

is _____

(b) Complete this table.

Value of x	Value of y	Expression x − y
16	10	16 − 10 is 6
100	5	
3	7	
8	0	

Notice that variables are used on your calculator keys. On the ⓨˣ key, *y* represents the base and *x* represents the exponent. You first evaluated y^x by entering 43 as the base and 3 as the exponent for the first fish. Then you evaluated y^x again by entering 37 as the base and 3 as the exponent for the second fish.

Some expressions (rules) involve several variables. For example, if you go bowling and want to know your average score, you can use this expression.

$$\frac{t}{g}$$ ← Total score for all games (variable)
← Number of games (variable)

EXAMPLE 5 **Evaluating Expressions with Two Variables**

(a) Find your average score if you bowl three games and your total score for all three games is 378.

Use the expression (rule) for finding your average score.

Replace *t* with your total score of 378.

$$\frac{t}{g} \longrightarrow \frac{378}{3}$$ } Follow the rule. Divide 378 by 3 to get 126.

Replace *g* with 3, the number of games.

Your average score is **126**

(b) Complete these tables to show how to evaluate each expression.

Value of x	Value of y	Expression (Rule) x + y
2	5	2 + 5 is 7
−6	4	__ + __ is __
0	16	__ + __ is __

Value of x	Value of y	Expression (Rule) xy
2	5	2 • 5 is 10
−6	4	__ • __ is __
0	16	__ • __ is __

The expression (rule) is to *add* the two variables. So the completed table is:

Value of x	Value of y	Expression (Rule) x + y
2	5	2 + 5 is 7
−6	4	−6 + 4 is −2
0	16	0 + 16 is 16

The expression (rule) is to *multiply* the two variables. We know that it's multiplication because there is no operation symbol between the *x* and *y*. So the completed table is:

Value of x	Value of y	Expression (Rule) xy
2	5	2 • 5 is 10
−6	4	−6 • 4 is −24
0	16	0 • 16 is 0

◀ **Work Problem ⑤ at the Side.**

Answers

5. **(a)** $\frac{532}{4}$; average score is 133.

(b) 100 − 5 is 95; 3 − 7 is −4;
8 − 0 is 8

OBJECTIVE ➤ **3** **Write properties of operations using variables.** Now you can use variables as a shorthand way to express the properties you learned earlier. We'll use the letters a and b to represent any two numbers.

Commutative Property of Addition	**Commutative Property of Multiplication**
$a + b = b + a$	$a \cdot b = b \cdot a$
	or, $ab = ba$

To get specific examples, you can pick values for a and b. For example, if a is -3, replace every a with -3. If b is 5, replace every b with 5.

$$a + b = b + a$$
$$-3 + 5 = 5 + (-3)$$
$$2 \;\; = \;\; 2$$

Both sums are 2

$$ab = ba$$
$$-3 \cdot 5 = 5 \cdot (-3)$$
$$-15 \;\; = \;\; -15$$

Both products are -15

Of course, you could pick many different values for a and b, because the commutative "rule" will always work for adding *any* two numbers or multiplying *any* two numbers.

EXAMPLE 6 **Writing Properties of Operations Using Variables**

Use the variable b to state this property: When any number is divided by 1, the quotient is the number.

Use the letter b to represent any number.

$$\frac{b}{1} = b$$

— **Work Problem 6 at the Side.** ▶

OBJECTIVE ➤ **4** **Use exponents with variables.** We have used an exponent as a quick way to write repeated multiplication. For example,

$$\underbrace{3 \cdot 3 \cdot 3 \cdot 3 \cdot 3}_{} \quad \text{can be written} \quad 3^5 \leftarrow \text{Exponent}$$
3 is used as a factor 5 times.   Base

The meaning of an exponent is the same when a variable is the base.

$$\underbrace{c \cdot c \cdot c \cdot c \cdot c}_{} \quad \text{can be written} \quad c^5 \leftarrow \text{Exponent}$$
c is used as a factor 5 times.   Base

m^2 means $m \cdot m$ Here m is used as a factor 2 times.

$x^4 y^3$ means $x^4 \cdot y^3$ or $\underbrace{x \cdot x \cdot x \cdot x}_{x^4} \cdot \underbrace{y \cdot y \cdot y}_{y^3}$

$7b^2$ means $7 \cdot b \cdot b$ The exponent applies *only* to b.

$-4xy^2z$ means $-4 \cdot x \cdot y \cdot y \cdot z$ The exponent applies *only* to y.

6 **(a)** Use the variable a to state this property: Multiplying any number by 0 gives a product of 0.

Any number	times	zero		
↓	↓	↓		
a	$\cdot$	__	=	__

(b) Use the variables a, b, and c to state the associative property of addition: Changing the grouping of addends does not change the sum.

Answers

6. **(a)** $a \cdot 0 = 0$
(b) $(a + b) + c = a + (b + c)$

7 Rewrite each expression without exponents.

GS **(a)** x^5

$$x \cdot \underline{} \cdot \underline{} \cdot \underline{} \cdot \underline{}$$

x is used as a factor 5 times.

GS **(b)** $4a^2b^2$

$$4 \cdot \underline{} \cdot \underline{} \cdot \underline{} \cdot \underline{}$$

(c) $-10xy^3$

(d) s^4tu^2

EXAMPLE 7 **Understanding Exponents Used with Variables**

Rewrite each expression without exponents.

(a) y^6 can be written as $\underbrace{y \cdot y \cdot y \cdot y \cdot y \cdot y}$

y is used as a factor 6 times.

(b) $12bc^3$ can be written as $12 \cdot b \cdot \underbrace{c \cdot c \cdot c}_{c^3}$

Coefficient is 12

The exponent applies *only* to c.

(c) $-2m^2n^4$ can be written as $-2 \cdot \underbrace{m \cdot m}_{m^2} \cdot \underbrace{n \cdot n \cdot n \cdot n}_{n^4}$

Coefficient is -2

◀ **Work Problem 7 at the Side.**

To evaluate an expression with exponents, multiply all the factors.

EXAMPLE 8 **Evaluating Expressions with Exponents**

Evaluate each expression.

(a) x^2 when x is -3

x^2 means $x \cdot x$ Replace each x with -3

$\underbrace{(-3) \cdot (-3)}$ Multiply -3 times -3

9

So x^2 becomes $(-3)^2$, which is $(-3)(-3)$, or **9**

(b) x^3y when x is -4 and y is -10

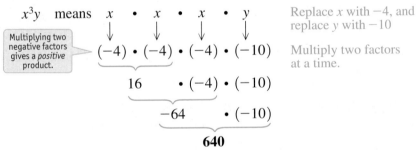

So x^3y becomes $(-4)^3(-10)$, which is $(-4)(-4)(-4)(-10)$, or **640**

8 Evaluate each expression.

GS **(a)** y^3 when y is -5

y^3 means $y \cdot y \cdot y$

$\underbrace{(-5) \cdot (-5)} \cdot (-5)$

$\underbrace{\underline{} \cdot (-5)}$

$\underline{}$

GS **(b)** r^2s^2 when r is 6 and s is 3

r^2s^2 means $r \cdot r \cdot s \cdot s$

$\underline{6} \cdot \underline{} \cdot \underline{3} \cdot \underline{}$

(c) $10xy^2$ when x is 4 and y is -3

(d) $-3c^4$ when c is 2

(c) $-5ab^2$ when a is 5 and b is 3

$-5ab^2$ means $-5 \cdot a \cdot b \cdot b$ Replace a with 5, and replace b with 3

Multiplying two factors with *different* signs gives a *negative* product. ➤ $-5 \cdot 5 \cdot 3 \cdot 3$ Multiply two factors at a time.

$\underbrace{-25} \cdot 3 \cdot 3$

$\underbrace{-75} \cdot 3$

-225

So $-5ab^2$ becomes $-5(5)(3)^2$, which is $-5(5)(3)(3)$, or **-225**

Coefficient is -5

Answers

7. **(a)** $x \cdot x \cdot x \cdot x \cdot x$ **(b)** $4 \cdot a \cdot a \cdot b \cdot b$
 (c) $-10 \cdot x \cdot y \cdot y \cdot y$
 (d) $s \cdot s \cdot s \cdot s \cdot t \cdot u \cdot u$
8. **(a)** $25 \cdot (-5)$; -125
 (b) $6 \cdot 6 \cdot 3 \cdot 3$ is 324
 (c) $10(4)(-3)^2$ is 360
 (d) $-3(2)^4$ is -48

◀ **Work Problem 8 at the Side.**

2.1 Exercises

FOR EXTRA HELP

Go to MyMathLab *for worked-out, step-by-step solutions to exercises enclosed in a square* ▢ *and video solutions to* ▶ *exercises.*

CONCEPT CHECK   *Identify the parts of each expression. Choose from these labels:* variable, constant, coefficient.

1. $c + 4$
GS c is the variable;
  4 is the _____ .

2. $d + 6$
GS   _____ is the variable;
  6 is the constant.

3. $-3 + m$

4. $-4 + n$

5. $5h$

6. $3s$

7. $10 + 2c$

8. $1 + 6b$

9. $x - y$

10. xy

11. $-6g + 9$

12. $-10k + 15$

Evaluate each expression. **See Examples 2–5.**

13. The expression (rule) for ordering robes for the graduation ceremony at West Community College is $g + 10$, where g is the number of graduates. Evaluate the expression when

(a) there are 654 graduates.

GS $g + 10$
   ↓   ↓
  $654 + 10$
  ⎵⎵⎵⎵⎵
  _____ robes

(b) there are 208 graduates.

(c) there are 95 graduates.

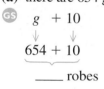

14. Crickets chirp faster as the weather gets hotter. The expression (rule) for finding the approximate temperature (in degrees Fahrenheit) is $c + 37$ where c is the number of chirps made in 15 seconds by a field cricket. (Data from AccuWeather.com) Evaluate the expression when the cricket

(a) chirps 45 times.

GS $c + 37$
   ↓   ↓
  $45 + 37$
  ⎵⎵⎵⎵⎵
  _____ degrees

(b) chirps 33 times.

(c) chirps 58 times.

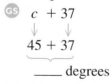

15. The expression for finding the perimeter of a triangle with sides of equal length is $3s$, where s is the length of one side. Evaluate the expression when

(a) the length of one side is 11 inches.

(b) the length of one side is 3 feet.

16. The expression for finding the perimeter of a pentagon with sides of equal length is $5s$, where s is the length of one side. Evaluate the expression when

(a) the length of one side is 25 meters.

(b) the length of one side is 8 inches.

17. The expression for ordering brushes for an art class is $3c - 5$, where c is the class limit. Evaluate the expression when

(a) the class limit is 12.

(b) the class limit is 16.

18. The expression for ordering doughnuts for the office staff is $2n - 4$, where n is the number of people at work. Evaluate the expression when

(a) there are 13 people at work.

(b) there are 18 people at work.

19. The expression for figuring a student's average test score is $\frac{p}{t}$, where p is the total points earned on all the tests and t is the number of tests. Evaluate the expression when

(a) 332 points were earned on 4 tests.

(b) there were 7 tests and 637 points were earned.

20. The expression for deciding how many buses are needed for a group trip is $\frac{p}{b}$, where p is the total number of people and b is the number of people that one bus will hold. Evaluate the expression when

(a) 176 people are going on a trip and one bus holds 44 people.

(b) a bus holds 36 people and 72 people are going on a trip.

Complete each table by evaluating the expressions. **See Example 5.**

21.

Value of x	Expression $x + x + x + x$	Expression $4x$
-2	$(-2) + (-2) + (-2) + (-2)$ is -8	$4 \cdot (-2)$ is -8
12		
0		
-5		

22.

Value of y	Expression $3y$	Expression $y + 2y$
-6	$3(-6)$ is -18	$-6 + 2(-6)$ is $-6 + (-12)$, or -18
10		
-3		
0		

23.

Value of x	Value of y	Expression $-2x + y$
3	7	$-2(3) + 7$ is $-6 + 7$, or 1
-4	5	
-6	-2	
0	-8	

24.

Value of x	Value of y	Expression $-2xy$
3	7	$-2 \cdot 3 \cdot 7$ is -42
-4	5	
-6	-2	
0	-8	

25. Explain the words *variable* and *expression*.

26. Explain the words *coefficient* and *constant*.

Use the variable b to express each of these properties. **See Example 6.**

27. Multiplying a number by 1 leaves the number unchanged.

28. Adding 0 to any number leaves the number unchanged.

29. Any number divided by 0 is undefined.

30. Multiplication distributes over addition. (Use *a*, *b*, and *c* as the variables.)

Rewrite each expression without exponents. **See Example 7.**

31. c^6

32. d^7

33. x^4y^3

34. c^2d^5

35. $-3a^3b$

36. $-8m^2n$

37. $9xy^2$

38. $5ab^4$

39. $-2c^5d$

40. $-4x^3y$

41. a^3bc^2

42. x^2yz^6

In Exercises 43–54, evaluate each expression when r is -3, s is 2, and t is -4. **See Example 8.**

43. t^2

44. r^2

45. rs^3

46. s^4t

47. $3rs$

48. $6st$

49. $-2s^2t^2$

50. $-4rs^4$

51. $r^2s^5t^3$

52. $r^3s^4t^2$

53. $-10r^5s^7$

54. $-5s^6t^5$

Evaluate each expression when x is 4, y is −2, and z is −6.

55. $|xy| + |xyz|$

56. $x + |y^2| + |xz|$

57. $\dfrac{z^2}{-3y + z}$

58. $\dfrac{y^2}{x + 2y}$

Relating Concepts (Exercises 59–60) For Individual or Group Work

At the beginning of this chapter, we said there is an expression to estimate your distance from a thunderstorm. First, count the number of seconds from the time you see a lightning flash until you start to hear the thunder. Then, to estimate the distance (in miles), use the expression $\dfrac{s}{5}$, where s is the number of seconds.

*Use this expression as you **work Exercises 59 and 60 in order.** (The expression is based on the fact that the light from the lightning travels faster than the sound from the thunder.)*

59. Evaluate the thunderstorm expression for each number of seconds. How far away is the storm?

  (a) 15 seconds

  (b) 10 seconds

  (c) 5 seconds

60. Explain how you can use your answers from **Exercise 59** to:

  (a) Estimate the distance when the time is $2\frac{1}{2}$ seconds.

  (b) Find the number of seconds when the distance is $1\frac{1}{2}$ miles.

  (c) Find the number of seconds when the distance is $2\frac{1}{2}$ miles.

Study Skills
USING STUDY CARDS

Y ou may have used "flash cards" in other classes before. In math, study cards can be helpful, too. However, they are different because the main things to remember in math are *not* necessarily terms and definitions; they are *sets of steps to follow* to solve problems (and how to know which set of steps to follow) and *concepts about how math works* (principles). So, the cards will look different but will be just as useful.

In this two-part activity, you will find four types of study cards to use in math. Look carefully at what kinds of information to put on them and where to put it. Then use them the way you would any flash card:

▶ to quickly review when you have a few minutes,

▶ to do daily reviews,

▶ to review before a test.

Remember, the most helpful thing about study cards is making them. It is in the making of them that you have to do the kind of thinking that is most brain friendly and will improve your neural network of dendrites. After each card description you will find an assignment to try. It is marked "**NOW TRY THIS.**"

New Vocabulary Cards

For **new vocabulary cards,** put the word (spelled correctly) and the page number where it is found on the front of the card. On the back, write:

▶ the definition (in your own words if possible),

▶ an example, an exception (if there are any),

▶ any related words, and

▶ a sample problem (if appropriate).

variable *page*____	Front of Card

Definition: A letter representing a number that varies (changes) depending on the situation.

Example: $c + 5$ *where c represents the class limit which varies for different classes.*

variable   *constant*   *varies ⟷ variable*

—Use any letter you like: a, b, c, d, h, k, m, n, o, p, x, y, etc.
—Related words: constant, coefficient

Back of Card

OBJECTIVES

1 Create study cards for all new terms.

2 Create study cards for new procedures.

Why Are Study Cards Brain Friendly?

• Making cards is **active.**

• Cards are **visually** appealing.

• **Repetition** is good for your brain.

For more details, see ***Using Study Cards Revisited*** following the next section.

Study Skills

Procedure ("Steps") Cards

For **procedure cards,** write the name of the procedure at the top on the front of the card. Then write each step **in words.** If you need to know abbreviations for some words, include them along with the whole words written out. On the back, put an example of the procedure, showing each step you need to take. You can review by looking at the front and practicing a new worked example, or by looking at the back and remembering what the procedure is called and what the steps are.

Evaluating an expression

Step 1: Replace each variable with the value you are given.

Step 2: Do the calculations, following the order of operations.

Front of Card

Example: Evaluate $3c - 5$ when c is 20
                        must be given the replacement value

$3c - 5$
$\downarrow$
$3(20) - 5$ Replace c with 20.
           Do the calculations.

$60 - 5$ Follow order of operations; multiply first, then subtract.

55 _No variables in final answer._

Back of Card

2.2   Simplifying Expressions

OBJECTIVE ▶ ❶ Combine like terms using the distributive property. In the previous section, the expression for ordering math textbooks was $c + 5$. This expression was simple and easy to use. Sometimes expressions are *not* written in the simplest possible way. For example:

OBJECTIVES

❶ Combine like terms using the distributive property.

❷ Simplify expressions.

❸ Use the distributive property to multiply.

Evaluate this expression when c is 20.

$$c + 5 \quad \text{Replace } c \text{ with 20}$$
$$\downarrow$$
$$\underbrace{20 + 5} \quad \text{Add } 20 + 5$$
$$25$$

Evaluate this expression when c is 20.

$$2c - \quad 10 - \quad c \quad + 15 \quad \text{Replace } c \text{ with 20}$$
$$\downarrow \qquad\qquad \downarrow$$
$$2 \cdot 20 - \quad 10 - \quad 20 \quad + 15 \quad \text{Multiply } 2 \cdot 20$$
$$\underbrace{40} - \quad 10 - \quad 20 \quad + 15 \quad \text{Change subtraction to adding the opposite.}$$
$$\downarrow \qquad \downarrow \qquad \downarrow \qquad \downarrow$$
$$40 \quad + (-10) + (-20) + 15 \quad \text{Add from left to right.}$$
$$\underbrace{30} \qquad\qquad + (-20) + 15$$
$$\underbrace{10} \qquad\qquad\qquad\qquad + 15$$
$$25$$

These two expressions are actually equivalent. When you evaluate them, the final result is the same, but it takes a lot more work when you use the righthand expression. Once you know how to *simplify expressions,* you can rewrite $2c - 10 - c + 15$ as $c + 5$.

The basic idea in **simplifying expressions** is to *combine,* or *add,* **like terms.** Each addend in an expression is a **term.** Here are two examples.

VOCABULARY TIPS

Constant term   Think about the meaning of the words **constant** and **term.** A constant term is a type of term (or addend) that is a number, not a variable. An example is the number 5 in the expression $2x + 5$.

Variable term   Think about the meaning of the words **variable** and **term.** A variable term is a type of term (or addend) that contains a variable. An example is the $2x$ in the expression $2x + 5$.

┌─── Separates the terms

$c + 5$

└──┴── Two terms

┌────── Separates the terms

$6x^2 + (-2xy) + 8$

└────┴────── Three terms

In $6x^2 + (-2xy) + 8$, the 8 is the **constant term.** There are also two *variable terms* in the expression: $6x^2$ is a variable term, and $-2xy$ is a variable term. A **variable term** has a number part (coefficient) and a letter part (variable).

$6x^2$

Variable part is x^2

Coefficient is 6

$-2xy$

Variable part is xy

Coefficient is -2

If no coefficient is shown, it is assumed to be 1. Remember that multiplying any number by 1 does *not* change the number.

c can be written $1 \cdot c$ or just $1c$

Coefficient is understood to be 1

Variable part is c

Also, $-c$ can be written $-1 \cdot c$. The coefficient of $-c$ is understood to be -1.

Like Terms

Like terms are terms with exactly the same variable parts (the same letters and exponents). The coefficients do *not* have to match.

1 List the like terms in each expression. Then identify the coefficients of the like terms.

(a) $3b^2 + (-3b) + 3 + b^3 + b$

Like terms are $-3b$ and b.

Coefficients are _____ and _____ .

(b) $-4xy + 4x^2y + (-4xy^2) + (-4) + 4$

(c) $5r^2 + 2r + (-2r^2) + 5 + 5r^3$

(d) $-10 + (-x) + (-10x) + (-x^2) + (-10y)$

	Like Terms			Unlike Terms
$5x$ and $3x$	Variable parts match; both are x.	$3x$ and $3x^2$	Variable parts do *not* match; exponents are different.
$-6y^3$ and y^3	Variable parts match; both are y^3.	$-2x$ and $-2y$	Variable parts do *not* match; letters are different.
$4a^2b$ and $5a^2b$	Variable parts match; both are a^2b.	a^3b and a^2b	Variable parts do *not* match; exponents are different.
-8 and 4	There are no variable parts; numbers are like terms.	$-8c$ and 4	Variable parts do *not* match; one term has a variable part, but the other term does not.

EXAMPLE 1	Identifying Like Terms and Their Coefficients

List the like terms in each expression. Then identify the coefficients of the like terms.

(a) $-5x + (-5x^2) + 3xy + x + (-5)$
The like terms are $-5x$ and x.
The coefficient of $-5x$ is -5, and the coefficient of x is understood to be 1.

(b) $2yz^2 + 2y^2z + (-3y^2z) + 2 + (-6yz)$
The like terms are $2y^2z$ and $-3y^2z$.
The coefficients are 2 and -3.

(c) $10ab + 12 + (-10a) + 12b + (-6)$
The like terms are 12 and -6.
The like terms are **constants** ◁ Numbers alone (no variable parts) are like terms.
(there are no variable parts).

◀ **Work Problem 1** at the Side.

The distributive property can be used "in reverse" to combine like terms. Here is an example.

$$\underbrace{3 \bullet x}_{3x} + \underbrace{4 \bullet x}_{4x} \quad \text{can be written as} \quad \underbrace{(3 + 4)}_{7} \bullet x$$

$$7x \quad ◁ \; \substack{7x \text{ is written} \\ \text{in simplified form.}}$$

Thus, $3x + 4x$ can be written in *simplified form* as $7x$. To check that $3x + 4x$ is the same as $7x$, evaluate each expression when x is 2.

$$\underset{\downarrow}{3x} + \underset{\downarrow}{4x} \qquad\qquad \underset{\downarrow}{7x}$$
$$\underbrace{3 \bullet 2}_{6} + \underbrace{4 \bullet 2}_{8} \qquad\qquad \underbrace{7 \bullet 2}_{14}$$
$$\underbrace{\qquad 6 + 8 \qquad}_{14}$$

These expressions are equivalent, so both results are 14. But you can see how much easier it is to work with $7x$, the simplified expression.

Answers

1. (a) Coefficients are -3 and 1.
  (b) -4 and 4; they are constants.
  (c) $5r^2$ and $-2r^2$; the coefficients are 5 and -2.
  (d) $-x$ and $-10x$; the coefficients are -1 and -10.

❶ CAUTION

Notice that $3x + 4x$ is simplified to $7x$, **not** to $7x^2$.

Variable part is unchanged.　　　Do *not* change x to x^2.

Combining Like Terms

Step 1 If there are any variable terms with no coefficient, write in the understood 1.

Step 2 If there are any subtractions, change each one to adding the opposite.

Step 3 Find *like* terms (the variable parts match).

Step 4 Add the coefficients (number parts) of like terms. ***The variable part stays the same.***

EXAMPLE 2　Combining Like Terms

Combine like terms.

(a) $2x + 4x + x$

$2x + 4x + x$ 　　No coefficient; write understood 1.
　　　　　　　　There are no subtractions to change.

$2x + 4x + 1x$ 　　Find like terms: $2x$, $4x$, and $1x$ are like terms.
　　　　　　　　Add the coefficients, $2 + 4 + 1$

$(2 + 4 + 1)x$ 　　The variable part, x, stays the same.

$7x$

Therefore, $2x + 4x + x$ can be written as **$7x$** 　Do **not** write $7x^3$. Keep x as x.

(b) $-3y^2 - 8y^2$

$-3y^2 - 8y^2$ 　　Both coefficients are shown.
　　　　　　　　Change subtraction to adding the opposite.

$-3y^2 + (-8y^2)$ 　　Find like terms: $-3y^2$ and $-8y^2$ are like terms.
　　　　　　　　Add the coefficients, $-3 + (-8)$

$[-3 + (-8)]y^2$ 　　Brackets show what to do first: $-3 + (-8)$ is -11
　　　　　　　　The variable part, y^2, stays the same.

$-11y^2$

Therefore, $-3y^2 - 8y^2$ can be written as **$-11y^2$** 　Do **not** write $-11y^4$. Keep y^2 as y^2.

────────── **Work Problem ❷ at the Side.** ▶

OBJECTIVE ❷ Simplify expressions. When simplifying expressions, be careful to combine only *like* terms—those having variable parts that match. You *cannot* combine terms if the variable parts are different.

---

❷ Combine like terms.

ⓖⓢ (a) $10b + 4b + 10b$

$(10 + 4 + 10)b$

_____ b

(b) $y^3 + 8y^3$

(c) $-7n - n$

(d) $3c - 5c - 4c$

(e) $-9xy + xy$

(f) $-4p^2 - 3p^2 + 8p^2$

(g) $ab - ab$

Answers

2. (a) $24b$ **(b)** $9y^3$ **(c)** $-8n$ **(d)** $-6c$
(e) $-8xy$ **(f)** $1p^2$, or just p^2
(g) 0, because $1ab + (-1ab)$ is $0ab$, and 0 times anything is 0.

3 Simplify each expression by combining like terms.

(a) $3b^2 + 4d^2 + 7b^2$

Rewrite the expression so the like terms are next to each other.

$3b^2 + 7b^2 + 4d^2$

$(3 + 7)b^2 + 4d^2$

_____ b^2 + _____

(b) $4a + b - 6a + b$

(c) $-6x + 5 + 6x + 2$

(d) $2y - 7 - y + 7$

(e) $-3x - 5 + 12 + 10x$

EXAMPLE 3 Simplifying Expressions

Simplify each expression by combining like terms.

(a) $6xy + 2y + 3xy$

The *like* terms are $6xy$ and $3xy$. We can use the **commutative property** to rewrite the expression so that the like terms are next to each other. This helps to organize our work.

$6xy + 3xy + 2y$	Combine *like* terms only.
$(6 + 3)xy + 2y$	Add the coefficients, $6 + 3$. The variable part, xy, stays the same.
$9xy + 2y$	Keep writing $2y$, the term that was not combined; it is still part of the expression.

The simplified expression is **$9xy + 2y$** — Remember to include $2y$ in your answer.

(b) Here is the expression from the first page in this section.

$2c - 10 - c + 15$	Write the understood 1 as the coefficient of c.
$2c - 10 - 1c + 15$	Change subtractions to adding the opposite.
$2c + (-10) + (-1c) + 15$	We can add in any order, so rewrite the expression so that like terms are next to each other.
$2c + (-1c) + (-10) + 15$	Combine $2c + (-1c)$, and combine $-10 + 15$
$[2 + (-1)]c + 5$	Brackets show what to do first: $2 + (-1)$ is 1. The variable part, c, stays the same.
$1c + 5$	

The simplified expression is $1c + 5$ or just **$c + 5$**

$1c$ is the same as c.

Note

In this text, when combining like terms we will usually write the variable terms in *alphabetical order*. A constant term (number only) will be written last. So, in **Examples 3(a) and 3(b)** above, the preferred and alternative ways of writing the expressions are as follows.

The simplified expression is $9xy + 2y$ (alphabetical order). However, using the commutative property of addition, $2y + 9xy$ is also correct.

The simplified expression is $c + 5$ (constant written last). However, using the commutative property of addition, $5 + c$ is also correct.

◀ **Work Problem 3 at the Side.**

We can use the **associative property of multiplication** to simplify an expression such as $4(3x)$.

$4(3x)$ can be written as $4 \cdot (3 \cdot x)$

Understood multiplications

Answers

3. **(a)** $10b^2 + 4d^2$ **(b)** $-2a + 2b$
   (c) 7 **(d)** y **(e)** $7x + 7$

Using the associative property, we can regroup the factors.

$4 \cdot (3 \cdot x)$ can be written as $(4 \cdot 3) \cdot x$ To simplify, multiply $4 \cdot 3$

$\underbrace{12}_{} \cdot x$ Write $12 \cdot x$ without the multiplication dot.

$12x$

The simplified expression is $12x$.

EXAMPLE 4   **Simplifying Multiplication Expressions**

Simplify.

(a) $5(10y)$

Use the associative property.

$5 \cdot (10 \cdot y)$ can be written as $(5 \cdot 10) \cdot y$ Multiply $5 \cdot 10$

$\underbrace{50}_{} \cdot y$ Write $50 \cdot y$ without the multiplication dot.

$50y$

So, $5(10y)$ simplifies to **$50y$**

(b) $-6(3b)$

Use the associative property.

$-6(3b)$ can be written as $\underbrace{(-6 \cdot 3)}_{} b$

$-18b$

So, $-6(3b)$ simplifies to **$-18b$**

(c) $-4(-2x^2)$

Use the associative property.

> Brackets show what to do first: $-4 \cdot (-2)$ is 8.

$-4(-2x^2)$ can be written as $\underbrace{[-4 \cdot (-2)]}_{} x^2$

$8x^2$

So, $-4(-2x^2)$ simplifies to **$8x^2$**

─────── **Work Problem ④ at the Side.** ▶

OBJECTIVE ❸ **Use the distributive property to multiply.** The distributive property can also be used to simplify expressions such as $3(x + 5)$. You *cannot* add the terms inside the parentheses because x and 5 are *not* like terms. But notice the understood multiplication dot between the 3 and the parentheses.

$$3(x + 5)$$
$$3 \cdot (x + 5)$$

Thus you can *distribute* multiplication over addition, as you have done before. That is, multiply 3 times each term inside the parentheses.

$3 \cdot (x + 5)$ can be written as $\underbrace{3 \cdot x}_{3x} + \underbrace{3 \cdot 5}_{15}$

So, $3(x + 5)$ simplifies to $3x + 15$.

└── Stays as addition ──┘

---

④ Simplify.

GS **(a)** $7(4c)$ means $7 \cdot (4 \cdot c)$

Using the associative property, it can be rewritten as

$\underbrace{(7 \cdot 4)}_{} \cdot c$

$\underbrace{\quad\quad}_{} \cdot c$

$\underline{\quad\quad}$ Rewrite without the multiplication dot.

(b) $-3(5y^3)$

(c) $20(-2a)$

(d) $-10(-x)$

5 Simplify.

(a) $7(a + 10)$

$7(a + 10)$

$7 \cdot a + 7 \cdot 10$

$\underline{\quad} + \underline{\quad}$

(b) $3(x - 3)$

$3(x - 3)$

$3 \cdot x - 3 \cdot 3$

$\underline{\quad} - \underline{\quad}$

(c) $4(2y + 6)$

(d) $-5(3b + 2)$

(e) $-8(c + 4)$

Answers

5. **(a)** $7a + 70$ **(b)** $3x - 9$ **(c)** $8y + 24$
   (d) $-15b - 10$ **(e)** $-8c - 32$

EXAMPLE 5   **Using the Distributive Property**

Simplify.

(a) $5(3x + 2)$ can be written as $5 \cdot 3x + 5 \cdot 2$ ← Remember to multiply $5 \cdot 3x$ and $5 \cdot 2$.

$$5 \cdot 3 \cdot x + 10$$
$$15 \cdot x + 10$$
$$15x + 10$$

So, $5(3x + 2)$ simplifies to **$15x + 10$**

(b) $-2(4a + 3)$ can be written as $-2 \cdot 4a + (-2) \cdot 3$

$$-2 \cdot 4 \cdot a + (-6)$$
$$-8 \cdot a + (-6)$$
$$-8a + (-6)$$

Now we will use the definition of subtraction "in reverse" to rewrite $-8a + (-6)$.

Change -6 to its opposite, 6

Write $-8a + (-6)$ as $-8a - 6$

Change addition to subtraction.

Think back to the way we changed subtraction to adding the opposite. Here we are "working backward." From now on, whenever addition is followed by a negative number, we will change it to subtracting a positive number.

$$-8a + (-6)$$
$$-8a - 6$$

Equivalent expressions

So, $-2(4a + 3)$ simplifies to **$-8a - 6$**

(c) $6(y - 4)$

$6[y + (-4)]$ Change subtraction to adding the opposite.

$6[y + (-4)]$ Multiply 6 times each term inside the brackets.

$6 \cdot y + 6(-4)$ Do the multiplications.

$6y + (-24)$ Use the definition of subtraction "in reverse." Change addition to subtraction.

$6y - 24$ Change -24 to its opposite, 24

This is an example of how *multiplication distributes over subtraction*, in the same way that *multiplication distributes over addition* in **Example 5(a)** above. You can show fewer steps, like this:

$6(y - 4)$ can be written as $6 \cdot y - 6(4)$

$$6y - 24$$

So, $6(y - 4)$ simplifies to **$6y - 24$**

◀ **Work Problem** **5** at the Side.

Sometimes you need to do several steps to simplify an expression.

EXAMPLE 6 Simplifying a More Complex Expression

Simplify: $8 + 3(x - 2)$

$8 + 3(x - 2)$	Do *not* add $8 + 3$. Use the distributive property first because multiplying is done *before* adding.
$8 + 3 \cdot x - 3 \cdot 2$	Do the multiplications.
$8 + 3x - 6$	Rewrite so that like terms are next to each other.
$3x + 8 - 6$	Subtract to find $8 - 6$ or change to adding $8 + (-6)$
$3x + 2$	

The simplified expression is $3x + 2$

---

❶ CAUTION

Do ***not*** add $8 + 3$ as the first step in **Example 6** above. Remember that the order of operations tells you to do multiplying *before* adding.

Working with Expressions

- An expression expresses, or tells, a rule for doing something.

- An expression is a combination of operations on variables and numbers. Here are some examples:

$$c + 5 \qquad 4s \qquad 2 + \frac{d}{60}$$

$$4 + 3x - 10 + x \qquad -8a^2b^3 \qquad 7 + 9(2y - 1)$$

- You can **evaluate an expression** by replacing each variable with specific numbers. Then do all possible calculations, following the order of operations. There are **no variables** in the final answer.

  For example, if x is 5, then **evaluating** the expression $4 + 3x - 10 + x$ becomes $4 + 3(5) - 10 + 5$, or 14.

- You can **simplify an expression** by combining (adding) all the like terms. When you simplify an expression, you write it in a simpler way.

  For example, the expression $4 + 3x - 10 + x$ **simplifies** to $4x - 6$.

- **IMPORTANT NOTE:** An expression **does not have an = sign in it**. The next section shows you equations, which **do** have = signs. You will learn how to **solve equations.** So look carefully at the directions for exercises. You can *evaluate an expression,* or *simplify an expression,* but only equations can be *solved.*

Work Problem **6** at the Side. ▶

---

6 Simplify.

(a) $-4 + 5(y + 1)$

$$-4 + 5 \cdot y + 5 \cdot 1$$

$$-4 + \underline{} + \underline{}$$

$$5y - 4 + 5 \qquad \text{Combine like terms.}$$

(b) $2(3w + 4) - 5$

(c) $5(6x - 2) + 3x$

(d) $21 + 7(a^2 - 3)$

(e) $-y + 3(2y + 5) - 18$

Answers

6. **(a)** $-4 + 5y + 5$ is $5y + 1$ **(b)** $6w + 3$
(c) $33x - 10$ **(d)** $7a^2$ **(e)** $5y - 3$

2.2 Exercises

FOR EXTRA HELP

Go to MyMathLab for worked-out, step-by-step solutions to exercises enclosed in a square ▢ and video solutions to ▶ exercises.

CONCEPT CHECK *Circle the like terms in each expression. Then identify the coefficients of the like terms.* **See Example 1.**

1. $2b^2 + 2b + 2b^3 + b^2 + 6$

ⓖ The like terms are $2b^2$ and b^2.

The coefficient of $2b^2$ is _____ and
the coefficient of b^2 is understood to be _____ .

2. $3x + x^3 + 3x^2 + 3 + 2x^3$

3. $-x^2y + (-xy) + 2xy + (-2xy^2)$

4. $ab^2 + (-a^2b) + 2ab + (-3a^2b)$

5. $7 + 7c + 3 + 7c^3 + (-4)$

6. $4d + (-5) + 1 + (-5d^2) + 4$

Simplify each expression. **See Example 2.**

7. $6r + 6r$

ⓖ $(6 + 6)r$

_____ r

8. $4t + 10t$

ⓖ $(4 + 10)t$

_____ t

9. $x^2 + 5x^2$

10. $9y^3 + y^3$

11. $p - 5p$

12. $n - 3n$

13. $-2a^3 - a^3$

14. $-10x^2 - x^2$

15. $c - c$

16. $b^2 - b^2$

17. $9xy + xy - 9xy$

18. $r^2s - 7r^2s + 7r^2s$

19. $5t^4 + 7t^4 - 6t^4$
▶

20. $10mn - 9mn + 3mn$

21. $y^2 + y^2 + y^2 + y^2$

22. $a + a + a$

23. $-x - 6x - x$

24. $-y - y - 3y$

Simplify by combining like terms. Write each answer with the variables in alphabetical order and any constant term last. **See Example 3.**

25. $8a + 4b + 4a$

26. $6x + 5y + 4y$

27. $6 + 8 + 7rs$

28. $10 + 2c^2 + 15$

29. $a + ab^2 + ab^2$

30. $n + mn + n$

31. $6x + y - 8x + y$

32. $d + 3c - 7c + 3d$

33. $8b^2 - a^2 - b^2 + a^2$

34. $5ab - ab + 3a^2b - 4ab$

35. $-x^3 + 3x - 3x^2 + 2$

36. $a^2b - 2ab - ab^3 + 3a^3b$

37. $-9r + 6t - s - 5r + s + t - 6t + 5s - r$

38. $-x - 3y + 4z + x - z + 5y - 8x - y$

Simplify by using the associative property of multiplication. **See Example 4.**

39. $3(10a)$

40. $8(4b)$

41. $-4(2x^2)$

42. $-7(3b^3)$

43. $5(-4y^3)$

44. $2(-6x)$

45. $-9(-2cd)$

46. $-6(-4rs)$

47. $7(3a^2bc)$

48. $4(2xy^2z^2)$

49. $-12(-w)$

50. $-10(-k)$

Use the distributive property to simplify each expression. ***See Example 5.***

51. $6(b + 6)$

GS

$6(b + 6)$

$6 \cdot b + 6 \cdot 6$

_____ + _____

52. $5(a + 3)$

GS

$5(a + 3)$

$5 \cdot a + 5 \cdot 3$

_____ + _____

53. $7(x - 1)$

▶

54. $4(y - 4)$

55. $3(7t + 1)$

▶

56. $8(2c + 5)$

57. $-2(5r + 3)$

▶

58. $-5(6z + 2)$

59. $-9(k + 4)$

60. $-3(p + 7)$

61. $50(m - 6)$

62. $25(n - 1)$

Simplify each expression. ***See Example 6.***

63. $10 + 2(4y + 3)$ Part of the work is done. Now finish it.

GS

$10 + 2(4y + 3)$

$10 + \underbrace{2 \cdot 4} \cdot y + \underbrace{2 \cdot 3}$

64. $4 + 7(x^2 + 3)$ Part of the work is done. Now finish it.

GS

$4 + 7(x^2 + 3)$

$4 + \underbrace{7 \cdot x^2} + \underbrace{7 \cdot 3}$

65. $6(a^2 - 2) + 15$

66. $5(b - 4) + 25$

67. $2 + 9(m - 4)$

▶

68. $6 + 3(n - 8)$

69. $-5(k + 5) + 5k$

70. $-7(p + 2) + 7p$

71. $4(6x - 3) + 12$

72. $6(3y - 3) + 18$

73. $5 + 2(3n + 4) - n$

74. $8 + 8(4z + 5) - z$

75. $-p + 6(2p - 1) + 5$

76. $-k + 3(4k - 1) + 2$

77. Explain the difference between *simplifying* an expression and *evaluating* an expression.

78. Simplify each expression. Are the answers equivalent? Explain why or why not.

$$5(3x + 2) \qquad 5(2 + 3x)$$

79. Explain what makes two terms *like* terms. Include several examples in your explanation.

80. Explain how to combine like terms. Include an example in your explanation.

81. CONCEPT CHECK   The work shown below has a mistake in it.

What Went Wrong?   First write a sentence explaining what the mistake is. Then fix the mistake and find the correct solution.

Joaquin simplified the expression this way:

$$\underbrace{-2x + 7x}_{} + 8$$
$$5x^2 + 8$$

82. CONCEPT CHECK   The work shown below has a mistake in it.

What Went Wrong?   First write a sentence explaining what the mistake is. Then fix the mistake and find the correct solution.

Paetyn simplified the expression this way:

$$\underbrace{-10a + 6a}_{} \underset{\downarrow}{-} \underset{\downarrow}{7} + 2$$
$$-4a \quad + \underbrace{(-7) + 2}_{}$$
$$-4a \quad + \quad (-5)$$
$$4a - 5$$

Simplify.

83. $-4(3y) - 5 + 2(5y + 7)$

84. $6(-3x) - 9 + 3(-2x + 6)$

85. $-10 + 4(-3b + 3) + 2(6b - 1)$

86. $12 + 2(4a - 4) + 4(-2a - 1)$

87. $-5(-x + 2) + 8(-x) + 3(-2x - 2) + 16$

88. $-7(-y) + 6(y - 1) + 3(-2y) + 6 - y$

Study Skills

OBJECTIVES

1. Create study cards for difficult problems.
2. Create study cards of quiz problems.

This is the second part of the Study Cards activity. As you get further into a chapter, you can choose particular problems that will serve as a good test review. Here are two more types of study cards that will help you.

Tough Problem Cards

When you are doing your homework and find yourself saying, "This is really hard" or "I'm worried I'll make a mistake," make a **tough problem** study card! On the front, write out the procedure to work the type of problem *in words*. Include any special notes, like what *not* to do. On the back, work at least one example; make sure you label what you are doing.

Simplifying Expressions   *page____*   Front of Card

    Combine (add) <u>like</u> terms only.

    Like terms have variable parts that match.

    Use the distributive property if possible.

Example   Simplify this expression.   Back of Card

$7 + 3(2n - 4)$	*Do <u>not</u> add 7 + 3!*
$7 + 3(2n - 4)$	*Use the distributive property.*
$7 + 3 \cdot 2n - 3 \cdot 4$	*Do the multiplications.*
$7 + 6n - 12$	*Rewrite so like terms are next to each other.*
$6n + 7 + (-12)$	*Change subtraction to adding the opposite.*
$6n + (-5)$	*Do <u>not</u> stop here!*
$6n - 5$	*Use definition of subtraction "in reverse."*

Now Try This

Choose three types of difficult problems or problems that have given you trouble, and work them out on study cards. Be sure to put the words for solving the problem on one side and the worked problem on the other side.

Practice Quiz Cards

Make up a few **quiz cards** for each type of problem you learn, and use them to prepare for a test. Choose two or three problems from the different sections of the chapter. Be sure you don't just choose the easiest problems! Put the problem

with the direction words (such as *solve, simplify, evaluate*) on the front, and work the problem on the back. If you like, put the page number from the text there, too. When you review, you work the problem on a separate paper, and check it by looking at the back.

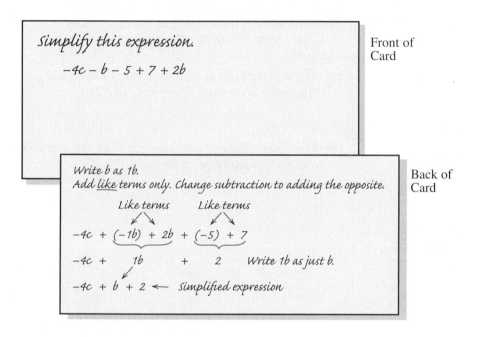

simplify this expression.

$-4c - b - 5 + 7 + 2b$

Front of Card

Write b as 1b.
Add like terms only. Change subtraction to adding the opposite.

Like terms ⟋⟍ Like terms ⟋⟍

$-4c + (-1b) + 2b + (-5) + 7$

$-4c + 1b + 2$ *Write 1b as just b.*

$-4c + b + 2$ ← *simplified expression*

Back of Card

Why Are Study Cards Brain Friendly?

▶ Making the study cards is an **active technique** that really gets your dendrites growing. You have to make decisions about what is most important and how to put it on the card. This kind of thinking is more in depth than just memorizing, and as a result, you will understand the concepts better and remember them longer.

▶ The cards are **visually appealing** (if you write neatly and use some color). Your brain responds to pleasant visual images, and again, you will remember longer and may even be able to "picture in your mind" how your cards look. This will help you during tests.

▶ Because study cards are small and portable, you can review them easily whenever you have a few minutes. Even while you're waiting for a bus or have a few minutes between classes you can take out your cards and read them to yourself. Your **brain really benefits from repetition;** each time you review your cards your dendrites are growing thicker and stronger. After a while, the information will become automatic and you will remember it for a long time.

Summary Exercises *Variables and Expressions*

CONCEPT CHECK   *Identify the parts of each expression. Choose from these labels:*
variable, constant, coefficient.

1. $-10 - m$

2. $-8cd$

3. $6 + 4x$

4. The expression for finding the perimeter of an octagon with sides of equal length is $8s$. Evaluate the expression when

   (a) the length of one side is 4 yards.

   (b) the length of one side is 15 inches.

5. The expression for the total cost of a car is $d + mt$ where d is the down payment, m is the monthly payment, and t is the number of months you must make payments. Evaluate the expression for each situation.

   (a) The down payment on a used car is $3000 and you make 36 monthly payments of $280.

   (b) The down payment on a new car is $1750 and you make 48 monthly payments of $429.

CONCEPT CHECK   *Rewrite each expression without exponents.*

6. ad^4

7. b^3cd

8. $-7ab^5c^2$

Evaluate each expression when w is 5, x is −2, y is −6, and z is 0.

9. w^4

10. $5xz$

11. yz^2

12. wxy

13. x^3

14. $-4wy$

15. $3xy^2$

16. w^2x^5

17. $-7w\,x^4y^3$

Simplify each expression.

18. $10b + 4b + 10b$

19. $-3x - 5 + 12 + 10x$

20. $-8(c + 4)$

21. $-9xy + 9xy$

22. $-4(-3c^2d)$

23. $3f - 5f - 4f$

24. $2(3w + 4)$

25. $-a - 6b - a$

26. $-10(-5x^3y^2)$

27. $5r^3 + 2r^2 - 2r^2 + 5r^3$

28. $21 + 7(h^2 - 3)$

29. $-3(m + 3) + 3m$

30. $4(8y - 5) + 5$

31. $2 + 12(3x - 1)$

32. $-n + 5(4n - 2) + 11$

33. CONCEPT CHECK   The work shown below has **two** mistakes in it.

What Went Wrong?   First write a sentence explaining what each mistake is. Then fix the mistakes and find the correct solution.

Kalley simplified two expressions.

She simplified the first expression this way:
$$6(n + 2)$$
$$6n + 2$$

She simplified the second expression this way:
$$-5(-4a)$$
$$-20a$$

34. CONCEPT CHECK   The work shown below has a mistake in it.

What Went Wrong?   First write a sentence explaining what the mistake is. Then fix the mistake and find the correct solution.

Hunter simplified the expression this way:
$$\underbrace{-8x + x}_{-7x} - \underbrace{12 + 3}_{15}$$

2.3 | Solving Equations Using Addition

Now you are ready for a look at the "heart" of algebra, writing and solving **equations.** *Writing* an equation is a way to show the relationship between what you *know* about a problem and what you *don't* know. Then, *solving* the equation is a way to figure out the part that you didn't know and answer your question.

The questions you can answer by writing and solving equations are as varied as the careers people choose. A zookeeper can solve an equation that answers the question of how long to incubate the egg of a particular tropical bird. A fitness instructor can solve an equation that answers the question of how hard a certain person should exercise for maximum benefit.

OBJECTIVE ▶ 1 Determine whether a given number is a solution of an equation. Let's start with the example from the beginning of this chapter: ordering textbooks for math classes. The expression we used to order books was $c + 5$, where c was the class limit (the maximum number of students allowed in the class). Suppose that 30 prealgebra books were ordered. What is the class limit for prealgebra? To answer this question, write an equation showing the relationship between what you know and what you don't know.

You don't know   You do know the total
the class limit.   ↓   ↓   number of books ordered.

$$c + 5 = 30$$

            ↑——— You do know that 5 extra books were ordered.

> **Note**
>
> An equation has an equal sign. Notice the similarity in the words equa**tion** and equa**l.** An *expression* does *not* have an equal sign.

The equal sign in an equation is like the balance point on a playground teeter-totter, or seesaw. To have a true equation, the two sides must balance.

These equations balance, so we can use the = sign.

$$6 + 8 = 14$$

$$10 = 5 \cdot 2$$

$$3 \cdot 2 = 5 + 1$$

These equations do *not* balance, so we write ≠ to mean "is not equal to."

$$6 + 8 \neq 15$$

$$10 \neq 4 \cdot 2$$

$$4 + 5 \neq 5 \cdot 4$$

When an equation has a variable, we **solve the equation** by finding a number that can replace the variable and make the equation balance. For the example about ordering prealgebra textbooks:

$$c + 5 = 30$$

What number can replace c so that the equation balances?

Try replacing c with 15. $15 + 5 \neq 30$ Does *not* balance: $15 + 5$ is only 20

Try replacing c with 40. $40 + 5 \neq 30$ Does *not* balance: $40 + 5$ is more than 30

Try replacing c with 25. $25 + 5 = 30$ Balances: $25 + 5$ is 30

The **solution** is 25 because 25 is the *only* number that makes the equation balance. By solving the equation, you have answered the question about the class limit for prealgebra. The class limit is 25.

Note

Most of the equations that you will solve in this text have only one solution, that is, one number that makes the equation balance. Later in this text, and in other algebra courses, you will solve equations that have two or more solutions.

EXAMPLE 1 Identifying the Solution of an Equation

Which of these numbers, 70, 40, or 60, is the solution of the equation $c - 10 = 50$?

Replace c with each of the numbers. The one that makes the equation balance is the solution.

$70 - 10 \neq 50$

$40 - 10 \neq 50$

The solution is 60.
$60 - 10 = 50$

Does *not* balance: $70 - 10$ is more than 50

Does *not* balance: $40 - 10$ is only 30

Balances: $60 - 10$ is 50

The solution is **60** because, when c is 60, the equation balances.

──────── **Work Problem 1 at the Side.** ▶

1 **(a)** Which of these numbers, 95, 65, 80, or 70, is the solution of the equation $c + 15 = 80$?

Try replacing c with 95.
$95 + 15 \neq 80$

Does **not** balance: $95 + 15$ is more than 80.

Now try the other numbers. Which one makes the equation balance? ____

(b) Which of these numbers, 28, 20, 24, or 32, is the solution of the equation $28 = c - 4$?

Answers

1. **(a)** The solution is 65.
   (b) The solution is 32.

OBJECTIVE 2 **Solve equations using the addition property of equality.**
When solving the book-ordering equation, $c + 5 = 30$, you could just look at the equation and think, "What number, plus 5, would balance with 30?" You could easily see that c had to be 25. Not all equations can be solved this easily, so you'll need some tools for the harder ones. The first tool, called the **addition property of equality,** allows us to add the *same* number to *both* sides of an equation.

Addition Property of Equality

If $a = b$, then $a + c = b + c$.

In other words, you may add the *same* number to *both* sides of an equation and still keep it balanced.

Think of the teeter-totter. If there are 3 children of the same weight on each side, it will balance. If 2 more children climb onto the left side, the only way to keep the balance is to have 2 more children of the same weight climb onto the right side as well.

> Keep the equation balanced by adding the same number to both sides.

2 more   2 more

$$3 = 3 \qquad\qquad 3 + 2 = 3 + 2$$

All the tools you will learn to use with equations have one goal.

Goal in Solving an Equation

The goal is to end up with the variable (the letter) on one side of the equal sign balancing a number on the other side.

We work on the original equation until we get:

$$\text{variable} = \text{number} \qquad \text{or} \qquad \text{number} = \text{variable}$$

Once we arrive at that point, the number balancing the variable is the solution to the original equation.

EXAMPLE 2	**Using the Addition Property of Equality**

Solve each equation and check the solution.

(a) $c + 5 = 30$

We want to get the variable, c, by itself on the left side of the equal sign. To do that, we add the *opposite* of 5, which is -5. Then $5 + (-5)$ will be 0.

$$c + 5 = 30$$

Add -5 to the left side. $\longrightarrow$ $\underline{\quad -5 \qquad -5 }$ $\longleftarrow$ To keep the balance, add -5 to the right side also.

$$c + 0 = 25 \longleftarrow 30 + (-5) \text{ is } 25$$

$5 + (-5)$ is 0 $\longrightarrow$

Adding 0 to any number leaves the number unchanged, so $c + 0$ is c.

$$\underbrace{c + 0}_{} = 25$$
$$c \quad = 25$$

Because c *balances* with 25, the *solution* is **25**

Check the solution by replacing c with 25 in the *original equation*.

$$c + 5 = 30 \qquad \text{Original equation}$$
$$\underbrace{25 + 5}_{} = 30 \qquad \text{Replace } c \text{ with } 25 \quad \boxed{\text{The solution is 25, \textbf{not} 30.}}$$
$$30 \quad = 30 \quad \checkmark \quad \text{Balances}$$

Because the equation balances when we use 25 to replace the variable, we know that **25 is the correct solution**. If it had *not* balanced, we would need to rework the problem, find our error, and correct it.

(b) $-5 = x - 3$

We want the variable, x, by itself on the right side of the equal sign. (Remember, it doesn't matter which side of the equal sign the variable is on, just so it ends up by itself.) To see what number to add, we change the subtraction to adding the opposite.

$$-5 = x - \quad 3 \qquad \text{Change subtraction to adding the opposite.}$$
$$\qquad \downarrow \quad \downarrow$$
$$-5 = x + (-3) \qquad \text{To get } x \text{ by itself on the right side, add the opposite of } -3, \text{ which}$$

To keep the balance, $\qquad \underline{3 \qquad\qquad 3}$ is 3. Then $-3 + 3$ is 0
add 3 to the left side also.
$-5 + 3$ is -2 $\qquad -2 = \underbrace{x +}_{} \quad 0 \qquad$ Adding 0 to x leaves x unchanged.

$$-2 = \quad x \qquad \text{x balances with -2, so -2 is the solution.}$$

Because -2 *balances* with x, the solution is -2

We check the solution by replacing x with -2 in the *original equation*. If the equation balances when we use -2, we know that it is the correct solution. If the equation does *not* balance when we use -2, we made an error and need to try solving the equation again.

Continued on Next Page

2 Solve each equation and check each solution.

GS **(a)** $12 = y + 5$

To get y by itself, add the opposite of 5, which is -5. To keep the balance, add -5 to *both* sides.

$$12 = y + 5$$
$$\underline{-5 \qquad -5}$$
$$7 = y + \_\_$$
$$7 = \_\_\_$$

CHECK

$12 = y + 5$ Original equation

$\downarrow$

$12 = \underbrace{7 + 5}$ Replace y with 7.

$12 = \_\_\_$

Does it balance? ____

So the solution is ____

(b) $b - 2 = -6$

CHECK

CHECK $-5 = x - 3$ Original equation

$\qquad\qquad\quad \downarrow$

$-5 = -2 - 3$ Replace x with -2

$-5 = \underbrace{-2 + (-3)}$ Change subtraction to adding the opposite.

$-5 = \quad -5$ ✓ Balances; this means -2 is the correct solution.

When x is replaced with -2, the equation balances, so -2 **is the correct solution**.

> **❗ CAUTION**
>
> When checking the solution to **Example 2(b)** above, we ended up with $-5 = -5$. Notice that -5 is *not* the solution. The solution is -2, the number used to replace x in the original equation.

◀ **Work Problem ② at the Side.**

OBJECTIVE ▶ ③ Simplify equations before using the addition property of equality. Sometimes you can simplify the expression on one or both sides of the equal sign. Doing so will make it easier to solve the equation.

EXAMPLE 3 **Simplifying before Solving Equations**

Solve each equation and check each solution.

(a) $y + 8 = 3 - 7$

You cannot simplify the left side because y and 8 are *not* like terms.

$y + 8 = 3 - 7$ Simplify the right side by changing subtraction to adding the opposite.

$y + 8 = \underbrace{3 + (-7)}$ Add $3 + (-7)$ to get -4

To get y by itself on the left side, add the opposite of 8, which is -8

$y + 8 = \quad -4$

$\underline{-8 \qquad\quad -8}$ To keep the balance, add -8 to the right side also.

$8 + (-8)$ is 0

$\underbrace{y + 0} = -12$ $-4 + (-8)$ is -12

$y \quad = -12$

The solution is -12. Now check the solution.

CHECK $y + 8 = 3 - 7$ Go back to the *original* equation and replace y with -12

$\quad \downarrow$

Add $-12 + 8$ $\underbrace{-12 + 8} = 3 - 7$ Change $3 - 7$ to $3 + (-7)$

$-4 \quad = \underbrace{3 + (-7)}$ Add $3 + (-7)$ to get -4

$-4 \quad = \quad -4$ ✓ Balances; so -12 is the correct solution.

When y is replaced with -12, the equation balances, so -12 **is the correct solution** (not -4).

─── **Continued on Next Page**

Answers

2. **(a)** $7 = y + 0; 7 = y$

CHECK $12 = y + 5$

$\qquad\qquad \downarrow$

$12 = \underbrace{7 + 5}$

Balances $12 = 12$ ✓

So the solution is 7.

(b) $b = -4$

CHECK $b - 2 = -6$

$\quad \downarrow$

$\underbrace{-4 + (-2)} = -6$

Balances $-6 = -6$ ✓

So the solution is -4.

(b) $-2 + 2 = -4b - 6 + 5b$

Simplify the left side by adding $-2 + 2$

$$-2 + 2 = -4b - 6 + 5b$$

Simplify the right side by changing subtraction to adding the opposite.

$$0 = -4b + (-6) + 5b$$

$$0 = \underbrace{-4b + 5b} + (-6)$$ Find like terms.

$$0 = 1b + (-6)$$ Combine $-4b + 5b$.

To keep the balance, add 6 to the left side also.

$$\underline{6} \qquad \qquad \underline{6}$$ ← To get $1b$ by itself, add the opposite of -6, which is 6

$$6 = 1b + 0$$

$$6 = 1b$$ $1b$ is equivalent to b.

$$\mathbf{6} = \mathbf{b}$$

The solution is **6**.

CHECK $\qquad -2 + 2 = -4b - 6 + 5b$

Go back to the *original* equation and replace each b with 6

Add $-2 + 2$ $\quad -2 + 2 = -4 \cdot 6 - 6 + 5 \cdot 6$

On the right side, do multiplications first.

$$0 = \underbrace{-24 + (-6)} + 30$$ Change subtraction to adding the opposite.

$$0 = \underbrace{-30 + 30}$$ Add from left to right.

$$0 = 0 \qquad ✓ \text{ Balances}$$

When b is replaced with 6, the equation balances, so **6 is the correct solution (not 0)**.

---

❶ CAUTION

When checking a solution, always go back to the *original* equation. That way, you will catch any errors you made when simplifying each side of the equation.

— **Work Problem ③ at the Side.** ▶

---

③ Simplify each side of the equation when possible. Then solve the equation and check the solution.

ᴳˢ (a) $2 - 8 = k - 2$

$$2 + (-8) = k + (-2) \text{ Simplify.}$$

$$-6 = k + (-2)$$

$$\underline{+2} \qquad \qquad \underline{+2} \text{ Add +2 to both sides.}$$

$$\underline{} = k + \underline{}$$

$$\underline{} = \underline{}$$

CHECK

$$2 - 8 = k - 2 \quad \text{Original equation}$$

$$2 - 8 = -4 - 2 \quad \text{Replace } k \text{ with } -4.$$

$$2 + (-8) = -4 + (-2)$$

$$\underline{} = \underline{}$$

Does it balance? ____

So the solution is ____.

(b) $4r + 1 - 3r = -8 + 11$

CHECK

Answers

3. (a) $-4 = k + 0; -4 = k$

  CHECK $\quad 2 - 8 = k - 2$

$$2 + (-8) = -4 + (-2)$$

  Balances $\quad -6 = -6$ ✓
  So the solution is -4.

(b) $r = 2$

  CHECK $\quad 4r + 1 - 3r = -8 + 11$

$$4 \cdot 2 + 1 - 3 \cdot 2 = -8 + 11$$

$$8 + 1 - 6 = 3$$

$$\underbrace{9 + (-6)} = 3$$

  Balances $\quad 3 = 3$ ✓
  So the solution is 2.

2.3 Exercises

FOR EXTRA HELP

Go to MyMathLab *for worked-out, step-by-step solutions to exercises enclosed in a square* ▢ *and video solutions to* ▶ *exercises.*

CONCEPT CHECK *In each list of numbers, find the one that is a solution of the given equation. See Example 1.*

1. $n - 50 = 8$

58, 42, 60, 8

2. $r - 20 = 5$

5, 15, 30, 25

3. $-6 = y + 10$
▶
$-4, -16, 16, -6$

4. $-4 = x + 13$

$-4, 17, -17, -9$

CONCEPT CHECK *To solve each equation, state the number you would add to both sides of the equation. Then explain why you picked that number. Do **not** finish the solution.*

5. (a) $m - 8 = 1$

(b) $-7 = w + 5$

6. (a) $n + 2 = -9$

(b) $10 = b - 6$

*Solve each equation and check each solution. **See Example 2.***

7. $p + 5 = 9$
GS

$$\begin{array}{r} p + 5 = 9 \\ \underline{-5 \quad -5} \\ p + 0 = 4 \\ \underbrace{} \\ \underline{} = \underline{} \end{array}$$

CHECK $p + 5 = 9$
↓

8. $a + 3 = 12$
GS

$$\begin{array}{r} a + 3 = 12 \\ \underline{-3 \quad -3} \\ a + 0 = 9 \\ \underbrace{} \\ \underline{} = \underline{} \end{array}$$

CHECK $a + 3 = 12$
↓

9. $8 = r - 2$
GS

$$\begin{array}{r} 8 = r + (-2) \\ \underline{+2 \qquad +2} \\ 10 = r + 0 \\ \underbrace{} \\ \underline{} = \underline{} \end{array}$$

CHECK $8 = r - 2$
↓

10. $3 = b - 5$
GS

$$\begin{array}{r} 3 = b + (-5) \\ \underline{+5 \quad + \quad 5} \\ 8 = b + 0 \\ \underbrace{} \\ \underline{} = \underline{} \end{array}$$

CHECK $3 = b - 5$
↓

11. $-5 = n + 3$
▶

CHECK

12. $-1 = a + 8$

CHECK

13. $-4 + k = 14$

CHECK

14. $-9 + y = 7$

CHECK

15. $y - 6 = 0$

CHECK

16. $k - 15 = 0$

CHECK

17. $7 = r + 13$ **CHECK**

18. $12 = z + 19$ **CHECK**

19. $x - 12 = -12$ **CHECK**

20. $-3 = m - 3$ **CHECK**

21. $-5 = -2 + t$ **CHECK**

22. $-1 = -10 + w$ **CHECK**

A solution is given for each equation. Show how to check the solution. If the solution is correct, leave it. If the solution is not correct, solve the equation and check your new solution. ***See Example 2.***

23. $z - 5 = 3$
The solution is -2.   **CHECK** $z - 5 = 3$

24. $x - 9 = 4$
The solution is 13.   **CHECK** $x - 9 = 4$

25. $7 + x = -11$
The solution is -18.   **CHECK**

26. $2 + k = -7$
The solution is -5.   **CHECK**

27. $-10 = -10 + b$
The solution is 10.   **CHECK**

28. $0 = -14 + a$
The solution is 0.   **CHECK**

Simplify each side of the equation when possible. Then solve the equation and check the solution as you did in Exercises 7–28. Show your work. **See Example 3.**

29. $c - 4 = -8 + 10$ **CHECK**

30. $b - 8 = 10 - 6$ **CHECK**

31. $-1 + 4 = y - 2$ **CHECK**

32. $2 + 3 = k - 4$ **CHECK**

33. $10 + b = -14 - 6$ **CHECK**

34. $1 + w = -8 - 8$ **CHECK**

35. $t - 2 = 3 - 5$ **CHECK**

36. $p - 8 = -10 + 2$ **CHECK**

37. $10z - 9z = -15 + 8$ **CHECK**

38. $2r - r = 5 - 10$ **CHECK**

39. $-5w + 2 + 6w = -4 + 9$ **CHECK**

40. $-2t + 4 + 3t = 6 - 7$ **CHECK**

Solve each equation. Show your work. ***See Examples 2 and 3.***

41. $-3 - 3 = 4 - 3x + 4x$

42. $-5 - 5 = -2 - 6b + 7b$

43. $-3 + 7 - 4 = -2a + 3a$

44. $6 - 11 + 5 = -8c + 9c$

45. $y - 75 = -100$

46. $a - 200 = -100$

47. $-x + 3 + 2x = 18$

48. $-s + 2s - 4 = 13$

49. $82 = -31 + k$

50. $-5 = 72 + w$

51. $-2 + 11 = 2b - 9 - b$

52. $-6 + 7 = 2h - 1 - h$

53. $r - 6 = 7 - 10 - 8$

54. $m - 5 = 2 - 9 + 1$

55. $-14 = n + 91$

56. $66 = x - 28$

57. $-9 + 9 = 5 + h$

58. $18 - 18 = 6 + p$

59. CONCEPT CHECK A student did this work when solving an equation. Do you agree that the solution is -7? Explain why or why not.

$$\underbrace{-8 + 1}_{} = x + \;\; 7$$
$$-7 \;\;\; = x + \;\; 7$$
$$\underline{-7 -7}$$
$$-14 \;\; = \underbrace{x + \;\; 0}_{}$$
$$-14 \;\; = \;\;\;\; x$$

CHECK

$$-8 + 1 = \;\;\; x + 7$$
$$\downarrow$$
$$\underbrace{-8 + 1}_{} = \underbrace{-14 + 7}_{}$$
$$-7 \;\;\; = \;\;\;\; -7 \; \checkmark$$

Balances, so
-7 is the solution.

60. CONCEPT CHECK A student did this work when solving an equation. Show how to check the solution. If the solution does not check, find and correct the errors.

$$-3 - 6 = n - 5$$
$$\underbrace{-3 + 6}_{} = n - 5$$
$$3 \;\;\;\; = n - 5$$
$$\underline{-5 -5}$$
$$-2 \;\;\; = \underbrace{n + 0}_{}$$
$$-2 \;\;\; = \;\;\; n$$

61. West Community College always orders 10 extra robes for the graduation ceremony. The college ordered 305 robes this year. Solving the equation $g + 10 = 305$ will give the number of graduates (g) this year. Solve the equation.

62. Refer to **Exercise 61.** The college ordered 278 robes last year. Solve the equation $g + 10 = 278$ to find the number of graduates last year.

63. The warmer the temperature, the faster a field cricket chirps. Solving the equation $92 = c + 37$ will give you the number of chirps (in 15 seconds) when the temperature is 92 degrees. Solve the equation.

64. Refer to **Exercise 63.** Solve the equation $77 = c + 37$ to find the number of times a field cricket chirps (in 15 seconds) when the temperature is 77 degrees.

65. During the summer months, Ernesto spends an average of only $45 per month on parking fees by riding his bike to work on nice days. This is $65 less per month than what he spends for parking in the winter. Solving the equation $p - 65 = 45$ will give you his monthly parking fees in the winter. Solve the equation.

66. By walking to work several times a week in the summer, Aimee spends an average of $56 less per month on parking fees. If she spends $98 per month on parking in the summer, solve the equation $p - 56 = 98$ to find her monthly parking fees in the winter.

Solve each equation. Show your work.

67. $-17 - 1 + 26 - 38 = -3 - m - 8 + 2m$

68. $19 - 38 - 9 + 11 = -t - 6 + 2t - 6$

69. $-6x + 2x + 6 + 5x = |0 - 9| - |-6 + 5|$

70. $-h - |-9 - 9| + 8h - 6h = -12 - |-5 + 0|$

Relating Concepts (Exercises 71–72) For Individual or Group Work

*Use what you have learned about solving equations to **work Exercises 71 and 72 in order.***

71. **(a)** Write two *different* equations that have -2 as the solution. Be sure that you have to use the *addition property of equality* to solve the equations. Show how to solve each equation. Use **Exercises 7 to 22** as models.

(b) Follow the directions in part (a), but this time write two equations that have 0 as the solution.

72. Not all equations have solutions that are integers. Try solving these equations.

(a) $x + 1 = 1\frac{1}{2}$

(b) $\frac{1}{4} = y - 1$

(c) $\$2.50 + n = \$3.35$

(d) Write two more equations that have fraction or decimal solutions.

2.4 | Solving Equations Using Division

OBJECTIVES

1. Solve equations using the division property of equality.

2. Simplify equations before using the division property of equality.

3. Solve equations such as −x = 5.

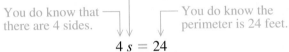

OBJECTIVE ▶ ① Solve equations using the division property of equality.
Earlier in this chapter you worked with the expression for finding the perimeter of a square-shaped garden, that is, finding the total distance around all four sides of the garden:

$$4s, \text{ where } s \text{ is the length of one side of the square}$$

Suppose you know that 24 feet of fencing was used around a square-shaped garden. What was the length of one side of the garden? To answer this question, write an equation showing the relationship between what you know and what you don't know.

You don't know the length of one side.

You do know that there are 4 sides.   You do know the perimeter is 24 feet.

$$4\,s = 24$$

To solve the equation, what number can replace *s* so that the equation balances? You can see that *s* is 6 feet.

$$4 \cdot 6 = 24$$

> Balances:
> 4 • 6 is exactly 24.

The **solution** is **6 feet** because 6 is the *only* number that makes the equation balance. You have answered the question about the length of one side: the length is 6 feet.

There is a tool that you can use to solve equations such as $4s = 24$ called the **division property of equality.** It allows you to *divide* both sides of an equation by the *same* number. (The only exception is that you cannot divide by 0.)

Division Property of Equality

If $a = b$, then $\dfrac{a}{c} = \dfrac{b}{c}$ as long as *c* is *not* 0.

In other words, you may divide both sides of an equation by the same nonzero number and still keep it balanced.

In the previous section, you saw that *adding* the same number to both sides of an equation kept it balanced. We could also have *subtracted* the same number from both sides because subtraction is defined as adding the opposite. Now we're saying that you can *divide* both sides by the same number. Later, we'll *multiply* both sides by the same number.

Equality Principle for Solving an Equation

As long as you do the *same* thing to *both* sides of an equation, the balance is maintained. (The only exception is that you cannot divide by 0.)

EXAMPLE 1 Using the Division Property of Equality

Solve each equation and check each solution.

(a) $4s = 24$

As with any equation, the goal is to get the variable by itself on one side of the equal sign. On the left side we have $4s$, which means $4 \cdot s$. The variable is multiplied by 4. Division is the inverse of multiplication, so dividing by 4 can be used to "undo" multiplying by 4.

Divide $4s$ by 4. The fraction bar indicates division: $\dfrac{4s}{4} = \dfrac{24}{4}$ To keep the balance, divide the right side by 4 also: $24 \div 4$ is 6
$4s \div 4$ is s

$$s = 6$$

So, as we already knew, the solution is **6**. Check the solution by replacing s with 6 in the original equation.

CHECK $4s = 24$ Original equation

$4 \cdot 6 = 24$ Replace s with 6

$24 = 24$ ✓   Balances   *The solution is 6, not 24.*

When s is replaced with 6, the equation balances, so **6 is the correct solution**.

(b) $42 = -6w$

On the right side of the equation, the variable is *multiplied* by -6. To undo the multiplication, *divide* by -6.

To keep the balance, divide by -6 on the left side also. $\dfrac{42}{-6} = \dfrac{-6w}{-6}$ Use division to undo multiplication: $-6w \div (-6)$ is w

$$-7 = w$$

The solution is -7

CHECK $42 = -6w$ Original equation

$42 = -6 \cdot (-7)$ Replace w with -7

$42 = 42$ ✓   Balances   *The solution is -7, not 42.*

When w is replaced with -7, the equation balances, so -7 **is the correct solution**.

❗ CAUTION

Be careful to divide both sides by the *same* number as the coefficient of the variable term. In **Example 1(b)** above, the coefficient of $-6w$ is -6, so divide both sides by -6. (Do **not** divide by the *opposite* of -6, which is 6. Use the opposite only when you're *adding* the same number to both sides.)

——— Work Problem ❶ at the Side. ▶

❶ Solve each equation and check each solution.

(a) $4s = 44$

$\dfrac{4s}{4} = \dfrac{44}{4}$ Divide both sides by 4.

$s = \underline{\quad}$

CHECK

$4s = 44$ Original equation

$4 \cdot \underline{\quad} = 44$ Replace s with ____.

$\underline{\quad} = 44$

Does it balance? ____

(b) $27 = -9p$

CHECK

(c) $-40 = -5x$

CHECK

(d) $7t = -70$

CHECK

Answers

1. **(a)** $s = 11$
   CHECK $4s = 44$
   $4 \cdot 11 = 44$ Replace s with 11.
   Balances $44 = 44$ ✓
   (b) $p = -3$
   CHECK $27 = -9p$
   $27 = -9 \cdot (-3)$
   Balances $27 = 27$ ✓
   (c) $x = 8$
   CHECK $-40 = -5x$
   $-40 = -5 \cdot 8$
   Balances $-40 = -40$ ✓
   (d) $t = -10$
   CHECK $7t = -70$
   $7 \cdot (-10) = -70$
   Balances $-70 = -70$ ✓

2 Simplify each side of the equation when possible. Then solve the equation and check the solution.

(a) $-28 = -6n + 10n$

$-28 = \underbrace{-6n + 10n}$ Simplify the right side.

$-28 = \quad 4n$ Divide both sides by ____.

Find the solution.

CHECK

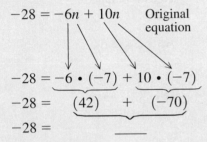

$-28 = -6n + 10n$ Original equation

$-28 = \underbrace{-6 \cdot (-7)} + \underbrace{10 \cdot (-7)}$

$-28 = \quad (42) \quad + \quad (-70)$

$-28 = \quad \underline{\qquad}$

Does it balance? _____

(b) $p - 14p = -2 + 18 - 3$

CHECK

Answers

2. **(a)** $\dfrac{-28}{4} = \dfrac{4n}{4}; -7 = n$

    CHECK $-28 = -28$ ✓

    Balances

(b) $p = -1$

    CHECK $p - \quad 14p = -2 + 18 - 3$

       $-1 - 14(-1) = \underbrace{16} \ -3$

       $-1 - (-14) = \quad 13$

       $\underbrace{-1 + (+14)} = \quad 13$

    Balances $13 \quad = \quad 13$ ✓

OBJECTIVE **2** **Simplify equations before using the division property of equality.** You can sometimes simplify the expression on one or both sides of the equal sign, as you did in the previous section.

EXAMPLE 2   **Simplifying before Solving Equations**

Solve each equation and check each solution.

(a) $4y - 7y = -12$

Simplify the left side by combining like terms. $4y - \quad 7y = -12$ The right side cannot be simplified.

Change subtraction to adding the opposite. $4y + (-7y) = -12$

Divide by the coefficient, which is -3 $\dfrac{-3y}{-3} = \dfrac{-12}{-3}$ To keep the balance, divide by -3 on the right side also.

    $y \quad = \quad 4$ $-12 \div (-3)$ is 4

The solution is **4**

CHECK $4y - \quad 7y \quad = -12$ Go back to the *original* equation and replace each y with 4

Do multiplications first. $4 \cdot 4 - 7 \cdot 4 = -12$

Change subtraction to adding the opposite. $16 \ - \ 28 \ = -12$

    $16 + (-28) = -12$

    $-12 \quad = -12$ ✓ Balances

When y is replaced with 4, the equation balances, so **4 is the correct solution.**

(b) $3 - 10 + 7 = h + 7h$

Change subtraction to adding the opposite. $3 - \quad 10 + 7 = h + 7h$ Write the understood 1 as the coefficient of h.

Add from left to right. $3 + (-10) + 7 = \underbrace{1h + 7h}$ Combine like terms on the right side.

    $\underbrace{-7 \quad + 7} = \quad 8h$

To keep the balance, divide by 8 on the left side also $\dfrac{0}{8} = \dfrac{8h}{8}$ Divide by the coefficient, which is 8

$0 \div 8$ is 0 $0 \quad = \quad h$

The solution is **0**

CHECK $3 - \quad 10 + 7 = h + \quad 7h$ Go back to the *original* equation and replace each h with 0.

    $3 + (-10) + 7 = 0 + 7 \cdot 0$

    $\underbrace{-7 \quad + 7} = \underbrace{0 + \quad 0}$

    $0 \quad = \quad 0$ ✓ Balances

When h is replaced with 0, the equation balances, so **0 is the correct solution.**

◀ **Work Problem** **2** at the Side.

OBJECTIVE 3 **Solve equations such as** $-x = 5$**.** When solving equations, do *not* leave a negative sign in front of the variable.

EXAMPLE 3 **Solving an Equation of the Type** $-x = 5$

Solve $-x = 5$ and check the solution.

It may look as if there is nothing more we can do to the equation $-x = 5$, but $-x$ is *not* the same as x. To see this, we write in the understood -1 as the coefficient of $-x$.

Coefficient is understood to be -1

$-x = 5$ can be written $-1x = 5$

We want the coefficient of x to be positive 1, not -1. To accomplish that, we can divide both sides by the coefficient of x, which is -1.

$$\frac{-1x}{-1} = \frac{5}{-1}$$ Divide both sides by -1

On the left side,
$-1 \div (-1)$ is 1 $1x = -5$ On the right side, $5 \div (-1)$ is -5

Now x is by itself on one side of the equal sign and has a coefficient of 1. The solution is **-5**

CHECK $-x = 5$ Go back to the *original* equation.

$-\mathbf{1}x = 5$ Write in the understood -1 as the coefficient of $-x$.

$-\mathbf{1} \cdot (-\mathbf{5}) = 5$ Replace x with -5

> The solution is -5, **not** 5.

$5 \quad = 5$ ✓ Balances

When x is replaced with -5, the equation balances, so **-5 is the correct solution.**

⊗ CAUTION

As the last step in solving an equation, do *not* leave a negative sign in front of a variable. For example, do *not* leave $-y = -8$. Write in the understood -1 as the coefficient, so that

$-y = -8$ is written as $-1y = -8$

Then divide both sides by -1 to get $y = 8$. The solution is 8.

—————— Work Problem **3** at the Side. ▶

---

3 Solve each equation and check each solution.

(a) $-k = -12$

$-k = -12$ $-k$ can be written as $-1k$.

$-1k = -12$ Divide both sides by ____.

         Find the solution.

CHECK

$-1k = -12$

$-1 \cdot 12 = -12$

$\rule{1cm}{0.4pt} = -12$ Does it balance? ____

(b) $7 = -t$

CHECK

(c) $-m = -20$

CHECK

Answers

3. **(a)** Divide both sides by -1; $k = 12$
     CHECK $-12 = -12$ ✓
           Balances
(b) $t = -7$
     CHECK $7 = -1t$

           $7 = -1 \cdot (-7)$
     Balances $7 = \quad 7$ ✓
(c) $m = 20$

     CHECK $-1m = -20$

           $-1 \cdot 20 = -20$
     Balances $-20 = -20$ ✓

2.4 Exercises

FOR EXTRA HELP

Go to MyMathLab for worked-out, step-by-step solutions to exercises enclosed in a square ☐ and video solutions to ▶ exercises.

CONCEPT CHECK *Solve each equation and check each solution.* ***See Example 1.***

1. $6z = 12$
GS Divide both sides by ____.
$$\frac{6z}{6} = \frac{12}{6}$$
$z =$ ____

CHECK $\quad 6z = 12$
$\quad 6 \cdot 2 = 12$
____ = ____

2. $8k = 24$
GS Divide both sides by ____.
$$\frac{8k}{8} = \frac{24}{8}$$
$k =$ ____

CHECK $\quad 8k = 24$
$\quad 8 \cdot 3 = 24$
____ = ____

3. $48 = 12r$

CHECK $\quad 48 = 12r$

4. $99 = 11m$

CHECK $\quad 99 = 11m$

5. $3y = 0$

CHECK

6. $5a = 0$

CHECK

Solve each equation. Show your work. ***See Example 1.***

7. $-7k = 70$

8. $-6y = 36$

9. $-54 = -9r$
▶

10. $-36 = -4p$

11. $-25 = 5b$

12. $-70 = 10x$

Simplify where possible. Then solve each equation and check each solution.
See Example 2.

13. $2r = -7 + 13$

CHECK $\quad 2r = \underbrace{-7 + 13}$
$\qquad \downarrow$

14. $6y = 28 - 4$

CHECK $\quad 6y = \underbrace{28 - 4}$
$\qquad \downarrow$

15. $-12 = 5p - p$

CHECK

16. $20 = z - 11z$

CHECK

Solve each equation. Show your work. ***See Examples 1 and 2.***

17. $3 - 28 = 5a$

18. $-55 + 7 = 8n$

19. $x - 9x = 80$

20. $4c - c = -27$

21. $13 - 13 = 2w - w$

22. $-11 + 11 = 8t - 7t$

23. $3t + 9t = 20 - 10 + 26$
▶

24. $6m + 6m = 40 + 20 - 12$

25. $0 = -9t$

26. $-10 = 10b$

27. $-14m + 8m = 6 - 60$
▶

28. $7w - 14w = 1 - 50 + 49$

29. $100 - 96 = 31y - 35y$

30. $150 - 139 = 20x - 9x$

Use multiplication to simplify the side of the equation with the variable. Then solve each equation.

31. $3(2z) = -30$

32. $2(4k) = 16$

33. $50 = -5(5p)$

34. $60 = 4(-3a)$

35. $-2(-4k) = 56$

36. $-5(4r) = -80$

37. $-90 = -10(-3b)$

38. $-90 = -5(-2y)$

Solve each equation. See Example 3.

39. $-x = 32$

40. $-c = 23$

41. $-2 = -w$

42. $-75 = -t$

43. $-n = -50$

44. $-x = -1$

45. $10 = -p$

46. $100 = -k$

47. Look again at the solutions to **Exercises 39–46.** Describe the pattern you see. Then write a rule for solving equations with a negative sign in front of the variable, such as $-x = 5$.

48. Write two *different* equations that have -4 as the solution. Be sure that you have to use the division property of equality to solve the equations. Show how to solve each equation.

49. CONCEPT CHECK The work shown below has a mistake in it.

What Went Wrong? First write a sentence explaining what the mistake is. Then fix the mistake and find the correct solution.

Keng solved the equation this way:

$$3x = \underbrace{16 - 1}$$
$$\frac{3x}{-3} = \frac{15}{-3}$$
$$x = -5$$

50. CONCEPT CHECK The work shown below has a mistake in it.

What Went Wrong? First write a sentence explaining what the mistake is. Then fix the mistake and find the correct solution.

Maynou solved the equation this way:

$$\underbrace{2x - 3x} = \underbrace{10 - 18}$$
$$-x = -8$$

51. The perimeter of a triangle with sides of equal length is 3 times the length of one side (s). If the perimeter is 45 feet, solving the equation $3s = 45$ will give the length of one side. Solve the equation.

52. Refer to **Exercise 51.** If the perimeter of the triangle is 63 inches, solve the equation $3s = 63$ to find the length of one side.

53. The perimeter of a pentagon with sides of equal length is 5 times the length of one side (s). If the perimeter is 120 meters, solve the equation $120 = 5s$ to find the length of one side.

54. Refer to **Exercise 53.** If the pentagon has a perimeter of 335 yards, solving the equation $335 = 5s$ will give the length of one side. Solve the equation.

Solve each equation. Show your work.

55. $89 - 116 = -4(-4y) - 9(2y) + y$

56. $58 - 208 = -b + 8(-3b) + 5(-5b)$

57. $-37(14x) + 28(21x) = |72 - 72| + |-166 + 96|$

58. $6a - 10a - 3(2a) = |-25 - 25| - 5(8)$

2.5 | Solving Equations with Several Steps

OBJECTIVES

1. Solve equations using the addition and division properties of equality.

2. Solve equations using the distributive, addition, and division properties.

OBJECTIVE ▶ ① Solve equations using the addition and division properties of equality. To solve some equations, you need to use both the addition and division properties of equality. Here are the steps.

Solving an Equation Using the Addition and Division Properties

Step 1   Add the same amount to both sides of the equation so that the variable term (the variable and its coefficient) ends up by itself on one side of the equal sign.

Step 2   Divide both sides by the coefficient of the variable term to find the solution.

Step 3   Check the solution by going back to the *original* equation.

① Solve; check each solution.

GS **(a)** $2r + 7 = 13$

$$\begin{array}{r} -7 = -7 \\ \hline 2r + 0 = 6 \end{array} \quad \text{Add } -7 \text{ to both sides.}$$

$$2r = 6$$

$$\frac{2r}{2} = \frac{6}{2} \quad \text{Divide both sides by 2.}$$

$$r = \underline{\quad}$$

CHECK $2(3) + 7 = 13$ Replace r with 3.

$$6 + 7 = 13$$

$$\underline{\quad} = 13$$

(b) $-10z - 9 = 11$

CHECK

EXAMPLE 1   Solving an Equation with Several Steps

Solve this equation and check the solution: $5m + 1 = 16$

Step 1   Get the variable term by itself on one side of the equal sign. The variable term is $5m$. Adding -1 to the left side of the equation will leave $5m$ by itself. To keep the balance, add -1 to the right side also.

$$5m + 1 = 16$$

$$\begin{array}{r} -1 = -1 \\ \hline 5m + 0 = 15 \end{array} \quad \text{First } add \text{ the same thing to } both \text{ sides.}$$

$$5m = 15$$

Step 2   Divide both sides by the coefficient of the variable term. In $5m$, the coefficient is 5, so divide both sides by 5.

$$\frac{5m}{5} = \frac{15}{5} \quad \text{Divide both sides by the coefficient.}$$

$$m = 3$$

Step 3   Check the solution by going back to the *original* equation.

$$5m + 1 = 16 \quad \text{Use the } original \text{ equation and replace } m \text{ with 3}$$

The solution is 3.

$$5(3) + 1 = 16$$

$$15 + 1 = 16$$

$$16 = 16 \quad ✓ \text{ Balances}$$

When m is replaced with 3, the equation balances, so **3 is the correct solution**.

◀ **Work Problem ① at the Side.**

So far, variable terms have appeared on just one side of the equal sign. But some equations start with variable terms on both sides. In that case, you can use the addition property of equality to add the same *variable term* to both sides of the equation, just as you have added the same *number* to both sides.

Answers

1. (a) $r = 3$

CHECK $13 = 13$ ✓
Balances

(b) $z = -2$

CHECK $-10(-2) - 9 = 11$

$$20 - 9 = 11$$

Balances $11 = 11$ ✓

First decide whether to keep the variable term on the left side, or to keep the variable term on the right side. It doesn't matter which one you keep; just pick one side or the other. Then use the addition property to "get rid of" the variable term on the *other* side by adding its opposite.

EXAMPLE 2 Solving an Equation with Variable Terms on Both Sides

Solve this equation and check the solution: $2x - 2 = 5x - 11$

First let's keep $2x$, the variable term on the *left* side. Then we need to "get rid of" $5x$ on the *right* side. Add the opposite of $5x$, which is $-5x$.

To keep the balance, add $-5x$ to the left side also. Write $-5x$ under $2x$, *not* under 2

$$
\begin{aligned}
2x - \quad 2 &= \quad 5x - \quad 11 \\
\underline{-5x \qquad\quad} &= \underline{-5x \qquad\quad} \\
-3x - \quad 2 &= \quad 0 - \quad 11 \\
\downarrow \quad\ \downarrow \quad & \qquad \downarrow \quad\ \downarrow
\end{aligned}
$$

Write $-5x$ under $5x$, *not* under 11

$5x + (-5x)$ is $0x$, or 0

Change subtractions to adding the opposite.

To get $-3x$ by itself, add 2 to both sides.

$$
\begin{aligned}
-3x + (-2) &= 0 + (-11) \\
\underline{\quad 2 \qquad} & \quad \underline{\quad 2 \qquad} \\
-3x + \quad 0 &= 0 + (-9)
\end{aligned}
$$

Divide both sides by -3, the coefficient of the variable term.

$$
\frac{-3x}{-3} = \frac{-9}{-3}
$$

$$
x = 3
$$

The solution is **3**.

Suppose that, in the first step, we decided to keep $5x$ on the *right* side and "get rid of" $2x$ on the *left* side. Let's see what happens.

Add $-2x$ to both sides.

$$
\begin{aligned}
2x - \quad 2 &= \quad 5x - \quad 11 \\
\underline{-2x \qquad\quad} &= \underline{-2x \qquad\quad} \\
0 - \quad 2 &= \quad 3x - \quad 11 \\
\downarrow \quad\ \downarrow \quad & \qquad \downarrow \quad\quad \downarrow \\
0 + (-2) &= \quad 3x + (-11)
\end{aligned}
$$

Change subtractions to adding the opposite.

To get $3x$ by itself, add 11 to both sides.

$$
\begin{aligned}
\underline{\quad 11 \qquad} &= \underline{\quad 11 \qquad} \\
0 + \quad 9 &= 3x + \quad 0 \\
\frac{9}{3} &= \frac{3x}{3}
\end{aligned}
$$

Divide both sides by 3

$$
3 = x
$$

The solution is **3**.

The two solutions are the same. In both cases, x balances with 3.

Notice that we used the addition principle *twice:* once to "get rid of" the variable term $2x$ and once to "get rid of" the number -11. We could have done those steps in the reverse order without changing the result.

Note

More than one sequence of steps will work. The basic approach is:

- Simplify each side of the equation, if possible.

- Get the variable term by itself on one side of the equal sign and a number by itself on the other side.

- Divide both sides by the coefficient of the variable term.

---

② Solve each equation *two* ways. First keep the variable term on the *left* side when you solve. Then solve again, keeping the variable term on the *right* side. Compare the solutions.

(a) Keep variable on *left* side.

$$
\begin{aligned}
3y - \quad 1 &= \quad 2y + 7 \\
\underline{-2y \qquad\quad} & \quad \underline{-2y \qquad\quad} \\
1y - \quad 1 &= \quad 0 + 7 \\
\downarrow \quad\ \downarrow \quad & \\
y + (-1) &= \qquad 7 \\
\underline{\quad +1 \qquad} & \quad \underline{\quad +1 \qquad}
\end{aligned}
$$

Now find the solution.

Keep variable on *right* side.

$$
\begin{aligned}
3y - \quad 1 &= \quad 2y + 7 \\
\underline{-3y \qquad\quad} & \quad \underline{-3y \qquad\quad} \\
0 - \quad 1 &= -1y + 7 \\
\downarrow \quad\ \downarrow \quad & \\
0 + (-1) &= -1y + 7 \\
\underline{\quad -7 \qquad} & \quad \underline{\quad -7 \qquad} \\
0 + (-8) &= -1y + 0 \\
-8 \quad &= \quad -1y
\end{aligned}
$$

Now find the solution.

(b) $3p - 2 = p - 6$

$$
3p - 2 = p - 6
$$

Answers

2. **(a)** $y = 8$ and $8 = y$
   (b) $p = -2$ and $-2 = p$

Work Problem ② at the Side. ▶

❸ Solve each equation.

ⓖⓢ **(a)** $-12 = 4(y - 1)$

$-12 = 4(y - 1)$ Use the distributive property.

$-12 = 4 \cdot y - 4 \cdot 1$
$-12 = 4y - 4$

$-12 = 4y + (-4)$ Add 4 to both sides.

$\underline{+ 4 \qquad\qquad +4}$

$-8 = 4y$ Divide both sides by ____ and find the solution.

(b) $5(m + 4) = 20$

(c) $6(t - 2) = 18$

OBJECTIVE ▶ ❷ Solve equations using the distributive, addition, and division properties. If an equation contains parentheses, check to see whether you can use the distributive property to remove them.

EXAMPLE 3 Solving an Equation Using the Distributive Property

Solve this equation and check the solution: $-6 = 3(y - 2)$

We can use the distributive property to simplify the right side of the equation. Recall that

$$3(y - 2) \quad \text{can be written as} \quad \underbrace{3 \cdot y} - \underbrace{3 \cdot 2}$$
$$\qquad\qquad\qquad\qquad\qquad\qquad 3y \quad - \quad 6$$

So the original equation $-6 = 3(y - 2)$ becomes $-6 = 3y - 6$.

$-6 = 3y - 6$ Change subtraction to adding the opposite.

$-6 = 3y + (-6)$

$\underline{6 \qquad\qquad\quad 6}$ To get $3y$ by itself, add 6 to both sides.

$\dfrac{0}{3} = \dfrac{3y}{3}$ Divide both sides by 3, the coefficient of $3y$.

$\mathbf{0 = y}$

The solution is **0**

CHECK $-6 = 3(y - 2)$ Go back to the *original* equation and replace y with 0

$-6 = 3(0 - 2)$ Follow the order of operations; work inside parentheses first.

$-6 = 3[0 + (-2)]$ Change subtraction to addition. Brackets show what to do first: $0 + (-2)$ is -2

$-6 = \quad 3(-2)$

$-6 = \quad -6$ ✓ Balances

When y is replaced with 0, the equation balances, so **0 is the correct solution**.

◀ **Work Problem ❸ at the Side.**

Here is a summary of all the steps you can use to solve an equation.

Solving an Equation

Step 1 If possible, use the **distributive property** to remove parentheses.

Step 2 **Combine** any like terms on the left side of the equation. **Combine** any like terms on the right side of the equation.

Step 3 **Add** the same amount to both sides of the equation so that the variable term ends up by itself on one side of the equal sign and a number is by itself on the other side. You may have to do this step more than once.

Step 4 **Divide** both sides by the coefficient of the variable term to find the solution.

Step 5 **Check** your solution by going back to the *original* equation. Replace the variable with your solution. Follow the order of operations to complete the calculations. If the two sides of the equation balance, your solution is correct.

Answers

3. **(a)** Divide both sides by 4; $\dfrac{-8}{4} = \dfrac{4y}{4}$; $-2 = y$

    (b) $m = 0$ **(c)** $t = 5$

Sometimes you will use only two or three of the solution steps and sometimes you will need all five steps. The example below uses all five steps.

EXAMPLE 4 Solving an Equation

Solve this equation and check the solution: $8 + 5(m + 2) = 6 + 2m$

Step 1 Use the distributive property on the left side.

$$8 + 5(m + 2) = 6 + 2m$$

Step 2 Combine like terms on the left side.

$$8 + 5m + 10 = 6 + 2m \qquad \text{No like terms on}$$

$$5m + 18 = 6 + 2m \qquad \text{the right side}$$

Step 3 Add $-2m$ to both sides.

$$\underline{\quad -2m \qquad\qquad -2m \quad}$$

$$3m + 18 = 6 + 0$$

$$3m + 18 = \quad 6$$

Step 3 To get $3m$ by itself, add -18 to both sides.

$$\underline{\quad -18 \qquad -18 \quad}$$

$$3m + 0 = -12$$

Step 4 Divide both sides by 3, the coefficient of the variable term $3m$.

$$\frac{3m}{3} = \frac{-12}{3}$$

$$\boldsymbol{m = -4}$$

The solution is $\boldsymbol{-4}$

Step 5 **CHECK**

$$8 + 5(m + 2) = 6 + 2m \qquad \text{Replace each } m \text{ with } -4$$

$$8 + 5(-4 + 2) = 6 + 2(-4)$$

$$8 + \quad 5(-2) \quad = 6 + \quad (-8)$$

$$8 + \quad (-10) \quad = \qquad -2$$

$$-2 \qquad = \qquad -2 \qquad \checkmark \text{ Balances}$$

When m is replaced with -4, the equation balances, so $\boldsymbol{-4}$ **is the correct solution (not** -2**).**

—————— **Work Problem 4 at the Side.** ▶

---

4 Solve each equation.

(a) $3(b + 7) = 2b - 1$

$$3(b + 7) = 2b - 1$$

$$3b + 21 = 2b + (-1)$$

$$\underline{-2b \qquad\qquad -2b \quad}$$

$$1b + 21 = 0 + (-1) \quad \text{Find the solution.}$$

(b) $6 - 2n = 14 + 4(n - 5)$

Answers

4. **(a)** $b + 21 = -1$
$$\underline{\quad -21 \qquad -21 \quad}$$
$$b + 0 = -22$$
$$b \quad = -22$$

(b) $n = 2$

2.5 Exercises

FOR EXTRA HELP Go to MyMathLab for worked-out, step-by-step solutions to exercises enclosed in a square ▮ and video solutions to ▶ exercises.

CONCEPT CHECK *Solve each equation and check each solution. See Example 1.*

1. $7p + 5 = 12$

$\ \underline{-5\ \ -5}$

$\underbrace{7p + 0}_{} = 7$ Find the solution.

CHECK $7p\ + 5 = 12$

$7(1) + 5 = 12$

$\underbrace{7}_{}\ + 5 = 12$

$\underline{\quad}\ \ \underline{\quad}$

2. $6k + 3 = 15$

$\ \underline{-3\ \ -3}$

$\underbrace{6k + 0}_{} = 12$ Find the solution.

CHECK $6k\ + 3 = 15$

$6(2) + 3 = 15$

$\underbrace{12}_{}\ + 3 = 15$

$\underline{\quad}\ \ \underline{\quad}$

3. $2 = 8y - 6$

$2 = 8y + (-6)$

$\underline{+6\ \ +6}$

$8 = \underbrace{8y + \ 0}_{}$ Find the solution.

CHECK $2 = 8y - 6$

4. $10 = 11p - 12$ **CHECK**

5. $28 = -9a + 10$ **CHECK**

6. $-4k + 5 = 5$ **CHECK**

Solve each equation. Show your work. See Example 1.

7. $-3m + 1 = 1$

8. $75 = -10w + 25$

9. $-5x - 4 = 16$

10. $-12b - 3 = 21$

*In Exercises 11–14, solve each equation **two** ways. First keep the variable term on the left side when you solve it. Then solve it again, keeping the variable term on the right side. Finally, check your solution. **See Example 2.***

11. $6p - 2 = 4p + 6$ $\qquad\qquad$ $6p - 2 = 4p + 6$ $\qquad\qquad$ **CHECK** $\quad 6p - 2 = 4p + 6$

12. $5y - 5 = 2y + 10$ $\qquad\qquad$ $5y - 5 = 2y + 10$ $\qquad\qquad$ **CHECK** $\quad 5y - 5 = 2y + 10$

13. $-2k - 6 = 6k + 10$ $\qquad\qquad$ $-2k - 6 = 6k + 10$ $\qquad\qquad$ **CHECK**

14. $5x + 4 = -3x - 4$ $\qquad\qquad$ $5x + 4 = -3x - 4$ $\qquad\qquad$ **CHECK**

First simplify where possible. Then solve each equation. Show your work.

15. $-18 + 7a = 2a + 3 + 4$ $\qquad$ **16.** $-10 + 5r = -7 - 12 - 1$ $\qquad$ **17.** $-3t = 8t$

18. $15z = -9z$

19. $4 + 16 - 2 = 2 - 2b$

20. $-9 + 2z = 9z - 1 + 13$

Use the distributive property to help you solve each equation. Show your work.
See Example 3.

21. $8(w - 2) = 32$

22. $9(b - 4) = 27$

23. $-10 = 2(y + 4)$

24. $-3 = 3(x + 6)$

25. $-4(t + 2) = 12$
$\blacktriangleright$

26. $-5(k + 3) = 25$

27. $6(x - 5) = -30$

28. $7(r - 7) = -49$

29. $-12 = 12(h - 2)$

30. $-11 = 11(c - 3)$

31. $0 = -2(y + 2)$

32. $0 = -9(b + 1)$

Solve each equation. Show your work. ***See Example 4.***

33. $6m + 18 = 0$

34. $8p - 40 = 0$

35. $6 = 9w - 12$

36. $8 = 8h + 24$

37. $5x = 3x + 10$

38. $7n = -2n - 36$

39. $2a + 11 = 8a - 7$

40. $r - 10 = 10r + 8$

41. $7 - 5b = 28 + 2b$

42. $1 - 8t = -9 - 3t$

43. $-20 + 2k = k - 4k$

44. $6y - y = -16 + y$

45. $10(c - 6) + 4 = 2 + c - 58$

46. $8(z + 7) - 6 = z + 60 - 10$

47. $-18 + 13y + 3 = 3(5y - 1) - 2$

48. $3 + 5h - 9 = 4(3h + 4) - 1$

49. $6 - 4n + 3n = 20 - 35$

50. $-19 + 8 = 6p - 7p - 5$

51. $6(c - 2) = 7(c - 6)$

52. $-3(5 + x) = 4(x - 2)$

53. $-5(2p + 2) - 7 = 3(2p + 5)$

54. $4(3m - 6) = 72 + 3(m - 8)$

55. $2(3b - 2) - 5b = 4(b - 1) + 8b$

56. $-3(w + 3) + 10 = -1(w + 14) + w$

57. CONCEPT CHECK Solve $-2t - 10 = 3t + 5$. Show each step you take while solving it. Next to each step, write a sentence that explains what you did in that step. Be sure to tell when you used the addition property of equality and when you used the division property of equality.

58. CONCEPT CHECK Explain the distributive property in your own words. Show two examples of using the distributive property to remove parentheses in an expression.

59. Here is one student's solution to an equation. Show how to check the solution. If the solution doesn't check, explain the error and correct it.

$$
\begin{array}{rcl}
-8 + 4a & = & 2a + 2 \\
\underline{-2a} & & \underline{-2a} \\
-10 + 4a & = & 0 + 2 \\
-10 + 4a & = & 2 \\
\underline{10} & & \underline{10} \\
0 + 4a & = & 12 \\
\dfrac{4a}{4} & = & \dfrac{12}{4} \\
a & = & 3
\end{array}
$$

60. Here is one student's solution to an equation. Show how to check the solution. If the solution doesn't check, explain the error and correct it.

$$
\begin{array}{rcl}
2(x + 4) & = & -16 \\
2x + 4 & = & -16 \\
\underline{-4} & & \underline{-4} \\
2x + 0 & = & -20 \\
\dfrac{2x}{2} & = & \dfrac{-20}{2} \\
x & = & -10
\end{array}
$$

Relating Concepts (Exercises 61–64) For Individual or Group Work

Work Exercises 61–64 in order.

61. (a) Simplify this expression.

$2(x + 4) + 3x$

(b) Solve this equation.

$2(x + 4) = 3x$

62. (a) Simplify this expression.

$11 + 3(w - 2) - 7w + 5 - 2w$

(b) Solve this equation.

$11 + 3(w - 2) = 7w + 5 - 2w$

63. (a) Simplify this expression.

$3y - 6(y + 3) + 4(3y + 2) - 11$

(b) Solve this equation.

$3y - 6(y + 3) = 4(3y + 2) - 11$

64. Exercises 61–63 each had two questions that were very similar. The only difference was that one had an **equal sign** and one did not. But that one symbol made a big difference in your answers! Look at your results from **Exercises 61–63**. Write a sentence or two explaining the difference between **simplifying an expression** and **solving an equation.**

Study Skills

Tips for Taking Math Tests

OBJECTIVES

1 Apply suggestions to tests and quizzes.

2 Develop a set of "best practices" to apply while testing.

Improving Your Test Score

To Improve Your Test Score	Comments
Come prepared with a pencil, eraser, and calculator, if allowed. If you are easily distracted, sit away from high-traffic areas.	**Working in pencil lets you erase,** keeping your work neat and readable.
Scan the entire test, note the point value of different problems, and plan your time accordingly. Allow at least five minutes to check your work at the end of the testing time.	If you have 50 minutes to do 20 problems, $50 \div 20 = 2.5$ minutes per problem. **Spend less time on easy ones,** more time on problems with higher point values.
Read directions carefully, and circle any significant words. When you finish a problem, read the directions again to make sure you did what was asked.	**Pay attention to announcements** written on the board or made by your instructor. Ask if you don't understand. You don't want to get problems wrong because you misread the directions!
Show your work. Most math teachers give partial credit when some of the steps in your work are correct, even if the final answer is wrong. **Write neatly.** If you like to scribble when first working or checking a problem, do it on scratch paper.	**If your teacher can't read your writing, you won't get credit for it.** If you need more space to work, ask if you can use extra pieces of paper that you hand in with your test paper.
Check that the **answer to an application problem is reasonable,** makes sense, and includes the units. Read the problem again to make sure you've answered the question.	**Use common sense.** Can the father really be seven years old? Would a month's rent be $32,140? Label your answer: $, years, inches, etc.
To check for careless errors, you need to **rework the problem again, without looking at your previous work.** Cover up your work with a piece of scratch paper, and pretend you are doing the problem for the first time. Then compare the two answers.	If you just "look over" your work, your mind can make the same mistake again without noticing it. Reworking the problem from the beginning **forces you to rethink it.** If possible, use a different method to solve the problem the second time.

Reducing Anxiety

To Reduce Anxiety	Comments
Do not try to review up until the last minute before the test. Instead, go for a walk, do some deep breathing, and arrive just in time for the test. Ignore other students.	Listening to anxious classmates before the test *may cause you to panic.* Moderate exercise and deep breathing will calm your mind.
Do a "knowledge dump" as soon as you get the test. Write important notes to yourself in a corner of the test paper: formulas, or common errors you want to watch out for.	Writing down tips and things that you've memorized *lets you relax;* you won't have to worry about forgetting those things and can refer to them as needed.
Do the easy problems first to build confidence. If you feel your anxiety starting to build, *immediately* stop for a minute, close your eyes, and take several slow, deep breaths.	Greater confidence helps you *get the easier problems correct.* Anxiety causes shallow breathing, which leads to confusion and reduced concentration. Deep breathing calms you.
As you work on more difficult problems, *notice your "inner voice."* You may have negative thoughts such as "I can't do it" or "Who cares about this test anyway?" In your mind, yell "STOP" and take several deep, slow breaths. Or replace the negative thought with a positive one.	Here are *examples of positive statements.* Try writing one of them on the top of your test paper. • I know I can do it. • I can do this one step at a time. • I've studied hard, and I'll do the best I can.
Read the harder problems twice. Write down *anything* that might help solve the problem: a formula, a picture, etc. If you still can't get it, circle the problem and *come back to it later.* Do *not* erase any of the things you wrote down.	If you know even a *little* bit about the problem, write it down. The *answer may come to you* as you work on it, or you may get partial credit. Don't spend too long on any one problem. Your subconscious mind will work on the tough problem while you go on with the test.
If you still can't solve a difficult problem when you come back to it the second time, *make a guess and do not change it.* In this situation, your first guess is your best bet. Do not change the answer just because you're a little unsure. *Change it only if you find an obvious mistake.*	If you are thinking about changing an answer, be sure you have a good reason for changing it. If you cannot find a specific error, leave your first answer alone. *When the tests are returned, check to see if changing answers helped or hurt you.*
Ignore students who finish early. Use the entire test time. *You do not get extra credit for finishing early.* Use the extra time to rework problems and correct careless errors.	Students who leave early are often the ones who didn't study or who are too anxious to continue working. If they bother you, *sit as far from the door as possible.*

Why Are These Suggestions Brain Friendly?

Several suggestions address anxiety. **Reducing anxiety** allows your brain to make the connections between dendrites; in other words, you can think clearly.

Remember that **your brain continues to work** on a difficult problem even if you skip it and go on to the next one. Your subconscious mind will come through for you if you are open to the idea!

Some of the suggestions ask you to **use your common sense.** Follow the directions, show your work, write neatly, and pay attention to whether your answers really make sense.

Key Terms

2.1

variable A variable is a letter that represents a number that varies or changes, depending on the situation.

constant A constant is a number that is added or subtracted in an expression. It does not vary. For example, 5 is the constant in the expression $c + 5$.

expression An expression expresses, or tells, the rule for doing something. It is a combination of operations on variables and numbers. An expression does *not* have an equal sign.

evaluate the expression To evaluate an expression, replace each variable with specific values (numbers). Then do all possible calculations, following the order of operations. There are no variables in the final answer.

coefficient The number part in a multiplication expression is the coefficient. For example, 4 is the coefficient in the expression $4s$.

2.2

simplifying expressions To simplify an expression, write it in a simpler way by combining all the like terms.

term Each addend in an expression is a term.

constant term A constant term is a type of term that is just a number. In the expression $4x + 8$, the constant term is 8.

variable term A variable term has a number part (called the coefficient) multiplied by a variable part (a letter). In the expression $4x + 8$, the variable term is $4x$.

like terms Like terms are terms with exactly the same variable parts (the same letters and exponents). The coefficients may be different.

2.3

equations An equation has an equal sign. It shows the relationship between what is known about a problem and what isn't known. The left side of the equation has the same value as the right side, so the two sides balance.

solve the equation To solve an equation, find a number that can replace the variable and make the equation balance.

solution A solution of an equation is a number that can replace the variable and make the equation balance.

addition property of equality The addition property of equality states that adding the same quantity to both sides of an equation will keep it balanced.

check the solution To check the solution of an equation, go back to the *original* equation and replace the variable with the solution. If the equation balances, the solution is correct.

2.4

division property of equality The division property of equality states that dividing both sides of an equation by the same nonzero number will keep it balanced.

Test Your Word Power

See how well you have learned the vocabulary in this chapter. The answers are on the next page.

1 A **variable**
  A. can only be the letter x
  B. is never an addend in an expression
  C. represents a number that varies.

2 Which expression has 2 as a **coefficient?**
  A. x^2
  B. $2x$
  C. $2 - x$

3 Which expression has 4 as a **constant** term?
  A. $4y$
  B. y^4
  C. $4 + y$

4 Which expression has four **terms?**
  A. $2 + 3x = -6 + x$
  B. $2 + 3x - 6 + x$
  C. $(2)(3x)(-6)(x)$

5 **Like terms**
  A. have the same coefficients
  B. have the same solutions
  C. have the same variable parts.

6 To **simplify an expression,**
  A. combine all the like terms
  B. multiply the exponents
  C. add the same quantity to both sides.

Quick Review

Concepts	Examples
2.1 Evaluating Expressions Replace each variable with the specified value. Then follow the order of operations to simplify the expression.	The expression for ordering textbooks for two prealgebra classes is $2c + 10$, where c is the class limit. Evaluate the expression when the class limit is 24. $\quad 2c + 10 \qquad$ Replace c with 24 $\quad 2 \cdot 24 + 10 \qquad$ Multiply first. $\quad\quad 48 + 10 \qquad$ Add last. $\quad\quad\quad \mathbf{58} \qquad$ Order 58 books.
2.1 Using Exponents with Variables An exponent next to a variable tells how many times to use the variable as a factor in multiplication.	Rewrite $-6x^4$ without exponents. $-6x^4$ can be written as $-6 \cdot x \cdot x \cdot x \cdot x$ Coefficient is -6 $\qquad$ x is used as a factor 4 times.
2.1 Evaluating Expressions with Exponents Rewrite the expression without exponents, replace each variable with the specified value, and multiply all the factors.	Evaluate x^3y when x is -4 and y is 5. x^3y means $x \cdot x \cdot x \cdot y$ $\quad$ Replace x with -4 and y with 5 $(-4) \cdot (-4) \cdot (-4) \cdot 5$ $\quad$ Multiply two factors at a time. $\quad 16 \quad \cdot (-4) \cdot 5$ $\quad\quad -64 \quad \cdot 5$ $\quad\quad\quad -320$
2.2 Identifying Like Terms Like terms have *exactly* the same letters and exponents. The coefficients may be different.	List the like terms in this expression. Then identify the coefficients of the like terms. $$-3b + (-3b^2) + 3ab + b + 3$$ The like terms are $-3b$ and b. The coefficient of $-3b$ is -3, and the coefficient of b is understood to be 1.

Concepts	Examples

2.2 Combining Like Terms

Step 1 If there are any variable terms with no coefficient, write in the understood 1.

Step 2 Change any subtractions to adding the opposite.

Step 3 Find like terms.

Step 4 Add the coefficients of like terms, keeping the variable part the same.

Simplify: $4x^2 - 10 + x^2 + 15$

$4x^2 - \quad 10 + \quad 1x^2 + 15$ Write understood 1

$4x^2 + (-10) + \quad 1x^2 + 15$ Change subtraction to adding the opposite.

$4x^2 + 1x^2 + (-10) + 15$ Combine $4x^2 + 1x^2$ The variable part stays the same.

$(4 + 1)x^2 + \quad 5$ Also combine $-10 + 15$

$5x^2 + \quad 5$

The simplified expression is $\mathbf{5x^2 + 5}$

2.2 Simplifying Multiplication Expressions

Use the associative property to rewrite the expression so that the two number parts can be multiplied. The variable part stays the same.

Simplify: $-7(5k)$

Use the associative property of multiplication.

$-7 \cdot (5 \cdot k)$ can be written as $(-7 \cdot 5) \cdot k$

$-35 \cdot k$

$-35k$

The simplified expression is $\mathbf{-35k}$

2.2 Using the Distributive Property

Multiplication distributes over addition and over subtraction. Be careful to multiply *every* term inside the parentheses by the number outside the parentheses.

Simplify.

(a) $6(w - 4)$ can be written as $6 \cdot w - 6 \cdot 4$

$6w - 24$

The simplified expression is $\mathbf{6w - 24}$

(b) $-3(2b + 5)$ can be written as $-3 \cdot 2b + (-3) \cdot 5$

$-6b + (-15)$

Use the definition of subtraction "in reverse" to write $-6b + (-15)$ as $-6b - 15$.

The simplified expression is $\mathbf{-6b - 15}$

2.3 Solving and Checking Equations Using the Addition Property of Equality

If possible, *simplify* the expression on one or both sides of the equal sign.

Next, to get the variable by itself on one side of the equal sign, *add* the same number to both sides.

Finally, *check* the solution by going back to the original equation and replacing the variable with the solution. If the equation balances, the solution is correct.

Solve this equation and check the solution.

$$-5 + 8 = \quad 9 + r \quad \text{Simplify the left side by adding } -5 + 8$$

$$3 = \quad 9 + r$$

$$\underline{-9 \qquad -9} \quad \text{To get } r \text{ by itself, add the opposite of 9, which is } -9, \text{ to both sides.}$$

$$-6 = \quad 0 + r$$

$$-6 = \quad r$$

The solution is $\mathbf{-6}$ *(Continued)*

Concepts	Examples

2.3 **Solving and Checking Equations Using the Addition Property of Equality** *(Continued)*

CHECK $-5 + 8 = 9 + \quad r$ — Use the original equation and replace r with -6

$$-5 + 8 = 9 + (-6)$$

$$3 = 3 \quad \checkmark \text{ Balances}$$

When r is replaced with -6, the equation balances, so **-6 is the correct solution (not** 3**).**

2.4 **Solving and Checking Equations Using the Division Property of Equality**

If possible, *simplify* the expression on one or both sides of the equal sign.

Next, to get the variable by itself on one side of the equal sign, *divide* both sides by the coefficient of the variable term.

Finally, *check* the solution by going back to the original equation and replacing the variable with the solution. If the equation balances, the solution is correct.

Solve this equation and check the solution.

Simplify the left side. $2h - 6h = 18 + 22$ Simplify the right side; add $18 + 22$

Change subtraction to adding the opposite. $2h + (-6h) = 40$

$-4h = 40$

Divide by -4, the coefficient of $-4h$. $\dfrac{-4h}{-4} = \dfrac{40}{-4}$ Also divide 40 by -4 to keep the balance.

$$h = -10$$

The solution is **-10**

CHECK $2h - 6h = 18 + 22$ Original equation; replace h with -10

$$2(-10) - 6(-10) = 40$$

$$-20 - (-60) = 40$$

$$-20 + (+60)$$

$$40 = 40 \quad \checkmark \text{ Balances}$$

When h is replaced with -10, the equation balances, so **-10 is the correct solution (not** 40**).**

2.4 **Solving Equations Such as** $-x = 5$

As the last step in solving an equation, do *not* leave a negative sign in front of the variable, such as $-x = 5$, because $-x$ is *not* the same as x. Divide both sides by -1, the understood coefficient of $-x$.

Solve this equation and check the solution.

$$9 = -n$$

Write the understood -1 as the coefficient of n.

$9 = -n$ can be written as $9 = -1n$

Now divide both sides by -1.

$$\frac{9}{-1} = \frac{-1n}{-1}$$

$$-9 = n$$

The solution is **-9**

(Continued)

Concepts	Examples

2.4 Solving Equations Such as $-x = 5$ (Continued)

CHECK

$$9 = -n \qquad \text{Original equation}$$
$$9 = -1n \qquad \text{Write understood } -1$$
$$9 = -1(-9) \qquad \text{Replace } n \text{ with } -9$$
$$9 = \quad 9 \quad \checkmark \text{ Balances}$$

When n is replaced with -9, the equation balances, so **-9 is the correct solution** (**not** 9).

2.5 Solving Equations with Several Steps

Solve this equation and check the solution.

Step 1 If possible, use the distributive property to remove parentheses.

Step 1 $\qquad 3 + 2(y + 8) = \quad 5y + 4$

Step 2 Combine any like terms on the left side of the equal sign. Combine any like terms on the right side of the equal sign.

Step 2 $\qquad 3 + 2y + 16 = \quad 5y + 4$

$$2y + 19 = \quad 5y + 4$$

Step 3 Add the same amount to both sides of the equation so that the variable term ends up by itself on one side of the equal sign, and a number is by itself on the other side. You may have to do this step more than once.

Step 3
$$\begin{array}{rcl} -2y & & -2y \\ \hline 0 + 19 & = & 3y + 4 \end{array}$$

Step 3
$$\begin{array}{rcl} -4 & & -4 \\ \hline 0 + 15 & = & 3y + 0 \end{array}$$

Step 4 Divide both sides by the coefficient of the variable term to find the solution.

Step 4
$$\frac{15}{3} = \frac{3y}{3}$$
$$5 = y$$

The solution is **5**

Step 5 Check the solution by going back to the original equation. Replace the variable with the solution. If the equation balances, the solution is correct.

CHECK $\qquad 3 + 2(y + 8) = \quad 5y + 4 \quad$ Original equation

Step 5
$$3 + 2(5 + 8) = 5(5) + 4$$
$$3 + 2(13) = 25 + 4$$
$$3 + 26 = 29$$
$$29 = 29 \quad \checkmark \text{ Balances}$$

When y is replaced with 5, the equation balances, so **5 is the correct solution** (**not** 29).

Chapter 2 *Review Exercises*

2.1

1. Identify the variable, the coefficient, and the constant in this expression: $-3 + 4k$

2. The expression for ordering test tubes for a chemistry lab is $4c + 10$, where c is the class limit. Evaluate the expression when the class limit is 24.

3. Rewrite each expression without exponents.
   (a) x^2y^4
   (b) $5ab^3$

4. Evaluate each expression when m is 2, n is -3, and p is 4.
   (a) n^3 (b) $-4mp^2$ (c) $5m^4n^2$

2.2 *Simplify.*

5. $ab + ab^2 + 2ab$

6. $-3x + 2y - x - 7$

7. $-8(-2g^3)$

8. $4(3r^2t)$

9. $5(k + 2)$

10. $-2(3b + 4)$

11. $3(2y - 4) + 12$

12. $-4 + 6(4x + 1) - 4x$

2.3 *Solve each equation and check each solution.*

13. $16 + n = 5$ **CHECK** $16 + \quad n = 5$

14. $-4 + 2 = 2a - 6 - a$ **CHECK**

$$-4 + 2 = 2a - \quad 6 - \quad a$$

2.4 *Solve each equation. Show your work.*

15. $48 = -6m$

16. $k - 5k = -40$

17. $-2p + 5p = 3 - 21$

18. $12 = -h$

2.5 *Solve each equation. Show your work.*

19. $12w - 4 = 8w + 12$

20. $0 = -4(c + 2)$

21. $5d + 4 = 3 + 2(d + 8)$

Chapter 2 Mixed Review Exercises

Practicing exercises in mixed-up order helps you prepare for tests.

1. Evaluate each expression when x is 5 and y is -1.

   (a) y^2 **(b)** $-2xy^3$ **(c)** $4x^2y$

2. Circle the expression that has 20 as the constant term and -9 as the coefficient.

   $20m - 9$ $-9 + 20x$ $-9y + 20$

3. Simplify: $-6(b + 3)$

4. Solve: $-17 + 5 + 12 = 4r$

5. Solve: $10(-3z) = -30$

6. Simplify: $-7a^3 - 10 + 6a^3 + 15$

7. Rewrite each expression without exponents.

   (a) $-4m^2$ **(b)** $60xyz^3$

8. The expression for making sandwiches at a homeless shelter is $c + 2p$, where c is the number of children and p is the number of parents. Evaluate the expression for 75 parents and 30 children.

Solve each equation. Show your work.

9. $12 + 7a = 4a - 3$

10. $-2(p + 3) = -14$

11. $10y = 6y + 20$

12. $2m - 7m = 5 - 20$

13. $20 = 3x - 7$

14. $b + 6 = 3b - 8$

15. $z + 3 = 0$

16. $3(2n - 1) = 3(n + 3)$

17. $-4 + 46 = 7(-3t + 6)$

18. $6 + 10d - 19 = 2(3d + 4) - 1$

19. $-35 = -m$

20. $-4(3b + 9) = 24 + 3(2b - 8)$

1. Identify the parts of this expression: $-7w + 6$ Choose from these labels: variable, constant, coefficient.

2. The expression for buying hot dogs for the company picnic is $3a + 2c$, where a is the number of adults and c is the number of children. Evaluate the expression when there are 45 adults and 21 children.

Rewrite each expression without exponents.

3. x^5y^3

4. $4ab^4$

5. Evaluate $-2s^2t$ when s is -5 and t is 4.

Simplify each expression.

6. $3w^3 - 8w^3 + w^3$

7. $xy - xy$

8. $-6c - 5 + 7c + 5$

9. $3m^2 - 3m + 3mn$

10. $-10\left(4b^2\right)$

11. $-5\left(-3k\right)$

12. $7\left(3t + 4\right)$

13. $-4\left(a + 6\right)$

14. $-8 + 6\left(x - 2\right) + 5$

15. $-9b - c - 3 + 9 + 2c$

Solve each equation and check each solution.

16. $-4 = x - 9$ **CHECK**

17. $-7w = 77$ **CHECK**

18. $-p = 14$ **CHECK**

19. $-15 = -3(a + 2)$ **CHECK**

Solve each equation. Show your work.

20. $6n + 8 - 5n = -4 + 4$

21. $5 - 20 = 2m - 3m$

22. $-2x + 2 = 5x + 9$

23. $3m - 5 = 7m - 13$

24. $2 + 7b - 44 = -3b + 12 + 9b$

25. $3c - 24 = 6(c - 4)$

26. Write an equation that requires the *addition property of equality* to solve it and has -4 as its solution. Then write a different equation that requires the *division property of equality* to solve it and has -4 as its solution. Show how to solve each equation.

Chapters 1-2 *Cumulative Review Exercises*

1. Write this number in words.
306,000,004,210

2. Write this number, using digits.
Eight hundred million, sixty-six thousand

3. Write $<$ or $>$ between each pair of numbers to make a true statement.

(a) -3 _____ -10 (b) -1 _____ 0

4. Name the property illustrated by each example.

(a) $-6 + 2 = 2 + (-6)$

(b) $0 \cdot 25 = 0$

5. (a) Round 9047 to the nearest hundred.

(b) Round 289,610 to the nearest thousand.

(c) $5(-6 + 4) = 5 \cdot (-6) + 5 \cdot 4$

Simplify.

6. $0 - 8$

7. $|-6| + |4|$

8. $-3(-10)$

9. $(-5)^2$

10. $\dfrac{-42}{-6}$

11. $-19 + 19$

12. $(-4)^3$

13. $\dfrac{-14}{0}$

14. $-5 \cdot 12$

15. $-20 - 20$

16. $\dfrac{45}{-5}$

17. $-50 + 25$

18. $-10 + 6(4 - 7)$

19. $\dfrac{-20 - 3(-5) + 16}{(-4)^2 - 3^3}$

First use front end rounding to estimate the answer to each application problem. Then find the exact answer.

20. One Siberian tiger was tracked for 22 days while it was searching for food. It traveled 616 miles. What was the average distance it traveled each day?

Estimate:

Exact:

21. The temperature in Siberia got down to -48 degrees one night. The next day the temperature rose 23 degrees. What was the daytime temperature?

Estimate:

Exact:

22. Doug owned 52 shares of Mathtronic stock that had a total value of $2132. Yesterday the value of each share dropped $8. What are his shares worth now?

Estimate:

Exact:

23. Ikuko's monthly rent is $758 plus $45 for parking. How much will she spend for rent and parking in one year?

Estimate:

Exact:

24. Rewrite $-4ab^3c^2$ without exponents.

25. Evaluate $3xy^3$ when x is -5 and y is -2.

Simplify.

26. $3h - 7h + 5h$

27. $c^2d - c^2d$

28. $4n^2 - 4n + 6 - 8 + n^2$

29. $-10(3b^2)$

30. $7(4p - 4)$

31. $3 + 5(-2w^2 - 3) + w^2$

Solve each equation and check each solution.

32. $3x = x - 8$ **CHECK**

33. $-44 = -2 + 7y$ **CHECK**

34. $2k - 5k = -21$ **CHECK**

35. $m - 6 = -2m + 6$ **CHECK**

Solve each equation. Show your work.

36. $4 - 4x = 18 + 10x$

37. $18 = -r$

38. $-8b - 11 + 7b = b - 1$

39. $-2(t + 1) = 4(1 - 2t)$

40. $5 + 6y - 23 = 5(2y + 8) - 10$

3

Solving Application Problems

In this chapter, you'll apply your skills for solving equations to real-world situations. Some examples are using formulas for perimeter and area to solve problems about soccer fields, volleyball courts, gymnastics floor mats, hockey rinks, and camping tents.

3.1 Problem Solving: Perimeter

OBJECTIVES

1 Use the formula for perimeter of a square to find the perimeter or the length of one side.

2 Use the formula for perimeter of a rectangle to find the perimeter, the length, or the width.

3 Find the perimeter of parallelograms, triangles, and irregular shapes.

OBJECTIVE ▶ 1 Use the formula for perimeter of a square to find the perimeter or the length of one side. If you have ever studied geometry, you probably used several different formulas such as $P = 2l + 2w$ and $A = lw$. A **formula** is just a shorthand way of writing a rule for solving a particular type of problem. A formula uses variables (letters) and it has an equal sign, so it is an equation. That means you can use the equation-solving techniques you have learned to work with formulas.

But let's start at the beginning. Geometry was developed centuries ago when people needed a way to measure land. The name *geometry* comes from the Greek *geo*, meaning earth, and *metron*, meaning measure. Today we still use geometry to measure land. It is also important in architecture, construction, navigation, art and design, physics, chemistry, and astronomy. You can use geometry at home when you buy carpet or wallpaper, hang a picture, or do home repairs. In this chapter you'll learn about two basic ideas, perimeter and area. Other geometry concepts will appear in later chapters.

Earlier, you found the *perimeter* of a square garden.

> **Perimeter**
>
> The distance around the outside edges of any flat shape is called the **perimeter** of the shape.

VOCABULARY TIP

Perimeter Remember, in **perimeter, peri** means "around," and **meter** means "measure." To find the perimeter, you measure around the outside edges of a flat shape. Write the perimeter using **linear units,** such as feet or yards; for example, 10 **ft** or 15 **yd**.

To review, a **square** has four sides that are all the same length. Also, the sides meet to form *right angles,* which measure 90° (90 degrees). This means that the sides form "square corners." Two examples of squares are shown below.

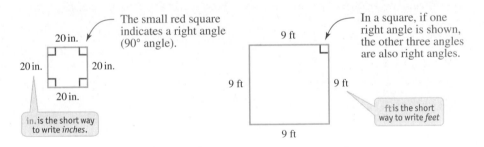

To find the *perimeter* of a square, we can "unfold" the shape so the four sides lie end-to-end, as shown below.

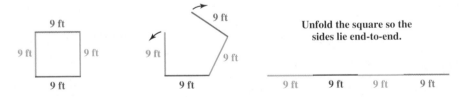

Now we can see the total length of the four sides. The total length is the *perimeter* of the square. We can find the perimeter by adding.

$$\text{Perimeter} = 9\text{ ft} + 9\text{ ft} + 9\text{ ft} + 9\text{ ft} = 36\text{ ft}$$

A shorter way is to multiply the length of one side times 4, because all 4 sides are the same length.

Finding the Perimeter of a Square

Perimeter of a square = side + side + side + side

or $P = 4 \cdot \text{side}$

$P = 4s$

EXAMPLE 1   **Finding the Perimeter of a Square**

Find the perimeter of the square on the previous page that measures 9 ft on each side.

Use the formula for perimeter of a square, $P = 4s$. You know that for this particular square, the value of s is 9 ft.

$P = 4s$	Formula for perimeter of a square
$P = 4 \cdot 9\text{ ft}$	Replace s with 9 ft. Multiply 4 times 9 ft
$P = \textbf{36 ft}$	Write 36 ft; ft is the unit of measure.

The perimeter of the square is **36 ft**. Notice that this answer matches the result we got by adding the four sides.

———— **Work Problem ❶ at the Side.** ▶

EXAMPLE 2   **Finding the Length of One Side of a Square**

If the perimeter of a square is 40 cm, find the length of one side.
(Note: **cm** is the short way to write *centimeters*.)

Use the formula for perimeter of a square, $P = 4s$. This time you know that the value of P (the perimeter) is 40 cm.

$P = 4s$	Formula for perimeter of a square
$40\text{ cm} = 4s$	Replace P with 40 cm
$\dfrac{40\text{ cm}}{4} = \dfrac{4s}{4}$	To get the variable by itself on the right side, divide both sides by 4
$\textbf{10 cm} = s$	

The length of one side of the square is **10 cm**   ◁ Write **cm** as part of your answer.

CHECK   Check the solution by drawing a square and labeling the length of each side as 10 cm. Then, the perimeter is 10 cm + 10 cm + 10 cm + 10 cm = 40 cm

10 cm
10 cm   10 cm
10 cm

Matches the perimeter given in the problem.

———— **Work Problem ❷ at the Side.** ▶

OBJECTIVE ▶ ❷ Use the formula for perimeter of a rectangle to find the perimeter, the length, or the width. A **rectangle** is a figure with four sides that meet to form right angles, which measure 90°. Each set of opposite sides is parallel and congruent (has the same length). Three examples of rectangles are shown on the next page.

---

❶ Find the perimeter of each square, using the appropriate formula.

GS **(a)** The 20 in. square shown on the previous page

$P = 4s$

$P = 4 \cdot \underline{\hspace{1cm}} = \underline{\hspace{1.5cm}}$

(b) A square measuring 14 miles (14 mi) on each side (*Hint:* Draw a sketch of the square and label each side with its length.)

❷ Use the perimeter of each square and the appropriate formula to find the length of one side. Then check your solution by drawing a square, labeling each side, and finding the perimeter.

GS **(a)** Perimeter is 28 in.

$P = 4s$ 7 in.

$28\text{ in.} = 4s$ 7 in. ⬜ 7 in.

$\dfrac{28\text{ in.}}{4} = \dfrac{4s}{4}$ 7 in.

$\underline{\hspace{1cm}} = s$

(b) Perimeter is 100 ft

Answers

1. (a) $P = 4 \cdot 20\text{ in.} = 80\text{ in.}$
  (b) $P = 56\text{ miles}$

14 mi
14 mi ⬜ 14 mi
14 mi

2. (a) 7 in. = s
CHECK $P = 7\text{ in.} + 7\text{ in.} + 7\text{ in.} + 7\text{ in.} = 28\text{ in.}$

  (b) $s = 25\text{ ft}$
CHECK $P = 25\text{ ft} + 25\text{ ft} + 25\text{ ft} + 25\text{ ft} = 100\text{ ft}$

25 ft
25 ft ⬜ 25 ft
25 ft

VOCABULARY TIP

Rect means "right." It reminds you that all the angles in a **rect**angle are "right" angles.

3 Find the perimeter of each rectangle using the appropriate formula. Check your solutions by adding the lengths of the four sides.

GS (a)

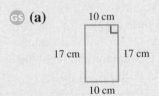

$P = 2l \qquad + 2w$

$P = 2 \cdot 17 \text{ cm} + 2 \cdot 10 \text{ cm}$

$P = \underline{\quad} + \underline{\quad}$

$P = \underline{\quad}$

CHECK

$P = \underline{\quad} + \underline{\quad} + \underline{\quad}$

$\quad + \underline{\quad} = \underline{\quad}$

(b)

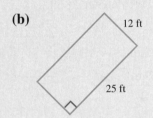

12 ft

25 ft

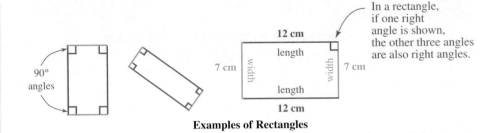

In a rectangle, if one right angle is shown, the other three angles are also right angles.

Examples of Rectangles

Each longer side of a rectangle is called the length (l) and each shorter side is called the width (w).

Look at the rectangle above with the lengths of the sides labeled. To find the perimeter (distance around), you could unfold the shape so the sides lie end-to-end. Then add the lengths of the sides.

$$P = 12 \text{ cm} + 7 \text{ cm} + 12 \text{ cm} + 7 \text{ cm} = 38 \text{ cm}$$

or $\quad P = 12 \text{ cm} + 12 \text{ cm} + 7 \text{ cm} + 7 \text{ cm} = 38 \text{ cm} \quad$ Commutative property

Because the two long sides are both 12 cm, and the two short sides are both 7 cm, you can also use the formula below.

Finding the Perimeter of a Rectangle

Perimeter of a rectangle = length + length + width + width

$$P = (2 \cdot \text{length}) + (2 \cdot \text{width})$$

$$P = 2l + 2w$$

EXAMPLE 3 Finding the Perimeter of a Rectangle

Find the perimeter of this rectangle.

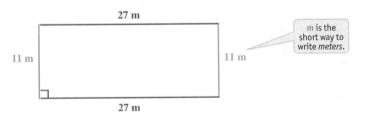

27 m

11 m

11 m

m is the short way to write *meters*.

27 m

The length is **27 m**, and the width is **11 m**

$P = \qquad 2l \quad + \quad 2w \qquad$ Replace l with 27 m and w with 11 m

$P = 2 \cdot 27 \text{ m} + 2 \cdot 11 \text{ m} \qquad$ Do the multiplications first.

$P = \qquad 54 \text{ m} + \quad 22 \text{ m} \qquad$ Add last.

$\mathbf{P = 76 \ m}$

The perimeter of the rectangle (the distance you would walk around the outside edges of the rectangle) is **76 m**

CHECK To check the solution, add the lengths of the four sides.

$$P = 27 \text{ m} + 27 \text{ m} + 11 \text{ m} + 11 \text{ m}$$

$$P = 76 \text{ m} \leftarrow \text{Matches the solution above}$$

◀ **Work Problem 3 at the Side.**

Answers

3. (a) $P = 34 \text{ cm} + 20 \text{ cm}; P = 54 \text{ cm}$
   CHECK
   $P = 17 \text{ cm} + 17 \text{ cm} + 10 \text{ cm}$
   $+ 10 \text{ cm} = 54 \text{ cm}$

   (b) $P = 74 \text{ ft}$
   CHECK
   $25 \text{ ft} + 25 \text{ ft} + 12 \text{ ft} + 12 \text{ ft} = 74 \text{ ft}$

EXAMPLE 4   Finding the Length or Width of a Rectangle

If the perimeter of a rectangle is 20 ft and the width is 3 ft, find the length.

First draw a sketch of the rectangle and label the widths as 3 ft.

Then use the formula for perimeter of a rectangle, $P = 2l + 2w$. The value of P is 20 ft and the value of w is 3 ft.

$$P = 2l + 2w$$ Formula for perimeter of a rectangle

$$20 \text{ ft} = 2l + 2 \cdot 3 \text{ ft}$$ Replace P with 20 ft and w with 3 ft
Simplify the right side by multiplying $2 \cdot 3$ ft

$$20 \text{ ft} = 2l + 6 \text{ ft}$$

$$\underline{-6 \text{ ft} \qquad\qquad -6 \text{ ft}}$$ To get $2l$ by itself, add -6 ft to both sides:
$$14 \text{ ft} = 2l + 0$$ $6 + (-6)$ is 0

$$\frac{14 \text{ ft}}{2} = \frac{2l}{2}$$ To get l by itself, divide both sides by 2

$$7 \text{ ft} = l$$

The length is **7 ft** ◁ Write **ft** as part of your answer.

CHECK To check the solution, put the length measurements on your sketch. Then add the four measurements.

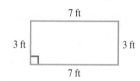

$$P = 7 \text{ ft} + 7 \text{ ft} + 3 \text{ ft} + 3 \text{ ft}$$

$$P = 20 \text{ ft}$$

Matches the perimeter given in the original problem, so the solution checks.

──────── **Work Problem 4 at the Side.** ▶

OBJECTIVE 3 Find the perimeter of parallelograms, triangles, and irregular shapes. A **parallelogram** is a four-sided figure in which opposite sides are both parallel and equal in length. Some examples are shown below. Notice that opposite sides have the same length.

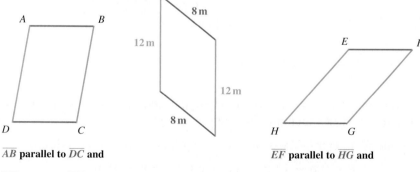

$\overline{AB}$ **parallel to** $\overline{DC}$ **and**

$\overline{AD}$ **parallel to** $\overline{BC}$

$\overline{EF}$ **parallel to** $\overline{HG}$ **and**

$\overline{EH}$ **parallel to** $\overline{FG}$

Perimeter is the distance around a flat shape, so the easiest way to find the perimeter of a parallelogram is to add the lengths of the four sides.

---

4 Use the perimeter of each rectangle and the appropriate formula to find the length or width. Draw a sketch of each rectangle and use it to check your solution.

(GS) **(a)** The perimeter of a rectangle is 36 in. and the width is 8 in. Find the length.

$$P = 2l + 2w$$

$$36 \text{ in.} = 2l + 2 \cdot 8 \text{ in.}$$

$$36 \text{ in.} = 2l + 16 \text{ in.}$$

$$\underline{-16 \text{ in.} \qquad\qquad -16 \text{ in.}}$$

$$20 \text{ in.} = 2l + 0 \qquad \text{Find the solution.}$$

(b) A rectangle has a length of 12 cm. The perimeter is 32 cm. Find the width.

Answers

4. **(a)** $\dfrac{20 \text{ in.}}{2} = \dfrac{2l}{2}$; 10 in. $= l$

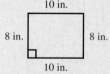

CHECK   10 in. + 10 in. + 8 in. + 8 in. = 36 in.

(b) $w = 4$ cm

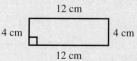

CHECK   12 cm + 12 cm + 4 cm + 4 cm = 32 cm

5 Find the perimeter of each parallelogram.

GS (a)

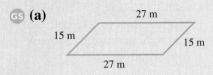

$P = 27 \text{ m} + \underline{\quad} + \underline{\quad}$

$+ \underline{\quad} = \underline{\quad}$

(b)

VOCABULARY TIP

Tri- means "three". Just like a **tri**cycle has three wheels, a **tri**angle has three sides.

6 Find the perimeter of each triangle.

GS (a)

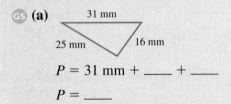

$P = 31 \text{ mm} + \underline{\quad} + \underline{\quad}$

$P = \underline{\quad}$

(b) A triangle with sides that each measure 5 in.

7 How much fencing is needed to go around a flower bed with the measurements shown below?

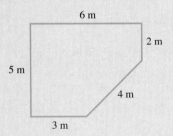

Answers

5. (a) $P = 27 \text{ m} + 27 \text{ m} + 15 \text{ m} + 15 \text{ m} = 84 \text{ m}$
   (b) $P = 18 \text{ ft}$
6. (a) $P = 31 \text{ mm} + 16 \text{ mm} + 25 \text{ mm} = 72 \text{ mm}$
   (b) $P = 15 \text{ in.}$
7. 20 m of fencing are needed.

EXAMPLE 5 Finding the Perimeter of a Parallelogram

Find the perimeter of the middle parallelogram on the previous page.

$$P = 12 \text{ m} + 12 \text{ m} + 8 \text{ m} + 8 \text{ m}$$

$$P = 40 \text{ m}$$

◀ **Work Problem 5 at the Side.**

A **triangle** is a figure with exactly three sides. Some examples are shown below.

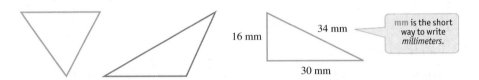

mm is the short way to write millimeters.

To find the perimeter of a triangle (the distance around the edges), add the lengths of the three sides.

EXAMPLE 6 Finding the Perimeter of a Triangle

Find the perimeter of the triangle above on the right.
To find the perimeter, add the lengths of the sides.

$$P = 16 \text{ mm} + 30 \text{ mm} + 34 \text{ mm}$$

$$P = 80 \text{ mm}$$

◀ **Work Problem 6 at the Side.**

As with any other shape, you can find the perimeter of an irregular shape by adding the lengths of the sides.

EXAMPLE 7 Finding the Perimeter of an Irregular Shape

The floor of a room has the shape shown here.

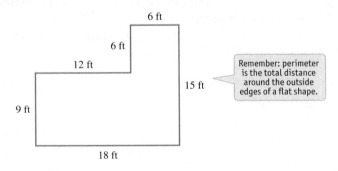

Remember: perimeter is the total distance around the outside edges of a flat shape.

You want to put in a new baseboard, where the floor meets the wall, around the entire room. How much material do you need?

Find the perimeter of the room by adding the lengths of the sides.

$$P = 9 \text{ ft} + 12 \text{ ft} + 6 \text{ ft} + 6 \text{ ft} + 15 \text{ ft} + 18 \text{ ft}$$

$$P = 66 \text{ ft}$$

You need **66 ft of baseboard.**

◀ **Work Problem 7 at the Side.**

3.1 Exercises

FOR EXTRA HELP

Go to MyMathLab *for worked-out, step-by-step solutions to exercises enclosed in a square* ▢ *and video solutions to* ▶ *exercises.*

1. **CONCEPT CHECK** Explain what is special about a square. Make a drawing of a square and label the length of each side.

2. **CONCEPT CHECK** Write the formula for finding the perimeter of a square. What do the letters in the formula stand for?

Find the perimeter of each square, using the appropriate formula. **See Example 1.**

3.

$P = 4s$

$P = 4 \cdot \underline{\hspace{1cm}}$

$P = \underline{\hspace{1cm}}$

4.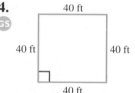

40 ft

40 ft   40 ft

40 ft

$P = 4s$

$P = 4 \cdot \underline{\hspace{1cm}}$

$P = \underline{\hspace{1cm}}$

5.

25 in.

25 in.

6.

3 cm

3 cm

Draw a sketch of each square and label the lengths of the sides. Then find the perimeter. (Sketches may vary; show your sketches to your instructor.)

7. A square park measuring 1 mile on each side

8. A square garden measuring 4 meters on each side

9. A 22 mm square postage stamp

10. A 10 in. square piece of cardboard

For the given perimeter of each square, find the length of one side using the appropriate formula. **See Example 2.**

11. The perimeter is 120 ft

$P = 4s$

$\dfrac{120 \text{ ft}}{4} = \dfrac{4s}{4}$

$\underline{\hspace{1cm}} = s$

12. The perimeter is 52 cm

$P = 4s$

$\dfrac{52 \text{ cm}}{4} = \dfrac{4s}{4}$

$\underline{\hspace{1cm}} = s$

13. A square parking lot with a perimeter of 92 yards

14. A square building with a perimeter of 144 meters

15. A square closet with a perimeter of 8 ft

16. A square bedroom with a perimeter of 44 ft

17. **CONCEPT CHECK** Explain what is special about a rectangle. Make a drawing of a rectangle and label the length of each side.

18. **CONCEPT CHECK** Write the formula for finding the perimeter of a rectangle. What do the letters in the formula stand for?

*Find the perimeter of each rectangle, using the appropriate formula. Check your solutions by adding the lengths of the four sides. See **Example 3**.*

19.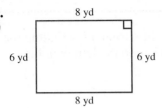

8 yd

6 yd   6 yd

8 yd

$P = 2l + 2w$

$P = 2 \cdot 8 \text{ yd} + 2 \cdot 6 \text{ yd}$

$P = \underline{\quad} + \underline{\quad}$

$P = \underline{\quad}$

20.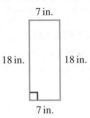

7 in.

18 in.   18 in.

7 in.

$P = 2l + 2w$

$P = 2 \cdot 18 \text{ in.} + 2 \cdot 7 \text{ in.}$

$P = \underline{\quad} + \underline{\quad}$

$P = \underline{\quad}$

21.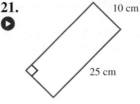

10 cm

25 cm

22.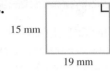

15 mm

19 mm

Draw a sketch of each rectangle and label the lengths of the sides. Then find the perimeter by using the appropriate formula. (Sketches may vary; show your sketches to your instructor.)

23. A rectangular living room 20 ft long by 16 ft wide

24. A rectangular place mat 45 cm long by 30 cm wide

25. An 8 in. by 5 in. rectangular piece of paper

26. A 2 ft by 3 ft rectangular window

*For each rectangle, you are given the perimeter and either the length or width. Find the unknown measurement using the appropriate formula. Check your solution by adding the lengths of the four sides. **See Example 4**.*

27. The perimeter is 30 cm and the width is 6 cm.

$P = 2l + 2w$

$30 \text{ cm} = 2l + 2 \cdot 6 \text{ cm}$

$30 \text{ cm} = 2l + 12 \text{ cm}$

$\underline{-12 \text{ cm} = \qquad\quad -12 \text{ cm}}$ Find the solution.

$18 \text{ cm} = 2l + 0$

28. The perimeter is 48 yards and the length is 14 yards.

$P = 2l + 2w$

$48 \text{ yd} = 2 \cdot 14 \text{ yd} + 2w$

$48 \text{ yd} = 28 \text{ yd} + 2w$

$\underline{-28 \text{ yd} \quad -28 \text{ yd}}$ Find the solution.

$20 \text{ yd} = 0 + 2w$

29. The length is 4 miles and the perimeter is 10 miles.

30. The width is 8 meters and the perimeter is 34 meters.

31. A 6 ft long rectangular table has a perimeter of 16 ft.

32. A 13 in. wide rectangular picture frame has a perimeter of 56 in.

33. A rectangular door 1 meter wide has a perimeter of 6 meters.

|←1 m→|

34. A rectangular house 33 ft long has a perimeter of 118 ft.

33 ft

*In Exercises 35–44, find the perimeter of each shape. The figures in Exercises 35–36 are parallelograms. **See Examples 5–7.***

35.

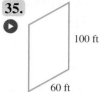
100 ft
60 ft

36.
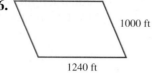
1000 ft
1240 ft

37.
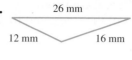
26 mm
12 mm   16 mm

38.

9 yd   7 yd
11 yd

39.
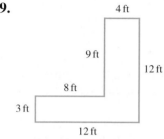
4 ft
9 ft
12 ft
8 ft
3 ft
12 ft

40.
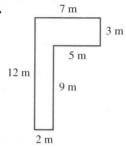
7 m
3 m
5 m
12 m
9 m
2 m

41.
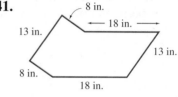
8 in.
18 in.
13 in.
13 in.
8 in.
18 in.

42.
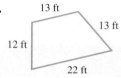
13 ft
13 ft
12 ft
22 ft

43.
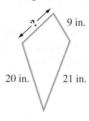
22 m
34 m
27 m
20 m
22 m

44.
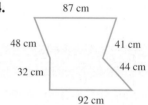
87 cm
48 cm   41 cm
32 cm   44 cm
92 cm

For each shape, you are given the perimeter and the lengths of all sides except one. Find the length of the unlabeled side.

45. The perimeter is 115 cm
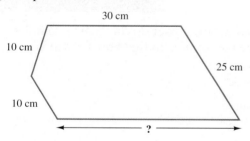
30 cm
10 cm
25 cm
10 cm
?

46. The perimeter is 63 in.
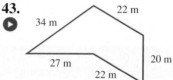
?
9 in.
20 in.   21 in.

47. The perimeter is 78 in.
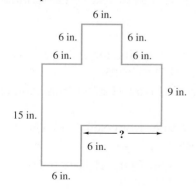
6 in.
6 in.   6 in.
6 in.   6 in.
9 in.
15 in.
?
6 in.
6 in.

48. The perimeter is 196 ft
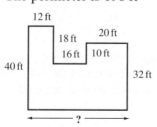
12 ft
18 ft   20 ft
16 ft   10 ft
40 ft
32 ft
?

49. In an *equilateral* triangle, all sides have the same length.

    (a) Draw sketches of four different equilateral triangles, label the lengths of the sides, and find the perimeters.

    (b) Write a "shortcut" rule (a formula) for finding the perimeter of an equilateral triangle.

    (c) Will your formula work for other kinds of triangles that are not equilateral? Explain why or why not.

50. Be sure that you have done **Exercise 49** first.

    (a) Draw a sketch of a figure with five sides of equal length. Write a "shortcut" rule (a formula) for finding the perimeter of this shape.

    (b) Draw a sketch of a figure with six sides of equal length. Write a formula for finding the perimeter of the shape.

    (c) Write a formula for finding the perimeter of a shape with 10 sides of equal length.

    (d) Write a formula for finding the perimeter of a shape with n sides of equal length.

Relating Concepts (Exercises 51–54) For Individual or Group Work

A formula that has many uses for drivers is $d = rt$, called the distance formula. If you are driving a car, then

    d is the *distance* you travel (how many miles),
    r is the *rate* (how fast you are driving in miles per hour),
    t is the *time* (how many hours you drive).

Use the distance formula as you **work Exercises 51–54** *in order.*

51. Suppose you are driving on an interstate highway at an average rate of 70 miles per hour. Use the distance formula to find out how far you will travel in

    (a) 2 hours.

    (b) 5 hours.

    (c) 8 hours.

52. An ice storm slows your driving rate to 35 miles per hour. Use the formula to find how far will you travel in

    (a) 2 hours.

    (b) 5 hours.

    (c) 8 hours.

    (d) Explain how to find each answer using the results from **Exercise 51** instead of the formula.

53. Use the distance formula to find out how many hours you would have to drive to travel the 3000 miles from Boston to San Francisco if your average rate is

    (a) 60 miles per hour.

    (b) 50 miles per hour.

    (c) 20 miles per hour (which was the speed limit 100 years ago).

54. Use the distance formula to find the average driving rate (speed) on each of these trips.

    (a) It took 11 hours to drive 671 miles from Atlanta to Chicago.

    (b) Sam drove 1539 miles from New York City to Dallas in 27 hours.

    (c) Carlita drove 16 hours to travel 1040 miles from Memphis to Denver.

3.2 | Problem Solving: Area

OBJECTIVE ▶ 1 Use the formula for area of a rectangle to find the area, the length, or the width.

OBJECTIVES

1 Use the formula for area of a rectangle to find the area, the length, or the width.

2 Use the formula for area of a square to find the area or the length of one side.

3 Use the formula for area of a parallelogram to find the area, the base, or the height.

4 Solve application problems involving perimeter and area of rectangles, squares, or parallelograms.

Understanding the Difference between Perimeter and Area

Perimeter is the *distance around the outside edges* of a flat shape.
Area is the amount of *surface inside* a flat shape.

The *perimeter* of a rectangle is the distance around the *outside edges*. Recall that we unfolded a shape and laid the sides end-to-end so we could see the total distance. The *area* of a rectangle is the amount of surface *inside* the rectangle. We measure area by finding the number of squares of a certain size needed to cover the surface inside the rectangle. Think of covering the floor of a living room with carpet. Carpet is measured in square yards, that is, square pieces that measure 1 yard along each side. Here is a drawing of a rectangular living room floor.

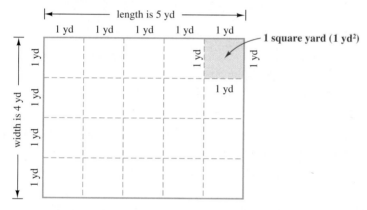

You can see from the drawing that it takes 20 squares to cover the floor. We say that the area of the floor is 20 *square yards*. A short way to write *square yards* is yd².

20 **square yards**   can be written as   20 **yd²**

To find the number of squares, you can count them, or you can multiply the number of squares in the length (5) times the number of squares in the width (4) to get 20. The formula is given below.

Finding the Area of a Rectangle

Area of a rectangle = length • width

$$A = lw$$

Use *square units* when measuring area, such as yd², in.², and so on.

Squares of many sizes can be used to measure area.

For smaller areas, you might use the ones shown at the right.

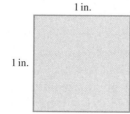

1 square inch (1 in.²)

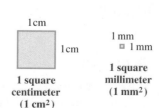

1 square centimeter (1 cm²)   1 square millimeter (1 mm²)

Actual-size drawings

VOCABULARY TIP

Area The **area** is the amount of surface **inside** a flat shape. Write the area using **square units,** such as square inches or square feet—for example, 10 **ft²** or 12 **in.²**.

1 Find the area of each rectangle, using the appropriate formula.

GS **(a)**

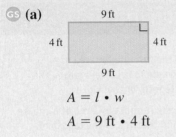

9 ft

4 ft ⌐ 4 ft

9 ft

$A = l \cdot w$

$A = 9 \text{ ft} \cdot 4 \text{ ft}$

$A = \underline{\hspace{1cm}}$

(b) A rectangle is 35 yd long and 20 yd wide. (First make a sketch of the rectangle and label the lengths of the sides.)

(c) A rectangular patio measures 3 m by 2 m. (First make a sketch of the patio and label the lengths of the sides.)

Answers

1. **(a)** $A = 36 \text{ ft}^2$
   (b)
   20 yd ⌐
   35 yd
   $A = 700 \text{ yd}^2$
   (c)
   2 m
   3 m
   $A = 6 \text{ m}^2$

Other sizes of squares that are often used to measure area are listed here, but they are too large to draw on this page.

1 square meter (1 m^2)
1 square kilometer (1 km^2)

1 square foot (1 ft^2)
1 square yard (1 yd^2)
1 square mile (1 mi^2)

> **! CAUTION**
>
> The raised 2 in 4^2 means that you multiply $4 \cdot 4$ to get 16. The raised 2 in cm^2 or yd^2 is a short way to write the word "square." It means that you multiplied cm times cm to get cm^2, or yd times yd to get yd^2. Recall that a short way to write $x \cdot x$ is x^2. Similarly, cm $\cdot$ cm is cm^2. When you see 5 cm^2, say "five square centimeters." Do *not* multiply $5 \cdot 5$, because the exponent applies only to the *first* thing to its immediate left. The exponent applies to cm, *not* to the number.

EXAMPLE 1   **Finding the Area of Rectangles**

Find the area of each rectangle.

(a)

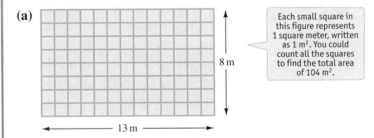

8 m

13 m

Each small square in this figure represents 1 square meter, written as 1 m^2. You could count all the squares to find the total area of 104 m^2.

The length of this rectangle is 13 m and the width is 8 m. Use the formula $A = lw$, which means $A = l \cdot w$.

$A = \quad l \quad \cdot \quad w$ — Replace l with 13 m and w with 8 m

$A = 13 \text{ m} \cdot 8 \text{ m}$ — Multiply 13 times 8 to get 104

$A = 104 \text{ m}^2$ — Multiply m times m to get m^2

The area of the rectangle is **104 m^2**

If you count the number of squares in the sketch, you will also get 104 m^2. Each square in the sketch represents 1 m by 1 m, which is 1 square meter (1 m^2).

(b) A rectangle measuring 7 cm by 21 cm

First make a sketch of the rectangle. The length is 21 cm (the longer measurement) and the width is 7 cm. Then use the formula $A = lw$.

7 cm

21 cm

$A = \quad l \quad \cdot \quad w$ — Replace l with 21 cm and w with 7 cm

$A = 21 \text{ cm} \cdot 7 \text{ cm}$ — Multiply $21 \cdot 7$ to get 147

$A = 147 \text{ cm}^2$ — Multiply cm $\cdot$ cm to get cm^2

The area of the rectangle is **147 cm^2**

◄ **Work Problem 1 at the Side.**

> **! CAUTION**
>
> The units for *area* will always be *square* units (cm^2, m^2, yd^2, mi^2, and so on). The units for *perimeter* will always be *linear* units (cm, m, yd, mi, and so on), *not* square units.

EXAMPLE 2   Finding the Length or Width of a Rectangle

If the area of a rectangular rug is 12 yd^2 and
the length is 4 yd, find the width.

4 yd

    Draw a sketch of the rug. Label the length as 4 yd.
Use the formula for area of a rectangle, $A = lw$.

$$A \quad = \quad l \quad \bullet \; w \qquad$$ Replace A with 12 yd^2 and
replace l with 4 yd

$$12 \text{ yd}^2 = 4 \text{ yd} \bullet w \qquad$$ To get w by itself, divide both sides by 4 yd

$$\frac{12 \text{ yd} \bullet \text{yd}}{4 \text{ yd}} = \frac{4 \text{ yd} \bullet w}{4 \text{ yd}} \qquad$$ On the left side, rewrite yd^2 as yd $\bullet$ yd. Then $\frac{\text{yd}}{\text{yd}}$ is 1, so they divide out.

On the right side, use division to "undo" multiplication:
(4 yd $\bullet$ w) $\div$ 4 yd is w

$$\textbf{3 yd} \quad = \quad w \qquad$$ On the left side, 12 yd $\div$ 4 is 3 yd

The width of the rug is **3 yd**

CHECK   To check the solution, put the width measurement on your sketch.
Then use the area formula.

$$A = \quad l \quad \bullet \quad w$$

$$A = 4 \text{ yd} \bullet 3 \text{ yd}$$

$$A = 12 \text{ yd}^2$$

3 yd

4 yd

An area of 12 yd^2 matches the information in the original problem.
So 3 yd is the correct width of the rug.

—————— **Work Problem** ❷ **at the Side.** ▶

OBJECTIVE ❷ **Use the formula for area of a square to find the area or
the length of one side.** As with a rectangle, you can multiply length times
width to find the area (surface inside) of a square. Because the length and the
width are the same in a square, the formula is written as shown below.

> **Finding the Area of a Square**
>
> $$\text{Area of a square} = \text{side} \bullet \text{side}$$
>
> $$A = s \bullet s$$
>
> $$A = s^2$$
>
> Remember to use *square units* when measuring area.

EXAMPLE 3   Finding the Area of a Square

Find the area of a square highway sign that is 4 ft on each side.
    Use the formula for area of a square, $A = s^2$.

$$A = \quad s^2 \qquad$$ Remember that s^2 means $s \bullet s$

$$A = s \quad \bullet \quad s \qquad$$ Replace s with 4 ft

$$A = 4 \text{ ft} \bullet 4 \text{ ft} \qquad$$ Multiply 4 $\bullet$ 4 to get 16

> Be careful!
> s^2 is 4 $\bullet$ 4 or 16
> (**not** 4 $\bullet$ 2).

$$A = \textbf{16 ft}^2 \qquad$$ Multiply ft $\bullet$ ft to get ft^2

The area of the sign is **16 ft^2**

—————— **Continued on Next Page**

---

❷ Use the area of each rectangle
and the appropriate formula to
find the length or width. Draw a
sketch of each rectangle and use
it to check your solution.

GS **(a)** The area of a microscope
slide is 12 cm^2, and the
length is 6 cm. Find the
width.

$$A = l \bullet w$$

$$12 \text{ cm}^2 = 6 \text{ cm} \bullet w$$

$$\frac{12 \text{ cm} \bullet \text{cm}}{6 \text{ cm}} = \frac{6 \text{ cm} \bullet w}{6 \text{ cm}}$$

$$\underline{\qquad\qquad} = w$$

(b) A child's play lot is 10 ft
wide and has an area of
160 ft^2. Find the length.

(c) A hallway floor is 31 m long
and has an area of 93 m^2.
Find the width of the floor.

Answers

2. (a) 2 cm $= w$

2 cm

6 cm

CHECK $A = 6$ cm $\bullet$ 2 cm
$A = 12$ cm^2
Matches original problem

(b) $l = 16$ ft

10 ft

16 ft

CHECK $A = 16$ ft $\bullet$ 10 ft
$A = 160$ ft^2
Matches original problem

(c) $w = 3$ m

3 m

31 m

CHECK $A = 31$ m $\bullet$ 3 m
$A = 93$ m^2
Matches original problem

3 Find the area of each square, using the appropriate formula. Make a sketch of each square.

GS **(a)** A 12 in. square piece of fabric

$$A = s^2$$

$$A = s \cdot s$$

$$A = 12 \text{ in.} \cdot 12 \text{ in.}$$

$$A = \underline{\hspace{2cm}}$$

(b) A square township 7 miles on a side

(c) A square earring measuring 20 mm on each side

4 Given the area of each square, find the length of one side by inspection.

GS **(a)** The area of a square-shaped nature center is 16 mi².

$$A = s^2$$ What number

$$16 \text{ mi}^2 = s \cdot s$$ times itself gives 16?

$$16 \text{ mi}^2 = \underline{\hspace{1cm}} \cdot \underline{\hspace{1cm}}$$

So s is _____.

(b) A square floor with an area of 100 m²

(c) A square clock face with an area of 81 in.²

Answers

3. **(a)** 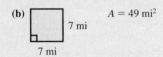 $A = 144$ in.²
12 in.
12 in.

(b) $A = 49$ mi²
7 mi
7 mi

(c) $A = 400$ mm²
20 mm
20 mm

4. **(a)** $16 \text{ mi}^2 = 4 \text{ mi} \cdot 4 \text{ mi}$; s is 4 mi
(b) $s = 10$ m   **(c)** $s = 9$ in.

⚠ CAUTION

Be careful! s^2 means $s \cdot s$. It does **not** mean $s \cdot 2$. In this example s is 4 ft, so $(4 \text{ ft})^2$ is $4 \text{ ft} \cdot 4 \text{ ft} = 16 \text{ ft}^2$. It is **not** $4 \text{ ft} \cdot 2 = 8 \text{ ft}$.

CHECK Check the solution by drawing a square and labeling each side as 4 ft. You can multiply length (4 ft) times width (4 ft), as you did for a rectangle. So the area is $4 \text{ ft} \cdot 4 \text{ ft}$, or 16 ft^2. This result matches the solution you got by using the formula $A = s^2$.

4 ft
4 ft

◀ **Work Problem 3 at the Side.**

EXAMPLE 4 **Finding the Length of One Side of a Square**

If the area of a square township is 49 mi², what is the length of one side of the township?

Use the formula for area of a square, $A = s^2$. The value of A is 49 mi².

$A = s^2$ Replace A with 49 mi²

$49 \text{ mi}^2 = s^2$ To get s by itself, we have to "undo" the squaring of s. This is called *finding the square root.* (More on this in a later chapter.)

$49 \text{ mi}^2 = s \cdot s$ For now, solve by inspection. Ask, what number times itself gives 49?

$49 \text{ mi}^2 = 7 \text{ mi} \cdot 7 \text{ mi}$ $7 \cdot 7$ is 49, so $7 \text{ mi} \cdot 7 \text{ mi}$ is 49 mi²

The value of s is **7 mi**, so the length of one side of the township is **7 mi**. Notice how this result matches the information about the township in **Margin Problem 3(b).**

◀ **Work Problem 4 at the Side.**

OBJECTIVE ▶ 3 **Use the formula for area of a parallelogram to find the area, the base, or the height.** To find the area of a parallelogram, first draw a dashed line inside the figure, as shown here.

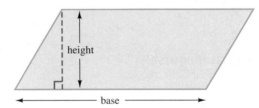
height
base

As an experiment, try this yourself by tracing this parallelogram onto a piece of paper.

The length of the dashed line is the *height* of the parallelogram. It forms a 90° angle (a right angle) with the base. The height is the shortest distance between the base and the opposite side.

Now cut off the triangle created on the left side of the parallelogram and move it to the right side, as shown on the next page.

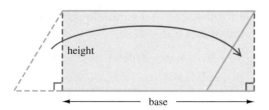

You have made the parallelogram into a rectangle. You can see that the area of the parallelogram and the rectangle are the same.

Equal areas
→ Area of the rectangle = length • width
→ Area of the parallelogram = base • height

Finding the Area of a Parallelogram

Area of a parallelogram = base • height

$$A = bh$$

Remember to use *square units* when measuring area.

EXAMPLE 5   **Finding the Area of Parallelograms**

Find the area of each parallelogram.

(a)

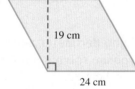

19 cm

24 cm

(b)

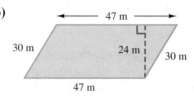

47 m

30 m   24 m

30 m

47 m

(a) The base is 24 cm and the height is 19 cm. The formula for the area of a parallelogram is $A = bh$.

$A = \quad b \quad • \quad h$ Replace *b* with 24 cm and *h* with 19 cm

$A = $ **24 cm • 19 cm**   Multiply 24 • 19 to get 456

$A = $ **456 cm²**   Multiply cm • cm to get cm²

The area of the parallelogram is **456 cm²**   ◁ Write **cm²** as part of your answer.

(b) Use the formula for area of a parallelogram, $A = bh$.

$A = \quad b \quad • \quad h$ Replace *b* with 47 m and *h* with 24 m

$A = $ **47 m • 24 m**   Multiply 47 • 24 to get 1128

$A = $ **1128 m²**   Multiply m • m to get m²   ◁ Write **m²** as part of your answer.

Notice that the 30 m sides are *not* used in finding the area. But you would use them when finding the *perimeter* of the parallelogram.

—————— **Work Problem ⑤ at the Side.** ▶

EXAMPLE 6   **Finding the Base or Height of a Parallelogram**

The area of a parallelogram is 24 ft² and the base is 6 ft. Find the height.

First draw a sketch of the parallelogram and label the base as 6 ft.

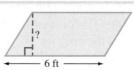

?

6 ft

————————— **Continued on Next Page**

---

⑤ Find the area of each parallelogram.

GS (a)

50 ft

44 ft   42 ft

44 ft

50 ft

$A = b • h$

$A = 50\ \text{ft} • 42\ \text{ft}$

$A = \underline{\hspace{2cm}}$

(b)   11 in.

18 in.   10 in.   18 in.

11 in.

(c) A parallelogram with base 8 cm and height 1 cm

Answers

5. (a) $A = 2100\ \text{ft}^2$ **(b)** $A = 180\ \text{in.}^2$
   (c) $A = 8\ \text{cm}^2$

6 Use the area of each parallelogram and the appropriate formula to find the base or height. Draw a sketch of each parallelogram and use it to check your solution.

(a) The area of a parallelogram is 140 in.² and the base is 14 in. Find the height.

$$A = b \cdot h$$

$$140 \text{ in.}^2 = 14 \text{ in.} \cdot h$$

$$\frac{140 \text{ in.} \cdot \text{in.}}{14 \text{ in.}} = \frac{14 \text{ in.} \cdot h}{14 \text{ in.}}$$

$$\underline{\hspace{1.5cm}} = h$$

(b) A parallelogram has an area of 4 yd². The height is 1 yd. Find the base.

7 If grass costs $3 per square yard, how much will the neighbors in **Example 7** spend to cover the playground with grass?

Answers

6. (a) 10 in. $= h$

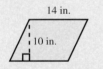
14 in.
10 in.

CHECK $A = 14$ in. $\cdot$ 10 in.
$A = 140$ in.²
Matches original problem

(b) $b = 4$ yd

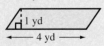
1 yd
4 yd

CHECK $A = 4$ yd $\cdot$ 1 yd
$A = 4$ yd²
Matches original problem

7. $1056 (Find the area by multiplying 22 yd $\cdot$ 16 yd to get 352 yd². Then multiply $3 $\cdot$ 352 to get $1056.)

Use the formula for the area of a parallelogram, $A = bh$. The value of A is 24 ft², and the value of b is 6 ft.

$$A = b \cdot h \quad \text{Replace } A \text{ with 24 ft}^2 \text{ and } b \text{ with 6 ft}$$

$$24 \text{ ft}^2 = 6 \text{ ft} \cdot h \quad \text{To get } h \text{ by itself, divide both sides by 6 ft}$$

On the left side, rewrite ft² as ft $\cdot$ ft. Then $\frac{ft}{ft}$ is 1, so they divide out.

$$\frac{24 \text{ ft} \cdot \text{ft}}{6 \text{ ft}} = \frac{6 \text{ ft} \cdot h}{6 \text{ ft}}$$

On the right side, use division to "undo" multiplication: $(6 \text{ ft} \cdot h) \div 6 \text{ ft}$ is h

$$4 \text{ ft} = h \quad \text{On the left side, 24 ft} \div 6 \text{ is 4 ft}$$

The height of the parallelogram is **4 ft**

CHECK To check the solution, put the height measurement on your sketch. Then use the area formula.

$$A = b \cdot h$$
$$A = 6 \text{ ft} \cdot 4 \text{ ft}$$
$$A = 24 \text{ ft}^2$$

4 ft
6 ft

An area of 24 ft² matches the information in the original problem. So 4 ft is the correct height of the parallelogram.

◀ **Work Problem 6 at the Side.**

OBJECTIVE ▶ **4** **Solve application problems involving perimeter and area of rectangles, squares, or parallelograms.** When you are solving problems, first decide whether you need to find the perimeter or the area.

EXAMPLE 7 **Solving an Application Problem Involving Perimeter or Area**

A group of neighbors is fixing up a playground for their children. The rectangular lot is 22 yd by 16 yd. If chain-link fencing costs $6 per yard, how much will they spend to put a fence around the lot?

First draw a sketch of the rectangular lot and label the length and width. The fence will go around the edges of the lot, so you need to find the *perimeter* of the lot.

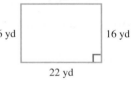
22 yd
16 yd
16 yd
22 yd

$$P = 2l + 2w \quad \text{Formula for perimeter of a rectangle}$$

$$P = 2 \cdot 22 \text{ yd} + 2 \cdot 16 \text{ yd} \quad \text{Replace } l \text{ with 22 yd and } w \text{ with 16 yd}$$

$$P = 44 \text{ yd} + 32 \text{ yd}$$

$$P = 76 \text{ yd}$$

The perimeter of the lot is 76 yd, so the neighbors need to buy 76 yd of fencing. The cost of the fencing is $6 *per yard,* which means $6 *for* 1 *yard.* To find the cost for 76 yd, multiply $6 $\cdot$ 76.
The neighbors will spend $456 on the fence.

◀ **Work Problem 7 at the Side.**

3.2 Exercises

FOR EXTRA HELP   *Go to* MyMathLab *for worked-out, step-by-step solutions to exercises enclosed in a square* [ ] *and video solutions to* ▶ *exercises.*

1. **CONCEPT CHECK**   Write the formula for finding the area of a square. What do the letters in the formula stand for?

2. **CONCEPT CHECK**   Write the formula for finding the area of a rectangle. What do the letters in the formula stand for?

CONCEPT CHECK   *Perform each calculation.*

3. (a) 10^2 (b) $2 \cdot 10$

   (c) 25^2 (d) $25 \cdot 2$

4. (a) 8^2 (b) $2 \cdot 8$

   (c) 30^2 (d) $30 \cdot 2$

Find the area of each rectangle, square, or parallelogram using the appropriate formula. **See Examples 1, 3, and 5.**

5.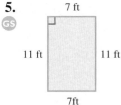
7 ft
11 ft   11 ft
7ft

$A = l \cdot w$

$A = 11 \text{ ft} \cdot 7 \text{ ft}$

$A = \underline{\hspace{1cm}}$

6.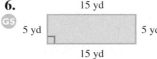
15 yd
5 yd   5 yd
15 yd

$A = l \cdot w$

$A = 15 \text{ yd} \cdot 5 \text{ yd}$

$A = \underline{\hspace{1cm}}$

7.
6 in.

$A = s^2$

$A = s \cdot s$

$A = \underline{\hspace{1cm}} \cdot \underline{\hspace{1cm}}$

$A = \underline{\hspace{1cm}}$

8.
20 cm

$A = s^2$

$A = s \cdot s$

$A = \underline{\hspace{1cm}} \cdot \underline{\hspace{1cm}}$

$A = \underline{\hspace{1cm}}$

9.
31 mm
31 mm
25 mm   31 mm
31 mm

Hint: $A = b \cdot h$

10.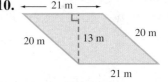
← 21 m →
20 m   13 m   20 m
21 m

Hint: $A = b \cdot h$

In Exercises 11–16, first draw a sketch of the shape and label the length and width or base and height. Then find the area. (Sketches may vary; show your sketches to your instructor.)

11. A rectangular calculator that measures 15 cm by 7 cm

12. A table 12 ft long by 3 ft wide

13. ▶ A parallelogram with height of 9 ft and base of 8 ft

14. A parallelogram measuring 18 mm on the base and 3 mm on the height

15. A fire burned a square-shaped forest 25 mi on a side.

16. A square piece of window glass 1 m on each side

In Exercises 17–22, use the area of each rectangle and either its length or width, and the appropriate formula, to find the other measurement. Draw a sketch of each rectangle and use it to check your solution. See Example 2. (Sketches may vary; show your sketches to your instructor.)

17. The area of a desk is 18 ft², and the width is 3 ft. Find its length.

$$A = l \cdot w$$
$$18 \text{ ft}^2 = l \cdot 3 \text{ ft}$$ Now divide both sides by 3 ft

18. The area of a classroom is 630 ft², and the length is 30 ft. Find its width.

$$A = l \cdot w$$
$$630 \text{ ft}^2 = 30 \text{ ft} \cdot w$$ Now divide both sides by 30 ft

19. A parking lot is 90 yd long and has an area of 7200 yd². Find its width.

20. A playground is 60 yd wide and has an area of 6000 yd². Find its length.

21. A 154 in.² photo has a width of 11 in. Find its length.

22. A 15 in.² note card has a width of 3 in. Find its length.

Given the area of each square, find the length of one side by inspection. See Example 4.

23. A square floor has an area of 36 m²

$$A = s^2$$
$$36 \text{ m}^2 = s \cdot s$$

What number times itself gives 36? _____

So, $s =$ _____

24. A square stamp has an area of 9 cm²

$$A = s^2$$
$$9 \text{ cm}^2 = s \cdot s$$

What number times itself gives 9? _____

So, $s =$ _____

25. The area of a square sign is 4 ft²

26. The area of a square piece of metal is 64 in.²

Use the area of each parallelogram and either its base or height, and the appropriate formula, to find the other measurement. Draw a sketch of each parallelogram and use it to check your solution. See Example 6. (Sketches may vary; show your sketches to your instructor.)

27. The area is 500 cm², and the base is 25 cm. Find the height.

$$A = b \cdot h$$
$$500 \text{ cm}^2 = 25 \text{ cm} \cdot h$$ Now divide both sides by 25 cm

28. The area is 1500 m², and the height is 30 m. Find the base.

$$A = b \cdot h$$
$$1500 \text{ m}^2 = b \cdot 30 \text{ m}$$ Now divide both sides by 30 m

29. The height is 13 in. and the area is 221 in.². Find the base.

30. The base is 19 cm, and the area is 114 cm². Find the height.

31. The base is 9 m, and the area is 9 m². Find the height.

32. The area is 25 mm², and the height is 5 mm. Find the base.

33. CONCEPT CHECK   The work shown below has **two** mistakes in it.

What Went Wrong?   First write a sentence explaining what each mistake is. Then fix the mistakes and find the correct solution.

25 cm   25 cm
24 cm
25 cm   25 cm

$P = 25$ cm $+ 24$ cm $+ 25$ cm $+ 25$ cm $+ 25$ cm

$P = 124$ cm^2

34. CONCEPT CHECK   The work shown below has **two** mistakes in it.

What Went Wrong?   First write a sentence explaining what each mistake is. Then fix the mistakes and find the correct solution.

7 ft

$A = s^2$
$A = 7$ ft $\cdot 2$
$A = 14$ ft

Name each figure and find its perimeter and area.

35.

45 in.
45 in.

36.
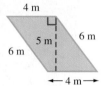
4 m
5 m   6 m
6 m
← 4 m →

37.
12 cm
18 cm
18 cm   10 cm
12 cm

38.
100 yd
80 yd

Solve each application problem to find the perimeter, the area, or one of the side measurements. In Exercises 39–44, draw a sketch, including the appropriate measurements. **See Example 7.** *(Sketches may vary; show sketches to your instructor.)*

39. Gymnastic floor exercises are performed on a square mat that is 12 meters on a side. Find the perimeter and area of a mat. (Data from nist.gov)

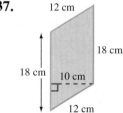

40. A regulation volleyball court is 18 meters by 9 meters. Find the perimeter and area of a regulation court. (Data from nist.gov)

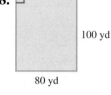

41. Tyra's kitchen is 4 m wide and 5 m long. She is pasting a decorative strip that costs $6 per meter around the top edge of all the walls. How much will she spend?

42. The Wang's family room measures 20 ft by 25 ft. They are covering the floor with square tiles that measure 1 ft on a side and cost $1 each. How much will they spend on tile?

43. Mr. and Mrs. Gomez are buying carpet for their square-shaped bedroom, which measures 5 yd along each wall. The carpet is $23 per square yard and padding and installation is another $6 per square yard. How much will they spend in all?

44. A standard piece of printer paper measures about 28 cm from top to bottom and about 22 cm from side to side. Find the perimeter and the area of the paper.

45. A regulation football field is 100 yd long (excluding end zones) and has an area of 5300 yd². Find the width of the field. (Data from NFL.)

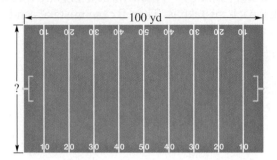

46. There are 14,790 ft² of ice on the rectangular playing area for a major league hockey game (excluding the area behind the goal lines). If the playing area is 85 ft wide, how long is it? (Data from NHL.)

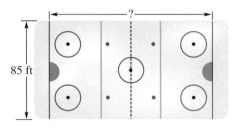

47. The table below shows information on two square tents for camping.

Tents	Coleman Evanston Tent	Mountain Trails Tent
Dimensions	12 ft × 12 ft	11 ft × 11 ft
Sleeps	8 campers	6 campers
Sale price	$175	$99

Data from www.amazon.com

For the Coleman tent, find the perimeter, area, and number of square feet of floor space for each camper. Round to the nearest whole number if necessary.

48. Look at the table in **Exercise 47.** For the Mountain Trails tent, find the perimeter, area, and number of square feet of floor space for each camper. Round to the nearest whole number if necessary.

Relating Concepts (Exercises 49–52) For Individual or Group Work

Use your knowledge of perimeter and area to **work Exercises 49–52 in order.**

49. Suppose you have 12 ft of fencing to make a square or rectangular garden plot. Draw sketches of *all* the possible plots that use exactly 12 ft of fencing and label the lengths of the sides. Use only *whole number* lengths. (*Hint:* There are three possibilities.)

50. (a) Find the area of each plot in **Exercise 49.**

(b) Which plot has the greatest area?

51. Repeat **Exercise 49** using 16 ft of fencing. Be sure to draw *all* possible square or rectangular plots that have whole number lengths for the sides.

52. (a) Find the area of each plot in **Exercise 51.**

(b) Compare your results to those from **Exercise 50.** What do you notice about the plots with the greatest area?

Summary Exercises *Perimeter and Area*

Name each figure and find its perimeter and area using the appropriate formulas.

1.
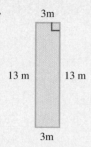
3m

13 m   13 m

3m

2.
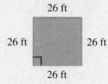
26 ft

26 ft   26 ft

26 ft

3.
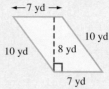
←7 yd→

10 yd   10 yd   8 yd

7 yd

4.
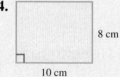
8 cm

10 cm

5.

9 in.

9 in.

6.
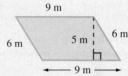
9 m

6 m   5 m   6 m

9 m

7.
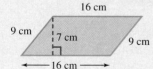
4 ft

9 ft

9 ft

4 ft

8.
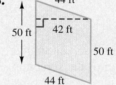
44 ft

50 ft   42 ft

50 ft

44 ft

9. CONCEPT CHECK   The work shown below has **two** mistakes in it.

What Went Wrong?   First write a sentence explaining what each mistake is. Then fix the mistakes and find the correct solution.

Rayvon found the area of the parallelogram this way:

16 cm

9 cm   7 cm   9 cm

16 cm

$A = 16 \text{ cm} \cdot 9 \text{ cm}$

$A = 144$

10. CONCEPT CHECK   The work shown below has **two** mistakes in it.

What Went Wrong?   First write a sentence explaining what each mistake is. Then fix the mistakes and find the correct solution.

Tommy found the perimeter of the rectangle this way:

55 yd

21 yd   21 yd

55 yd

$P = 21 \text{ yd} + 55 \text{ yd}$

$P = 76 \text{ yd}^2$

In Exercises 11–14, use the appropriate formula to find the unknown measurement. Draw a sketch of the figure and use it to check your solution. (Sketches may vary; show your instructor.)

11. If a square photograph has an area of 36 in.^{2}, what is the length of one side?

12. A sidewalk is 42 m long and has an area of 84 m^2. Find the width of the sidewalk.

13. The perimeter of a square map is 64 cm. What is the length of one side?

14. A rectangular patio is 9 ft wide and has a perimeter of 48 ft. What is the patio's length?

Solve each application problem using the appropriate formula. Show your work.

15. How much fencing is needed to enclose a triangular building site that measures 168 meters on each side?

16. Regulation soccer fields for teens and adults can measure 65 to 80 yards wide and 110 to 120 yards long, depending on the age and skill level of the players. Find the area of the smallest soccer field and the area of the largest soccer field. What is the difference in playing room between the smallest and largest fields? (Data from www.socceru.com)

17. A laptop computer has a rectangular screen with a perimeter of 44 inches. The length of the screen is 13 inches. Find the width of the screen.

18. Kari is decorating the cover of a square-shaped photo album that is 22 cm along each side. She is gluing braid along the edges of the front and back covers. How much braid will she need?

19. A school flag measures 3 feet by 5 feet. Find the amount of fabric needed to make seven flags, and the amount of binding needed to go around all the edges of all the flags.

20. The lobby floor of a new skyscraper is in the shape of a parallelogram, with an area of 3024 ft^2. If the base of the parallelogram measures 63 ft, what is its height?

3.3 Solving Application Problems with One Unknown Quantity

OBJECTIVE ▸ ① Translate word phrases into algebraic expressions. Earlier in this chapter, you worked with applications involving perimeter and area. You were able to use well-known rules (formulas) to write equations to solve. However, you will encounter many problems for which no formula is available. Then you need to analyze the problem and translate the words into an equation that fits the particular situation. We'll start by translating word phrases into algebraic expressions.

OBJECTIVES

① Translate word phrases into algebraic expressions.

② Translate sentences into equations.

③ Solve application problems with one unknown quantity.

EXAMPLE 1 Translating Word Phrases into Algebraic Expressions

Write each phrase as an algebraic expression. Use x as the variable.

Words	Algebraic Expression	
A number plus 2	$x + 2$ or $2 + x$	
The sum of 8 and a number	$8 + x$ or $x + 8$	Two correct ways to write each *addition* expression
5 more than a number	$x + 5$ or $5 + x$	
−35 added to a number	$-35 + x$ or $x + (-35)$	
A number increased by 6	$x + 6$ or $6 + x$	
9 less than a number	$x - 9$	
A number subtracted from 3	$3 - x$	
3 subtracted from a number	$x - 3$	Only *one* correct way to write each *subtraction* expression
7 fewer than a number	$x - 7$	
A number decreased by 4	$x - 4$	
10 minus a number	$10 - x$	

> **❶ CAUTION**
> Recall that addition can be done in any order, so $x + 2$ gives the same result as $2 + x$. This is *not* true in subtraction, so be careful. $10 - x$ does *not* give the same result as $x - 10$.

—————— Work Problem ① at the Side. ▶

EXAMPLE 2 Translating Word Phrases into Algebraic Expressions

Write each phrase as an algebraic expression. Use x as the variable.

Words	Algebraic Expression	
8 times a number	$8x$	
The product of 12 and a number	$12x$	
Double a number (meaning "2 times")	$2x$	
The quotient of −6 and a number	$\dfrac{-6}{x}$	Be careful when dividing! The order of the numbers *does* matter.
A number divided by 10	$\dfrac{x}{10}$	
15 subtracted from 4 times a number	$4x - 15$	
The result is	$=$	

—————— Work Problem ② at the Side. ▶

① Write each phrase as an algebraic expression. Use x as the variable.

(a) 15 less than a number

(b) 12 more than a number

(c) A number increased by 13

(d) A number minus 8

(e) A number subtracted from 6

(f) 6 subtracted from a number

② Write each phrase as an algebraic expression. Use x as the variable.

(a) Triple a number

(b) The product of −8 and a number

(c) The quotient of 15 and a number

(d) 5 times a number subtracted from 30

(e) A number divided by −9

Answers

1. (a) $x - 15$ (b) $x + 12$ or $12 + x$
   (c) $x + 13$ or $13 + x$ (d) $x - 8$
   (e) $6 - x$ (f) $x - 6$

2. (a) $3x$ (b) $-8x$ (c) $\dfrac{15}{x}$
   (d) $30 - 5x$ (*not* $5x - 30$)   (e) $\dfrac{x}{-9}$

③ Translate each sentence into an equation and solve it. Check your solution by going back to the words in the original problem.

GS (a) If 3 times a number is added to 4, the result is 19. Find the number.

Let x represent the unknown number.

$$\underbrace{\text{3 times}}_{} \quad \underset{\downarrow}{\text{added}} \quad \underset{\downarrow}{4} \quad \underbrace{\text{result}}_{} \quad \underset{\downarrow}{19}$$
$$\text{a number} \quad \text{to} \quad \quad \text{is}$$
$$3x \quad + \quad 4 \quad = \quad 19$$

(b) If 7 is subtracted from 6 times a number, the result is −25. Find the number.

Answers

3. (a) $3x + 4 = 19$
   $x = 5$

   CHECK $3 \cdot 5 + 4$ does equal 19
   $$\underbrace{15}_{} + 4$$
   $$19$$

   (b) $6x - 7 = -25$
   $x = -3$

   CHECK $6 \cdot (-3) - 7$ does equal -25
   $$\underbrace{-18}_{} + (-7)$$
   $$-25$$

OBJECTIVE ▶ ② Translate sentences into equations. The next example shows you how to translate a sentence into an equation that you can solve.

EXAMPLE 3 Translating a Sentence into an Equation

If 5 times a number is added to 11, the result is 26. Find the number.

Let x represent the unknown number. Use the information in the problem to write an equation.

$$\underbrace{\text{5 times a number}}_{} \underbrace{\text{added to}}_{} \underset{\downarrow}{11} \quad \underset{\downarrow}{\text{is}} \quad \underset{\downarrow}{26}$$
$$5x \quad\quad\quad + \quad 11 = 26$$

Next, solve the equation.

$$\begin{array}{rcl} 5x + 11 & = & 26 \\ \underline{-11} & & \underline{-11} \\ 5x + 0 & = & 15 \end{array}$$ To get $5x$ by itself, add -11 to both sides.

$$\frac{5x}{5} = \frac{15}{5}$$ To get x by itself, divide both sides by 5.

$$x = 3$$

The unknown number is **3**

CHECK Go back to the words of the *original* problem.

$$\underset{\downarrow}{\text{If 5}} \underset{\downarrow}{\text{times}} \underbrace{\text{a number}}_{} \underbrace{\text{is added to}}_{} \underset{\downarrow}{11}, \underbrace{\text{the result is}}_{} \underset{\downarrow}{26}$$
$$5 \quad \cdot \quad 3 \quad\quad + \quad 11 \quad = \quad 26$$

Does $5 \cdot 3 + 11$ really equal 26? Yes, $5 \cdot 3 + 11 = 15 + 11 = 26$.

So 3 is the correct solution because it "works" when you put it back into the original problem. That is, 3 fits all the facts in the problem.

──────────────────── ◀ **Work Problem ③ at the Side.**

OBJECTIVE ▶ ③ Solve application problems with one unknown quantity. Now you are ready to tackle application problems. Here are the steps.

Solving an Application Problem	
Step 1	**Read** the problem once to see what it is about. Read it carefully a second time. As you read, make a sketch or write word phrases that identify the known and the unknown parts of the problem.
Step 2(a)	If there is one unknown quantity, **assign a variable** to represent it. Write down what your variable represents.
Step 2(b)	If there is more than one unknown quantity, **assign a variable** to represent "the thing you know the least about." Then write variable expressions, using the same variable, to show the relationship of the other unknown quantities to the first one.
Step 3	**Write an equation using your variable.** Use your sketch or word phrases as the guide.
Step 4	**Solve the equation.**
Step 5	**State the answer** to the question in the problem and label your answer.
Step 6	**Check** whether your answer fits all the facts given in the *original* statement of the problem. If it does, you are done. If it doesn't, start again at *Step 1*.

EXAMPLE 4 Solving an Application Problem with One Unknown Quantity

Heather put some money aside in an envelope for small household expenses. Yesterday she took out $20 for groceries. Today a friend paid back a loan and Heather put the $34 in the envelope. Now she has $43 in the envelope. How much was in the envelope at the start?

Step 1 **Read** the problem once. It is about money in an envelope. Read it a second time and write word phrases.

> Unknown: amount of money in the envelope at the start
> Known: took out $20; put in $34; ended up with $43

Step 2(a) There is only one unknown quantity, so **assign a variable** to represent it. Let m represent the money at the start.

Step 3 **Write an equation using your variable.** Use the phrases you wrote as a guide.

Money at the start	Took out $20	Put in $34	Ended up with $43
m	$- \$20$	$+ \$34$	$= \$43$

Step 4 **Solve the equation.**

$$m - 20 + 34 = 43 \quad \text{Change subtraction to adding the opposite.}$$
$$m + (-20) + 34 = 43 \quad \text{Simplify the left side.}$$
$$m + 14 = 43$$
$$\underline{ -14 \qquad -14} \quad \text{To get } m \text{ by itself, add } -14 \text{ to both sides.}$$
$$m + 0 = 29$$
$$m = 29$$

Step 5 **State the answer** to the question, "How much was in the envelope at the start?" There was **$29 in the envelope.**

Step 6 **Check** the solution by going back to the *original* problem and inserting the solution.

Started with $29 in the envelope

Took out $20, so $29 − $20 = $9 in the envelope

Put in $34, so $9 + $34 = $43

Now has $43 ← Matches

Because $29 fits all the facts in the original problem, you know it is the correct solution.

> If your answer does **not** work in the original problem, start again at *Step 1.*

——— **Work Problem** 4 **at the Side.** ▶

---

4 **GS** Some people got on an empty bus at its first stop. At the second stop, 3 people got on. At the third stop, 5 more people got on. At the fourth stop, 10 people got off, but 4 people were still on the bus. How many people got on at the first stop? Show your work for each of the six problem-solving steps.

Step 1 **Read.**

Unknown: number of people who got on at first stop

Known: 3 got on; 5 got on; 10 got off; 4 people still on bus

Step 2(a) **Assign a variable.**

Let p be people who got on at first stop. (You may use any letter you like as the variable.)

Step 3 **Write an equation using your variable.**

$$p + \rule{0.6cm}{0.15mm} + \rule{0.6cm}{0.15mm} - \rule{0.6cm}{0.15mm} = \rule{0.6cm}{0.15mm}$$

Answer

4. *Step 3* **Write an equation.**
$$p + 3 + 5 - 10 = 4$$
Step 4 **Solve the equation.**
$$p + 3 + 5 + (-10) = 4$$
$$p + (-2) = 4$$
$$\underline{ +2 \qquad +2}$$
$$p + 0 = 6$$
$$p = 6$$

Step 5 **State the answer.**
6 people got on at the first stop.

Step 6 **Check.**
6 got on at first stop
3 got on at 2nd stop: 6 + 3 = 9
5 got on at 3rd stop: 9 + 5 = 14
10 got off at 4th stop: 14 − 10 = 4
4 people are left. ← Matches

⑤ **Five donors each gave the same amount of money to a college to use for scholarships. From the money, scholarships of $1250, $900, and $850 were given to students; $250 was left. How much money did each donor give to the college? Show your work for each of the six problem-solving steps.**

Step 1 **Read.**

Unknown: money given by each donor

Known: 5 donors; gave out $1250, $900, $850; $250 left

Step 2(a) **Assign a variable.**

Let m be each donor's money.

Step 3 **Write an equation. using your variable.**

$5 \cdot m - \underline{\hspace{0.6cm}} - \underline{\hspace{0.6cm}} - \underline{\hspace{0.6cm}}$

$= \underline{\hspace{0.6cm}}$

Answer

5. *Step 3* **Write an equation.**
  $5 \cdot m - \$1250 - \$900 - \$850 = \$250$

Step 4 **Solve the equation.**
$5m + (-1250) + (-900) + (-850) = 250$

$5m + (-3000) = 250$
$ 3000 3000$
$\overline{5m + 0 = 3250}$

$\dfrac{5m}{5} = \dfrac{3250}{5}$

$m = 650$

Step 5 **State the answer.**
Each donor gave $650.

Step 6 **Check.**
5 donors each gave $650, so
$5 \cdot \$650 = \$3250$
Gave out $1250, $900, $850, so
$\$3250 - 1250 - 900 - 850 = \$250$
Had $250 left ⟵ Matches ⟶

EXAMPLE 5 **Solving an Application Problem with One Unknown Quantity**

Three friends each put in the same amount of money to buy a gift. After they spent $2 for a card and $31 for the gift, they had $6 left. How much money did each friend put in originally?

Step 1 **Read** the problem. It is about 3 friends buying a gift.

Unknown: amount of money each friend contributed
Known: 3 friends put in money; spent $2 and $31; had $6 left

Step 2(a) There is only one unknown quantity. **Assign a variable,** m, to represent the amount of money each friend contributed.

Step 3 **Write an equation using your variable.**

Number of friends	Amount each friend put in	Spent on card	Spent on gift	Left over
3 •	m	− $2	− $31	= $6

To see why this is multiplication, think of an example. If each friend put in $10, how much money would there be? 3 • $10, or $30

Step 4 **Solve the equation.**

$3m - 2 - 31 = 6$ Change subtractions to adding the opposite.

$3m + (-2) + (-31) = 6$ Simplify the left side.

$3m + (-33) = 6$
$ 33 33$ To get $3m$ by itself, add 33 to both sides.
$\overline{3m + 0 = 39}$

$\dfrac{3m}{3} = \dfrac{39}{3}$ To get m by itself, divide both sides by 3.

$m = 13$

Write a $ as part of your answer.

Step 5 **State the answer.** Each friend put in **$13**

Step 6 **Check** the solution by putting it back into the *original* problem.

3 friends each put in $13, so 3 • $13 = $39

Spent $2, spent $31, so $39 − $2 − $31 = $6

Had $6 left ⟵ Matches ⟶

$13 is the correct solution because it fits the facts in the original problem.

◀ **Work Problem ⑤ at the Side.**

EXAMPLE 6   **Solving a More Complex Application Problem with One Unknown Quantity**

Michael has completed 5 less than three times as many lab experiments as David. If Michael has completed 13 experiments, how many experiments has David completed?

Step 1   **Read** the problem. It is about the number of experiments done by two students.

Unknown: number of experiments David did
Known: Michael did 5 less than 3 times the number David did; Michael did 13.

Step 2(a)   **Assign a variable.** Let n represent the number of experiments David did.

Step 3   **Write an equation using your variable.**

The number Michael did	is	5 less than 3 times David's number.
13	=	$3n - 5$

> Be careful with subtraction! $5 - 3n$ is **not** the same as $3n - 5$.

Step 4   **Solve the equation.**

$13 = 3n - 5$ Change subtraction to adding the opposite.

$13 = 3n + (-5)$ To get $3n$ by itself, add 5 to both sides.

$\dfrac{5}{18 = 3n + 0}\quad\dfrac{5}{}$

$\dfrac{18}{3} = \dfrac{3n}{3}$ To get n by itself, divide both sides by 3

$6 = n$

Step 5   **State the answer.** David did **6 experiments.**

> Label your answer. Write *6 experiments*, **not** just 6.

Step 6   **Check** the solution by putting it back into the *original* problem.

3 times David's number $3 \cdot 6 = 18$

Less 5 $18 - 5 = 13$

Michael did 13 ←——— Matches ———↑

6 experiments is the correct solution because it fits the facts in the original problem.

——————— Work Problem ⑥ at the Side. ▶

⑥ Susan donated $10 more than twice what LuAnn donated. If Susan donated $22, how much did LuAnn donate?

Show your work for each of the six problem-solving steps.

Step 1   Read.

Unknown: LuAnn's donation

Known: Susan donated $10 more than twice what LuAnn donated; Susan donated $22.

Step 2(a)   Assign a variable.

Let d be _____ donation.

3.3 Exercises

FOR EXTRA HELP Go to MyMathLab *for worked-out, step-by-step solutions to exercises enclosed in a square* ▢ *and video solutions to* ▶ *exercises.*

CONCEPT CHECK *Write* true *or* false *for each statement. If a statement is* false, *write the correct statement.*

1. (a) $14 - x$
means "14 subtracted from a number."

(b) $x - 7$
means "a number decreased by 7."

(c) $\dfrac{-20}{x}$
means "the product of -20 and a number."

2. (a) $3x - 6$
means "6 subtracted from 3 times a number."

(b) $18x$
means "the sum of 18 and a number."

(c) $\dfrac{x}{42}$
means "the quotient of 42 and a number."

Write an algebraic expression, using x as the variable. ***See Examples 1 and 2.***

3. 14 plus a number

4. The sum of a number and -8

5. -5 added to a number
▶

6. 16 more than a number

7. 20 minus a number
▶

8. A number decreased by 25

9. 9 less than a number

10. A number subtracted from -7

11. -6 times a number

12. The product of -3 and a number

13. Double a number

14. A number times 10

15. A number divided by 2

16. 4 divided by a number

17. Twice a number added to 8

18. Five times a number plus 5

19. 10 fewer than seven times a number
▶

20. 12 less than six times a number

21. The sum of twice a number and the number

22. Triple a number subtracted from the number

Translate each sentence into an equation and solve it. Check your solution by going back to the words in the original problem. ***See Example 3.***

23. If four times a number is decreased by 2, the result is 26. Find the number.

4 times a number	decreased by	2	result is	26
$4n$	$-$	2	$=$	26

24. The sum of 8 and five times a number is 53. Find the number.

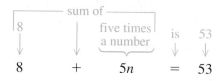

$$8 + 5n = 53$$

25. If a number is added to twice the number, the result is -15. What is the number?

26. If a number is subtracted from three times the number, the result is -8. What is the number?

27. If the product of some number and 5 is increased by 12, the result is seven times the number. Find the number.

28. If eight times a number is subtracted from eleven times the number, the result is -9. Find the number.

29. When three times a number is subtracted from 30, the result is 2 plus the number. What is the number?

30. When twice a number is decreased by 8, the result is the number increased by 7. Find the number.

Solve each application problem. Use the six problem-solving steps you learned in this section. ***See Examples 4–6.***

31. Ricardo gained 15 pounds over the winter. He went on a diet and lost 28 pounds. Then he regained 5 pounds and weighed 177 pounds. How much did he weigh originally?

Let *w* be Ricardo's original weight.

32. Mr. Chee deposited $200 into his bank account. Then, after paying $43 for gas and $190 for his child's day care, the balance in his account was $67. How much was in his account before he made the deposit?

Let *m* be the original amount in the account.

33. There were 18 cookies in Magan's cookie jar. While she was busy in another room, her children ate some of the cookies. Magan bought three dozen cookies and added them to the jar. At that point she had 49 cookies in the jar. How many cookies did her children eat?

34. The Greens had a 20-pound bag of bird seed in their garage. Mice got into the bag and ate some of it. The Greens then bought an 8-pound bag of seed and put all the seed in a metal container. They now have 24 pounds of seed. How much did the mice eat?

35. A college bookstore ordered six boxes of red pens. The store sold 32 red pens last week and 35 red pens this week. Five pens were left on the shelf. How many pens were in each box?

36. A local charity received a donation of eight cartons filled with cans of soup. The charity gave out 100 cans of soup yesterday and 92 cans today before running out. How many cans were in each carton?

37. The 14 music club members each paid the same amount for dues. The club also earned $340 selling garden plants. They spent $575 to organize a jazz festival. Now their bank account, which started at $0, is overdrawn by $25. How much did each member pay in dues?

38. The manager of an apartment complex had 11 packages of light bulbs on hand. He replaced 29 burned out bulbs in hallway lights and 7 bulbs in the party room. Eight bulbs were left. How many bulbs were in each package?

39. When 75 is subtracted from four times Tamu's age, the result is Tamu's age. How old is Tamu?

Let *a* be Tamu's age.

40. If three times Linda's age is decreased by 36, the result is twice Linda's age. How old is Linda?

Let *a* be Linda's age.

41. While shopping for clothes, Consuelo spent $3 less than twice what Brenda spent. Consuelo spent $81. How much did Brenda spend?

42. Dennis weighs 184 pounds. His weight is 2 pounds less than six times his child's weight. How much does his child weigh?

43. Paige bought five identical bags of candy for Halloween. Forty-eight children visited her home and she gave each child three pieces of candy. At the end of the night she still had one bag of candy. How many pieces of candy were in each bag?

44. A restaurant ordered four packages of paper napkins. Yesterday they used up one package, and today they used up 140 napkins. Two packages plus 60 napkins remain. How many napkins are in each package?

45. The recommended daily intake of vitamin C for an adult female is 15 mg more than four times the recommended amount for a young child. The amount for an adult female is 75 mg. How much should a young child receive? (Data from National Institutes of Health.)

46. A cheetah's sprinting speed is 61 miles per hour less than three times a zebra's running speed. A cheetah can sprint 68 miles per hour. Find the zebra's running speed. (Data from *Grolier Multimedia Encyclopedia*.)

3.4 Solving Application Problems with Two Unknown Quantities

OBJECTIVE

1 Solve application problems with two unknown quantities.

OBJECTIVE ▶ 1 Solve application problems with two unknown quantities. In the preceding section, the problems had only one unknown quantity. As a result, we used *Step 2(a)* rather than *Step 2(b)* in the problem-solving steps. For easy reference, we repeat the steps here.

Solving an Application Problem

Step 1	**Read** the problem once to see what it is about. Read it carefully a second time. As you read, make a sketch or write word phrases that identify the known and the unknown parts of the problem.
Step 2(a)	If there is one unknown quantity, **assign a variable** to represent it. Write down what your variable represents.
Step 2(b)	If there is more than one unknown quantity, **assign a variable** to represent "the thing you know the least about." Then write variable expressions, using the same variable, to show the relationship of the other unknown quantities to the first one.
Step 3	**Write an equation using your variable.** Use your sketch or word phrases as the guide.
Step 4	**Solve the equation.**
Step 5	**State the answer** to the question in the problem and label your answer.
Step 6	**Check** whether your answer fits all the facts given in the *original* statement of the problem. If it does, you are done. If it doesn't, start again at *Step 1*.

Now you are ready to solve problems with two unknown quantities.

EXAMPLE 1 Solving an Application Problem with Two Unknown Quantities

Last month, Sheila worked 72 hours more than Russell. Together they worked a total of 232 hours. Find the number of hours each person worked last month.

Step 1 **Read** the problem. It is about the number of hours worked by Sheila and by Russell.

Unknowns: hours worked by Sheila;
　　　　　hours worked by Russell
Known: Sheila worked 72 hours more than Russell;
　　　　232 hours total for Sheila and Russell

Step 2(b) There are *two* unknowns so **assign a variable** to represent "the thing you know the least about." You know the *least* about the hours worked by Russell, so let h represent Russell's hours.

Sheila worked 72 hours more than Russell, so her hours are $h + 72$, that is, Russell's hours (h) plus 72 more.

———— **Continued on Next Page**

Step 3   **Write an equation using your variable.**

Hours worked by Russell		Hours worked by Sheila		Total hours worked
h	$+$	$h + 72$	$=$	232

Step 4   **Solve the equation.**

$$h + h + 72 = 232 \qquad \text{Simplify the left side by combining like terms.}$$

$$2h + 72 = 232$$

$$\underline{ -72 \quad -72} \qquad \text{To get } 2h \text{ by itself, add } -72 \text{ to both sides.}$$

$$2h + 0 = 160$$

$$\frac{2h}{2} = \frac{160}{2} \qquad \text{To get } h \text{ by itself, divide both sides by 2}$$

$$h = 80$$

Step 5   **State the answer.**

Because h represents Russell's hours, and the solution of the equation is $h = 80$, Russell worked 80 hours.

$h + 72$ represents Sheila's hours. Replace h with 80.

$80 + 72 = 152$, so Sheila worked 152 hours.

The final answer is:
Russell worked 80 hours and **Sheila worked 152 hours.**   *Write hours as part of both answers.*

Step 6   **Check** the solution by putting both numbers back into the *original* problem.

"Sheila worked 72 hours more than Russell."

Sheila's 152 hours are 72 more than Russell's 80 hours, so the solution checks.

$$\begin{array}{r} 152 \\ -72 \\ \hline 80 \end{array}$$

"Together they worked a total of 232 hours."

Sheila's 152 hours + Russell's 80 hours = 232 hours, so the solution checks.

$$\begin{array}{r} 152 \\ +80 \\ \hline 232 \end{array}$$

You've answered the question correctly because 80 hours and 152 hours fit all the facts given in the problem.

—————— **Work Problem 1 at the Side.** ▶

EXAMPLE 2	**Solving an Application Problem with Two Unknown Quantities**

Riley cut a cord that was 83 ft long into two pieces. The second piece was 19 ft shorter than the first piece. How long was each piece?

Step 1   **Read** the problem. It is about a cord that is cut into two pieces.

    Unknowns: length of first piece; length of second piece
    Known: Second piece is 19 ft shorter than first piece; total length of both pieces is 83 ft

—————— **Continued on Next Page**

---

1 In a day of work, Keonda made $12 more than her daughter. Together they made $182. Find the amount that each person made. Use the six problem-solving steps.

Step 1   **Read.**

Unknowns: Keonda's amount; daughter's amount

Known: Keonda made $12 more than her daughter; $182 total for both.

Step 2(b)   **Assign a variable** to represent "the thing you know the *least* about." You know the *least* about the daughter's amount, so let m be the daughter's amount.

Keonda's amount is ___ + ___.

Step 3   **Write an equation using your variable.**

Answer

1. Keonda's amount is $m + 12$.
   Step 3 $m + m + 12 = 182$
   Step 4 $2m + 12 = 182$

$$\underline{ -12 \quad -12}$$
$$2m + 0 = 170$$
$$\frac{2m}{2} = \frac{170}{2}$$
$$m = 85$$

   Step 5   Daughter made $85.
           Keonda made $85 + 12 = $97.
   Step 6   **Check.** $97 - $85 = $12 and
                 $97 + $85 = $182

2 Charles had 175 yd of fishing line to put on two fishing reels. He put 25 yd less of line on the second reel than on the first reel. How much line did he put on each reel? Use the six problem-solving steps.

Hint: Which length of fishing line do you know the *least* about; the length on the first reel, or the length on the second reel? Let *r* represent that length.

Step 2(b) There are *two* unknowns so **assign a variable** to represent "the thing you know the least about." You know the *least* about the first piece, so let *p* represent the length of the first piece.

The second piece is 19 ft shorter than the first piece, so its length is $p - 19$.

Step 3 **Write an equation using your variable.**

Length of first piece		Length of second piece		Total length
p	$+$	$p - 19$	$=$	83

Step 4 **Solve the equation.**

$$p + p - 19 = 83 \quad \text{Simplify the left side by combining like terms.}$$

$$2p - 19 = 83 \quad \text{Change subtraction to adding the opposite.}$$

$$2p + (-19) = 83$$

$$\underline{\qquad\qquad 19 \qquad 19} \quad \text{To get } 2p \text{ by itself, add 19 to both sides.}$$

$$2p + 0 = 102$$

$$\frac{2p}{2} = \frac{102}{2} \quad \text{To get } p \text{ by itself, divide both sides by 2}$$

$$p = 51$$

Step 5 **State the answer.**

p represents the length of the first piece, and $p = 51$, so the first piece is 51 ft long.

$p - 19$ represents the second piece. Replace *p* with 51.

$51 - 19 = 32$, so the second piece is 32 ft long.

The final answer is: **First piece is 51 ft long** and the **second piece is 32 ft long**. ⟨Write **ft** as part of both answers.⟩

Step 6 **Check** the solution by putting both numbers back into the *original* problem. ⟨If your answers do **not** fit all the given facts, start over at *Step 1*.⟩

"Riley cut a cord that was 83 ft long."
51 ft + 32 ft = 83 ft, so the solution checks.

"The second piece was 19 ft shorter than the first piece."
32 ft is 19 ft shorter than 51 ft, so the solution checks.

The answers are correct because 51 ft and 32 ft fit all the facts in the problem.

◀ **Work Problem** **2** **at the Side.**

Answer

2. *Step 1* Unknowns: length on first reel, length on second reel
Known: total of 175 yd for both reels; second reel has 25 yd less.

Step 2 Length on first reel is *r*.
Length on second reel is $r - 25$.

Step 3 $r + r - 25 = 175$

Step 4 $2r + (-25) = 175$
$$\underline{\qquad 25 \qquad 25}$$
$$\frac{2r}{2} = \frac{200}{2}$$
$$r = 100$$

Step 5 First reel had 100 yd of line.
Second reel had $100 - 25 = 75$ yd of line.

Step 6 **Check.** 100 yd + 75 yd = 175 yd
and 100 yd − 75 yd = 25 yd

EXAMPLE 3 **Solving a Geometry Application with Two Unknown Quantities**

The length of a rectangle is 2 cm more than the width. The perimeter is 68 cm. Find the length and width.

Step 1 **Read** the problem. It is about a rectangle. Make a sketch of a rectangle.

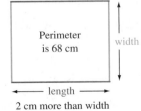

Perimeter is 68 cm

width

length

2 cm more than width

Continued on Next Page

Unknowns: length of the rectangle; width of the rectangle

Known: The length is 2 cm more than the width; the perimeter is 68 cm.

Step 2(b) There are *two* unknowns so **assign a variable** to represent "the thing you know the least about." You know the *least* about the width, so let *w* represent the **width**.

The length is 2 cm more than the width, so the **length** is *w* + 2.

A drawing will help you see these relationships.

width is *w*

length is *w* + 2

Step 3 **Write an equation using your variable.**

Use the formula for perimeter of a rectangle, $P = 2l + 2w$, to help you write the equation.

$$P = 2 \quad l \quad + 2 \quad w$$
$$\downarrow \quad \downarrow \quad \downarrow \quad \downarrow \quad \downarrow$$
$$68 = 2(w + 2) + 2 \cdot w$$

Replace *P* with 68

Replace *l* with (*w* + 2)

Step 4 **Solve the equation.**

$$68 = 2(w + 2) + 2w \quad \text{Use the distributive property.}$$
$$68 = 2w + 4 + 2w \quad \text{Combine like terms.}$$

$$68 = 4w + 4$$
$$\underline{-4 \qquad -4} \qquad \text{To get } 4w \text{ by itself, add } -4 \text{ to}$$
$$64 = 4w + 0 \qquad \text{both sides.}$$

$$\frac{64}{4} = \frac{4w}{4} \qquad \text{To get } w \text{ by itself, divide both sides by 4}$$

$$16 = w$$

Step 5 **State the answer.**

w represents the width, and *w* = 16, so the width is 16 cm.

w + 2 represents the length. Replace *w* with 16.
↓

16 + 2 = 18, so the length is 18 cm.

> Write **cm** as the label for both width and length.

The final answer is: The **width is 16 cm** and the **length is 18 cm**

Step 6 **Check** the solution by putting the measurements on your sketch and going back to the original problem.

"The length of a rectangle is 2 cm more than the width."

18 cm is 2 cm more than 16 cm, so the solution checks.

"The perimeter is 68 cm."

$$P = 2 \cdot 18 \text{ cm} + 2 \cdot 16 \text{ cm}$$
$$P = 36 \text{ cm} + 32 \text{ cm}$$
$$P = 68 \text{ cm}$$

> Matches the perimeter given in the original problem, so the solution checks.

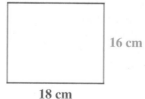
16 cm

18 cm

Work Problem ③ at the Side. ▶

---

③ Make a sketch to help solve this **GS** problem.

The length of Ann's rectangular garden plot is 3 yd more than the width. She used 22 yd of fencing to go around the entire garden. Find the length and the width of the garden, using the six problem-solving steps.

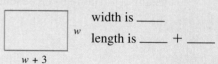

width is _____

length is _____ + _____

w + 3

Answer

3.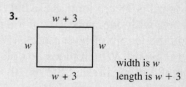

w + 3

w   *w*

w + 3

width is *w*

length is *w* + 3

$$22 = 2(w + 3) + 2 \cdot w$$
width is 4 yd
length is 7 yd

Check. 7 yd is 3 yd more than 4 yd.
$$P = 2 \cdot 7 \text{ yd} + 2 \cdot 4 \text{ yd} = 22 \text{ yd}$$
Matches perimeter given in the original problem

3.4 Exercises

Go to MyMathLab *for worked-out, step-by-step solutions to exercises enclosed in a square* ☐ *and video solutions to* ▶ *exercises.*

CONCEPT CHECK *Fill in each blank to complete the statement about one of the six steps in solving an application problem.*

1. (a) As you read the problem, identify the _____ and the _____.

    (b) When there is more than one unknown quantity, assign a variable to represent the thing you know the _____ about.

2. (a) When you state the answer, be sure to _____ your answer.

    (b) To check the answer, make sure it fits _____.

Solve each application problem using the six problem-solving steps you learned in this section. ***See Examples 1 and 2.***

3. My sister is 9 years older than I am. The sum of our ages is 51. Find our ages.

    Hint: For *Step 2,* you know the *least* about my age. So, *a* is my age; $a + 9$ is my sister's age.

4. Ed and Marge were candidates for city council. Marge won with 93 more votes than Ed. The total number of votes cast in the election was 587. Find the number of votes received by each candidate.

    Hint: For *Step 2,* you know the *least* about Ed's votes. So, *v* is Ed's votes; $v + 93$ is Marge's votes.

5. Last year, Lien earned $1500 more than her husband. Together they earned $37,500. How much did each of them earn?

6. A $149,000 estate is divided between two charities so that one charity receives $18,000 less than the other. How much does each charity receive?

7. Jason paid eight times as much for his laptop computer as he did for his printer. He paid a total of $396 for both items. What did each item cost?

8. The attendance at the Saturday night baseball game was three times the attendance at Sunday's game. In all, 56,000 fans attended the games. How many fans were at each game?

9. A board is 78 cm long. Rosa cut the board into two pieces, with one piece 10 cm longer than the other. Find the length of both pieces. (*Hint:* Make a sketch of the board. Which piece do you know the least about? Let *x* represent the length of that piece.)

    Longer piece | Shorter piece

10. A rope is 21 yd long. Marcos cut it into two pieces so that one piece is 3 yd longer than the other. Find the length of each piece.

    Longer piece | Shorter piece

11. A wire is cut into two pieces, with one piece 7 ft shorter than the other. The wire was 31 ft long before it was cut. How long is each piece?

12. A 90 cm pipe is cut into two pieces so that one piece is 6 cm shorter than the other. Find the length of each piece.

13. In the U.S. Congress, the number of Representatives is 65 less than five times the number of Senators. There are a total of 535 members of Congress. Find the number of Senators and the number of Representatives. (Data from *The World Almanac and Book of Facts*.)

14. Florida's record low temperature is 68 degrees higher than Montana's record low. The sum of the two record lows is −72 degrees. What is the record low for each state? (Data from National Climatic Data Center.)

15. A fence is 706 m long. It is cut into three parts. Two parts are the same length, and the third part is 25 m longer than either of the other two. Find the length of each part.

16. A wooden railing is 82 m long. It is divided into four pieces. Three pieces are the same length, and the fourth piece is 2 m longer than each of the other three. Find the length of each piece.

In Exercises 17–22, use the formula for the perimeter of a rectangle, $P = 2l + 2w$. Make a sketch for each problem. **See Example 3.** *(Sketches may vary; show your sketches to your instructor.)*

17. The perimeter of a rectangle is 48 yd. The width is 5 yd. Find the length.

18. The length of a rectangle is 27 cm, and the perimeter is 74 cm. Find the width of the rectangle.

19. A rectangular dog pen is twice as long as it is wide. The perimeter of the pen is 36 ft. Find the length and the width of the pen.

20. A new city park is a rectangular shape. The length is triple the width. It will take 240 meters of fencing to go around the park. Find the length and width of the park.

21. The length of a rectangular jewelry box is 3 in. more than twice the width. The perimeter is 36 in. Find the length and the width.

22. The perimeter of a rectangular house is 122 ft. The width is 5 ft less than the length. Find the length and the width.

23. A photograph measures 8 in. by 10 in. Earl put it in a frame that added 2 in. to every side. Find the outside perimeter and total area of the photograph and frame.

24. Barb had a 16 in. by 20 in. photograph. She cropped 3 in. off every side of the photo. What are the perimeter and the area of the cropped photo?

Study Skills

TIPS FOR IMPROVING TEST SCORES

OBJECTIVES

1. Restate the importance of sleep and good nutrition to learning.
2. Explain the effect of anxiety and stress on learning.

Many things besides studying can improve your test scores. You may not realize that eating the right foods, as well as getting enough exercise and sleep, can also improve your scores. Your brain (and therefore your ability to think) is affected by the condition of your whole body. So part of your preparation for tests includes keeping yourself in good physical shape as well as spending time on the actual course material. Try these suggestions and see the difference.

Performance Health Tips

Performance Health Tips to Improve Your Test Score	Explanation
Get **seven to eight hours of sleep** the night before the exam. (It's helpful to get that much sleep *every* night.)	**Fatigue and exhaustion** reduce efficiency. They also cause poor memory and recall. If you didn't sleep much the night before a test, 20 minutes of relaxation or meditation can help. (Also see the comments below about eating carbohydrates to help you sleep.)
Eat a **small, high-energy meal** about two hours before the test. Start the meal with a small amount of protein such as fish, chicken, or nonfat yogurt. Include carbohydrates if you like, but no high-fat foods.	Just 3 to 4 ounces of protein increases the amount of a chemical in the brain called tyrosine, which **improves your alertness, accuracy, and motivation.** High-fat foods dull your mind and slow down your brain.
Drink plenty of water. Don't wait until you feel thirsty; your body is already dehydrated by the time you feel thirst.	Research suggests that staying well hydrated improves the electrochemical communications in your brain.
Give your brain the time it needs to grow dendrites!	**Cramming doesn't work;** your brain cannot grow dendrites that quickly. **Studying every day** using these study skills techniques is the way to give your brain the time it needs.

Anxiety Prevention Tips

To Prevent Anxiety	Explanation
Practice slow, deep breathing for five minutes each day. Then do a minute or two of deep breathing right before the test. Also, if you feel your anxiety building during the test, stop for a minute, close your eyes, and do some deep breathing.	When **test anxiety** hits, you breathe more quickly and shallowly, which causes hyperventilation. Symptoms may include confusion, inability to concentrate, shaking, dizziness, and more. Slow, deep breathing will **calm you and prevent panic.**

To Prevent Anxiety	Explanation
Do 15 to 20 minutes of **moderate exercise** (like walking) shortly before the test. Daily exercise is even better!	**Exercise reduces stress** and will help prevent "blanking out" on a test. Exercise also increases your alertness, clear thinking, and energy.
To help you sleep the night before the test, or any time you need to calm down, **eat high-carbohydrate foods** such as popcorn, bread, rice, crackers, muffins, bagels, pasta, corn, baked potatoes (not fries or chips), and cereals.	Carbohydrates increase the level of a chemical in the brain called serotonin, which has a *calming effect on the mind*. It reduces feelings of tension and stress and improves your ability to concentrate. You only need to eat a small amount, like half a bagel, to get this effect.
Before the test, **go easy on caffeinated beverages** such as coffee, tea, and soft drinks. Do not eat candy bars or other sugary snacks.	Extra caffeine can **make you jittery,** "hyper," and shaky for the test. It can increase the tendency to panic. Too much sugar causes negative emotional reactions in some people.

Now Try This

What will you do to improve your next test score? List the three or four tips you think will help you the most.

1 _____

2 _____

3 _____

4 _____

What changes will you have to make in order to try the tips you chose?

_____

_____

_____

_____

See *Tips for Taking Math Tests* **and** *Preparing for Your Final Exam* **for more ideas about managing anxiety.** (Check the Contents to find their locations.)

Chapter 3 *Summary*

Key Terms

3.1

formula Formulas are well-known rules for solving common types of problems. They are written in a shorthand form that uses variables.

perimeter Perimeter is the distance around the outside edges of a flat shape. It is measured in linear units such as in., ft, yd, mm, cm, m, and so on.

square A square is a figure with four sides that are all the same length and meet to form right angles, which measure 90°. See example at the right.

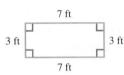

rectangle A rectangle is a four-sided figure in which all sides meet to form right angles, which measure 90°. The opposite sides are the same length. See example at the right.

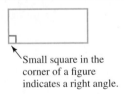

parallelogram A parallelogram is a four-sided figure in which opposite sides are both parallel and equal in length. See example at the right.

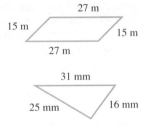

triangle A triangle is a figure with exactly three sides. See example at the right.

3.2

area Area is the surface inside a two-dimensional (flat) shape. It is measured by determining the number of squares of a certain size needed to cover the surface inside the shape. Some of the commonly used units for measuring area are square inches (in.^{2}), square feet (ft^2), square yards (yd^2), square centimeters (cm^2), and square meters (m^2).

New Symbols

Right angle:
(90° angle)

Small square in the corner of a figure indicates a right angle.

Square units for measuring area: in.^{2} ft^2 yd^2 mi^2 mm^2 cm^2 m^2 km^2

New Formulas

Perimeter of a square: $P = 4s$

Perimeter of a rectangle: $P = 2l + 2w$

Area of a square: $A = s^2$

Area of a rectangle: $A = lw$

Area of a parallelogram: $A = bh$

Test Your Word Power

See how well you have learned the vocabulary in this chapter.

1 The **perimeter** of a flat shape is
A. measured in square units
B. found by adding the lengths of the sides
C. found by multiplying length times width.

2 The **area** of a flat shape is
A. found by adding length plus width
B. measured in linear units
C. measured in square units.

3 When working with a **square shape,**
A. the perimeter formula is $P = s^2$
B. the area formula is $A = \frac{1}{2}bh$
C. all sides have the same length.

4 When working with a **rectangular shape,**
A. the area formula is $A = lw$
B. opposite sides have different lengths
C. the perimeter formula is $P = bh$.

5 In **triangles**
A. all the sides meet at 90° angles
B. there are exactly three sides
C. the area formula is $A = s^2$.

6 In *all* **parallelograms**
A. there are exactly six sides
B. all sides have the same length
C. opposite sides are parallel and equal in length.

Answers to Test Your Word Power

1. B; *Example:* If a triangle has sides measuring 4 ft, 10 ft, and 8 ft, then
$P = 4$ ft $+ 10$ ft $+ 8$ ft $= 22$ ft.

2. C; *Example:* Area is measured in square units such as in.², ft², yd², cm², and km².

3. C; *Example:* If one side of a square is 5 in. long, all the other sides are also 5 in. long.

4. A; *Example:* If the length of a rectangle is 12 cm and the width is 8 cm, then $A = 12$ cm $\cdot$ 8 cm $= 96$ cm².

5. B; *Examples:*

6. C; *Example:* In parallelogram *ABCD*, sides *AB* and *DC* are parallel and equal in length. Also, sides *AD* and *BC* are parallel and equal in length.

Quick Review

Concepts	Examples

3.1 Finding Perimeter

To find the perimeter of *any* flat shape with straight sides, add the lengths of the sides. Perimeter is measured in linear units (cm, m, ft, yd, and so on).

Find the perimeter of each figure.

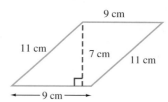

$P = 9$ cm $+ 11$ cm $+ 9$ cm $+ 11$ cm

$P = 40$ cm

Or, for squares and rectangles, you can use the formulas shown below.

Perimeter of a square: $P = 4s$,
where *s* is the length of one side.

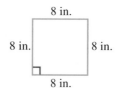

$P = 4s$

$P = 4 \cdot 8$ in.

$P = 32$ in.

Perimeter of a rectangle: $P = 2l + 2w$,
where *l* is the length and *w* is the width.

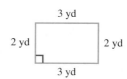

$P = 2 \cdot l \quad + 2 \cdot w$

$P = 2 \cdot 3$ yd $+ 2 \cdot 2$ yd

$P = \quad 6$ yd $\quad + \quad 4$ yd

$P = 10$ yd

Concepts	**Examples**

3.1 Finding the Length of One Side of a Square

If you know the perimeter of a square, use the formula $P = 4s$. Replace P with the value for the perimeter and solve the equation for s.

If the perimeter of a square room is 44 ft, find the length of one side.

$$P = 4s \qquad \text{Replace } P \text{ with 44 ft}$$

$$44 \text{ ft} = 4s$$

$$\frac{44 \text{ ft}}{4} = \frac{4s}{4} \qquad \text{Divide both sides by 4}$$

$$\mathbf{11 \text{ ft}} = s$$

The length of one side is **11 ft**

CHECK Check the solution by drawing a sketch of the room. The perimeter is
11 ft + 11 ft + 11 ft + 11 ft = 44 ft.
The result matches the perimeter given in the problem.

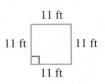

3.1 Finding the Length or Width of a Rectangle

If you know the perimeter of a rectangle and either its width or length, use the formula $P = 2l + 2w$. Replace P and either l or w with the values that you know. Then solve the equation.

The width of a rectangular rug is 8 ft. The perimeter is 36 ft. Find the length.

$$P = 2l + 2w \qquad \text{Replace } P \text{ with 36 ft and } w \text{ with 8 ft}$$

$$36 \text{ ft} = 2l + 2 \cdot 8 \text{ ft}$$

$$36 \text{ ft} = 2l + 16 \text{ ft}$$

$$\underline{-16 \text{ ft} \qquad\qquad -16 \text{ ft}} \qquad \text{Add } -16 \text{ ft to both sides}$$

$$20 \text{ ft} = 2l + 0$$

$$\frac{20 \text{ ft}}{2} = \frac{2l}{2} \qquad \text{Divide both sides by 2}$$

$$\mathbf{10 \text{ ft}} = l \qquad \text{The length is 10 ft}$$

CHECK To check the solution, draw a sketch of the rectangle and label the lengths of the sides. Then add the four measurements.

10 ft + 10 ft + 8 ft + 8 ft = 36 ft

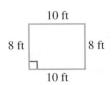

The result matches the perimeter given in the problem.

Concepts	**Examples**

3.2 Finding Area

Use the appropriate formula. Remember to measure area in *square* units (cm², m², ft², yd², and so on).

Area of a rectangle: $A = lw$,
  where l is the length and w is the width.

Area of a square: $A = s^2$, which means $s \cdot s$,
  where s is the length of one side.

Area of a parallelogram: $A = bh$,
  where b is the base and h is the height.

Find the area of each figure.

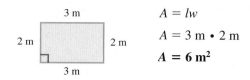

$A = lw$
$A = 3 \text{ m} \cdot 2 \text{ m}$
$A = 6 \text{ m}^2$

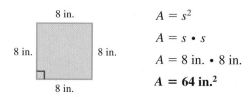

$A = s^2$
$A = s \cdot s$
$A = 8 \text{ in.} \cdot 8 \text{ in.}$
$A = 64 \text{ in.}^2$

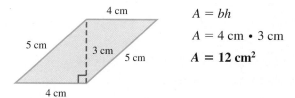

$A = bh$
$A = 4 \text{ cm} \cdot 3 \text{ cm}$
$A = 12 \text{ cm}^2$

3.2 Finding the Unknown Length in a Rectangle or Parallelogram

If you know the area of a rectangle or parallelogram and one of the other measurements, use the appropriate area formula (see above). Replace A and one of the other variables with the values that you know. Then solve the equation.

The area of a parallelogram is 72 yd², and its height is 9 yd. Find the base.

$$A = bh$$

Replace A with 72 yd² and h with 9 yd

$$72 \text{ yd}^2 = b \cdot 9 \text{ yd}$$

$$\frac{72 \text{ yd} \cdot \text{yd}}{9 \text{ yd}} = \frac{b \cdot 9 \text{ yd}}{9 \text{ yd}}$$

Divide both sides by 9 yd

$$8 \text{ yd} = b$$

The base of the parallelogram is **8 yd**

3.2 Finding the Length of One Side of a Square

If you know the area of a square, use the formula $A = s^2$. Replace A with the value that you know. Then solve the equation by asking, "What number, times itself, gives the value of A?"

A square ceiling has an area of 100 ft². What is the length of each side of the ceiling?

$$A = s^2$$

Replace A with 100 ft²

$$100 \text{ ft}^2 = s^2$$

Rewrite s^2 as $s \cdot s$

$$100 \text{ ft}^2 = s \cdot s$$

Ask, "What number times itself gives 100?"

$$100 \text{ ft}^2 = 10 \text{ ft} \cdot 10 \text{ ft}$$

The value of s is 10 ft, so each side of the ceiling is **10 ft** long.

| **Concepts** | **Examples** |

3.3 Translating Sentences into Equations

Translate word phrases into symbols using x (or any other letter) as the variable. Then solve the equation. Check the solution by putting it back in the original problem.

If 10 is subtracted from three times a number, the result is 14. Find the number.

Let x represent the unknown number.

$$3x - \quad 10 \ = 14 \qquad \text{Change subtraction to adding the opposite.}$$

$$3x + (-10) = 14$$
$$\underbrace{\ \ 10 \qquad 10}_{3x + \quad 0\ = 24} \qquad \text{Add 10 to both sides.}$$

$$\frac{3x}{3} \ = \frac{24}{3} \qquad \text{Divide both sides by 3}$$

$$x \quad = 8$$

The unknown number is **8**

CHECK If 10 is subtracted from three times 8, do you get 14? Yes, $3 \cdot 8 - 10 = 24 - 10 = 14$.

3.3 Solving Application Problems with One Unknown Quantity

Use the six problem-solving steps outlined in this chapter. They are listed below in abbreviated form.

Step 1 **Read** the problem and identify what is known and what is unknown.

Step 2 **Assign a variable** to represent the unknown quantity.

Step 3 **Write an equation using your variable.**

Step 4 **Solve the equation.**

Step 5 **State the answer.**

Step 6 **Check** whether your answer fits all the facts given in the *original* statement of the problem.

Denise had some money in her purse this morning. She gave $15 to her daughter and paid $4 to park in the lot at work. At that point she still had $27. How much was in her purse this morning?

Step 1 Unknown: money in purse this morning
Known: took out $15; took out $4; still had $27.

Step 2 Let m represent the money in her purse this morning.

Step 3 $m - \$15 - \$4 = \$27$

Step 4 $m - \quad 15 - \quad 4 = 27$

$$m + \underbrace{(-15) + (-4)}_{} = 27 \qquad \text{Change subtractions.}$$
$$m + \quad (-19) \quad = 27 \qquad \text{Combine like terms.}$$
$$\underbrace{ \quad 19 \qquad\quad 19}_{m + \quad 0 \quad = 46} \qquad \text{Add 19 to both sides.}$$

$$m = \ 46$$

Step 5 She had **$46 in her purse** this morning.

Step 6 Started with $46.
Took out $15, so $46 − $15 = $31
Took out $4, so $31 − $4 = $27
Had $27 left ⟵ Matches ⟶

Concepts	Examples

3.4 Solving Application Problems with Two Unknown Quantities

Use the six problem-solving steps outlined in this chapter.

Last week, Brian earned $50 more than twice what Dan earned. How much did each person earn if the total for both of them was $254?

Step 1 Unknowns: Brian's earnings;
Dan's earnings

Known: Brian earned $50 more than twice what Dan earned;
the sum of their earnings was $254.

In **Step 2,** there are *two* unknown quantities, so assign a variable to represent "the thing you know the least about." Then write a variable expression, using the same variable, to show the relationship of the other unknown quantity to the first one.

Step 2 You know the *least* about Dan's earnings, so let m represent Dan's earnings.

Brian earned $50 more than twice what Dan earned, so Brian's earnings are $2m + \$50$.

Step 3 $\underbrace{m + 2m}_{} + 50 = 254$

Step 4
$$
\begin{array}{rcl}
3m + 50 &=& 254 \\
\underline{-50 -50} & & \quad \text{Add } -50 \text{ to both sides.}\\
\underbrace{3m + 0}_{} &=& 204
\end{array}
$$

$$\frac{3m}{3} = \frac{204}{3} \qquad \text{Divide both sides by 3}$$

$$m = 68$$

Step 5 m represents Dan's earnings, so Dan earned $68.

$2m + \$50$ represents Brian's earnings,

and $2 \cdot \$68 + \$50$ is $\$136 + \$50 = \$186$.

Dan earned $68; Brian earned $186.

Step 6 Is $186 actually $50 more than twice $68?
Yes, the solution checks.

Does $\$68 + \$186 = \$254$?
Yes, the solution checks.

Chapter 3 Review Exercises

3.1 *In Exercises 1–4, find the perimeter of each figure. Also, name each figure in Exercises 1–3.*

1.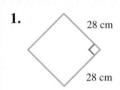
28 cm
28 cm

2.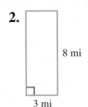
8 mi
3 mi

3.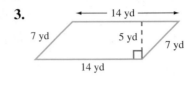
14 yd
7 yd   5 yd   7 yd
14 yd

4.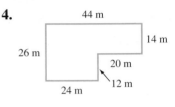
44 m
26 m   14 m
20 m
24 m   12 m

In Exercises 5–7, use the appropriate formula to find the unknown measurement.

5. A square card table has a perimeter of 12 ft. Find the length of one side of the table.

6. A rectangular playground has a perimeter of 128 yd. Find its length if it is 31 yd wide.

7. A rectangular watercolor painting that is 21 in. long has a perimeter of 72 in. What is the width of the painting?

3.2 *In Exercises 8–10, draw a sketch of each shape with the appropriate measurements on it. (Sketches may vary. Show sketches to your instructor.) Then find the area, using the appropriate formula.*

8. A rectangular tablecloth that measures 5 ft by 8 ft

9. A 25 m square dance floor

10. A parallelogram-shaped lot with a base of 16 yd and a height of 13 yd

In Exercises 11–13, use the appropriate formula to find the unknown measurement.

11. A rectangular patio that is 14 ft long has an area of 126 ft². Find its width.

12. A parallelogram has an area of 88 cm². If the base is 11 cm, what is the height?

13. The area of a square piece of land is 100 mi². What is the length of one side?

3.3 *Write each phrase as an algebraic expression. Use x as the variable.*

14. A number subtracted from 57

15. The sum of 15 and twice a number

16. The product of −9 and a number

Translate each sentence into an equation and solve it. Show your work.

17. The sum of four times a number and 6 is −30. What is the number?

18. When twice a number is subtracted from 10, the result is 4 plus the number. Find the number.

3.3–3.4 *Use the six problem-solving steps to solve each problem. Show your work.*

19. Grace wrote a $600 check for her rent. Then she deposited her $750 paycheck and a $75 tax refund into her account. The new balance was $309. How much was in her account before she wrote the rent check?

20. Yoku ordered four boxes of candles for his restaurant. One candle was put on each of the 25 tables. There were 23 candles left. How many candles were originally in each box?

21. A school store ordered 20 boxes of pencils. The first month they sold 83 pencils, and the second month they sold 107 pencils. At the end of the second month there were two pencils still on the shelf and four unopened boxes in the back. How many pencils were in each box?

22. A soup kitchen had 12 pounds of hamburger in the freezer. They used some of the hamburger to make chili to serve for dinner. The next day they received another 30 pounds of hamburger. Now there are 36 pounds of hamburger in the freezer. How much hamburger was used to make the chili?

23. $1000 in prize money is being split between Reggie and Donald. Donald should get $300 more than Reggie. How much will each man receive?

24. A rectangular photograph is twice as long as it is wide. The perimeter of the photograph is 84 cm. Find the length and the width of the photograph.

25. A 36-inch ribbon is cut into three pieces. Two pieces are the same length. The third piece is 3 inches shorter than the other two pieces. Find the length of each piece.

26. Enrique spent $750 on a new sofa and coffee table. The sofa cost four times as much as the coffee table. How much did each item cost?

Chapter 3 Mixed Review Exercises

Practicing exercises in mixed-up order helps you prepare for tests.

In Exercises 1–3, name each figure and find its perimeter and area.

1.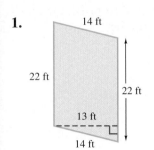

14 ft

22 ft

22 ft

13 ft

14 ft

2.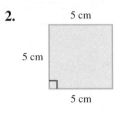

5 cm

5 cm

5 cm

5 cm

3.

60 yd

18 yd

18 yd

60 yd

Solve Exercises 4–6 using the appropriate formulas. Show your work.

4. A rectangular classroom that is 12 m wide has an area of 204 m². Find the length of one side.

5. The area of a parallelogram is 33 in.² and the height is 3 in. How long is the base?

6. A square nature preserve has a perimeter of 64 miles. What is the length of one side?

Use the information in the advertisement and the appropriate formulas to solve Exercises 7–10. Show your work.

Build Your Own Dog Pen

Kit #1 includes 20 feet of fencing.

Kit #2 includes 36 feet of fencing.

7. Anthony made a square dog pen using Kit #2.
 (a) What was the length of each side of the pen?

 (b) What was the area of the pen?

8. First draw sketches of two *different* rectangular dog pens that you could build using all the fencing in Kit #1. Label the lengths of the sides. Then find the area of each pen.

9. Timotha bought Kit #2. But she used some of the fencing around her garden, so she went back and bought Kit #1. Now she has 41 ft of fencing for a dog pen. How much fencing did she use around her garden? Use the six problem-solving steps.

10. Diana bought Kit #2. The pen she built had a length that was 2 ft more than the width. Find the length and width of the pen. Use the six problem-solving steps.

Find the perimeter of each shape.

1.

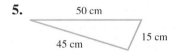

72 m
59 m   46 m   59 m
← 72 m →

2.

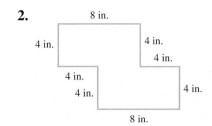

8 in.
4 in.   4 in.
             4 in.
4 in.
4 in.   4 in.
8 in.

3. A square wetland 3 miles on a side

4. A rectangular mirror that measures 2 ft by 4 ft

5.

50 cm
15 cm
45 cm

Find the area of each shape.

6.

27 mm
18 mm   18 mm
27 mm

7.

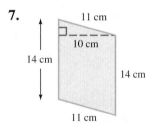

11 cm
10 cm
14 cm
14 cm
11 cm

8. A rectangular animal preserve is 55 mi wide and 68 mi long.

9. The floor of a square room measures 6 m on each side.

Solve Exercises 10–14 using the appropriate formulas. Show your work.

10. A square table has a perimeter of 12 ft. Find the length of one side.

11. The Mercado family has 34 ft of fencing to put around a garden plot. The plot is rectangular in shape. If it is 6 ft wide, find the length of the plot.

12. The area of a parallelogram is 65 in.^{2}, and the base is 13 in. What is the height of the parallelogram?

13. A rectangular postage stamp has a length of 4 cm and an area of 12 cm^2. Find its width.

14. A square bulletin board has an area of 16 ft². How long is each side of the bulletin board?

15. Explain the difference between ft and ft². For which types of problems might you use each of these units?

Translate each sentence into an equation and solve it. Show your work.

16. If 40 is added to four times a number, the result is zero. Find the number.

17. When 7 times a number is decreased by 23, the result is the number plus 7. What is the number?

Solve each application problem, using the six problem-solving steps.

18. Josephine had $43 in her wallet. Her son used some of it when he bought groceries. Josephine found $16 in her desk drawer and put it in her wallet. She counted $44 in the wallet. How much money did her son spend on groceries?

19. Ray is 39 years old. His age is 4 years more than five times his daughter's age. How old is his daughter?

20. A board is 118 cm long. Karin cut it into two pieces, with one piece 4 cm longer than the other. Find the length of both pieces.

21. The perimeter of a rectangular building is 420 ft. The length is four times as long as the width. Find the length and the width. Draw a sketch to help you solve this problem.

22. Marcella and her husband, Tim, spent a total of 19 hours redecorating their living room. Tim spent 3 hours less time than Marcella. How long did each person work on the room?

Chapters 1–3 *Cumulative Review Exercises*

1. Write this number in words.
   4,000,206,300

2. Write this number, using digits.
   seventy million, five thousand, four hundred eighty-nine

3. Write $<$ or $>$ between each pair of numbers to make a true statement.
   -7 ____ -1 0 ____ -5

4. Name the property illustrated by each equation.

   (a) $1(97) = 97$

   (b) $-10 + 0 = -10$

   (c) $(3 \cdot 7) \cdot 6 = 3 \cdot (7 \cdot 6)$

5. **(a)** Round 3795 to the nearest ten.

   (b) Round 493,662 to the nearest ten-thousand.

Simplify.

6. $-12 - 12$

7. $-3(-9)$

8. $|7| - |-10|$

9. $-40 \div 2 \cdot 5$

10. $3 - 8 + 10$

11. $\dfrac{0}{-6}$

12. $-8 + 5(2 - 3)$

13. $(-3)^2 + 4^2$

14. $4 - 3(-6 \div 3) + 7(0 - 6)$

15. $\dfrac{4 - 2^3 + 5^2 - 3(-2)}{-1(3) - 6(-2) - 9}$

16. Rewrite $10w^2xy^4$ without exponents.

17. Evaluate $-6cd^3$ when c is 5 and d is -2.

Simplify.

18. $-4k + k + 5k$

19. $m^2 + 2m + 2m^2$

20. $xy^3 - xy^3$

21. $5(-4a)$

22. $-8 + x + 5 - 2x^2 - x$

23. $-3(4n + 3) + 10$

Solve each equation and check each solution.

24. $6 - 20 = 2x - 9x$ **CHECK**

25. $-5y = y + 6$ **CHECK**

Solve each equation. Show your work.

26. $3b - 9 = 19 - 4b$

27. $-16 - h + 2 = h - 10$

28. $-5(2x + 4) = 3x - 20$

29. $6 + 4(a + 8) = -8a + 5 + a$

Find the perimeter and area of each shape. The figure in Exercise 30 is a parallelogram.

30.
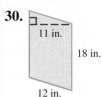
11 in.

18 in.

12 in.

31. A square plaza is 15 m on a side.

32. A rectangular piece of plywood is 4 ft wide and 8 ft long.

Translate each sentence into an equation and solve it. Show your work.

33. When -50 is added to five times a number, the result is 0. What is the number?

34. When three times a number is subtracted from 10, the result is two times the number. Find the number.

Solve each application problem, using the six problem-solving steps.

35. Some people were waiting in line at Walmart. Three people were checked out and left. Six more people got in line. Then two more people were helped, but five people were still in line. How many were in the line originally?

36. Twelve soccer players each paid the same amount toward a team trip. Expenses for the trip were $2200, and now the team bank account, which started at $0, is overdrawn by $40. How much did each player originally pay?

37. A group of 192 students was divided into two smaller groups. One smaller group was three times the size of the other. How many students were in each smaller group?

38. A rectangular swimming pool is 14 ft longer than it is wide. If the perimeter of the pool is 92 ft, find the length and the width.

Rational Numbers: Positive and Negative Fractions

In this chapter, you'll see how fractions are used when measuring things for remodeling projects, cooking, sewing, medical jobs, or any time you want to talk about part of a whole thing.

4.1 Introduction to Signed Fractions

OBJECTIVES

1 Use a fraction to name part of a whole.

2 Identify numerators, denominators, proper fractions, and improper fractions.

3 Graph positive and negative fractions on a number line.

4 Find the absolute value of a fraction.

5 Write equivalent fractions.

OBJECTIVE **1** **Use a fraction to name part of a whole.** So far you have worked with integers. Recall that a list of the integers can be written as follows.

$$\ldots, -6, -5, -4, -3, -2, -1, 0, 1, 2, 3, 4, 5, 6, \ldots$$

The dots show that the list goes on forever in both directions.

Now we will work with *fractions*.

> **Fractions**
>
> A **fraction** is a number of the form $\dfrac{a}{b}$ where a and b are integers and b is not 0.

One use for fractions is situations in which we need a number that is between two integers. Here is an example.

$\frac{2}{3}$ cup

A recipe uses $\frac{2}{3}$ cup of milk.

$\frac{2}{3}$ is between 0 and 1.

$\frac{2}{3}$ is a fraction because it is of the form $\dfrac{a}{b}$ and 2 and 3 are integers.

The number $\frac{2}{3}$ is a fraction that represents 2 of 3 equal parts. In this example, the cup is divided into 3 equal parts and we use enough milk to fill 2 of the parts.

We read $\frac{2}{3}$ as "two-thirds."

VOCABULARY TIP

Fract- means "to break." Take a whole and "break" it into equal-sized parts. Each part is a **fraction** of the whole.

1 Write fractions for the shaded portion and the unshaded portion of each figure.

GS **(a)**

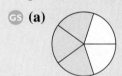

Number of shaded parts → 3

Number of equal parts → ☐

Number of unshaded parts → ☐

Number of equal parts → ☐

(b)

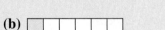

EXAMPLE 1 **Using Fractions to Represent Part of One Whole**

Use fractions to represent the shaded portion and the unshaded portion of each figure.

(a)

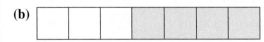

The whole circle must be divided into parts that are the *same* size.

The figure has 3 equal parts. The 2 shaded parts are represented by the fraction $\frac{2}{3}$. The *un*shaded part is $\frac{1}{3}$ of the figure.

(b)

The figure has 7 equal parts. The 4 shaded parts are represented by the fraction $\frac{4}{7}$. The *un*shaded part is $\frac{3}{7}$ of the figure.

◀ **Work Problem 1 at the Side.**

Answers

1. (a) $\dfrac{3}{5}; \dfrac{2}{5}$ (b) $\dfrac{1}{6}; \dfrac{5}{6}$

Fractions can also be used to represent more than one whole object.

EXAMPLE 2 **Using Fractions to Represent More Than One Whole**

Use a fraction to represent the shaded parts.

(a)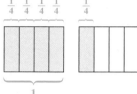

An area equal to 5 of the $\frac{1}{4}$ parts is shaded, so $\frac{5}{4}$ is shaded.

(b)

An area equal to 5 of the $\frac{1}{3}$ parts is shaded, so $\frac{5}{3}$ is shaded.

—————— **Work Problem ② at the Side.** ▶

OBJECTIVE ▶ **②** **Identify numerators, denominators, proper fractions, and improper fractions.** In the fraction $\frac{2}{3}$, the number 2 is the *numerator* and 3 is the *denominator*. The bar between the numerator and the denominator is the *fraction bar*.

$$\text{Fraction bar} \rightarrow \frac{2}{3} \begin{array}{l} \leftarrow \text{Numerator} \\ \leftarrow \text{Denominator} \end{array}$$

Numerator and Denominator

The **denominator** of a fraction shows the number of equal parts in the whole, and the **numerator** shows how many parts are being considered.

Note

Recall that a fraction bar, —, is a symbol for division and division by 0 is *undefined*. Therefore a fraction with a denominator of 0 is also *undefined*.

EXAMPLE 3 **Identifying Numerators and Denominators**

Identify the numerator and denominator in each fraction. Then state the number of equal parts in the whole.

(a) $\frac{5}{9}$ $\begin{array}{l} 5 \leftarrow \text{Numerator} \\ 9 \leftarrow \text{Denominator} \end{array}$

9 equal parts in the whole

(b) $\frac{11}{7}$ $\begin{array}{l} 11 \leftarrow \text{Numerator} \\ 7 \leftarrow \text{Denominator} \end{array}$

7 equal parts in the whole

—————— **Work Problem ③ at the Side.** ▶

Fractions are sometimes called *proper* or *improper* fractions.

Proper and Improper Fractions

If the numerator of a fraction is *less* than the denominator, the fraction is a **proper fraction.** A proper fraction is less than 1.

If the numerator is *greater than or equal to* the denominator, the fraction is an **improper fraction.** An improper fraction is greater than or equal to 1.

---

② Write fractions for the shaded portions.

(a)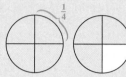

(b)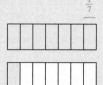

VOCABULARY TIP

De- means "down," so you can remember that the **denominator** is the bottom number in a fraction.

③ Identify the numerator and the denominator. Draw a picture with shaded parts to show each fraction. Your drawings may vary, but they should have the correct number of shaded parts.

(a) $\frac{2}{3}$

(b) $\frac{1}{4}$

(c) $\frac{8}{5}$

(d) $\frac{5}{2}$

Answers

2. (a) $\frac{8}{7}$ (b) $\frac{7}{4}$

3. (a) N: 2; D: 3

(b) N: 1; D: 4

(c) N: 8; D: 5

(d) N: 5; D: 2

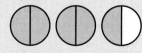

4 From this group of fractions:

$$\frac{3}{4} \quad \frac{8}{7} \quad \frac{5}{7} \quad \frac{6}{6} \quad \frac{1}{2} \quad \frac{2}{1}$$

GS **(a)** list all proper fractions.

Hint: In a proper fraction, the numerator is *less* than the denominator.

Proper Fractions	Improper Fractions
$\dfrac{1}{2} \quad \dfrac{5}{11} \quad \dfrac{35}{36}$	$\dfrac{9}{7} \quad \dfrac{126}{125} \quad \dfrac{7}{7}$

EXAMPLE 4 **Classifying Types of Fractions**

(a) Identify all proper fractions in this list.

$$\frac{3}{4} \quad \frac{5}{9} \quad \frac{17}{5} \quad \frac{9}{7} \quad \frac{3}{3} \quad \frac{12}{25} \quad \frac{1}{9} \quad \frac{5}{3}$$

Proper fractions have a numerator that is *less* than the denominator. The **proper fractions** in the list are shown below.

$$\frac{3}{4} \leftarrow \text{3 is less than 4.} \qquad \frac{5}{9} \quad \frac{12}{25} \quad \frac{1}{9}$$

(b) Identify all improper fractions in the list in part (a).

Improper fractions have a numerator that is *equal to or greater* than the denominator. The **improper fractions** in the list are shown below.

$$\frac{17}{5} \leftarrow \text{17 is greater than 5.} \quad \frac{9}{7} \quad \frac{3}{3} \quad \frac{5}{3}$$

◀ **Work Problem 4 at the Side.**

> **! CAUTION**
>
> So far, you have seen only positive fractions. For negative fractions, the negative sign will be written in front of the fraction bar: for example, $-\frac{3}{4}$. As with integers, the negative sign tells you that a fraction is *less than 0*; it is to the *left* of 0 on the number line. When there is *no* sign in front of a fraction, the fraction is assumed to be positive. For example, $\frac{3}{4}$ is understood to be $+\frac{3}{4}$. It is to the *right* of 0 on the number line.
>
> When the negative sign might be confused with the symbol for subtraction, we will write parentheses around the negative number. Here is an example.
>
> $$3 - \left(-\tfrac{1}{2}\right) \quad \text{means} \quad 3 \text{ minus } \left(\text{negative } \tfrac{1}{2}\right)$$

(b) list all improper fractions.

OBJECTIVE **3** **Graph positive and negative fractions on a number line.**
Sometimes we need *negative* numbers that are between two integers. For example, $-\frac{3}{4}$ is between 0 and -1. Graphing numbers on a number line helps us see the difference between $\frac{3}{4}$ and $-\frac{3}{4}$. Both represent 3 out of 4 equal parts, but they are in opposite directions from 0 on the number line. For $\frac{3}{4}$, divide the distance from 0 to 1 into 4 equal parts. Then start at 0, count over 3 parts, and make a dot. For $-\frac{3}{4}$, repeat the same process between 0 and -1.

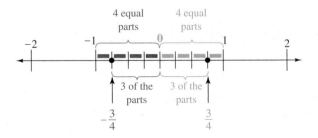

Answers

4. (a) $\dfrac{3}{4}, \dfrac{5}{7}, \dfrac{1}{2}$ **(b)** $\dfrac{8}{7}, \dfrac{6}{6}, \dfrac{2}{1}$

EXAMPLE 5 **Graphing Positive and Negative Fractions**

EXAMPLE 5 **Graphing Positive and Negative Fractions**

Graph each fraction on the number line.

(a) $\dfrac{2}{5}$

There is *no* sign in front of $\frac{2}{5}$, so it is *positive*. Because $\frac{2}{5}$ is between 0 and 1, we divide that space into 5 equal parts. Then we start at 0 and count to the right 2 parts.

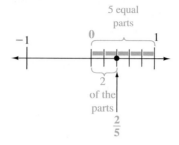

(b) $-\dfrac{4}{5}$

The fraction is *negative,* so it is between 0 and -1. We divide that space into 5 equal parts. Then we start at 0 and count to the left 4 parts.

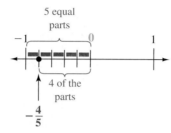

———— **Work Problem** 5 **at the Side.** ▶

 4 Find the absolute value of a fraction. Earlier you learned that the *absolute value* of a number is its distance from 0 on the number line. Two vertical bars indicate absolute value, as shown below.

$$\left|-\dfrac{3}{4}\right| \text{ is read "the absolute value of negative three-fourths."}$$

As with integers, the absolute value of fractions will *always* be positive (or 0) because it is the *distance* from 0 on the number line.

EXAMPLE 6 **Finding the Absolute Value of Fractions**

Find each absolute value: $\left|\dfrac{1}{2}\right|$ and $\left|-\dfrac{1}{2}\right|$

The distance from 0 to $\dfrac{1}{2}$ on the number line is $\dfrac{1}{2}$ space, so $\left|\dfrac{1}{2}\right| = \mathbf{\dfrac{1}{2}}$

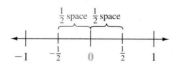

The distance from 0 to $-\dfrac{1}{2}$ is also $\dfrac{1}{2}$ space, so $\left|-\dfrac{1}{2}\right| = \mathbf{\dfrac{1}{2}}$

———— **Work Problem** 6 **at the Side.** ▶

5 Graph each fraction on the number line.

GS **(a)** $\dfrac{2}{4}$ A *positive* fraction is to the *right* of 0.

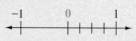

(b) $\dfrac{1}{2}$

(c) $-\dfrac{2}{3}$

6 Find each absolute value.

(a) $\left|-\dfrac{3}{4}\right|$ **(b)** $\left|\dfrac{5}{8}\right|$

(c) $|0|$

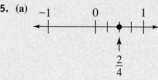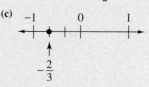

OBJECTIVE ▶ **5** **Write equivalent fractions.** You may have noticed in **Margin Problems 5(a)** and **5(b)** on the previous page that $\frac{2}{4}$ and $\frac{1}{2}$ were at the same point on the number line. Both of them were halfway between 0 and 1. There are actually *many* different names for this point. We illustrate some of them below.

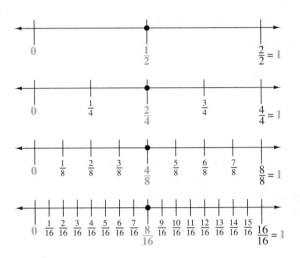

That is, $\frac{8}{16} = \frac{4}{8} = \frac{2}{4} = \frac{1}{2}$. If you have used a standard ruler with inches divided into sixteenths, you probably already noticed that these distances are the same. Although the fractions look different, they all name the same point that is halfway between 0 and 1. In other words, they have the same value. We say that they are *equivalent fractions*.

Equivalent Fractions

Fractions that represent the same number (the same point on a number line) are **equivalent fractions.**

Drawing number lines is tedious, so we usually find equivalent fractions by multiplying or dividing both the numerator and the denominator by the same number. We can use some of the fractions that we just graphed to illustrate this method.

$$\frac{1}{2} = \frac{1 \cdot 2}{2 \cdot 2} = \frac{2}{4} \qquad\qquad \frac{8}{16} = \frac{8 \div 4}{16 \div 4} = \frac{2}{4}$$

Multiply both numerator and denominator by 2.   Divide both numerator and denominator by 4.

Writing Equivalent Fractions

If *a*, *b*, and *c* are numbers (and *b* and *c* are not 0), then:

$$\frac{a}{b} = \frac{a \cdot c}{b \cdot c} \quad \text{or} \quad \frac{a}{b} = \frac{a \div c}{b \div c}$$

In other words, if the numerator and denominator of a fraction are multiplied or divided by the *same* nonzero number, the result is an *equivalent* fraction.

EXAMPLE 7	**Writing Equivalent Fractions**

(a) Write $-\dfrac{1}{2}$ as an equivalent fraction with a denominator of 16.

In other words: $\quad -\dfrac{1}{2} = -\dfrac{?}{16}$

The original denominator is 2. *Multiplying* 2 times 8 gives 16, the new denominator. To write an equivalent fraction, multiply *both* the numerator and denominator by 8.

$$-\underset{\uparrow\, 2}{\dfrac{1}{2}} = -\underset{\uparrow\, 2\cdot 8}{\dfrac{1\cdot 8}{2\cdot 8}} = -\underset{\uparrow\, 16}{\dfrac{8}{16}}$$

> Multiply numerator and denominator by *the same* number.

Keep the negative sign.

So, $-\dfrac{1}{2}$ is equivalent to $-\dfrac{\mathbf{8}}{\mathbf{16}}$

(b) Write $\dfrac{12}{15}$ as an equivalent fraction with a denominator of 5.

In other words: $\quad \dfrac{12}{15} = \dfrac{?}{5}$

The original denominator is 15. *Dividing* 15 by 3 gives 5, the new denominator. To write an equivalent fraction, divide *both* the numerator and denominator by 3.

$$\dfrac{12}{15} = \dfrac{12 \div 3}{15 \div 3} = \dfrac{4}{5}$$

> Divide numerator and denominator by the *same* number.

So, $\dfrac{12}{15}$ is equivalent to $\dfrac{\mathbf{4}}{\mathbf{5}}$

──────── **Work Problem 7 at the Side.** ▶

Look back at the set of four number lines on the previous page. Notice that there are many different names for 1.

$$\dfrac{2}{2} = 1 \qquad \dfrac{4}{4} = 1 \qquad \dfrac{8}{8} = 1 \qquad \dfrac{16}{16} = 1$$

Because a fraction bar is a symbol for division, you can think of $\frac{2}{2}$ as $2 \div 2$, which equals 1. Similarly, $\frac{4}{4}$ is $4 \div 4$, which also is 1, and so on. These examples illustrate one of the division properties.

Division Properties

If a is any number (except 0), then $\dfrac{a}{a} = 1$. In other words, when a nonzero number is divided by itself, the result is 1.

For example: $\quad \dfrac{6}{6} = 1 \quad$ and $\quad \dfrac{-4}{-4} = 1$

Also recall that when any number is divided by 1, the result is the number. That is, $\dfrac{a}{1} = a$.

For example: $\quad \dfrac{6}{1} = 6 \quad$ and $\quad -\dfrac{12}{1} = -12$

7 **(a)** Write $\frac{2}{5}$ as an equivalent fraction with a denominator of 20.

$$\dfrac{2}{5} = \dfrac{?}{20}$$

The original denominator is 5. *Multiplying* 5 times ____ gives 20, the new denominator.

$$\dfrac{2}{5} = \dfrac{2 \cdot \square}{5 \cdot \square} = \dfrac{\square}{20}$$

(b) Write $-\frac{21}{28}$ as an equivalent fraction with a denominator of 4.

Answers

7. **(a)** multiplying 5 times 4; $\dfrac{2 \cdot 4}{5 \cdot 4} = \dfrac{8}{20}$

(b) $-\dfrac{21}{28} = -\dfrac{21 \div 7}{28 \div 7} = -\dfrac{3}{4}$

8 Simplify each fraction by dividing the numerator by the denominator.

GS **(a)** $\dfrac{10}{10}$

Think of $\frac{10}{10}$ as $10 \div 10$.

The result is _____

So $\frac{10}{10} =$ _____

(b) $-\dfrac{3}{1}$

(c) $\dfrac{8}{2}$

(d) $-\dfrac{25}{5}$

EXAMPLE 8 **Using Division to Simplify Fractions**

Simplify each fraction by dividing the numerator by the denominator.

(a) $\dfrac{5}{5}$ Think of $\dfrac{5}{5}$ as $5 \div 5$. The result is 1, so $\dfrac{5}{5} = \mathbf{1}$

> A fraction bar is a symbol for division.

(b) $-\dfrac{12}{4}$ Think of $-\dfrac{12}{4}$ as $-12 \div 4$. The result is -3, so $-\dfrac{12}{4} = \mathbf{-3}$

Keep the negative sign.

(c) $\dfrac{6}{1}$ Think of $\dfrac{6}{1}$ as $6 \div 1$. The result is 6, so $\dfrac{6}{1} = \mathbf{6}$

◀ **Work Problem** 8 **at the Side.**

Note

The title of this chapter is "Rational Numbers: Positive and Negative Fractions." *Rational numbers* are numbers that can be written in the form $\dfrac{a}{b}$, where a and b are integers and b is not 0. In **Example 8(c)** above, you saw that an integer can be written in the form $\dfrac{a}{b}$ (6 can be written as $\frac{6}{1}$).

So rational numbers include all the integers and all the fractions. Later you'll work with rational numbers that are in decimal form.

Answers

8. **(a)** $1; \frac{10}{10} = 1$ **(b)** -3 **(c)** 4 **(d)** -5

4.1 Exercises

FOR EXTRA HELP   *Go to* MyMathLab *for worked-out, step-by-step solutions to exercises enclosed in a square ▢ and video solutions to* ▶ *exercises.*

CONCEPT CHECK   *Circle the correct word or phrase to answer the question.*

1. **(a)** When using fractions, the whole must be cut into ⟨ parts of equal size. / an odd number of parts.

   (b) Which number in a fraction tells how many equal parts are in the whole? ⟨ denominator / numerator

2. **(a)** Which number in a fraction tells how many equal parts are being considered? ⟨ numerator / denominator

   (b) In a fraction, what separates the numerator from the denominator? ⟨ an integer / a fraction bar

Write the fractions that represent the shaded and unshaded portions of each figure.
See Examples 1 and 2.

3.

Number of shaded parts →
Number of equal parts ⟶

4.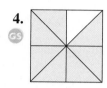

Number of shaded parts →
Number of equal parts ⟶

5.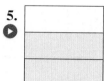

6.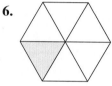

7.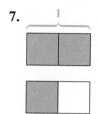

8.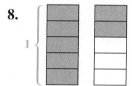

9.

10.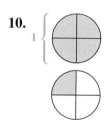

11. What fraction of these coins are dimes? What fraction are pennies? What fraction are nickels?

12. What fraction of these musicians are men? What fraction are women? What fraction are wearing something black?

The circle graph shows the results of a survey on where women would like to have flowers delivered on Valentine's Day. Use the graph to answer Exercises 13–14.

13. **(a)** What fraction of the women would like flowers delivered at work?

    (b) Delivered at home or at work?

14. **(a)** What fraction picked a location other than home or work?

    (b) What fraction picked at home or other?

Delivering Flowers on Valentine's Day

Out of every 20 women surveyed, the number who would like flowers delivered at home, at work, or elsewhere

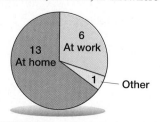

CONCEPT CHECK *Identify the numerator and denominator in each fraction. Then state the number of equal parts in the whole. **See Example 3.***

15. $\frac{3}{4}$

16. $\frac{5}{8}$

17. $\frac{12}{7}$

18. $\frac{8}{3}$

*List the proper and improper fractions in each group of numbers. **See Example 4.***

	Proper	Improper

19. $\frac{8}{5}, \frac{1}{3}, \frac{5}{8}, \frac{6}{6}, \frac{12}{2}, \frac{7}{16}$ _____ _____

	Proper	Improper

20. $\frac{1}{6}, \frac{5}{8}, \frac{15}{14}, \frac{11}{9}, \frac{7}{7}, \frac{3}{4}$ _____ _____

*Graph each pair of fractions on the number line. **See Example 5.***

21. $\frac{1}{4}, -\frac{1}{4}$

22. $-\frac{1}{3}, \frac{1}{3}$

23. $-\frac{3}{5}, \frac{3}{5}$

24. $\frac{5}{6}, -\frac{5}{6}$

25. $\frac{7}{8}, -\frac{7}{8}$

26. $-\frac{3}{4}, \frac{3}{4}$

CONCEPT CHECK *Write a positive or negative fraction to describe each situation.*

27. The baby lost $\frac{3}{4}$ pound in weight while she was sick.

28. Greta needed $\frac{1}{3}$ cup of brown sugar for the cookie recipe.

29. Barb Brown's driveway is $\frac{3}{10}$ mile long.

30. The oil level in my car is $\frac{1}{2}$ quart below normal.

*Find each absolute value. **See Example 6.***

31. $\left| -\frac{2}{5} \right|$

32. $\left| -\frac{1}{4} \right|$

33. $|0|$

34. $\left| \frac{9}{10} \right|$

35. Rewrite each fraction as an equivalent fraction with a denominator of 24. **See Example 7.**

(a) $\frac{1}{2} = \frac{\square}{24}$

Multiplying 2 times _____ gives 24.

(b) $\frac{1}{3} =$

(c) $\frac{2}{3} =$

(d) $\frac{1}{4} =$

(e) $\frac{3}{4} =$

(f) $\frac{1}{6} =$

(g) $\frac{5}{6} =$

(h) $\frac{1}{8} =$

(i) $\frac{3}{8} =$

(j) $\frac{5}{8} =$

36. Rewrite each fraction as an equivalent fraction with a denominator of 36.
See Example 7.

(a) $\dfrac{1}{2} = \dfrac{\square}{36}$

Multiplying
2 times ____
gives 36.

(b) $\dfrac{1}{3} =$

(c) $\dfrac{2}{3} =$

(d) $\dfrac{1}{4} =$

(e) $\dfrac{3}{4} =$

(f) $\dfrac{1}{6} =$

(g) $\dfrac{5}{6} =$

(h) $\dfrac{1}{9} =$

(i) $\dfrac{4}{9} =$

(j) $\dfrac{8}{9} =$

37. Rewrite each fraction as an equivalent fraction with a denominator of 3.
See Example 7.

(a) $-\dfrac{2}{6} = -\dfrac{\square}{3}$

(b) $-\dfrac{4}{6} =$

(c) $-\dfrac{12}{18} =$

(d) $-\dfrac{6}{18} =$

(e) $-\dfrac{200}{300} =$

(f) Write two more fractions that are equivalent to $-\frac{1}{3}$ and two more fractions equivalent to $-\frac{2}{3}$.

38. Rewrite each fraction as an equivalent fraction with a denominator of 4.
See Example 7.

(a) $-\dfrac{2}{8} = -\dfrac{\square}{4}$

(b) $-\dfrac{6}{8} =$

(c) $-\dfrac{15}{20} =$

(d) $-\dfrac{50}{200} =$

(e) $-\dfrac{150}{200} =$

(f) Write two more fractions that are equivalent to $-\frac{1}{4}$ and two more fractions equivalent to $-\frac{3}{4}$.

39. CONCEPT CHECK   Can you write $\frac{3}{5}$ as an equivalent fraction with a denominator of 18? Explain why or why not. If not, what denominators could you use instead of 18?

40. CONCEPT CHECK   Can you write $\frac{3}{4}$ as an equivalent fraction with a denominator of 0? Explain why or why not.

Simplify each fraction by dividing the numerator by the denominator. See Example 8.

41. $\dfrac{10}{1}$ means $10 \div 1 =$ ____

42. $\dfrac{9}{9}$ means $9 \div 9 =$ ____

43. $-\dfrac{16}{16}$

44. $-\dfrac{7}{1}$

45. $-\dfrac{18}{3}$

46. $-\dfrac{40}{4}$

47. $\dfrac{24}{8}$

48. $\dfrac{42}{6}$

49. $\dfrac{14}{7}$

50. $\dfrac{8}{2}$

51. $-\dfrac{90}{10}$

52. $-\dfrac{45}{9}$

53. $\dfrac{150}{150}$

54. $\dfrac{55}{5}$

55. $-\dfrac{32}{4}$

56. $-\dfrac{200}{200}$

There are many correct ways to draw the answers for Exercises 57–64,
so ask your instructor to check your work.

57. Shade $\frac{3}{5}$ of this figure. What fraction is unshaded?

58. Shade $\frac{5}{6}$ of this figure. What fraction is unshaded?

59. Shade $\frac{3}{8}$ of this figure. What fraction is unshaded?

60. Shade $\frac{1}{3}$ of this figure. What fraction is unshaded?

61. Draw a group of figures. Make $\frac{1}{10}$ of the figures circles, $\frac{6}{10}$ of the figures squares, and $\frac{3}{10}$ of the figures triangles.

62. Write a group of capital letters. Make $\frac{4}{9}$ of the letters A's, $\frac{2}{9}$ of the letters B's, and $\frac{3}{9}$ of the letters C's.

63. Draw a group of punctuation marks. Make $\frac{5}{12}$ of them exclamation points, $\frac{1}{12}$ of them commas, $\frac{3}{12}$ of them periods, and add enough question marks to make a full $\frac{12}{12}$ in all. Then circle $\frac{2}{5}$ of the exclamation points.

64. Draw a group of symbols. Make $\frac{2}{15}$ of them addition symbols, $\frac{4}{15}$ of them subtraction symbols, $\frac{2}{15}$ of them division symbols, and add enough equal signs to make a full $\frac{15}{15}$ in all. Then circle $\frac{3}{4}$ of the subtraction symbols.

Relating Concepts (Exercises 65–68) For Individual or Group Work

▦ *Use your calculator as you* **work Exercises 65–68 in order.**

65. (a) Write $\frac{3}{8}$ as an equivalent fraction with a denominator of 3912.

(b) Explain how you solved part (a).

66. (a) Write $\frac{7}{9}$ as an equivalent fraction with a denominator of 5472.

(b) Explain how you solved part (a).

67. (a) Is $-\dfrac{697}{3485}$ equivalent to $-\dfrac{1}{2}, -\dfrac{1}{3}$, or $-\dfrac{1}{5}$?

(b) Explain how you solved part (a).

68. (a) Is $-\dfrac{817}{4902}$ equivalent to $-\dfrac{1}{4}, -\dfrac{1}{6}$, or $-\dfrac{1}{8}$?

(b) Explain how you solved part (a).

4.2 Writing Fractions in Lowest Terms

OBJECTIVE ▶ **1** **Identify fractions written in lowest terms.** You can see from these drawings that $\frac{1}{2}$ and $\frac{4}{8}$ are different names for the same amount of pizza.

$\frac{1}{2}$ of the pizza has pepperoni on it.

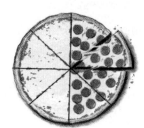

$\frac{4}{8}$ of the pizza has pepperoni on it.

You saw in the last section that $\frac{1}{2}$ and $\frac{4}{8}$ are equivalent fractions. But we say that the fraction $\frac{1}{2}$ is in *lowest terms*. That means that 1 is the *only* number that divides evenly into both 1 and 2. However, the fraction $\frac{4}{8}$ is *not* in lowest terms because its numerator and denominator have a common factor of 4. That means 4 will divide evenly into both 4 and 8.

> **Note**
>
> Recall that *factors* are numbers being multiplied to give a product. For example,
>
> $1 \cdot 4 = 4$, so 1 and 4 are factors of 4.
>
> $2 \cdot 4 = 8$, so 2 and 4 are factors of 8.
>
> 4 is a factor of both 4 and 8, so 4 is a *common factor* of those numbers.

> **Writing a Fraction in Lowest Terms**
>
> A fraction is written in **lowest terms** when the numerator and denominator have no common factor other than 1. Examples are $\frac{1}{3}, \frac{3}{4}, \frac{2}{5},$ and $\frac{7}{10}$. When you are working with fractions, always write the final answer in lowest terms.

EXAMPLE 1 **Identifying Fractions Written in Lowest Terms**

Are the following fractions in lowest terms? If not, find a common factor of the numerator and denominator (other than 1).

(a) $\frac{3}{8}$ 「3 and 8 have no common factor (other than 1)」

The numerator and denominator have no common factor other than 1, so the fraction $\frac{3}{8}$ **is in lowest terms.**

(b) $\frac{21}{36}$ 「Both 21 and 36 are divisible by 3」

The numerator and denominator have a common factor of 3, so the fraction $\frac{21}{36}$ is *not* **in lowest terms.**

───── **Work Problem** ① **at the Side.** ▶

OBJECTIVES

1 Identify fractions written in lowest terms.

2 Write a fraction in lowest terms using common factors.

3 Write a number as a product of prime factors.

4 Write a fraction in lowest terms using prime factorization.

5 Write a fraction with variables in lowest terms.

---

① Are the following fractions in lowest terms? If not, find a common factor of the numerator and denominator (other than 1).

GS **(a)** $\frac{2}{3}$ 2 and 3 have no common factor other than 1.

Is $\frac{2}{3}$ in lowest terms? _____

(b) $-\frac{8}{10}$

(c) $-\frac{9}{11}$

(d) $\frac{15}{20}$

Answers

1. **(a)** Yes, $\frac{2}{3}$ is in lowest terms
 (b) No; 2 is a common factor.
 (c) Yes   **(d)** No; 5 is a common factor.

2 Divide by a common factor to write each fraction in lowest terms.

GS (a) $\dfrac{5}{10}$ What number is a common factor for 5 and 10? ____

$$\dfrac{5}{10} = \dfrac{5 \div \square}{10 \div \square} =$$

(b) $\dfrac{9}{12}$

(c) $-\dfrac{24}{30}$

(d) $\dfrac{15}{40}$

(e) $-\dfrac{50}{90}$

Answers

2. **(a)** 5; $\dfrac{5 \div 5}{10 \div 5} = \dfrac{1}{2}$ **(b)** $\dfrac{3}{4}$ **(c)** $-\dfrac{4}{5}$

(d) $\dfrac{3}{8}$ **(e)** $-\dfrac{5}{9}$

OBJECTIVE 2 Write a fraction in lowest terms using common factors.
We will show you two methods for writing a fraction in lowest terms. The first method, dividing by a common factor, works best when the numerator and denominator are small numbers.

EXAMPLE 2 Using Common Factors to Write Fractions in Lowest Terms

Divide by a common factor to write each fraction in lowest terms.

(a) $\dfrac{20}{24}$

The *greatest* common factor of 20 and 24 is 4. Divide both numerator and denominator by 4.

$$\dfrac{20}{24} = \dfrac{20 \div 4}{24 \div 4} = \dfrac{5}{6}$$ ◁ Now the fraction is in lowest terms.

(b) $\dfrac{30}{50} = \dfrac{30 \div 10}{50 \div 10} = \dfrac{3}{5}$ Divide both numerator and denominator by 10.

(c) $-\dfrac{24}{42} = -\dfrac{24 \div 6}{42 \div 6} = -\dfrac{4}{7}$ Divide both numerator and denominator by 6. Keep the negative sign.

(d) $\dfrac{60}{72}$

Suppose we made an error and thought that 4 was the greatest common factor of 60 and 72. Dividing by 4 gives the following.

$$\dfrac{60}{72} = \dfrac{60 \div 4}{72 \div 4} = \dfrac{15}{18}$$

But $\frac{15}{18}$ is *not* in lowest terms because 15 and 18 have a common factor of 3. Therefore, divide the numerator and denominator by 3.

$$\dfrac{15}{18} = \dfrac{15 \div 3}{18 \div 3} = \dfrac{5}{6}$$ ← Lowest terms

The fraction $\frac{60}{72}$ could have been written in lowest terms in one step by dividing by 12, the *greatest* common factor of 60 and 72.

$$\dfrac{60}{72} = \dfrac{60 \div 12}{72 \div 12} = \dfrac{5}{6}$$ { Same answer as above

Either way works. Just keep dividing until the fraction is in lowest terms.

This method of writing a fraction in lowest terms by dividing by a common factor is summarized below.

Dividing by a Common Factor to Write a Fraction in Lowest Terms

Step 1 Find the *greatest* number that will divide evenly into both the numerator and denominator. This number is a **common factor.**

Step 2 **Divide** both numerator and denominator by the common factor.

Step 3 **Check** to see if the new numerator and denominator have any common factors (besides 1). If they do, repeat *Steps 2 and 3*. If the only common factor is 1, the fraction is in lowest terms.

◁ **Work Problem 2** at the Side.

OBJECTIVE ▶ **3** **Write a number as a product of prime factors.** In Example 2(d) on the previous page, the greatest common factor of 60 and 72 was difficult to see quickly. You can handle a problem like that by writing the numerator and denominator as a product of *prime numbers*.

> **Prime Numbers**
>
> A **prime number** is a natural number that has exactly *two different* factors, itself and 1. (Natural numbers are 1, 2, 3, 4, 5, and so on.)

The number 3 is a prime number because it can be divided evenly only by itself and 1. The number 8 is *not* a prime number. The number 8 is a *composite number* because it can be divided evenly by 2 and 4, as well as by itself and 1.

> **Composite Numbers**
>
> A number with a factor other than itself or 1 is called a **composite number.**

> ⊘ **CAUTION**
>
> A prime number has *only two* different factors, itself and 1. The number 1 is *not* a prime number because it does not have *two different* factors; the only factor of 1 is 1. Therefore, 1 is *neither* prime nor composite.

EXAMPLE 3 **Finding Prime Numbers**

Label each number as *prime* or *composite* or *neither*.

$$1 \quad 2 \quad 5 \quad 10 \quad 11 \quad 15$$

First, 1 is *neither* prime nor composite. Next, 2, 5, and 11, are *prime*. Each of these numbers is divisible only by itself and 1.

The number 10 can be divided by 5 and 2, so it is *composite*. Also, 15 is a *composite* number because 15 can be divided by 5 and 3.

─────────── **Work Problem** ③ **at the Side.** ▶

For reference, here are the prime numbers smaller than 100.

2	3	5	7	11
13	17	19	23	29
31	37	41	43	47
53	59	61	67	71
73	79	83	89	97

> ⊘ **CAUTION**
>
> All prime numbers are odd numbers except the number 2. Be careful though, because *not all odd numbers are prime numbers*. For example, 9, 15, and 21 are odd numbers but they are *not* prime numbers.

The *prime factorization* of a number can be especially useful when working with fractions.

> **Prime Factorization**
>
> A **prime factorization** of a number shows the number as the product of prime numbers.

3 Label each number as *prime* or *composite* or *neither*.

$$1, 2, 3, 4, 7, 9, 13, 19, 25, 29$$

Answers

3. 2, 3, 7, 13, 19, and 29 are *prime*.
4, 9, and 25 are *composite*.
1 is *neither* prime nor composite.

4 Find the prime factorization of each number using division.

(GS) (a) 8

Keep
dividing. $2\overline{)4}$ ← Divide 4 by 2

Start
here. $\Big\}$ → $2\overline{)8}$ ← Divide 8 by 2, the first prime.

8 = ____ • ____ • ____

(b) 42

(c) 90

(d) 100

(e) 81

EXAMPLE 4   Factoring Using the Division Method

(a) Find the prime factorization of 48. Start by dividing 48 by 2 and work your way up the chain of divisions.

```
                  1  ← Continue to divide until the quotient is 1
   All the      3)3  ← Divide 3 by 3; quotient is 1
   divisors
   are prime    2)6  ← Divide 6 by 2; quotient is 3
   factors.    2)12  ← Divide 12 by 2; quotient is 6
               2)24  ← Divide 24 by 2; quotient is 12
Start here
and work up. → 2)48  ← Divide 48 by 2 (the first prime number); quotient is 24
```

Because all the factors (divisors) are prime, the prime factorization of 48 is

$$2 \cdot 2 \cdot 2 \cdot 2 \cdot 3$$

Check by multiplying the factors to see if the product is 48.
Yes, $2 \cdot 2 \cdot 2 \cdot 2 \cdot 3$ does equal 48.

> **Note**
>
> You may write the factors in any order because multiplication is commutative. So you could write the factorization of 48 as $3 \cdot 2 \cdot 2 \cdot 2 \cdot 2$. We will show the factors from least to greatest in our examples.

(b) Find the prime factorization of 225. Start by dividing 225 by 3.

```
                   1  ← Quotient is 1
   All the       5)5  ← Divide 5 by 5
   divisors
   are prime    5)25  ← 25 is not divisible by 3; use 5
   factors.     3)75  ← Divide 75 by 3; quotient is 25
Start here
and work up. → 3)225 ← 225 is not divisible by 2 (first prime) so use 3 (next prime)
```

$225 = 3 \cdot 3 \cdot 5 \cdot 5$ CHECK: 3 • 3 is 9; 9 • 5 is 45; 45 • 5 is **225**

> **⚠ CAUTION**
>
> When you're using the division method of factoring, the last quotient is 1. Do **not** list 1 as a prime factor because 1 is not a prime number.

◀ **Work Problem 4 at the Side.**

Another method of factoring uses what is called a *factor tree*.

EXAMPLE 5   Factoring Using a Factor Tree

Find the prime factorization of each number.

(a) 60

Try to divide 60 by the first prime number, 2. The quotient is 30. Write the factors 2 and 30 under the 60. Circle the 2, because it is a prime.

$60 = 2 \cdot 30$

Continued on Next Page

Try dividing 30 by 2. The quotient is 15. Write the factors 2 and 15 under the 30.

60
② 30
  ② 15

Because 15 cannot be divided evenly by 2, try dividing 15 by the next prime number, 3. The quotient is 5. Write the factors 3 and 5 under the 15.

60
② 30
  ② 15
    ③ ⑤ ← Circle 3 and 5 because both are primes.

No uncircled factors remain, so you have found the prime factorization (the circled factors).

$$60 = 2 \cdot 2 \cdot 3 \cdot 5$$ CHECK: 2 · 2 is 4;
4 · 3 is 12; 12 · 5 is **60**

(b) 72

Divide 72 by 2, the first prime number.

72 ← Divide by 2; 72 = 2 · 36
② 36 ← Divide by 2 again; 36 = 2 · 18
  ② 18 ← Divide by 2 again; 18 = 2 · 9
    ② 9 ← Divide by 3; 9 = 3 · 3
      ③ ③

$$72 = 2 \cdot 2 \cdot 2 \cdot 3 \cdot 3$$

(c) 45

Because 45 cannot be divided evenly by 2, try dividing by the next prime, 3.

45 ← Divide by 3; 45 = 3 · 15
③ 15 ← Divide by 3 again; 15 = 3 · 5
  ③ ⑤

$$45 = 3 \cdot 3 \cdot 5$$ CHECK: 3 · 3 is 9;
9 · 5 is **45**

Note

Here is a reminder about the quick way to see whether a number is *divisible* by 2, 3, or 5; in other words, there is no remainder when you do the division.

A number is divisible by 2 if the ones digit is 0, 2, 4, 6, or 8.
For example, 30, 512, 76, and 3018 are all divisible by 2.

A number is divisible by 3 if the *sum* of the digits is divisible by 3.
For example, 129 is divisible by 3 because $1 + 2 + 9 = 12$ and 12 is divisible by 3.

A number is divisible by 5 if the ones digit is 0 or 5.
For example, 85, 610, and 1725 are all divisible by 5.

Work Problem **5** at the Side. ▶

---

5 Complete each factor tree and write the prime factorization.

(a)   28 ← Divide by 2;

             28 = 2 · 14
② 14 ← Divide by 2 again;
             14 = 2 · ____
○ ○

(b)   35
⑤ ○

(c)   90

Answers

5. (a)   28
   ② 14
     ② ⑦
   28 = 2 · 2 · 7
(b)   35
   ⑤ ⑦
   35 = 5 · 7
(c)   90
   ② 45
     ③ 15
       ③ ⑤
   90 = 2 · 3 · 3 · 5

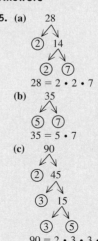

▦ **Calculator Tip**

You can use your calculator to find the prime factorization of a number. Here is an example that uses 539.

Try dividing 539 by the first prime number, 2.

539 ÷ 2 = **269.5**   Does not divide evenly

Try dividing 539 by the next prime number, 3.

539 ÷ 3 = **179.6666667**   Does not divide evenly

Keep trying the next prime numbers until you find one that divides evenly.

539 ÷ 5 = **107.8**   Does not divide evenly

539 ÷ 7 = **77**   Divides evenly

Once you have found that 7 works, try using it again on the quotient 77.

77 ÷ 7 = **11**   Divides evenly

Because 11 is prime, you're finished. The prime factorization of 539 is 7 • 7 • 11.

Now try factoring 2431 using your calculator. (The answer is at the bottom left of this page.)

OBJECTIVE ▶ 4   **Write a fraction in lowest terms using prime factorization.**
Now you can use the second method for writing fractions in lowest terms: prime factorization. This is a good method to use when the numerator and denominator are larger numbers or include variables.

EXAMPLE 6   **Using Prime Factorization to Write Fractions in Lowest Terms**

(a) Write $\frac{20}{35}$ in lowest terms.

20 can be written as 2 • 2 • 5 ⟵ Prime factors

35 can be written as 5 • 7 ⟵ Prime factors

$$\frac{20}{35} = \frac{2 \cdot 2 \cdot 5}{5 \cdot 7}$$

The numerator and denominator have 5 as a common factor. Dividing both numerator and denominator by 5 will give an equivalent fraction.

Any number divided by itself is 1.

$$\frac{20}{35} = \frac{2 \cdot 2 \cdot 5}{5 \cdot 7} = \frac{2 \cdot 2 \cdot 5}{7 \cdot 5} = \frac{2 \cdot 2 \cdot \boxed{5 \div 5}}{7 \cdot \boxed{5 \div 5}} = \frac{2 \cdot 2 \cdot 1}{7 \cdot 1} = \frac{4}{7}$$

Multiplication is commutative.

$\frac{20}{35}$ is written in lowest terms as $\frac{4}{7}$

Calculator Tip Answer

2431 = 11 • 13 • 17

Continued on Next Page

To shorten the work, you may use slashes to indicate the divisions. For example, the work on $\frac{20}{35}$ can be shown as follows.

$$\frac{20}{35} = \frac{2 \cdot 2 \cdot \overset{1}{\cancel{5}}}{\underset{1}{\cancel{5}} \cdot 7}$$

Slashes indicate $5 \div 5$, and the result is 1.

(b) Write $\frac{60}{72}$ in lowest terms.

Use the prime factorizations of 60 and 72 from **Examples 5(a) and 5(b)**.

$$\frac{60}{72} = \frac{2 \cdot 2 \cdot 3 \cdot 5}{2 \cdot 2 \cdot 2 \cdot 3 \cdot 3}$$

This time there are three common factors. Use slashes to show the three divisions.

2 ÷ 2 is 1

2 ÷ 2 is 1 ⟶   ⟵ 3 ÷ 3 is 1

$$\frac{60}{72} = \frac{\overset{1}{\cancel{2}} \cdot \overset{1}{\cancel{2}} \cdot \overset{1}{\cancel{3}} \cdot 5}{\underset{1}{\cancel{2}} \cdot \underset{1}{\cancel{2}} \cdot 2 \cdot \underset{1}{\cancel{3}} \cdot 3} = \frac{5}{6}$$

⟵ Multiply 1 • 1 • 1 • 5 to get 5
⟵ Multiply 1 • 1 • 2 • 1 • 3 to get 6

(c) $\frac{18}{90}$

$$\frac{18}{90} = \frac{\overset{1}{\cancel{2}} \cdot \overset{1}{\cancel{3}} \cdot \overset{1}{\cancel{3}}}{\underset{1}{\cancel{2}} \cdot \underset{1}{\cancel{3}} \cdot \underset{1}{\cancel{3}} \cdot 5} = \frac{1}{5}$$

⟵ Multiply 1 • 1 • 1 to get 1
⟵ Multiply 1 • 1 • 1 • 5 to get 5

> **⊘ CAUTION**
>
> In **Example 6(c)** above, all factors of the numerator divided out. But 1 • 1 • 1 is still 1, so the final answer is $\frac{1}{5}$ (**not** 5).

This method of using prime factorization to write a fraction in lowest terms is summarized below.

Using Prime Factorization to Write a Fraction in Lowest Terms

Step 1 Write the **prime factorization** of both numerator and denominator.

Step 2 Use slashes to show where you are **dividing** the numerator and denominator by any common factors.

Step 3 **Multiply** the factors in the numerator and in the denominator.

Work Problem ⑥ at the Side. ▶

OBJECTIVE ▶ 5 **Write a fraction with variables in lowest terms.** Fractions may have variables in the numerator or denominator. Examples are shown below.

$$\frac{6}{2x} \qquad \frac{3xy}{9xy} \qquad \frac{4b^3}{8ab} \qquad \frac{7ab^2}{n^2}$$

You can use prime factorization to write these fractions in lowest terms.

---

⑥ Use the method of prime factorization to write each fraction in lowest terms.

GS (a) $\frac{16}{48}$

$$\frac{16}{48} = \frac{2 \cdot 2 \cdot 2 \cdot 2}{2 \cdot 2 \cdot 2 \cdot 2 \cdot 3} =$$

Divide out all the common factors.

(b) $\frac{28}{60}$

(c) $\frac{74}{111}$

(d) $\frac{124}{340}$

Answers

6. (a) $\dfrac{\overset{1}{\cancel{2}} \cdot \overset{1}{\cancel{2}} \cdot \overset{1}{\cancel{2}} \cdot \overset{1}{\cancel{2}}}{\underset{1}{\cancel{2}} \cdot \underset{1}{\cancel{2}} \cdot \underset{1}{\cancel{2}} \cdot \underset{1}{\cancel{2}} \cdot 3} = \dfrac{1}{3}$

(b) $\dfrac{\overset{1}{\cancel{2}} \cdot \overset{1}{\cancel{2}} \cdot 7}{\underset{1}{\cancel{2}} \cdot \underset{1}{\cancel{2}} \cdot 3 \cdot 5} = \dfrac{7}{15}$

(c) $\dfrac{2 \cdot \overset{1}{\cancel{37}}}{3 \cdot \underset{1}{\cancel{37}}} = \dfrac{2}{3}$

(d) $\dfrac{\overset{1}{\cancel{2}} \cdot \overset{1}{\cancel{2}} \cdot 31}{\underset{1}{\cancel{2}} \cdot \underset{1}{\cancel{2}} \cdot 5 \cdot 17} = \dfrac{31}{85}$

7 Write each fraction in lowest terms.

(a) $\dfrac{5c}{15}$

$$\dfrac{5c}{15} = \dfrac{5 \cdot c}{3 \cdot 5} = \dfrac{\square}{\square}$$

Divide out all the common factors.

(b) $\dfrac{10x^2}{8x^2}$

(c) $\dfrac{9a^3}{11b^3}$

(d) $\dfrac{6m^2n}{9n^2}$

EXAMPLE 7 Writing Fractions with Variables in Lowest Terms

Write each fraction in lowest terms.

(a) $\dfrac{6}{2x}$ ← Prime factors of 6 are 2 · 3
 ← 2x means 2 · x

$$\dfrac{6}{2x} = \dfrac{\overset{1}{\cancel{2}} \cdot 3}{\underset{1}{\cancel{2}} \cdot x} = \dfrac{3}{x} \begin{array}{l} \leftarrow 1 \cdot 3 \text{ is } 3 \\ \leftarrow 1 \cdot x \text{ is } x \end{array}$$

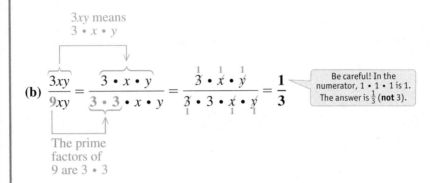

3xy means 3 · x · y

(b) $\dfrac{3xy}{9xy} = \dfrac{3 \cdot x \cdot y}{3 \cdot 3 \cdot x \cdot y} = \dfrac{\overset{1}{\cancel{3}} \cdot \overset{1}{\cancel{x}} \cdot \overset{1}{\cancel{y}}}{\underset{1}{\cancel{3}} \cdot 3 \cdot \underset{1}{\cancel{x}} \cdot \underset{1}{\cancel{y}}} = \dfrac{1}{3}$

> Be careful! In the numerator, 1 · 1 · 1 is 1. The answer is $\frac{1}{3}$ (**not** 3).

The prime factors of 9 are 3 · 3

b^3 means $b \cdot b \cdot b$

(c) $\dfrac{4b^3}{8ab} = \dfrac{2 \cdot 2 \cdot b \cdot b \cdot b}{2 \cdot 2 \cdot 2 \cdot a \cdot b} = \dfrac{\overset{1}{\cancel{2}} \cdot \overset{1}{\cancel{2}} \cdot \overset{1}{\cancel{b}} \cdot b \cdot b}{\underset{1}{\cancel{2}} \cdot \underset{1}{\cancel{2}} \cdot 2 \cdot a \cdot \underset{1}{\cancel{b}}} = \dfrac{b^2}{2a} \begin{array}{l} \leftarrow b \cdot b \text{ is } b^2 \\ \leftarrow 2 \cdot a \text{ is } 2a \end{array}$

The prime factors of 8 are 2 · 2 · 2

(d) $\dfrac{7ab^2}{n^2} = \dfrac{7 \cdot a \cdot b \cdot b}{n \cdot n}$

> There are no common factors.

$\dfrac{7ab^2}{n^2}$ is **already** in lowest terms.

◄ **Work Problem 7 at the Side.**

⚙ **Study Skills Reminder**

Are you getting the most out of your text? This text is full of helpful features such as caution boxes, margin exercises, chapter summaries, review exercises, and more. These features are here to help you succeed. See the Study Skills activity "Using Your Text" for more information on these helpful features and how to use them.

Answers

7. **(a)** $\dfrac{\overset{1}{\cancel{5}} \cdot c}{3 \cdot \underset{1}{\cancel{5}}} = \dfrac{1c}{3}$ or $\dfrac{c}{3}$

(b) $\dfrac{\overset{1}{\cancel{2}} \cdot 5 \cdot \overset{1}{\cancel{x}} \cdot \overset{1}{\cancel{x}}}{\underset{1}{\cancel{2}} \cdot 2 \cdot 2 \cdot \underset{1}{\cancel{x}} \cdot \underset{1}{\cancel{x}}} = \dfrac{5}{4}$

(c) already in lowest terms

(d) $\dfrac{2 \cdot \overset{1}{\cancel{3}} \cdot m \cdot m \cdot \overset{1}{\cancel{n}}}{\underset{1}{\cancel{3}} \cdot 3 \cdot n \cdot \underset{1}{\cancel{n}}} = \dfrac{2m^2}{3n}$

4.2 Exercises

FOR EXTRA HELP

Go to MyMathLab *for worked-out, step-by-step solutions to exercises enclosed in a square* ☐ *and video solutions to* ▶ *exercises.*

CONCEPT CHECK *Are the following fractions in lowest terms? If not, find a common factor of the numerator and denominator (other than 1).* **See Example 1.**

1. (a) $-\dfrac{3}{10}$ **(b)** $\dfrac{10}{15}$ **(c)** $\dfrac{9}{16}$ **(d)** $-\dfrac{4}{21}$ **(e)** $\dfrac{6}{9}$ **(f)** $-\dfrac{7}{28}$

2. (a) $\dfrac{10}{12}$ **(b)** $\dfrac{3}{18}$ **(c)** $-\dfrac{8}{15}$ **(d)** $-\dfrac{22}{33}$ **(e)** $\dfrac{2}{25}$ **(f)** $-\dfrac{14}{15}$

Write each fraction in lowest terms. Use the method of dividing by a common factor. **See Example 2.**

3. (a) $\dfrac{10}{15}$ **(b)** $\dfrac{6}{9}$ **(c)** $-\dfrac{7}{28}$ **(d)** $-\dfrac{25}{50}$ **(e)** $\dfrac{16}{18}$ **(f)** $-\dfrac{8}{20}$

4. (a) $\dfrac{10}{12}$ **(b)** $\dfrac{3}{18}$ **(c)** $-\dfrac{22}{33}$ **(d)** $\dfrac{12}{16}$ **(e)** $-\dfrac{15}{20}$ **(f)** $-\dfrac{9}{15}$

CONCEPT CHECK *Label each number as* prime *or* composite *or* neither. **See Example 3.**

5. 9   2   8   1   5   11   10   21

6. 12   3   7   6   1   15   13   25

Find the prime factorization of each number. **See Examples 4 and 5.**

7. 6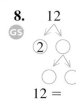
6 =

8. 12
12 =

9. 20

10. 30

11. 25

12. 18

13. 36

14. 56

15. (a) 44

(b) 88

16. (a) 45

(b) 135

17. (a) 75

(b) 68

(c) 189

18. (a) 80

(b) 64

(c) 385

Write each numerator and denominator as a product of prime factors. Then use the prime factorization to write the fraction in lowest terms. **See Example 6.**

19. $\frac{8}{16}$ $8 = 2 \cdot 2 \cdot 2$
$16 = 2 \cdot 2 \cdot 2 \cdot 2$

$\frac{8}{16} = \dfrac{\overset{1}{\cancel{2}} \cdot \overset{1}{\cancel{2}} \cdot \overset{1}{\cancel{2}}}{\underset{1}{\cancel{2}} \cdot \underset{1}{\cancel{2}} \cdot \underset{1}{\cancel{2}} \cdot 2} =$

20. $\frac{6}{8}$ $6 = 2 \cdot 3$
$8 = 2 \cdot 2 \cdot 2$

$\frac{6}{8} = \dfrac{\overset{1}{\cancel{2}} \cdot 3}{\underset{1}{\cancel{2}} \cdot 2 \cdot 2} =$

21. $\frac{32}{48}$

22. $\frac{9}{27}$

23. $\frac{14}{21}$

24. $\frac{20}{32}$

25. $\frac{36}{42}$

26. $\frac{22}{33}$

27. $\frac{50}{63}$

28. $\frac{72}{80}$

29. $\frac{27}{45}$

30. $\frac{36}{63}$

31. $\frac{12}{18}$

32. $\frac{63}{90}$

33. $\frac{35}{40}$

34. $\frac{36}{48}$

35. $\frac{90}{180}$

36. $\frac{16}{64}$

37. $\frac{210}{315}$

38. $\frac{96}{192}$

39. $\frac{429}{495}$

40. $\frac{135}{182}$

Write your answers to Exercises 41–46 in lowest terms.

41. There are 60 minutes in an hour. What fraction of an hour is

   (a) 15 minutes?   **(b)** 30 minutes?

   (c) 6 minutes?   **(d)** 60 minutes?

42. There are 24 hours in a day. What fraction of a day is

   (a) 8 hours?   **(b)** 18 hours?

   (c) 12 hours?   **(d)** 3 hours?

43. SueLynn's monthly income is $2400.

   (a) She spends $800 on rent. What fraction of her income is spent on rent?

   (b) She spends $400 on food. What fraction of her income is spent on food?

   (c) What fraction of her income is left for other expenses?

44. There are 15,000 students at Fox River College.

   (a) 10,500 of the students receive some form of financial aid. What fraction of the students receive financial aid?

   (b) 7500 of the students are going to school full time. What fraction are full-time students?

   (c) What fraction of the students are going to school part time?

45. The pictograph below shows the average number of hours spent on household chores each month by men and by women.

Of the average 140 hours spent monthly on household chores, here are the amounts of time spent by men and by women.

Men:

Women:

Each broom represents 10 hours.

What fraction of the time spent on household chores is done by men? What fraction is done by women?

46. A survey asked 100 smartphone users how often they checked their devices. The circle graph shows the number who gave each answer.

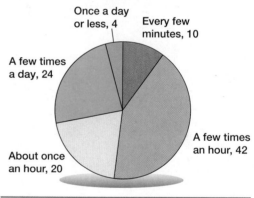

How Often Do Smartphone Users Check Their Device?

Once a day or less, 4
Every few minutes, 10
A few times a day, 24
A few times an hour, 42
About once an hour, 20

Data from Gallup.

What fraction of the smartphone users gave each answer?

47. CONCEPT CHECK The work shown below has a mistake in it.

What Went Wrong? First write a sentence explaining what the mistake is. Then fix the mistake and find the correct solution.

Marcus wrote $\frac{9}{36}$ in lowest terms by doing the following calculations:

$$\frac{9}{36} = \frac{3 \cdot 3}{2 \cdot 2 \cdot 3 \cdot 3} = 4$$

48. CONCEPT CHECK The work shown below has a mistake in it.

What Went Wrong? First write a sentence explaining what the mistake is. Then fix the mistake and find the correct solution.

Ginny wrote $\frac{9}{16}$ in lowest terms by doing these divisions:

$$\frac{9}{16} = \frac{9 \div 3}{16 \div 4} = \frac{3}{4}$$

*Write each fraction in lowest terms. **See Example 7.***

49. $\dfrac{16c}{40}$

50. $\dfrac{36}{54a}$

51. $\dfrac{20x}{35x}$

52. $\dfrac{21n}{28n}$

53. $\dfrac{18r^2}{15rs}$

54. $\dfrac{18ab}{48b^2}$

55. $\dfrac{6m}{42mn^2}$

56. $\dfrac{10g^2}{90g^2h}$

57. $\dfrac{9x^2}{16y^2}$

58. $\dfrac{5rst}{8st}$

59. $\dfrac{7xz}{9xyz}$

60. $\dfrac{6a^3}{23b^3}$

61. $\dfrac{21k^3}{6k^2}$

62. $\dfrac{16x^3}{12x^4}$

63. $\dfrac{13a^2bc^3}{39a^2bc^3}$

64. $\dfrac{22m^3n^4}{55m^3n^4}$

65. $\dfrac{14c^2d}{14cd^2}$

66. $\dfrac{19rs}{19s^3}$

67. $\dfrac{210ab^3c}{35b^2c^2}$

68. $\dfrac{81w^4xy^2}{300wy^4}$

69. $\dfrac{25m^3rt^2}{36n^2s^3w^2}$

70. $\dfrac{42a^5b^4c^3}{7a^4b^3c^2}$

71. $\dfrac{33e^2fg^3}{11efg}$

72. $\dfrac{21xy^2z^3}{17ab^2c^3}$

4.3 Multiplying and Dividing Signed Fractions

OBJECTIVE ▶ **1 Multiply signed fractions.** Suppose that you give $\frac{1}{3}$ of your candy bar to your friend Ann. Then Ann gives $\frac{1}{2}$ of her share to Tim. How much of the bar does Tim get to eat?

OBJECTIVES

1 Multiply signed fractions.

2 Multiply fractions that involve variables.

3 Divide signed fractions.

4 Divide fractions that involve variables.

5 Solve application problems involving multiplying and dividing fractions.

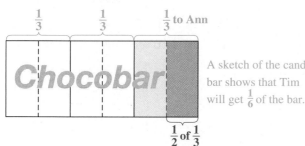

A sketch of the candy bar shows that Tim will get $\frac{1}{6}$ of the bar.

$\frac{1}{2}$ of $\frac{1}{3}$ to Tim

Tim's share is $\frac{1}{2}$ of $\frac{1}{3}$ candy bar. When used with fractions, the word **of** indicates multiplication.

$$\frac{1}{2} \text{ of } \frac{1}{3} \quad \text{means} \quad \frac{1}{2} \cdot \frac{1}{3}$$

Tim's share is $\frac{1}{6}$ bar, so $\frac{1}{2} \cdot \frac{1}{3} = \frac{1}{6}$.

This example illustrates the rule for multiplying fractions.

Multiplying Fractions

If a, b, c, and d are numbers (but b and d are not 0), then

$$\frac{a}{b} \cdot \frac{c}{d} = \frac{a \cdot c}{b \cdot d}$$

In other words, multiply the numerators and multiply the denominators.

When we apply this rule to find Tim's part of the candy bar, we get

$$\frac{1}{2} \cdot \frac{1}{3} = \frac{1 \cdot 1}{2 \cdot 3} = \frac{1}{6} \quad \begin{array}{l} \leftarrow \text{Multiply numerators.} \\ \leftarrow \text{Multiply denominators.} \end{array}$$

EXAMPLE 1 Multiplying Signed Fractions

Find each product.

(a) $-\frac{5}{8} \cdot -\frac{3}{4}$ Recall that the product of two negative numbers is a positive number.

Multiply the numerators and multiply the denominators.

$$-\frac{5}{8} \cdot -\frac{3}{4} = \frac{5 \cdot 3}{8 \cdot 4} = \frac{15}{32} \leftarrow \text{Lowest terms}$$

The product of two negative numbers is positive.

The answer is in lowest terms because 15 and 32 have no common factor other than 1.

Continued on Next Page

1 Find each product.

(a) $-\dfrac{3}{4} \cdot \dfrac{1}{2}$

$= -\dfrac{3 \cdot 1}{4 \cdot 2} = -\dfrac{\square}{\square}$

Product of a negative number and a positive number is negative.

(b) $\left(-\dfrac{2}{5}\right)\left(-\dfrac{2}{3}\right)$

The product of two negative numbers is _____ .

(c) $\dfrac{3}{4}\left(\dfrac{3}{8}\right)$

(b) $\left(\dfrac{4}{7}\right)\left(-\dfrac{2}{5}\right) = -\dfrac{4 \cdot 2}{7 \cdot 5} = -\dfrac{8}{35}$

The product of a negative number and a positive number is negative.

◀ **Work Problem 1 at the Side.**

Sometimes the result won't be in lowest terms. For example, find $\frac{3}{10}$ of $\frac{5}{6}$.

$$\frac{3}{10} \text{ of } \frac{5}{6} \quad \text{means} \quad \frac{3}{10} \cdot \frac{5}{6} = \frac{3 \cdot 5}{10 \cdot 6} = \frac{15}{60} \quad \left\{ \begin{array}{l} \text{Not in lowest} \\ \text{terms} \end{array} \right.$$

Now write $\frac{15}{60}$ in lowest terms.

$$\frac{15}{60} = \frac{\overset{1}{\cancel{3}} \cdot \overset{1}{\cancel{5}}}{2 \cdot 2 \cdot \underset{1}{\cancel{3}} \cdot \underset{1}{\cancel{5}}} = \frac{1}{4} \leftarrow \text{Lowest terms}$$

You used prime factorization earlier in this chapter to write fractions in lowest terms. You can also use it when multiplying fractions. Writing the prime factors of the original fractions and dividing out common factors *before* multiplying usually saves time. If you divide out *all* the common factors, the result will automatically be in lowest terms. Let's see how that works when finding $\frac{3}{10}$ of $\frac{5}{6}$.

3 and 5 are already prime.

$$\frac{3}{10} \cdot \frac{5}{6} = \frac{3 \cdot 5}{2 \cdot 5 \cdot 2 \cdot 3} = \frac{\overset{1}{\cancel{3}} \cdot \overset{1}{\cancel{5}}}{2 \cdot \underset{1}{\cancel{5}} \cdot 2 \cdot \underset{1}{\cancel{3}}} = \frac{1}{4} \quad \left\{ \begin{array}{l} \text{Same result} \\ \text{as above} \end{array} \right.$$

Write 10 as 2 • 5

Write 6 as 2 • 3

Divide out the common factors.

> **! CAUTION**
>
> When you are working with fractions, always write the final result in lowest terms. Visualizing $\frac{15}{60}$ is hard to do. But when $\frac{15}{60}$ is written as $\frac{1}{4}$, working with it is much easier. A fraction is *simplified* when it is written in lowest terms.

EXAMPLE 2 Using Prime Factorization to Multiply Fractions

(a) $-\dfrac{8}{5}\left(\dfrac{5}{12}\right)$

Multiplying a negative number times a positive number gives a negative product.

Write 8 as 2 • 2 • 2

$$-\frac{8}{5}\left(\frac{5}{12}\right) = -\frac{2 \cdot 2 \cdot 2 \cdot 5}{5 \cdot 2 \cdot 2 \cdot 3} = -\frac{\overset{1}{\cancel{2}} \cdot \overset{1}{\cancel{2}} \cdot 2 \cdot \overset{1}{\cancel{5}}}{\underset{1}{\cancel{5}} \cdot \underset{1}{\cancel{2}} \cdot \underset{1}{\cancel{2}} \cdot 3} = -\frac{2}{3} \quad \left\{ \begin{array}{l} \text{Lowest} \\ \text{terms} \end{array} \right.$$

5 is already prime.

Write 12 as 2 • 2 • 3

Negative product

Answers

1. **(a)** $-\dfrac{3}{8}$ **(b)** positive; $\dfrac{4}{15}$ **(c)** $\dfrac{9}{32}$

Continued on Next Page

(b) Find $\dfrac{2}{9}$ of $\dfrac{15}{16}$

When used with fractions, *of* indicates multiplication.

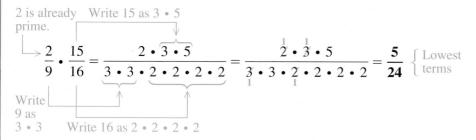

2 is already prime.

Write 15 as 3 • 5

$$\dfrac{2}{9} \cdot \dfrac{15}{16} = \dfrac{2 \cdot 3 \cdot 5}{3 \cdot 3 \cdot 2 \cdot 2 \cdot 2 \cdot 2} = \dfrac{\overset{1}{2} \cdot \overset{1}{3} \cdot 5}{\underset{1}{3} \cdot 3 \cdot \underset{1}{2} \cdot 2 \cdot 2 \cdot 2} = \dfrac{5}{24} \left\{ \begin{array}{l} \text{Lowest} \\ \text{terms} \end{array} \right.$$

Write 9 as 3 • 3

Write 16 as 2 • 2 • 2 • 2

—————— **Work Problem 2 at the Side.** ▶

▦ **Calculator Tip**

If your calculator has a fraction key ⓐᵇ/ᶜ, or Ⓐᵇ/ᶜ, you can do calculations with fractions. You'll also need the change of sign key ⊕/⊖, or the negative sign key ⊖, to enter negative fractions, just as you did to enter negative integers.

Start by entering several different fractions. Clear your calculator after each one. Different calculator models may vary, so check your calculator's instructions.

To enter $\frac{3}{4}$, press 3 ⓐᵇ/ᶜ 4. The display will show **3⌐4**.

↑ Fraction bar

To enter $-\frac{9}{10}$, press ⊖ 9 ⓐᵇ/ᶜ 10. The display will show **−9⌐10**.

Let's check the result of **Example 2(a)** on the previous page: Multiply $-\frac{8}{5}\left(\frac{5}{12}\right)$ by pressing the following keys.

⊖ 8 ⓐᵇ/ᶜ 5 ⊗ 5 ⓐᵇ/ᶜ 12 ⊜   The display shows **−2⌐3**

$-\dfrac{8}{5}$ $\dfrac{5}{12}$ $-\dfrac{2}{3}$

Now try **Example 2(b)** above: Find $\frac{2}{9}$ of $\frac{15}{16}$. (Did you get $\frac{5}{24}$?)

❗ **CAUTION**

The fraction key on a calculator is useful for *checking* your work. But knowing the rules for fraction computation is important because you'll need them when fractions involve variables. (See **Example 4** on the next page).

---

2 Use prime factorization to multiply these fractions.

ɢꜱ (a) $\dfrac{15}{28}\left(-\dfrac{6}{5}\right)$ Product will be negative.

$$= -\dfrac{3 \cdot 5 \cdot 2 \cdot 3}{2 \cdot 2 \cdot 7 \cdot 5} = -\dfrac{\square}{\square}$$

Divide out the common factors.

(b) $\dfrac{12}{7} \cdot \dfrac{7}{24}$ Product will be _____

(c) $\left(-\dfrac{11}{18}\right)\left(-\dfrac{9}{20}\right)$

Product will be _____

Answers

2. **(a)** $-\dfrac{3 \cdot \overset{1}{5} \cdot \overset{1}{2} \cdot 3}{2 \cdot \underset{1}{2} \cdot 7 \cdot \underset{1}{5}} = -\dfrac{9}{14}$

(b) positive; $\dfrac{\overset{1}{2} \cdot \overset{1}{2} \cdot \overset{1}{3} \cdot \overset{1}{7}}{\underset{1}{7} \cdot \underset{1}{2} \cdot \underset{1}{2} \cdot 2 \cdot \underset{1}{3}} = \dfrac{1}{2}$

(c) positive; $\dfrac{11 \cdot \overset{1}{3} \cdot \overset{1}{3}}{2 \cdot \underset{1}{3} \cdot \underset{1}{3} \cdot 2 \cdot 2 \cdot 5} = \dfrac{11}{40}$

3 Use prime factorization to find these products.

GS (a) $\dfrac{3}{4}$ of 36

$$\dfrac{3}{4} \downarrow \quad \dfrac{36}{\downarrow}$$

$$\dfrac{3}{4} \cdot \dfrac{36}{1} = \dfrac{3 \cdot 2 \cdot 2 \cdot 3 \cdot 3}{2 \cdot 2 \cdot 1}$$

$$= \dfrac{\square}{\square} =$$

(b) $-10 \cdot \dfrac{2}{5}$ Product will be _____

(c) $\left(-\dfrac{7}{8}\right)(-24)$ Product will be _____

4 Use prime factorization to find these products.

GS (a) $\dfrac{2c}{5} \cdot \dfrac{c}{4} = \dfrac{2 \cdot c \cdot c}{5 \cdot 2 \cdot 2}$ Divide out common factors.

$$= \dfrac{\square}{\square}$$

(b) $\left(\dfrac{m}{6}\right)\left(\dfrac{9}{m^2}\right)$

(c) $\left(\dfrac{w^2}{y}\right)\left(\dfrac{x^2y}{w}\right)$

Answers

3. (a) $\dfrac{3 \cdot \overset{1}{\cancel{2}} \cdot \overset{1}{\cancel{2}} \cdot 3 \cdot 3}{\underset{1}{\cancel{2}} \cdot \underset{1}{\cancel{2}} \cdot 1} = \dfrac{27}{1} = 27$

(b) negative; $-\dfrac{2 \cdot \overset{1}{\cancel{5}} \cdot 2}{1 \cdot \underset{1}{\cancel{5}}} = -\dfrac{4}{1} = -4$

(c) positive; $\dfrac{7 \cdot \overset{1}{\cancel{2}} \cdot \overset{1}{\cancel{2}} \cdot \overset{1}{\cancel{2}} \cdot 3}{\underset{1}{\cancel{2}} \cdot \underset{1}{\cancel{2}} \cdot \underset{1}{\cancel{2}}} = \dfrac{21}{1} = 21$

4. (a) $\dfrac{\overset{1}{\cancel{2}} \cdot c \cdot c}{5 \cdot \underset{1}{\cancel{2}} \cdot 2} = \dfrac{c^2}{10}$

(b) $\dfrac{\overset{1}{\cancel{m}} \cdot \overset{1}{\cancel{3}} \cdot 3}{2 \cdot \underset{1}{\cancel{3}} \cdot \underset{1}{\cancel{m}} \cdot m} = \dfrac{3}{2m}$

(c) $\dfrac{\overset{1}{\cancel{y}} \cdot w \cdot x \cdot x \cdot \overset{1}{\cancel{y}}}{\underset{1}{\cancel{y}} \cdot \underset{1}{\cancel{w}}} = \dfrac{wx^2}{1} = wx^2$

EXAMPLE 3   **Multiplying a Fraction and an Integer**

Find $\frac{2}{3}$ of 6.

We can write 6 in fraction form as $\frac{6}{1}$. Recall that $\frac{6}{1}$ means $6 \div 1$, which is 6. So we can write any integer a as $\frac{a}{1}$.

$$\dfrac{2}{3} \text{ of } 6 \text{ means } \dfrac{2}{3} \cdot \dfrac{6}{1} = \dfrac{2 \cdot 2 \cdot \overset{1}{\cancel{3}}}{\underset{1}{\cancel{3}} \cdot 1} = \dfrac{4}{1} = 4$$

$4 \div 1$ is 4

◀ **Work Problem 3 at the Side.**

OBJECTIVE ▸ 2 Multiply fractions that involve variables. The multiplication method that uses prime factors also works when there are variables in the numerators and/or denominators of the fractions.

EXAMPLE 4   **Multiplying Fractions with Variables**

Find each product.

(a) $\dfrac{3x}{5} \cdot \dfrac{2}{9x}$

$3x$ means $3 \cdot x$ $\dfrac{x}{x}$ is 1

$$\dfrac{3x}{5} \cdot \dfrac{2}{9x} = \dfrac{3 \cdot x \cdot 2}{5 \cdot 3 \cdot 3 \cdot x} = \dfrac{\overset{1}{\cancel{3}} \cdot \overset{1}{\cancel{x}} \cdot 2}{5 \cdot \underset{1}{\cancel{3}} \cdot 3 \cdot \underset{1}{\cancel{x}}} = \dfrac{2}{15}$$

The prime factors of 9 are $3 \cdot 3$, so $9x$ is $3 \cdot 3 \cdot x$

(b) $\left(\dfrac{3y}{4x}\right)\left(\dfrac{2x^2}{y}\right)$

$2x^2$ means $2 \cdot x \cdot x$

$$\left(\dfrac{3y}{4x}\right)\left(\dfrac{2x^2}{y}\right) = \dfrac{3 \cdot y \cdot 2 \cdot x \cdot x}{2 \cdot 2 \cdot x \cdot y} = \dfrac{3 \cdot \overset{1}{\cancel{y}} \cdot \overset{1}{\cancel{2}} \cdot \overset{1}{\cancel{x}} \cdot x}{2 \cdot \underset{1}{\cancel{2}} \cdot \underset{1}{\cancel{x}} \cdot \underset{1}{\cancel{y}}} = \dfrac{3x}{2}$$

The prime factors of 4 are $2 \cdot 2$, so $4x$ is $2 \cdot 2 \cdot x$

◀ **Work Problem 4 at the Side.**

OBJECTIVE ▸ 3 Divide signed fractions. To divide fractions, you will rewrite division problems as multiplication problems. For division, you will leave the first number (the dividend) as it is, but change the second number (the divisor) to its *reciprocal*.

Reciprocal of a Fraction

Two numbers are **reciprocals** of each other if their product is 1.

The reciprocal of the fraction $\dfrac{a}{b}$ is $\dfrac{b}{a}$ because

$$\frac{a}{b} \cdot \frac{b}{a} = \frac{\overset{1}{\cancel{a}} \cdot \overset{1}{\cancel{b}}}{\cancel{b} \cdot \cancel{a}} = \frac{1}{1} = 1$$

Notice that you "flip" or "invert" a fraction to find its reciprocal. Here are some examples.

Number	Reciprocal	Reason
$\dfrac{1}{6}$	$\dfrac{6}{1}$	Because $\dfrac{1}{6} \cdot \dfrac{6}{1} = \dfrac{6}{6} = 1$
$-\dfrac{2}{5}$	$-\dfrac{5}{2}$	Because $\left(-\dfrac{2}{5}\right)\left(-\dfrac{5}{2}\right) = \dfrac{10}{10} = 1$
4 Think of 4 as $\frac{4}{1}$	$\dfrac{1}{4}$	Because $4 \cdot \dfrac{1}{4} = \dfrac{4}{1} \cdot \dfrac{1}{4} = \dfrac{4}{4} = 1$

VOCABULARY TIP

Vert- means "to turn." When you invert a fraction, you switch the position of the numerator and the denominator. For example, $\frac{1}{2}$ becomes $\frac{2}{1}$.

Note

Every number has a reciprocal except 0. Why not 0? Recall that a number times its reciprocal equals 1. But that doesn't work for 0.

$$0 \cdot (\text{reciprocal}) \neq 1$$

Put any number here. When you multiply it by 0, you get 0, never 1.

Dividing Fractions

If a, b, c, and d are numbers (but b, c, and d are not 0), then we have the following.

$$\frac{a}{b} \div \frac{c}{d} = \frac{a}{b} \cdot \frac{d}{c}$$

Reciprocals

In other words, change division to multiplying by the reciprocal of the divisor.

Use this method to find the quotient for $\frac{2}{3} \div \frac{1}{6}$. Rewrite it as a multiplication problem and then use the steps for multiplying fractions.

Change division to multiplication.

$$\frac{2}{3} \div \frac{1}{6} = \frac{2}{3} \cdot \frac{6}{1} = \frac{2 \cdot 2 \cdot \overset{1}{\cancel{3}}}{\cancel{3} \cdot 1} = \frac{4}{1} = \mathbf{4}$$

The reciprocal of $\frac{1}{6}$ is $\frac{6}{1}$

Does it make sense that $\frac{2}{3} \div \frac{1}{6} = 4$? Let's compare dividing fractions to dividing whole numbers.

$15 \div 3$ is asking, "How many 3s are in 15?"

$\frac{2}{3} \div \frac{1}{6}$ is asking, "How many $\frac{1}{6}$s are in $\frac{2}{3}$?"

The figure below illustrates $\frac{2}{3} \div \frac{1}{6}$.

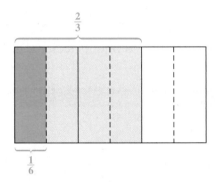

How many $\frac{1}{6}$s are in $\frac{2}{3}$?

There are 4 of the $\frac{1}{6}$ pieces in $\frac{2}{3}$.

So $\frac{2}{3} \div \frac{1}{6} = 4$

As a final check on this method of dividing, try changing $15 \div 3$ into a multiplication problem. You know that the quotient should be 5.

$$15 \div 3 = 15 \cdot \frac{1}{3} = \frac{15}{1} \cdot \frac{1}{3} = \frac{3 \cdot 5 \cdot 1}{1 \cdot 3} = \frac{5}{1} = 5 \quad \left\{ \begin{array}{l} \text{The quotient} \\ \text{we expected} \end{array} \right.$$

Reciprocals

So, you can see that *dividing* by a fraction is the same as *multiplying* by the *reciprocal* of the fraction.

EXAMPLE 5 | **Dividing Signed Fractions**

Rewrite each division problem as a multiplication problem. Then multiply.

(a) $\dfrac{3}{10} \div \dfrac{4}{5} = \dfrac{3}{10} \cdot \dfrac{5}{4} = \dfrac{3 \cdot 5}{2 \cdot 5 \cdot 2 \cdot 2} = \dfrac{3}{8}$

Reciprocals

Do **not** change $\frac{3}{10}$.

Change the *divisor* $\left(\frac{4}{5}\right)$ to its reciprocal, $\frac{5}{4}$.

Note

When multiplying fractions, you don't always have to factor the numerator and denominator completely into prime numbers. In **Example 5(a)** above, if you notice that 5 is a common factor of the numerator and denominator, you can write

$$\frac{3}{10} \cdot \frac{5}{4} = \frac{3 \cdot 5}{2 \cdot 5 \cdot 4} = \frac{3}{8}$$

Factor 10 into $2 \cdot 5$

Leave 4 as it is.

If no common factors are obvious to you, then write out the complete prime factorization to help find the common factors.

—— **Continued on Next Page**

(b) $2 \div \left(-\dfrac{1}{3}\right)$

First notice that the numbers have different signs. In a division problem, different signs mean that the quotient is negative. Then write 2 in fraction form as $\frac{2}{1}$.

$$2 \div \left(-\frac{1}{3}\right) = \frac{2}{1} \cdot \left(-\frac{3}{1}\right) = -\frac{2 \cdot 3}{1 \cdot 1} = -\frac{6}{1} = -6$$

Reciprocals   Negative product

(c) $-\dfrac{3}{4} \div (-8) = -\dfrac{3}{4} \cdot \left(-\dfrac{1}{8}\right) = \dfrac{3 \cdot 1}{4 \cdot 8} = \dfrac{3}{32}$

Reciprocals

No common factor to divide out

> Both numbers in the problem were negative. When signs *match*, the quotient is *positive*.

(d) $\dfrac{9}{16} \div 0$ Dividing by 0 **cannot** be done.

This is **undefined,** just as dividing by 0 was undefined for integers. Recall that 0 does *not* have a reciprocal, so you can't change the division to multiplying by the reciprocal of the divisor.

(e) $0 \div \dfrac{9}{16} = 0 \cdot \dfrac{16}{9} = 0$ Recall that 0 divided by any nonzero number gives a result of 0

Reciprocals

—— **Work Problem 5 at the Side.** ▶

OBJECTIVE ▶ 4 Divide fractions that involve variables. The method for dividing fractions also works when there are variables in the numerators and/or denominators of the fractions.

EXAMPLE 6   **Dividing Fractions with Variables**

Divide. (Assume that none of the variables represent zero.)

(a) $\dfrac{x^2}{y} \div \dfrac{x}{3y} = \dfrac{x^2}{y} \cdot \dfrac{3y}{x} = \dfrac{\overset{1}{x} \cdot x \cdot 3 \cdot \overset{1}{y}}{\underset{1}{y} \cdot \underset{1}{x}} = \dfrac{3x}{1} = 3x$

Reciprocals

(b) $\dfrac{8b}{5} \div b^2 = \dfrac{8b}{5} \cdot \dfrac{1}{b^2} = \dfrac{8 \cdot \overset{1}{b} \cdot 1}{5 \cdot \underset{1}{b} \cdot b} = \dfrac{8}{5b}$

Think of b^2 as $\dfrac{b^2}{1}$ so the reciprocal is $\dfrac{1}{b^2}$

—— **Work Problem 6 at the Side.** ▶

5 Rewrite each division problem as a multiplication problem. Then multiply.

GS (a) $-\dfrac{3}{4} \div \dfrac{5}{8} = -\dfrac{3}{4} \cdot \dfrac{8}{5} =$

Reciprocals

$-\dfrac{3 \cdot 2 \cdot \overset{1}{4}}{\underset{1}{4} \cdot 5} = -\dfrac{\square}{\square}$

(b) $0 \div \left(-\dfrac{7}{12}\right)$

(c) $\dfrac{5}{6} \div 10$

(d) $-9 \div \left(-\dfrac{9}{16}\right)$

(e) $\dfrac{2}{5} \div 0$

6 Divide. (Assume none of the variables represent zero.)

GS (a) $\dfrac{c^2d^2}{4} \div \dfrac{c^2d}{4} = \dfrac{c^2d^2}{4} \cdot \dfrac{4}{c^2d} =$

$\dfrac{\overset{1}{c} \cdot \overset{1}{c} \cdot \overset{1}{d} \cdot d \cdot \overset{1}{2} \cdot \overset{1}{2}}{\underset{1}{2} \cdot \underset{1}{2} \cdot \underset{1}{c} \cdot \underset{1}{c} \cdot \underset{1}{d}} =$

$\dfrac{\square}{1} =$

(b) $\dfrac{20}{7h} \div \dfrac{5h}{7}$

(c) $\dfrac{n}{8} \div mn$

Answers

5. (a) $-\dfrac{6}{5}$ **(b)** $0 \cdot \left(-\dfrac{12}{7}\right) = 0$

     (c) $\dfrac{5}{6} \cdot \dfrac{1}{10} = \dfrac{\overset{1}{5} \cdot 1}{6 \cdot 2 \cdot \underset{1}{5}} = \dfrac{1}{12}$

     (d) $-\dfrac{9}{1} \cdot \left(-\dfrac{16}{9}\right) = \dfrac{\overset{1}{9} \cdot 16}{1 \cdot \underset{1}{9}} = \dfrac{16}{1} = 16$

     (e) undefined; can't be written as multiplication because 0 doesn't have a reciprocal

6. (a) $\dfrac{d}{1} = d$ **(b)** $\dfrac{4}{h^2}$ **(c)** $\dfrac{1}{8m}$

7 Look for indicator words or draw sketches to help you with these problems.

GS (a) How many times can a $\frac{2}{3}$-cup spray bottle be filled from an 18-cup bottle of window cleaner?

$$18 \div \frac{2}{3} = \frac{\square}{\square} \cdot \frac{\square}{\square} =$$

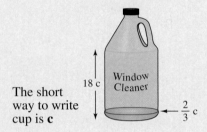

The short way to write cup is **c**

18 c Window Cleaner

$\frac{2}{3}$ c

GS (b) A retiring police officer will receive $\frac{5}{8}$ of her highest annual salary as retirement income. If her highest annual salary is $64,000, how much will she receive as retirement income?

$\frac{5}{8}$ of her highest salary

$\frac{5}{8} \cdot \frac{}{1} =$

Answers

7. **(a)** $\dfrac{18}{1} \cdot \dfrac{3}{2} = \dfrac{\overset{1}{\cancel{2}} \cdot 9 \cdot 3}{1 \cdot \cancel{2}} = \dfrac{27}{1} = 27$

The bottle can be filled 27 times.

(b $\dfrac{5}{8} \cdot \dfrac{64,000}{1} = \dfrac{5 \cdot \overset{1}{\cancel{8}} \cdot 8000}{\cancel{8} \cdot 1} = \dfrac{40,000}{1}$

$= 40,000$

The officer will receive $40,000.

OBJECTIVE ► 5 **Solve application problems involving multiplying and dividing fractions.** When you're solving application problems, some indicator words are used to suggest multiplication and some are used to suggest division.

INDICATOR WORDS FOR MULTIPLICATION	INDICATOR WORDS FOR DIVISION
product	per
double	each
triple	goes into
times	divided by
twice	divided into
of (when *of* follows a fraction)	divided equally

Look for these indicator words in the following examples. However, *you won't always find an indicator word.* Then, you need to think through the problem to decide what to do. Sometimes, drawing a sketch of the situation described in the problem will help you decide which operation to use.

> **EXAMPLE 7** **Using Indicator Words and Sketches to Solve Application Problems**
>
> **(a)** Lois gives $\frac{1}{10}$ of her income to her church. Last month she earned $1980. How much of that did she give to her church?
>
> Notice the word **of**. Because the word **of** *follows the fraction* $\frac{1}{10}$, it indicates multiplication.
>
> $$\frac{1}{10} \text{ of } 1980 = \frac{1}{10} \cdot \frac{1980}{1} = \frac{1 \cdot \overset{1}{\cancel{10}} \cdot 198}{\underset{1}{\cancel{10}} \cdot 1} = \frac{198}{1} = 198$$
>
> Lois gave **$198** to her church.
>
> **(b)** The apparel design class is making infant snowsuits to give to a local shelter. A fabric store donated a 12 yd length of fabric for the project. If one snowsuit needs $\frac{2}{3}$ yd of fabric, how many suits can the class make?
>
> The word **of** appears in the second sentence: "A fabric store donated a 12 yd length **of** fabric." But the word **of** does *not follow a fraction,* so it is *not* an indicator to multiply. Let's try a sketch. There is a piece of fabric 12 yd long. One snowsuit will use $\frac{2}{3}$ yd. The question is, how many $\frac{2}{3}$ yd pieces can be cut from the 12 yards?

12 yd of fabric

Cutting 12 yards into equal size pieces indicates division. How many $\frac{2}{3}$s are in 12?

$\frac{2}{3}$ yd

$$12 \div \frac{2}{3} = \frac{12}{1} \cdot \frac{3}{2} = \frac{\overset{1}{\cancel{2}} \cdot 6 \cdot 3}{1 \cdot \cancel{2}} = \frac{18}{1} = 18$$

Reciprocals

The class can make **18 snowsuits.**

◄ **Work Problem 7 at the Side.**

4.3 Exercises

FOR EXTRA HELP

Go to MyMathLab for worked-out, step-by-step solutions to exercises enclosed in a square ▢ *and video solutions to* ▶ *exercises.*

CONCEPT CHECK *Circle the correct word or phrase to complete each statement.*

1. To multiply fractions, you should $\lessgtr$ divide / multiply the numerators and multiply the $\lessgtr$ denominators / factors.

2. When you multiply two negative fractions, the product is $\lessgtr$ positive / negative. When you multiply a positive fraction by a negative fraction, the product is $\lessgtr$ positive / negative.

Multiply. Write the products in lowest terms. See **Examples 1–4.**

3. $-\dfrac{3}{8} \cdot \dfrac{1}{2} = -\dfrac{3 \cdot 1}{8 \cdot 2} =$

4. $\left(\dfrac{2}{3}\right)\left(-\dfrac{5}{7}\right) = -\dfrac{2 \cdot 5}{3 \cdot 7} =$

5. $\left(-\dfrac{3}{8}\right)\left(-\dfrac{12}{5}\right)$

6. $\dfrac{4}{9} \cdot \dfrac{12}{7}$

7. $\dfrac{21}{30}\left(\dfrac{5}{7}\right)$

8. $\left(-\dfrac{6}{11}\right)\left(-\dfrac{22}{15}\right)$

9. $10\left(-\dfrac{3}{5}\right)$

10. $-20\left(\dfrac{3}{4}\right)$

11. $\dfrac{4}{9}$ of 81

12. $\dfrac{2}{3}$ of 48

13. $\left(\dfrac{3x}{4}\right)\left(\dfrac{5}{xy}\right)$

14. $\left(\dfrac{2}{5a^2}\right)\left(\dfrac{a}{8}\right)$

Divide. Write the quotients in lowest terms. (Assume that none of the variables represent zero.) **See Examples 5 and 6.**

15. $\dfrac{1}{6} \div \dfrac{1}{3} = \dfrac{1}{6} \cdot \dfrac{3}{1} =$
Reciprocals

16. $-\dfrac{1}{2} \div \dfrac{2}{3} = -\dfrac{1}{2} \cdot \dfrac{3}{2} =$
Reciprocals

17. $-\dfrac{3}{4} \div \left(-\dfrac{5}{8}\right)$

18. $\dfrac{7}{10} \div \dfrac{2}{5}$

19. $6 \div \left(-\dfrac{2}{3}\right)$

20. $-7 \div \left(-\dfrac{1}{4}\right)$

21. $-\dfrac{2}{3} \div 4$

22. $\dfrac{5}{6} \div (-15)$

23. $\dfrac{11c}{5d} \div 3c$

24. $8x^2 \div \dfrac{4x}{7}$

25. $\dfrac{ab^2}{c} \div \dfrac{ab}{c}$

26. $\dfrac{mn}{6} \div \dfrac{n}{3m}$

Find each product or quotient. Write all answers in lowest terms. **See Examples 1–6.**

27. $\dfrac{4}{5} \div 3 = \dfrac{4}{5} \cdot \dfrac{1}{3} =$
Reciprocals

28. $\left(-\dfrac{20}{21}\right)\left(-\dfrac{14}{15}\right) = \dfrac{2 \cdot 2 \cdot \overset{1}{5} \cdot 2 \cdot \overset{1}{7}}{3 \cdot 7 \cdot 3 \cdot \underset{1}{5}} =$

The product of two negative numbers is positive.

29. $-\dfrac{3}{8}\left(\dfrac{3}{4}\right)$

30. $-\dfrac{8}{17} \div \dfrac{4}{5}$

31. $\dfrac{3}{5}$ of 35

32. $\dfrac{2}{3} \div (-6)$

33. $-9 \div \left(-\dfrac{3}{5}\right)$

34. $\dfrac{7}{8} \cdot \dfrac{25}{21}$

35. $\dfrac{12}{7} \div 0$

36. $\dfrac{5}{8}$ of (-48)

37. $\left(\dfrac{11}{2}\right)\left(-\dfrac{5}{6}\right)$

38. $\dfrac{3}{4} \div \dfrac{3}{16}$

39. $\dfrac{4}{7}$ of $14b$

40. $\dfrac{ab}{6} \div \dfrac{b}{9}$

41. $\dfrac{12}{5} \div 4d$

42. $\dfrac{18}{7} \div 2t$

43. $\dfrac{x^2}{y} \div \dfrac{w}{2y}$

44. $\dfrac{5}{6}$ of $18w$

Solve each application problem. See Example 7.

45. Al is helping Tim make a mahogany lamp table for Jill's birthday. Find the area of the rectangular top of the table if it is $\frac{4}{5}$ yd long by $\frac{3}{8}$ yd wide.

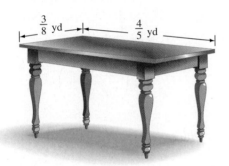

46. A rectangular dog bed is $\frac{4}{5}$ yd by $\frac{9}{10}$ yd. Find its area.

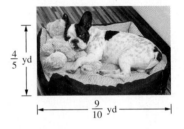

47. How many $\frac{1}{8}$-ounce eyedrop dispensers can be filled from a container that holds 10 ounces of eyedrops?

48. Ms. Shaffer has a piece of property with an area that is $\frac{9}{10}$ acre. She wishes to divide it into three equal parts for her children. How many acres of land will each child get?

49. Todd estimates that it will cost him $12,400 to attend a community college for one year. He thinks he can earn $\frac{3}{4}$ of the cost and borrow the balance. Find the amount he must earn and the amount he must borrow.

50. Joyce Chen wants to make children's vests to sell at a craft fair. Each vest requires $\frac{3}{4}$ yd of material. She has 36 yd of material. Find the number of vests she can make.

51. There are about 300 baseball players in the Baseball Hall of Fame. About one-fourth of them played at infield positions (1st base, 2nd base, 3rd base, or shortstop). About how many infield players are in the Hall of Fame? (Data from www.baseball-almanac.com)

52. At the Garlic Festival Fun Run, $\frac{5}{12}$ of the runners are women. If there are 780 runners, how many are women? How many are men?

53. Pam Trizlia has a small pickup truck that can carry $\frac{2}{3}$ cord of firewood. Find the number of trips needed to deliver 6 cords of wood.

54. Parking lot A is $\frac{1}{4}$ mile long and $\frac{3}{16}$ mile wide, and parking lot B is $\frac{3}{8}$ mile long and $\frac{1}{8}$ mile wide. Which parking lot has the larger area?

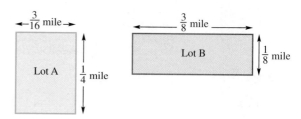

55. An adult male alligator may weigh 400 pounds. A newly hatched alligator weighs only $\frac{1}{8}$ pound. The adult weighs how many times the hatchling? (Data from St. Marks NWR.)

56. A female alligator lays about 35 eggs in a nest of marsh grass and mud. But $\frac{4}{5}$ of the eggs or hatchlings will fall prey to raccoons, wading birds, or larger alligators. How many of the 35 eggs will hatch and survive? (Data from St. Marks NWR.)

A survey of 1600 students asked them where they got the money for college expenses. The circle graph shows what fraction of the students gave each answer. Use this information to work Exercises 57–60.

57. How many students in the survey borrowed money for college expenses?

58. What number of students used money from their parents' income and savings?

59. How many more students used their own income and savings rather than money from relatives and friends?

60. How many fewer students used money that their parents borrowed rather than money that they borrowed themselves?

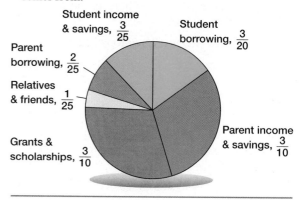

Paying for College

A survey of 1600 students shows where the money comes from:

Student income & savings, $\frac{3}{25}$

Student borrowing, $\frac{3}{20}$

Parent borrowing, $\frac{2}{25}$

Relatives & friends, $\frac{1}{25}$

Grants & scholarships, $\frac{3}{10}$

Parent income & savings, $\frac{3}{10}$

Data from Sallie Mae.

There are a total of about 175 million companion pets in the United States. This table shows the fraction of pets that are dogs, cats, birds, and horses. Use the table to answer Exercises 61–64.

COMPANION PETS IN THE UNITED STATES

Type of Pet	Fraction of All Pets
Dog	$\frac{2}{5}$
Cat	$\frac{12}{25}$
Bird	$\frac{3}{50}$
Horse	$\frac{1}{25}$

Data from American Veterinary Medical Association.

61. How many U.S. pets are horses?

62. How many U.S. pets are dogs?

63. How many dogs and cats are pets?

64. How many birds are pets?

65. CONCEPT CHECK The work shown below has a mistake in it.

What Went Wrong? First write a sentence explaining what the mistake is. Then fix the mistake and find the correct solution.

Evan did the following calculation:

$$\frac{3}{4} \cdot \frac{8}{9} = \frac{3 \cdot 8}{4 \cdot 9} = \frac{24}{36}$$

66. CONCEPT CHECK The work shown below has a mistake in it.

What Went Wrong? First write a sentence explaining what the mistake is. Then fix the mistake and find the correct solution.

Haddi did the following calculation.

$$8 \cdot \frac{2}{3} = \frac{8}{1} \cdot \frac{3}{2} = \frac{\overset{1}{\cancel{2}} \cdot 4 \cdot 3}{1 \cdot \underset{1}{\cancel{2}}} = \frac{12}{1} = 12$$

67. CONCEPT CHECK The work shown below has **two** mistakes in it.

What Went Wrong? First write a sentence explaining what each mistake is. Then fix the mistakes and find the correct solutions.

Bailey did two test questions like this:

$$\frac{2}{3} \div 4 = \frac{2}{3} \cdot \frac{4}{1} = \frac{2 \cdot 4}{3 \cdot 1} = \frac{8}{3}$$

$$\frac{1}{2} \div 0 = 0$$

68. CONCEPT CHECK The work shown below has **two** mistakes in it.

What Went Wrong? First write a sentence explaining what each mistake is. Then fix the mistakes and find the correct solutions.

Jillian did these two calculations on a quiz:

$$\frac{3}{14} \cdot \frac{7}{9} = \frac{\cancel{3} \cdot \cancel{7}}{2 \cdot \cancel{7} \cdot \cancel{3} \cdot 3} = 6$$

$$\frac{2}{5} \cdot \frac{3}{8} = \frac{\overset{1}{\cancel{2}}}{5} \cdot \frac{3}{\underset{2}{\cancel{8}}} = \frac{3}{10}$$

4.4 Adding and Subtracting Signed Fractions

OBJECTIVES

1. Add and subtract like fractions.
2. Find the lowest common denominator for unlike fractions.
3. Add and subtract unlike fractions.
4. Add and subtract unlike fractions that contain variables.

OBJECTIVE ▶ **1 Add and subtract like fractions.** You probably remember learning something about "common denominators" in other math classes. When fractions have the *same* denominator, we say that they have a *common* denominator, which makes them **like fractions**. When fractions have different denominators, they are called **unlike fractions**. Here are examples.

Like Fractions	**Unlike Fractions**

$$\frac{3}{4} \quad \text{and} \quad -\frac{7}{4}$$

Common denominator

$$\frac{2}{9} \quad \text{and} \quad \frac{2}{8}$$

Different denominators

$$\frac{6}{x} \quad \text{and} \quad \frac{y}{x}$$

Common denominator

$$\frac{a^2}{3} \quad \text{and} \quad \frac{3}{a}$$

Different denominators

You can add or subtract fractions *only* when they have a common denominator. To see why, let's look at more pizzas.

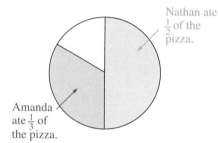

Nathan ate $\frac{1}{2}$ of the pizza.

Amanda ate $\frac{1}{3}$ of the pizza.

What fraction of the pizza has been eaten? We can't write a fraction until the pizza is cut into pieces of *equal* size. That's what the denominator of a fraction tells us: the number of *equal* size pieces in the pizza.

Now the pizza is cut into 6 *equal* pieces, and we can find out how much was eaten.

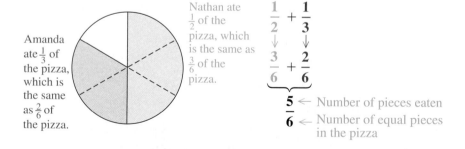

Amanda ate $\frac{1}{3}$ of the pizza, which is the same as $\frac{2}{6}$ of the pizza.

Nathan ate $\frac{1}{2}$ of the pizza, which is the same as $\frac{3}{6}$ of the pizza.

$$\frac{1}{2} + \frac{1}{3}$$
$$\downarrow \qquad \downarrow$$
$$\frac{3}{6} + \frac{2}{6}$$

$\frac{5}{6}$ ← Number of pieces eaten

$\frac{5}{6}$ ← Number of equal pieces in the pizza

Adding and Subtracting Like Fractions

You can add or subtract fractions **only** when they have a common denominator. If a, b, and c are numbers (and b is not 0), then

$$\frac{a}{b} + \frac{c}{b} = \frac{a+c}{b} \quad \text{and} \quad \frac{a}{b} - \frac{c}{b} = \frac{a-c}{b}$$

In other words, add or subtract the numerators and write the result over the common denominator. Then check to be sure that the answer is in lowest terms.

1 Write each sum or difference in lowest terms.

(a) $\dfrac{1}{6} + \dfrac{5}{6}$

$$\dfrac{1}{6} + \dfrac{5}{6} = \dfrac{\boxed{} + \boxed{}}{6} = \dfrac{\boxed{}}{\boxed{}} =$$

Common denominator

(b) $-\dfrac{11}{12} + \dfrac{5}{12}$

$$-\dfrac{11}{12} + \dfrac{5}{12} = \dfrac{-11 + \boxed{}}{12} =$$

$$\dfrac{\boxed{}}{12} =$$

Write your answer in lowest terms.

(c) $-\dfrac{2}{9} - \dfrac{3}{9}$

(d) $\dfrac{8}{ab} + \dfrac{3}{ab}$

Answers

1. **(a)** $\dfrac{1+5}{6} = \dfrac{6}{6} = 1$

(b) $\dfrac{-11+5}{12} = \dfrac{-6}{12} = -\dfrac{1}{2}$

(c) $-\dfrac{5}{9}$ **(d)** $\dfrac{11}{ab}$

EXAMPLE 1 Adding and Subtracting Like Fractions

Find each sum or difference.

(a) $\dfrac{1}{8} + \dfrac{3}{8}$

These are *like* fractions because they have a *common denominator of 8.* So they are ready to be added. Add the numerators and write the sum over the common denominator.

$$\dfrac{1}{8} + \dfrac{3}{8} = \dfrac{1+3}{8} = \dfrac{4}{8} \qquad \text{Now write } \tfrac{4}{8} \text{ in} \qquad \dfrac{4}{8} = \dfrac{\overset{1}{2} \cdot \overset{1}{2}}{2 \cdot 2 \cdot 2} = \dfrac{1}{2}$$

Common denominator   lowest terms.

The sum, in lowest terms, is $\tfrac{1}{2}$

> **⚠ CAUTION**
>
> Add *only* the numerators. ***Do not add the denominators.*** In **Example 1a** above we ***kept the common denominator.***
>
> $$\dfrac{1}{8} + \dfrac{3}{8} = \dfrac{1+3}{8} \qquad \text{not} \qquad \dfrac{1}{8} + \dfrac{3}{8} = \dfrac{1+3}{8+8} \; \cancel{= \dfrac{4}{16}} \quad \text{Incorrect}$$
>
> To help you understand why we add *only* the numerators, think of $\tfrac{1}{8}$ as $1\left(\tfrac{1}{8}\right)$ and $\tfrac{3}{8}$ as $3\left(\tfrac{1}{8}\right)$. Then we use the distributive property.
>
> $$\dfrac{1}{8} + \dfrac{3}{8} = 1\left(\dfrac{1}{8}\right) + 3\left(\dfrac{1}{8}\right) = (1+3)\left(\dfrac{1}{8}\right) = 4\left(\dfrac{1}{8}\right) = \dfrac{4}{8} = \dfrac{1}{2}$$
>
> Use the distributive property.
>
> Or think about a pie cut into 8 equal pieces. If you eat 1 piece and your friend eats 3 pieces, together you've eaten 4 of the 8 pieces or $\tfrac{4}{8}$ of the pie (***not*** $\tfrac{4}{16}$ of the pie, which would be 4 out of 16 pieces).

(b) $-\dfrac{3}{5} + \dfrac{4}{5} = \dfrac{-3+4}{5} = \dfrac{1}{5}$ ← Lowest terms

Common denominator

Rewrite subtraction as adding the opposite.

(c) $\dfrac{3}{10} - \dfrac{7}{10} = \dfrac{3-7}{10} = \dfrac{3+(-7)}{10} = \dfrac{-4}{10} \quad \text{or} \quad -\dfrac{4}{10}$

Common denominator

Now write $-\tfrac{4}{10}$ $-\dfrac{4}{10} = -\dfrac{\overset{1}{2} \cdot 2}{2 \cdot 5} = -\dfrac{2}{5}$ *Always write fraction answers in lowest terms.*

in lowest terms.

(d) $\dfrac{5}{x^2} - \dfrac{2}{x^2} = \dfrac{5-2}{x^2} = \dfrac{3}{x^2}$

Common denominator

◀ **Work Problem 1** at the Side.

OBJECTIVE ❷ Find the lowest common denominator for unlike fractions. When we first tried to add the pizza eaten by Nathan and Amanda, we could *not* do so because the pizza was *not cut into pieces of the same size*. So we rewrote $\frac{1}{2}$ and $\frac{1}{3}$ as equivalent fractions that both had 6 as the common denominator.

$$\frac{1}{2} = \frac{1 \cdot 3}{2 \cdot 3} = \frac{3}{6} \overset{\longleftarrow}{\underset{\longleftarrow}{}} \quad \begin{array}{l} \text{Common} \\ \text{denominator of 6} \end{array}$$

$$\frac{1}{3} = \frac{1 \cdot 2}{3 \cdot 2} = \frac{2}{6}$$

Then we could add the fractions, because the pizza was cut into pieces of *equal* size.

$$\frac{1}{2} + \frac{1}{3} = \frac{3}{6} + \frac{2}{6} = \frac{3 + 2}{6} = \frac{5}{6} \leftarrow \text{Lowest terms}$$

In other words, when you want to add or subtract unlike fractions, the first thing you must do is rewrite them so that they have a common denominator.

> **A Common Denominator for Unlike Fractions**
>
> To find a common denominator for two unlike fractions, find a number that is divisible by *both* of the original denominators.
>   For example, a common denominator for $\frac{1}{2}$ and $\frac{1}{3}$ is 6 because 2 goes into 6 evenly and 3 goes into 6 evenly.

Notice that 12 is also a common denominator for $\frac{1}{2}$ and $\frac{1}{3}$ because 2 and 3 both go into 12 evenly.

$$\frac{1}{2} = \frac{1 \cdot 6}{2 \cdot 6} = \frac{6}{12} \overset{\longleftarrow}{\underset{\longleftarrow}{}} \quad \begin{array}{l} \text{Common} \\ \text{denominator of 12} \end{array}$$

$$\frac{1}{3} = \frac{1 \cdot 4}{3 \cdot 4} = \frac{4}{12}$$

Now that the fractions have a common denominator, we can add them.

$$\frac{1}{2} + \frac{1}{3} = \frac{6}{12} + \frac{4}{12} = \frac{6 + 4}{12} = \frac{10}{12} \left\{ \begin{array}{l} \text{Not in} \\ \text{lowest} \\ \text{terms} \end{array} \right. \text{but} \quad \frac{10}{12} = \frac{\overset{1}{\cancel{2}} \cdot 5}{\underset{1}{\cancel{2}} \cdot 6} = \frac{5}{6} \left\{ \begin{array}{l} \text{Same} \\ \text{result as} \\ \text{above} \end{array} \right.$$

Both 6 and 12 worked as common denominators for adding $\frac{1}{2}$ and $\frac{1}{3}$, but using the smaller number saved some work. You should always try to find the smallest common denominator. If you don't for some reason, you can still work the problem—but it may take you longer. You'll have to divide out some common factors at the end in order to write the answer in lowest terms.

> **Least Common Denominator (LCD)**
>
> The **least common denominator** (LCD) for two fractions is the *smallest* positive number divisible by both denominators of the original fractions. For example, both 6 and 12 are common denominators for $\frac{1}{2}$ and $\frac{1}{3}$, but 6 is smaller, so it is the LCD.

2 Find the LCD for each pair of fractions by inspection.

(a) $\frac{3}{5}$ and $\frac{3}{10}$

10 is the larger denominator.

Is 10 divisible by 5? _____

So the LCD is _____

(b) $\frac{1}{2}$ and $\frac{2}{5}$ LCD is _____

(c) $\frac{3}{4}$ and $\frac{1}{6}$ LCD is _____

(d) $\frac{5}{6}$ and $\frac{7}{18}$ LCD is _____

3 Use prime factorization to find the LCD for each pair of fractions.

(a) $\frac{1}{10}$ and $\frac{13}{14}$ $10 = 2 \cdot 5$

                      $14 = 2 \cdot 7$

LCD = _____ • _____ • _____

LCD = _____

(b) $\frac{5}{12}$ and $\frac{17}{20}$

(c) $\frac{7}{15}$ and $\frac{7}{9}$

Answers

2. **(a)** Yes; LCD is 10   **(b)** 10   **(c)** 12   **(d)** 18

3. **(a)** LCD = $2 \cdot 5 \cdot 7 = 70$

(b) $12 = 2 \cdot 2 \cdot 3$ LCD = $2 \cdot 2 \cdot 3 \cdot 5 = 60$
     $20 = 2 \cdot 2 \cdot 5$

(c) $15 = 3 \cdot 5$ LCD = $3 \cdot 3 \cdot 5 = 45$
     $9 = 3 \cdot 3$

There are several ways to find the LCD. When the original denominators are small numbers, you can often find the LCD by inspection. **Hint: Always check to see if the larger denominator will work as the LCD.**

EXAMPLE 2   Finding the LCD by Inspection

(a) Find the LCD for $\frac{2}{3}$ and $\frac{1}{9}$ by inspection.

Check to see if 9 (the larger denominator) will work as the LCD. Is 9 divisible by 3 (the other denominator)? Yes, so **9** is the LCD for $\frac{2}{3}$ and $\frac{1}{9}$.

(b) Find the LCD for $\frac{5}{8}$ and $\frac{5}{6}$ by inspection.

Check to see if 8 (the larger denominator) will work. No, 8 is not divisible by 6. So start checking numbers that are multiples of 8, that is, 16, 24, and 32. Notice that 24 will work because it is divisible by 8 and by 6. The LCD for $\frac{5}{8}$ and $\frac{5}{6}$ is **24**.

◀ **Work Problem 2 at the Side.**

For large denominators, you can use prime factorization to find the LCD. Factor each denominator completely into prime numbers. Then use the factors to build the LCD.

EXAMPLE 3   Using Prime Factors to Find the LCD

(a) What is the LCD for $\frac{7}{12}$ and $\frac{13}{18}$?

Write 12 and 18 as the product of prime factors. Then use just enough prime factors in the LCD so you "cover" both 12 and 18.

$$12 = 2 \cdot 2 \cdot 3$$

Factors of 12

$$LCD = 2 \cdot 2 \cdot 3 \cdot 3 = 36$$

$$18 = 2 \cdot 3 \cdot 3$$

Factors of 18

Check whether 36 is divisible by 12 (yes) and by 18 (yes). The LCD for $\frac{7}{12}$ and $\frac{13}{18}$ is **36**.

> **! CAUTION**
> When finding the LCD, notice that we did *not* have to repeat the factors that 12 and 18 have in common. If we had used *all* the 2s and 3s, we would get a common denominator, but not the *smallest* one.

(b) What is the LCD for $\frac{11}{15}$ and $\frac{9}{70}$?

Factors of 15

$$15 = 3 \cdot 5$$

$$LCD = 3 \cdot 5 \cdot 2 \cdot 7 = 210$$

$$70 = 2 \cdot 5 \cdot 7$$

Factors of 70

Check whether 210 is divisible by 15 (yes) and divisible by 70 (yes). The LCD for $\frac{11}{15}$ and $\frac{9}{70}$ is **210**.

◀ **Work Problem 3 at the Side.**

OBJECTIVE ▶ ❸ **Add and subtract unlike fractions.** Here are the steps for adding or subtracting unlike fractions. The key idea is that you must rewrite the fractions so that they have a common denominator before you can add or subtract them.

Adding and Subtracting Unlike Fractions

Step 1 Find the LCD, the smallest number divisible by both denominators in the problem.

Step 2 Rewrite each original fraction as an equivalent fraction whose denominator is the LCD.

Step 3 Add or subtract the numerators of the like fractions. Keep the common denominator.

Step 4 Write the sum or difference in lowest terms.

EXAMPLE 4   **Adding and Subtracting Unlike Fractions**

Find each sum or difference.

(a) $\dfrac{1}{5} + \dfrac{3}{10}$ First check to see if the larger denominator is the LCD.

Step 1 The larger denominator (10) is the LCD.

Step 2 $\dfrac{1}{5} = \dfrac{1 \cdot 2}{5 \cdot 2} = \dfrac{2}{10} \leftarrow \text{LCD}$ and $\dfrac{3}{10}$ already has the LCD.

Step 3 Add the numerators. Write the sum over the common denominator.

$$\dfrac{1}{5} + \dfrac{3}{10} = \dfrac{2}{10} + \dfrac{3}{10} = \dfrac{2+3}{10} = \dfrac{5}{10}$$

Step 4 Write $\frac{5}{10}$ in lowest terms.

$$\dfrac{5}{10} = \dfrac{\overset{1}{5}}{2 \cdot \underset{1}{5}} = \dfrac{1}{2} \leftarrow \text{Lowest terms}$$

(b) $\dfrac{3}{4} - \dfrac{5}{6}$

Step 1 The LCD is 12.

Step 2 $\dfrac{3}{4} = \dfrac{3 \cdot 3}{4 \cdot 3} = \dfrac{9}{12} \leftarrow \text{LCD}$ and $\dfrac{5}{6} = \dfrac{5 \cdot 2}{6 \cdot 2} = \dfrac{10}{12} \leftarrow \text{LCD}$

Step 3 Subtract the numerators. Write the difference over the common denominator.

$9 + (-10)$ is -1

$$\dfrac{3}{4} - \dfrac{5}{6} = \dfrac{9}{12} - \dfrac{10}{12} = \dfrac{9-10}{12} = \dfrac{-1}{12} \quad \text{or} \quad -\dfrac{1}{12}$$

Step 4 $-\frac{1}{12}$ is in lowest terms.

Continued on Next Page

4 Find each sum or difference. Write all answers in lowest terms.

(a) $\dfrac{2}{3} + \dfrac{1}{6}$ 6 is the larger denominator. Is 6 divisible by 3? ____ The LCD is ____

$$\dfrac{2}{3} = \dfrac{2 \cdot \square}{3 \cdot \square} = \dfrac{\square}{6}$$

$\frac{1}{6}$ already has the LCD.

$$\dfrac{\square}{6} + \dfrac{1}{6} = \dfrac{\square + \square}{6} =$$

(b) $\dfrac{1}{12} - \dfrac{5}{6}$

(c) $3 - \dfrac{4}{5}$

(d) $-\dfrac{5}{12} + \dfrac{9}{16}$

Answers

4. **(a)** Yes; LCD is 6

$$\dfrac{2 \cdot 2}{3 \cdot 2} = \dfrac{4}{6}; \dfrac{4}{6} + \dfrac{1}{6} = \dfrac{4+1}{6} = \dfrac{5}{6}$$

(b) $-\dfrac{3}{4}$ **(c)** $\dfrac{11}{5}$ **(d)** $\dfrac{7}{48}$

(c) $-\dfrac{5}{12} + \dfrac{5}{9}$

Step 1 Use prime factorization to find the LCD.

Factors of 12

$$12 = 2 \cdot 2 \cdot 3$$
$$9 = 3 \cdot 3$$
$$LCD = 2 \cdot 2 \cdot 3 \cdot 3 = 36$$

Factors of 9

Step 2 $-\dfrac{5}{12} = -\dfrac{5 \cdot 3}{12 \cdot 3} = -\dfrac{15}{36}$ and $\dfrac{5}{9} = \dfrac{5 \cdot 4}{9 \cdot 4} = \dfrac{20}{36}$

Step 3 Add the numerators. Keep the common denominator.

$$-\dfrac{5}{12} + \dfrac{5}{9} = -\dfrac{15}{36} + \dfrac{20}{36} = \dfrac{-15 + 20}{36} = \dfrac{5}{36}$$

Step 4 $\frac{5}{36}$ is in lowest terms.

(d) $4 - \dfrac{2}{3}$

Step 1 Think of 4 as $\frac{4}{1}$. The LCD for $\frac{4}{1}$ and $\frac{2}{3}$ is 3, the larger denominator.

Step 2 $\dfrac{4}{1} = \dfrac{4 \cdot 3}{1 \cdot 3} = \dfrac{12}{3}$ and $\dfrac{2}{3}$ already has the LCD.

Step 3 Subtract the numerators. Keep the common denominator.

$$\dfrac{4}{1} - \dfrac{2}{3} = \dfrac{12}{3} - \dfrac{2}{3} = \dfrac{12 - 2}{3} = \dfrac{10}{3}$$

Step 4 $\frac{10}{3}$ is in lowest terms.

◀ **Work Problem 4 at the Side.**

OBJECTIVE 4 Add and subtract unlike fractions that contain variables. We use the same steps to add or subtract unlike fractions with variables in the numerators or denominators.

EXAMPLE 5 Adding and Subtracting Unlike Fractions with Variables

Find each sum or difference.

(a) $\dfrac{1}{4} + \dfrac{b}{5}$

Step 1 The LCD is 20.

Step 2 $\dfrac{1}{4} = \dfrac{1 \cdot 5}{4 \cdot 5} = \dfrac{5}{20}$ and $\dfrac{b}{5} = \dfrac{b \cdot 4}{5 \cdot 4} = \dfrac{4b}{20}$

Continued on Next Page

Step 3 $\dfrac{1}{4}+\dfrac{b}{5}=\dfrac{5}{20}+\dfrac{4b}{20}=\dfrac{5+4b}{20}$ ← Add the numerators.
← Keep the common denominator.

Step 4 $\dfrac{5+4b}{20}$ is in lowest terms.

> **! CAUTION**
>
> In *Step 4* above, we could **not** add $5+4b$ in the numerator of the answer because 5 and $4b$ are **not** like terms. We *could* add $5b+4b$ but **not** $5+4b$.
>
> Variable parts match.

(b) $\dfrac{2}{3}-\dfrac{6}{x}$

Step 1 The LCD is $3 \cdot x$, or $3x$

Step 2 $\dfrac{2}{3}=\dfrac{2 \cdot x}{3 \cdot x}=\dfrac{2x}{3x}$ and $\dfrac{6}{x}=\dfrac{6 \cdot 3}{x \cdot 3}=\dfrac{18}{3x}$

Step 3 $\dfrac{2}{3}-\dfrac{6}{x}=\dfrac{2x}{3x}-\dfrac{18}{3x}=\dfrac{2x-18}{3x}$ ← Subtract the numerators.
← Keep the common denominator.

Step 4 $\dfrac{2x-18}{3x}$ is in lowest terms.

> **Note**
>
> Notice in **Example 5(b)** above that we found the LCD for $\frac{2}{3}-\frac{6}{x}$ by multiplying the two denominators. The LCD is $3 \cdot x$ or $3x$.
>   Multiplying the two denominators will *always* give you a common denominator, but it may not be the *least* common denominator. Here are more examples.
>
> $\dfrac{1}{3}-\dfrac{2}{5}$ If you multiply the denominators, $3 \cdot 5 = 15$, and 15 is the LCD.
>
> $\dfrac{5}{6}+\dfrac{3}{4}$ If you multiply the denominators, $6 \cdot 4 = 24$, and 24 will work. But you'll save some time by using the least common denominator, which is 12.
>
> $\dfrac{7}{y}+\dfrac{a}{4}$ If you multiply the denominators, $y \cdot 4 = 4y$, and $4y$ is the LCD.

——— **Work Problem 5 at the Side.** ▶

---

5 Find each sum or difference.

GS (a) $\dfrac{5}{6}-\dfrac{h}{2}$ The LCD is 6

$\dfrac{5}{6}$ already has the LCD.

$\dfrac{h}{2}=\dfrac{h \cdot \boxed{}}{2 \cdot \boxed{}}=\dfrac{\boxed{}}{6}$

$\dfrac{5}{6}-\dfrac{\boxed{}}{6}=$

Is the answer in lowest terms? _____

(b) $\dfrac{7}{t}+\dfrac{3}{5}$

(c) $\dfrac{4}{x}-\dfrac{8}{3}$

Answers

5. **(a)** $\dfrac{h \cdot 3}{2 \cdot 3}=\dfrac{3h}{6};\dfrac{5}{6}-\dfrac{3h}{6}=\dfrac{5-3h}{6}$; Yes

(b) $\dfrac{35+3t}{5t}$ **(c)** $\dfrac{12-8x}{3x}$

4.4 Exercises

FOR EXTRA HELP   Go to MyMathLab for worked-out, step-by-step solutions to exercises enclosed in a square [   ] and video solutions to ▶ exercises.

CONCEPT CHECK *Circle the correct word or phrase to complete each statement.*

1. **(a)** When two fractions have the same denominator, they are called $\Big\langle$ fractions in lowest terms. / like fractions.

   (b) Two fractions need to have a common denominator before you can $\Big\langle$ multiply them. / add them.

2. **(a)** If two fractions are unlike fractions, they have $\Big\langle$ different denominators. / different numerators.

   (b) When adding two fractions with a common denominator $\Big\langle$ add only the denominators. / add only the numerators.

Find each sum or difference. Write all answers in lowest terms. **See Examples 1–5.**

3. $\dfrac{3}{4} + \dfrac{1}{8}$ $\begin{cases} \text{The LCD is 8.} \\ \dfrac{3}{4} = \dfrac{3 \cdot 2}{4 \cdot 2} = \dfrac{6}{8} \end{cases}$

$\dfrac{6}{8} + \dfrac{1}{8} =$

4. $\dfrac{1}{3} + \dfrac{1}{2}$ $\begin{cases} \text{The LCD is 6.} \\ \dfrac{1}{3} = \dfrac{1 \cdot 2}{3 \cdot 2} = \dfrac{2}{6} \\ \dfrac{1}{2} = \dfrac{1 \cdot 3}{2 \cdot 3} = \dfrac{3}{6} \end{cases}$

$\dfrac{2}{6} + \dfrac{3}{6} =$

5. $-\dfrac{1}{14} + \left(-\dfrac{3}{7}\right)$

6. $-\dfrac{2}{9} + \dfrac{2}{3}$

7. $\dfrac{2}{3} - \dfrac{1}{6}$

8. $\dfrac{5}{12} - \dfrac{1}{4}$

9. $\dfrac{3}{8} - \dfrac{3}{5}$

10. $\dfrac{1}{3} - \dfrac{3}{5}$

11. $-\dfrac{5}{8} + \dfrac{1}{12}$

12. $-\dfrac{13}{16} + \dfrac{13}{16}$

13. $-\dfrac{7}{20} - \dfrac{5}{20}$

14. $-\dfrac{7}{9} - \dfrac{5}{6}$

15. $0 - \dfrac{7}{18}$

16. $-\dfrac{7}{8} + 3$

17. $2 - \dfrac{6}{7}$

18. $5 - \dfrac{2}{5}$

19. $-\dfrac{1}{2} + \dfrac{3}{24}$

20. $\dfrac{7}{10} + \dfrac{7}{15}$

21. $\dfrac{1}{5} + \dfrac{c}{3}$

22. $\dfrac{x}{4} + \dfrac{2}{3}$

23. $\dfrac{5}{m} - \dfrac{1}{2}$

Hint: The LCD is $2m$

24. $\dfrac{2}{9} - \dfrac{4}{y}$

Hint: The LCD is $9y$

25. $\dfrac{3}{b^2} + \dfrac{5}{b^2}$

26. $\dfrac{10}{xy} - \dfrac{7}{xy}$

27. $\dfrac{c}{7} + \dfrac{3}{b}$

28. $\dfrac{2}{x} - \dfrac{y}{5}$

29. $-\dfrac{4}{c^2} - \dfrac{d}{c}$

30. $-\dfrac{1}{n} + \dfrac{m}{n^2}$

31. $-\dfrac{11}{42} - \dfrac{11}{70}$

32. $\dfrac{7}{45} - \dfrac{7}{20}$

33. $\dfrac{b}{3} + \dfrac{4}{a^2}$

34. $\dfrac{16}{y^3} - \dfrac{x^2}{y^3}$

35. $-\dfrac{w}{10} + \dfrac{5}{w^2}$

36. $-\dfrac{r^2}{t^2} - \dfrac{8}{t^3}$

37. $\dfrac{m^4}{n^2} - 0$

38. $\dfrac{6c + 1}{h^3} + 0$

39. CONCEPT CHECK The work shown below has **two** mistakes in it.

What Went Wrong? First write a sentence explaining what each mistake is. Then fix the mistakes and find the correct solutions.

Katelyn solved the following test questions.

She did the first question this way:

$$\frac{3}{4} + \frac{2}{5} = \frac{3+2}{4+5} = \frac{5}{9}$$

She did the second question this way:

$$\frac{5}{6} - \frac{4}{9} = \frac{5}{18} - \frac{4}{18} = \frac{5-4}{18} = \frac{1}{18}$$

40. CONCEPT CHECK The work shown below has **two** mistakes in it.

What Went Wrong? First write a sentence explaining what each mistake is. Then fix the mistakes and find the correct solutions.

Sean solved the following test questions.

Here is his work on the first question:

$$-\frac{1}{4} + \frac{7}{12} = -\frac{3}{12} + \frac{7}{12} = \frac{-3+7}{12} = \frac{4}{12}$$

Here is his work on the second question:

$$\frac{3}{10} - \frac{1}{4} = \frac{3-1}{10-4} = \frac{2}{6} = \frac{1}{3}$$

Solve each application problem. Write all answers in lowest terms.

41. When assembling shelving units, Pam Phelps must use the proper type and size of hardware. Find the total length of the bolt shown.

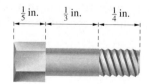

$\frac{1}{5}$ in. $\frac{1}{3}$ in. $\frac{1}{4}$ in.

42. How much fencing will be needed to enclose this rectangular wildflower preserve?

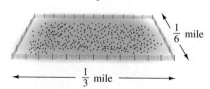

$\frac{1}{6}$ mile

$\frac{1}{3}$ mile

43. LaSonya needs $\frac{1}{8}$ cup of milk for a muffin recipe and $\frac{1}{2}$ cup of milk to make pudding. She also needs $\frac{1}{3}$ cup of milk to make instant mashed potatoes. How much milk does she need in all?

44. A flower grower purchased $\frac{9}{10}$ acre of land one year and $\frac{3}{10}$ acre the next year. She then sold $\frac{7}{10}$ acre of land. How much land does she now have?

45. Samar works out every morning. She jogs $\frac{3}{4}$ mile from her house to the park. Then she walks $\frac{2}{5}$ mile through the park to the library. Then she jogs $\frac{1}{2}$ mile back to her house.

(a) What is the total distance that Samar travels while working out?

(b) How much farther does Samar jog than walk?

46. Swiss cheese is full of holes. In the United States today, the holes must measure from $\frac{3}{8}$ inch to $\frac{13}{16}$ inch across the center. Before 2001, the holes had to measure between $\frac{11}{16}$ inch and $\frac{13}{16}$ inch across the center.

(a) Before 2001, what was the difference in the measurements of the smaller and larger holes?

(b) Find the size of the smaller holes before 2001 and the size of the smaller holes today. How much less do they measure today?

The circle graph shows the fraction of U.S. electricity generated by each source. Use the graph to answer Exercises 47–50.

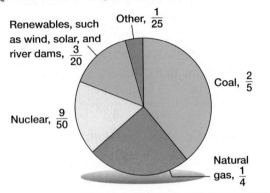

U.S. Electricity Production

Renewables, such as wind, solar, and river dams, $\frac{3}{20}$

Other, $\frac{1}{25}$

Coal, $\frac{2}{5}$

Nuclear, $\frac{9}{50}$

Natural gas, $\frac{1}{4}$

Data from www.eia.gov

47. What fraction of U.S. electricity is generated by coal plants and nuclear plants?

48. What fraction of U.S. electricity is generated by natural gas and "other" sources?

49. What is the difference between the fraction of U.S. electricity generated by coal plants and by renewables?

50. What is the difference between the fraction of U.S. electricity generated by coal plants and by natural gas?

Use the photo to answer Exercises 51–52. A nut driver is like a screwdriver but is used to tighten nuts instead of screws. The ends of the drivers are sized from $\frac{3}{16}$ inch to $\frac{1}{2}$ inch to fit smaller or larger nuts. (The symbol " is for inches.)

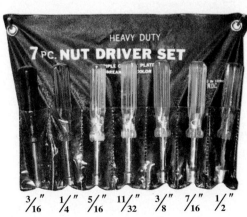

HEAVY DUTY
7 PC. NUT DRIVER SET

$\frac{3}{16}"$ $\frac{1}{4}"$ $\frac{5}{16}"$ $\frac{11}{32}"$ $\frac{3}{8}"$ $\frac{7}{16}"$ $\frac{1}{2}"$

Data from author's tool collection.

51. The rightmost driver fits a nut that is how much larger than the nut for the leftmost driver?

52. The nut size for the yellow-handled driver is how much less than the nut size for the blue-handled driver?

53. A hazardous waste dump site needs $\frac{7}{8}$ mile of security fencing. The site has four sides with three of the sides measuring $\frac{1}{4}$ mile, $\frac{1}{6}$ mile, and $\frac{3}{8}$ mile. Find the length of the fourth side.

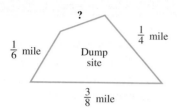

54. Chakotay is fitting a turquoise stone into a bear claw pendant. Find the length of the red dashed line.

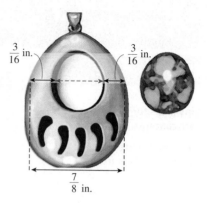

Relating Concepts (Exercises 55–56) For Individual or Group Work

As you **work Exercises 55 and 56 in order,** *think about the properties you learned when working with integers. Then explain the property that each pair of problems illustrates.*

55. (a) $-\dfrac{2}{3} + \dfrac{3}{4} =$ _____ $\dfrac{3}{4} + \left(-\dfrac{2}{3}\right) =$ _____

56. (a) $-\dfrac{7}{12} + \dfrac{7}{12} =$ _____ $\dfrac{3}{5} + \left(-\dfrac{3}{5}\right) =$ _____

(b) $\dfrac{5}{6} - \dfrac{1}{2} =$ _____ $\dfrac{1}{2} - \dfrac{5}{6} =$ _____

(b) $-\dfrac{13}{16} \div \left(-\dfrac{13}{16}\right) =$ _____ $\dfrac{1}{8} \div \dfrac{1}{8} =$ _____

(c) $\left(-\dfrac{2}{3}\right)\left(\dfrac{9}{10}\right) =$ _____ $\left(\dfrac{9}{10}\right)\left(-\dfrac{2}{3}\right) =$ _____

(c) $\dfrac{5}{6} \cdot 1 =$ _____ $1\left(-\dfrac{17}{20}\right) =$ _____

(d) $\dfrac{2}{5} \div \dfrac{1}{15} =$ _____ $\dfrac{1}{15} \div \dfrac{2}{5} =$ _____

(d) $\left(-\dfrac{4}{5}\right)\left(-\dfrac{5}{4}\right) =$ _____ $7 \cdot \dfrac{1}{7} =$ _____

Study Skills
MAKING A MIND MAP

OBJECTIVES

1 Create mind maps for appropriate concepts.

2 Visually show how concepts are related to each other using arrows or lines.

Why Is Mapping Brain Friendly?

Remember that your brain grows dendrites when you are **actively thinking** about and working with information. Making a map requires you to think hard about **how to place the information, how to show connections** between parts of the map, and **how color will be useful.** It also takes a lot of thinking to fill in all related details and **show how those details connect to the larger concept.** All that thinking will let your brain grow a complex, many-branched neural network of interconnected dendrites. It is time well spent.

Mind mapping is a visual way to show information that you have learned. It is an excellent way to review. Mapping is flexible and can be personalized, which is helpful for your memory. Your brain likes to see things that are **pleasing** to look at, are **colorful,** and **show connections** between ideas. Take advantage of this by creating maps that

► are easy to read,

► use color in a systematic way, and

► clearly show you how different concepts are related (using arrows or dotted lines, for example).

Directions for Making a Mind Map

Below are some general directions for making a map. After you read them, go to the next page and work on completing the map that has been started for you. It is from the two previous sections in this chapter.

► To begin a mind map, write the concept in the center of a piece of paper and either circle it or draw a box around it.

► Make a line out from the center concept, and draw a box large enough to write the definition of the concept.

► Think of the other aspects (subpoints) of the concept that you have learned, such as procedures to follow or formulas. Make a separate line and box connecting each subpoint to the center.

► From each of the new boxes, add the information you've learned. You can continue making new lines and boxes and circles, or you can list items below the new information.

► Use color to highlight the major points. For example, everything related to one subpoint might be the same color. That way you can easily see related ideas.

► You may also use arrows, underlining, or small drawings to help yourself remember.

Try This Fractions Mind Map

On a separate paper, make a map that summarizes computation with fractions. Follow the directions below. Use the starter map.

▶ The longest rectangles are instructions for all four operations. (The first one starts "Rewrite all numbers as fractions . . . " and the second one is at the bottom of the map.)

▶ Notice the wavy dividing lines that separate the map into two sides.

▶ Your job is to complete the map by writing the steps used in multiplying fractions and the steps used in adding and subtracting fractions.

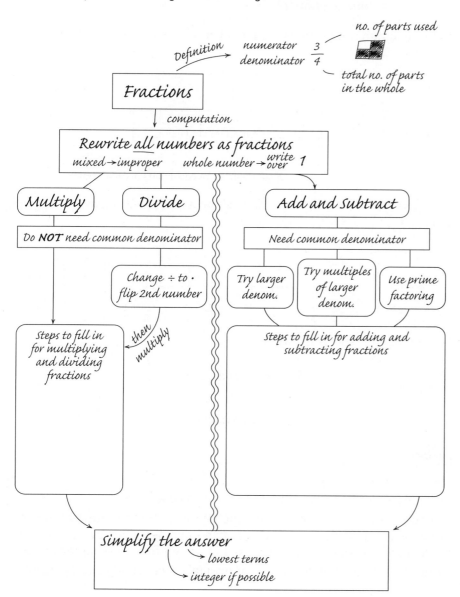

4.5 Problem Solving: Mixed Numbers and Estimating

OBJECTIVES

1. Identify mixed numbers and graph them on a number line.

2. Rewrite mixed numbers as improper fractions, or the reverse.

3. Estimate the answer and multiply or divide mixed numbers.

4. Estimate the answer and add or subtract mixed numbers.

5. Solve application problems containing mixed numbers.

OBJECTIVE ▶ 1 **Identify mixed numbers and graph them on a number line.** When a fraction and a whole number are written together, the result is a **mixed number.** For example, the mixed number

$$3\frac{1}{2} \quad \text{represents} \quad 3 + \frac{1}{2}$$

or 3 wholes and $\frac{1}{2}$ of a whole. Read $3\frac{1}{2}$ as "three and one half."

One common use of mixed numbers is to measure things. Examples are shown below.

Juan worked $5\frac{1}{2}$ hours.   The box weighs $2\frac{3}{4}$ pounds.

The park is $1\frac{7}{10}$ miles long.   Add $1\frac{2}{3}$ cups of flour.

EXAMPLE 1   **Illustrating a Mixed Number with a Diagram and a Number Line**

As this diagram shows, the mixed number $3\frac{1}{2}$ is equivalent to the improper fraction $\frac{7}{2}$.

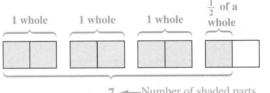

We can also use a number line to show mixed numbers, as in this graph of $3\frac{1}{2}$ and $-3\frac{1}{2}$.

The number line shows the following.

$$3\frac{1}{2} \quad \text{is equivalent to} \quad \frac{7}{2}$$

$$-3\frac{1}{2} \quad \text{is equivalent to} \quad -\frac{7}{2}$$

Note

$3\frac{1}{2}$ represents $3 + \frac{1}{2}$.

$-3\frac{1}{2}$ represents $-3 + \left(-\frac{1}{2}\right)$, which can also be written as $-3 - \frac{1}{2}$.

In algebra we usually work with the improper fraction form of mixed numbers, especially for negative mixed numbers. However, positive mixed numbers are frequently used in daily life, so it's important to know how to work with them. For example, we usually say $3\frac{1}{2}$ inches rather than $\frac{7}{2}$ inches.

---

1. (a) Use this diagram to write $1\frac{2}{3}$ as an improper fraction.

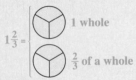

(b) Now graph $1\frac{2}{3}$ and $-1\frac{2}{3}$.

(c) Use this diagram to write $2\frac{1}{4}$ as an improper fraction.

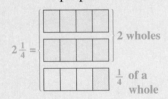

(d) Now graph $2\frac{1}{4}$ and $-2\frac{1}{4}$.

Answers

1. (a) $\frac{5}{3}$ (b) $-1\frac{2}{3}$ $1\frac{2}{3}$

(c) $\frac{9}{4}$ (d) $-2\frac{1}{4}$ $2\frac{1}{4}$

◀ **Work Problem 1 at the Side.**

OBJECTIVE ▶ 2 Rewrite mixed numbers as improper fractions, or the reverse. You can use the following steps to write $3\frac{1}{2}$ as an improper fraction without drawing a diagram or a number line.

Step 1 Multiply 2 times 3 and add 1 to the product.

$$3\frac{1}{2} \qquad 2 \cdot 3 = 6 \qquad \text{Then } 6 + 1 = 7$$

Step 2 Use 7 (from *Step 1*) as the numerator and 2 as the denominator.

$$3\frac{1}{2} = \frac{7}{2} \leftarrow (2 \cdot 3) + 1$$

Same denominator

To see why this method works, recall that $3\frac{1}{2}$ represents $3 + \frac{1}{2}$. Let's add $3 + \frac{1}{2}$.

$$3 + \frac{1}{2} = \frac{3}{1} + \frac{1}{2} = \frac{6}{2} + \frac{1}{2} = \frac{6+1}{2} = \frac{7}{2} \quad \left\{ \begin{array}{l} \text{Same result} \\ \text{as above} \end{array} \right.$$

Common denominator

In summary, use the following steps to *write a mixed number as an improper fraction*.

Writing a Mixed Number as an Improper Fraction

Step 1 ***Multiply*** the denominator of the fraction times the whole number and ***add*** the numerator of the fraction to the product.

Step 2 Write the result of *Step 1* as the ***numerator*** and keep the original ***denominator***.

EXAMPLE 2 **Writing a Mixed Number as an Improper Fraction**

Write $7\frac{2}{3}$ as an improper fraction (numerator greater than denominator).

Step 1 $7\frac{2}{3} \qquad 3 \cdot 7 = 21 \qquad \text{Then } 21 + 2 = 23$

Step 2 $7\frac{2}{3} = \frac{23}{3} \leftarrow (3 \cdot 7) + 2$

Same denominator

So, $7\frac{2}{3} = \frac{22}{3}$ ◄ Keep the same denominator.

──────── **Work Problem 2 at the Side.** ▶

2 Write each mixed number as an equivalent improper fraction.

GS **(a)** $4\frac{2}{3}$

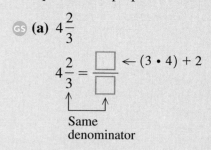

$$4\frac{2}{3} = \frac{\square}{\square} \leftarrow (3 \cdot 4) + 2$$

Same denominator

(b) $4\frac{7}{10}$

(c) $-5\frac{3}{4}$

(d) $8\frac{5}{6}$

Answers

2. **(a)** $\frac{14}{3}$ **(b)** $\frac{47}{10}$ **(c)** $-\frac{23}{4}$ **(d)** $\frac{53}{6}$

We used *multiplication* for the first step in writing a mixed number as an improper fraction. To work in *reverse*, writing an improper fraction as a mixed number, we use *division*. Recall that the fraction bar is a symbol for division.

Writing an Improper Fraction as a Mixed Number

To write an *improper fraction* as a mixed number, divide the numerator by the denominator. The quotient is the whole number part (of the mixed number), the remainder is the numerator of the fraction part, and the denominator remains the same.

Always check to be sure that the fraction part of the mixed number is in lowest terms. Then the mixed number is in *simplest form*.

EXAMPLE 3 **Writing Improper Fractions as Mixed Numbers**

Write each improper fraction as an equivalent mixed number in simplest form.

(a) $\dfrac{17}{5}$

Divide 17 by 5.

$$
\begin{array}{r}
\mathbf{3} \quad \longleftarrow \text{Whole number part} \\
5\overline{)17} \\
\underline{15} \\
\mathbf{2} \quad \longleftarrow \text{Remainder}
\end{array}
$$

The quotient **3** is the whole number part of the mixed number. The remainder **2** is the numerator of the fraction, and the denominator stays as **5**.

$$\frac{17}{5} = 3\frac{2}{5} \longleftarrow \text{Remainder}$$

$\uparrow \qquad \uparrow \qquad$ Same denominator

Let's look at a drawing of $\frac{17}{5}$ to check our work.

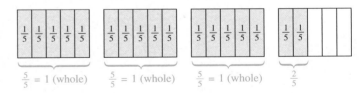

$\frac{5}{5} = 1$ (whole) $\qquad \frac{5}{5} = 1$ (whole) $\qquad \frac{5}{5} = 1$ (whole) $\qquad \frac{2}{5}$

So, $\frac{17}{5} = 3\frac{2}{5}$

Continued on Next Page

In other words,

$$\frac{17}{5} = \frac{5}{5} + \frac{5}{5} + \frac{5}{5} + \frac{2}{5}$$

$$= \underbrace{1 + 1 + 1}_{} + \frac{2}{5}$$

$$= \qquad 3 \qquad + \frac{2}{5}$$

$$= \mathbf{3\frac{2}{5}}$$

(b) $\dfrac{26}{4}$

Divide 26 by 4.

$$\begin{array}{r} 6 \\ 4\overline{)26} \\ 24 \\ \hline 2 \end{array} \quad \text{so} \quad \frac{26}{4} = 6\frac{2}{4} = 6\frac{1}{2}$$

Write $\frac{2}{4}$ in
lowest terms.

You could write $\frac{26}{4}$ in lowest terms first.

$$\frac{26}{4} = \frac{\overset{1}{\cancel{2}} \cdot 13}{\underset{1}{\cancel{2}} \cdot 2} = \frac{13}{2} \quad \text{Then} \quad \begin{array}{r} 6 \\ 2\overline{)13} \\ 12 \\ \hline 1 \end{array} \quad \text{so} \quad \frac{13}{2} = 6\frac{1}{2} \; \left\{ \begin{array}{l} \text{Same result} \\ \text{as above} \end{array} \right.$$

So, $\frac{26}{4} = \mathbf{6\frac{1}{2}}$

—————————— **Work Problem ③ at the Side.** ▶

OBJECTIVE ▶ **③ Estimate the answer and multiply or divide mixed numbers.**
Rewrite mixed numbers as improper fractions. Then use the steps you
learned earlier for multiplying and dividing. However, it's a good idea to
estimate the answer before you start any other work.

EXAMPLE 4   Rounding Mixed Numbers to the Nearest Whole Number

To estimate answers, first round each mixed number to the *nearest whole
number.* If the numerator is *half* of the denominator or *more,* round up
the whole number part. If the numerator is *less* than half the denominator,
leave the whole number as it is.

(a) Round $1\dfrac{5}{8}$ ← 5 is more than 4
            ← Half of 8 is 4 $1\dfrac{5}{8}$ rounds up to **2**

(b) Round $3\dfrac{2}{5}$ ← 2 is less than $2\frac{1}{2}$
            ← Half of 5 is $2\frac{1}{2}$ $3\dfrac{2}{5}$ rounds to **3**

—————————— **Work Problem ④ at the Side.** ▶

③ Write each improper fraction as
an equivalent mixed number in
simplest form.

GS **(a)** $\dfrac{7}{2}$ $\begin{array}{r} 3 \\ 2\overline{)7} \\ 6 \\ \hline 1 \end{array}$ ← Whole number part

        ← Remainder

$$\frac{7}{2} = 3\frac{\square}{\square} \quad \begin{array}{l} \leftarrow \text{Remainder} \\ \\ \leftarrow \text{Same denominator} \end{array}$$

(b) $\dfrac{14}{4}$

(c) $\dfrac{33}{5}$

(d) $-\dfrac{58}{10}$

④ Round each mixed number to
the nearest whole number.

GS **(a)** $7\dfrac{3}{4}$ ← 3 is more than 2
        ← Half of 4 is 2 ↗

$7\dfrac{3}{4}$ rounds up to _____

(b) $6\dfrac{3}{8}$

(c) $4\dfrac{2}{3}$

(d) $1\dfrac{7}{10}$

(e) $3\dfrac{1}{2}$

(f) $5\dfrac{4}{9}$

Answers

3. (a) $\dfrac{7}{2} = 3\dfrac{1}{2}$ (b) $3\dfrac{1}{2}$ (c) $6\dfrac{3}{5}$ (d) $-5\dfrac{4}{5}$

4. (a) 8 (b) 6 (c) 5 (d) 2 (e) 4 (f) 5

5 First, round the numbers and estimate each answer. Then find the exact answer. Write exact answers in simplest form.

GS (a) $2\dfrac{1}{4} \cdot 7\dfrac{1}{3}$

$$\dfrac{\downarrow}{2} \cdot \dfrac{\downarrow}{7} = \underline{\quad}$$

Estimate

Step 1 $2\dfrac{1}{4} = \dfrac{9}{4}$ and $7\dfrac{1}{3} = \dfrac{22}{3}$

$$\dfrac{9}{4} \cdot \dfrac{22}{3} = \dfrac{\overset{}{3} \cdot 3 \cdot \overset{}{2} \cdot 11}{\underset{1}{2} \cdot 2 \cdot \underset{1}{3}} =$$

(b) $\left(4\dfrac{1}{2}\right)\left(1\dfrac{2}{3}\right)$

$$(\underline{\quad})(\underline{\quad}) = \underline{\quad}$$

Estimate

(c) $3\dfrac{3}{5} \cdot 4\dfrac{4}{9}$

$$\underline{\quad} \cdot \underline{\quad} = \underline{\quad}$$

Estimate

(d) $\left(3\dfrac{1}{5}\right)\left(5\dfrac{3}{8}\right)$

$$(\underline{\quad})(\underline{\quad}) = \underline{\quad}$$

Estimate

Answers

5. **(a)** *Estimate:* $2 \cdot 7 = 14$; *Exact:* $\dfrac{33}{2}$ or $16\dfrac{1}{2}$

(b) *Estimate:* $(5)(2) = 10$; *Exact:* $\dfrac{15}{2}$ or $7\dfrac{1}{2}$

(c) *Estimate:* $4 \cdot 4 = 16$; *Exact:* $\dfrac{16}{1}$ simplifies to 16

(d) *Estimate:* $(3)(5) = 15$; *Exact:* $\dfrac{86}{5}$ or $17\dfrac{1}{5}$

Multiplying and Dividing Mixed Numbers

Step 1 **Rewrite** each mixed number as an improper fraction.

Step 2 **Multiply** or **divide** the improper fractions.

Step 3 Write the answer in lowest terms. Then the answer is in simplest form. If desired, change an improper fraction answer to a mixed number.

EXAMPLE 5 | **Estimating the Answer and Multiplying Mixed Numbers**

First, round the numbers and estimate each answer. Then find the exact answer. Write exact answers in simplest form.

(a) $2\dfrac{1}{2} \cdot 3\dfrac{1}{5}$

Estimate the answer by rounding the mixed numbers.

$$2\dfrac{1}{2} \text{ rounds to } 3 \quad \text{and} \quad 3\dfrac{1}{5} \text{ rounds to } 3$$

$$3 \cdot 3 = \mathbf{9} \leftarrow \text{Estimated answer}$$

To find the exact answer, first rewrite each mixed number as an improper fraction.

$$\text{Step 1} \qquad 2\dfrac{1}{2} = \dfrac{5}{2} \quad \text{and} \quad 3\dfrac{1}{5} = \dfrac{16}{5}$$

Next, multiply.

$$2\dfrac{1}{2} \cdot 3\dfrac{1}{5} = \dfrac{5}{2} \cdot \dfrac{16}{5} = \dfrac{\overset{1}{5} \cdot 2 \cdot 8}{\underset{1}{2} \cdot \underset{1}{5}} = \dfrac{8}{1} = 8$$

> **Careful!** $\dfrac{8}{1}$ can be simplified to a whole number.

The **estimate was 9,** so an **exact answer of 8 is reasonable.**

> Estimating helps you catch errors.

(b) $\left(3\dfrac{5}{8}\right)\left(4\dfrac{4}{5}\right)$

First, round each mixed number and estimate the answer.

$$3\dfrac{5}{8} \text{ rounds to } 4 \quad \text{and} \quad 4\dfrac{4}{5} \text{ rounds to } 5$$

$$4 \cdot 5 = \mathbf{20} \leftarrow \text{Estimated answer}$$

Now find the exact answer.

$$\left(3\dfrac{5}{8}\right)\left(4\dfrac{4}{5}\right) = \overset{\text{Step 1}}{\left(\dfrac{29}{8}\right)\left(\dfrac{24}{5}\right)} = \overset{\text{Step 2}}{\dfrac{29 \cdot 3 \cdot \overset{1}{8}}{\underset{1}{8} \cdot 5}} = \dfrac{87}{5} \text{ or } 17\dfrac{2}{5}$$

The **estimate was 20,** so an **exact answer of $17\dfrac{2}{5}$ is reasonable.**

◀ **Work Problem 5** at the Side.

EXAMPLE 6 Estimating the Answer and Dividing Mixed Numbers

First, round the numbers and estimate each answer. Then find the exact answer. Write exact answers in simplest form.

(a) $3\frac{3}{5} \div 1\frac{1}{2}$

To estimate the answer, round each mixed number to the nearest whole number.

$$3\frac{3}{5} \quad \div \quad 1\frac{1}{2}$$
$$\downarrow \qquad \text{Rounded} \qquad \downarrow$$
$$4 \quad \div \quad 2 \quad = \quad \mathbf{2} \leftarrow \text{Estimate}$$

To find the exact answer, first rewrite each mixed number as an improper fraction.

$$3\frac{3}{5} \div 1\frac{1}{2} = \frac{18}{5} \div \frac{3}{2}$$

> You do **not** need a common denominator when multiplying or dividing fractions.

Now rewrite the problem as multiplying by the reciprocal of $\frac{3}{2}$.

$$\frac{18}{5} \div \frac{3}{2} = \frac{18}{5} \cdot \frac{2}{3} = \frac{\overset{1}{3} \cdot 6 \cdot 2}{5 \cdot \underset{1}{3}} = \frac{\mathbf{12}}{\mathbf{5}} \text{ or } \mathbf{2\frac{2}{5}}$$

Reciprocals

The **estimate was 2,** so an **exact answer of $2\frac{2}{5}$ is reasonable.**

(b) $4\frac{3}{8} \div 5$

First, round the numbers and estimate the answer.

$$4\frac{3}{8} \quad \div \quad 5$$
$$\downarrow \qquad \text{Rounded} \qquad \downarrow$$
$$4 \quad \div \quad 5 \qquad \text{Write } 4 \div 5 \text{ using a fraction bar.} \Big\} \frac{\mathbf{4}}{\mathbf{5}} \leftarrow \text{Estimate}$$

Now find the exact answer.

Write 5 as $\frac{5}{1}$

$$4\frac{3}{8} \div 5 = \frac{35}{8} \div \frac{5}{1} = \frac{35}{8} \cdot \frac{1}{5} = \frac{\overset{1}{5} \cdot 7 \cdot 1}{8 \cdot \underset{1}{5}} = \frac{\mathbf{7}}{\mathbf{8}} \quad \Big\{ \begin{array}{l} \text{Simplest} \\ \text{form} \end{array}$$

Reciprocals

The **estimate was $\frac{4}{5}$,** so an **exact answer of $\frac{7}{8}$ is reasonable.** They are both less than 1.

—— Work Problem **6** at the Side. ▶

---

6 First, round the numbers and estimate each answer. Then find the exact answer. Write exact answers in simplest form.

GS (a) $6\frac{1}{4} \div 3\frac{1}{3}$
$$\downarrow \qquad\qquad \downarrow$$
$$\underline{}6 \div \underline{}3 = \underline{}$$
$$\qquad\qquad\qquad\qquad \textit{Estimate}$$

$$6\frac{1}{4} = \frac{\square}{4} \quad \text{and} \quad 3\frac{1}{3} = \frac{\square}{3}$$

$$\frac{25}{4} \div \frac{10}{3} = \frac{25}{4} \cdot \frac{3}{10} = \frac{\overset{1}{5} \cdot 5 \cdot 3}{4 \cdot 2 \cdot \underset{1}{5}} =$$

Reciprocals

(b) $3\frac{3}{8} \div 2\frac{4}{7}$
$$\downarrow \qquad\qquad \downarrow$$
$$\underline{} \div \underline{} = \underline{}$$
$$\qquad\qquad\qquad \textit{Estimate}$$

(c) $8 \div 5\frac{1}{3}$
$$\downarrow \qquad\qquad \downarrow$$
$$\underline{} \div \underline{} = \underline{}$$
$$\qquad\qquad\qquad \textit{Estimate}$$

(d) $4\frac{1}{2} \div 6$
$$\downarrow \qquad\qquad \downarrow$$
$$\underline{} \div \underline{} = \underline{}$$
$$\qquad\qquad\qquad \textit{Estimate}$$

Answers

6. (a) *Estimate:* $6 \div 3 = 2$
Exact: $6\frac{1}{4} = \frac{25}{4}; 3\frac{1}{3} = \frac{10}{3}; \frac{15}{8} = 1\frac{7}{8}$

(b) *Estimate:* $3 \div 3 = 1$; *Exact:* $\frac{21}{16}$ or $1\frac{5}{16}$

(c) *Estimate:* $8 \div 5 = 1\frac{3}{5}$; *Exact:* $\frac{3}{2}$ or $1\frac{1}{2}$

(d) *Estimate:* $5 \div 6 = \frac{5}{6}$; *Exact:* $\frac{3}{4}$

OBJECTIVE ▶ **4** **Estimate the answer and add or subtract mixed numbers.**
The steps you learned earlier for adding and subtracting fractions will also work for mixed numbers. Just rewrite the mixed numbers as equivalent improper fractions. Again, it is a good idea to estimate the answer before you start any other work.

EXAMPLE 7 **Estimating the Answer and Adding or Subtracting Mixed Numbers**

First, estimate each answer. Then add or subtract to find the exact answer. Write exact answers in simplest form.

(a) $2\frac{3}{8} + 3\frac{3}{4}$

To estimate the answer, round each mixed number to the nearest whole number.

$$2\frac{3}{8} + 3\frac{3}{4}$$
$$\downarrow \qquad \downarrow$$
$$2 \;+\; 4 \;\;=\;\; 6 \leftarrow \text{Estimate}$$

To find the exact answer, first rewrite each mixed number as an equivalent improper fraction.

$$2\frac{3}{8} + 3\frac{3}{4} = \frac{19}{8} + \frac{15}{4}$$

> You **do** need a common denominator to add or subtract fractions.

You can't add fractions until they have a common denominator. The LCD for $\frac{19}{8}$ and $\frac{15}{4}$ is 8. Rewrite $\frac{15}{4}$ as an equivalent fraction with a denominator of 8.

$$\frac{19}{8} + \frac{15}{4} = \frac{19}{8} + \frac{30}{8} = \frac{19 + 30}{8} = \frac{49}{8} \text{ or } 6\frac{1}{8}$$

Common denominator

The **estimate was 6,** so an **exact answer of $6\frac{1}{8}$ is reasonable.**

(b) $4\frac{2}{3} - 2\frac{4}{5}$

Round each number and estimate the answer.

$$4\frac{2}{3} - 2\frac{4}{5}$$
$$\downarrow \qquad \downarrow$$
$$5 \;-\; 3 \;=\; 2 \leftarrow \text{Estimate}$$

To find the exact answer, rewrite the mixed numbers as improper fractions and subtract.

$$4\frac{2}{3} - 2\frac{4}{5} = \frac{14}{3} - \frac{14}{5} = \frac{70}{15} - \frac{42}{15} = \frac{70 - 42}{15} = \frac{28}{15} \text{ or } 1\frac{13}{15}$$

LCD is 15

The **estimate was 2,** so an **exact answer of $1\frac{13}{15}$ is reasonable.**

Continued on Next Page

(c) $5 - 1\frac{3}{8}$

$$5 - 1\frac{3}{8}$$
$$\downarrow \qquad \downarrow$$
$$5 - 1 = \mathbf{4} \leftarrow \text{Estimate}$$

Write 5 as $\frac{5}{1}$

$$5 - 1\frac{3}{8} = \frac{5}{1} - \frac{11}{8} = \frac{40}{8} - \frac{11}{8} = \frac{40 - 11}{8} = \frac{\mathbf{29}}{\mathbf{8}} \text{ or } \mathbf{3\frac{5}{8}}$$

LCD is 8

The **estimate was 4,** so an **exact answer of $3\frac{5}{8}$ is reasonable.**

——————————————— Work Problem ⑦ at the Side. ▶

Note

In some situations the method of rewriting mixed numbers as improper fractions may result in very large numerators. Here is an example.

Last year Hue's child was $48\frac{3}{8}$ in. tall. This year the child is $51\frac{1}{4}$ in. tall. How much has the child grown?

First, estimate the answer by rounding each mixed number to the nearest whole number.

$$51\frac{1}{4} - 48\frac{3}{8}$$
$$\downarrow \qquad \downarrow$$
$$51 - 48 = 3 \text{ in.} \leftarrow \text{Estimate}$$

To find the exact answer, rewrite the mixed numbers as improper fractions.

Rewrite $\frac{205}{4}$ as $\frac{410}{8}$

$$51\frac{1}{4} - 48\frac{3}{8} = \frac{205}{4} - \frac{387}{8} = \frac{410}{8} - \frac{387}{8} = \frac{410 - 387}{8} = \frac{23}{8} = 2\frac{7}{8} \text{ in.}$$

LCD is 8

You can also use the fraction key $\boxed{a^{b/c}}$ on your *scientific* calculator to solve this problem. Either way the exact answer is $2\frac{7}{8}$ in., which is close to the estimate of 3 in.

Another efficient method for handling large mixed numbers is to rewrite them in decimal form. You will learn how to do that soon.

OBJECTIVE ▶ ⑤ **Solve application problems containing mixed numbers.** Rounding mixed numbers to the nearest whole number can also help you decide whether to solve an application problem by adding, subtracting, multiplying, or dividing.

⑦ First, round the numbers and estimate each answer. Then add or subtract to find the exact answer.

GS **(a)** $5\frac{1}{3} - 2\frac{5}{6}$
$$\downarrow \qquad \downarrow$$
$$\underline{\;5\;} - \underline{} = \underline{}$$
$$\textit{Estimate}$$

$$5\frac{1}{3} = \frac{16}{3} \quad \text{and} \quad 2\frac{5}{6} = \frac{17}{6}$$

$$\frac{16}{3} - \frac{17}{6} = \frac{32}{6} - \frac{17}{6} = \frac{\square - \square}{6}$$

$$= \frac{\square}{6} = \underline{}$$

Write the answer in simplest form.

(b) $\frac{3}{4} + 3\frac{1}{8}$
$$\downarrow \qquad \downarrow$$
$$\underline{} + \underline{} = \underline{}$$
$$\textit{Estimate}$$

(c) $6 - 3\frac{4}{5}$
$$\downarrow \qquad \downarrow$$
$$\underline{} - \underline{} = \underline{}$$
$$\textit{Estimate}$$

Answers

7. **(a)** *Estimate:* $5 - 3 = 2$
   Exact: $\frac{32 - 17}{6} = \frac{15}{6} = \frac{5}{2}$ or $2\frac{1}{2}$

   (b) *Estimate:* $1 + 3 = 4$; *Exact:* $\frac{31}{8}$ or $3\frac{7}{8}$

   (c) *Estimate:* $6 - 4 = 2$; *Exact:* $\frac{11}{5}$ or $2\frac{1}{5}$

8 First, round the numbers and estimate the answer to each problem. Then find the exact answer. Write exact answers in simplest form and, when possible, as mixed numbers.

GS **(a)** Richard's son grew $3\frac{5}{8}$ inches last year and $2\frac{1}{4}$ inches this year. How much has his height increased over the two years?

Estimate: $3\frac{5}{8}$ rounds to _____

$2\frac{1}{4}$ rounds to _____

Read the problem using the rounded numbers. Do you need to add, subtract, multiply, or divide? _____

Estimate is _____

Now find the exact answer.

$3\frac{5}{8}$ is $\frac{\square}{8}$ and $2\frac{1}{4}$ is $\frac{\square}{4}$

Exact: _____

(b) Ernestine used $2\frac{1}{2}$ packages of chocolate chips in her cookie recipe. Each package has $5\frac{1}{2}$ ounces of chips. How many ounces of chips did she use in the recipe?

Estimate:

Exact:

Answers

8. **(a)** *Estimate:* 4; 2; add; $4 + 2 = 6$ inches

Exact: $\frac{29}{8}; \frac{9}{4}; \frac{47}{8} = 5\frac{7}{8}$ inches

(b) *Estimate:* $3 \cdot 6 = 18$ ounces

Exact: $\frac{55}{4} = 13\frac{3}{4}$ ounces

EXAMPLE 8 **Solving Application Problems with Mixed Numbers**

First, estimate the answer to each application problem. Then find the exact answer. Write exact answers in simplest form and, when possible, as mixed numbers.

(a) Gary needs to haul $15\frac{3}{4}$ **tons** of sand to a construction site. His truck can carry $2\frac{1}{4}$ **tons**. How many trips will he need to make?

First, round each mixed number to the nearest whole number.

$$15\frac{3}{4} \text{ rounds to } \boxed{16} \text{ and } 2\frac{1}{4} \text{ rounds to } \boxed{2}$$

Now read the problem again, *using the rounded numbers.*

Gary needs to haul **16 tons** of sand to a construction site. His truck can carry **2 tons**. How many trips will he need to make?

Using the rounded numbers in the problem makes it easier to see that you need to *divide*.

$$16 \div 2 = \textbf{8 trips} \leftarrow \text{Estimate}$$

To find the exact answer, use the original mixed numbers and divide.

$$15\frac{3}{4} \div 2\frac{1}{4} = \frac{63}{4} \div \frac{9}{4} = \frac{63}{4} \cdot \frac{4}{9} = \frac{7 \cdot \overset{1}{9} \cdot \overset{1}{4}}{\underset{1}{4} \cdot \underset{1}{9}} = \frac{7}{1} = 7 \quad \left\{ \begin{array}{l} \text{Simplest} \\ \text{form} \end{array} \right.$$

Reciprocals

Gary needs to make **7 trips to haul all the sand.** This result is **close to the estimate of 8 trips.**

(b) Zenitia worked $3\frac{5}{6}$ hours on Monday and $6\frac{1}{2}$ hours on Tuesday. How much longer did she work on Tuesday than on Monday?

First, round each mixed number to the nearest whole number.

$$3\frac{5}{6} \text{ rounds to } \boxed{4} \text{ and } 6\frac{1}{2} \text{ rounds to } \boxed{7}$$

Now read the problem again, *using the rounded numbers.*

Zenitia worked **4 hours** on Monday and **7 hours** on Tuesday. How much longer did she work on Tuesday than on Monday?

Using the rounded numbers in the problem makes it easier to see that you need to *subtract*.

$$7 - 4 = \textbf{3 hours} \leftarrow \text{Estimate}$$

To find the exact answer, use the original mixed numbers and subtract.

Simplest form

$$6\frac{1}{2} - 3\frac{5}{6} = \frac{13}{2} - \frac{23}{6} = \frac{39}{6} - \frac{23}{6} = \frac{39 - 23}{6} = \frac{16}{6} = \frac{8}{3} = 2\frac{2}{3} \quad \left\{ \begin{array}{l} \text{Changed} \\ \text{to a mixed} \\ \text{number} \end{array} \right.$$

LCD is 6

Zenitia worked $2\frac{2}{3}$ **hours longer on Tuesday.** This result is **close to the estimate of 3 hours.**

— **◄ Work Problem 8 at the Side.**

4.5 Exercises

FOR EXTRA HELP

Go to MyMathLab *for worked-out, step-by-step solutions to exercises enclosed in a square* ▢ *and video solutions to* ▶ *exercises.*

CONCEPT CHECK *For Exercises 1–4, first graph the mixed numbers or improper fractions on the number line. Then shade the shapes to show the positive mixed number or improper fraction. See Example 1.*

1. ▶ Graph $2\frac{1}{3}$ and $-2\frac{1}{3}$

Shade $2\frac{1}{3}$ of the circles.

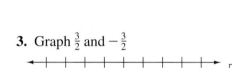

2. Graph $1\frac{3}{4}$ and $-1\frac{3}{4}$

Shade $1\frac{3}{4}$ of the rectangles.

3. Graph $\frac{3}{2}$ and $-\frac{3}{2}$

Shade $\frac{3}{2}$ of the squares.

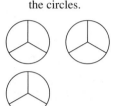

4. Graph $\frac{11}{3}$ and $-\frac{11}{3}$

Shade $\frac{11}{3}$ of the circles.

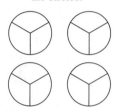

Write each mixed number as an improper fraction. See Example 2.

5. $4\frac{1}{2}$

6. $2\frac{1}{4}$

7. $-1\frac{3}{5}$

8. $-1\frac{5}{6}$

9. $2\frac{3}{8}$

10. $3\frac{4}{9}$

11. $-5\frac{7}{10}$

12. $-4\frac{5}{7}$

13. $10\frac{11}{15}$

14. $12\frac{9}{11}$

Write each improper fraction as a mixed number in simplest form. See Example 3.

15. $\frac{13}{3}$

16. $\frac{11}{2}$

17. $-\frac{10}{4}$

18. $-\frac{14}{5}$

19. $\frac{22}{6}$

20. $\frac{28}{8}$

21. $-\frac{51}{9}$

22. $-\frac{44}{10}$

23. ▶ $\frac{188}{16}$

24. $\frac{200}{15}$

First, round each mixed number to the nearest whole number and estimate the answer. Then find the exact answer. Write exact answers in simplest form. See Examples 4–7.

25. ▶ $2\frac{1}{4} \cdot 3\frac{1}{2}$

Estimate:

____ • ____ = ____

Exact:

26. $\left(1\frac{1}{2}\right)\left(3\frac{3}{4}\right)$

Estimate:

(____) (____) = ____

Exact:

27. ▶ $3\frac{1}{4} \div 2\frac{5}{8}$

Estimate:

____ ÷ ____ = ____

Exact:

28. $2\frac{1}{4} \div 1\frac{1}{8}$

Estimate:

____ ÷ ____ = ____

Exact:

29. ▶ $3\frac{2}{3} + 1\frac{5}{6}$

Estimate:

____ + ____ = ____

Exact:

30. $4\frac{4}{5} + 2\frac{1}{3}$

Estimate:

____ + ____ = ____

Exact:

31. $4\dfrac{1}{4} - \dfrac{7}{12}$

Estimate:

____ − ____ = ____

Exact:

32. $10\dfrac{1}{3} - 6\dfrac{5}{6}$

Estimate:

____ − ____ = ____

Exact:

33. $5\dfrac{2}{3} \div 6$

Estimate:

____ ÷ ____ = ____

Exact:

34. $1\dfrac{7}{8} \div 6\dfrac{1}{4}$

Estimate:

____ ÷ ____ = ____

Exact:

35. $8 - 1\dfrac{4}{5}$

Estimate:

____ − ____ = ____

Exact:

36. $7 - 3\dfrac{3}{10}$

Estimate:

____ − ____ = ____

Exact:

Find the perimeter and the area of each square, rectangle, or parallelogram. Write all answers in simplest form and, when possible, as mixed numbers.

37.

$1\frac{3}{4}$ in.

$1\frac{3}{4}$ in.

$1\frac{3}{4}$ in.

38.

$2\frac{1}{3}$ ft

$2\frac{1}{3}$ ft

39.

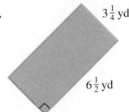

$3\frac{1}{4}$ yd

$6\frac{1}{2}$ yd

40.

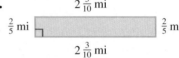

$2\frac{3}{10}$ mi

$\frac{2}{5}$ mi

$\frac{2}{5}$ mi

$2\frac{3}{10}$ mi

41.

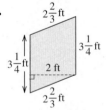

$2\frac{2}{3}$ ft

$3\frac{1}{4}$ ft

$3\frac{1}{4}$ ft

2 ft

$2\frac{2}{3}$ ft

42.

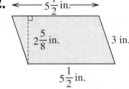

$5\frac{1}{2}$ in.

$2\frac{5}{8}$ in.

3 in.

$5\frac{1}{2}$ in.

First, estimate the answer to each application problem. Then find the exact answer. Write all answers in simplest form and, when possible, as mixed numbers. ***See Example 8.***

43. A carpenter has two pieces of oak trim. One piece of trim is $12\frac{1}{2}$ ft long and the other is $8\frac{2}{3}$ ft long. How many feet of oak trim does he have in all?

Estimate:

Exact:

44. On Monday, $5\frac{3}{4}$ tons of cans were recycled, and on Tuesday, $9\frac{3}{5}$ tons were recycled. How many tons were recycled on these two days?

Estimate:

Exact:

45. The directions for mixing an insect spray say to use $1\frac{3}{4}$ ounces of chemical in each gallon of water. How many ounces of chemical should be mixed with $5\frac{1}{2}$ gallons of water?

Estimate:

Exact:

46. Shirley Cicero wants to make 16 holiday wreaths to sell at a craft fair. Each wreath requires $2\frac{1}{4}$ yd of ribbon. How many yards does she need in all?

Estimate:

Exact:

47. The Boy Scout troop has volunteered to pick up trash along a 4-mile stretch of highway. So far they have done $1\frac{7}{10}$ miles. How much do they have left to do?

Estimate:

Exact:

48. Marv bought a 10 yd length of Italian silk fabric. He used $3\frac{7}{8}$ yd to make a jacket. What length of fabric is left for other sewing projects?

Estimate:

Exact:

49. Suppose that a bridesmaid's floor-length dress requires a piece of material $3\frac{3}{4}$ yd in length. What length of material is needed to make dresses for five bridesmaids?

Estimate:

Exact:

50. A cookie recipe uses $\frac{2}{3}$ cup brown sugar. How much brown sugar is needed to make $2\frac{1}{2}$ times the original recipe?

Estimate:

Exact:

Solve each application problem. Write all answers in simplest form and, when possible, as mixed numbers. Use the information on the Oregon state park sign for Exercises 51–54. The distances on the sign are in miles.

51. What is the distance from Devils Kitchen to the beach parking?

52. How far is it from Face Rock to the beach parking?

53. Suppose you bicycle from the park sign to the picnic parking, then bicycle back to Face Rock, and finally bicycle to the beach parking. How far will you have traveled?

54. If you drive back and forth from the sign to the picnic parking each day for a week, how many miles will you drive?

55. Find the length of the arrow shaft.

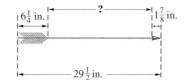

56. Find the length of the indented section on this board.

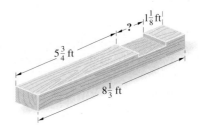

First, estimate the answer to each application problem. Then find the exact answer.
Write exact answers in simplest form and, when possible, as mixed numbers.

57. A craftsperson must attach a lead strip around all four sides of a rectangular stained glass window before it is installed. Find the length of lead stripping needed for the window shown.

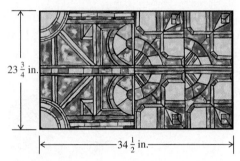

$23\frac{3}{4}$ in.

$34\frac{1}{2}$ in.

Estimate:

Exact:

58. To complete a custom order, Zak Morten of Home Depot must find the number of inches of brass trim needed to go around the four sides of the lamp base plate shown. Find the length of brass trim needed.

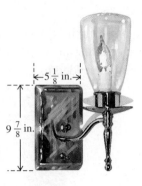

$5\frac{1}{8}$ in.

$9\frac{7}{8}$ in.

Estimate:

Exact:

59. Alex bought a 126-ounce container of powdered laundry detergent. It takes $1\frac{1}{2}$ ounces of detergent to wash a load of clothes. How many loads can Alex wash?

Estimate:

Exact:

60. The smallest post office in the United States was in Wheeler Springs, California. It was a rectangular building measuring $5\frac{3}{4}$ ft wide by $7\frac{1}{3}$ ft long. Find the area of the post office floor.

Estimate:

Exact:

61. Claire and Deb create custom hat bands that people can put around the crowns of their hats. The finished bands are the lengths shown in the table. The strip of fabric for each band must include the finished length plus an extra $\frac{3}{4}$ in. for the seam.

Band Size	Finished Length
Small	$21\frac{7}{8}$ in.
Medium	$22\frac{5}{8}$ in.
Large	$23\frac{1}{2}$ in.

What length of fabric strip is needed to make 4 small bands, 5 medium bands, and 3 large bands, including the seam allowance?

Estimate:

Exact:

62. Three sides of a parking lot are $108\frac{1}{4}$ ft, $162\frac{3}{8}$ ft, and $143\frac{1}{2}$ ft. If the distance around the lot is $518\frac{3}{4}$ ft, find the length of the fourth side.

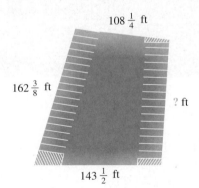

$108\frac{1}{4}$ ft

$162\frac{3}{8}$ ft

? ft

$143\frac{1}{2}$ ft

Estimate:

Exact:

Summary Exercises *Computation with Fractions*

1. Write fractions that represent the shaded and unshaded portion of each figure.

(a)

(b)

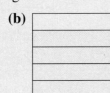

2. Graph $\dfrac{2}{3}$ and $-\dfrac{2}{3}$ on the number line.

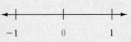

3. Rewrite each fraction with the indicated denominator.

(a) $-\dfrac{4}{5} = -\dfrac{}{30}$

(b) $\dfrac{2}{7} = \dfrac{}{14}$

4. Simplify.

(a) $\dfrac{15}{15}$

(b) $-\dfrac{24}{6}$

(c) $\dfrac{9}{1}$

5. Write the prime factorization of each number.

(a) 72

(b) 105

6. Write each fraction in lowest terms.

(a) $\dfrac{24}{30}$

(b) $\dfrac{175}{200}$

Simplify.

7. $\left(-\dfrac{3}{4}\right)\left(-\dfrac{2}{3}\right)$

8. $-\dfrac{7}{8} + \dfrac{2}{3}$

9. $\dfrac{7}{16} + \dfrac{5}{8}$

10. $\dfrac{5}{8} \div \dfrac{3}{4}$

11. $\dfrac{2}{3} - \dfrac{4}{5}$

12. $\dfrac{7}{12}\left(-\dfrac{9}{14}\right)$

13. $-21 \div \left(-\dfrac{3}{8}\right)$

14. $\dfrac{7}{8} - \dfrac{5}{12}$

15. $-\dfrac{35}{45} \div \dfrac{10}{15}$

16. $-\dfrac{5}{6} - \dfrac{3}{4}$

17. $\dfrac{7}{12} + \dfrac{5}{6} + \dfrac{2}{3}$

18. $\dfrac{5}{8}$ of 56

First round the numbers and estimate each answer. Then find the exact answer.

19. $4\dfrac{3}{4} + 2\dfrac{5}{6}$

Estimate:

___ + ___ = ___

Exact:

20. $2\dfrac{2}{9} \cdot 5\dfrac{1}{7}$

Estimate:

___ • ___ = ___

Exact:

21. $6 - 2\dfrac{7}{10}$

Estimate:

___ − ___ = ___

Exact:

22. $1\dfrac{3}{5} \div 3\dfrac{1}{2}$

Estimate:

___ ÷ ___ = _____

Exact:

23. $4\dfrac{2}{3} \div 1\dfrac{1}{6}$

Estimate:

___ ÷ ___ = ___

Exact:

24. $3\dfrac{5}{12} - \dfrac{3}{4}$

Estimate:

___ − ___ = ___

Exact:

Solve each application problem. Write all answers in simplest form and, when possible, as mixed numbers.

25. When installing cabinets, Cecil Feathers must be certain that the proper type and size of mounting screw is used.

    (a) Find the total length of the screw shown.

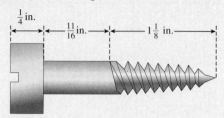

    (b) If the screw is put into a board that is $1\frac{3}{4}$ in. thick, how much of the screw will stick out the back of the board?

26. A stamp collector needs to make room in his album for this postage stamp. Find the perimeter and the area of the stamp.

27. A batch of cookies requires $\frac{3}{4}$ pound of chocolate chips. If you have nine pounds of chocolate chips, how many batches of cookies can you make?

28. The Municipal Utility District says that the cost of operating a treadmill is $\frac{1}{5}$ ¢ per minute. Find the cost of operating the treadmill for half an hour.

29. A survey asked 1500 adults if UFOs (unidentified flying objects) are real. The circle graph shows the fraction of the adults who gave each answer. How many adults gave each answer?

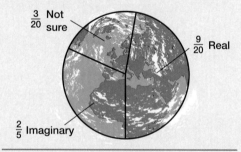

Out of This World?

When asked in a survey if UFOs are real, respondents answered:

$\frac{3}{20}$ Not sure

$\frac{9}{20}$ Real

$\frac{2}{5}$ Imaginary

Data from Yankelovich Partners for *Life* magazine.

30. Find the diameter of the hole in the rectangular mounting bracket shown. (The diameter is the distance across the center of the hole.) Then find the perimeter of the bracket.

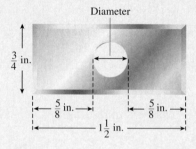

31. A travel-size bottle of contact lens cleaning solution holds $\frac{1}{3}$ fluid ounce. How many bottles can be filled with $15\frac{1}{3}$ fluid ounces? (Data from Alcon Laboratories, Inc.)

32. The first experimental TV, built in 1928, had a tiny square screen that was only $1\frac{1}{2}$ in. on each side. Find the perimeter and area of the screen.

4.6 Exponents, Order of Operations, and Complex Fractions

OBJECTIVE ▶ ① Simplify fractions with exponents. We have used exponents as a quick way to write repeated multiplication of integers and variables. Here are two examples as a review.

$$
\overbrace{(-3)^2}^{\text{Exponent}} = \underbrace{(-3)(-3)}_{\text{Two factors of } -3} = 9
\qquad \text{and} \qquad
\overbrace{x^3}^{\text{Exponent}} = \underbrace{x \cdot x \cdot x}_{\text{Three factors of } x}
$$

Base   Two factors of -3 Base   Three factors of x

The meaning of an exponent remains the same when a fraction is the base.

OBJECTIVES

① Simplify fractions with exponents.

② Use the order of operations with fractions.

③ Simplify complex fractions.

EXAMPLE 1   Simplifying Fractions with Exponents

Simplify.

(a) $\left(-\dfrac{1}{2}\right)^3$

The base is $-\dfrac{1}{2}$. The exponent indicates that there are three factors of $-\dfrac{1}{2}$.

$$
\left(-\frac{1}{2}\right)^3 = \overbrace{\left(-\frac{1}{2}\right)\left(-\frac{1}{2}\right)\left(-\frac{1}{2}\right)}^{\text{Three factors of } -\frac{1}{2}}
\quad \text{Multiply } \left(-\frac{1}{2}\right)\left(-\frac{1}{2}\right) \text{ to get } \frac{1}{4}
$$

> Watch the signs! Multiply *two* factors at a time.

$$
\underbrace{\frac{1}{4} \quad \left(-\frac{1}{2}\right)}
\qquad \text{Now multiply } \frac{1}{4}\left(-\frac{1}{2}\right) \text{ to get } -\frac{1}{8}
$$

$$
-\frac{1}{8}
$$

> The product is *negative*.

(b) $\left(\dfrac{3}{4}\right)^2 \left(\dfrac{2}{3}\right)^3$

$$
\left(\frac{3}{4}\right)^2 \left(\frac{2}{3}\right)^3 =
\underbrace{\left(\frac{3}{4}\right)\left(\frac{3}{4}\right)}_{\text{Two factors of } \frac{3}{4}}
\underbrace{\left(\frac{2}{3}\right)\left(\frac{2}{3}\right)\left(\frac{2}{3}\right)}_{\text{Three factors of } \frac{2}{3}}
$$

$$
= \frac{\overset{1}{\cancel{3}} \cdot \overset{1}{\cancel{3}} \cdot \overset{1}{\cancel{2}} \cdot \overset{1}{\cancel{2}} \cdot \overset{1}{\cancel{2}}}{\underset{1}{\cancel{2}} \cdot \underset{1}{\cancel{2}} \cdot \underset{1}{\cancel{2}} \cdot \underset{1}{\cancel{2}} \cdot \underset{1}{\cancel{3}} \cdot \underset{1}{\cancel{3}} \cdot 3}
\qquad \text{Divide out all common factors.}
$$

$$
= \frac{1}{6}
$$

> Remember the 1 in the numerator. The answer is $\frac{1}{6}$, **not** 6.

Work Problem ① at the Side. ▶

---

① Simplify.

(a) $\left(-\dfrac{3}{5}\right)^2$

$$
\underbrace{\left(-\frac{3}{5}\right)\left(-\frac{3}{5}\right)}
$$

Two factors of $-\frac{3}{5}$

The product of two negative numbers is *positive*.

(b) $\left(\dfrac{1}{3}\right)^4$

(c) $\left(-\dfrac{2}{3}\right)^3 \left(\dfrac{1}{2}\right)^2$

(d) $\left(-\dfrac{1}{2}\right)^2 \left(\dfrac{1}{4}\right)^2$

OBJECTIVE ▶ ② Use the order of operations with fractions. The order of operations used for integers also applies to fractions.

Answers

1. (a) $\dfrac{9}{25}$ (b) $\dfrac{1}{81}$ (c) $-\dfrac{2}{27}$ (d) $\dfrac{1}{64}$

2 Simplify.

(a) $\dfrac{1}{3} - \dfrac{5}{9}\left(\dfrac{3}{4}\right)$

$\dfrac{1}{3} - \dfrac{\square}{\square}$ Multiply first.

$\dfrac{\square}{36} - \dfrac{\square}{36}$ Now subtract. The LCD is 36

$\dfrac{\square - \square}{36}$

$-\dfrac{\square}{36}$ Now write the answer in lowest terms.

(b) $-\dfrac{3}{4} + \left(-\dfrac{1}{2}\right)^2 \div \dfrac{2}{3}$

(c) $\dfrac{12}{5} - \dfrac{1}{6}\left(3 - \dfrac{3}{5}\right)$

Answers

2. **(a)** $\dfrac{1}{3} - \dfrac{15}{36}; \dfrac{12}{36} - \dfrac{15}{36}; \dfrac{12 - 15}{36}; -\dfrac{3}{36}; -\dfrac{1}{12}$

 (b) $-\dfrac{3}{8}$ **(c)** 2

Order of Operations

Step 1 Work inside *parentheses* or *other grouping symbols.*

Step 2 Simplify expressions with *exponents.*

Step 3 Do the remaining *multiplications and divisions* as they occur **from left to right.**

Step 4 Do the remaining *additions and subtractions* as they occur **from left to right.**

EXAMPLE 2   **Using the Order of Operations with Fractions**

Simplify.

(a) $-\dfrac{1}{3} + \dfrac{1}{2}\left(\dfrac{4}{5}\right)$ There is no work to be done inside the parentheses. There are no exponents, so start with *Step 3*, multiplying and dividing.

$-\dfrac{1}{3} + \dfrac{1 \cdot 4}{2 \cdot 5}$ Multiply.   Multiply $\frac{1}{2} \cdot \frac{4}{5}$ first. Do **not** start by adding $-\frac{1}{3} + \frac{1}{2}$

$-\dfrac{1}{3} + \dfrac{4}{10}$ Now add. The LCD is 30
$-\dfrac{1}{3} = -\dfrac{10}{30}$ and $\dfrac{4}{10} = \dfrac{12}{30}$

$\dfrac{-10 + 12}{30}$ Add the numerators; keep the common denominator.

$\dfrac{2}{30}$ Write $\dfrac{2}{30}$ in lowest terms: $\dfrac{\overset{1}{\cancel{2}}}{\underset{1}{\cancel{2}} \cdot 15} = \dfrac{1}{15}$

$\dfrac{1}{15}$ ⟵ The answer is in lowest terms.

(b) $-2 + \left(\dfrac{1}{4} - \dfrac{3}{2}\right)^2$ Work inside parentheses.
The LCD for $\dfrac{1}{4}$ and $\dfrac{3}{2}$ is 4

$-2 + \left(\dfrac{1 - 6}{4}\right)^2$ Rewrite $\dfrac{3}{2}$ as $\dfrac{6}{4}$ and subtract; $\dfrac{1}{4} - \dfrac{6}{4}$ is $\dfrac{1-6}{4}$

$-2 + \left(\dfrac{-5}{4}\right)^2$ Simplify the term with the exponent.
Multiply $\left(-\dfrac{5}{4}\right)\left(-\dfrac{5}{4}\right)$. *Signs match*, so the product is *positive.*

$-2 + \left(\dfrac{25}{16}\right)$ Add last.
Write -2 as $-\dfrac{2}{1}$

$-\dfrac{2}{1} + \dfrac{25}{16}$ The LCD is 16. Rewrite $-\dfrac{2}{1}$ as $-\dfrac{32}{16}$

$\dfrac{-32 + 25}{16}$ Add the numerators; keep the common denominator.

$-\dfrac{7}{16}$ ⟵ The answer is in lowest terms.

◀ **Work Problem 2 at the Side.**

OBJECTIVE **3** **Simplify complex fractions.** We have used both the ÷ symbol and a fraction bar to indicate division. For example,

$$\text{Indicates division} \to \frac{6}{2} \quad \text{can be written as} \quad 6 \div 2$$

Indicates division

That means we could write $-\frac{4}{5} \div \left(-\frac{3}{10}\right)$ using a fraction bar instead of ÷

$$-\frac{4}{5} \div \left(-\frac{3}{10}\right) \quad \text{can be written as} \quad \frac{-\dfrac{4}{5}}{-\dfrac{3}{10}} \leftarrow \text{Indicates division}$$

Indicates division

The result looks a bit complicated, and its name reflects that fact. We call it a *complex fraction*.

Complex Fractions

A **complex fraction** is a fraction in which the numerator and/or the denominator contain one or more fractions.

To simplify a complex fraction, rewrite it in horizontal format using the ÷ symbol for division.

EXAMPLE 3   Simplifying a Complex Fraction

Simplify: **(a)** $\dfrac{-\dfrac{4}{5}}{-\dfrac{3}{10}}$ **(b)** $\dfrac{\left(\dfrac{2}{3}\right)^2}{6}$

(a) Rewrite the complex fraction using the ÷ symbol for division. Then follow the steps for dividing fractions.

$$\frac{-\dfrac{4}{5}}{-\dfrac{3}{10}} = -\frac{4}{5} \div -\frac{3}{10} = -\frac{4}{5} \cdot -\frac{10}{3} = \frac{4 \cdot 2 \cdot \overset{1}{\cancel{5}}}{\underset{1}{\cancel{5}} \cdot 3} = \frac{8}{3} \quad \text{or} \quad 2\frac{2}{3}$$

Reciprocals

The quotient is *positive* because both numbers in the problem had the *same* sign (both were negative).

(b) Rewrite the complex fraction using the ÷ symbol for division. Then follow the order of operations.

$$\left(\tfrac{2}{3}\right)\left(\tfrac{2}{3}\right) \text{ is } \tfrac{4}{9}$$

$$\frac{\left(\dfrac{2}{3}\right)^2}{6} = \left(\frac{2}{3}\right)^2 \div 6 = \frac{4}{9} \div 6 = \frac{4}{9} \cdot \frac{1}{6} = \frac{\overset{1}{\cancel{2}} \cdot 2 \cdot 1}{9 \cdot \underset{1}{\cancel{2}} \cdot 3} = \frac{2}{27}$$

Reciprocals

──────── **Work Problem** **3** **at the Side.** ▶

3 Simplify.

GS **(a)** $\dfrac{-\dfrac{3}{5}}{\dfrac{9}{10}}$ Rewrite this complex fraction using the ÷ symbol.

$$-\frac{3}{5} \div \frac{9}{10} = -\frac{3}{5} \cdot \frac{10}{9} =$$

Reciprocals

$$-\frac{3 \cdot \Box \cdot \Box}{5 \cdot \Box \cdot \Box} =$$

Divide out common factors.

(b) $\dfrac{6}{\dfrac{3}{4}}$ ← *Hint:* Write 6 as $\frac{6}{1}$

(c) $\dfrac{-\dfrac{15}{16}}{-5}$

(d) $\dfrac{-3}{\left(-\dfrac{3}{4}\right)^2}$

Answers

3. **(a)** $-\dfrac{\overset{1}{\cancel{3}} \cdot 2 \cdot \overset{1}{\cancel{5}}}{\underset{1}{\cancel{5}} \cdot 3 \cdot \underset{1}{\cancel{3}}} = -\dfrac{2}{3}$ **(b)** 8 **(c)** $\dfrac{3}{16}$

(d) $-\dfrac{16}{3}$ or $-5\dfrac{1}{3}$

4.6 Exercises

FOR EXTRA HELP Go to MyMathLab for worked-out, step-by-step solutions to exercises enclosed in a square ▢ and video solutions to ▶ exercises.

CONCEPT CHECK

1. (a) What part of the expression $\left(-\frac{3}{8}\right)^2$ is called the base? _____

   (b) What is the small raised "2" called? _____

   (c) Circle the correct factored form.

   $\left(-\frac{3}{8}\right)(2)$ or $\left(-\frac{3}{8}\right)\left(-\frac{3}{8}\right)$ or $\left(-\frac{3}{8}\right)\left(-\frac{8}{3}\right)$

CONCEPT CHECK

2. (a) When two factors are both negative numbers, is the product positive or negative? _____

   (b) Circle all the multiplications that give a negative product.

   $5\left(-\frac{3}{10}\right)$ or $\left(-\frac{1}{2}\right)(-8)$ or $\left(-\frac{5}{6}\right)\left(\frac{3}{5}\right)$

*Simplify. **See Example 1.***

3. $\left(-\frac{3}{4}\right)^2 = \left(-\frac{3}{4}\right)\left(-\frac{3}{4}\right)$

   $= \dfrac{\square}{\square}$ Signs match so the product is _____.

4. $\left(-\frac{4}{5}\right)^2 = \left(-\frac{4}{5}\right)\left(-\frac{4}{5}\right)$

   $= \dfrac{\square}{\square}$ Signs match so the product is _____.

5. $\left(\frac{2}{5}\right)^3$

6. $\left(\frac{1}{4}\right)^3$

7. $\left(-\frac{1}{3}\right)^3$

8. $\left(-\frac{3}{5}\right)^3$

9. $\left(\frac{1}{2}\right)^5$

10. $\left(\frac{1}{3}\right)^4$

11. $\left(\frac{7}{10}\right)^2$

12. $\left(\frac{8}{9}\right)^2$

13. $\left(-\frac{6}{5}\right)^2$

14. $\left(-\frac{8}{7}\right)^2$

15. $\frac{15}{16}\left(\frac{4}{5}\right)^3$

16. $-8\left(-\frac{3}{8}\right)^2$

17. $\left(\frac{1}{3}\right)^4\left(\frac{9}{10}\right)^2$

18. $\left(\dfrac{4}{5}\right)^2\left(\dfrac{1}{2}\right)^6$

19. $\left(-\dfrac{3}{2}\right)^3\left(-\dfrac{2}{3}\right)^2$

20. $\left(\dfrac{5}{6}\right)^2\left(-\dfrac{2}{5}\right)^3$

*Simplify. **See Example 2.***

21. $\dfrac{1}{5} - \underbrace{6\left(\dfrac{7}{10}\right)}$ Multiply first!

$\dfrac{1}{5} - \dfrac{\Box}{\Box}$ Now subtract.
The LCD is 10.

$\dfrac{\Box - \Box}{10} =$

22. $\dfrac{2}{9} - \underbrace{4\left(\dfrac{5}{6}\right)}$ Multiply first!

$\dfrac{2}{9} - \dfrac{\Box}{\Box}$ Now subtract.
The LCD is 18.

$\dfrac{\Box - \Box}{18} =$

23. $\left(\dfrac{4}{3} \div \dfrac{8}{3}\right) + \left(-\dfrac{3}{4} \cdot \dfrac{1}{4}\right)$

24. $\left(-\dfrac{1}{3} \cdot \dfrac{3}{5}\right) + \left(\dfrac{3}{4} \div \dfrac{1}{4}\right)$

25. $-\dfrac{3}{10} \div \dfrac{3}{5}\left(-\dfrac{2}{3}\right)$

26. $5 \div \left(-\dfrac{10}{3}\right)\left(-\dfrac{4}{9}\right)$

27. $\dfrac{8}{3}\left(\dfrac{1}{4} - \dfrac{1}{2}\right)^2$

28. $\dfrac{1}{3}\left(\dfrac{4}{5} - \dfrac{3}{10}\right)^3$

29. $-\dfrac{3}{8} + \dfrac{2}{3}\left(-\dfrac{2}{3} + \dfrac{1}{6}\right)$

30. $\dfrac{1}{6} + 4\left(\dfrac{2}{5} - \dfrac{7}{10}\right)$

31. $2\left(\dfrac{1}{3}\right)^3 - \dfrac{2}{9}$

32. $8\left(-\dfrac{3}{4}\right)^2 + \dfrac{3}{2}$

33. $\left(-\dfrac{2}{3}\right)^3\left(\dfrac{1}{8} - \dfrac{1}{2}\right) - \dfrac{2}{3}\left(\dfrac{1}{8}\right)$

34. $\left(\dfrac{3}{5}\right)^2\left(\dfrac{5}{9} - \dfrac{2}{3}\right) \div \left(-\dfrac{1}{5}\right)^2$

Solve each application problem. Write all answers in simplest form and, when possible, as mixed numbers.

35. A square operation key on a calculator is $\frac{3}{8}$ inch on each side. What is the area of the key? Use the formula $A = s^2$. (Data from Texas Instruments.)

$\frac{3}{8}$ in.

36. A square lot for sale in the country is $\frac{3}{10}$ mile on a side. Find the area of the lot by using the formula $A = s^2$.

$\frac{3}{10}$ mi

37. A rectangular parking lot at the megamall is $\frac{7}{10}$ mile long and $\frac{1}{4}$ mile wide. How much fencing is needed to enclose the lot? Use the formula $P = 2l + 2w$.

38. A rectangular computer chip is $\frac{7}{8}$ inch long and $\frac{5}{16}$ inch wide. An insulating strip must be put around all sides of the chip. Find the length of the strip. Use the formula $P = 2l + 2w$.

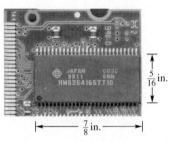

$\frac{5}{16}$ in.

$\frac{7}{8}$ in.

Simplify. ***See Example 3.***

39. $\dfrac{-\dfrac{7}{9}}{-\dfrac{7}{36}}$ Rewrite using the ÷ symbol.

$-\dfrac{7}{9} \div \left(-\dfrac{7}{36} \right) =$

40. $\dfrac{\dfrac{15}{32}}{-\dfrac{5}{64}}$ Rewrite using the ÷ symbol.

$\dfrac{15}{32} \div \left(-\dfrac{5}{64} \right) =$

41. $\dfrac{-15}{\dfrac{6}{5}}$

42. $\dfrac{-6}{-\dfrac{5}{8}}$

43. $\dfrac{\dfrac{4}{7}}{8}$

44. $\dfrac{-\dfrac{11}{5}}{3}$

45. $\dfrac{-\dfrac{2}{3}}{-2\dfrac{2}{5}}$

46. $\dfrac{\dfrac{1}{2}}{3\dfrac{1}{3}}$

47. $\dfrac{-4\dfrac{1}{2}}{\left(\dfrac{3}{4}\right)^2}$

48. $\dfrac{1\dfrac{2}{3}}{\left(-\dfrac{1}{6}\right)^2}$

49. $\dfrac{\left(\dfrac{2}{5}\right)^2}{\left(-\dfrac{4}{3}\right)^2}$

50. $\dfrac{\left(-\dfrac{5}{6}\right)^2}{\left(\dfrac{1}{2}\right)^3}$

Relating Concepts (Exercises 51 and 52)   For Individual or Group Work

Use your knowledge of exponents as you **work Exercises 51 and 52 in order.**

51. (a) Evaluate this series of examples.

$\left(-\dfrac{1}{2}\right)^2 = $ _____ $\left(-\dfrac{1}{2}\right)^6 = $ _____

$\left(-\dfrac{1}{2}\right)^3 = $ _____ $\left(-\dfrac{1}{2}\right)^7 = $ _____

$\left(-\dfrac{1}{2}\right)^4 = $ _____ $\left(-\dfrac{1}{2}\right)^8 = $ _____

$\left(-\dfrac{1}{2}\right)^5 = $ _____ $\left(-\dfrac{1}{2}\right)^9 = $ _____

(b) Explain the pattern in the sign of the answers.

52. Several drops of ketchup fell on Ron's homework. Explain how he can figure out what number is covered by each drop. Be careful. More than one number may work.

(a) $\left(\text{●} \right)^2 = \dfrac{4}{9}$

(b) $\left(\text{●} \right)^3 = -\dfrac{1}{27}$

(c) $\left(\text{●} \right)^4 = \dfrac{1}{16}$

4.7 | Problem Solving: Equations Containing Fractions

OBJECTIVE ▶ **1** **Use the multiplication property of equality to solve equations containing fractions.** You already used the division property of equality to solve an equation such as $4s = 24$. The division property says that you may divide both sides of an equation by the same nonzero number and the equation will still be balanced. Now that you have some experience with fractions, let's look again at how the division property works.

$$4s = 24$$

$4s$ means $4 \cdot s$

$$\frac{4 \cdot s}{4} = \frac{24}{4} \qquad \text{Divide both sides by 4.}$$

Divide out the common factor of 4

$$\frac{\overset{1}{\cancel{4}} \cdot s}{\underset{1}{\cancel{4}}} = 6$$

$$s = 6$$

Because multiplication and division are related to each other, we can also *multiply* both sides of an equation by the same nonzero number and keep it balanced.

Multiplication and Division Properties of Equality

The **multiplication property of equality** says, if $a = b$, then $a \cdot c = b \cdot c$

The **division property of equality** says, if $a = b$, then $\dfrac{a}{c} = \dfrac{b}{c}$ as long as c is not 0. In other words, you may multiply or divide both sides of an equation by the same nonzero number and it will still be balanced.

EXAMPLE 1   Using the Multiplication Property of Equality

Solve each equation and check each solution.

(a) $\dfrac{1}{2}b = 5$

We want the variable by itself on one side of the equal sign. In this example, we want $1b$, not $\frac{1}{2}b$, on the left side. (Recall that $1b$ is equivalent to b.)

Earlier in this chapter you learned that the product of a number and its reciprocal is 1. Thus, multiplying $\frac{1}{2}$ by $\frac{2}{1}$ will give the desired coefficient of 1.

$$\frac{1}{2}b = 5$$

$$\frac{2}{1}\left(\frac{1}{2}b\right) = \frac{2}{1}(5) \qquad \text{Multiply both sides by } \tfrac{2}{1} \text{ (the reciprocal of } \tfrac{1}{2}\text{)}$$

On the left side, use the associative property to regroup the factors.

$$\left(\frac{2}{1} \cdot \frac{1}{2}\right)b = \frac{2}{1}\left(\frac{5}{1}\right) \qquad \text{On the right side, 5 is equivalent to } \tfrac{5}{1}$$

$$\left(\frac{\overset{1}{\cancel{2}}}{1} \cdot \frac{1}{\underset{1}{\cancel{2}}}\right)b = \frac{10}{1}$$

$1b$ is equivalent to b

$$1b = 10$$

$$b = 10 \qquad \boxed{\text{The solution is 10}}$$

Continued on Next Page

Once you understand the process, you don't have to show every step. Here is a shorthand solution of the same problem.

$$\frac{1}{2}b = 5$$

$$\frac{\overset{1}{\cancel{2}}}{1}\left(\frac{1}{\underset{1}{\cancel{2}}}b\right) = \frac{2}{1}\left(\frac{5}{1}\right)$$

$$b = 10 \quad \boxed{\text{The solution is 10}}$$

The solution is **10**. Check the solution by going back to the *original* equation.

CHECK $\frac{1}{2}b = 5$ Replace b with 10 in the *original* equation.

$$\frac{1}{2}(10) = 5 \qquad \text{Multiply on the left side: } \frac{1}{2}(10)$$
$$\text{is } \frac{1}{2} \cdot \frac{10}{1} \text{ or } \frac{1}{2} \cdot \frac{2 \cdot 5}{1}$$

$$\frac{1 \cdot \overset{1}{\cancel{2}} \cdot 5}{\underset{1}{\cancel{2}} \cdot 1} = 5$$

$$5 = 5 \checkmark \quad \text{Balances}$$

When b is 10, the equation balances, so **10** is the correct solution (***not*** 5).

(b) $12 = -\dfrac{3}{4}x$

$$12 = -\frac{3}{4}x \qquad \begin{array}{l}\text{Multiply both sides by } -\frac{4}{3} \text{ (the reciprocal of } -\frac{3}{4}\text{)} \\ \text{The reciprocal of a negative number} \\ \text{is also negative.}\end{array}$$

$$-\frac{4}{3}(12) = -\frac{\overset{1}{\cancel{4}}}{\underset{1}{\cancel{3}}}\left(-\frac{\overset{1}{\cancel{3}}}{\underset{1}{\cancel{4}}}x\right) \qquad \begin{array}{l}\text{On the left side } -\frac{4}{3}(12) \text{ is } -\frac{4}{3} \cdot \frac{12}{1} \\ \text{or } -\frac{4 \cdot \overset{1}{\cancel{3}} \cdot 4}{\underset{1}{\cancel{3}} \cdot 1} = -16\end{array}$$

$$-16 = x \quad \boxed{\begin{array}{c}\text{The solution} \\ \text{is } -16\end{array}}$$

The solution is $-\mathbf{16}$. Check the solution by going back to the *original* equation.

CHECK $12 = -\dfrac{3}{4}x$ Replace x with -16 in the *original* equation.

$$12 = -\frac{3}{4}(-16) \qquad \begin{array}{l}\text{The product of two negative} \\ \text{numbers is positive.} \\ -\frac{3}{4}(-16) \text{ is } \frac{3}{4} \cdot \frac{16}{1} \text{ or } \frac{3}{4} \cdot \frac{4 \cdot 4}{1}\end{array}$$

$$12 = \frac{3 \cdot \overset{1}{\cancel{4}} \cdot 4}{\underset{1}{\cancel{4}} \cdot 1}$$

$$12 = 12 \checkmark \quad \text{Balances} \quad \boxed{\begin{array}{l}\text{If the equation does \textbf{not} balance,} \\ \text{your solution is wrong. Start over.}\end{array}}$$

When x is -16, the equation balances, so $-\mathbf{16}$ is the correct solution (***not*** 12).

Continued on Next Page

1 Solve each equation. Check each solution.

(a) $\dfrac{1}{6}m = 3$ — Multiply both sides by $\dfrac{6}{1}$ (reciprocal of $\dfrac{1}{6}$).

$$\dfrac{\overset{1}{6}}{1}\left(\dfrac{1}{\underset{1}{6}}m\right) = \dfrac{6}{1}\left(\dfrac{3}{1}\right) \quad \text{Write 3 as } \dfrac{3}{1}$$

$$m = \underline{\hphantom{000}}$$

CHECK

$$\dfrac{1}{6}\,m = 3 \qquad \begin{array}{l}\text{Replace } m \\ \text{with } \underline{\hphantom{00}}\end{array}$$

$$\dfrac{1}{6}(\underline{\hphantom{00}}) = 3$$

$$\underline{\hphantom{00}} = 3 \quad \text{Does it balance?}$$

(b) $\dfrac{3}{2}a = -9$

CHECK

(c) $\dfrac{3}{14} = -\dfrac{2}{7}x$

CHECK

Answers

1. (a) $m = 18$ Replace m with 18.

    CHECK $\dfrac{1}{6}m = 3$

$$\dfrac{1}{6}(18) = 3$$

    Balances $3 = 3$ ✓

(b) $a = -6$ **CHECK** $\dfrac{3}{2}a = -9$

$$\dfrac{3}{2}(-6) = -9$$

    Balances $-9 = -9$ ✓

(c) $x = -\dfrac{3}{4}$ **CHECK** $\dfrac{3}{14} = -\dfrac{2}{7}x$

$$\dfrac{3}{14} = -\dfrac{2}{7}\left(-\dfrac{3}{4}\right)$$

    Balances $\dfrac{3}{14} = \dfrac{3}{14}$ ✓

(c) $-\dfrac{2}{5}n = -\dfrac{1}{3}$

$$-\dfrac{\overset{1}{\cancel{5}}}{\underset{1}{\cancel{2}}}\left(-\dfrac{\cancel{2}}{\cancel{5}}n\right) = \left(-\dfrac{5}{2}\right)\left(-\dfrac{1}{3}\right) \quad \begin{array}{l}\text{Multiply both sides by } -\dfrac{5}{2} \\ \text{(the reciprocal of } -\dfrac{2}{5})\end{array}$$

$$n = \underbrace{\dfrac{5 \cdot 1}{2 \cdot 3}} \qquad \begin{array}{l}\text{The product of two negative} \\ \text{numbers is positive.}\end{array}$$

$$n = \dfrac{5}{6} \quad \boxed{\text{The solution is } \tfrac{5}{6}}$$

CHECK $\quad -\dfrac{2}{5}n = -\dfrac{1}{3} \qquad$ Original equation

$$\left(-\dfrac{2}{5}\right)\left(\dfrac{5}{6}\right) = -\dfrac{1}{3} \qquad \text{Replace } n \text{ with } \tfrac{5}{6}$$

$$-\dfrac{\overset{1}{2} \cdot \overset{1}{\cancel{5}}}{\underset{1}{\cancel{5}} \cdot \underset{1}{2} \cdot 3} = -\dfrac{1}{3} \qquad \text{Multiply on the left side.}$$

$$-\dfrac{1}{3} = -\dfrac{1}{3} \;✓ \quad \text{Balances}$$

When n is $\tfrac{5}{6}$, the equation balances, so $\tfrac{5}{6}$ is the correct solution (**not** $-\tfrac{1}{3}$).

◀ **Work Problem 1 at the Side.**

OBJECTIVE 2 Use both the addition and multiplication properties of equality to solve equations containing fractions.

EXAMPLE 2   **Using the Addition and Multiplication Properties of Equality**

Solve each equation and check each solution.

(a) $\dfrac{1}{3}c + 5 = 7$

The first step is to get the variable term $\tfrac{1}{3}c$ by itself on the left side of the equal sign. Recall that to "get rid of" the 5 on the left side, add the opposite of 5, which is -5, to both sides.

$$\dfrac{1}{3}c + 5 = 7$$

$$\underline{\phantom{\dfrac{1}{3}c +} -5 \quad -5} \qquad \text{Add } -5 \text{ to both sides.}$$

$$\underbrace{\dfrac{1}{3}c + 0} = 2$$

$$\dfrac{1}{3}c = 2$$

$$\dfrac{\overset{1}{\cancel{3}}}{1}\left(\dfrac{1}{\underset{1}{\cancel{3}}}c\right) = \dfrac{3}{1}\left(\dfrac{2}{1}\right) \qquad \begin{array}{l}\text{Multiply both sides by } \tfrac{3}{1} \\ \text{(the reciprocal of } \tfrac{1}{3})\end{array}$$

$$c = 6 \quad \boxed{\text{The solution is 6}}$$

Continued on Next Page

The solution is **6**. Check the solution by going back to the *original* equation.

CHECK $\dfrac{1}{3}c + 5 = 7$ Original equation

$\dfrac{1}{3}(6) + 5 = 7$ Replace c with 6

$\underbrace{2} + 5 = 7$

$7 = 7$ ✓   Balances

When c is 6, the equation balances, so **6** is the correct solution (*not* 7).

(b) $-3 = \dfrac{2}{3}y + 7$

To get the variable term $\frac{2}{3}y$ by itself on the right side, add -7 to both sides.

$-3 = \dfrac{2}{3}y + 7$

$\underline{-7 \qquad\quad -7}$ Add -7 to both sides.

$-10 = \dfrac{2}{3}y + 0$

$\dfrac{3}{2}(-10) = \dfrac{\cancel{3}^{1}}{\cancel{2}_{1}}\left(\dfrac{\cancel{2}^{1}}{\cancel{3}_{1}}y\right)$ Multiply both sides by $\frac{3}{2}$ (the reciprocal of $\frac{2}{3}$)

$-15 = y$ ◁ The solution is −15

CHECK $-3 = \dfrac{2}{3}y + 7$ Original equation

$-3 = \dfrac{2}{3}(-15) + 7$ Replace y with −15

$-3 = \underbrace{-10} + 7$

$-3 = -3$ ✓   Balances

When y is -15, the equation balances, so -15 is the correct solution (*not* −3).

—————— **Work Problem ② at the Side.** ▶

OBJECTIVE ▶ ❸ Solve application problems using equations containing fractions. Use the six problem-solving steps to solve application problems.

---

❷ Solve each equation. Check each solution.

GS (a) $18 = \dfrac{4}{5}x + 2$

$\underline{-2 \qquad\quad -2}$ Add -2 to both sides.

$16 = \dfrac{4}{5}x + 0$

$16 = \underbrace{\dfrac{4}{5}x}$ Multiply both sides by the reciprocal of $\frac{4}{5}$

CHECK

(b) $\dfrac{1}{4}h - 5 = 1$

CHECK

(c) $\dfrac{4}{3}r + 4 = -8$

CHECK

Answers

2. **(a)** $\dfrac{5}{\cancel{4}_{1}}\left(\dfrac{\cancel{16}^{4}}{1}\right) = \dfrac{\cancel{5}^{1}}{\cancel{4}_{1}}\left(\dfrac{\cancel{4}^{1}}{\cancel{5}_{1}}x\right)$; $20 = x$

CHECK $18 = \dfrac{4}{5}x + 2$

$18 = \dfrac{4}{5}(20) + 2$

$18 = 16 + 2$

Balances $18 = 18$ ✓

(b) $h = 24$ **CHECK** $\dfrac{1}{4}h - 5 = 1$

$\dfrac{1}{4}(24) - 5 = 1$

$6 - 5 = 1$

Balances $1 = 1$ ✓

(c) $r = -9$ **CHECK** $\dfrac{4}{3}r + 4 = -8$

$\dfrac{4}{3}(-9) + 4 = -8$

$-12 + 4 = -8$

Balances $-8 = -8$ ✓

③ ⑤ On the second day, Riley's grandparents were on the road for 12 hours. Find how far they drove on the second day using the expression from **Example 3** and the six problem-solving steps.

Step 1

Unknown: distance traveled

Known: expression to find time on the road is $2 + \dfrac{\text{distance}}{60}$; time on the road was 12 hours.

Step 2 Let d represent the distance the grandparents drove.

Step 3 Write an equation.

$2 + \dfrac{\text{distance}}{60}$ is time on the road

$2 + \dfrac{d}{60} =$ _____

Answer

3. *Step 3* $2 + \dfrac{d}{60} = 12$

Step 4 $2 + \dfrac{d}{60} = 12$

$\underline{-2 -2}$

$0 + \dfrac{d}{60} = 10$

$\dfrac{\overset{1}{\cancel{60}}}{1}\left(\dfrac{1}{\underset{1}{\cancel{60}}}d\right) = \dfrac{60}{1}(10)$

$d = 600$

Step 5 Riley's grandparents drove 600 miles on the second day.

Step 6

Check. $2 + \frac{600}{60} = 2 + 10 = 12$ ← Matches

Time on the road is 12 hours. ←

EXAMPLE 3 **Solving an Application Problem Using an Equation with Fractions**

Riley's grandparents drive from Minnesota to Florida every winter. It takes them several days. Each day they use the expression $2 + \dfrac{\text{distance}}{60}$ to figure out how many hours they will be on the road. This expression includes 2 hours to stop for meals and buy gas. It also includes the distance they want to drive that day and their average driving speed of 60 miles per hour. The first day, the grandparents were on the road for 10 hours. How far did they drive that day?

Step 1 **Read** the problem. It is about how long the grandparents were on the road (time) and how far they drove (distance).

Unknown: distance traveled

Known: expression to find time on the road is $2 + \dfrac{\text{distance}}{60}$; time on the road was 10 hours.

Step 2 **Assign a variable.** Let d represent the distance the grandparents drove.

Step 3 **Write an equation** using your variable.

$2 + \dfrac{\text{distance}}{60}$ is time on the road

$2 + \dfrac{d}{60} = 10$

Step 4 **Solve the equation.**

$2 + \dfrac{d}{60} = 10$

$\underline{-2 -2}$ Add -2 to both sides.

$0 + \dfrac{d}{60} = 8$

$\frac{1}{60}d$ is equivalent to $\frac{d}{60}$ because $\frac{1}{60}d$ is $\frac{1}{60} \cdot \frac{d}{1}$ $\left\{ \begin{array}{l} \dfrac{d}{60} = 8 \\[2mm] \dfrac{1}{60}d = 8 \end{array} \right.$

$\dfrac{\overset{1}{\cancel{60}}}{1}\left(\dfrac{1}{\underset{1}{\cancel{60}}}d\right) = \dfrac{60}{1}(8)$ Multiply both sides by $\frac{60}{1}$ (the reciprocal of $\frac{1}{60}$)

$d = 480$ *Remember to write the units (miles) in your answer.*

Step 5 **State the answer.** Riley's grandparents drove **480 miles** on the first day.

Step 6 **Check** the solution by putting it back into the *original* problem.

Time on the road is $2 + \frac{d}{60}$.

If the distance traveled is 480 miles,

then $2 + \frac{480}{60} = 2 + 8 = 10$. ←

Time on the road is 10 hours. ←   Matches   *Your answer must fit all the information in the original problem.*

A distance of 480 miles is the correct solution because it fits the facts of the original problem.

◄ Work Problem **③** at the Side.

4.7 Exercises

FOR EXTRA HELP
Go to MyMathLab *for worked-out, step-by-step solutions to exercises enclosed in a square* ⬜ *and video solutions to* ▶ *exercises.*

CONCEPT CHECK

1. (a) To solve $\frac{2}{5}c = 6$, multiply both sides by which number? Circle your answer.

$$\frac{2}{5} \qquad \frac{1}{6} \qquad \frac{5}{2} \qquad \frac{6}{1}$$

   (b) Explain why you picked that number.

CONCEPT CHECK

2. (a) To solve $12 = \frac{1}{4}x$, multiply both sides by which number? Circle your answer.

$$\frac{1}{12} \qquad \frac{1}{4} \qquad \frac{12}{1} \qquad \frac{4}{1}$$

   (b) Explain why you picked that number.

Solve each equation and check each solution. ***See Examples 1 and 2.***

3. $\frac{1}{3}a = 10$ **CHECK**

   Multiply both sides by $\frac{3}{1}$, the reciprocal of $\frac{1}{3}$

   $$\frac{\overset{1}{\cancel{3}}}{1}\left(\frac{1}{\underset{1}{\cancel{3}}}a\right) = \frac{3}{1}\left(\frac{10}{1}\right)$$

   $$a = \underline{\quad}$$

4. $7 = \frac{1}{5}y$ **CHECK**

   Multiply both sides by $\frac{5}{1}$, the reciprocal of $\frac{1}{5}$

   $$\frac{5}{1}(7) = \frac{\overset{1}{\cancel{5}}}{1}\left(\frac{1}{\underset{1}{\cancel{5}}}y\right)$$

   $$\underline{\quad} = y$$

5. $-20 = \frac{5}{6}b$ **CHECK**

6. $-\frac{4}{9}w = 16$ **CHECK**

7. $-\frac{7}{2}c = -21$ **CHECK**

8. $-25 = \frac{5}{3}x$ **CHECK**

9. $\frac{3}{10} = -\frac{1}{4}d$ **CHECK**

10. $-\frac{7}{8}h = -\frac{1}{6}$ **CHECK**

11. $\frac{1}{6}n + 7 = 9$ **CHECK**

    Hint: Start by adding -7 to both sides.

12. $3 + \frac{1}{4}p = 5$ **CHECK**

    Hint: Start by adding -3 to both sides.

13. $-10 = \dfrac{5}{3}r + 5$ **CHECK**

14. $0 = 6 + \dfrac{3}{2}t$ **CHECK**

15. $\dfrac{3}{8}x - 9 = 0$ **CHECK**

16. $\dfrac{1}{3}s - 10 = -5$ **CHECK**

Solve each equation. Show your work.

17. $\underbrace{7 - 2}_{} = \dfrac{1}{5}y - 4$ First simplify the left side.

GS

$\underline{} = \dfrac{1}{5}y - 4$ Now add 4 to both sides.

     Finish the work.

18. $0 - 8 = \dfrac{1}{10}k - 3$ First simplify the left side.

GS

$\underbrace{0 + (-8)}_{} = \dfrac{1}{10}k - 3$

$\underline{} = \dfrac{1}{10}k - 3$ Now add 3 to both sides.

     Finish the work.

19. $4 + \dfrac{2}{3}n = -10 + 2$

20. $-\dfrac{2}{5}m - 3 = -9 + 0$

21. $3x + \dfrac{1}{2} = \dfrac{3}{4}$

22. $\dfrac{6}{21} = \dfrac{2}{7} - 6m$

23. $\dfrac{3}{10} = -4b - \dfrac{1}{5}$

24. $\dfrac{5}{6} = -3c - \dfrac{2}{3}$

25. $-1 - \dfrac{3}{8}n = 5 - 6$

26. $4y + \dfrac{1}{3} = \dfrac{7}{9}$

CONCEPT CHECK *Check the solution given for each equation in Exercises 27 and 28. If a solution doesn't balance, show how to find the correct solution.*

27. (a) $\dfrac{1}{6}x + 1 = -2$

$x = 18$

(b) $-\dfrac{3}{2} = \dfrac{9}{4}k$

$-\dfrac{2}{3} = k$

28. (a) $-\dfrac{3}{4}y = -\dfrac{5}{8}$

$y = \dfrac{5}{6}$

(b) $16 = -\dfrac{7}{3}w + 2$

$6 = w$

Riley's grandparents drive from Florida to Minnesota every spring. It takes them several days. Each day they use the expression $2 + \dfrac{\text{distance}}{60}$ *to figure out how many hours they will be on the road. Use this expression to solve Exercises 29 and 30.* ***See Example 3.***

29. The first day they were on the road for 8 hours. How far did they travel on the first day?

30. They were on the road for 11 hours on the second day. How far did they travel on the second day?

Robb is moving from Nevada to Wisconsin. He rented a truck to drive his furniture to his new home. Each day he uses the expression $2 + \dfrac{\text{distance}}{50}$ *to figure out how long he will be on the road. This expression includes 2 hours to stop for meals and buy gas, the distance he wants to drive that day, and his average driving speed of 50 miles per hour. In Exercises 31–34, find the distance Robb travels each day using the six problem-solving steps and the expression* $2 + \dfrac{\text{distance}}{50}$. ***See Example 3.***

31. The first day Robb was on the road for 12 hours. How far did he travel on the first day?

32. Robb was on the road for 10 hours on the second day. How far did he travel on the second day?

33. How far did Robb travel on the third day, when he was on the road for only 9 hours?

34. Find the distance Robb traveled on the final day, when he was on the road for 13 hours.

Nails for home remodeling projects are classified by the "penny" system, which is a number that indicates the nail's length. The expression for finding the length of a nail, in inches, is $\dfrac{\text{penny size}}{4} + \dfrac{1}{2}$ inch. In Exercises 35–38, find the penny size for each nail using this expression and the six problem-solving steps.
(Data from Season by Season Home Maintenance.*)*

35. The length of a common nail is 3 inches. What is its penny size?

|← ——————— 3 in. ——————— →|

36. The length of a drywall nail is 2 inches. Find the nail's penny size.

|← ——— 2 in. ——— →|

37. The length of a box nail is $2\frac{1}{2}$ inches. What penny size would you ask for when buying these nails?

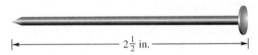

|← ——— $2\frac{1}{2}$ in. ——— →|

38. The length of a finishing nail is $1\frac{1}{2}$ inches. What is its penny size?

|← —— $1\frac{1}{2}$ in. —— →|

Relating Concepts (Exercises 39–40)   For Individual or Group Work

Use your skills in solving equations as you **work Exercises 39 and 40 in order.**

39. Write two different equations that have 8 as a solution. Write your equations with a fraction as the coefficient of the variable term. Use **Exercises 3–10** as models.

40. Write two different equations that have -12 as the solution. Write your equations with a fraction as the coefficient of the variable term. Use **Exercises 3–10** as models.

4.8 | Geometry Applications: Area and Volume

OBJECTIVE ▶ 1 Find the area of a triangle. Earlier, you worked with triangles, which are flat shapes that have exactly three sides. You found the perimeter of a triangle by adding the lengths of the three sides. Now you are ready to find the area of a triangle (the amount of surface inside the triangle).

You can find the *height* of a triangle by measuring the distance from one vertex of the triangle to the opposite side (the base). The height must be *perpendicular* to the base, that is, it must form a right angle with the base. Sometimes you have to extend the base before you can draw the height perpendicular to it, as shown below in the figure on the right.

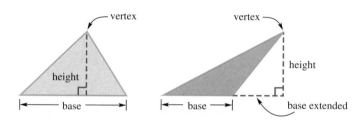

If you cut out two identical triangles and turn one upside down, you can fit them together to form a parallelogram, as shown below.

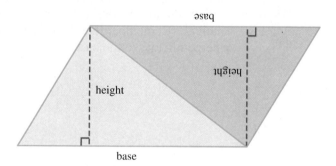

Recall that the area of the parallelogram is *base* times *height*. Because each triangle is *half* of the parallelogram, the area of one triangle is

$$\frac{1}{2} \text{ of base times height}$$

Use the following formula to find the area of a triangle.

Finding the Area of a Triangle

$$\text{Area of triangle} = \frac{1}{2} \cdot \text{base} \cdot \text{height}$$

$$A = \frac{1}{2}bh$$

Remember to use **square units** when measuring area.

1 Find the area of each triangle. Write all answers in simplest form and, when possible, as mixed numbers.

GS **(a)**

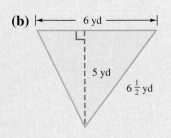

$$A = \frac{1}{2} \cdot b \cdot h$$

$$A = \frac{1}{\overset{\cancel{2}}{1}} \cdot \overset{13}{\cancel{26}} \text{ m} \cdot 20 \text{ m}$$

$$A = \underline{\hspace{2cm}}$$

(b)

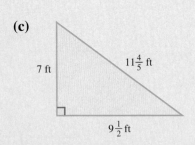

(c)

7 ft $11\frac{4}{5}$ ft

$9\frac{1}{2}$ ft

Hint: The perpendicular sides are the base and height. (Look for the small red square.)

Answers

1. **(a)** 260 m² **(b)** 15 yd²

(c) $33\frac{1}{4}$ ft²

EXAMPLE 1 **Finding the Area of Triangles**

Find the area of each triangle. Write all answers in simplest form and, when possible, as mixed numbers.

(a)

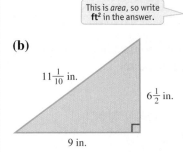

26 ft $39\frac{3}{4}$ ft   22 ft   47 ft

The base is 47 ft and the height is 22 ft. You do *not* need the 26 ft or $39\frac{3}{4}$ ft sides to find the area.

$$A = \frac{1}{2} \cdot b \cdot h$$

Replace *b* with 47 ft and *h* with 22 ft

$$A = \frac{1}{2} \cdot 47 \text{ ft} \cdot 22 \text{ ft}$$

$$A = \frac{1}{2} \cdot \frac{47 \text{ ft}}{1} \cdot \frac{22 \text{ ft}}{1}$$

Divide out the common factor of 2

$$A = \frac{1 \cdot 47 \text{ ft} \cdot \overset{1}{\cancel{2}} \cdot 11 \text{ ft}}{\underset{1}{\cancel{2}} \cdot 1 \cdot 1}$$

Multiply 47 • 11 to get 517

This is area, so write ft² in the answer. → $A = 517 \text{ ft}^2$ ← Multiply ft • ft to get ft²

(b)

$11\frac{1}{10}$ in. $6\frac{1}{2}$ in.   9 in.

Here is a right triangle. Because two sides of a right triangle are perpendicular to each other, use those sides as the base and the height. (Remember that the height must be perpendicular to the base.)

$$A = \frac{1}{2}bh$$

Formula for area of a triangle

$$A = \frac{1}{2} \cdot 9 \text{ in.} \cdot 6\frac{1}{2}\text{in.}$$

Replace *b* with 9 in. and *h* with $6\frac{1}{2}$ in.

$$A = \frac{1}{2} \cdot \frac{9 \text{ in.}}{1} \cdot \frac{13 \text{ in.}}{2}$$

Write 9 in. and $6\frac{1}{2}$ in. as improper fractions.

$$A = \frac{1 \cdot 9 \text{ in.} \cdot 13 \text{ in.}}{2 \cdot 1 \cdot 2}$$

← Multiply 9 • 13 to get 117 and in. • in. to get in.²

$$A = \frac{117}{4} \text{ in.}^2 = 29\frac{1}{4} \text{ in.}^2$$

This is area, so write in.²

◄ **Work Problem** **1** **at the Side.**

EXAMPLE 2 **Using the Concept of Area**

Find the area of the shaded part in this figure.

First find the area of the entire figure. Then *subtract* the area of the *unshaded* part.

The *entire* figure is a rectangle. Find its area.

$$A = lw$$

$$A = 30 \text{ cm} \cdot 40 \text{ cm}$$

$$A = 1200 \text{ cm}^2$$

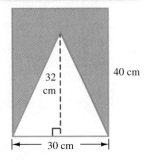

32 cm   40 cm   30 cm

Continued on Next Page

The *unshaded* part of the figure is a triangle.

$$A = \frac{1}{2} bh$$

$$A = \frac{1}{2} \cdot \frac{30 \text{ cm}}{1} \cdot \frac{32 \text{ cm}}{1}$$

$$A = \frac{1 \cdot \overset{1}{2} \cdot 15 \text{ cm} \cdot 32 \text{ cm}}{\underset{1}{2} \cdot 1 \cdot 1}$$

$$A = 480 \text{ cm}^2$$

Subtract to find the area of the shaded part.

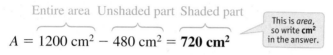

Entire area Unshaded part Shaded part

$$A = 1200 \text{ cm}^2 - 480 \text{ cm}^2 = \textbf{720 cm}^2$$

This is *area*, so write **cm²** in the answer.

——————— **Work Problem ➋ at the Side.** ▶

OBJECTIVE ➋ Find the volume of a rectangular solid. A shoe box is an example of a three-dimensional (or solid) figure. The three dimensions are length, width, and height. If we want to know how much a shoe box will hold, we find its *volume*. We measure volume by seeing how many cubes of a certain size will fill the space inside the box. Three sizes of *cubic units* are shown here. Notice that all the edges of a cube have the same length and all sides meet at right angles.

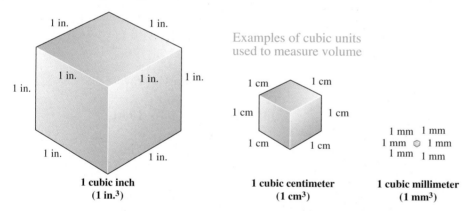

Examples of cubic units used to measure volume

1 cubic inch
(1 in.³)

1 cubic centimeter
(1 cm³)

1 cubic millimeter
(1 mm³)

Some larger sizes of cubes that are used to measure volume are 1 cubic foot (1 ft³), 1 cubic yard (1 yd³), and 1 cubic meter (1 m³).

❶ CAUTION

The raised 3 in 4^3 means you multiply 4 • 4 • 4 to get 64. The raised 3 in cm³ or ft³ is a short way to write the word "cubic." It means that you multiplied cm times cm times cm to get cm³, or ft times ft times ft to get ft³. Recall that a short way to write $x \cdot x \cdot x$ is x^3. Similarly, cm • cm • cm is cm³.

When you see 5 cm³, say "five cubic centimeters." Do *not* multiply 5 • 5 • 5 because the exponent applies only to the *first* thing to its left. The exponent applies to cm, *not* to the 5.

Volume

Volume is a measure of the space inside a solid shape. The volume of a solid is the number of cubic units it takes to fill the solid.

➋ Find the area of the shaded part in this figure.

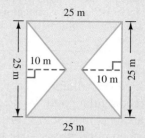

The entire figure is a square, so the area of the square is:

$$A = 25 \text{ m} \cdot 25 \text{ m}$$

$$A = \underline{\hspace{2cm}} \text{ for entire figure.}$$

Now find the area of the white triangles. Then subtract to find the shaded area.

Answer

2. $A = 625 \text{ m}^2$ for entire figure. For shaded portion, $A = 625 \text{ m}^2 - 125 \text{ m}^2 - 125 \text{ m}^2$; $A = 375 \text{ m}^2$

Use the formula below to find the volume of *rectangular solids*.

Finding the Volume of Rectangular Solids (Box-like Shapes)

Volume of a rectangular solid = length • width • height

$$V = lwh$$

Remember to use *cubic units* when measuring volume.

---

3 Find the volume of each rectangular solid. Write all answers in simplest form and, when possible, as mixed numbers.

GS (a)

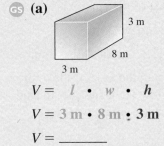

3 m

8 m

3 m

$V = \ l \ \bullet \ w \ \bullet \ h$

$V = 3 \text{ m} \bullet 8 \text{ m} \bullet 3 \text{ m}$

$V = \underline{}$

EXAMPLE 3 Finding the Volume of Rectangular Solids

Find the volume of each box.

(a)

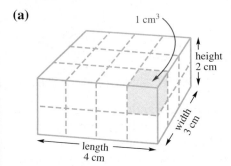

1 cm³

height 2 cm

width 3 cm

length 4 cm

Each cube that fits in the box is 1 cubic centimeter (1 cm³). To find the volume, you can count the number of cubes.

Bottom layer has 12 cubes.

Top layer has 12 cubes.

Total of 24 cubes (24 cm³)

Or you can use the formula for rectangular solids.

$$V = \ l \ \bullet \ w \ \bullet h$$

$$V = 4 \text{ cm} \bullet 3 \text{ cm} \bullet 2 \text{ cm} \quad \text{Multiply } 4 \bullet 3 \bullet 2 \text{ to get } 24$$

$$V = 24 \text{ cm}^3 \quad \text{Multiply cm} \bullet \text{cm} \bullet \text{cm to get cm}^3$$

(b)

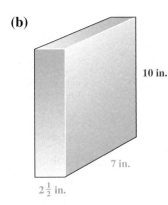

10 in.

7 in.

$2\frac{1}{2}$ in.

Use the formula $V = lwh$.

$$V = 7 \text{ in.} \bullet 2\frac{1}{2} \text{ in.} \bullet 10 \text{ in.} \quad \begin{array}{l}\text{Write each} \\ \text{measurement} \\ \text{as an improper} \\ \text{fraction.}\end{array}$$

$$V = \frac{7 \text{ in.}}{1} \bullet \frac{5 \text{ in.}}{2} \bullet \frac{10 \text{ in.}}{1}$$

$$V = \frac{7 \text{ in.} \bullet 5 \text{ in.} \bullet \overset{1}{2} \bullet 5 \text{ in.}}{1 \bullet \underset{1}{2} \bullet 1} \quad \begin{array}{l}\text{Divide out} \\ \text{the common} \\ \text{factor of 2.}\end{array}$$

$$V = 175 \text{ in.}^3 \quad \boxed{\begin{array}{l}\text{This is } volume \text{ so write} \\ \textbf{in.}^3 \text{ in the answer.}\end{array}}$$

(b) Length $6\frac{1}{4}$ ft,

width $3\frac{1}{2}$ ft, height 2 ft

◀ **Work Problem 3 at the Side.**

OBJECTIVE 3 Find the volume of a pyramid. A pyramid is a solid shape like the one below. We will study pyramids with square or rectangular bases.

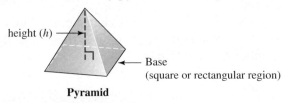

height (*h*)

Base (square or rectangular region)

Pyramid

The height is the distance from the base to the highest point of the pyramid. The height must be perpendicular to the base.

Answers

3. (a) 72 m³ (b) $43\frac{3}{4}$ ft³

Note

In this text we will work only with pyramids that have a square or rectangular base. In later math courses you may work with pyramids that have a base with three sides (triangle), five sides (pentagon), six sides (hexagon), and so on.

Use this formula to find the volume of a pyramid.

Finding the Volume of a Pyramid

$$\text{Volume of a pyramid} = \frac{1}{3} \cdot B \cdot h$$

$$V = \frac{1}{3}Bh$$

where B is the area of the base of the pyramid and h is the height of the pyramid.

Remember to use **cubic units** when measuring volume.

EXAMPLE 4 **Finding the Volume of a Pyramid**

Find the volume of this pyramid with a rectangular base. Write the answers in simplest form and, when possible, as mixed numbers.

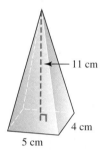

11 cm

4 cm

5 cm

First find the value of B in the formula, which is the *area of the rectangular base*. Recall that the area of a rectangle is found by multiplying length times width.

$$B = 5 \text{ cm} \cdot 4 \text{ cm}$$

$$B = 20 \text{ cm}^2 \quad \boxed{\text{First find the area of the rectangular base.}}$$

Now find the volume.

$$V = \frac{1}{3}Bh \qquad \text{Formula for volume of pyramid}$$

$$V = \frac{1}{3} \cdot 20 \text{ cm}^2 \cdot 11 \text{ cm} \qquad \text{Replace } B \text{ with 20 cm}^2 \text{ and } h \text{ with 11 cm}$$

$$V = \frac{1}{3} \cdot \frac{20 \text{ cm}^2}{1} \cdot \frac{11 \text{ cm}}{1} \qquad \text{There are no common factors to divide out.}$$

$$V = \frac{1 \cdot 20 \text{ cm}^2 \cdot 11 \text{ cm}}{3 \cdot 1 \cdot 1}$$

$$V = \frac{220}{3} \text{ cm}^3 = 73\frac{1}{3} \text{ cm}^3 \qquad \boxed{\text{This is } volume, \text{ so write cm}^3 \text{ in the answer.}}$$

— **Work Problem 4 at the Side.** ▶

4 Find the volume of a pyramid
GS with a square base measuring 10 ft by 10 ft. The height of the pyramid is 6 ft.

First find the area of the square base.

$$B = 10 \text{ ft} \cdot 10 \text{ ft}$$

$$B = \underline{\qquad}$$

Now find the volume of the pyramid.

$$V = \frac{1}{3}Bh$$

$$V = \frac{1}{3} \cdot \underline{\qquad} \cdot \underline{\qquad}$$

$$V =$$

Answer

4. $B = 100 \text{ ft}^2$; $V = \dfrac{1}{\cancel{3}} \cdot 100 \text{ ft}^2 \cdot \cancel{6}^{2} \text{ ft}$;

$V = 200 \text{ ft}^3$

4.8 Exercises

FOR EXTRA HELP   Go to MyMathLab for worked-out, step-by-step solutions to exercises enclosed in a square ▢ and video solutions to ▶ exercises.

1. **CONCEPT CHECK**   Look at **Exercise 3** below. Which measurements will you use to find the area of the triangle? Will the units in your answer be m of m²?

2. **CONCEPT CHECK**   Look at **Exercise 3** below. Which measurements will you use to find the perimeter of the triangle? Will the units in your answer be m or m²?

Find the perimeter and area of each triangle. Write all answers in simplest form and, when possible, as mixed numbers. See Example 1.

3.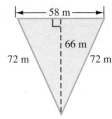
58 m · 66 m · 72 m · 72 m

4.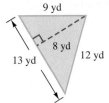
9 yd · 8 yd · 13 yd · 12 yd

5. ▶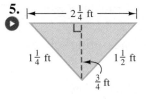
$2\frac{1}{4}$ ft · $1\frac{1}{4}$ ft · $1\frac{1}{2}$ ft · $\frac{3}{4}$ ft

6.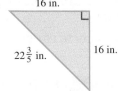
16 in. · $22\frac{3}{5}$ in. · 16 in.

7.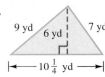
9 yd · 6 yd · 7 yd · $10\frac{1}{4}$ yd

8.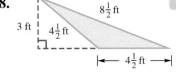
3 ft · $4\frac{1}{2}$ ft · $8\frac{1}{2}$ ft · $4\frac{1}{2}$ ft

9.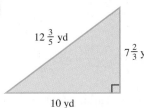
$12\frac{3}{5}$ yd · $7\frac{2}{3}$ yd · 10 yd

10.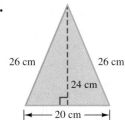
26 cm · 26 cm · 24 cm · 20 cm

Find the shaded area in each figure. See Example 2.

11. GS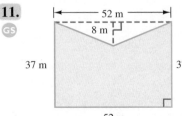
52 m · 8 m · 37 m · 37 m · 52 m

Find the area of the entire figure, which is a rectangle.

$A = 52$ m • _____

$A =$ _____

Now find the area of the white triangle and subtract it from the rectangle.

12. GS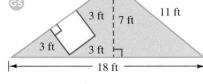
3 ft · 7 ft · 11 ft · 3 ft · 3 ft · 18 ft

Find the area of the entire figure, which is a triangle.

$A = \dfrac{1}{2} \cdot 18$ ft • ____

$A =$ ____

Now find the area of the white square and subtract it from the triangle.

Solve each application problem. Write all answers in simplest form and, when possible, as mixed numbers.

13. A triangular tent flap measures $3\frac{1}{2}$ ft along the base and has a height of $4\frac{1}{2}$ ft. How much canvas is needed to make the flap?

14. A wooden sign in the shape of a right triangle has perpendicular sides measuring $1\frac{1}{2}$ yd and 1 yd. How much surface area does the front of the sign have?

15. A triangular space between three streets has the measurements shown below.

   (a) How much new curbing is needed to go around the space?

   (b) How much grass is needed to cover the space?

16. A city lot with an unusual shape is shown below.

   (a) How much frontage (distance along streets) does the lot have?

   (b) What is the area of the lot?

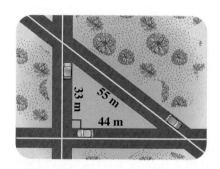

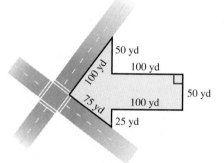

17. CONCEPT CHECK The work shown below has **two** mistakes in it.

What Went Wrong? First write a sentence explaining what each mistake is. Then fix the mistakes and find the correct solution. Pedro found the area of the triangle this way:

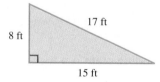

$A = 15 \text{ ft} + 8 \text{ ft} + 17 \text{ ft}$

$A = 40$

18. CONCEPT CHECK The work shown below has **two** mistakes in it.

What Went Wrong? First write a sentence explaining what each mistake is. Then fix the mistakes and find the correct solution. Rosario found the perimeter of the triangle this way:

$$P = \frac{1}{2} \cdot 10 \text{ cm} \cdot 12 \text{ cm}$$

$$P = 60 \text{ cm}^2$$

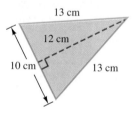

*In Exercises 19–24, name each solid and find its volume. Write all answers in simplest form and, when possible, as mixed numbers. **See Examples 3 and 4.***

19.
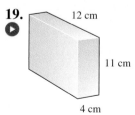

20.
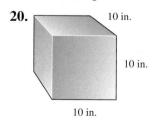

21.
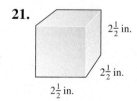

22.

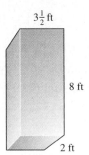

$3\frac{1}{2}$ ft

8 ft

2 ft

23.

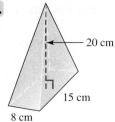

20 cm

15 cm

8 cm

24.

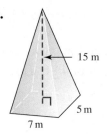

15 m

5 m

7 m

Solve each application problem.

25. Find the volume of a pyramid with a height of 5 ft and a square base that is 8 ft on each side.

26. A pyramid has a height of 4 yd. The rectangular base measures 10 yd by 4 yd. What is the volume of the pyramid?

27. A box to hold pencils measures 3 in. by 8 in. by $\frac{3}{4}$ in. high. Find the volume of the box. (Data from Faber Castell.)

$\frac{3}{4}$ in.

8 in.

3 in.

Hint: A pencil box is a rectangular solid.

28. A train is being loaded with shipping crates. Each crate is 12 ft long, 8 ft wide, and $2\frac{1}{4}$ ft high. How much space will each crate take?

$2\frac{1}{4}$ ft

8 ft

12 ft

Hint: A shipping crate is a rectangular solid.

29. One of the ancient stone pyramids in Egypt has a square base that measures 145 m on each side. The height is 93 m. What is the volume of the pyramid? (Data from *Columbia Encyclopedia*.)

30. A cardboard model of an ancient stone pyramid has a square base that is $10\frac{3}{8}$ in. on each side. The height is $6\frac{1}{2}$ in. Find the volume of the model.

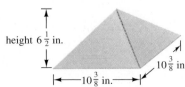

height $6\frac{1}{2}$ in.

$10\frac{3}{8}$ in.

$10\frac{3}{8}$ in.

Study Skills
ANALYZING YOUR TEST RESULTS

OBJECTIVES

1. Determine the reason for errors.
2. Develop a plan to avoid test-taking errors.
3. Review material to correct misunderstandings.

After taking a test, many students heave a big sigh of relief and try to forget it ever happened. Don't fall into this trap! An exam is a learning opportunity. It gives you clues about *what your instructor thinks is important,* what *concepts and skills are valued* in mathematics, and *if you are on the right track.*

Immediately After the Test

Jot down problems that caused you trouble. Find out how to solve them by checking your text, reviewing your notes, or asking your instructor or tutor (if available). You might see those same problems again on a final exam.

After the Test Is Returned

Find out what you got wrong and why you had points deducted. Write down the problem so you can learn how to do it correctly. Sometimes you have only a short time in class to review your test. *If you need more time,* ask your instructor if you can look at the test in his or her office.

Find Out Why You Made the Errors You Made

Here is a list of typical reasons for making errors on math tests.

1. You read the directions wrong.
2. You read the question wrong or skipped over something.
3. You made a computation error (maybe even an easy one).
4. Your answer is not accurate.
5. Your answer is not complete.
6. You labeled your answer wrong. For example, you labeled it "ft" and it should have been ft^2.
7. You didn't show your work.
8. *You didn't understand the concept.
9. *You were unable to go from words (in an application problem) to setting up the problem.
10. *You were unable to apply a procedure to a new situation.
11. You were so anxious that you made errors even when you knew the material.

The first seven errors are **test-taking errors.** They are easy to correct if you decide to carefully read test questions and directions, proofread or rework your problems, show all your work, and double check units and labels every time.

The three starred errors (*) are **test preparation errors.** Remember that to grow a complex neural network, you need to practice the kinds of problems that you will see on the tests. So, for example, if application problems are difficult for you, you must *do more application problems!* If you have practiced the study skills techniques, however, you are less likely to make these kinds of errors on tests because you will have a deeper understanding of course concepts and you will be able to remember them better.

The last error isn't really an error. **Anxiety** can play a big part in your test results. Go back to the *Preparing for Tests* activity and read the suggestions about exercise and deep breathing. Recall from the *Your Brain* Can *Learn Mathematics* activity that when you are anxious, your body produces adrenaline. The presence of *adrenaline in the brain blocks connections* between dendrites. If you can *reduce the adrenaline* in your system, you will be able to *think more clearly* during your test.

Study Skills

Just five minutes of brisk walking right before your test can help do that. Also, *practicing a relaxation technique while you do your homework* will make it more likely that you can benefit from using the technique during a test. *Deep breathing* is helpful because it gets *oxygen into your brain*. When you are anxious you tend to breathe more shallowly, which can make you feel confused and easily distracted.

Make a Plan for the Next Test

Make a plan for your next test based on your results from this test. You might review the Chapter Summary and work the problems in the Chapter Review Exercises or the Chapter Test. Ask your instructor or a tutor (if available) for more help if you are confused about any of the problems.

Now Try This

Below is a record sheet to track your progress in test taking. Use it to find out if you make particular kinds of errors. Then you can work specifically on correcting them. Just check in the box when you made one of the errors. If you take more than four tests, make your own grid on separate paper.

Test Taking Errors

Test #	Read directions wrong	Read question wrong	Computation error	Not exact or accurate	Not complete	Labeled wrong	Didn't show work
1							
2							
3							
4							

Test Preparation Errors

Test #	Didn't understand concept	Didn't set up problem correctly	Couldn't apply concept to new situation
1			
2			
3			
4			

Anxiety

Test #	Felt anxious *before* the exam	Felt anxious *during* the exam	Blanked out on questions	Got questions wrong that I knew how to do
1				
2				
3				
4				

What will you do to avoid test-taking errors?

_____
_____
_____

What will you do to avoid test preparation errors?

_____
_____
_____

What will you do to reduce anxiety?

_____
_____
_____

Chapter 4   *Summary*

Key Terms

4.1

fraction A fraction is a number of the form $\frac{a}{b}$, where a and b are integers and b is not 0.

numerator The top number in a fraction is the numerator. It shows how many of the equal parts are being considered.

denominator The bottom number in a fraction is the denominator. It shows how many equal parts are in the whole.

proper fraction In a proper fraction, the numerator is less than the denominator. The fraction is less than 1.

improper fraction In an improper fraction, the numerator is greater than or equal to the denominator. The fraction is greater than or equal to 1.

equivalent fractions Equivalent fractions have the same value even though they look different. When graphed on a number line, they are names for the same point.

4.2

lowest terms A fraction is written in lowest terms when its numerator and denominator have no common factor other than 1.

prime number A prime number is a natural number that has exactly two different factors, itself and 1. The first few prime numbers are 2, 3, 5, 7, 11, 13, and 17.

composite number A composite number has at least one factor other than itself and 1. Examples are 4, 6, 9, and 10. The number 1 is neither prime nor composite.

prime factorization In a prime factorization, every factor is a prime number. For example, the prime factorization of 24 is $2 \cdot 2 \cdot 2 \cdot 3$.

4.3

reciprocal Two numbers are reciprocals of each other if their product is 1. The reciprocal of $\frac{a}{b}$ is $\frac{b}{a}$ because $\frac{a}{b} \cdot \frac{b}{a} = 1$.

4.4

like fractions Like fractions have the same denominator.

unlike fractions Unlike fractions have different denominators.

least common denominator The least common denominator (LCD) for two fractions is the smallest positive number that can be divided evenly by both denominators.

4.5

mixed number A mixed number is a fraction and a whole number written together. It represents the sum of the whole number and the fraction. For example, $5\frac{1}{3}$ represents $5 + \frac{1}{3}$.

4.6

complex fraction A complex fraction is a fraction in which the numerator and/or denominator contain one or more fractions.

4.7

multiplication property of equality The multiplication property of equality states that you may multiply both sides of an equation by the same nonzero number and it will still be balanced.

division property of equality The division property of equality states that you may divide both sides of an equation by the same nonzero number and it will still be balanced.

4.8

volume Volume is a measure of the space inside a solid shape. Volume is measured in cubic units, such as in.^{3}, ft^3, yd^3, mm^3, cm^3, and so on.

New Symbols

Cubic units (for measuring volume)   in.^{3} ft^3 yd^3 mm^3 cm^3 m^3

New Formulas

Area of a triangle: $A = \frac{1}{2}bh$

Volume of a rectangular solid: $V = lwh$

Volume of a pyramid: $V = \frac{1}{3}Bh$

Test Your Word Power

See how well you have learned the vocabulary in this chapter.

① A **fraction** is in lowest terms if
A. it has a value less than 1
B. its numerator and denominator have no common factor other than 1
C. it is rewritten as a mixed number.

② The **denominator** of a fraction
A. is written above the fraction bar
B. is a prime number
C. shows how many equal parts are in the whole.

③ The **LCD** of two fractions is the
A. smallest number divisible by both denominators
B. largest factor common to both denominators
C. smallest number divisible by both numerators.

④ Two numbers are **reciprocals** of each other if
A. their sum is 0
B. they are written in lowest terms
C. their product is 1.

⑤ **Volume** is
A. measured in square units
B. the space inside a solid shape
C. found by multiplying base times height.

⑥ A **mixed number**
A. has a value equal to 1
B. is the sum of a whole number and a fraction
C. has a value less than 1.

⑦ A whole number is **prime** if
A. it is divisible by itself and 1
B. it can be divided by two
C. it has exactly two different factors, itself and 1.

⑧ Equivalent fractions
A. have the same denominators
B. are written in lowest terms
C. name the same point on a number line.

⑨ An **improper fraction**
A. has a value greater than or equal to 1
B. is always a negative number
C. has a value less than 1.

Answers to Test Your Word Power

1. B; *Example:* $\frac{2}{5}$ is in lowest terms, but $\frac{4}{10}$ is not.

2. C; *Example:* In $\frac{3}{4}$ the denominator, 4, shows that the whole is divided into 4 equal parts.

3. A; *Example:* The LCD of $\frac{1}{4}$ and $\frac{5}{6}$ is 12, because 12 is the smallest number that can be divided evenly by 4 and by 6.

4. C; *Example:* The reciprocal of $\frac{3}{8}$ is $\frac{8}{3}$ because $\frac{3}{8} \cdot \frac{8}{3} = \frac{24}{24} = 1$.

5. B; *Example:* The volume of a rectangular solid (box-like shape) is the space inside the box, measured in cubic units.

6. B; *Example:* The mixed number $2\frac{3}{4}$ represents $2 + \frac{3}{4}$.

7. C; *Example:* 7 is prime because it has *exactly* two different factors, 7 and 1.

8. C; *Example:* $\frac{1}{2}$ and $\frac{2}{4}$ are equivalent fractions because they both name the point halfway between 0 and 1.

9. A; *Examples:* $\frac{3}{2}$ and $\frac{8}{8}$ are improper fractions.

Quick Review

Concepts	**Examples**

4.1 **Understanding Fraction Terminology**

The *numerator* is the top number. The *denominator* is the bottom number.

In a *proper fraction,* the numerator is less than the denominator, so the fraction is less than 1. In an *improper fraction,* the numerator is greater than or equal to the denominator, so the fraction is greater than or equal to 1.

Proper fractions $\dfrac{2}{3}, \dfrac{3}{4}, \dfrac{15}{16}, \dfrac{1}{8}$ ⟵ Numerator / ⟵ Denominator

Improper fractions $\dfrac{17}{8}, \dfrac{19}{12}, \dfrac{11}{2}, \dfrac{5}{3}, \dfrac{7}{7}$

4.1 **Writing Equivalent Fractions**

Multiply or divide the numerator and denominator by the same nonzero number. The result is an equivalent fraction.

$$\frac{1}{2} = \frac{1 \cdot 8}{2 \cdot 8} = \frac{8}{16} \longleftarrow \text{Equivalent to } \tfrac{1}{2}$$

$$-\frac{12}{15} = -\frac{12 \div 3}{15 \div 3} = -\frac{4}{5} \longleftarrow \text{Equivalent to } -\tfrac{12}{15}$$

Concepts	Examples

4.2 Finding Prime Factorizations

A prime factorization of a number shows the number as the product of prime numbers. The first few prime numbers are 2, 3, 5, 7, 11, 13, and 17. You can use a division method or a factor tree to find the prime factorization.

Find the prime factorization of 24.

Division Method: Start at the bottom and work up.

$$1 \longleftarrow \text{Continue to divide until the quotient is 1}$$
$$3\overline{)3} \longleftarrow \text{Divide 3 by 3; quotient is 1}$$
$$2\overline{)6} \longleftarrow \text{Divide 6 by 2; quotient is 3}$$
$$2\overline{)12} \longleftarrow \text{Divide 12 by 2; quotient is 6}$$

Start here; divide up. $\quad 2\overline{)24} \longleftarrow$ Divide 24 by 2, the first prime; quotient is 12

$$24 = 2 \cdot 2 \cdot 2 \cdot 3$$

Factor Tree Method:

Circle each prime number.

$$24 = 2 \cdot 2 \cdot 2 \cdot 3$$

4.2 Writing Fractions in Lowest Terms

Write the prime factorization of both numerator and denominator. Divide out all common factors, using slashes to show the division. Multiply the factors in the numerator and in the denominator.

$$\frac{18}{90} = \frac{\overset{1}{\cancel{2}} \cdot \overset{1}{\cancel{3}} \cdot \overset{1}{\cancel{3}}}{\underset{1}{\cancel{2}} \cdot \underset{1}{\cancel{3}} \cdot \underset{1}{\cancel{3}} \cdot 5} = \frac{1}{5}$$

$$\frac{2b^3}{8ab} = \frac{\overset{1}{\cancel{2}} \cdot \overset{1}{\cancel{b}} \cdot b \cdot b}{\underset{1}{\cancel{2}} \cdot 2 \cdot 2 \cdot a \cdot \underset{1}{\cancel{b}}} = \frac{b^2}{4a}$$

4.3 Multiplying Fractions

Multiply the numerators and multiply the denominators. The product must be written in lowest terms. One way to do this is to write each original numerator and each original denominator as the product of primes and divide out any common factors before multiplying.

$$\left(-\frac{7}{10}\right)\left(\frac{5}{6}\right) = -\frac{7 \cdot \overset{1}{\cancel{5}}}{2 \cdot \underset{1}{\cancel{5}} \cdot 2 \cdot 3} = -\frac{7}{12}$$

$$\frac{3x^2}{5} \cdot \frac{2}{9x} = \frac{\overset{1}{\cancel{3}} \cdot \overset{1}{\cancel{x}} \cdot x \cdot 2}{5 \cdot \underset{1}{\cancel{3}} \cdot 3 \cdot \underset{1}{\cancel{x}}} = \frac{2x}{15}$$

4.3 Dividing Fractions

Rewrite the division problem as multiplying by the reciprocal of the divisor. In other words, the first number (dividend) stays the same and the second number (divisor) is changed to its reciprocal. Then use the steps for multiplying fractions. The quotient must be in lowest terms.

Division by 0 is undefined.

$$2 \div \left(-\frac{1}{3}\right) = \frac{2}{1} \cdot \left(-\frac{3}{1}\right) = -\frac{2 \cdot 3}{1 \cdot 1} = -6$$

Reciprocals

$$\frac{x^2}{y^2} \div \frac{x}{3y} = \frac{x^2}{y^2} \cdot \frac{3y}{x} = \frac{\overset{1}{\cancel{x}} \cdot x \cdot 3 \cdot \overset{1}{\cancel{y}}}{\underset{1}{\cancel{y}} \cdot y \cdot \underset{1}{\cancel{x}}} = \frac{3x}{y}$$

Reciprocals

Concepts	Examples

4.4 Adding and Subtracting Like Fractions

You can add or subtract fractions *only* when they have the *same* denominator. Add or subtract the numerators and write the result over the common denominator. Be sure that the final result is in lowest terms.

$$\frac{3}{10} - \frac{7}{10} = \frac{3 - 7}{10} = \frac{-4}{10} \quad \text{or} \quad -\frac{4}{10}$$

Now write $-\frac{4}{10}$ in lowest terms.

$$-\frac{4}{10} = -\frac{\overset{1}{2} \cdot 2}{2 \cdot 5} = -\frac{2}{5} \left.\begin{cases} \text{Lowest} \\ \text{terms} \end{cases}\right.$$

$$\frac{5}{a} + \frac{7}{a} = \frac{5 + 7}{a} = \frac{\mathbf{12}}{\mathbf{a}} \left.\begin{cases} \text{Lowest} \\ \text{terms} \end{cases}\right.$$

4.4 Finding the Lowest Common Denominator (LCD)

Write the prime factorization of each denominator. Then use just enough prime factors in the LCD so you "cover" both denominators.

What is the LCD for $\frac{5}{12}$ and $\frac{5}{18}$?

Factors of 12

$$12 = 2 \cdot 2 \cdot 3$$
$$18 = 2 \cdot 3 \cdot 3$$
$$\text{LCD} = 2 \cdot 2 \cdot 3 \cdot 3 = 36$$

Factors of 18

The LCD for $\frac{5}{12}$ and $\frac{5}{18}$ is **36**

4.4 Adding and Subtracting Unlike Fractions

Find the LCD. Rewrite each original fraction as an equivalent fraction whose denominator is the LCD. Then add or subtract the numerators and keep the common denominator. Be sure that the final result is in lowest terms.

$$-\frac{5}{12} + \frac{7}{9}$$

The LCD is 36.

Rewrite: $\quad -\dfrac{5}{12} = -\dfrac{5 \cdot 3}{12 \cdot 3} = -\dfrac{15}{36}$

Rewrite: $\quad \dfrac{7}{9} = \dfrac{7 \cdot 4}{9 \cdot 4} = \dfrac{28}{36}$

Add: $\quad -\dfrac{15}{36} + \dfrac{28}{36} = \dfrac{-15 + 28}{36} = \dfrac{\mathbf{13}}{\mathbf{36}} \left.\begin{cases} \text{Lowest} \\ \text{terms} \end{cases}\right.$

$$\frac{2}{3} - \frac{6}{x}$$

The LCD is $3 \cdot x$ or $3x$.

Rewrite: $\quad \dfrac{2}{3} = \dfrac{2 \cdot x}{3 \cdot x} = \dfrac{2x}{3x}$

Rewrite: $\quad \dfrac{6}{x} = \dfrac{6 \cdot 3}{x \cdot 3} = \dfrac{18}{3x}$

Subtract: $\quad \dfrac{2x}{3x} - \dfrac{18}{3x} = \dfrac{\mathbf{2x - 18}}{\mathbf{3x}} \left.\begin{cases} \text{Lowest} \\ \text{terms} \end{cases}\right.$

Concepts	Examples

4.5 Mixed Numbers and Improper Fractions

Changing Mixed Numbers to Improper Fractions
Multiply the denominator by the whole number, add the numerator, and place the result over the original denominator.

Changing Improper Fractions to Mixed Numbers
Divide the numerator by the denominator and place the remainder over the original denominator.

Mixed to improper $\quad 7\dfrac{2}{3} = \dfrac{23}{3} \leftarrow (3 \cdot 7) + 2$

Same denominator

Improper to mixed $\quad \dfrac{17}{5} = 3\dfrac{2}{5}$

Same denominator

4.5 Multiplying Mixed Numbers

First, round the numbers and estimate the answer. Then follow these steps to find the exact answer.

Step 1 Rewrite each mixed number as an improper fraction.

Step 2 Multiply.

Step 3 Write the answer in lowest terms. Then the answer is in simplest form. If desired, change an improper fraction answer to a mixed number.

Estimate:

$$1\dfrac{3}{5} \quad \bullet \quad 3\dfrac{1}{3}$$

$\downarrow$ Rounded $\downarrow$

$$2 \quad \bullet \quad 3 = 6$$

$\uparrow$ Estimate

Exact:

$$1\dfrac{3}{5} \cdot 3\dfrac{1}{3} = \dfrac{8}{5} \cdot \dfrac{10}{3}$$

$$\dfrac{8 \cdot 2 \cdot \overset{1}{\cancel{5}}}{\underset{1}{\cancel{5}} \cdot 3}$$

$$\dfrac{16}{3} \text{ or } 5\dfrac{1}{3} \leftarrow$$

Close to estimate of 6

4.5 Dividing Mixed Numbers

First, round the numbers and estimate the answer. Then follow these steps to find the exact answer.

Step 1 Rewrite each mixed number as an improper fraction.

Step 2 Divide. (Rewrite as multiplication using the reciprocal of the divisor.)

Step 3 Write the answer in lowest terms. Then the answer is in simplest form. If desired, change an improper fraction answer to a mixed number.

Estimate:

$$3\dfrac{3}{4} \quad \div \quad 2\dfrac{2}{5}$$

$\downarrow$ Rounded $\downarrow$

$$4 \quad \div \quad 2 = 2$$

$\uparrow$ Estimate

Exact:

$$3\dfrac{3}{4} \div 2\dfrac{2}{5} = \dfrac{15}{4} \div \dfrac{12}{5}$$

Reciprocal of $\frac{12}{5}$ is $\frac{5}{12}$

$$\dfrac{15}{4} \cdot \dfrac{5}{12}$$

$$\dfrac{\overset{1}{\cancel{3}} \cdot 5 \cdot 5}{4 \cdot \underset{1}{\cancel{3}} \cdot 4}$$

$$\dfrac{25}{16} \text{ or } 1\dfrac{9}{16} \leftarrow$$

Close to estimate of 2

4.5 Adding and Subtracting Mixed Numbers

First round the numbers and estimate the answer. Then rewrite the mixed numbers as improper fractions and follow the steps for adding and subtracting fractions. Write the answer in simplest form.

Estimate:

$$2\dfrac{3}{8} + 3\dfrac{3}{4}$$

$\downarrow \qquad \downarrow$

$$2 \quad + \quad 4 = 6 \leftarrow \text{Estimate}$$

Exact:

$$2\dfrac{3}{8} + 3\dfrac{3}{4} = \dfrac{19}{8} + \dfrac{15}{4} = \dfrac{19}{8} + \dfrac{30}{8} = \dfrac{19 + 30}{8} =$$

$$\dfrac{49}{8} \text{ or } 6\dfrac{1}{8} \leftarrow \text{Close to estimate of 6}$$

Concepts	**Examples**

4.6 Simplifying Fractions with Exponents

The meaning of an exponent is the same for fractions as it is for integers. An exponent is a way to write repeated multiplication.

$$\left(-\frac{2}{3}\right)^2 \quad \text{means} \quad \left(-\frac{2}{3}\right)\left(-\frac{2}{3}\right) = \frac{2 \cdot 2}{3 \cdot 3} = \frac{4}{9}$$

The product of two negative numbers is positive.

4.6 Order of Operations

The order of operations is the same for fractions as for integers.

1. Work inside *parentheses* or *other grouping symbols*.

2. Simplify expressions with *exponents*.

3. Do the remaining *multiplications and divisions* as they occur **from left to right**.

4. Do the remaining *additions and subtractions* as they occur **from left to right**.

Simplify.

$$-\frac{2}{3} + 3\left(\frac{1}{4}\right)^2 \qquad \text{Cannot work inside parentheses.} \\ \text{Apply the exponent: } \frac{1}{4} \cdot \frac{1}{4} \text{ is } \frac{1}{16}$$

$$-\frac{2}{3} + 3\left(\frac{1}{16}\right) \qquad \text{Multiply next: } 3\left(\frac{1}{16}\right) \text{ is } \frac{3}{1} \cdot \frac{1}{16} = \frac{3}{16}$$

$$-\frac{2}{3} + \frac{3}{16} \qquad \text{Add last. The LCD is 48}$$

$$-\frac{32}{48} + \frac{9}{48} \qquad \text{Rewrite } -\frac{2}{3} \text{ as } -\frac{32}{48} \\ \text{Rewrite } \frac{3}{16} \text{ as } \frac{9}{48}$$

$$\frac{-32 + 9}{48} \qquad \text{Add the numerators. Keep the common denominator.}$$

$$-\frac{\mathbf{23}}{\mathbf{48}} \qquad \text{The answer is in lowest terms.}$$

4.6 Simplifying Complex Fractions

Recall that the fraction bar indicates division. Rewrite the complex fraction using the ÷ symbol for division. Then follow the steps for dividing fractions.

Simplify. $\dfrac{-\dfrac{4}{5}}{10}$

Rewrite: $-\dfrac{4}{5} \div 10$

$$-\frac{4}{5} \div 10 = -\frac{4}{5} \cdot \frac{1}{10} = -\frac{\overset{1}{\cancel{2}} \cdot 2 \cdot 1}{5 \cdot \underset{1}{\cancel{2}} \cdot 5} = -\frac{\mathbf{2}}{\mathbf{25}}$$

Reciprocals

Concepts	**Examples**

4.7 **Solving Equations Containing Fractions**

Solve the equation. Check the solution.

$$\frac{1}{3}b + 6 = 10$$

Step 1 If necessary, add the same number to both sides of the equation so that the variable term is by itself on one side of the equal sign.

$$\underline{\quad -6 \quad -6 \quad} \qquad \text{Add } -6 \text{ to both sides.}$$

$$\frac{1}{3}b + 0 = 4$$

$$\underbrace{\qquad\qquad}$$

$$\frac{1}{3}b = 4$$

Step 2 Multiply both sides by the reciprocal of the coefficient of the variable term.

$$\frac{\cancel{3}}{1}\left(\frac{1}{\cancel{3}}b\right) = \frac{3}{1}(4) \qquad \begin{array}{l}\text{Multiply both sides by } \frac{3}{1}\\ \text{(the reciprocal of } \frac{1}{3}\text{)}\end{array}$$

$$\boldsymbol{b = 12} \qquad \text{The solution is 12}$$

The solution is **12**

Step 3 To check your solution, go back to the *original* equation and replace the variable with your solution. If the equation balances, your solution is correct. If it does not balance, rework the problem.

CHECK

$$\frac{1}{3}b + 6 = 10 \qquad \text{Original equation}$$

$$\frac{1}{3}(12) + 6 = 10 \qquad \text{Replace } b \text{ with 12}$$

$$\underbrace{\quad 4 \quad} + 6 = 10$$

$$10 = 10 \checkmark \quad \text{Balances}$$

When b is 12, the equation balances, so **12** is the correct solution (**not** 10).

4.8 **Finding the Area of a Triangle**

Use this formula to find the area of a triangle.

$$\text{Area} = \frac{1}{2} \cdot \text{base} \cdot \text{height}$$

$$A = \frac{1}{2}bh$$

Remember that area is measured in **square units**.

Find the area of this triangle.

$$A = \frac{1}{2}bh$$

$$A = \frac{1}{2} \cdot 20 \text{ ft} \cdot 5 \text{ ft}$$

$$A = \frac{1}{2} \cdot \frac{20 \text{ ft}}{1} \cdot \frac{5 \text{ ft}}{1}$$

$$A = \frac{1 \cdot \overset{1}{2} \cdot 10 \text{ ft} \cdot 5 \text{ ft}}{\underset{1}{2} \cdot 1 \cdot 1}$$

$$\boldsymbol{A = 50} \text{ ft}^2 \quad \text{Measure area in square units.}$$

4.8 **Finding the Volume of a Rectangular Solid**

Use this formula to find the volume of box-like solids.

$$\text{Volume} = \text{length} \cdot \text{width} \cdot \text{height}$$

$$V = lwh$$

Volume is measured in **cubic units**.

Find the volume of this box.

$$V = l \cdot w \cdot h$$

$$V = 5 \text{ cm} \cdot 3 \text{ cm} \cdot 6 \text{ cm}$$

$$\boldsymbol{V = 90} \text{ cm}^3$$

Measure volume in cubic units.

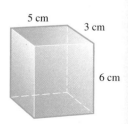

Concepts

4.8   Finding the Volume of a Pyramid

Use this formula to find the volume of a pyramid.

$$\text{Volume} = \frac{1}{3} \cdot B \cdot h$$

$$V = \frac{1}{3} Bh$$

where B is the area of the base and h is the height of the pyramid.

Volume is measured in **cubic units**.

Examples

Find the volume of a pyramid with a square base 2 cm by 2 cm and a height of 6 cm.

$$\text{Area of square base} = 2 \text{ cm} \cdot 2 \text{ cm}$$

$$B = 4 \text{ cm}^2$$

$$V = \frac{1}{3} \cdot B \cdot h$$

$$V = \frac{1}{3} \cdot 4 \text{ cm}^2 \cdot 6 \text{ cm}$$

$$V = \frac{1}{3} \cdot \frac{4 \text{ cm}^2}{1} \cdot \frac{6 \text{ cm}}{1}$$

$$V = \frac{1 \cdot 4 \text{ cm}^2 \cdot \overset{1}{3} \cdot 2 \text{ cm}}{\underset{1}{3} \cdot 1 \cdot 1}$$

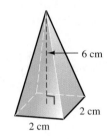

$$V = 8 \text{ cm}^3 \quad \text{Measure volume in cubic units.}$$

Chapter 4 Review Exercises

4.1 **1.** What fraction of these figures are squares? What fraction are circles?

2. Write fractions to represent the shaded and unshaded portions of this figure.

3. Graph $-\frac{1}{2}$ and $1\frac{1}{2}$ on the number line.

$$-3 \quad -2 \quad -1 \quad 0 \quad 1 \quad 2 \quad 3$$

4. Simplify each fraction.

(a) $-\dfrac{20}{5}$ (b) $\dfrac{8}{1}$ (c) $-\dfrac{3}{3}$

4.2 *Write each fraction in lowest terms.*

5. $\dfrac{28}{32}$

6. $\dfrac{54}{90}$

7. $\dfrac{16}{25}$

8. $\dfrac{15x^2}{40x}$

9. $\dfrac{7a^3}{35a^3b}$

10. $\dfrac{12mn^2}{21m^3n}$

4.3 *Multiply or divide. Simplify all answers.*

11. $-\dfrac{3}{8} \div (-6)$

12. $\dfrac{2}{5}$ of (-30)

13. $\dfrac{4}{9}\left(\dfrac{2}{3}\right)$

14. $\left(\dfrac{7}{3x^3}\right)\left(\dfrac{x^2}{14}\right)$

15. $\dfrac{ab}{5} \div \dfrac{b}{10a}$

16. $\dfrac{18}{7} \div 3k$

4.4 *Add or subtract. Simplify all answers.*

17. $-\dfrac{5}{12} + \dfrac{5}{8}$

18. $\dfrac{2}{3} - \dfrac{4}{5}$

19. $4 - \dfrac{5}{6}$

20. $\dfrac{7}{9} + \dfrac{13}{18}$

21. $\dfrac{n}{5} + \dfrac{3}{4}$

22. $\dfrac{3}{10} - \dfrac{7}{y}$

4.5 *First, round each mixed number to the nearest whole number and estimate the answer. Then find the exact answer.*

23. $2\frac{1}{4} \div 1\frac{5}{8}$

Estimate:

_____ ÷ _____ = _____

Exact:

24. $7\frac{1}{3} - 4\frac{5}{6}$

Estimate:

_____ − _____ = _____

Exact:

25. $1\frac{3}{4} + 2\frac{3}{10}$

Estimate:

_____ + _____ = _____

Exact:

4.6 *Simplify.*

26. $\left(-\frac{3}{4}\right)^3$

27. $\left(\frac{2}{3}\right)^2 \left(-\frac{1}{2}\right)^4$

28. $\frac{2}{5} + \frac{3}{10}(-4)$

29. $-\frac{5}{8} \div \left(-\frac{1}{2}\right)\left(\frac{14}{15}\right)$

30. $\dfrac{\frac{5}{8}}{\frac{1}{16}}$

31. $\dfrac{\frac{8}{9}}{-6}$

4.7 *Solve each equation. Show your work.*

32. $-12 = -\frac{3}{5}w$

33. $18 + \frac{6}{5}r = 0$

34. $3x - \frac{2}{3} = \frac{5}{6}$

4.8 *Find the area of the triangle. In Exercises 36 and 37, name each solid and find its volume. Write all answers in simplest form and, when possible, as mixed numbers.*

35.
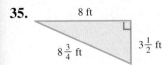
8 ft

$8\frac{3}{4}$ ft $3\frac{1}{2}$ ft

36.
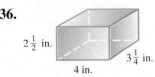
$2\frac{1}{2}$ in.

$3\frac{1}{4}$ in.

4 in.

37.
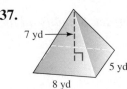
7 yd

5 yd

8 yd

Chapter 4   *Mixed Review Exercises*

Practicing exercises in mixed-up order helps you prepare for tests.

Simplify.

1. $-\dfrac{3}{10} + \dfrac{5}{6}$

2. $\dfrac{-\dfrac{2}{5}}{-4}$

3. $5 - \dfrac{7}{9}$

Solve. Show your work.

4. $-\dfrac{2}{5}y = 6$

5. $-1 = \dfrac{5}{3}d - 11$

6. $\dfrac{3}{10} = -\dfrac{1}{12}x$

Simplify.

7. $\left(\dfrac{3}{t}\right)\left(\dfrac{2t^3}{3}\right)$

8. $\dfrac{5}{b} - \dfrac{3}{4}$

9. $\left(-\dfrac{1}{3}\right)^2\left(-\dfrac{1}{2}\right)^3$

Solve. Show your work.

10. $-2 + \dfrac{3}{8}c = 0$

11. $3 - 10 = -\dfrac{1}{3}h - 4$

12. $-2x - \dfrac{4}{5} = \dfrac{7}{10}$

Solve each application problem. Write all answers in simplest form and, when possible, as mixed numbers.

13. A chili recipe that makes 10 servings uses $2\frac{1}{2}$ pounds of meat.

    (a) How much meat is in each serving?

    (b) How much meat would be needed to make 30 servings?

    (c) How much meat would be needed to make 6 servings?

14. Yanli worked as a math tutor for $4\frac{1}{2}$ hours on Monday, $2\frac{3}{4}$ hours on Tuesday, and $3\frac{2}{3}$ hours on Friday.

    (a) How much longer did she work on Monday than on Friday?

    (b) How many hours did she work in all?

    (c) How much less did she work on Tuesday than on Friday?

15. There are 60 children in the day care center. If $\frac{1}{5}$ of the children are preschoolers, $\frac{2}{3}$ of the children are toddlers, and the rest are infants, find the number of children in each age group.

16. A rectangular city park is $\frac{3}{4}$ mile long and $\frac{3}{10}$ mile wide. Find the perimeter and area of the park.

17.

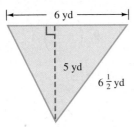

    (a) The perimeter of this triangle is $18\frac{1}{3}$ yd. Find the length of the third side.

    (b) If the perimeter was 19 yd, what would be the length of the third side?

    (c) Explain how you could check your answers in parts (a) and (b) above.

18.

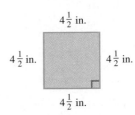

    (a) Use addition to find the perimeter of this square.

    (b) Now use multiplication to find the perimeter.

    (c) Suppose the perimeter of the square was $9\frac{3}{4}$ in. How long would each side be?

Chapter 4

The Chapter Test Prep Videos with step-by-step solutions are available in MyMathLab or on YouTube at **https://goo.gl/c3befo**

1. Write fractions to represent the shaded and unshaded portions of the figure.

2. Graph $-\frac{2}{3}$ and $2\frac{1}{3}$ on the number line below.

Write each fraction in lowest terms.

3. $\dfrac{21}{84}$

4. $\dfrac{25}{54}$

5. $\dfrac{6a^2b}{9b^2}$

Simplify.

6. $\dfrac{1}{6} + \dfrac{7}{10}$

7. $-\dfrac{3}{4} \div \dfrac{3}{8}$

8. $\dfrac{5}{8} - \dfrac{4}{5}$

9. $(-20)\left(-\dfrac{7}{10}\right)$

10. $\dfrac{\frac{4}{9}}{-6}$

11. $4 - \dfrac{7}{8}$

12. $-\dfrac{2}{9} + \dfrac{2}{3}$

13. $\dfrac{21}{24}\left(\dfrac{9}{14}\right)$

14. $\dfrac{12x}{7y} \div 3x$

15. $\dfrac{6}{n} - \dfrac{1}{4}$

16. $\dfrac{2}{3} + \dfrac{a}{5}$

17. $\left(\dfrac{5}{9b^2}\right)\left(\dfrac{b}{10}\right)$

18. $\left(-\dfrac{1}{2}\right)^3\left(\dfrac{2}{3}\right)^2$

19. $\dfrac{1}{6} + 4\left(\dfrac{2}{5} - \dfrac{7}{10}\right)$

First, round the numbers and estimate each answer. Then find the exact answer. Write exact answers in simplest form.

20. $4\dfrac{4}{5} \div 1\dfrac{1}{8}$ Estimate: Exact:

21. $3\dfrac{2}{5} - 1\dfrac{9}{10}$ Estimate: Exact:

Solve each equation. Show your work.

22. $7 = \dfrac{1}{5}d$

23. $-\dfrac{3}{10}t = \dfrac{9}{14}$

24. $0 = \dfrac{1}{4}b - 2$

25. $\dfrac{4}{3}x + 7 = -13$

Find the area of each triangle. Write all answers in simplest form and, when possible, as mixed numbers.

26.

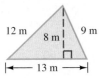

12 m   8 m   9 m

13 m

27.

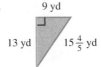

9 yd

13 yd $15\frac{4}{5}$ yd

Name each solid and find its volume.

28.

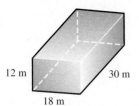

12 m   30 m

18 m

29.

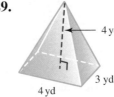

4 yd

3 yd

4 yd

Solve each application problem. Show your work. Write all answers in simplest form and, when possible, as mixed numbers.

30. Ann-Marie Sargent is training for an upcoming wheelchair race. She rides $4\frac{5}{6}$ hours on Monday, $6\frac{2}{3}$ hours on Tuesday, and $3\frac{1}{4}$ hours on Wednesday. How many hours did she spend in all? How many more hours did she train on Tuesday than on Monday?

31. A new vaccine is synthesized at the rate of $2\frac{1}{2}$ ounces per day. How long will it take to synthesize $8\frac{3}{4}$ ounces?

32. There are 8448 students at the Metro Community College campus. If $\frac{7}{8}$ of the students work either full time or part time, find the total number of students who work.

Chapters 1–4 *Cumulative Review Exercises*

1. Write this number in words. 505,008,238

2. Write this number using digits.
thirty-five billion, six hundred million,
nine hundred sixteen

3. (a) Round 60,719 to the nearest hundred.

   (b) Round 99,505 to the nearest thousand.

   (c) Round 3206 to the nearest ten.

4. Name the property illustrated by each example.

   (a) $-3 \cdot 6 = 6 \cdot (-3)$

   (b) $(7 + 18) + 2 = 7 + (18 + 2)$

   (c) $5(-10 + 7) = 5 \cdot (-10) + 5 \cdot 7$

Simplify.

5. $9\,(-6)$

6. $-10 - 10$

7. $\dfrac{-14}{0}$

8. $6 + 3\,(2 - 7)^2$

9. $|-8| + |2|$

10. $-8 + 24 \div 2$

11. $(-4)^2 - 2^5$

12. $\dfrac{-45 \div (-5)(3)}{-5 - 4(0 - 8)}$

13. $-3 \div \dfrac{3}{8}$

14. $-\dfrac{5}{6}\left(-42\right)$

15. $\left(\dfrac{5a^2}{12}\right)\left(\dfrac{18}{a}\right)$

16. $\dfrac{7}{x^2} \div \dfrac{7y^2}{3x}$

17. $\dfrac{3}{10} - \dfrac{5}{6}$

18. $-\dfrac{3}{8} + \dfrac{11}{16}$

19. $\dfrac{2}{3} - \dfrac{b}{7}$

20. $\dfrac{8}{5} + \dfrac{3}{n}$

21. $3\dfrac{1}{4} \div 2\dfrac{1}{4}$

22. $2\dfrac{2}{5} - 1\dfrac{3}{4}$

23. $\left(\dfrac{1}{2}\right)^3 (-2)^3$

24. $\dfrac{\dfrac{7}{12}}{-\dfrac{14}{15}}$

Solve each equation. Show your work.

25. $-5a + 2a = 12 - 15$

26. $y = -7y - 40$

27. $-10 = 6 + \dfrac{4}{9}k$

28. $20 + 5x = -2x - 8$

29. $3(-2m + 5) = -4m + 9 + m$

Name each figure and find its perimeter and area. Write all answers in simplest form and, when possible, as mixed numbers.

30.

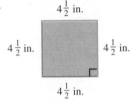

31.

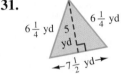

32.

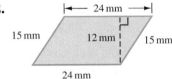

Solve each application problem using the six problem-solving steps.

33. Three new bags of birdseed each weighed the same amount. Sixteen pounds of seed were used from one bag and 25 pounds from another. There were still 79 pounds of seed left. How much did each bag weigh originally?

34. A rectangular parking lot is twice as long as it is wide. If the perimeter of the lot is 102 yd, find the length and width of the parking lot.

5

Rational Numbers: Positive and Negative Decimals

Veterinary professionals treat some of the 312.1 million pets in the United States. They use decimals as they diagnose sick pets and prescribe medications. (Data from www.americanpetproducts.org)

5.1 | Reading and Writing Decimal Numbers

VOCABULARY TIP

Deci- The prefix **deci** means **tenth**. For example, a **deci**meter is one **tenth** of a meter.

---

1 There are 10 dimes in one dollar. Each dime is $\frac{1}{10}$ of a dollar.

Write a fraction, a decimal, and the words that name the yellow shaded portion of each dollar.

(a)

(b)

Earlier, you worked with rational numbers written in fraction form to represent parts of a whole. In this chapter, we will use rational numbers written as **decimals** to show parts of a whole. For example, our money system is based on decimals. One dollar is divided into 100 equivalent parts. One cent ($0.01) is one of the parts, and a dime ($0.10) is 10 of the parts.

OBJECTIVE 1 Write parts of a whole using decimals. Decimals are used when a whole is divided into 10 equivalent parts or into 100 or 1000 or 10,000 equivalent parts. In other words, decimals are fractions with denominators that are a power of 10. For example, the square at the right is cut into 10 equivalent parts. Written as a fraction, each part is $\frac{1}{10}$ of the whole. Written as a decimal, each part is **0.1**. Both $\frac{1}{10}$ and 0.1 are read as "*one tenth*."

One tenth of the square is shaded.

The dot in 0.1 is called the **decimal point.**

$$0.1$$
↑
Decimal point

The square at the right has **7** of its 10 parts shaded.
Written as a *fraction*, $\frac{7}{10}$ of the square is shaded.
Written as a *decimal*, **0.7** of the square is shaded.
Both $\frac{7}{10}$ and **0.7** are read as "*seven tenths*."

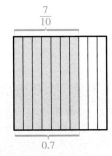

Seven tenths of the square is shaded.

◀ **Work Problem 1 at the Side.**

Each square below is cut into 100 equivalent parts.
Written as a fraction, each part is $\frac{1}{100}$ of the whole.
Written as a decimal, each part is **0.01** of the whole.
Both $\frac{1}{100}$ and **0.01** are read as "*one hundredth*."

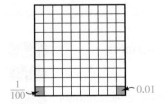

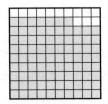

Eighty-seven hundredths of the square is shaded.

The square above on the right has 87 of its 100 parts shaded.
Written as a fraction, $\frac{87}{100}$ of the total area is shaded.
Written as a decimal, **0.87** of the total area is shaded.
Both $\frac{87}{100}$ and **0.87** are read as "*eighty-seven hundredths*."

Answers

1. (a) $\frac{1}{10}$; 0.1; one tenth

   (b) $\frac{9}{10}$; 0.9; nine tenths

Work Problem ② at the Side. ▶

The example below shows several numbers written as fractions, as decimals, and in words.

EXAMPLE 1 Using the Decimal Forms of Fractions

Fraction	Decimal	Read as
(a) $\dfrac{4}{10}$	**0.4**	four tenths
(b) $-\dfrac{9}{100}$	**−0.09**	**negative** nine hundredths
(c) $\dfrac{71}{100}$	**0.71**	seventy-one hundredths
(d) $\dfrac{8}{1000}$	**0.008**	eight thousandths
(e) $-\dfrac{45}{1000}$	**−0.045**	**negative** forty-five thousandths
(f) $\dfrac{832}{1000}$	**0.832**	eight hundred thirty-two thousandths

Work Problem ③ at the Side. ▶

OBJECTIVE ② Identify the place value of a digit. The decimal point separates the *whole number part* from the *fractional part* in a decimal number. In the chart below, you see that the **place value** names for fractional parts are similar to those on the whole number side but end in "*ths*."

Decimal Place Value Chart

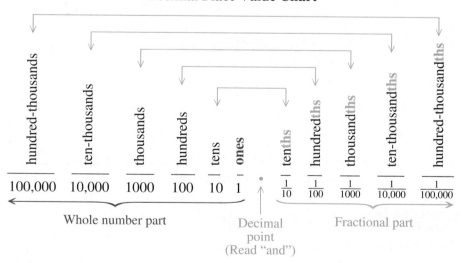

hundred-thousands	ten-thousands	thousands	hundreds	tens	ones	tenths	hundredths	thousandths	ten-thousandths	hundred-thousandths
100,000	10,000	1000	100	10	1	$\frac{1}{10}$	$\frac{1}{100}$	$\frac{1}{1000}$	$\frac{1}{10,000}$	$\frac{1}{100,000}$

← Whole number part   Decimal point (Read "and")   Fractional part

Note

Notice that the **ones** place is at the center of the place value chart above. There is no "oneths" place.

Also notice that each place is 10 times the value of the place to its right.

Finally, be sure to write a **hyphen** (dash) in ten–thousand**ths** and hundred–thousand**ths**.

② Write the shaded portion of each square as a fraction, as a decimal, and in words.

(GS) (a)

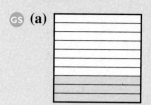

$$\frac{3}{10} = 0.\underline{\quad} = \underline{\quad} \text{ tenths}$$

(b)

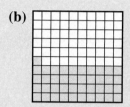

③ Write each decimal as a fraction.

(GS) (a) $-0.7 = -\dfrac{7}{\boxed{}}$

(b) 0.2

(c) −0.03

(d) 0.69

(e) 0.047

(f) −0.351

Answers

2. (a) $\dfrac{3}{10} = 0.3 = $ three tenths

   (b) $\dfrac{41}{100}$; 0.41; forty-one hundredths

3. (a) $-\dfrac{7}{10}$ (b) $\dfrac{2}{10}$ (c) $-\dfrac{3}{100}$

   (d) $\dfrac{69}{100}$ (e) $\dfrac{47}{1000}$ (f) $-\dfrac{351}{1000}$

4 Identify the place value of each digit.

(a) 971.54

(b) 0.4

(c) 5.60

(d) 0.0835

5 Tell how to read each decimal in words.

GS (a) 0.6   six _____

GS (b) 0.46   forty-six _____

(c) 0.05

(d) 0.409

(e) 0.0003

(f) 0.2703

(g) 0.088

Answers

4. (a)
hundreds | tens | ones | tenths | hundredths
9 7 1 . 5 4

(b)
ones | tenths
0 . 4

(c)
ones | tenths | hundredths
5 . 6 0

(d)
ones | tenths | hundredths | thousandths | ten-thousandths
0 . 0 8 3 5

5. (a) six tenths
   (b) forty-six hundredths
   (c) five hundredths
   (d) four hundred nine thousandths
   (e) three ten-thousandths
   (f) two thousand seven hundred three ten-thousandths
   (g) eighty-eight thousandths

! CAUTION

If a number does *not* have a decimal point, it is an *integer*. An integer has no fractional part. If you want to show the decimal point in an integer, it is just to the **right** of the digit in the ones place. Here are three examples.

$$8 = 8. \qquad 306 = 306. \qquad -42 = -42.$$

Decimal point · · · Decimal point · · · Decimal point

EXAMPLE 2 **Identifying the Place Value of a Digit**

Identify the place value of each digit.

> Notice the **hyphen** (dash) in ten-thousandths and in hundred-thousandths.

(a) 178.36

(b) 0.00935

(a)
hundreds | tens | ones | tenths | hundredths
1 7 8 . 3 6

(b)
ones | tenths | hundredths | thousandths | ten-thousandths | hundred-thousandths
0 . 0 0 9 3 5

Notice in **Example 2(b)** that we do *not* use commas on the right side of the decimal point.

◀ **Work Problem 4 at the Side.**

OBJECTIVE ▶ 3 Read decimal numbers. A decimal number is read according to its form as a fraction.

ones | tenths
0 . 9

We read 0.9 as "nine tenths" because 0.9 is the same as $\frac{9}{10}$. Notice that 0.9 ends in the tenths place.

ones | tenths | hundredths
0 . 0 2

We read 0.02 as "two hundredths" because 0.02 is the same as $\frac{2}{100}$. Notice that 0.02 ends in the hundredths place.

EXAMPLE 3 **Reading Decimal Numbers**

Tell how to read each decimal in words.

(a) 0.3   Because $0.3 = \frac{3}{10}$, read the decimal as: three <u>tenths</u>

(b) 0.49   Read it as: forty-nine <u>hundredths</u>

> Think: $0.08 = \frac{8}{100}$ so write *hundredths*.

(c) 0.08   Read it as: eight <u>hundredths</u>

(d) 0.918   Read it as: nine hundred eighteen <u>thousandths</u>

(e) 0.0106   Read it as: one hundred six <u>ten-thousandths</u>

> Think: $0.0106 = \frac{106}{10,000}$

◀ **Work Problem 5 at the Side.**

Reading a Decimal Number

Step 1 Read any whole number part to the **left** of the decimal point as you normally would.

Step 2 Read the decimal point as "**and**."

Step 3 Read the part of the number to the **right** of the decimal point as if it were an ordinary whole number.

Step 4 Finish with the place value name of the rightmost digit; these names all end in "**ths**."

When *the whole number part is zero,* use only *Steps 3 and 4*.

EXAMPLE 4   **Reading Decimal Numbers**

Read each decimal.

(a)

> 9 is in tenths place

$$16.9$$

sixteen **and** nine tenths

> Remember to say or write "and" when you see a decimal point.

16.9 is read "**sixteen and nine tenths**."

(b)

> 5 is in hundredths place

$$482.35$$

four hundred eighty-two **and** thirty-five hundredths

482.35 is read "**four hundred eighty-two and thirty-five hundredths**."

> 3 is in thousandths place

(c) 0.063 is "**sixty-three** thousandths." (Whole number part is zero.)

(d) 11.1085 is "**eleven and one thousand eighty-five** ten-thousandths."

❗ CAUTION

Use "and" when reading a decimal point. A common mistake is to read the whole number 405 as "four hundred *and* five." But there is *no decimal point* shown in 405, so it is read "four hundred five."

—————— **Work Problem ⑥ at the Side.** ▶

OBJECTIVE ▶ ④ Write decimals as fractions or mixed numbers. Knowing how to read decimals will help you when writing decimals as fractions or mixed numbers.

Writing a Decimal as a Fraction or Mixed Number

Step 1 The digits to the right of the decimal point are the numerator of the fraction.

Step 2 The denominator is 10 for tenths, 100 for hundredths, 1000 for thousandths, 10,000 for ten-thousandths, and so on.

Step 3 If the decimal has a whole number part (other than zero), it is written as a mixed number with the same whole number part.

---

⑥ Tell how to read each decimal in words.

GS (a) 3.8 is read

three **and** eight _____

(b) 15.001

(c) 0.0073

(d) 764.309

Answers

6. (a) three and eight tenths
   (b) fifteen and one thousandth
   (c) seventy-three ten-thousandths
   (d) seven hundred sixty-four and three hundred nine thousandths

7 Write each decimal as a fraction or mixed number.

GS (a) 0.9 $\dfrac{9}{\boxed{}}$

GS (b) 12.21 $12\dfrac{21}{\boxed{}}$

(c) 0.101

(d) 0.007

(e) 1.3717

8 Write each decimal as a fraction or mixed number in lowest terms.

GS (a) 0.5 Write in lowest terms.

$$\dfrac{5}{10} = \dfrac{5 \div 5}{10 \div \boxed{}} = \dfrac{1}{\boxed{}}$$

GS (b) 12.6

$$12.6 = 12\dfrac{6}{10} = 12\dfrac{6 \div \boxed{}}{10 \div \boxed{}} =$$

(c) 0.85

(d) 3.05

(e) 0.225

(f) 420.0802

Answers

7. (a) $\dfrac{9}{10}$ (b) $12\dfrac{21}{100}$ (c) $\dfrac{101}{1000}$

   (d) $\dfrac{7}{1000}$ (e) $1\dfrac{3717}{10,000}$

8. (a) $\dfrac{5 \div 5}{10 \div 5} = \dfrac{1}{2}$ (b) $12\dfrac{6 \div 2}{10 \div 2} = 12\dfrac{3}{5}$

   (c) $\dfrac{17}{20}$ (d) $3\dfrac{1}{20}$ (e) $\dfrac{9}{40}$ (f) $420\dfrac{401}{5000}$

EXAMPLE 5 **Writing Decimals as Fractions or Mixed Numbers**

Write each decimal as a fraction or mixed number.

(a) 0.19

The digits to the right of the decimal point, 19, are the numerator of the fraction. The denominator is 100 for hundredths because the rightmost digit is in the hundredths place.

$$0.19 = \dfrac{19}{100} \leftarrow 100 \text{ for hundredths}$$

Hundredths place

(b) 0.863

$$0.863 = \dfrac{863}{1000} \leftarrow 1000 \text{ for thousandths}$$

Thousandths place

(c) 4.0099

The whole number part stays the same.

$$4.0099 = 4\dfrac{99}{10,000} \leftarrow 10,000 \text{ for ten-thousandths}$$

Ten-thousandths place

◀ **Work Problem 7 at the Side.**

EXAMPLE 6 **Writing Decimals as Fractions or Mixed Numbers**

Write each decimal as a fraction or mixed number in lowest terms.

(a) $0.4 = \dfrac{4}{10} \leftarrow 10 \text{ for tenths}$

Write $\dfrac{4}{10}$ in lowest terms. $\dfrac{4}{10} = \dfrac{4 \div 2}{10 \div 2} = \dfrac{2}{5} \leftarrow$ Lowest terms

(b) $0.75 = \dfrac{75}{100} = \dfrac{75 \div 25}{100 \div 25} = \dfrac{3}{4} \leftarrow$ Lowest terms

(c) $18.105 = 18\dfrac{105}{1000} = 18\dfrac{105 \div 5}{1000 \div 5} = 18\dfrac{21}{200} \leftarrow$ Lowest terms

The whole number part stays the same.

⚠ CAUTION

Always check that your fraction answers are in lowest terms.

◀ **Work Problem 8 at the Side.**

▦ Calculator Tip

In this text, we will write **0**.45 instead of just .45, to emphasize that the whole number is zero. Your *scientific* calculator may show these zeros also. See how your calculator displays the number.

5.1 Exercises

FOR EXTRA HELP   Go to MyMathLab *for worked-out, step-by-step solutions to exercises enclosed in a square* ▢ *and video solutions to* ▶ *exercises.*

CONCEPT CHECK   *Circle the correct answer in Exercises 1–4.*

1. The number 11.084 has how many decimal places?
2   3   5

2. The number 0.7185 has how many decimal places?
4   5   1

3. In 22.85, the 8 means what?
8 tenths   8 hundredths   8 ones

4. In 57.213, the 3 means what?
3 thousandths   3 tenths   3 hundredths

Identify the digit that has the given place value. **See Example 2.**

5. 70.489
tens
ones
tenths

6. 135.296
ones
tenths
tens

7. 0.83472
thousandths
ten-thousandths
tenths

8. 0.51968
tenths
ten-thousandths
hundredths

9. 149.0832
hundreds
hundredths
ones

10. 3458.712
hundreds
hundredths
tenths

11. 6285.7125
thousands
thousandths
hundredths

12. 5417.6832
thousands
thousandths
ones

Write the decimal number that has the specified place values. **See Example 2.**

13. Fill in the blanks for the rest of this decimal number:

0 ones, 5 hundredths, 1 ten, 4 hundreds, 2 tenths   __4__ __1__ __0__ . ____ ____

14. Fill in the blanks for the rest of this decimal number:

7 tens, 9 tenths, 3 ones, 6 hundredths, 8 hundreds   __8__ __7__ __3__ . ____ ____

15. 3 thousandths, 4 hundredths, 6 ones, 2 ten-thousandths, 5 tenths

16. 8 ten-thousandths, 4 hundredths, 0 ones, 2 tenths, 6 thousandths

17. 4 hundredths, 4 hundreds, 0 tens, 0 tenths, 5 thousandths, 5 thousands, 6 ones

18. 7 tens, 7 tenths, 6 thousands, 6 thousandths, 3 hundreds, 3 hundredths, 2 ones

Write each decimal as a fraction or mixed number in lowest terms.
See Examples 5 and 6.

19. 0.7

20. 0.1

21. 13.4

22. 9.8

23. 0.35

24. 0.85

25. 0.66

26. 0.33

27. 10.17

28. 31.99

29. 0.06

30. 0.08

31. 0.205

32. 0.805

33. 5.002

34. 4.008

35. 0.686

36. 0.492

Tell how to read each decimal in words. See Examples 1, 3, and 4.

37. 0.5

38. 0.2

39. 0.78

40. 0.55

41. 0.105

42. 0.609

43. 12.04

44. 86.09

45. 1.075

46. 4.025

For Exercises 47–56, rewrite the words as a decimal number. See Examples 3 and 4.

47. Fill in the blanks to write six and seven tenths as a decimal number.

——— **.** ———

48. Fill in the blanks to write eight and twelve hundredths as a decimal number.

——— **.** ——— ———

49. Thirty-two hundredths

50. One hundred eleven thousandths

51. (a) Four hundred twenty and eight thousandths

(b) Four hundred twenty-eight thousandths

52. (a) Two hundred twenty-four thousandths

(b) Two hundred and twenty-four thousandths

53. (a) Seven hundred three ten-thousandths

(b) Seven hundred and three ten-thousandths

54. (a) Eighty-six and six hundredths

(b) Eighty-six hundredths

55. Seventy-five and thirty thousandths

56. Sixty and fifty hundredths

57. CONCEPT CHECK The work shown below has a mistake in it.

What Went Wrong? First write a sentence explaining what the mistake is. Then fix the mistake and find the correct solution.

Anne read the number 4302 as:
"four thousand three hundred and two."

58. CONCEPT CHECK The work shown below has a mistake in it.

What Went Wrong? First write a sentence explaining what the mistake is. Then fix the mistake and find the correct solution.

Jerry read the number 9.0106 as:
"nine and one hundred and six ten-thousandths."

A grandfather is selecting fishing line for his grandson's fishing reel. Fishing line is sold according to its "test strength," which is how many pounds of "pull" the line can withstand before breaking. Use the table to answer Exercises 59–62. Write all fractions in lowest terms. (Note: The diameter of the fishing line is its thickness.)

FISHING LINE

Test Strength (pounds)	Average Diameter (inches)
4	0.008
8	0.010
12	0.013
14	0.014
17	0.015
20	0.016

Data from Berkely Outdoor Technologies Group.

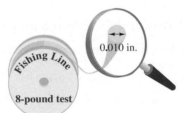

The diameter is the distance across the end of the line (or its thickness).

59. Write the diameter of 8-pound test line in words and as a fraction.

60. Write the diameter of 17-pound test line in words and as a fraction.

61. (a) What is the test strength of the line with a diameter of $\frac{13}{1000}$ inch?

(b) A fishing line has a diameter of eight thousandths inch. What is its test strength?

62. (a) What is the test strength of the line with a diameter of sixteen thousandths inch?

(b) A fishing line has a diameter of $\frac{14}{1000}$ inch. What is its test strength?

CONCEPT CHECK   *Suppose your job is to take phone orders for precision parts. Use the table below. In Exercises 63–66, write the correct part number that matches what you hear the customer say over the phone. In Exercises 67–68, write the words you would say to the customer.*

Part Number	Size in Centimeters
3-A	0.06
3-B	0.26
3-C	0.6
3-D	0.86
4-A	1.006
4-B	1.026
4-C	1.06
4-D	1.6
4-E	1.602

63. "Please send the six tenths centimeter bolt."

Part number _____

64. "The part missing from our order was the one and six hundredths size."

Part number _____

65. "The size we need is one and six thousandths centimeters."

Part number _____

66. "Do you still stock the twenty-six hundredths centimeter bolt?"

Part number _____

67. "What size is part number 4-E?" Write your answer in words.

68. "What size is part number 4-B?" Write your answer in words.

69. Look back at the decimal place value chart earlier in this section.

   (a) What do you think would be the names of the next four places to the *right* of hundred-thousandths?

   (b) What information did you use to come up with these names?

70. A common mistake is to think that the first place to the right of the decimal point is "oneths" and the second place is "tenths."

   (a) Why might someone make that mistake?

   (b) How would you explain why there is no "oneths" place?

Relating Concepts (Exercises 71–78) For Individual or Group Work

*Use your knowledge of place value to **work Exercises 71–78 in order.***

*Decimal points are sometimes used to write large numbers. For example, a website says there are **85.9 million** pet cats in the United States. The number **85.9 million** is the same as **85,900,000**. The ones place in **85.9 million** is actually referring to the millions place. The **tenths** place in **85.9 million** is referring to the place value to the right of the millions place, which is the **hundred-thousands place**.*

*Use this information and your knowledge of place value to complete **Exercises 71–78 in order.** (Data from www.americanpetproducts.org)*

71. Write 20.5 billion without the decimal point.

GS __ __, __ __ __, __ __ __, __ __ __

72. Write 14.6 thousand without the decimal point.

GS __ __, __ __ __

*In **Exercises 73–74,** rewrite each number without the decimal point.*

73. **(a)** 7.82 million

(b) 9.6 billion

(c) 42.7 thousand

(d) 704.8 trillion

74. **(a)** 123.2 thousand

(b) 503.42 million

(c) 36.51 billion

(d) 9.032 trillion

75. Write 81,030,000,000,000 in trillions

GS ____ ____ . ____ __3__ trillion

76. Write 6,450,000 in millions

GS ____ . ____ __5__ million

*In **Exercises 77–78,** rewrite each number using a decimal point.*

77. **(a)** Write 3,079,000 in millions.

(b) Write 25,600 in thousands.

(c) Write 590,800,000,000 in billions.

(d) Write 4,312,000,000,000 in trillions.

78. **(a)** Write 608,900,000,000 in billions.

(b) Write 42,387,000 in millions.

(c) Write 85,200 in thousands.

(d) Write 8,740,000,000,000 in trillions.

5.2 Rounding Decimal Numbers

Earlier, you learned how to round integers. For example, 89 rounded to the nearest ten is 90, and 8512 rounded to the nearest hundred is 8500.

OBJECTIVE ▶ 1 Learn the rules for rounding decimals. It is also important to be able to **round** decimals. For example, a store is selling 2 candy mints for $0.75 but you want only one mint. The price of each mint is $0.75 ÷ 2, which is $0.375, but you cannot pay part of a cent. Is $0.375 closer to $0.37 or to $0.38? Actually, it's exactly halfway between. When this happens in everyday situations, the rule is to round *up*. The store will charge you $0.38 for the mint.

OBJECTIVES

1 Learn the rules for rounding decimals.

2 Round decimals to any given place.

3 Round money amounts to the nearest cent or nearest dollar.

Rounding a Decimal Number

Step 1 Find the place to which the rounding is being done. Draw a "cut-off" line *after* that place to show that you are cutting off and dropping the rest of the digits.

Step 2 Look *only* at the *first* digit you are cutting off.

Step 3A If this digit is **4 or less,** the part of the number you are keeping *stays the same.*

Step 3B If this digit is **5 or more,** you must **round up** the part of the number you are keeping.

Step 4 You can use the ≈ symbol or the ≐ symbol to indicate that the rounded number is now an approximation (close, but *not exact*). Both symbols mean "is approximately equal to." (In this text we use the ≈ symbol.)

---

❶ CAUTION

Do *not* move the decimal point when rounding.

OBJECTIVE ▶ 2 Round decimals to any given place. These examples show you how to round decimals.

EXAMPLE 1 Rounding a Decimal Number

Round 14.39652 to the nearest thousandth. (Is it closer to 14.396 or to 14.397?)

Step 1 Draw a "cut-off" line after the thousandths place.

$$1\ 4\ .\ 3\ 9\ 6\ \big|\ 5\ 2$$

You are cutting off the 5 and 2
They will be dropped.

Thousandths ⬏

Step 2 Look *only* at the *first* digit you are cutting off. Ignore the other digits you are cutting off.

Look *only* at the 5
Ignore the 2

$$1\ 4\ .\ 3\ 9\ 6\ \big|\ 5\ 2$$

Continued on Next Page

1 Round to the nearest thousandth.

(a) 0.33492

$$0.3\,3\,\underline{4}\,\vdots\,9\,2$$

Thousandths ⟶

First digit cut is *5 or more.*

(b) 8.07029

$$8.0\,7\,\underline{0}\,\vdots\,2\,9$$

Thousandths ⟶

First digit cut is *4 or less.*

(c) 265.42068

(d) 10.70180

(e) 0.95745

Step 3 If the first digit you are cutting off is *5 or more*, round up the part of the number you are keeping.

$$\begin{array}{r} 1\,4\,.\,3\,9\,6\;\;5\,2 \\ +\;\;0\,.\,0\,0\,1 \\ \hline 1\,4\,.\,3\,9\,7 \end{array}$$

First digit cut is *5 or more*, so round up by adding 1 thousandth to the part you are keeping.

So, 14.39652 rounded to the nearest thousandth is 14.397
You can write **14.39652 ≈ 14.397**

> *Rounding to* **thousandths** *means three decimal places.*

> **❶ CAUTION**
>
> When rounding **integers** you keep all the digits but change some to zeros. With **decimals**, you cut off and *drop the extra digits*. In the example above, 14.39652 rounds to 14.397 (*not* 14.39700).

◀ **Work Problem ❶ at the Side.**

In **Example 1** above, the rounded number 14.397 had *three decimal places*. **Decimal places** are the number of digits to the *right* of the decimal point. The first decimal place is tenths, the second is hundredths, the third is thousandths, and so on.

> **EXAMPLE 2** Rounding Decimals to Different Places

Round to the place indicated.

(a) Round 5.3496 to the nearest tenth. (Is it closer to 5.3 or to 5.4?)

Step 1 Draw a cut-off line after the tenths place. ◀ *Tenths is one decimal place.*

$$5\,.\,3\;\;4\,9\,6$$

Tenths ⟶

You are cutting off the 4, 9, and 6

Step 2 $5\,.\,3\;\;\underline{4}\,9\,6$

Look *only* at the 4

Ignore these digits.

Step 3 $5\,.\,3\;\;4\,9\,6$

First digit cut is *4 or less*, so the part you are keeping stays the same.

$5\,.\,3$ ← Stays the same

5.3496 rounded to the nearest tenth is 5.3 (*one decimal place for tenths*).
You can write **5.3496 ≈ 5.3**

Notice that 5.3496 does *not* round to 5.3000 (which would be ten-thousandths instead of tenths).

(b) Round 0.69738 to the nearest hundredth. (Is it closer to 0.69 or to 0.70?)

Step 1 $0\,.\,6\,9\,\vert\,7\,3\,8$

Draw a cut-off line after the hundredths place.

Hundredths ⟶

Look *only* at the 7

Step 2 $0\,.\,6\,9\,\vert\,\underline{7}\,3\,8$

Answers

1. (a) 0.335 (b) 8.070 (c) 265.421
   (d) 10.702 (e) 0.957

Continued on Next Page

Step 3 $\underbrace{0 \cdot 6\,9}\,|\,\overset{\downarrow}{7}\,3\,8$ First digit cut is *5 or more*, so round up by adding 1 hundredth to the part you are keeping.

$$\begin{aligned}\overset{1}{}0 \cdot 6\,9 &\leftarrow \text{Keep this part.}\\ +\,0 \cdot 0\,1 &\leftarrow \text{To round up, add 1 hundredth.}\\ \hline 0 \cdot 7\,\mathbf{0} &\leftarrow 9 + 1 \text{ is } 10; \text{ write } 0 \text{ and regroup } 1 \text{ to the tenths place.}\end{aligned}$$

So, 0.69738 rounded to the nearest hundredth is 0.70. Hundredths is *two* decimal places so you *must* write the 0 in the hundredths place. You can write **0.69738 ≈ 0.70**

> ❗ **CAUTION**
>
> If a *rounded* number has a 0 in the rightmost place, you *must* keep the 0. As shown above, 0.69738 rounded to the nearest hundredth is 0.7**0**. Do **not** write 0.7, which is rounded to tenths instead of hundredths.

(c) Round 0.01806 to the nearest thousandth. (Is it closer to 0.018 or to 0.019?)

$$\underbrace{0 \cdot 0\,1\,8}\,|\,\overset{\downarrow}{0}\,6$$

First digit cut is *4 or less*, so the part you are keeping stays the same.

$$0 \cdot 0\,1\,8 \leftarrow \text{Stays the same}$$

> Rounding to *thousandths* means *three* decimal places.

So, 0.01806 rounded to the nearest thousandth is 0.018
You can write **0.01806 ≈ 0.018**

(d) Round 57.976 to the nearest tenth. (Is it closer to 57.9 or to 58.0?)

$$\underbrace{57.9}\,|\,\overset{\downarrow}{7}\,6$$

First digit cut is *5 or more*, so round up by adding 1 tenth to the part you are keeping.

$$\begin{aligned}\overset{1}{}57.9 &\leftarrow \text{Keep this part.}\\ +\ 0.1 &\leftarrow \text{To round up, add 1 tenth.}\\ \hline 58.0 &\leftarrow 9 + 1 \text{ is } 10; \text{ write } 0 \text{ and}\\ & \text{regroup } 1 \text{ to the ones place.}\end{aligned}$$

> Be sure to write the 0 in the *tenths* place.

So, 57.976 rounded to the nearest tenth is 58.0. You can write **57.976 ≈ 58.0**
You *must* write the 0 in the tenths place to show that the number was rounded to the nearest tenth.

> ❗ **CAUTION**
>
> Check that your rounded answer shows *exactly* the number of decimal places asked for in the problem. Be sure your answer shows *one decimal place* if you rounded to *tenths, two decimal places* for *hundredths, three decimal places* for *thousandths,* and so on.

— **Work Problem ② at the Side.** ▶

OBJECTIVE ③ Round money amounts to the nearest cent or nearest dollar. When you are shopping in a store, money amounts are usually rounded to the nearest cent. There are 100 cents in a dollar.

$$\text{Each cent is } \frac{1}{100} \text{ of a dollar.}$$

Another way to write $\frac{1}{100}$ is 0.01. So **rounding to the nearest cent is the same as rounding to the nearest hundredth of a dollar.**

---

② Round to the place indicated.

(a) 0.8988 to the nearest hundredth

$$\underbrace{0 \cdot 8\quad 9}\,|\,\overset{\downarrow}{8}\,8$$

First digit cut is *5 or more*, so round up.

$$\begin{aligned}0 \cdot 8\quad 9 &\leftarrow \text{Keep this part.}\\ +\,0 \cdot 0\quad 1 &\leftarrow \text{To round up, add}\\ \hline 0 \cdot \_\ \_ & \text{1 hundredth.}\end{aligned}$$

(b) 5.8903 to the nearest hundredth

(c) 11.0299 to the nearest thousandth

(d) 0.545 to the nearest tenth

Answers

2. **(a)** 0.90   **(b)** 5.89
   (c) 11.030   **(d)** 0.5

3 Round each money amount to the nearest cent.

GS **(a)** $14.595

$14 . 5 \underset{\displaystyle \smile}{9} \Big| 5$ ——— First digit cut is *5 or more*, so round up.

$\begin{array}{r} \$14 . 5\ 9 \\ +\ \ 0 . 0\ 1 \\ \hline \$14 . \_\_\_ \end{array}$

— To round up, add 1 hundredth (1 cent).

— You pay

(b) $578.0663

You pay _____

(c) $0.849

You pay _____

(d) $0.0548

You pay _____

Answers

3. **(a)** $14.60 **(b)** $578.07 **(c)** $0.85
(d) $0.05

| EXAMPLE 3 | **Rounding to the Nearest Cent** |

Round each money amount to the nearest cent.

(a) $2.4238 (Is it closer to $2.42 or to $2.43?)

$\$2.42 \Big| 38$ ——— First digit cut is *4 or less*, so the part you are keeping stays the same.

$\$2.42$ ——— You pay $2.42

$2.4238 rounded to the nearest cent is **$2.42**

> Rounding to the *nearest cent* is rounding to *hundredths*.

(b) $0.695 (Is it closer to $0.69 or to $0.70?)

$\$0.69 \Big| 5$ ——— *5 or more*; round up.

$\begin{array}{r} \$0.69 \\ +\ \$0.01 \\ \hline \$0.70 \end{array}$ ——— To round up, add 1 hundredth (1 cent).

——— You pay $0.70

$0.695 rounded to the nearest cent is **$0.70**

◀ **Work Problem 3 at the Side.**

Note

Some stores round *all* money amounts up to the next higher cent, even if the next digit is *4 or less*. In **Example 3(a)** above, some stores would round $2.4238 *up* to $2.43, even though it is closer to $2.42.

It is also common to round money amounts to the nearest dollar. For example, you can do that on your federal and state income tax returns to make the calculations easier.

| EXAMPLE 4 | **Rounding to the Nearest Dollar** |

Round to the nearest dollar.

(a) $48.69 (Is it closer to $48 or to $49?)

Draw a cut-off line after the ones place. ——— $\$48 \Big| .69$ ——— First digit cut is *5 or more*, so round up by adding $1

$\begin{array}{r} \$48 \\ +\ \ 1 \\ \hline \$49 \end{array}$

> Write $49 not $49.00

$48.69 rounded to the nearest dollar is **$49**

⊘ CAUTION

$48.69 rounded to the nearest dollar is $49. Be careful to write the answer as **$49** to show that the rounding is to the *nearest dollar*. Writing $49.00 would show rounding to the *nearest cent*.

——— **Continued on Next Page**

(b) $594.36 (Is it closer to $594 or to $595?)

Draw cut-off line after the ones place.

First digit cut is *4 or less,* so the part you are keeping stays the same.

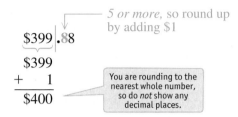

$594 .36

$594

$594.36 rounded to the nearest dollar is **$594** — Write $594 **not** $594.00

(c) $399.88 (Is it closer to $399 or to $400?)

5 or more, so round up by adding $1

$399 .88

$399
+ 1
_____
$400

You are rounding to the nearest whole number, so do *not* show any decimal places.

$399.88 rounded to the nearest dollar is **$400**

(d) $2689.50 (Is it closer to $2689 or to $2690?)

5 or more, so round up by adding $1

$2689 .50

$2689
+ 1
_____
$2690

Careful! Write $2690 **not** $2690.00

$2689.50 rounded to the nearest dollar is **$2690**

> **Note**
>
> When rounding $2689.50 to the nearest dollar, above, notice that it is exactly halfway between $2689 and $2690. When this happens in everyday situations, the rule is to round *up*. (Scientists working with technical data may use a more complicated rule when rounding numbers that are exactly in the middle.)

(e) $0.61 (Is it closer to $0 or to $1?)

5 or more, so round up.

$0 .61

$0.61 rounded to the nearest dollar is **$1** — Write $1 **not** $1.00

> ▦ **Calculator Tip**
>
> Accountants and other people who work with money amounts often set their calculators to automatically round to two decimal places (nearest cent) or to round to zero decimal places (nearest dollar). Your calculator may have this feature.

——————— **Work Problem ④ at the Side.** ▶

---

④ Round to the nearest dollar.

ⓖⓢ **(a)** $29.10

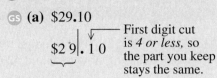

$2 9 . 1 0

First digit cut is *4 or less,* so the part you keep stays the same.

_____

(b) $136.49

(c) $990.91

(d) $5999.88

(e) $49.60

(f) $0.55

(g) $1.08

Answers

4. (a) $29 **(b)** $136 **(c)** $991
   (d) $6000 **(e)** $50 **(f)** $1 **(g)** $1

5.2 Exercises

FOR EXTRA HELP

Go to MyMathLab *for worked-out, step-by-step solutions to exercises enclosed in a square* ⬛ *and video solutions to* ▶ *exercises.*

1. **CONCEPT CHECK** Which digit would you look at when deciding how to round 4.8073 to the nearest tenth? _____

2. **CONCEPT CHECK** Which digit would you look at when deciding how to round 875.639 to the nearest hundredth? _____

3. **CONCEPT CHECK** Explain how to round 5.70961 to the nearest thousandth.

4. **CONCEPT CHECK** Explain how to round 10.028 to the nearest tenth.

Round each number to the place indicated. ***See Examples 1 and 2.***

┌── First digit cut is *5 or more.*

5. 16.8|974 to the nearest tenth
GS

┌── First digit cut is *5 or more.*

6. 193.84|5 to the nearest hundredth
GS

┌── First digit cut is *4 or less.*

7. 0.956|47 to the nearest thousandth
GS
▶

┌── First digit cut is *4 or less.*

8. 96.8158|4 to the nearest ten-thousandth
GS

9. 0.799 to the nearest hundredth

10. 0.952 to the nearest tenth

11. 3.66062 to the nearest thousandth

12. 1.5074 to the nearest hundredth

13. 793.988 to the nearest tenth
▶

14. 476.1196 to the nearest thousandth

15. 0.09804 to the nearest ten-thousandth

16. 176.004 to the nearest tenth

17. 9.0906 to the nearest hundredth
▶

18. 30.1290 to the nearest thousandth

19. 82.000151 to the nearest ten-thousandth

20. 0.400594 to the nearest ten-thousandth

Nardos is grocery shopping. The store will round the amount she pays for each item to the nearest cent. Write the rounded amounts. ***See Example 3.***

21. Soup is three cans for $2.45, so one can is $0.81666. Nardos pays _____
▶

22. Orange juice is two cartons for $3.89, so one carton is $1.945. Nardos pays _____

23. Facial tissue is four boxes for $4.89, so one box is $1.2225. Nardos pays _____
▶

24. Ramen noodles are twelve packages for $3.16, so one package is $0.26333. Nardos pays _____

25. Candy bars are six for $4.19, so one bar is $0.6983. Nardos pays _____

26. Boxes of spaghetti are four for $4.39, so one box is $1.0975. Nardos pays _____

As she gets ready to do her income tax return, Ms. Chen rounds each amount to the nearest dollar. Write the rounded amounts. ***See Example 4.***

27. Income from job, $48,649.60

28. Income from interest on bank account, $69.58

29. Donations to charity, $840.08

30. Federal withholding, $6064.49

Round each money amount as indicated. ***See Examples 3 and 4.***

31. $499.98 to the nearest dollar
▶

32. $9899.59 to the nearest dollar

33. $0.996 to the nearest cent
▶

34. $0.09929 to the nearest cent

35. $999.73 to the nearest dollar

36. $9999.80 to the nearest dollar

The table lists speed records for various types of transportation. Use the table to answer Exercises 37–40.

Record	Speed (miles per hour)
Land speed record (specially built car)	763.04
Motorcycle speed record (conventional motorcycle)	311.945
Fastest rollercoaster	149.1
Fastest military jet	4473.873
Boeing 737NG airplane (regular passenger service)	522
Indianapolis 500 auto race (fastest average winning speed)	185.981
Daytona 500 auto race (fastest average winning speed)	177.602

Data from *Guinness World Records*, www.visordown.com and www.migflug.com

37. Round these speed records to the nearest whole number.

   (a) Motorcycle

   (b) Rollercoaster

38. Round these speed records to the nearest hundredth.

   (a) Daytona 500 average winning speed

   (b) Indianapolis 500 average winning speed

39. Round these speed records to the nearest tenth.

   (a) Indianapolis 500 average winning speed

   (b) Land speed record

40. Round these speed records to the nearest hundred.

   (a) military jet

   (b) Boeing 737NG airplane

Relating Concepts (Exercises 41–44)   For Individual or Group Work

Use your knowledge about rounding money amounts to **work Exercises 41–44 in order.**

41. Explain what happens when you round $0.499 to the nearest dollar. Why does this happen?

42. Look again at **Exercise 41.** How else could you round $0.499 that would be more helpful? What kind of guideline does this suggest about rounding to the nearest dollar?

43. Explain what happens when you round $0.0015 to the nearest cent. Why does this happen?

44. You make purchasing decisions for your company and round all money amounts to the nearest cent before comparing them. Round each of these amounts to the nearest cent and compare them.

      $0.5968   and $0.6014

Explain what happens. What could you do instead of rounding to the nearest cent?

5.3 Adding and Subtracting Signed Decimal Numbers

OBJECTIVES

1. Add and subtract positive decimals.
2. Add and subtract negative decimals.
3. Estimate the answer when adding or subtracting decimals.

OBJECTIVE ▶ 1 Add and subtract positive decimals. When adding or subtracting *whole* numbers, you line up the numbers in columns so that you are adding ones to ones, tens to tens, and so on. A similar idea applies to adding or subtracting *decimal* numbers. With decimals, you line up the decimal points to be sure that you are adding tenths to tenths, hundredths to hundredths, and so on.

Adding and Subtracting Decimal Numbers

Step 1   Write the numbers in columns with the decimal points lined up.

Step 2   If necessary, write in zeros so both numbers have the same number of decimal places. Then add or subtract as if they were whole numbers.

Step 3   Line up the decimal point in the answer directly below the decimal points in the problem.

1 Find each sum.

(a) $2.86 + 7.09$

$$
\begin{array}{r}
2.86 \\
+\,7.09 \\
\end{array}
$$

└─ Decimal points are lined up.

(b) $13.761 + 8.325$

(c) $0.319 + 56.007 + 8.252$

(d) $39.4 + 0.4 + 177.2$

EXAMPLE 1   Adding Decimal Numbers

Find each sum.

(a) 16.92 and 48.34

Step 1   Write the numbers in columns with the decimal points lined up.

$$
\begin{array}{r}
\text{tens}\;\text{ones}\;.\;\text{tenths}\;\text{hundredths} \\
16.92 \\
+\,48.34 \\
\end{array}
$$

└─ Decimal points are lined up.

Step 2   Add as if these were whole numbers.

$$
\begin{array}{r}
\overset{1\,1}{16}.92 \\
+\,48.34 \\
\hline
65.26 \\
\end{array}
$$

Step 3

└─ Decimal point in answer is lined up under decimal points in problem.

(b) $5.897 + 4.632 + 12.174$

Write the numbers in columns with the decimal points lined up. Then add.

$$
\begin{array}{r}
\overset{1\,1\;\;\;2\,1}{5}.897 \\
4.632 \\
+\,12.174 \\
\hline
22.703 \\
\end{array}
$$

When you rewrite the numbers in columns, be careful to line up the decimal points.

◀ **Work Problem 1 at the Side.**

In **Example 1(a)** above, both numbers had *two* decimal places (two digits to the right of the decimal point). In **Example 1(b),** all the numbers had *three decimal places* (three digits to the right of the decimal point). That made it easy to add tenths to tenths, hundredths to hundredths, and so on.

Answers

1. (a) 9.95   **(b)** 22.086   **(c)** 64.578
   (d) 217.0

If the numbers of decimal places do *not* match, you can write in zeros as placeholders to make them match. This is shown in **Example 2** below.

EXAMPLE 2 **Writing Zeros as Placeholders before Adding**

Find each sum.

(a) 7.3 + 0.85

There are two decimal places in 0.85 (tenths and hundredths), so write a 0 in the hundredths place in 7.3 so that it has two decimal places also.

$$7.30 \leftarrow \text{One 0 is}$$
$$\underline{+\,0.85} \quad \text{written in.}$$
$$8.15$$

7.30 is equivalent to 7.3 because

$$7\frac{30}{100} \text{ in lowest terms is } 7\frac{3}{10}$$

(b) 6.42 + 9 + 2.576

Write in zeros so that all the addends have three decimal places. Notice how the whole number 9 is written with the decimal point on the *right* side. (If you put the decimal point on the *left* side of the 9, you would turn it into the decimal fraction 0.9.)

Write the decimal point on the *right* side of the 9.

$$6.420 \leftarrow \text{One 0 is written in.}$$
$$9.000 \leftarrow \text{9 is a whole number; decimal point}$$
$$\quad\quad\quad\quad\text{and three zeros are written in.}$$
$$\underline{+\,2.576} \leftarrow \text{No zeros are needed.}$$
$$17.996$$

Note

Writing zeros to the right of a *decimal* number does *not* change the value of the number, as shown in **Example 2(a)** above.

──── **Work Problem 2 at the Side.** ▶

EXAMPLE 3 **Subtracting Decimal Numbers**

Find each difference.

(a) 15.82 from 28.93

Step 1

$$28.93$$
$$-\,15.82$$

Line up decimal points. Then you will be subtracting hundredths from hundredths and tenths from tenths.

Watch the order of the numbers when you see "from." Subtraction is **not** commutative like addition. So the number you are subtracting "from" goes first.

Step 2

$$28.93$$
$$\underline{-\,15.82}$$
$$13\ 11 \leftarrow$$

Both numbers have two decimal places; no need to write in zeros.

Subtract as if they were whole numbers.

Step 3

$$28.93$$
$$\underline{-\,15.82}$$
$$13.11$$

Decimal point in answer is lined up.

──── **Continued on Next Page**

2 Find each sum.

(a) 6.54 + 9.8

$$6.5\,4$$
$$\underline{+\ 9.8\ 0} \leftarrow \text{One 0 is}$$
$$\quad\quad\quad\quad\quad \text{written in.}$$
$$\_\_ \_\_ . \_\_ \_\_$$

(b) 0.831 + 222.2 + 10

(c) 8.64 + 39.115 + 3.0076

(d) 5 + 429.823 + 0.76

Answers

2. **(a)** 16.34 **(b)** 233.031 **(c)** 50.7626
(d) 435.583

3 Find each difference.

(a) 22.7 from 72.9

$$\begin{array}{r} 7\ 2\ .\ 9 \\ -2\ 2\ .\ 7 \\ \hline \rule{1.5em}{0.4pt}\ \rule{0.8em}{0pt}.\ \rule{0.8em}{0pt} \end{array}$$

(b) 6.425 from 11.813

(c) $20.15 − $19.67

4 Find each difference.

(a) 18.651 from 25.3

(b) 5.816 − 4.98

(c) 40 less 3.66

(d) 1 − 0.325

(b) 146.35 minus 58.98

Regrouping is needed here.

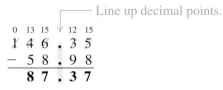

Line up decimal points.

$$\begin{array}{r} \overset{0\ \ 13\ \ 15\ \ \ \ \ 12\ \ 15}{\cancel{1}\ \cancel{4}\ \cancel{6}\ .\ \cancel{3}\ 5} \\ -\ \ 5\ 8\ .\ 9\ 8 \\ \hline 8\ 7\ .\ 3\ 7 \end{array}$$

◄ Work Problem **3** at the Side.

EXAMPLE 4 Writing Zeros as Placeholders before Subtracting

Find each difference.

(a) 16.5 from 28.362

Use the same steps as in **Example 3** above. Remember to write in zeros so both numbers have three decimal places.

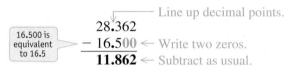

Line up decimal points.

16.500 is equivalent to 16.5

$$\begin{array}{r} 28.362 \\ -\ 16.500 \\ \hline 11.862 \end{array}$$ ← Write two zeros.
← Subtract as usual.

(b) 59.7 − 38.914

$$\begin{array}{r} 59.700 \\ -\ 38.914 \\ \hline 20.786 \end{array}$$ ← Write two zeros.
← Subtract as usual.

(c) 12 less 5.83

12.00 is equivalent to 12

$$\begin{array}{r} 12.00 \\ -\ 5.83 \\ \hline 6.17 \end{array}$$ ← Write a decimal point and two zeros.
← Subtract as usual.

◄ Work Problem **4** at the Side.

Calculator Tip

If you are *adding* decimal numbers, you can enter them in any order on your calculator. Try these; jot down the answers.

9.82 ⊕ 1.86 ⊜ _____   1.86 ⊕ 9.82 ⊜ _____

The answers are the same because addition is *commutative*. But subtraction is *not* commutative. It *does* matter which number you enter first. Try these:

9.82 ⊖ 1.86 ⊜ _____   1.86 ⊖ 9.82 ⊜ _____

The answers are 7.96 and −7.96. As you know, positive numbers are *greater* than 0, but negative numbers are *less* than 0. So it is important to do subtraction in the correct order, particularly for your bank account!

OBJECTIVE ▶ **2** **Add and subtract negative decimals.** The rules that you used to add integers will also work for positive and negative decimal numbers.

Adding Signed Numbers

To add two numbers with the *same* sign, add the absolute values of the numbers. Use the common sign as the sign of the sum.

To add two numbers with *unlike* signs, subtract the lesser absolute value from the greater absolute value. Use the sign of the number with the greater absolute value as the sign of the sum.

Answers

3. (a) 50.2   (b) 5.388   (c) $0.48
4. (a) 6.649   (b) 0.836   (c) 36.34
    (d) 0.675

EXAMPLE 5   **Adding Positive and Negative Decimal Numbers**

Find each sum.

(a) $-3.7 + (-16)$

Both addends are negative, so the sum will be negative. To begin, $|-3.7|$ is 3.7 and $|-16|$ is 16. Then add the absolute values.

$$
\begin{array}{r}
3.7 \\
+\ 16.0 \\
\hline
19.7
\end{array} \leftarrow \text{Write a decimal point and one 0}
$$

$$
\underset{\text{Both negative}}{-3.7} + \underset{}{(-16)} = \underset{\text{Negative sum}}{-\,\mathbf{19.7}}
$$

Note

We will continue to write parentheses around negative numbers when the negative sign might be confused with other symbols. Thus in **Example 5(a)** above,

$-3.7 + (-16)$ means   **negative** 3.7   plus   **negative** 16

(b) $-5.23 + 0.792$

The addends have different signs. To begin, $|-5.23|$ is 5.23 and $|0.792|$ is 0.792. Then subtract the lesser absolute value from the greater.

$$
\begin{array}{r}
5.230 \\
-\ 0.792 \\
\hline
4.438
\end{array} \leftarrow \text{Write one 0}
$$

$$
\underset{\substack{\text{Number with}\\\text{greater absolute}\\\text{value is negative.}}}{-5.23} + 0.792 = \underset{\substack{\text{Answer is}\\\text{negative.}}}{-\,\mathbf{4.438}}
$$

Work Problem ⑤ **at the Side.** ▶

Earlier in this course, you rewrote subtraction of integers as addition of the first number to the opposite of the second number. This same strategy works with positive and negative decimal numbers.

EXAMPLE 6   **Subtracting Positive and Negative Decimal Numbers**

Find each difference.

(a) $4.3 - 12.73$

Rewrite subtraction as adding the opposite.

$$
4.3\ \underset{\downarrow}{-}\ \underset{\downarrow}{12.73} \quad \boxed{\text{The opposite of } 12.73 \text{ is } -12.73}
$$

$$
4.3 + (-12.73)
$$

-12.73 has the greater absolute value and is negative, so the answer will be negative.

$$
4.3 + (-12.73) = \underset{\substack{\uparrow\\\text{Answer is}\\\text{negative.}}}{-\,\mathbf{8.43}} \quad \begin{array}{l}\text{Subtract the absolute values.}\\ \begin{array}{r}12.73\\ -\ 4.30\\ \hline 8.43\end{array}\end{array}
$$

Continued on Next Page

⑤ Find each sum.

(a) $13.245 + (-18)$

$|13.245|$ is _____

$|-18|$ is _____

Now subtract the lesser absolute value from the greater.

$$
\begin{array}{r}
1\ 8\ .\ 0\ \ 0\ \ 0 \\
-\ 1\ 3\ .\ 2\ \ 4\ \ 5 \\
\hline
\text{\_\_ \_\_ . \_\_ \_\_ \_\_}
\end{array} \begin{array}{l}\leftarrow\text{Write a}\\ \text{decimal point}\\ \text{and three 0s.}\end{array}
$$

The number with the greater absolute value is negative, so the answer is _____ .

$13.245 + (-18) =$ _____

(b) $-0.7 + (-0.33)$

(c) $-6.02 + 100.5$

6 Find each difference.

(a) $-0.37 - (-6)$

$\quad\quad\downarrow\quad\quad\downarrow$ Rewrite

$\quad -0.37 + \quad 6$ subtraction as adding the opposite.

(b) $5.8 - \quad 10.03$

$\quad\quad\downarrow\quad\quad\quad\downarrow$

$\quad 5.8 + (-10.03)$

(c) $-312.72 - 65.7$

(d) $0.8 - (6 - 7.2)$ Work inside parentheses first.

(b) $-3.65 - (-4.8)$

Rewrite subtraction as adding the opposite.

$$-3.65 - (-4.8)$$
$$\quad\quad\downarrow\quad\quad\downarrow$$
$$-3.65 + 4.8$$

The opposite of -4.8 is 4.8

4.8 has the greater absolute value and is positive, so the answer will be positive.

$$-3.65 + 4.8 = \mathbf{1.15} \quad\quad \text{Subtract the absolute values.}$$
$$\uparrow \quad\quad\quad\quad\quad 4.80$$
$$\text{Answer is} \quad\quad\quad -3.65$$
$$\text{positive.} \quad\quad\quad\quad 1.15$$

(c) $14.2 - (1.69 + 0.48)$ Work inside parentheses first.

$\quad 14.2 - \quad\quad (2.17)$ Change subtraction to adding the opposite.

$\quad 14.2 + (-2.17)$ 14.2 has the greater absolute value and is positive, so the answer will be positive.

$\quad\quad\quad \mathbf{12.03}$

◀ **Work Problem 6 at the Side.**

OBJECTIVE **3** **Estimate the answer when adding or subtracting decimals.** A common error when working decimal problems by hand is to misplace the decimal point in the answer. Or, when using a calculator, you might press the wrong key. Using *front end rounding* to estimate the answer first will help you avoid these mistakes. Start by rounding each number to the highest possible place. Here are several examples. In the rounded numbers, only the leftmost digit is something other than 0.

3.25	rounds to	3	0.812	rounds to	1
532.6	rounds to	500	26.397	rounds to	30
7094.2	rounds to	7000	351.24	rounds to	400

EXAMPLE 7   **Estimating Decimal Answers**

First, use front end rounding to round each number. Then add or subtract the rounded numbers to get an estimated answer. Finally, find the exact answer.

(a) Find the sum of 194.2 and 6.825

$$\begin{array}{ll} \textit{Estimate:} & \textit{Exact:} \\ \quad 200 \;\xleftarrow{\text{Rounds to}}\; & 194.200 \\ + \quad 7 \;\xleftarrow{\text{Rounds to}}\; & + \quad 6.825 \\ \hline \quad \mathbf{207} & \mathbf{201.025} \end{array}$$

The estimate goes out to the hundreds place (three places to the *left* of the decimal point), and so does the exact answer. Therefore, the decimal point is probably in the correct place in the exact answer.

(b) Subtract $13.78 from $69.42

$$\begin{array}{ll} \textit{Estimate:} & \textit{Exact:} \\ \quad \$70 \;\xleftarrow{\text{Rounds to}}\; & \$69.42 \\ - \quad 10 \;\xleftarrow{\text{Rounds to}}\; & - \quad 13.78 \\ \hline \quad \mathbf{\$60} & \mathbf{\$55.64} \xleftarrow{} \text{Exact answer is close to estimate,} \\ & \quad\quad\quad\quad \text{so it is reasonable.} \end{array}$$

Answers

6. (a) 5.63   **(b)** −4.23   **(c)** −378.42   **(d)** 2

Continued on Next Page

(c) $-1.861 - 7.3$

Rewrite subtraction as adding the opposite. Then use front end rounding to get an estimated answer.

$$-1.861 \quad - \quad 7.3$$
$$\downarrow \qquad \quad \downarrow$$
$$-1.861 \quad + \quad (-7.3)$$
$$\downarrow \quad \text{Rounded} \quad \downarrow$$
$$-2 \quad + \quad (-7) \; = \; \mathbf{-9} \leftarrow \text{Estimate}$$

To find the exact answer, add the absolute values.

$$\begin{array}{l} 1.861 \leftarrow \text{Absolute value of } -1.861 \\ \underline{+\ 7.300} \leftarrow \text{Absolute value of } -7.3 \text{ with two zeros written in} \\ 9.161 \end{array}$$

The answer will be negative because both numbers are negative.

$$-1.861 + (-7.3) = \mathbf{-9.161} \leftarrow \text{Exact}$$

The exact answer of -9.161 is reasonable because it is close to the estimated answer of -9.

───────── **Work Problem 7 at the Side.** ▶

Study Skills Reminder

Are you having trouble remembering all the material from this chapter? If so, you may need to change the way you do your homework. To learn and remember new math skills, you have to practice them. That's why instructors assign homework! See the Study Skills activity "Homework: How, Why, and When" to learn ways to get the most out of your homework.

7 Use front end rounding to estimate each answer. Then find the exact answer.

(a) $2.83 + 5.009 + 76.1$

$$\begin{array}{ll} \textit{Estimate:} & \textit{Exact:} \\ 3 \longleftarrow & 2.830 \\ 5 \longleftarrow & 5.009 \\ +\ 80 \longleftarrow & +76.100 \\ \hline \end{array}$$

(b) 19.28 less 1.53

Estimate: *Exact:*

(c) $11.365 - 38$

Estimate: *Exact:*

(d) $-214.6 + 300.72$

Estimate: *Exact:*

Answers

7. (a) *Estimate:* $3 + 5 + 80 = 88$
     Exact: 83.939
(b) *Estimate:* $\$20 - \$2 = \$18$
     Exact: $\$17.75$
(c) *Estimate:* $10 + (-40) = -30$
     Exact: -26.635
(d) *Estimate:* $-200 + 300 = 100$
     Exact: 86.12

5.3 Exercises

FOR EXTRA HELP Go to MyMathLab for worked-out, step-by-step solutions to exercises enclosed in a square ▢ and video solutions to ▶ exercises.

CONCEPT CHECK *Rewrite each addition in columns. Write in any decimal points or zeros, as needed. Do **not** complete the calculation.*

1. $6.42 + 10.163$

2. $7 + 9.204$

3. $20 - 9.1263$

4. $137.06 - 12$

Find each sum. See Examples 1 and 2.

5.
$$\begin{array}{r} 5.69 \\ 0.24 \\ + 11.79 \\ \hline \end{array}$$

6.
$$\begin{array}{r} 372.1 \\ 33.7 \\ + 42.3 \\ \hline \end{array}$$

7. ▶ $0.38 + 7 + 4.6$

8. $3.7 + 0.812 + 55$

9. $14.23 + 8 + 74.63 + 18.715 + 0.286$

10. $197.4 + 0.72 + 17.43 + 25 + 1.4$

11. ▶ $27.65 + 18.714 + 9.749 + 3.21$

12. $58.546 + 19.2 + 8.735 + 14.58$

13. CONCEPT CHECK The work shown below has a mistake in it.

What Went Wrong? First write a sentence explaining what the mistake is. Then fix the mistake and find the correct solution.

Kaneya added $0.72 + 6 + 39.5$ this way:

$$\begin{array}{r} 0.72 \\ 6 \\ + 39.50 \\ \hline 40.28 \end{array}$$

14. CONCEPT CHECK The work shown below has a mistake in it.

What Went Wrong? First write a sentence explaining what the mistake is. Then fix the mistake and find the correct solution.

Caleb added $7.21 + 65 + 13.15$ this way:

$$\begin{array}{r} 7.21 \\ .65 \\ + 13.15 \\ \hline 21.01 \end{array}$$

15. CONCEPT CHECK Explain why 0.3 is equivalent to 0.3000.

16. CONCEPT CHECK Explain why 7 may be written as 7.0 but not as 0.7.

Find each difference. See Examples 3 and 4.

17. $90.5 - 0.8$

18. $303.72 - 0.68$

19. 0.4 less 0.291

20. 0.35 less 0.088

21. GS
$$\begin{array}{r} 6.00 \\ - 5.09 \\ \hline \end{array}$$
6 minus 5.09

22. GS
$$\begin{array}{r} 80.0 \\ - 16.3 \\ \hline \end{array}$$
80 minus 16.3

23. ▶ Subtract 8.339 from 15

24. Subtract 0.08 from 44

25. CONCEPT CHECK The work shown below has a mistake in it.

What Went Wrong? First write a sentence explaining what the mistake is. Then fix the mistake and find the correct solution.

Keegan subtracted 7.45 from 15.32 this way:

$$\begin{array}{r} 7.45 \\ - 15.32 \\ \hline 12.13 \end{array}$$

26. Explain the difference between saying "subtract 2.9 from 8" and saying "2.9 minus 8."

This drawing of a human skeleton shows the average length of the longest bones, in inches. Use the drawing to answer Exercises 27–30.

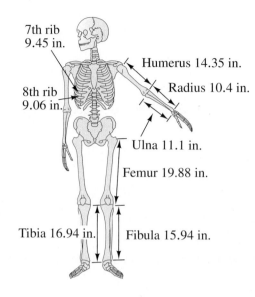

7th rib
9.45 in.

Humerus 14.35 in.

8th rib
9.06 in.

Radius 10.4 in.

Ulna 11.1 in.

Femur 19.88 in.

Tibia 16.94 in.

Fibula 15.94 in.

27. (a) What is the combined length of the humerus and radius bones?

(b) What is the difference in the lengths of these two bones?

28. (a) What is the total length of the femur and tibia bones?

(b) How much longer is the femur than the tibia?

29. (a) Find the sum of the lengths of the humerus, ulna, femur, and tibia.

(b) How much shorter is the 8th rib than the 7th rib?

30. (a) What is the difference in the lengths of the two bones in the lower arm?

(b) What is the difference in the lengths of the two bones in the lower leg?

Find each sum or difference. ***See Examples 5 and 6.***

31. $24.008 + (-0.995)$

32. $0.77 + 3.06$

33. $-6.05 + (-39.7)$

34. $-6.409 + 8.224$

35. $0.9 - 7.59$
  $\downarrow \quad \searrow$
  $0.9 + (-7.59)$

Change subtraction to adding the opposite.

36. $-489.7 - 38$
  $\downarrow \quad \searrow$
  $-489.7 + (-38)$

Change subtraction to adding the opposite.

37. $-2 - 4.99$

38. $2.068 - (-32.7)$

39. $-5.009 + 0.73$

40. $-0.33 - 65$

41. $-1.7035 - (5 - 6.7)$

42. $60 + (-0.9345 + 1.4)$

43. $8000 - (8002.63 - 8)$

44. $-210 - (-0.7306 + 0.5)$

CONCEPT CHECK   *Use front end rounding to estimate each answer. The first two estimates are done for you. Then use your estimate to select the correct exact answer. Circle your choice.* **See Example 7.**

45. $18 - 11.725$
Ⓖ **Estimate:** $20 - 10 = 10$
Exact: 29.725   6.275   −11.545

46. $20 - 1.37$
Ⓖ **Estimate:** $20 - 1 = 19$
Exact: −21.37   −1.863   18.63

47. $-6.5 + 0.7$
Estimate:
Exact: −5.8   7.2   −0.58

48. $-9.67 + 3.09$
Estimate:
Exact: −12.76   −6.58   6.58

49. $-0.671 - 9$
Estimate:
Exact: 1.571   8.329   −9.671

50. $-803 - 0.6$
Estimate:
Exact: −803.6   803.6   −203

51. $8.4 - (-50.83)$
Estimate:
Exact: −42.43   −59.23   59.23

52. $14.98 - (-6.506)$
Estimate:
Exact: −8.474   21.486   8.004

Use front end rounding to estimate each sum or difference. Then find the exact answer to each application problem. Use the information in the table for Exercises 53–56.

INTERNET USERS IN SELECTED COUNTRIES	
Country	**Number of Users**
China	642 million
United States	279.84 million
India	243.2 million
Nigeria	67 million
Mexico	50.9 million
Philippines	39.47 million
Iran	22.2 million
WORLD TOTAL	3130.8 million

Data from internetlivestats.com

53. How many fewer Internet users are there in Mexico than in Nigeria?
Estimate:
Exact:

54. How many more users are there in China than in the Philippines?
Estimate:
Exact:

55. How many Internet users are there in all the countries listed in the table?
Estimate:
Exact:

56. Using the exact answer from **Exercise 55,** calculate the number of worldwide Internet users in countries other than the ones in the table.
Estimate:
Exact:

57. The tallest known land mammal, a prehistoric ancestor of the rhino, was 6.4 m tall. Compare the rhino's height to the combined heights of these three NBA basketball players: Roy Hibbert at 2.18 meters, Dwayne Wade at 1.93 meters, and Carmelo Anthony at 2.03 meters. Is their combined height greater or less than the rhino's height? By how much? (Data from NBA.com/players)

6.4 m

Estimate:
Exact:

58. Sammy works in a veterinarian's office. He weighed two kittens. One was 3.9 ounces and the other was 4.05 ounces. What was the difference in the weight of the two kittens?
Estimate:
Exact:

59. Steven One Feather gave the cashier a $20 bill to pay for $9.12 worth of groceries. How much change did he get?
Estimate:
Exact:

60. The cost of Julie's tennis racket, with tax, was $41.09. She gave the clerk two $20 bills and a $10 bill. What amount of change did Julie receive?
Estimate:
Exact:

A mother is taking her children fishing. She is buying fishing equipment for them.
Use the information below on the sale prices of equipment to answer Exercises 61–64.
When estimating, round to the nearest whole number (nearest dollar).

Bobbers 3 for $3.74

8-pound test fishing line:
regular $4.99
clear $7.49
fluorescent $6.99

No-See Line

Environmentally safe
split shot $2.99

Leaded split shot 89¢

Tackle box: 2-trays $11.39
Tackle box: 3-trays $20.24

Spinning reels:
$17.49, $29.99, $39.99, $49.99

Spinning rods:
$9.99, $14.99, $19.99, $29.99

Data from www.Cabelas.com; www.Amazon.com

61. What is the difference in price between the clear and the fluorescent fishing lines?

Estimate:

Exact:

62. How much more does the three-tray tackle box cost than the second-most-expensive spinning rod?

Estimate:

Exact:

63. Find the total cost of the second-lowest-priced spinning reel, two packages of environmentally safe split shot, and a three-tray tackle box. Sales tax for all the items was $3.93.

Estimate:

Exact:

64. The mother bought leaded split shot on sale. She also bought some SPF 15 sunscreen for $7.53 and a flotation vest for $44.96. Sales tax was $3.74. How much did she spend in all?

Estimate:

Exact:

Olivia Sanchez kept track of her expenses for one month. Use her list to answer Exercises 65–68.

65. What were Olivia's total expenses for the month?

66. How much did Olivia pay for her cell phone, cable TV, and Internet?

67. What was the difference in the amounts spent for groceries and for the car payment?

68. How much more did Olivia spend on rent than on all her car expenses?

Monthly Expenses	
Rent	$1022
Car payment	$330.44
Car repairs, gas	$202
Cable TV	$39.99
Internet	$29.99
Electricity	$65.43
Cell phone	$73.75
Groceries	$258.70
Entertainment	$159.51
Clothing, laundry	$68

Find the length of the dashed line in each rectangle or circle.

69.

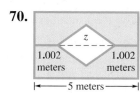

0.91 cm   0.7 cm b

3 cm

70.

1.002 meters z 1.002 meters

5 meters

71.

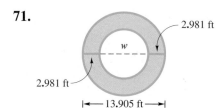

2.981 ft

w

2.981 ft

13.905 ft

5.4 Multiplying Signed Decimal Numbers

OBJECTIVES

1. Multiply positive and negative decimals.

2. Estimate the answer when multiplying decimals.

OBJECTIVE 1 Multiply positive and negative decimals. The decimals 0.3 and 0.07 can be multiplied by writing them as fractions.

$$0.\underline{3} \times 0.\underline{07} = \frac{3}{10} \times \frac{7}{100} = \frac{3 \times 7}{10 \times 100} = \frac{21}{1000} = 0.\underline{021}$$

1 decimal place + 2 decimal places → 3 decimal places

Can you see a way to multiply decimals without writing them as fractions? Use these steps. Remember that each number in a multiplication problem is called a *factor*, and the answer is called the *product*.

Multiplying Two Decimal Numbers

Step 1 Multiply the factors (the numbers being multiplied) as if they were whole numbers.

Step 2 Find the *total* number of decimal places in *both* factors.

Step 3 Write the decimal point in the product (the answer) so it has the same number of decimal places as the total from *Step 2*. You may need to write in extra zeros on the *left side* of the product in order to get the correct number of decimal places.

Step 4 If two factors have the *same sign,* the product is *positive.* If two factors have *different signs,* the product is *negative.*

Note

When multiplying decimals, you do **not** need to line up decimal points. (You **do** need to line up decimal points when adding or subtracting.)

EXAMPLE 1 Multiplying Decimal Numbers

Find the product of 8.34 and (−4.2)

Step 1 Multiply the numbers as if they were whole numbers.

```
    8.3 4
  ×   4.2
    1 6 6 8
  3 3 3 6
  3 5 0 2 8
```

You do *not* have to line up decimal points when multiplying.

Step 2 Count the total number of decimal places in both factors.

```
    8.3 4  ← 2 decimal places
  ×   4.2  ← 1 decimal place
    1 6 6 8     3 total decimal places
  3 3 3 6
  3 5 0 2 8
```

Continued on Next Page

Step 3 Count over 3 places in the product and write the decimal point. Count from *right to left*.

$$
\begin{array}{r}
8.3\,4 \leftarrow \text{2 decimal places} \\
\times\ \ 4.2 \leftarrow \text{1 decimal place} \\
\hline
1\,6\,6\,8 \quad \text{3 total decimal places} \\
3\,3\,3\,6 \quad\quad \\
\hline
3\,5.0\,2\,8 \leftarrow \text{3 decimal places in product}
\end{array}
$$

Count over 3 places from right to left
to position the decimal point.

Step 4 The factors have *different* signs, so the product is *negative:*
8.34 times (-4.2) is **−35.028**

———— **Work Problem ① at the Side.** ▶

EXAMPLE 2 **Writing Zeros as Placeholders in the Product**

Find the product: $(-0.042)(-0.03)$

Start by multiplying, then count decimal places.

$$
\begin{array}{r}
0.0\,4\,2 \leftarrow \text{3 decimal places} \\
\times\ \ 0.0\,3 \leftarrow \text{2 decimal places} \\
\hline
1\,2\,6 \leftarrow \text{5 decimal places needed in product}
\end{array}
$$

After multiplying, the answer has only three decimal places, but five are needed. So write two zeros on the *left* side of the answer.

$$
\begin{array}{r}
0.0\,4\,2 \\
\times\ \ 0.0\,3 \\
\hline
0\,0\,1\,2\,6
\end{array}
\qquad
\begin{array}{r}
0.0\,4\,2 \leftarrow \text{3 decimal places} \\
\times\ \ 0.0\,3 \leftarrow \text{2 decimal places} \\
\hline
.0\,0\,1\,2\,6 \leftarrow \text{5 decimal places}
\end{array}
$$

Write two zeros
on *left* side of answer. | Now count over 5 places
and write in the decimal point.

The final product is **0.00126**, which has five decimal places. The product is positive because the factors have the *same* sign (both factors are negative).

———— **Work Problem ② at the Side.** ▶

🖩 **Calculator Tip**

When working with money amounts, you may need to write a 0 in your answer. For example, try multiplying $3.54 × 5 on your calculator. Write down the result.

$$3.54\ \otimes\ 5\ \ominus\ \underline{\qquad}$$

Notice that the result is **17.7**, which is not the way to write a money amount. You have to write the 0 in the hundredths place: **$17.70** is correct. The calculator does not show the "extra" 0 because:

$$17.70 \text{ or } 17\frac{70}{100} \quad \text{simplifies to} \quad 17\frac{7}{10} \text{ or } 17.7$$

So keep an eye on your calculator—it doesn't know when you're working with money amounts.

---

① Find each product.

(a) $-2.6\,(0.4)$

$$
\begin{array}{r}
2.6 \leftarrow \text{1 decimal place} \\
\times\ 0.4 \leftarrow \text{1 decimal place} \\
\hline
1\,0\,4 \leftarrow \text{2 decimal places}
\end{array}
$$

Write in the decimal point. Is the product positive or negative? _____

(b) $(45.2)\,(0.25)$

(c)
$$
\begin{array}{r}
0.104 \leftarrow \text{3 decimal places} \\
\times\quad\quad 7 \leftarrow \text{0 decimal places} \\
\hline
\leftarrow \text{3 decimal places} \\
\text{in the product}
\end{array}
$$

(d) $(-3.18)^2$

Hint: Recall that squaring a number means multiplying the number by itself, so this is $(-3.18)(-3.18)$

② Find each product.

(a) $0.04\,(-0.09)$

$$
\begin{array}{r}
0.04 \leftarrow \text{2 decimal places} \\
\times\ 0.09 \leftarrow \text{2 decimal places} \\
\hline
\leftarrow \text{4 decimal places}
\end{array}
$$

Is the product positive or negative? _____

(b) $(0.2)\,(0.008)$

(c) $(-0.063)\,(-0.04)$

(d) $(0.003)^2$

Answers

1. **(a)** Product is negative; −1.04
   (b) 11.300 or 11.3 **(c)** 0.728 **(d)** 10.1124
2. **(a)** Product is negative; −0.0036
   (b) 0.0016 **(c)** 0.00252 **(d)** 0.000009

3 First, use front end rounding to estimate the answer. Then find the exact answer.

GS **(a)** $(11.62)(4.01)$

Estimate: *Exact:*

$$\begin{array}{r} 10 \longleftarrow \quad 11.62 \\ \times\ 4 \longleftarrow\ \times\ 4.01 \\ \hline \end{array}$$

(b) $(-5.986)(-33)$

Estimate: *Exact:*

(c) $8\,(\$4.35)$

Estimate: *Exact:*

(d) $58.6\,(-17.4)$

Estimate: *Exact:*

OBJECTIVE ▶ 2 **Estimate the answer when multiplying decimals.** If you are doing multiplication problems by hand, estimating the answer helps you check that the decimal point is in the right place. When you are using a calculator, estimating helps you catch an error like pressing the ⊝ key instead of the ⊗ key.

EXAMPLE 3   **Estimating before Multiplying**

First, use front end rounding to estimate $(76.34)(12.5)$. Then find the exact answer.

Estimate: *Exact:*

$$\begin{array}{r} 80 \quad \xleftarrow{\text{Rounds to}} \\ \times\ 10 \quad \xleftarrow{\text{Rounds to}} \\ \hline 800 \end{array} \qquad \begin{array}{r} 76.34 \leftarrow \text{2 decimal places} \\ \times\ \ 12.5 \leftarrow \text{1 decimal place} \\ \hline 38170 \\ 15268 \\ 7634 \\ \hline \mathbf{954.250} \end{array} \begin{array}{l} \text{3 decimal places are} \\ \text{in the product.} \end{array}$$

Both the estimate and the exact answer go out to the hundreds place, so the decimal point in 954.250 is probably in the correct place.

◀ **Work Problem** **3** **at the Side.**

Study Skills Reminder

Math classes are taught in many different ways. Regardless of how your class is taught, knowing how to take good notes can help you understand and remember the material covered. See the Study Skills activity "Taking Lecture Notes" for ways to get the most out of your notes.

Answers

3. **(a)** *Estimate:* $(10)(4) = 40$
    Exact: 46.5962
  (b) *Estimate:* $(-6)(-30) = 180$
    Exact: 197.538
  (c) *Estimate:* $8\,(\$4) = \$32$
    Exact: $\$34.80$
  (d) *Estimate:* $60\,(-20) = -1200$
    Exact: -1019.64

5.4 Exercises

FOR EXTRA HELP Go to MyMathLab *for worked-out, step-by-step solutions to exercises enclosed in a square* ⬚ *and video solutions to* ▶ *exercises.*

CONCEPT CHECK *Fill in the blanks for Exercises 1–4.*

1. In 4.2×3.46, how many decimal places will the product have? _____

2. In 8.071×2.79, how many decimal places will the product have? _____

3. In $(0.12)(0.03)$, how many zeros do you have to write in the product as placeholders? _____

4. In $(0.0006)(0.07)$, how many zeros do you have to write in the product as placeholders? _____

Find each product. **See Example 1.**

5. ▶
$$\begin{array}{r} 0.042 \\ \times\ \ 3.2 \\ \hline \end{array}$$

6.
$$\begin{array}{r} 0.571 \\ \times\ \ 2.9 \\ \hline \end{array}$$

7. ▶ $-21.5\,(7.4)$

8. $-85.4\,(-3.5)$

9. $(-23.4)\,(-0.66)$

10. $0.896\,(-0.7)$

11.
$$\begin{array}{r} \$51.88 \\ \times\ \ \ 665 \\ \hline \end{array}$$

12.
$$\begin{array}{r} \$736.75 \\ \times\ \ \ \ \ 118 \\ \hline \end{array}$$

CONCEPT CHECK *Use the fact that* $(72)(6) = 432$ *to solve Exercises 13–18 by simply counting decimal places and writing the decimal point in the correct location. Be sure to indicate the sign of the product.*

13. $72\,(-0.6) =$ 4 3 2

14. $7.2\,(-6) =$ 4 3 2

15. $(7.2)\,(0.06) =$ 4 3 2

16. $(0.72)\,(0.6) =$ 4 3 2

17. $-0.72\,(-0.06) =$ 4 3 2

18. $-72\,(-0.0006) =$ 4 3 2

Find each product. **See Example 2.**

19. ▶ $(0.006)\,(0.0052)$

20. $(0.0052)\,(0.009)$

21. $(-0.003)^2$

22. $(0.0004)^2$

First, use front end rounding to estimate the answer. Then find the exact answer.
See Example 3.

23. The *Estimate* is already done. Now find the *Exact* answer.

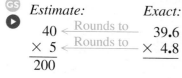

Estimate: *Exact:*
$$\begin{array}{r} 40 \\ \times\ 5 \\ \hline 200 \end{array} \quad \begin{array}{r} 39.6 \\ \times\ 4.8 \\ \hline \end{array}$$

24. The *Estimate* is already done. Now find the *Exact* answer.

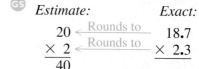

Estimate: *Exact:*
$$\begin{array}{r} 20 \\ \times\ 2 \\ \hline 40 \end{array} \quad \begin{array}{r} 18.7 \\ \times\ 2.3 \\ \hline \end{array}$$

25. *Estimate:*   *Exact:*
$$\begin{array}{r} \\ \times\ \ \ \\ \hline \end{array} \quad \begin{array}{r} 37.1 \\ \times\ 42 \\ \hline \end{array}$$

26. *Estimate:*   *Exact:*
$$\begin{array}{r} \\ \times\ \ \ \\ \hline \end{array} \quad \begin{array}{r} 5.08 \\ \times\ 71 \\ \hline \end{array}$$

27. *Estimate:*   *Exact:*
$$\begin{array}{r} \\ \times\ \ \ \\ \hline \end{array} \quad \begin{array}{r} 6.53 \\ \times\ 4.6 \\ \hline \end{array}$$

28. *Estimate:*   *Exact:*
$$\begin{array}{r} \\ \times\ \ \ \\ \hline \end{array} \quad \begin{array}{r} 7.51 \\ \times\ 8.2 \\ \hline \end{array}$$

29. *Estimate:*   *Exact:*
$$\begin{array}{r} \\ \times\ \ \ \\ \hline \end{array} \quad \begin{array}{r} 2.809 \\ \times\ 6.85 \\ \hline \end{array}$$

30. *Estimate:*   *Exact:*
$$\begin{array}{r} \\ \times\ \ \ \\ \hline \end{array} \quad \begin{array}{r} 73.52 \\ \times\ 22.34 \\ \hline \end{array}$$

CONCEPT CHECK   *Even with most of the problem missing, you can tell whether or not these answers are* reasonable. *Circle* reasonable *or* unreasonable. *If the answer is* unreasonable, *move the decimal point or insert a decimal point to make the answer* reasonable.

31. How much was his car payment? $28.90

reasonable

unreasonable, should be _____

32. How many hours did she work today? 25 hours

reasonable

unreasonable, should be _____

33. How tall is her son? 60.5 inches

reasonable

unreasonable, should be _____

34. How much does he pay for rent now? $6.92

reasonable

unreasonable, should be _____

35. What is the price of one gallon of milk? $419

reasonable

unreasonable, should be _____

36. How long is the living room? 16.8 feet

reasonable

unreasonable, should be _____

37. How much did the baby weigh? 0.095 pound

reasonable

unreasonable, should be _____

38. What was the sale price of the jacket? $1.49

reasonable

unreasonable, should be _____

First use front end rounding to round each number and estimate the answer. Then find the exact answer. Round money answers to the nearest cent when necessary.

39. LaTasha worked 50.5 hours over the last two weeks. She earns $18.73 per hour. How much did she make?

Estimate:

Exact:

40. Michael's time card shows 42.2 hours at $10.03 per hour. What are his earnings?

Estimate:

Exact:

41. Sid needs a piece of canvas material 0.6 meter in length to make a carry-all bag that fits on his wheelchair. If canvas is $4.09 per meter, how much will Sid spend? (*Note:* $4.09 *per* meter means $4.09 for *one* meter.)

Estimate:

Exact:

42. How much will Mrs. Nguyen pay for 3.5 yards of lace trim that costs $2.87 per yard?

Estimate:

Exact:

Solve each application problem. Round final money answers to the nearest cent when necessary.

43. Michelle is taking a road trip with friends. Her car's owner's manual recommends that she use only premium unleaded gasoline. Michelle is running out of gas money. How much would it cost Michelle to fill her 15.5-gallon tank with the recommended premium unleaded gas? How much would she save if she filled up with regular unleaded gas instead?

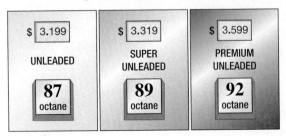

44. Ground beef and chicken legs are on sale. Juma bought 1.7 pounds of legs. Use the information in the ad to find the amount she paid.

45. Ms. Rolack is a real estate broker who helps people sell their homes. Her fee is 0.07 times the price of the home. What was her fee for selling a $289,500 home?

46. Jose Altuve of the Houston Astros baseball team had a batting average of 0.341 in the 2014 season. He went to bat 660 times. How many hits did he make? (*Hint:* Multiply the number of times at bat by his batting average.) Round to the nearest whole number. (Data from www.baseballreference.com)

Paper money in the United States has not always been the same size. Shown below are the measurements of bills printed before 1929 and the measurements from 1929 on. Use this information to answer Exercises 47–50. (Data from moneyfactory.com)

Before 1929

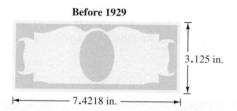

3.125 in.

7.4218 in.

From 1929 on

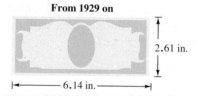

2.61 in.

6.14 in.

47. (a) Find the area of each bill. Round your final answers to the nearest tenth. (*Hint:* Multiply to find area.)

(b) What is the difference in the rounded areas?

48. (a) Find the perimeter of each bill. Round your final answers to the nearest hundredth. (*Hint:* Add to find perimeter.)

(b) How much less is the perimeter of today's bills than the bills printed before 1929?

49. The thickness of one piece of today's paper money is 0.0043 inch.

(a) If you had a pile of 100 bills, how high would it be?

(b) How high would a pile of 1000 bills be?

50. (a) Use your answers from **Exercise 49** to find the number of bills in a pile that is 43 inches high.

(b) How much money would you have if the pile is all $20 bills?

51. Judy is switching to a new cable and Internet provider. With a two-year agreement she gets an introductory price of $69.99 per month for the first 12 months. For the next 12 months the price increases to $89.99 per month. After the two-year agreement is over, the price increases to $119.94 per month. There is also a one-time installation fee of $90. (Data from Xfinity.)

(a) How much will Judy pay for the first year of cable and Internet and the installation fee? How much will she pay for the second year?

(b) How much will Judy pay for a year of cable and Internet after the two-year agreement is over?

52. Chuck is buying a car and has two options for paying it off. The first option is to pay a one-time down payment of $2000 plus monthly payments of $170.06 per month for four years. The second option is to pay zero down (no down payment) and pay monthly payments of $216.91 per month for four years.

(a) How much will Chuck pay altogether with the first option? How much will he pay altogether with the second option?

(b) Which option costs less? How much would Chuck save?

53. Barry bought 16.5 meters of rope at $0.47 per meter and three meters of wire at $1.05 per meter. How much change did he get from three $5 bills?

54. Susan bought a 46-inch LED Smart HDTV that cost $1249.97. She paid $66.59 per month for 24 months. How much could she have saved by paying for the HDTV when she bought it?

The minimum wage where Monique works is $7.75 per hour now. A new law will increase the minimum wage to $8.65 per hour next year.

55. (a) Monique worked 25 hours last week. How much did she make with the current $7.75 per hour minimum wage?

(b) Next year, how much will Monique make for working 25 hours in one week with the new $8.65 minimum wage?

56. (a) Monique worked 25 hours each week for 50 weeks this year. How much did she make with the current $7.75 per hour minimum wage?

(b) How much will Monique make next year for working 25 hours each week for 50 weeks with the new $8.65 minimum wage?

Relating Concepts (Exercises 57–58)   For Individual or Group Work

*Use your knowledge of multiplying decimal numbers as you **work Exercises 57–58 in order.***

57. (a) The average month-old puppy will grow up to be about ten times heavier as an adult. Here are the weights of six month-old puppies. Multiply each weight by 10 to predict the adult weight. (*Note:* **lb** is the short way to write **pounds.**)

$(0.538)(10) = $ _____ lb $(4.95)(10) = $ _____ lb

$(10.5)(10) = $ _____ lb $(0.9)(10) = $ _____ lb

$(7)(10) = $ _____ lb $(14.34)(10) = $ _____ lb

(b) What pattern do you see? Write a "rule" for multiplying by 10.

(c) What do you think the rule is for multiplying by 100? by 1000? Write the rules and try them out on the numbers above.

58. (a) Smaller dogs need less medication than larger dogs. A Chihuahua weighing 6 pounds needs about $\frac{1}{10}$ or 0.1 times the medication given to a Labrador retriever weighing 60 pounds. Here are some dosages for a 60-pound Labrador retriever. Multiply each dosage by 0.1 to find the amount of medication for a 6-pound Chihuahua. (*Note:* **mg** is the short way to write **milligrams.**)

$(0.538)(0.1) = $ _____ mg $(4.95)(0.1) = $ _____ mg

$(10.5)(0.1) = $ _____ mg $(0.9)(0.1) = $ _____ mg

$(58)(0.1) = $ _____ mg $(300)(0.1) = $ _____ mg

(b) What pattern do you see? Write a "rule" for multiplying by 0.1.

(c) What do you think the rule is for multiplying by 0.01? by 0.001? Write the rules and try them out on the numbers above.

5.5 | Dividing Signed Decimal Numbers

There are two kinds of decimal division problems: those in which a decimal is divided by an integer, and those in which a number is divided by a decimal. First recall the parts of a division problem.

OBJECTIVES

1. Divide a decimal by an integer.
2. Divide a number by a decimal.
3. Estimate the answer when dividing decimals.
4. Use the order of operations with decimals.

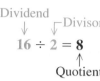

$$\text{Divisor} \rightarrow 2\overline{)16} \begin{array}{l} \leftarrow \text{Quotient} \\ \leftarrow \text{Dividend} \end{array}$$

$$\underset{\uparrow}{\overset{\text{Dividend}}{16}} \div \underset{}{\overset{\text{Divisor}}{2}} = \underset{}{8}$$
$$\text{Quotient}$$

$$\begin{array}{l} \text{Dividend} \rightarrow \dfrac{16}{2} = 8 \\ \text{Divisor} \rightarrow \phantom{\dfrac{16}{2}} \underset{\uparrow}{} \\ \text{Quotient} \end{array}$$

OBJECTIVE ▶ 1 Divide a decimal by an integer. When the divisor is an integer, use these steps.

Dividing a Decimal Number by an Integer

Step 1 Write the decimal point in the quotient (answer) directly above the decimal point in the dividend.

Step 2 Divide as if both numbers were whole numbers.

Step 3 If both numbers have the *same sign*, the quotient is *positive*. If they have *different signs*, the quotient is *negative*.

EXAMPLE 1 | **Dividing Decimals by Integers**

Find each quotient. Check the quotients by multiplying.

(a) $\underset{\text{Dividend}}{\underline{21.93}} \div \underset{\text{Divisor}}{(-3)}$

First consider $21.93 \div 3$. $\qquad\qquad 3\overline{)21.93}$

Step 1 Write the decimal point in the quotient directly above the decimal point in the dividend.

Decimal points lined up

$$3\overline{)21.\overset{\bullet}{9}3}$$

Step 2 Divide as if the numbers were whole numbers.

Check by multiplying the quotient times the divisor.

$$\dfrac{7.31}{3\overline{)21.93}}$$

Matches, so 7.31 is correct.

CHECK

$$\begin{array}{r} 7.31 \\ \times \quad 3 \\ \hline 21.93 \end{array}$$

Step 3 The quotient is -7.31 because the numbers have *different* signs.

$$\underset{\text{Different signs}}{21.93} \div \underset{}{(-3)} = \underset{\text{Negative quotient}}{-7.31}$$

Different signs, negative quotient.

The **quotient is** -7.31

— **Continued on Next Page**

1 Divide. Check the quotients by multiplying.

(a) 4)93.6
```
 23._
4)93.6
 8
 13
 12
  1
```
Finish the division. Then multiply the quotient by 4. If you get 93.6, the quotient is correct.

(b) 6)6.804

(c) $\dfrac{278.3}{11}$

(d) $-0.51835 \div 5$

Different signs; *negative* quotient

(e) $-213.45 \div (-15)$

Same sign; *positive* quotient

Answers

1. (a) 23.4
   CHECK (23.4)(4) = 93.6
   (b) 1.134
   CHECK (1.134)(6) = 6.804
   (c) 25.3
   CHECK (25.3)(11) = 278.3
   (d) −0.10367
   CHECK (−0.10367)(5) = −0.51835
   (e) 14.23
   CHECK (14.23)(−15) = −213.45

(b)
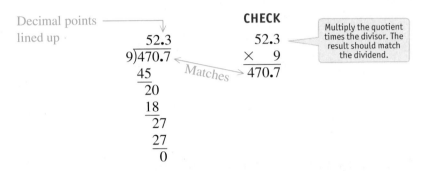

Write the decimal point in the quotient directly above the decimal point in the dividend. Then divide as if they were whole numbers.

Decimal points lined up
```
    52.3
9)470.7
  45
   20
   18
    27
    27
     0
```
Matches

CHECK
```
  52.3
×    9
470.7
```
Multiply the quotient times the divisor. The result should match the dividend.

The **quotient is 52.3** and is *positive* because both numbers have the *same* sign.

◄ **Work Problem 1 at the Side.**

EXAMPLE 2 **Writing Extra Zeros to Complete a Division**

Divide 1.5 by 8. Use multiplying to check the quotient.

Keep dividing until the remainder is 0, or until the digits in the quotient begin to repeat in a pattern. In **Example 1(b)** above, you ended up with a remainder of 0. But sometimes you run out of digits in the dividend before that happens. If so, write extra zeros on the right side of the dividend so you can continue dividing.

```
  0.1
8)1.5   ← All digits have been used.
  8
  7   ← Remainder is not yet 0
```

Write a 0 after the 5 in the dividend so you can continue dividing. Keep writing more zeros in the dividend if needed. Recall that writing zeros to the *right* of a decimal number does *not* change its value.

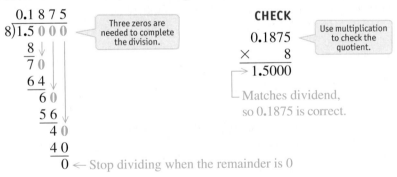

```
  0.1 8 7 5
8)1.5 0 0 0
  8
  7 0
  6 4
    6 0
    5 6
      4 0
      4 0
       0   ← Stop dividing when the remainder is 0
```
Three zeros are needed to complete the division.

CHECK
```
 0.1875
×     8
1.5000
```
Use multiplication to check the quotient.

Matches dividend, so 0.1875 is correct.

The **quotient is 0.1875**

❶ CAUTION

Notice that in decimals the dividend might *not* be the *greater* number. In **Example 2** above, the dividend is 1.5, which is *less* than 8.

——— **Continued on Next Page**

▦ **Calculator Tip**

When *multiplying* numbers, you can enter them in any order because multiplication is commutative. But division is *not* commutative. It *does* matter which number you enter first. Try **Example 2** on the previous page both ways; jot down your answers.

1.5 ⊝ 8 ⊜ _____　　8 ⊝ 1.5 ⊜ _____

Notice that the first answer, 0.1875, matches the result from **Example 2**. But the second answer is much different: 5.333333333. Be careful to enter the dividend first.

────── **Work Problem ② at the Side.** ▶

In the next example the remainder is never 0, even if we keep dividing.

EXAMPLE 3　**Rounding a Decimal Quotient**

Divide 4.6 by 3. Round the quotient to the nearest thousandth.

Write extra zeros in the dividend so that you can continue dividing.

$$
\begin{array}{r}
1.5\,3\,3\,3 \\
3\overline{)4.6\,0\,0\,0} \quad \leftarrow \text{Three zeros added so far}\\
\underline{3} \\
1\,6 \\
\underline{1\,5} \\
1\,0 \\
\underline{9} \\
1\,0 \\
\underline{9} \\
1\,0 \\
\underline{9} \\
1 \quad \leftarrow \text{Remainder is still not 0}
\end{array}
$$

Notice that the digit 3 in the quotient is repeating. It will continue to do so. The remainder will *never be 0*. There are two ways to show that an answer is a **repeating decimal** that goes on forever. You can write three dots after the answer, or you can write a bar above the digits that repeat (in this case, the 3).

1.5**333** ...　or　1.5$\overline{3}$　← Bar above repeating digit

　　　　Three dots

❗ **CAUTION**

Do not use *both* the dots *and* the bar at the same time. Use three dots *or* the bar.

When repeating decimals occur, write the quotient according to the directions in the problem. In this example, to round to thousandths, divide out one *more* place, to ten-thousandths.

4.6 ÷ 3 = 1.5333...　rounds to　**1.533** ◁ Nearest **thousandth** is **three** decimal places.

Check the answer by multiplying 1.533 by 3. Because 1.533 is a rounded answer, the check will *not* give exactly 4.6, but it should be very close.

$(1.533)(3) = 4.599$ ◁ Does not equal exactly 4.6 because 1.533 was rounded

────── **Continued on Next Page**

---

② Divide. Check the quotients by multiplying.

(a) $\dfrac{6.4}{5}$ ➝ $5\overline{)6\,.\,4\,\,0}$

Finish the division.

(b) $30.87 \div (-14)$

(c) $\dfrac{-259.5}{-30}$

(d) $0.3 \div 8$

Answers

2. **(a)** 1.28
　　CHECK $(1.28)(5) = 6.40$ or 6.4
　(b) −2.205
　　CHECK $(-2.205)(-14) = 30.870$
　　　or 30.87
　(c) 8.65
　　CHECK $(8.65)(-30) = -259.50$
　　　or −259.5
　(d) 0.0375
　　CHECK $(0.0375)(8) = 0.3000$ or 0.3

③ Divide. Round quotients to the nearest thousandth. If it is a repeating decimal, also write the answer using a bar. Check your answers by multiplying.

GS (a) $13\overline{)267.01}$

Use your calculator.

$267.01 \div 13 \approx$ _____

Is the quotient a repeating decimal? _____

So round your answer to the nearest thousandth. _____

(b) $6\overline{)20.5}$

(c) $\dfrac{17.02}{15}$

(d) $16.15 \div 3$

(e) $116.3 \div 11$

! CAUTION

When checking quotients that you've rounded, the check will *not* match the dividend exactly, but it should be very close.

◀ **Work Problem ③ at the Side.**

OBJECTIVE ② Divide a number by a decimal. To divide by a *decimal* divisor, first change the divisor to a whole number. Then divide as before. To see how this is done, write the problem in fraction form. Here is an example.

$$1.2\overline{)6.36} \quad \text{can be written} \quad \frac{6.36}{1.2}$$

You already learned that multiplying the numerator and denominator by the same number gives an equivalent fraction. We want the divisor (1.2) to be a whole number. Multiplying by 10 will accomplish that.

$$\text{Decimal divisor} \rightarrow \frac{6.36}{1.2} = \frac{(6.36)(10)}{(1.2)(10)} = \frac{63.6}{12} \leftarrow \text{Whole number divisor}$$

The short way to multiply by 10 is to move the decimal point *one place* to the *right* in both the divisor and the dividend.

$$1.2\overline{)6.36} \quad \text{is equivalent to} \quad 12\overline{)63.6}$$

Note

Moving the decimal points the **same** number of places to the right in **both** the divisor and dividend will **not** change the answer.

Dividing by a Decimal Number

Step 1 Count the number of decimal places in the divisor and move the decimal point that many places to the *right*. (This changes the divisor to a whole number.)

Step 2 Move the decimal point in the dividend the *same* number of places to the *right*. (Write in extra zeros if needed.)

Step 3 Write the decimal point in the quotient directly above the decimal point in the dividend. Then divide as usual.

Step 4 If both numbers have the *same sign,* the quotient is *positive.* If they have *different signs,* the quotient is *negative.*

Answers

3. (a) 20.5392; no repeating digits visible on calculator; 20.539 (rounded)
   CHECK $(20.539)(13) = 267.007$
(b) 3.417; $3.41\overline{6}$
   CHECK $(3.417)(6) = 20.502$
(c) 1.135; $1.134\overline{6}$
   CHECK $(1.135)(15) = 17.025$
(d) 5.383; $5.38\overline{3}$
   CHECK $(5.383)(3) = 16.149$
(e) 10.573; $10.5\overline{72}$
   CHECK $(10.573)(11) = 116.303$

EXAMPLE 4 **Dividing by Decimal Numbers**

(a) $\dfrac{27.69}{0.003}$

Move the decimal point in the divisor *three* places to the *right* so that 0.003 becomes the whole number 3. In order to move the decimal point in the dividend the same number of places, write in an extra 0.

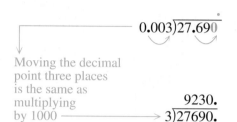

Move decimal points in divisor and dividend. Then line up the decimal point in the quotient.

Moving the decimal point three places is the same as multiplying by 1000 —→ $3\overline{)27690.}$ Divide as usual.

$9230.$

The **quotient is 9230**

(b) Divide −5 by −4.2 and round the quotient to the nearest hundredth.

First consider 5 ÷ 4.2. Move the decimal point in the divisor one place to the right so that 4.2 becomes the whole number 42. The decimal point in the dividend starts on the right side of 5 and is also moved one place to the right.

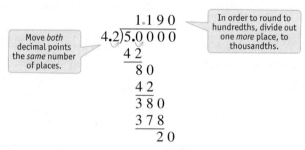

Move *both* decimal points the *same* number of places.

In order to round to hundredths, divide out one *more* place, to thousandths.

Rounding the quotient to the nearest hundredth gives 1.19. The quotient is *positive* because both the divisor and dividend have the *same* sign (both are negative).

$$-5 \div (-4.2) \approx 1.19$$

Same sign — Positive quotient

The **quotient is 1.19 (rounded)**

———— **Work Problem ④ at the Side.** ▶

OBJECTIVE ▶ 3 **Estimate the answer when dividing decimals.** Estimating the answer to a division problem helps you catch errors. Compare the estimate to your exact answer. If they are very different, do the division again.

④ Divide. If the quotient does not come out even, round to the nearest hundredth.

GS **(a)** $0.2\overline{)1.0\,4}$

(b) $0.06\overline{)1.8072}$

(c) $0.005\overline{)32}$

(d) $-8.1 \div 0.025$

(e) $\dfrac{7}{1.3}$

(f) $-5.3091 \div (-6.2)$

5 Decide whether each answer is reasonable by using front end rounding to estimate the answer. If the exact answer is *not* reasonable, find and correct the error.

GS **(a)** $42.75 \div 3.8 = 1.125$

Estimate: $40 \div 4 = 10$

Answer of 1.125 is **not** reasonable. Rework.

$3.8\overline{)42.75}$

(b) $807.1 \div 1.76 = 458.580$
to nearest thousandth

Estimate:

(c) $48.63 \div 52 = 93.519$
to nearest thousandth

Estimate:

(d) $9.0584 \div 2.68 = 0.338$

Estimate:

Answers

5. **(a)** Exact answer should be 11.25
   (b) Estimate is $800 \div 2 = 400$
   exact answer is reasonable.
   (c) Estimate is $50 \div 50 = 1$
   exact answer is not reasonable,
   should be 0.935
   (d) Estimate is $9 \div 3 = 3$
   exact answer is not reasonable,
   should be 3.38

EXAMPLE 5 **Estimating before Dividing**

First, use front end rounding to estimate the answer. Then divide to find the exact answer.

$$580.44 \div 2.8$$

Here is how one student solved this problem. She rounded 580.44 to 600 and 2.8 to 3 to estimate the answer.

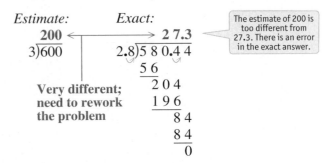

Estimate:
$$\begin{array}{r} 200 \\ 3\overline{)600} \end{array}$$

Exact:
$$\begin{array}{r} 27.3 \\ 2.8\overline{)580.44} \\ \underline{56} \\ 204 \\ \underline{196} \\ 84 \\ \underline{84} \\ 0 \end{array}$$

The estimate of 200 is too different from 27.3. There is an error in the exact answer.

Very different; need to rework the problem

Notice that the estimate, which is in the hundreds, is very different from the exact answer, which is only in the tens. This tells the student that she needs to rework the problem. Can you find the error? (The **exact answer should be 207.3**, which fits with the estimate of 200.)

◀ **Work Problem 5 at the Side.**

OBJECTIVE ▶ 4 **Use the order of operations with decimals.** Use the order of operations when a decimal problem involves more than one operation.

Order of Operations

Step 1 Work inside *parentheses* or *other grouping symbols*.

Step 2 Simplify expressions with *exponents*.

Step 3 Do the remaining *multiplications and divisions* as they occur from left to right.

Step 4 Do the remaining *additions and subtractions* as they occur from left to right.

EXAMPLE 6 **Using the Order of Operations**

Simplify by using the order of operations.

(a) $2.5 + (-6.3)^2 + 9.62$ Apply the exponent: $(-6.3)(-6.3)$ is 39.69

$ 2.5 + 39.69 + 9.62$ Add from left to right.

$ 42.19 + 9.62$

$ 51.81$

(b) $1.82 + (5.8)(5.2 - 6.7)$ Work inside parentheses.

$1.82 + (5.8)(-1.5)$ Multiply next.

$1.82 + (-8.7)$ Add last.

$ -6.88$

Continued on Next Page

(c) $\underline{3.7^2} - 1.8 \div 5\,(1.5)$ Apply the exponent.

$13.69 - \underline{\mathbf{1.8 \div 5}}\,(1.5)$ Multiply and divide *from left to right,* so first divide 1.8 by 5

$13.69 - \underline{\mathbf{0.36\,(1.5)}}$ Then multiply 0.36 by 1.5

> Careful!
> Do **not**
> subtract yet.

$13.69 - \underline{0.54}$ Subtract last.

$\mathbf{13.15}$

———————— **Work Problem ⑥ at the Side.** ▶

🔲 **Calculator Tip**

You may want to use a calculator to check your work. Most calculators that have parentheses keys ⓛ ⓛ can handle calculations like those in **Example 6(b)** on the previous page just by entering the numbers in the order given. You may need to enter any understood multiplications. For example, the keystrokes for **Example 6(b)** are:

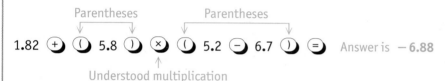

Parentheses   Parentheses

1.82 ⊕ ⓛ 5.8 ⓛ ⓧ ⓛ 5.2 ⊖ 6.7 ⓛ ⌷   Answer is -6.88

↑
Understood multiplication

    Standard, four-function calculators generally do *not* give the correct answer if you enter the numbers in the order given. Check the instruction manual that came with your calculator or look it up online. See if your model has the rules for order of operations built into it. For a quick check, try entering this problem:

2 ⊕ 2 ⓧ 2 ⌷

If the result is 6, the calculator follows the order of operations. If the result is 8, it does *not* have the rules built into it. Use the space below to explain how this test works.

Answer: The test works because a calculator that follows the order of operations will automatically do the multiplication first. If the calculator does *not* have the rules built into it, it will work from left to right.

Following Order of Operations	**Working from Left to Right**
$2 + \underbrace{2 \times 2}$ Multiply before adding.	$\underbrace{2 + 2} \times 2$
$2 + 4$	4×2
$6 \leftarrow$ Correct	✖ $8 \leftarrow$ Incorrect

⚙️ **Study Skills Reminder**

How do you learn and remember a new skill, like dividing decimals? Your brain has to grow new dendrites! You can help them grow and make strong connections so you'll remember longer. See the Study Skills activity "Your Brain **Can** Learn Mathematics" for more help.

---

⑥ Simplify. The blue brace shows you where to start.

(a) $-4.6 - 0.79 + \underbrace{1.5^2}$

(b) $\underbrace{3.64 \div 1.3}\,(3.6)$

(c) $0.08 + 0.6\,(\underbrace{2.99 - 3})$

(d) $10.85 - \underbrace{2.3\,(5.2)} \div 3.2$

Answers

6. **(a)** -3.14 **(b)** 10.08 **(c)** 0.074
    (d) 7.1125

5.5 Exercises

FOR EXTRA HELP Go to MyMathLab for worked-out, step-by-step solutions to exercises enclosed in a square ▢ and video solutions to ▶ exercises.

1. **CONCEPT CHECK** Which problem below has a whole number for the divisor? Circle it and then rewrite the problem using the $\overline{)}$ symbol.

   (a) Divide 0.25 by 0.05

   (b) $25 \div 0.5$

   (c) $25.5 \div 5$

2. **CONCEPT CHECK** Which problem below has a whole number for the divisor? Circle it and then rewrite the problem using the $\overline{)}$ symbol.

   (a) $0.423 \div 0.07$

   (b) $42.3 \div 7$

   (c) Divide 423 by 0.7

3. **CONCEPT CHECK** In $2.2\overline{)8.24}$ how do you make 2.2 a whole number, and how does that change 8.24? Use arrows to show how to move the decimal points.

4. **CONCEPT CHECK** In $5.1\overline{)10.5}$ how do you make 5.1 a whole number, and how does that change 10.5? Use arrows to show how to move the decimal points.

*Find each quotient. **See Examples 1, 2, and 4.***

5. ▶ $27.3 \div (-7)$

6. $-50.4 \div 8$

7. $\dfrac{4.23}{9}$

8. $\dfrac{1.62}{6}$

9. $-20.01 \div (-0.05)$

10. $-16.04 \div (-0.08)$

11. $1.5\overline{)54.0}$
GS

12. $2.4\overline{)132.0}$
GS

CONCEPT CHECK Use the fact that $108 \div 18 = 6$ to work Exercises 13–20 simply by moving decimal points. **See Examples 1, 2, and 4.**

13. $1.8\overline{)0.108}$

14. $18\overline{)10.8}$

15. $0.018\overline{)108}$

16. $0.18\overline{)1.08}$

17. $0.18\overline{)10.8}$

18. $0.18\overline{)108}$

19. $18\overline{)0.0108}$

20. $1.8\overline{)0.0108}$

*Divide. Round quotients to the nearest hundredth when necessary. **See Examples 3 and 4.***

21. $4.6\overline{)116.38}$

22. $2.6\overline{)4.992}$

23. ▶ $\dfrac{-3.1}{-0.006}$

24. $\dfrac{-1.7}{0.09}$

*Divide. Round quotients to the nearest thousandth. **See Examples 3 and 4.***

25. $-240.8 \div 9$

26. $-76.43 \div (-7)$

27. $0.034\overline{)342.81}$
🖩

28. $0.043\overline{)1748.4}$
🖩

CONCEPT CHECK Decide whether each answer is reasonable or unreasonable by using front end rounding to estimate the answer. If the exact answer is unreasonable, find and correct the error. **See Example 5.**

29. ▶ $37.8 \div 8 = 47.25$

   Estimate:

30. $345.6 \div 3 = 11.52$

   Estimate:

31. $54.6 \div 48.1 \approx 1.135$

Estimate:

32. $2428.8 \div 4.8 = 56$

Estimate:

33. $307.02 \div 5.1 = 6.2$

Estimate:

34. $395.415 \div 5.05 = 78.3$

Estimate:

Solve each application problem. Round money answers to the nearest cent when necessary.

35. Rob has discovered that his daughter's favorite brand of tights are on sale. He decided to buy one pair as a surprise for her. How much did he pay?

36. The bookstore has a special price on notepads. How much did Randall pay for one notepad?

37. It will take 21 months for Aimee to pay off her credit card balance of $1408.68. How much is she paying each month?

38. Marcella Anderson bought a 2.6 meter length of microfiber woven suede fabric for $33.77. How much did she pay per meter? (Cost per meter means the cost for *one* meter.)

39. Darren Jackson earned $476.80 for 40 hours of work. Find his earnings per hour. (Earnings *per* hour means how much he earned for *one* hour.)

40. Adrian Webb bought 108 patio blocks to build a backyard patio. He paid $237.60. Find the cost per block.

41. It took 12.3 gallons of gas to fill the gas tank of Kim's car. She had driven 344.1 miles since her last fill-up. How many miles per gallon did she get? Round to the nearest tenth.

42. Mr. Rodriquez pays $53.19 each month to Household Finance. How many months will it take him to pay off $1436.13?

43. Soup is on sale at six cans for $3.85, or you can buy individual cans for $0.67. How much will you save per can if you buy six cans? Round your final answer to the nearest cent.

44. Nadia's diet allows her to eat 3.5 ounces of chicken nuggets. The package weighs 10.5 ounces and contains 15 nuggets. How many nuggets can Nadia eat? (*Hint:* First find the weight of one nugget.)

Use the table of world records for the women's longest long jump (through the year 2014) to answer Exercises 45–50. The records are listed from longest to shortest. To find an average, add up the values you are interested in and then divide the sum by the number of values. Round your final answers to the nearest hundredth.

Athlete	Country	Year	Length (meters)
Galina Christyakova	USSR	1988	7.52
Jackie Joyner-Kersee	U.S.	1994	7.49
Heike Drechsler	Germany	1992	7.48
Jackie Joyner-Kersee	U.S.	1987	7.45
Jackie Joyner-Kersee	U.S.	1988	7.40
Jackie Joyner-Kersee	U.S.	1991	7.32
Jackie Joyner-Kersee	U.S.	1996	7.20
Chioma Ajunwa	Nigeria	1996	7.12
Fiona May	Italy	2000	7.09
Tatyana Lebedeva	Russia	2004	7.07

Data from www.CNNSI.com

45. Find the average length of the long jumps made by Jackie Joyner-Kersee.

46. Find the average length of all the long jumps listed in the table.

47. How much longer was the fifth-longest jump than the sixth-longest jump?

48. If the longest jump was made five times, what would be the total distance jumped?

49. What was the total length jumped by the top three athletes in the table?

50. How much less was the shortest jump in the table than the next-to-shortest jump?

51. CONCEPT CHECK The work shown below has **two** mistakes in it.

What Went Wrong? First write a sentence explaining what each mistake is. Then fix the mistakes and find the correct solution.

Ketruah simplified $4.4^2 - (2.5 + 3.1)$ this way:

$4.4^2 - (2.5 + 3.1)$
$8.8 - 2.5 + 3.1$
$6.3 + 3.1$
9.4

52. CONCEPT CHECK The work shown below has **two** mistakes in it.

What Went Wrong? First write a sentence explaining what each mistake is. Then fix the mistakes and find the correct solution.

Richlieu simplified $60.41 - 3(0.4 + 5.07)$ this way:

$60.41 - 3(0.4 + 5.07)$
$57.41(0.4 + 5.07)$
$57.41 + 5.47$
62.88

Simplify. See Example 6.

53. $7.2 - 5.2 + 3.5^2$

54. $6.2 + 4.3^2 - 9.72$

55. $38.6 + 11.6(10.4 - 13.4)$

56. $2.25 - 1.06(0.85 - 3.95)$

57. $-8.68 - 4.6(10.4) \div 6.4$

58. $25.1 + 11.4 \div 7.5(-3.75)$

59. $33 - 3.2(0.68 + 9) + (-1.3)^2$

60. $0.6 + (-1.89 + 0.11) \div 0.004(0.5)$

61. The U.S. Treasury prints about 38,000,000 pieces of paper money each day. The printing presses run 24 hours a day. How many pieces of money are printed each hour, each minute, and each second, to the nearest whole number?

38,000,000 pieces of paper money are printed each day. (Data from www.moneyfactory.com)

   (a) each hour

   (b) each minute

   (c) each second

62. Mach 1 is the speed of sound. Dividing a vehicle's speed by the speed of sound gives its speed on the Mach scale. In 1997, a specially built car with two 110,000-horsepower engines broke the world land speed record by traveling 763.035 miles per hour. The speed of sound changes slightly with the weather. That day it was 748.11 miles per hour. What was the car's Mach speed, to the nearest hundredth? (Data from Associated Press.)

General Mills will give a school 10¢ for each box top logo collected from its cereals and other products. A school can earn up to $20,000 per year. Use this information to answer Exercises 63–66. Round answers to the nearest whole number. (Data from General Mills.)

EARN CASH FOR YOUR SCHOOL!

BOX TOPS FOR EDUCATION

63. How many box tops would a school need to collect in one year to earn the maximum amount?

64. (Complete **Exercise 63** first.) If a school has 550 children, how many box tops would each child need to collect to reach the maximum?

65. How many box tops would need to be collected during each of the 38 weeks in the school year to reach the maximum amount?

66. How many box tops would each of the 550 children need to collect during each of the 38 weeks of school to reach the maximum amount?

Relating Concepts (Exercises 67–68)   For Individual or Group Work

*Look for patterns as you **work Exercises 67 and 68 below**.*

67. Do these division problems.

$3.77 \div 10 =$ _____ $9.1 \div 10 =$ _____

$0.886 \div 10 =$ _____ $30.19 \div 10 =$ _____

$406.5 \div 10 =$ _____ $6625.7 \div 10 =$ _____

   (a) What pattern do you see? Write a "rule" for dividing by 10. What do you think the rule is for dividing by 100? by 1000? Write the rules and try them out on the numbers above.

   (b) Compare your rules to the ones you wrote in the **previous section** for **Exercise 57**.

68. Do these division problems.

$40.2 \div 0.1 =$ _____ $7.1 \div 0.1 =$ _____

$0.339 \div 0.1 =$ _____ $15.77 \div 0.1 =$ _____

$46 \div 0.1 =$ _____ $873 \div 0.1 =$ _____

   (a) What pattern do you see? Write a "rule" for dividing by 0.1. What do you think the rule is for dividing by 0.01? by 0.001? Write the rules and try them out on the numbers above.

   (b) Compare your rules to the ones you wrote in the **previous section** for **Exercise 58**.

Summary Exercises *Computation with Decimal Numbers*

CONCEPT CHECK *Write each decimal as a fraction or mixed number in lowest terms.*

1. 0.8

2. 6.004

3. 0.35

Write each decimal in words.

4. 94.5

5. 2.0003

6. 0.706

Write each decimal in numbers.

7. five hundredths

8. three hundred nine ten-thousandths

9. ten and seven tenths

Round to the place indicated.

10. 6.1873 to the nearest hundredth

11. 0.95 to the nearest tenth

12. 0.42025 to the nearest thousandth

13. $0.893 to the nearest cent

14. $3.0017 to the nearest cent

15. $99.64 to the nearest dollar

Simplify.

16. $-0.27\,(3.5)$

17. $50 - 0.3801$

18. $0.35 \div (-0.007)$

19. $\dfrac{-90.18}{-6}$

20. $(0.004)\,(1.22)$

21. $1.55 - 3.7$

22. $-0.95 + 10.005$

23. $3.6 + 0.718 + 9 + 5.0829$

24. $32.305 - 40 + 0.7$

25. $-8.9 + 4^2 \div (-0.02)$

26. $(-0.18 + 2.5) + 4\,(-0.05)$

27. $0.64 \div 16.3$ Round your answer to the nearest hundredth.

Find the perimeter of each figure.

28.

19.75 in.

6.3 in.   6.3 in.

19.75 in.

29.

2 meters   1 meter

0.9 meter

1.7 meters

1.18 meters

0.86 meter

2.095 meters

Solve each application problem.

30. At a bakery, Sue Chee bought $7.42 worth of muffins and $10.09 worth of croissants for a staff party and a $0.69 cookie for herself. How much change did she receive from two $10 bills?

31. A craft cooperative paid $60.32 for enough fabric to make eight baby blankets for their store. They sold the blankets for $18.95 each. How much profit was made on the blankets?

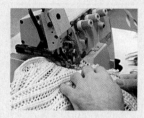

The table below shows information on two tents for camping. Use the table to answer Exercises 32 and 33.

Tents	Coleman Evanston Tent	Mountain Trails Tent
Dimensions	12 ft × 12 ft	11 ft × 11 ft
Sleeps	8 campers	6 campers
Sale price	$170	$99

Data from www.Amazon.com

32. For the Coleman tent, find:

   (a) the area of the tent floor

   (b) the cost per square foot of floor space. Round your answer to the nearest cent.

33. Find the same information for the Mountain Trails tent as you did for the Coleman tent.

Use the information in the table to answer Exercises 34–36.

Animal	Average Weight of Animal (ounces)	Average Weight of Food Eaten Each Day (ounces)
Hamster	3.5	0.4
Queen bee	0.004	?
Hummingbird	?	0.07

Data from *NCTM News Bulletin.*

34. (a) In how many days will a hamster eat enough food to equal its body weight? Round to the nearest whole number of days.

   (b) If a 140-pound woman ate her body weight of food in the same number of days as the hamster, how much would she eat each day? Round to the nearest tenth.

35. While laying eggs, a queen bee eats eighty times her weight each day. Use this information to fill in one of the missing values in the table.

36. A hummingbird's body weight is about 1.6 times the weight of its daily food intake. Find its body weight.

5.6 Fractions and Decimals

OBJECTIVES

1 Write fractions as equivalent decimals.

2 Compare the size of fractions and decimals.

Writing fractions as equivalent decimals can help you do calculations or compare the size of two numbers more easily.

OBJECTIVE 1 Write fractions as equivalent decimals. Recall that a fraction is one way to show division. For example, $\frac{3}{4}$ means $3 \div 4$. If you are doing the division by hand, write it as $4\overline{)3}$. When you do the division, the result is 0.75, the decimal equivalent of $\frac{3}{4}$.

> **Writing a Fraction as an Equivalent Decimal**
>
> **Step 1** Divide the numerator of the fraction by the denominator.
>
> **Step 2** If necessary, round the answer to the place indicated.

◀ **Work Problem 1 at the Side.**

1 Rewrite each fraction so you could do the division by hand. Do **not** complete the division.

GS (a) $\frac{1}{9}$ is written $9\overline{)}$

(b) $\frac{2}{3}$ is written $\overline{)}$

(c) $\frac{5}{4}$ is written $\overline{)}$

(d) $\frac{3}{10}$ is written $\overline{)}$

(e) $\frac{21}{16}$ is written $\overline{)}$

(f) $\frac{1}{50}$ is written $\overline{)}$

EXAMPLE 1 Writing Fractions or Mixed Numbers as Decimals

(a) Write $\frac{1}{8}$ as a decimal.

$\frac{1}{8}$ means $1 \div 8$. Write it as $8\overline{)1}$. The decimal point in the dividend is on the *right* side of the 1. Write extra zeros in the dividend so you can continue dividing until the remainder is 0.

$$\frac{1}{8} \rightarrow 1 \div 8 \rightarrow 8\overline{)1} \rightarrow \begin{array}{r} 0.125 \\ 8\overline{)1.000} \\ \underline{8} \\ 20 \\ \underline{16} \\ 40 \\ \underline{40} \\ 0 \end{array}$$

Decimal points lined up

← Three extra zeros needed

Be careful to divide $8\overline{)1}$, **not** $1\overline{)8}$

← Remainder is 0

Therefore, $\frac{1}{8} = \mathbf{0.125}$

To check, write 0.125 as a fraction, then write it in lowest terms.

$$0.125 = \frac{125}{1000} \quad \text{In lowest terms:} \quad \frac{125 \div 125}{1000 \div 125} = \frac{1}{8} \left\{ \begin{array}{l} \text{Original} \\ \text{fraction} \end{array} \right.$$

> **Calculator Tip**
>
> When using your calculator to write fractions as decimals, enter the numbers from the top down. Remember that the *order* in which you enter the numbers *does* matter in division. **Example 1(a)** above works like this:
>
> $\frac{1}{8}$ ↓ Top down   Enter 1 ÷ 8 =   Answer is 0.125
>
> What happens if you enter 8 ÷ 1 =? Do you see why that cannot possibly be correct? (Answer: $8 \div 1 = 8$. A proper fraction like $\frac{1}{8}$ *cannot* be equivalent to a whole number.)

Answers

1. **(a)** $9\overline{)1}$ **(b)** $3\overline{)2}$ **(c)** $4\overline{)5}$ **(d)** $10\overline{)3}$
   (e) $16\overline{)21}$ **(f)** $50\overline{)1}$

——— **Continued on Next Page**

(b) Write $2\frac{3}{4}$ as a decimal.

Recall that $2\frac{3}{4}$ is $2 + \frac{3}{4}$. So one method is to divide 3 by 4 to get 0.75 for the fraction part. Then add the whole number part to 0.75.

$$\frac{3}{4} \longrightarrow \begin{array}{r} 0.75 \\ 4\overline{)3.00} \\ \underline{2\ 8} \\ 20 \\ \underline{20} \\ 0 \end{array}$$

Fraction part $\longrightarrow$

$$\begin{array}{r} 2.00 \leftarrow \text{Whole number part} \\ +\ 0.75 \\ \hline 2.75 \end{array}$$

So, $2\frac{3}{4} = \mathbf{2.75}$

Whole number parts match.

CHECK $2.75 = 2\frac{75}{100} = 2\frac{3}{4} \leftarrow$ Lowest terms

A second method is to write $2\frac{3}{4}$ as an improper fraction before dividing numerator by denominator.

$$2\frac{3}{4} = \frac{11}{4}$$

$$\frac{11}{4} \longrightarrow 11 \div 4 \longrightarrow 4\overline{)11} \longrightarrow \begin{array}{r} 2.75 \\ 4\overline{)11.0\ 0} \\ \underline{8} \\ 3\ 0 \\ \underline{2\ 8} \\ 2\ 0 \\ \underline{2\ 0} \\ 0 \end{array} \leftarrow \text{Two extra zeros needed}$$

Whole number parts match.

So, $2\frac{3}{4} = \mathbf{2.75}$

$\frac{3}{4}$ is equivalent to $\frac{75}{100}$ or 0.75

─────── **Work Problem ❷ at the Side.** ▶

Continued on Next Page

| EXAMPLE 2 | **Writing a Fraction as a Decimal with Rounding** |

Write $\frac{2}{3}$ as a decimal and round to the nearest thousandth.

$\frac{2}{3}$ means $2 \div 3$. To round to thousandths, divide out one *more* place, to ten-thousandths.

$$\frac{2}{3} \longrightarrow 2 \div 3 \longrightarrow 3\overline{)2} \longrightarrow \begin{array}{r} 0.6666 \\ 3\overline{)2.0000} \\ \underline{1\ 8} \\ 20 \\ \underline{18} \\ 20 \\ \underline{18} \\ 20 \\ \underline{18} \\ 2 \end{array}$$

$\left\{ \begin{array}{l} \text{Four zeros needed} \\ \text{for ten-thousandths} \end{array} \right.$

Be careful to divide $3\overline{)2}$, **not** $2\overline{)3}$

Written as a repeating decimal, $\frac{2}{3} = 0.\overline{6} \leftarrow$ Bar above repeating digit

Rounded to the nearest thousandth, $\dfrac{2}{3} \approx \mathbf{0.667}$

❷ Write each fraction or mixed number as a decimal.

GS (a) $\dfrac{1}{4} \longrightarrow \begin{array}{r} 0.\underline{}\ \underline{} \\ 4\overline{)1.\ 0\ 0} \end{array}$ Finish the division.

GS (b) $2\dfrac{1}{2} = \dfrac{5}{2} \longrightarrow 2\overline{)5.0}$

(c) $\dfrac{5}{8}$

(d) $4\dfrac{3}{5}$

(e) $\dfrac{7}{8}$

Answers

2. (a) 0.25 (b) 2.5 (c) 0.625
   (d) 4.6 (e) 0.875

③ **▦** Write as decimals. Round to the nearest thousandth.

(a) $\dfrac{1}{3}$

GS (b) $2\dfrac{7}{9} = \dfrac{25}{9} = 25 \div 9$

Now use your calculator to do the division. Round the answer to the nearest thousandth.

(c) $\dfrac{10}{11}$

(d) $\dfrac{3}{7}$

(e) $3\dfrac{5}{6}$

Answers

3. (a) $\dfrac{1}{3} \approx 0.333$ (b) $2\dfrac{7}{9} \approx 2.778$

(c) $\dfrac{10}{11} \approx 0.909$ (d) $\dfrac{3}{7} \approx 0.429$

(e) $3\dfrac{5}{6} \approx 3.833$

▦ Calculator Tip

Try **Example 2** on your calculator. Enter 2 **÷** 3. Which answer do you get?

| **0.666666667** | or | **0.666666666** |

Many scientific calculators will show a 7 as the last digit. Because the 6s keep on repeating forever, the calculator automatically rounds in the last decimal place it has room to show. If you have a 10-digit display space, the calculator is rounding as shown below.

0.6666666666 (11 digits)   rounds to   0.666666667

Next digit is 5 or more, so 6 rounds to 7

Other calculators, especially standard, four-function ones, may *not* round. They cut off, or *truncate,* the extra digits. Such a calculator would show 0.6666666 in the display.

Would this difference in calculators show up when changing $\frac{1}{3}$ to a decimal? Why not? (Answer: No. The repeating digit is a 3, which is *4 or less,* so it stays as a 3 whether it's rounded or not.)

◀ **Work Problem ③ at the Side.**

OBJECTIVE ▸ ② **Compare the size of fractions and decimals.** You can use a number line to compare fractions and decimals. For example, the number line below shows the space between 0 and 1. The locations of some commonly used fractions are marked, along with their decimal equivalents.

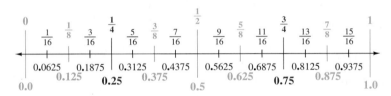

The next number line shows the locations of some commonly used fractions between 0 and 1 that are equivalent to repeating decimals. The decimal equivalents use a bar above repeating digits.

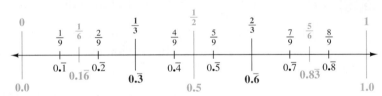

EXAMPLE 3 **Using a Number Line to Compare Numbers**

Use the number lines above to decide whether to write $>$, $<$, or $=$ in the blank between each pair of numbers.

(a) 0.6875 _____ 0.625

You have learned that the number farther to the right on the number line is the greater number. On the first number line, 0.6875 is to the *right* of 0.625, so use the $>$ symbol.

0.6875 is greater than 0.625   can be written as   **0.6875 > 0.625**

> The larger end of the $>$ symbol faces the greater number.

Continued on Next Page

(b) $\dfrac{3}{4}$ _____ 0.75

On the first number line, $\frac{3}{4}$ and 0.75 are at the same point on the number line. They are equivalent, so use the $=$ symbol.

$$\dfrac{3}{4} = \mathbf{0.75}$$

(c) 0.5 _____ $0.\overline{5}$

On the second number line, 0.5 is to the *left* of $0.\overline{5}$ (which is actually $0.555\dots$), so use the $<$ symbol.

0.5 is less than $0.\overline{5}$ can be written as $\mathbf{0.5 < 0.\overline{5}}$

> The smaller end of the $<$ symbol points to the lesser number.

(d) $\dfrac{2}{6}$ _____ $0.\overline{3}$

Write $\frac{2}{6}$ in lowest terms as $\frac{1}{3}$.
On the second number line you can see that $\frac{1}{3}$ and $0.\overline{3}$ are equivalent.

$$\dfrac{1}{3} = \mathbf{0.\overline{3}}$$

— **Work Problem ④ at the Side.** ▶

You can also compare fractions by first writing each one as a decimal.

EXAMPLE 4 **Arranging Numbers in Order**

Write each group of numbers in order, from least to greatest.

(a) 0.49   0.487   0.4903

It is easier to compare decimals if they are all tenths, or all hundredths, and so on. Because 0.4903 has four decimal places (ten-thousandths), write zeros to the right of 0.49 and 0.487 so they also have four decimal places. Recall that writing zeros to the right of a decimal number does *not* change its value. Then find the least and greatest number of ten-thousandths.

0.49 = 0.4900 = **4900** ten-thousandths ← 4900 is in the middle.
0.487 = 0.4870 = **4870** ten-thousandths ← 4870 is the least.
0.4903 = **4903** ten-thousandths ← 4903 is the greatest.

From least to greatest, the correct order is shown below.

0.487   0.49   0.4903

(b) $2\dfrac{5}{8}$ 2.63   2.6

Write $2\frac{5}{8}$ as $\frac{21}{8}$ and divide $8\overline{)21}$ to get the decimal form, 2.625. Then, because 2.625 has three decimal places, write zeros so all the numbers have three decimal places.

$$2\dfrac{5}{8} = 2.625 = 2 \text{ and } \mathbf{625} \text{ thousandths} \leftarrow 625 \text{ is in the middle.}$$

2.63 = 2.630 = 2 and **630** thousandths ← 630 is the greatest.
2.6 = 2.600 = 2 and **600** thousandths ← 600 is the least.

From least to greatest, the correct order is shown below.

2.6 $2\dfrac{5}{8}$ **2.63**

— **Work Problem ⑤ at the Side.** ▶

---

④ Use the number lines on the previous page to help you decide whether to write $<$, $>$, or $=$ in each blank.

(a) 0.4375 _____ 0.5

(b) 0.75 _____ 0.6875

(c) 0.625 _____ 0.0625

(d) $\dfrac{2}{8}$ _____ 0.375

(e) $0.8\overline{3}$ _____ $\dfrac{5}{6}$

(f) $\dfrac{1}{2}$ _____ $0.\overline{5}$

(g) $0.\overline{1}$ _____ $0.1\overline{6}$

(h) $\dfrac{8}{9}$ _____ $0.\overline{8}$

(i) $0.\overline{7}$ _____ $\dfrac{4}{6}$

(j) $\dfrac{1}{4}$ _____ 0.25

⑤ Arrange each group in order from least to greatest.

(a) 0.7   0.703   0.7029

(b) 6.39   6.309   6.401   6.4

(c) 1.085 $1\dfrac{3}{4}$ 0.9

(d) $\dfrac{1}{4}$ $\dfrac{2}{5}$ $\dfrac{3}{7}$ 0.428

Answers

4. (a) $<$ (b) $>$ (c) $>$ (d) $<$ (e) $=$
   (f) $<$ (g) $<$ (h) $=$ (i) $>$ (j) $=$
5. (a) 0.7   0.7029   0.703
   (b) 6.309   6.39   6.4   6.401
   (c) 0.9   1.085 $1\dfrac{3}{4}$
   (d) $\dfrac{1}{4}$ $\dfrac{2}{5}$ 0.428 $\dfrac{3}{7}$

5.6 Exercises

FOR EXTRA HELP

Go to MyMathLab for worked-out, step-by-step solutions to exercises enclosed in a square ▢ and video solutions to ▶ exercises.

CONCEPT CHECK *In Exercises 1 and 2, circle all the divisions that are correctly set up to convert the given fraction to a decimal. If a division is **not** set up correctly, fix it.*

1. (a) $\dfrac{2}{5} \rightarrow 2\overline{)5}$ (b) $\dfrac{1}{7} \rightarrow 7\overline{)1}$ (c) $\dfrac{3}{8} \rightarrow 8\overline{)3}$ 2. (a) $\dfrac{4}{9} \rightarrow 9\overline{)4}$ (b) $\dfrac{1}{3} \rightarrow 1\overline{)3}$ (c) $\dfrac{5}{6} \rightarrow 6\overline{)5}$

CONCEPT CHECK *In Exercises 3 and 4, show the correct division set-up for converting the given fraction to a decimal. Place the decimal points in the dividend and quotient. Then write the first digit in the quotient. Finally, explain what you will do next in order to continue dividing.*

3. $\dfrac{3}{4} \rightarrow$

4. $\dfrac{1}{8} \rightarrow$

CONCEPT CHECK *In Exercises 5 and 6, write* greater than *or* less than *between each pair of numbers.*

5. (a) $\dfrac{2}{5}$ is _____ 0.5

   (b) $\dfrac{3}{4}$ is _____ 0.6

   (c) 0.2 is _____ $\dfrac{5}{8}$

6. (a) 0.1 is _____ $\dfrac{1}{5}$

   (b) $\dfrac{2}{3}$ is _____ 0.5

   (c) 0.7 is _____ $\dfrac{1}{2}$

Write each fraction or mixed number as a decimal. Round to the nearest thousandth when necessary. ***See Examples 1 and 2.***

7. $\dfrac{1}{2} \longrightarrow 2\overline{)1.\,0}^{\,0.\,\underline{}}$
   GS

8. $\dfrac{1}{4} \longrightarrow 4\overline{)1.\,0\ 0}^{\,0.\,\underline{}\,\underline{}}$
   GS

9. $\dfrac{3}{4}$
   ▶

10. $\dfrac{1}{10}$

11. $\dfrac{3}{10}$

12. $\dfrac{7}{10}$

13. $\dfrac{9}{10}$

14. $\dfrac{4}{5}$

15. $\dfrac{3}{5}$

16. $\dfrac{2}{5}$

17. $\dfrac{7}{8}$
    ▶

18. $\dfrac{3}{8}$

19. $2\dfrac{1}{4}$
    ▶

20. $1\dfrac{1}{2}$

21. $14\dfrac{7}{10}$

22. $23\dfrac{3}{5}$

23. $3\dfrac{5}{8}$

24. $2\dfrac{7}{8}$

25. $6\dfrac{1}{3}$

26. $5\dfrac{2}{3}$

27. $\dfrac{5}{6}$
    ▶
    ▦

28. $\dfrac{1}{6}$
    ▦

29. $1\dfrac{8}{9}$
    ▦

30. $5\dfrac{4}{7}$
    ▦

Find the decimal or fraction equivalent for each number. Write fractions in lowest terms.

	Fraction	Decimal		Fraction	Decimal
31.	_____	0.4	**32.**	_____	0.75
33.	_____	0.625	**34.**	_____	0.111
35.	_____	0.35	**36.**	_____	0.9
37.	$\dfrac{7}{20}$	_____	**38.**	$\dfrac{1}{40}$	_____
39.	_____	0.04	**40.**	_____	0.52
41.	$\dfrac{1}{5}$	_____	**42.**	$\dfrac{1}{8}$	_____
43.	_____	0.09	**44.**	_____	0.02

Solve each application problem.

45. The average length of a newborn baby at Central Hospital is 20.8 inches. Charlene's baby is 20.08 inches long. Is her baby longer or shorter than the average? By how much?

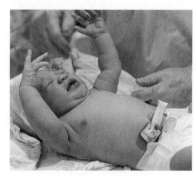

46. The patient in room 830 is supposed to get 8.3 milligrams of medicine. She was actually given 8.03 milligrams. Did she get too much or too little medicine? What was the difference?

47. Ginny Brown hoped her crops would get $3\frac{3}{4}$ inches of rain this month. The newspaper said the area received 3.8 inches of rain. Was that more or less than Ginny hoped for? By how much?

48. The rats in a medical experiment gained $\frac{3}{8}$ ounce. They were expected to gain 0.3 ounce. Was their actual gain more or less than expected? By how much?

49. The label on the bottle of vitamins says that each capsule contains 0.5 gram of calcium. When tested, each capsule had 0.505 gram of calcium. Was there too much or too little calcium? What was the difference?

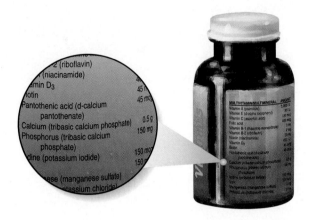

50. The glass mirror of the Hubble telescope had to be repaired in space because it would not focus properly. The problem was that the mirror's outer edge had a thickness of 0.6248 cm when it was supposed to be 0.625 cm. Was the edge too thick or too thin? By how much? (Data from NASA.)

51. **CONCEPT CHECK** Precision Medical Parts makes an artificial heart valve that must measure between 0.998 centimeter and 1.002 centimeters. Circle the lengths that are acceptable.

   1.01 cm   0.9991 cm   1.0007 cm   0.99 cm

52. **CONCEPT CHECK** The white rats in a medical experiment must start out weighing between 2.95 ounces and 3.05 ounces. Circle the weights that can be used.

   3.0 ounces   2.995 ounces 3.055 ounces

   3.005 ounces

Use the number lines above Example 3 in the text to decide whether to write >, <, or = between each pair of numbers. ***See Example 3.***

53. (a) 0.3125 ____ 0.375 (b) $\dfrac{6}{8}$ ____ 0.75 (c) $0.\overline{8}$ ____ $0.8\overline{3}$ (d) 0.5 ____ $\dfrac{5}{9}$

54. (a) 0.125 ____ 0.0625 (b) $\dfrac{4}{9}$ ____ $\dfrac{3}{6}$ (c) 1.0 ____ $\dfrac{4}{4}$ (d) $\dfrac{1}{6}$ ____ $0.\overline{1}$

Arrange each group of numbers in order from least to greatest. ***See Example 4.***

55. 0.54, 0.5455, 0.5399

56. 0.76, 0.7, 0.7006

57. 5.8, 5.79, 5.0079, 5.804
   ▶

58. 12.99, 12.5, 13.0001, 12.77

59. 0.628, 0.62812, 0.609, 0.6009

60. 0.27, 0.281, 0.296, 0.3

61. 5.8751, 4.876, 2.8902, 3.88

62. 0.98, 0.89, 0.904, 0.9

63. 0.043, 0.051, 0.006, $\dfrac{1}{20}$

64. 0.629, $\dfrac{5}{8}$, 0.65, $\dfrac{7}{10}$

65. $\dfrac{3}{8}$, $\dfrac{2}{5}$, 0.37, 0.4001

66. 0.1501, 0.25, $\dfrac{1}{10}$, $\dfrac{1}{5}$

CONCEPT CHECK *Four boxes of fishing line are in a sale bin. The thicker the line, the stronger it is. The diameter of the fishing line is its thickness. Use the information on the boxes to answer Exercises 67–68.*

67. **(a)** Which color box has the strongest line? The weakest line?

   (b) What is the difference in line diameter between the weakest and strongest lines?

68. **(a)** Which color box has the line that is $\frac{1}{125}$ inch in diameter?

   (b) What is the difference in line diameter between the blue and purple boxes?

🖩 *Some rulers for technical drawings show each inch divided into tenths. Use this scale drawing for Exercises 69–74. Change the measurements on the drawing to decimals and round them to the nearest tenth of an inch.*

69. Length **(a)** is _____

70. Length **(b)** is _____

71. Length **(c)** is _____

72. Length **(d)** is _____

73. Length **(e)** is _____

74. Length **(f)** is _____

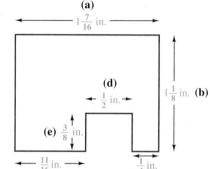

Relating Concepts (Exercises 75–78) For Individual or Group Work

*Use your knowledge of fractions and decimals as you **work Exercises 75–78 in order.***

75. **(a)** Explain how you can tell that Keith made an error *just by looking at his final answer.* Here is his work.

$$\frac{5}{9} \quad 5)\overline{9.0}^{\,1.8} \quad \text{so} \quad \frac{5}{9} = 1.8$$

   (b) Show the correct way to change $\frac{5}{9}$ to a decimal. Explain why your answer makes sense.

76. **(a)** How can you prove to Sandra that $2\frac{7}{20}$ is *not* equivalent to 2.035? Here is her work.

$$2\frac{7}{20} \quad 20)\overline{7.00}^{\,0.35} \quad \text{so} \quad 2\frac{7}{20} = 2.035$$

   (b) What is the correct answer? Show how to prove it.

77. Ving knows that $\frac{3}{8} = 0.375$. How can he write $1\frac{3}{8}$ as a decimal *without* having to do a division? How can he write $3\frac{3}{8}$ as a decimal? $295\frac{3}{8}$ as a decimal? Explain your answer.

78. Iris has a shortcut for writing mixed numbers as decimals.

$$2\frac{7}{10} = 2.7 \qquad 1\frac{13}{100} = 1.13$$

   Does her shortcut work for all mixed numbers? Explain when it works and why it works.

5.7 Problem Solving with Statistics: Mean, Median, and Mode

OBJECTIVES

1 Find the mean of a list of numbers.

2 Find a weighted mean.

3 Find the median.

4 Find the mode.

The word *statistics* originally came from words that mean *state numbers*. State numbers refer to numerical information, or *data,* gathered by the government, such as the number of births, deaths, or marriages in a population. Today the word *statistics* has a much broader meaning; data from the fields of economics, social science, science, and business can all be organized and studied under the branch of mathematics called *statistics*.

OBJECTIVE ▶ 1 Find the mean of a list of numbers. Making sense of a long list of numbers can be difficult. So when you analyze data, one of the first things to look for is a *measure of central tendency*—a single number that you can use to represent the entire list of numbers. One such measure is the *average* or **mean.** The mean can be found using this formula.

VOCABULARY TIP

Average You may think of the word "average" as "okay" or "not good or bad, just average." When it refers to math, the definition is the mathematical average, or **mean.**

Finding the Mean (Average)

$$\text{mean} = \frac{\text{sum of all values}}{\text{number of values}}$$

EXAMPLE 1 Finding the Mean (Average)

David had test scores of 84, 90, 95, 98, and 88. Find his mean (average) score.

Use the formula for finding the mean. Add up all the test scores and then divide the sum by the number of tests.

$$\text{mean} = \frac{84 + 90 + 95 + 98 + 88}{5}$$

Add all the test scores.

Divide by the number of tests.

$$\text{mean} = \frac{455}{5}$$

$$\text{mean} = 91$$

David has a mean (average) score of **91**

◀ **Work Problem 1 at the Side.**

1 Tanya had test scores of 95, 91, 81, 78, 81, and 90. Find her mean (average) score.

$$\text{mean} = \frac{\leftarrow \text{Total of scores}}{\leftarrow \text{Number of tests}}$$

$$\text{mean} =$$

2 Find the mean for each list of numbers.

(a) Jorge's last five monthly electric bills: $56.12, $63.62, $74.53, $102.31, $136.29. Find his average monthly bill. Round your answer to the nearest cent.

(b) The sales for one year at eight different office supply stores: $749,820; $765,480; $643,744; $824,222; $485,886; $668,178; $702,294; $525,800

EXAMPLE 2 Applying the Mean (Average)

The cash donations to a school fund raiser for each day last week were $86, $103, $118, $117, $126, $158, and $149. Find the mean daily cash donation.

To find the mean, add all the daily donation amounts and then divide the sum by the number of days (7).

$$\text{mean} = \frac{\$86 + \$103 + \$118 + \$117 + \$126 + \$158 + \$149}{7}$$

$$\text{mean} = \frac{\$857}{7} \quad \begin{array}{l} \longleftarrow \text{Sum of donations} \\ \longleftarrow \text{Number of days} \end{array}$$

$$\text{mean} \approx \$122.43 \quad \text{Rounded to nearest cent}$$

◀ **Work Problem 2 at the Side.**

Answers

1. $\dfrac{516}{6}$; mean $= 86$

2. (a) $\dfrac{\$432.87}{5} \approx \$86.57$ (rounded to nearest cent)

(b) $\dfrac{\$5,365,424}{8} = \$670,678$

OBJECTIVE ► ❷ **Find a weighted mean.** Some items in a list of data might appear more than once. In this case, we find a **weighted mean,** in which each value is "weighted" by multiplying it by the number of times it occurs.

EXAMPLE 3 Finding a Weighted Mean

The table below shows the amount of contribution and the number of times the amount was given (frequency) to a food pantry. Find the weighted mean.

Contribution Value	Frequency
$ 3	4
$ 5	2
$ 7	1
$ 8	5
$ 9	3
$10	2
$12	1
$13	2

4 people each contributed $3.

Some amounts were given by more than one person: for example, $3 was given by four people, and $8 was given by five people. Other amounts, such as $12, were given by only one person.

To find the mean, multiply each contribution value by its frequency. Then add the products. Next, add the numbers in the *frequency* column to find the total number of values, that is, the total number of people who contributed money.

Value	Frequency	Product
$ 3	4	$3 • 4 = $12
$ 5	2	$5 • 2 = $10
$ 7	1	$7 • 1 = $ 7
$ 8	5	$8 • 5 = $40
$ 9	3	$9 • 3 = $27
$10	2	$10 • 2 = $20
$12	1	$12 • 1 = $12
$13	2	$13 • 2 = $26
Totals	20	$154

Finally, divide the totals.

$$\text{mean} = \frac{\$154}{20} = \$7.70$$

The **mean contribution to the food pantry was $7.70**

──────── **Work Problem ❸ at the Side.** ▶

VOCABULARY TIP

Frequency You may know that "frequent" means "often." You can think of **frequency** as "how often," that is, how often the same number appears in a list.

❸ Alison Nakano works downtown. Some days she can park in cheaper lots that charge $6, $7, or $8. Other days, she has to park in lots that charge $9 or $10. Last month she kept track of the amount she spent each day for parking and the number of days she spent that amount. Find her average daily parking cost.

Parking Fee	Frequency
$ 6	2
$ 7	6
$ 8	3
$ 9	4
$10	6

Answer

3. $\dfrac{\$174}{21} \approx \$8.29$ (to nearest cent)

4 Find the GPA (grade point average) for a student who earned the following grades. Round to the nearest hundredth.

Course	Credits	Grade
Mathematics	5	A (= 4)
English	3	C (= 2)
Biology	4	B (= 3)
History	3	B (= 3)

Credits • Grade

$5 \cdot 4 \quad = 20$

$3 \cdot 2 \quad = 6$

$4 \cdot \underline{\quad} = \underline{\quad}$

$\underline{\quad} \cdot \underline{\quad} = \underline{\quad}$

Total credits $\underline{\quad}$ Total of $\underline{\quad}$ products

$GPA = \dfrac{\boxed{}}{15} \approx \underline{\quad}$ To nearest $\underline{\quad}$ hundredth

VOCABULARY TIP

Median To remember that **median** means "middle," think of the **median** of a highway, which is the strip in the middle between the opposite lanes of traffic.

5 Find the median for the numbers of students helped each hour in the tutoring center.

35, 33, 27, 30, 39, 50, 59, 25, 30

First, list the numbers in order from least (25) to greatest (59).

A common use of the weighted mean is to find a student's *grade point average (GPA),* as shown in the next example.

EXAMPLE 4 Applying the Weighted Mean

Find the GPA (grade point average) for a student who earned the following grades last semester. Assume A = 4, B = 3, C = 2, D = 1, and F = 0. The number of credits determines how many times the grade is counted (the frequency).

Course	Credits	Grade	Credits • Grade
Mathematics	4	A (= 4)	$4 \cdot 4 = 16$
Speech	3	C (= 2)	$3 \cdot 2 = 6$
English	3	B (= 3)	$3 \cdot 3 = 9$
Computer science	2	A (= 4)	$2 \cdot 4 = 8$
Theater	2	D (= 1)	$2 \cdot 1 = 2$
Totals	14		41

It is common to round grade point averages to the nearest hundredth. So the grade point average for this student is rounded to 2.93.

$$GPA = \frac{41}{14} \approx 2.93$$

This GPA is close to a B.

◀ **Work Problem 4 at the Side.**

OBJECTIVE ❸ Find the median. Because it can be affected by extremely high or low numbers, the mean is often a poor indicator of central tendency for a list of numbers. In cases like this, another measure of central tendency, called the *median,* can be used. The **median** divides a group of numbers in half, so half the numbers lie above the median, and half lie below the median.

Find the median by listing the numbers *in order* from *least* to *greatest.* If the list contains an *odd* number of items, the median is the *middle number.*

EXAMPLE 5 Finding the Median (Odd Number of Items)

Find the median for this list of prices.

$7, $23, $15, $6, $18, $12, $24

First, arrange the numbers in numerical order from least to greatest.

Least ⟶ 6, 7, 12, 15, 18, 23, 24 ⟵ Greatest

Next, find the *middle* number in the list.

6, 7, 12, **15**, 18, 23, 24

List the numbers from least to greatest **before** finding the middle.

Three are below. ↓ Three are above.

Middle number

The **median price is $15**

◀ **Work Problem 5 at the Side.**

If a list contains an *even* number of items, there is no single middle number. In this case, the median is defined as the mean (average) of the *middle two* numbers.

| **EXAMPLE 6** | **Finding the Median (Even Number of Items)** |

Find the median for this list of ages, in years.

$$74, 7, 15, 13, 25, 28, 47, 59, 33, 68$$

First, arrange the numbers in numerical order from least to greatest. Because the list has an even number of ages, find the middle *two* numbers.

Least ⟶ 7, 13, 15, 25, $\underbrace{28, 33}_{\text{Middle two numbers}}$, 47, 59, 68, 74 ⟵ Greatest

The median age is the mean (average) of the two middle numbers.

$$\text{median} = \frac{28 + 33}{2} = \frac{61}{2} = \textbf{30.5 years}$$

──────── **Work Problem 6 at the Side.** ▶

OBJECTIVE ▶ **4** **Find the mode.** Another statistical measure is the **mode,** which is the number that occurs *most often* in a list of numbers. For example, the test scores for ten students are shown below.

↓   ↓   ↓

74, 81, 39, 74, 82, 80, 100, 92, 74, 85

The mode is 74. Three students earned a score of 74, so 74 appears more times on the list than any other score. It is *not* necessary to place the numbers in numerical order when looking for the mode, although that may help you find it more easily.

A list can have two modes; such a list is sometimes called *bimodal.* If no number occurs more frequently than any other number in a list, the list has *no mode.*

| **EXAMPLE 7** | **Finding the Mode** |

Find the mode for each list of numbers.

(a) 51, 32, 49, 51, 49, 90, 49, 60, 17, 60, 51, 49

The number 49 occurs four times, which is more often than any other number. Therefore, **49 is the mode.**

(b) 482, 485, 483, 485, 487, 487, 489, 486

Because **both 485 and 487 occur twice, each is a mode.** This list is *bimodal.*

(c) $10,708; $11,519; $10,972; $12,546; $13,905; $12,182

No price occurs more than once. This list has **no mode.**

──────── **Work Problem 7 at the Side.** ▶

Measures of Central Tendency

The **mean** is the sum of all the values divided by the number of values. It is the mathematical *average.*

The **median** is the *middle number* (or the average of the two middle numbers) in a group of values that are listed from least to greatest. It divides a group of numbers in half.

The **mode** is the value that occurs *most often* in a group of values.

---

6 Find the median for this list of measurements.

178 ft, 261 ft, 126 ft, 189 ft, 121 ft, 195 ft, 121 ft, 200 ft

VOCABULARY TIP

Mode The first two letters of the word **mode** are **mo-**, the same as the first two letters of the word **most**. The **mode** is the number that appears **most** often in a list of numbers.

7 Find the mode for each list of numbers.

(a) Total points on a screening exam: 312, 219, 782, 312, 219, 426, 507, 600

(b) Ages of part-time employees (in years): 28, 21, 16, 22, 28, 34, 22, 28, 19, 18

(c) Monthly starting salaries for different jobs: $2769, $1785, $1560, $1594, $1663, $1892, $2120, $1989

Answers

6. $\dfrac{178 + 189}{2} = 183.5$ ft

7. **(a)** bimodal, 219 points and 312 points (this list has two modes)
   (b) 28 years
   (c) no mode (no number occurs more than once)

5.7 Exercises

FOR EXTRA HELP

Go to MyMathLab *for worked-out, step-by-step solutions to exercises enclosed in a square* ▢ *and video solutions to* ▶ *exercises.*

CONCEPT CHECK *Fill in the blank to complete each statement.*

1. (a) Another name for "mean" is _____.

(b) To find the mean, add all the values and then divide by _____.

2. (a) To find a "weighted mean," multiply each value by _____.

(b) A common use of the weighted mean is finding a student's _____.

Find the mean for each list of numbers. Round answers to the nearest tenth when necessary. **See Examples 1 and 2.**

3. Final exam scores: 92, 51, 59, 86, 68, 73, 49, 80

$$\text{mean} = \frac{92 + 51 + 59 + 86 + 68 + 73 + 49 + 80}{\square}$$

$$\text{mean} = \frac{\square}{8} \approx \underline{\quad} \text{ (rounded)}$$

4. Quiz scores: 18, 25, 21, 8, 16, 13, 23

$$\text{mean} = \frac{18 + 25 + 21 + 8 + 16 + 13 + 23}{\square}$$

$$\text{mean} = \frac{\square}{7} \approx \underline{\quad} \text{ (rounded)}$$

5. Annual salaries: $31,900; $32,850; $34,930; $39,712; $38,340, $60,000

6. Numbers of people attending baseball games: 27,500; 18,250; 17,357; 14,298; 33,110

Find the weighted mean. Round answers to the nearest tenth when necessary. **See Example 3.**

7. ▶

Quiz Score	Frequency
3	4
5	2
6	5
8	5
9	2

8.

Credits per Student	Frequency
9	3
12	5
13	2
15	6
18	1

Find the GPA (grade point average) for students earning the following grades. Assume A = 4, B = 3, C = 2, D = 1, and F = 0. Round answers to the nearest hundredth when necessary. **See Example 4.**

9.

Course	Credits	Grade
Biology	4	B
Biology lab	2	A
Mathematics	5	C
Health	1	F
Psychology	3	B

10.

Course	Credits	Grade
Chemistry	3	A
English	3	B
Mathematics	4	B
Theater	2	C
Astronomy	3	C

11. Look again at the grades in **Exercise 9.** Find the student's GPA in each of these situations.

  (a) The student earned a B instead of an F in the 1-credit class.

  (b) The student earned a B instead of a C in the 5-credit class.

  (c) Both (a) and (b) happened.

12. List the credits for the courses you're taking at this time. List the lowest grades you think you will earn in each class and find your GPA. Then list the highest grades you think you will earn and find your GPA.

Find the median for each list of numbers. ***See Examples 5 and 6.***

13. Number of text messages received each hour:
9, 15, 23, 12, 14, 24, 28

List the numbers in order.

Least → 9, 12, 14, 15, 23, 24, 28 ← Greatest

Find the *middle* number in the list.
Median = ____ texts

14. Patients seen each day at a clinic:
108, 99, 123, 109, 126, 146, 129, 170, 168

List the numbers in order.

99, 108, 109, 123, 126, 129, 146, 168, 170
↑ Least   Greatest ↑

Find the *middle* number in the list.
Median = ____ patients

15. Students enrolled in algebra each semester:
328, 549, 420, 592, 715, 483

16. Number of cars in the parking lot each day:
520, 523, 513, 1283, 338, 509, 290, 420

17. Pounds of shrimp sold each day:
51, 48, 96, 40, 47, 40, 95, 56, 34, 49

18. Number of gallons of paint sold per week:
1072, 1068, 1093, 1042, 1056, 205, 1009, 1081

The table lists the cruising speed and distance flown without refueling for several types of larger airplanes used to carry passengers. Use the table to answer Exercises 19–22.

Type of Airplane	Cruising Speed (miles per hour)	Distance without Refueling (miles)
B757–200	530	4722
B737–900 ER	517	2870
B777–200 LR	560	10,840
B747–400	561	8357
A320	521	3009
MD 90	509	2400

Data from Delta.

19. What is the average distance flown without refueling, to the nearest mile?

20. Find the average cruising speed in miles per hour.

21. Find the median distance flown without refueling.

22. Find the median cruising speed.

Find the mode or modes for each list of numbers. See Example 7.

23. Number of samples taken each hour:
3, 8, 5, 1, 7, 6, 8, 4, 5, 8

24. Monthly water bills:
$21, $32, $46, $32, $49, $32, $49, $25, $32

25. Ages of senior residents (in years):
74, 68, 68, 68, 75, 75, 74, 74, 70, 77

26. Patients admitted to the hospital each week:
30, 19, 25, 78, 36, 20, 45, 85, 38

27. The number of boxes of candy sold by each child:
5, 9, 17, 3, 2, 8, 19, 1, 4, 20, 10, 6

28. The weights of soccer players (in pounds):
158, 161, 165, 162, 165, 157, 163, 162

The table lists monthly normal temperatures from November through April for two of the coldest U.S. cities. Use the table to answer Exercises 29–30. Round answers to the nearest whole degree.

Normal Monthly Temperatures (in Degrees Fahrenheit)						
City	Nov.	Dec.	Jan.	Feb.	Mar.	Apr.
Barrow, Alaska	-2	-11	-13	-18	-15	-2
Fairbanks, Alaska	3	-7	-10	-4	11	31

Data from National Climatic Data Center.

29. Find Barrow's mean temperature and Fairbanks' mean temperature for the six-month period. How much warmer is Fairbanks' mean than Barrow's?

30. Find Barrow's median temperature and Fairbanks' median temperature for the six-month period. How much cooler is Barrow's median than Fairbanks'?

31. **CONCEPT CHECK** The work shown below has a mistake in it.

What Went Wrong? First write a sentence explaining what the mistake is. Then fix the mistake and find the correct solution.

Magneilia was given the following list of numbers:
30, 19, 25, 78, 46, 20, 35, 85, 38

She said the median was 46 because 46 was the middle number.

32. **CONCEPT CHECK** The work shown below has a mistake in it.

What Went Wrong? First write a sentence explaining what the mistake is. Then fix the mistake and find the correct solution.

Leopoldo had the following test scores:
86, 72, 90, 86, 67

He found his test average this way:

$$\frac{86 + 72 + 90 + 67}{4} = \frac{315}{4} = 78.75$$

5.8 Geometry Applications: Pythagorean Theorem and Square Roots

Earlier, you used this formula for the area of a square, $A = s^2$. The blue square below has an area of 25 cm² because $(5 \text{ cm})(5 \text{ cm}) = 25 \text{ cm}^2$.

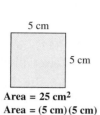

5 cm

5 cm

Area = 25 cm²
Area = (5 cm)(5 cm)

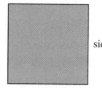

side = ? cm

Area = 49 cm²
Area = (? cm)(? cm)

OBJECTIVES

1. Find square roots using the square root key on a calculator.
2. Find the unknown length in a right triangle.
3. Solve application problems involving right triangles.

The red square above has an area of 49 cm². To find the length of a side, ask yourself, "What number can be multiplied by itself to give 49?" Because $(7)(7) = 49$, the length of each side is 7 cm. Also, because $(7)(7) = 49$ we say that 7 is the *square root* of 49, or $\sqrt{49} = 7$.

Square Root

The positive **square root** of a positive number is one of two identical positive factors of that number.

For example, $\sqrt{36} = 6$ because $(6)(6) = 36$.

Note

There is another square root for 36. We know that

$$(6)(6) = 36 \quad \text{and} \quad (-6)(-6) = 36$$

so the *positive* square root of 36 is 6 and the *negative* square root of 36 is −6. In this section we will work only with positive square roots. You will learn more about negative square roots in other math courses.

Work Problem ① at the Side. ▶

A number that has a whole number as its square root is called a *perfect square*. For example, 9 is a perfect square because $\sqrt{9} = 3$, and 3 is a whole number. Some of the perfect squares are listed below.

Some of the Perfect Squares

$\sqrt{1} = 1$	$\sqrt{16} = 4$	$\sqrt{49} = 7$	$\sqrt{100} = 10$
$\sqrt{4} = 2$	$\sqrt{25} = 5$	$\sqrt{64} = 8$	$\sqrt{121} = 11$
$\sqrt{9} = 3$	$\sqrt{36} = 6$	$\sqrt{81} = 9$	$\sqrt{144} = 12$

OBJECTIVE ① Find square roots using the square root key on a calculator. If a number is *not* a perfect square, then you can find its *approximate* square root by using a calculator with a square root key.

① Find each square root.

(a) $\sqrt{36}$

What number, multiplied by itself, gives 36? ____

So, $\sqrt{36} =$ ____

(b) $\sqrt{25}$

(c) $\sqrt{9}$

(d) $\sqrt{100}$

(e) $\sqrt{121}$

Answers

1. (a) 6; $\sqrt{36} = 6$ (b) 5 (c) 3
(d) 10 (e) 11

2 Use a calculator with a square root key to find each square root. Round to the nearest thousandth when necessary.

(a) $\sqrt{11}$

(b) $\sqrt{40}$

(c) $\sqrt{56}$

(d) $\sqrt{196}$

(e) $\sqrt{147}$

▦ **Calculator Tip**

To find a square root on a *scientific* calculator, use the ⎷ key, or on some calculators, the ⎷x key. Try these.

To find $\sqrt{16}$ press: ⎷ 16   Answer is 4

To find $\sqrt{7}$ press: ⎷ 7   Answer is 2.645751311

On some models you enter the number first, then press the ⎷ key. Check your model's operating instructions.

For $\sqrt{7}$, your calculator shows 2.645751311, an *approximate* answer. (Some calculators may show more or fewer digits.) We will round to the nearest thousandth, so $\sqrt{7} \approx 2.646$. To check, multiply 2.646 times 2.646. Do you get 7 as the result? No, you get 7.001316, which is very close to 7. The difference is due to rounding.

EXAMPLE 1   **Finding the Square Root of Numbers**

▦ Use a calculator to find each square root. Round to the nearest thousandth.

(a) $\sqrt{35}$ Calculator shows 5.916079783; round to **5.916**

(b) $\sqrt{124}$ Calculator shows 11.13552873; round to **11.136**

◀ **Work Problem 2 at the Side.**

OBJECTIVE ▶ 2 **Find the unknown length in a right triangle.** One place you will use square roots is when working with the *Pythagorean Theorem.* This theorem applies only to *right* triangles (triangles with a 90° angle). The longest side of a right triangle is called the **hypotenuse.** It is opposite the right angle. The other two sides are called *legs.* The legs form the right angle. Here are some right triangles.

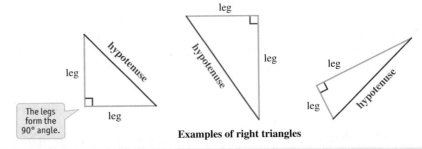

Examples of right triangles

Pythagorean Theorem

$$(\text{hypotenuse})^2 = (\text{leg})^2 + (\text{leg})^2$$

In other words, square the length of each side. After you have squared all the sides, the sum of the squares of the two legs will equal the square of the hypotenuse. An example is shown below.

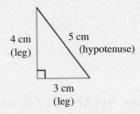

$$(\text{hypotenuse})^2 = (\text{leg})^2 + (\text{leg})^2$$
$$5^2 = 4^2 + 3^2$$
$$25 = 16 + 9$$
$$25 = 25$$

Answers

2. (a) $\sqrt{11} \approx 3.317$ **(b)** $\sqrt{40} \approx 6.325$
(c) $\sqrt{56} \approx 7.483$ **(d)** $\sqrt{196} = 14$
(e) $\sqrt{147} \approx 12.124$

The theorem is named after Pythagoras, a Greek mathematician who lived about 2500 years ago.

If you know the lengths of any two sides in a right triangle, you can use the Pythagorean Theorem to find the length of the third side.

Formulas Based on the Pythagorean Theorem
To find the hypotenuse: $\textbf{hypotenuse} = \sqrt{(\textbf{leg})^2 + (\textbf{leg})^2}$
To find a leg: $\textbf{leg} = \sqrt{(\textbf{hypotenuse})^2 - (\textbf{leg})^2}$

❗ CAUTION

Remember: A small square drawn in one angle of a triangle indicates a right angle (90°). You can use the Pythagorean Theorem **only** on triangles that have a right angle.

EXAMPLE 2   **Finding the Unknown Length in Right Triangles**

Find the unknown length in each right triangle. Round your answers to the nearest tenth when necessary.

(a)

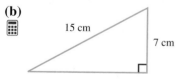

The **unknown length** is the side opposite the right angle, which is the **hypotenuse.** Use the formula for finding the hypotenuse.

$\text{hypotenuse} = \sqrt{(\text{leg})^2 + (\text{leg})^2}$ Formula to find the hypotenuse

$\text{hypotenuse} = \sqrt{(3)^2 + (4)^2}$ Legs are 3 and 4

$\text{hypotenuse} = \sqrt{9 + 16}$ (3)(3) is 9   and   (4)(4) is 16

$\text{hypotenuse} = \sqrt{25}$

$\text{hypotenuse} = 5$ This is the length of a side, so write **ft** in the answer (**not** ft²).

The **hypotenuse is 5 ft long.**

(b)

You *do* know the length of the hypotenuse (15 cm), so it is the **length of one of the legs that is unknown.** Use the formula for finding a leg.

$\text{leg} = \sqrt{(\text{hypotenuse})^2 - (\text{leg})^2}$ Formula to find a leg

$\text{leg} = \sqrt{(15)^2 - (7)^2}$ Hypotenuse is 15; one leg is 7

$\text{leg} = \sqrt{225 - 49}$ (15)(15) is 225   and   (7)(7) is 49

$\text{leg} = \sqrt{176}$ Use a calculator to find $\sqrt{176}$

$\text{leg} \approx 13.3$ Round 13.26649916 to 13.3

The length of the **leg is approximately 13.3 cm**   Write **cm** in the answer, **not** cm²

Work Problem 3 at the Side. ▶

---

3 Find the unknown length in each right triangle. Round your answers to the nearest tenth when necessary.

(a)

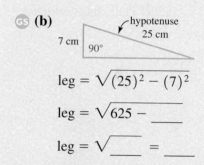

$\text{hypotenuse} = \sqrt{(5)^2 + (12)^2}$

$\text{hypotenuse} = \sqrt{25 + \underline{\quad}}$

$\text{hypotenuse} = \sqrt{\underline{\quad}} = \underline{\quad}$

(b)

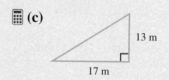

$\text{leg} = \sqrt{(25)^2 - (7)^2}$

$\text{leg} = \sqrt{625 - \underline{\quad}}$

$\text{leg} = \sqrt{\underline{\quad}} = \underline{\quad}$

(c)

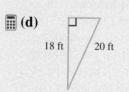

(d)

Answers

3. **(a)** $\sqrt{25 + 144} = \sqrt{169} = 13$ in.

(b) $\sqrt{625 - 49} = \sqrt{576} = 24$ cm

(c) hypotenuse $= \sqrt{458} \approx 21.4$ m

(d) leg $= \sqrt{76} \approx 8.7$ ft

4 These problems show ladders leaning against buildings. Find the unknown lengths. Round answers to the nearest tenth of a foot when necessary.

(a)

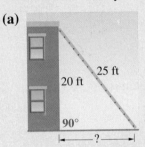

How far away from the building is the bottom of the ladder? (*Hint:* The ladder is the **hypotenuse.**)

(b)

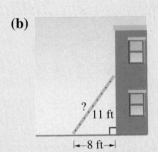

How long is the ladder?

(c) A 17 ft ladder is leaning against a building. The bottom of the ladder is 10 ft from the building. How high up on the building will the ladder reach? (*Hint:* Start by drawing a sketch of the building and the ladder.)

Answers

4. (a) leg = $\sqrt{225}$ = 15 ft

(b) hypotenuse = $\sqrt{185}$ ≈ 13.6 ft

(c) leg = $\sqrt{189}$ ≈ 13.7 ft

OBJECTIVE **3** **Solve application problems involving right triangles.** This example is an application of the Pythagorean Theorem.

EXAMPLE 3 **Using the Pythagorean Theorem**

A television antenna is on the roof of a house, as shown. Find the length of the support wire. Round your answer to the nearest tenth of a meter if necessary.

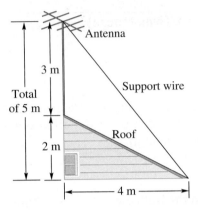

A right triangle is formed.
The total length of the leg on the left is 3 m + 2 m = 5 m

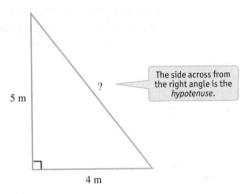

The side across from the right angle is the *hypotenuse.*

Notice that the support wire is opposite the right angle, so it is the *hypotenuse* of the right triangle.

$$\text{hypotenuse} = \sqrt{(\text{leg})^2 + (\text{leg})^2}$$ Formula to find the hypotenuse

$$\text{hypotenuse} = \sqrt{(5)^2 + (4)^2}$$ Legs are 5 and 4

$$\text{hypotenuse} = \sqrt{25 + 16}$$ 5^2 is 25 and 4^2 is 16

$$\text{hypotenuse} = \sqrt{41}$$ Use a calculator to find $\sqrt{41}$

$$\text{hypotenuse} \approx 6.4$$ ← Rounded

The **length of the support wire is approximately 6.4 m** This is **length** so write **m** in the answer (**not** m²).

! CAUTION

You use the Pythagorean Theorem to find the *length* of one side, *not* the area of the triangle. Your answer will be in linear units, such as ft, yd, cm, m, and so on (**not** ft², yd², cm², m²).

◄ **Work Problem** **4** at the Side.

5.8 Exercises

FOR EXTRA HELP   Go to MyMathLab *for worked-out, step-by-step solutions to exercises enclosed in a square* ▢ *and video solutions to* ▶ *exercises.*

Find each square root. Starting with Exercise 5, find the square root using a calculator.
Round your answers to the nearest thousandth when necessary. **See Example 1.**

1. $\sqrt{16}$

2. $\sqrt{4}$

3. $\sqrt{64}$

4. $\sqrt{81}$

5. $\sqrt{11}$

6. $\sqrt{23}$

7. $\sqrt{5}$

8. $\sqrt{2}$

9. $\sqrt{73}$

10. $\sqrt{80}$

11. $\sqrt{101}$

12. $\sqrt{125}$

13. $\sqrt{361}$

14. $\sqrt{729}$

15. $\sqrt{1000}$

16. $\sqrt{2000}$

17. CONCEPT CHECK   You know that $\sqrt{25} = 5$ and $\sqrt{36} = 6$. Using just that information (no calculator), describe how you could *estimate* $\sqrt{30}$. How would you estimate $\sqrt{26}$ or $\sqrt{35}$? Now check your estimates using a calculator.

18. CONCEPT CHECK   Explain the relationship between *squaring* a number and finding the *square root* of a number. Include two examples to illustrate your explanation.

▦ *Find the unknown length in each right triangle. Use a calculator to find square roots.*
Round your answers to the nearest tenth when necessary. **See Example 2.**

19.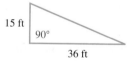

Hint: The unknown length is the **hypotenuse.**

hypotenuse = $\sqrt{(15)^2 + (36)^2}$

20.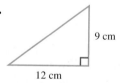

Hint: The unknown length is the **hypotenuse.**

hypotenuse = $\sqrt{(9)^2 + (12)^2}$

21.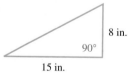

22.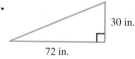

23.

The unknown length is a **leg.**

leg = $\sqrt{(20)^2 - (16)^2}$

24.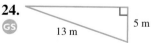

The unknown length is a **leg.**

leg = $\sqrt{(13)^2 - (5)^2}$

25.
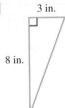
3 in.
8 in.

26.
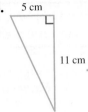
5 cm
11 cm

27.
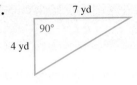
7 yd
90°
4 yd

28.
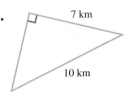
7 km
10 km

29.
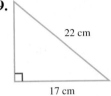
22 cm
17 cm

30.
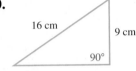
16 cm
9 cm
90°

31.
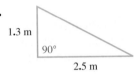
1.3 m
90°
2.5 m

32.
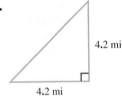
4.2 mi
4.2 mi

33.

11.5 cm
8.2 cm

34.
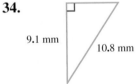
9.1 mm
10.8 mm

35.

13.2 km   90°
21.6 km

36.
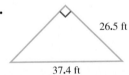
26.5 ft
37.4 ft

37. CONCEPT CHECK The work shown below has **two** mistakes in it.

What Went Wrong? First write a sentence explaining what each mistake is. Then fix the mistakes and find the correct solution.

Jenoveba had to find the unknown side length and round the answer to the nearest tenth. This is the work she did:

$$? = \sqrt{(13)^2 + (20)^2}$$
$$? = \sqrt{169 + 400}$$
$$? = \sqrt{569} \approx 23.9 \text{ m}^2$$

?   13 m
20 m

38. CONCEPT CHECK The work shown below has **three** mistakes in it.

What Went Wrong? First write a sentence explaining what each mistake is. Then fix the mistakes and find the correct solution.

Ibrahim had to find the hypotenuse and round his answer to the nearest tenth. This is the work he did:

$$? = \sqrt{(9)^2 + (7)^2}$$
$$? = \sqrt{18 + 14}$$
$$? = \sqrt{32} \approx 5.657 \text{ in.}$$

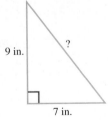
9 in.   ?
7 in.

*Solve each application problem. Round your answers to the nearest tenth when necessary. **See Example 3.***

39. Find the length of this loading ramp.
(*Hint:* The ramp is the **hypotenuse** in a right triangle.)

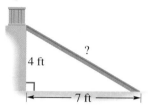

40. Find the unknown length in this window frame.

41. How high is the airplane above the ground?
(*Hint:* The height of the plane is a **leg** in the right triangle.)

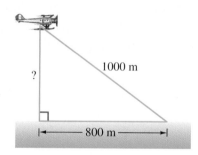

42. Find the height of this farm silo.

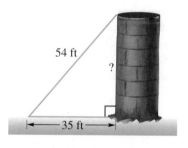

43. How long is the diagonal brace on this storage shed door?

44. Find the height of this rectangular television screen.

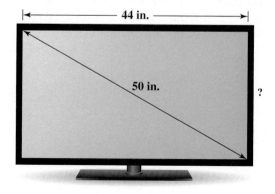

45. To reach his ladylove, a knight placed a 12 ft ladder against the castle wall. If the base of the ladder is 3 ft from the building, how high on the castle will the top of the ladder reach? Draw a sketch of the castle and ladder and solve the problem.

46. William drove his car 15 miles north, then made a 90° right turn and drove 7 miles east. How far is he, in a straight line, from his starting point? Draw a sketch to illustrate the problem and then solve it.

Relating Concepts (Exercises 47–50) For Individual or Group Work

▦ *Use your knowledge of the Pythagorean Theorem to* **work Exercises 47–50 in order.** *Round answers to the nearest tenth.*

47. A major league baseball diamond is a square shape measuring 90 ft on each side. If the catcher throws a ball from home plate to second base, how far is he throwing the ball? (Data from American League of Professional Baseball Clubs.)

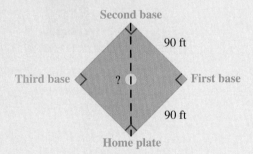

48. A softball diamond is 60 ft on each side. (Data from Amateur Softball Association.)

   (a) Draw a sketch of the softball diamond and label the bases and the lengths of the sides.

   (b) How far is it to throw a ball from home plate to second base?

49. Look back at your answer to **Exercise 47.** Explain how you can tell the distance from third base to first base without doing any further calculations.

50. (a) Look back at your answer to **Exercise 48.** Suppose you measured the distance from home plate to second base on a softball diamond and found it was 80 ft. What would this tell you about the length of each side of the diamond? (Assume the diamond is still a square.)

   (b) Bonus question: For the diamond in part (a), find the length of each side, to the nearest tenth.

5.9 | Problem Solving: Equations Containing Decimals

OBJECTIVE ▶ ① Solve equations containing decimals using the addition property of equality. You have used the addition property of equality to solve an equation like $c + 5 = 30$. The addition property says that you can add the *same* number to *both* sides of an equation and still keep it balanced. You can also use this property when an equation contains decimal numbers.

OBJECTIVES

① Solve equations containing decimals using the addition property of equality.

② Solve equations containing decimals using the division property of equality.

③ Solve equations containing decimals using both properties of equality.

④ Solve application problems involving equations with decimals.

EXAMPLE 1   Using the Addition Property of Equality

Solve each equation and check each solution.

(a) $w + 2.9 = -0.6$

The first step is to get the variable term (w) by itself on the left side of the equal sign. Use the addition property to "get rid of" the 2.9 on the left side by adding its opposite, -2.9, to both sides.

$$w + 2.9 = -0.6$$
$$\underline{-2.9 \quad -2.9} \quad \text{Add } -2.9 \text{ to both sides.}$$
$$w + 0 = -3.5$$
$$w \quad = -3.5$$

The solution is -3.5. To check the solution, go back to the *original* equation.

CHECK $w + 2.9 = -0.6$ Original equation

$-3.5 + 2.9 = -0.6$ Replace w with -3.5

$-0.6 = -0.6$ ✓   Balances

When w is replaced with -3.5, the equation balances, so -3.5 is the correct solution (*not* -0.6).

(b) $7 = -4.3 + x$

To get x by itself on the right side of the equal sign, add 4.3 to both sides.

$$7 = -4.3 + x$$
$$\underline{+4.3 \quad +4.3} \quad \text{Add 4.3 to both sides.}$$
$$11.3 = 0 + x$$
$$11.3 = x$$

The solution is 11.3. To check the solution, go back to the *original* equation.

CHECK $7 = -4.3 + x$ Original equation

$7 = -4.3 + 11.3$ Replace x with 11.3   — The solution is 11.3 (**not** 7)

$7 = 7$ ✓   Balances

When x is replaced with 11.3, the equation balances, so **11.3** is the correct solution (*not* 7).

Work Problem ① at the Side. ▶

---

① Solve each equation and check each solution.

GS (a) $8.1 = h + 9$ **CHECK**
$$\underline{-9 \qquad -9}$$
$$-0.9 = h + 0$$
$$\underline{\qquad} = \underline{\qquad}$$

(b) $-0.75 + y = 0$ **CHECK**

(c) $c - 6.8 = -4.8$ **CHECK**

Answers

1. (a) $-0.9 = h$ **CHECK** $8.1 = h + 9$
   $8.1 = -0.9 + 9$
   Balances $8.1 = 8.1$ ✓

(b) $y = 0.75$ **CHECK** $-0.75 + y = 0$
   $-0.75 + 0.75 = 0$
   Balances $0 = 0$ ✓

(c) $c = 2$ **CHECK** $c - 6.8 = -4.8$
   $2 - 6.8 = -4.8$
   Balances $-4.8 = -4.8$ ✓

2 Solve each equation and check each solution.

(a) $-3y = -0.63$

$$\frac{\overset{1}{\cancel{-3}} \cdot y}{\underset{1}{\cancel{-3}}} = \frac{-0.63}{-3}$$ Divide both sides by -3

$$y = \underline{\quad}$$ Same sign, positive quotient.

CHECK

$$-3y = -0.63$$

$$-3(\underline{\quad}) = -0.63$$

$$\underline{\quad} = -0.63$$

(b) $2.25r = -18$

CHECK

(c) $1.7 = 0.5n$

CHECK

Answers

2. **(a)** $y = 0.21$ **CHECK** $\quad -3y = -0.63$

$$-3(0.21) = -0.63$$

Balances $-0.63 = -0.63$ ✓

(b) $r = -8$ **CHECK** $\quad 2.25r = -18$

$$2.25(-8) = -18$$

Balances $-18 = -18$ ✓

(c) $3.4 = n$ **CHECK** $\quad 1.7 = 0.5n$

$$1.7 = 0.5(3.4)$$

Balances $1.7 = 1.7$ ✓

OBJECTIVE 2 **Solve equations containing decimals using the division property of equality.** You can also use the division property of equality when an equation contains decimals.

EXAMPLE 2 **Using the Division Property of Equality**

Solve each equation and check each solution.

(a) $5x = 12.4$

On the left side of the equation, the variable is multiplied by 5. To undo the multiplication, divide both sides by 5.

$5x$ means
$5 \cdot x$

$$5x = 12.4$$

$$\frac{5 \cdot x}{5} = \frac{12.4}{5}$$ Divide both sides by 5

On the left side, divide out the common factor of 5

$$\frac{\overset{1}{\cancel{5}} \cdot x}{\underset{1}{\cancel{5}}} = 2.48$$ On the right side, $12.4 \div 5$ is 2.48

$$x = 2.48$$

The solution is 2.48. To check the solution, go back to the original equation.

CHECK $\quad 5x = 12.4$ Original equation

$$5(2.48) = 12.4$$ Replace x with 2.48 ◁ The solution is 2.48 (**not** 12.4)

$$12.4 = 12.4 ✓$$ Balances

When x is replaced with 2.48, the equation balances, so **2.48** is the correct solution (**not** 12.4).

(b) $-9.3 = 1.5t$

Signs are different, so quotient is negative.

$$\frac{-9.3}{1.5} = \frac{\overset{1}{\cancel{1.5}}t}{\underset{1}{\cancel{1.5}}}$$ Divide both sides by the coefficient of the variable term, 1.5

$$-6.2 = t$$

The solution is -6.2. To check the solution, go back to the original equation.

CHECK $\quad -9.3 = 1.5t$ Original equation

$$-9.3 = 1.5(-6.2)$$ Replace t with -6.2

$$-9.3 = -9.3 ✓$$ Balances

When t is replaced with -6.2, the equation balances, so -6.2 is the correct solution (**not** -9.3).

◀ **Work Problem 2 at the Side.**

OBJECTIVE 3 **Solve equations containing decimals using both properties of equality.** Sometimes you need to use both the addition and division properties to solve an equation, as shown in **Example 3.**

EXAMPLE 3	Solving Equations with Several Steps

(a) $2.5b + 0.35 = -2.65$

The first step is to get the variable term, $2.5b$, by itself on the left side of the equal sign.

$$2.5b + 0.35 = -2.65$$
$$\underline{\quad -0.35 \quad -0.35 \quad} \quad \text{Add } -0.35 \text{ to both sides.}$$
$$\underbrace{2.5b + \quad 0} = -3.00$$
$$2.5b \quad\quad = -3$$

The next step is to divide both sides by the coefficient of the variable term. In $2.5b$, the coefficient is 2.5.

$$\frac{\overset{1}{2.5}b}{\underset{1}{2.5}} = \frac{-3}{2.5} \qquad \begin{array}{l}\text{On the right side, signs do}\\ \textit{not} \text{ match, so the quotient}\\ \text{is } \textit{negative.}\end{array}$$

$$b = -1.2$$

The solution is -1.2. To check the solution, go back to the original equation.

CHECK
$$2.5b \;\; + 0.35 = -2.65 \qquad \text{Original equation}$$
$$2.5(-1.2) + 0.35 = -2.65 \qquad \text{Replace } b \text{ with } -1.2$$
$$\underbrace{-3} \quad\;\; + 0.35 = -2.65$$
$$-2.65 \quad\quad\;\; = -2.65 \;\checkmark \qquad \begin{array}{l}\text{Balances, so } -1.2 \text{ is the}\\ \text{correct solution.}\end{array}$$

(b) $5x - 0.98 = 2x + 0.4$

There is a variable term on both sides of the equation. You can choose to keep the variable term on the left side, or to keep the variable term on the right side. Either way will work. Just pick the left side or the right side.

Suppose that you decide to keep the variable term, $5x$, on the left side. Use the addition property to "get rid of" $2x$ on the right side by adding its opposite, $-2x$, to both sides.

$$5x - 0.98 = \;\; 2x + 0.4$$

Change subtraction to adding the opposite.
$$\underline{-2x \qquad\qquad -2x \quad} \quad \text{Add } -2x \text{ to both sides.}$$
$$3x - 0.98 = \;\; 0 \; + 0.4$$
$$3x + (-0.98) = \qquad 0.4$$
$$\underline{\quad +0.98 \qquad\;\; +0.98 \quad} \quad \text{Add } 0.98 \text{ to both sides.}$$
$$3x + \quad 0 \quad = \qquad 1.38$$

$$\frac{\overset{1}{3}x}{\underset{1}{3}} \quad = \quad \frac{1.38}{3} \qquad \text{Divide both sides by 3}$$

$$x \quad\quad = \quad\quad 0.46$$

The solution is 0.46. To check the solution, go back to the original equation.

CHECK
$$5x - 0.98 = 2x + 0.4 \qquad \text{Original equation}$$
$$5(0.46) - 0.98 = 2(0.46) + 0.4 \qquad \text{Replace } x \text{ with } 0.46$$
$$\underbrace{2.3} \;\; - 0.98 = \underbrace{0.92} \; + 0.4$$
$$1.32 \quad = \quad\;\; 1.32 \;\checkmark \qquad \begin{array}{l}\text{Balances, so } 0.46 \text{ is}\\ \text{the correct solution.}\end{array}$$

Work Problem ➌ at the Side. ▶

➌ Solve each equation and check each solution.

GS (a) $\quad 4 = 0.2c - 2.6$
$$\underline{+2.6 \qquad\qquad +2.6 \quad}$$
$$6.6 = 0.2c \; +0$$

$$\frac{6.6}{0.2} = \frac{\overset{1}{0.2}c}{\underset{1}{0.2}}$$

$$\underline{\qquad} = c$$

CHECK

(b) $3.1k - 4 = 0.5k + 13.42$

CHECK

(c) $-2y + 3 = 3y - 6$

CHECK

Answers

3. **(a)** $33 = c$ **CHECK** $\quad 4 = 0.2c - 2.6$
$$4 = 0.2(33) - 2.6$$
$$4 = \underbrace{6.6} \;- 2.6$$
$$\text{Balances} \quad 4 = \quad 4 \;\checkmark$$

(b) $k = 6.7$ **CHECK**
$$3.1k - 4 = 0.5k + 13.42$$
$$3.1(6.7) - 4 = 0.5(6.7) + 13.42$$
$$\underbrace{20.77} - 4 = \underbrace{3.35} \; + 13.42$$
$$\text{Balances } 16.77 \; = \quad 16.77 \;\checkmark$$

(c) $y = 1.8$ **CHECK**
$$-2y + 3 = 3y - 6$$
$$-2(1.8) + 3 = 3(1.8) - 6$$
$$\underbrace{-3.6} \; + 3 = \underbrace{5.4} \; - 6$$
$$\text{Balances} \quad -0.6 \; = \quad -0.6 \;\checkmark$$

④ Maria found a cheaper calling
⑤ card. The new card charges a
$1 connection fee and $0.06 per
minute. How many minutes
long was a call that cost $4?
Use the six problem-solving
steps.

Step 1 Cost of a call

Unknown: number of
minutes the call lasted
Known: Costs are $0.06
per minute plus $ _____
connection fee.

Step 2 Only one unknown; let *m*
be the number of minutes.

Step 3 Write an equation using
your variable.

OBJECTIVE ④ **Solve application problems involving equations with decimals.** Use the six problem-solving steps.

EXAMPLE 4   Solving an Application Problem

Many international calling cards add a connection fee to the cost of each call. The card that Maria uses to call her sister's cell phone in Mexico charges $0.084 per minute plus a $1.25 connection fee for each call. The cost of the call Maria made today to her sister was $5.03. How many minutes did the call last?

Step 1   **Read** the problem. It is about the cost of a call to Mexico.

Unknown: number of minutes the call lasted
Known: Costs are $0.084 per minute plus $1.25 connection fee;
total cost was $5.03.

Step 2   **Assign a variable:** There is only one unknown, so let *m* be the number of minutes.

Step 3   **Write an equation using your variable.**

Cost per minute		Number of minutes		Connection fee		Total cost
0.084	•	*m*	+	1.25	=	5.03

Step 4   **Solve** the equation.

$$0.084m + 1.25 = 5.03$$
$$\underline{\qquad\quad -1.25 \quad -1.25} \quad \text{Add} -1.25 \text{ to both sides.}$$
$$0.084m + 0 = 3.78$$

$$\frac{\overset{1}{0.\cancel{0}84}m}{0.\cancel{0}84} = \frac{3.78}{0.084} \quad \text{Divide both sides by 0.084}$$
$$\underset{1}{}$$

$$m = 45$$

> Write **minutes** as part of your answer.

Step 5   **State the answer.** The call lasted **45 minutes**.

Step 6   **Check** the solution by putting it back into the original problem.

$0.084 per minute times 45 minutes = $3.78

$3.78 plus $1.25 connection fee = $5.03

Cost of the call was $5.03. ← Matches ⤴

Because 45 minutes fits all the facts in the original problem, it is the correct solution.

> If your answer does **not** fit the facts in the original problem, start again at *Step 1*.

◄ **Work Problem ④ at the Side.**

Answer

4. *Step 1:* $1 connection fee
   Step 3: 0.06*m* + 1 = 4
   Step 4: Add −1 to both sides;
   divide both sides by 0.06
   Step 5: The call lasted 50 minutes.
   Step 6: $0.06 per minute times
   50 minutes = $3.00, or $3
   Then, $3 + $1 = $4
   Maria's call ↑
   cost $4 ← Matches

Study Skills Reminder

You're coming to the end of this chapter and you've learned a lot! It is a good idea to spend time at the end of each chapter reviewing all the material that was covered. See the Study Skills activity "Reviewing a Chapter" for different ways to work through the material.

5.9 Exercises

FOR EXTRA HELP

Go to MyMathLab for worked-out, step-by-step solutions to exercises enclosed in a square ▢ and video solutions to ▶ exercises.

1. **CONCEPT CHECK** When solving this equation, state the number you would add to both sides. Then explain why you picked that number.
$$-0.3 = x - 6.2$$

2. **CONCEPT CHECK** When solving this equation, state the number you would divide both sides by. Then explain why you picked that number.
$$-4.5y = 2.25$$

Solve each equation and check each solution. **See Examples 1 and 2.**

3. $-20.6 + n = -22$ **CHECK**

GS
$$\underline{+20.6 \qquad\quad +20.6}$$
$$0 + n = \underline{\quad}$$
$$\underline{\quad} = \underline{\quad}$$

4. $g - 5 = 6.03$ **CHECK**

GS
$$\underline{+5 = +5}$$
$$g + 0 = \underline{\quad}$$
$$\underline{\quad} = \underline{\quad}$$

5. $0 = b - 0.008$ **CHECK**

6. $0.18 + m = -4.5$ **CHECK**

7. $2.03 = 7a$ **CHECK**
▶

8. $-6.2c = 0$ **CHECK**

9. $0.8p = -96$ **CHECK**

10. $-10.16 = -4r$ **CHECK**

11. $-3.3t = -2.31$ **CHECK**
▶

12. $8.3w = -49.8$ **CHECK**

Solve each equation. Show your work. **See Example 3.**

13. Finish the work.

GS
$$7.5x + 0.15 = -6$$
$$\underline{\quad -0.15 \quad\; -0.15}$$

14. Finish the work.

GS
$$0.8 = 0.2y + 3.4$$
$$\underline{-3.4 \qquad\quad -3.4}$$

15. $-7.38 = 2.05z - 7.38$

16. $6.2h - 0.4 = 2.7$

17. $-0.9 = 0.2 - 0.01h$

18. $3 = -0.3 - 2.2m$

19. $3c + 10 = 6c + 8.65$
▶

20. $2.1b + 5 = 1.6b + 10$

21. $0.8w - 0.4 = -6 + w$

22. $7r + 9.64 = -2.32 + 5r$ **23.** $-10.9 + 0.5p = 0.9p + 5.3$ **24.** $0.7x - 4.38 = x - 2.16$

Solve each application problem using the six problem-solving steps. **See Example 4.**

25. Most adult medication doses are for a person weighing 150 pounds. For a 45-pound child, the adult dose should be multiplied by 0.3. If the child's dose of a decongestant is 9 milligrams, what is the adult dose?

26. For a 30-pound child, an adult dose of medication should be multiplied by 0.2. If the child's dose of a cough suppressant is 3 milliliters, find the adult dose.

27. A storm blew down many trees in a neighborhood. Several neighbors rented a chain saw for $315.80 and helped each other cut up and stack the wood. The rental company charges $75.95 per day plus a $12 sharpening fee. How many days was the saw rented? (Data from Central Rental.)

28. A 20-inch chain saw can be rented for $32.99 for the first two hours, and $11 for each additional hour. Steve's rental charge was $65.99. How many hours did he rent the saw? (Data from Central Rental.)

Relating Concepts (Exercises 29–32) For Individual or Group Work

When doing aerobic exercises, it is important to increase your heart rate (the number of beats per minute) so that you get the maximum benefit from the exercise. But you don't want your heart rate to be so fast that it is dangerous. Here is an expression for finding a safe maximum heart rate for a healthy person with no heart disease: $0.7(220 - a)$, *where a is the person's age.* **Work Exercises 29–32 in order:** *Write an equation and solve it to find each person's age. Assume all the people are healthy.*

29. How old is a person who has a maximum safe heart rate of 140 beats per minute? *Hint:* Use the distributive property to simplify $0.7(220 - a)$.

30. How old is a person who has a maximum safe heart rate of 126 beats per minute?

31. If a person's maximum safe heart rate is 134 (rounded to the nearest whole number), how old is the person, to the nearest whole year?

32. If a person's maximum safe heart rate is 117 (rounded to the nearest whole number), how old is the person, to the nearest whole year?

5.10 | Geometry Applications: Circles, Cylinders, and Surface Area

OBJECTIVE ▶ **1** **Find the radius and diameter of a circle.** Suppose you start with one dot on a piece of paper. Then you draw many dots that are each 2 cm away from the first dot. If you draw enough dots (points) you'll end up with a circle, as shown below on the left. Each point on the circle is exactly 2 cm away from the *center* of the circle. The 2 cm distance is called the *radius, r,* of the circle. The distance across the circle (passing through the center) is called the *diameter, d,* of the circle. In this circle, the diameter is 4 cm.

 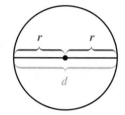

OBJECTIVES

1 Find the radius and diameter of a circle.

2 Find the circumference of a circle.

3 Find the area of a circle.

4 Find the volume of a cylinder.

5 Find the surface area of a rectangular solid.

6 Find the surface area of a cylinder.

Circle, Radius, and Diameter

A **circle** is a two-dimensional (flat) figure with all points the same distance from a fixed center point.

The **radius** (*r*) is the distance from the center of the circle to any point on the circle.

The **diameter** (*d*) is the distance across the circle passing through the center.

VOCABULARY TIPS

Diameter **Dia** means "through," and **meter** means "measure." So the **diameter** of a circle is a line segment through the circle, passing through the center.

Ray and Radius come from the same root word, which means "spoke," like the spoke of a bicycle wheel. So the **radius** is a line segment from the center of a circle out to the edge of the circle.

Using the circle above on the right as a model, you can see some relationships between the radius and diameter.

Finding the Diameter and Radius of a Circle

$$\text{diameter} = 2 \cdot \text{radius}$$

$$d = 2r$$

$$\text{and} \quad \text{radius} = \frac{\text{diameter}}{2} \quad \text{or} \quad r = \frac{d}{2}$$

EXAMPLE 1   **Finding the Diameter and Radius of a Circle**

Find the unknown length of the diameter or radius in each circle.

(a)

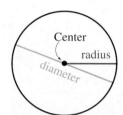

Because the radius is 9 cm, the diameter is twice as long.

$$d = 2 \cdot r$$

$$d = 2 \cdot 9 \text{ cm}$$

$$d = 18 \text{ cm}$$

Multiply the radius by 2 to get the diameter. Write **cm** in your answer, **not** cm².

Continued on Next Page

1 Find the unknown length of the diameter or radius in each circle.

(a)

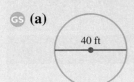

radius (r) is unknown

$$r = \frac{d}{2}$$

$$r = \frac{40 \text{ ft}}{2} = \underline{\hspace{1cm}}$$

(b)

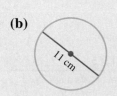

(c)

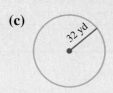

(d)

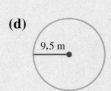

(b)

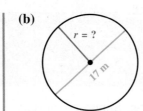

The radius is half the diameter.

$$r = \frac{d}{2} \quad \text{so} \quad r = \frac{17 \text{ m}}{2}$$

> $8.5 = 8\frac{1}{2}$ because $8\frac{5}{10}$ simplifies to $8\frac{1}{2}$.

$$r = 8.5 \text{ m} \quad \text{or} \quad 8\frac{1}{2} \text{ m}$$

◀ **Work Problem 1 at the Side.**

OBJECTIVE **2** **Find the circumference of a circle.** The perimeter of a circle is called its **circumference.** Circumference is the distance around the edge of a circle.

The diameter of the can in the drawing is about 10.6 cm, and the circumference of the can is about 33.3 cm. Dividing the circumference of the circle by the diameter gives an interesting result.

$$\frac{\text{Circumference}}{\text{diameter}} = \frac{33.3}{10.6} \approx 3.14 \quad \text{Rounded to the nearest hundredth}$$

Dividing the circumference of *any* circle by its diameter *always* gives an answer close to 3.14. This means that going around the edge of any circle is a little more than 3 times as far as going straight across the circle.

This ratio of circumference to diameter is called π (the Greek letter **pi,** pronounced PIE). There is no decimal that is exactly equal to π, but here is the *approximate* value.

$$\pi \approx 3.14159265359$$

Rounding the Value of *Pi* (π)

We usually round π to 3.14. Therefore, calculations involving π will give approximate answers and should be written using the $\approx$ symbol.

Use the following formulas to find the *circumference* of a circle.

Finding the Circumference (Distance around a Circle or its Perimeter)

$$\text{Circumference} = \pi \cdot \text{diameter}$$

$$C = \pi d$$

or, because $d = 2r$ then $C = \pi \cdot 2r$ usually written $C = 2\pi r$

Because circumference is the perimeter of a circle, remember to use linear units such as ft, yd, m, and cm (**not** square units).

EXAMPLE 2   **Finding the Circumference of Circles**

Find the circumference of each circle. Use 3.14 as the approximate value for π. Round answers to the nearest tenth.

(a)

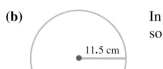

The *diameter* is 38 m, so use the formula with *d* in it.

$$C = \pi \cdot d$$

$$C \approx 3.14 \cdot 38 \text{ m}$$ 〔Write **m** in your answer, **not** m².〕

$$C \approx 119.3 \text{ m} \longleftarrow \text{Rounded}$$

(b)

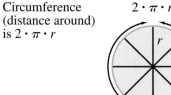

In this example, the length of the *radius* is labeled, so it is easier to use the formula with *r* in it.

$$C = 2 \cdot \pi \cdot r$$

$$C \approx 2 \cdot 3.14 \cdot 11.5 \text{ cm}$$ 〔Write **cm** in your answer, **not** cm².〕

$$C \approx 72.2 \text{ cm} \longleftarrow \text{Rounded}$$

⊞ **Calculator Tip**

Most *scientific* calculators have a (π) key. Try pressing it. With a 10-digit display, you'll see the value of π to the nearest billionth.

(π) (=)   ▮ **3.141592654**

But this is still an approximate value, although it is more precise than rounding π to 3.14. Try finding the circumference in **Example 2(a)** above using the (π) key.

(π) (×) 38 (=) **119.3805208**   Rounds to 119.4

When you used 3.14 as the approximate value of π, the result rounded to 119.3, so the answers are slightly different. In this text we will use 3.14 because some students may be using a calculator without a (π) key.

— **Work Problem ② at the Side.** ▶

OBJECTIVE ▶ ③ **Find the area of a circle.** To find the formula for the area of a circle, start by cutting two circles into many pie-shaped pieces.

Circumference (distance around) is $2 \cdot \pi \cdot r$

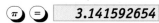

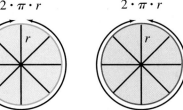

Unfold the circles, much as you might "unfold" a peeled orange, and put them together as shown here.

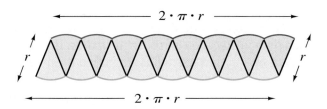

VOCABULARY TIP

Circumference **Circum-** means "around, as in a circle." So the **circumference** is the distance around the outside edge of a circle. It is the perimeter of a circle.

❷ Find the circumference of each circle. Use 3.14 as the approximate value for π. Round answers to the nearest tenth.

GS **(a)**

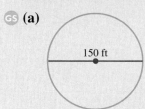

diameter is known; use

$$C = \pi \cdot d$$

$$C \approx 3.14 \cdot 150 \text{ ft}$$

$$C \approx \underline{\hspace{1cm}}$$

GS **(b)**

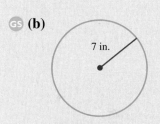

radius is known; use

$$C = 2 \cdot \pi \cdot r$$

(c) diameter 0.9 km

(d) radius 4.6 m

Answers

2. **(a)** $C \approx 471$ ft **(b)** $C \approx 44.0$ in. (rounded)
   (c) $C \approx 2.8$ km (rounded)
   (d) $C \approx 28.9$ m (rounded)

❸ Find the area of each circle. Use 3.14 for π. Round your answers to the nearest tenth.

ᴳˢ (a)

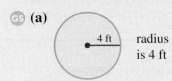

radius is 4 ft

$$A = \pi r^2$$

$$A = \pi \cdot r \cdot r$$

$$A \approx 3.14 \cdot 4\text{ ft} \cdot 4\text{ ft}$$

$$A \approx \underline{\qquad}$$

(b)

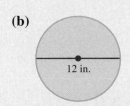

12 in.

Hint: The diameter is 12 in., so $r = \underline{\qquad}$ in.

▦ (c)

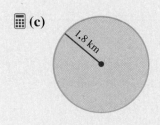

1.8 km

▦ (d)

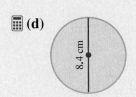

8.4 cm

Answers

3. All area answers are rounded.
   (a) $A \approx 50.2\text{ ft}^2$
   (b) $r = 6$ in.; $A \approx 113.0\text{ in.}^2$
   (c) $A \approx 10.2\text{ km}^2$ **(d)** $A \approx 55.4\text{ cm}^2$

The figure is approximately a parallelogram with height r (the radius of the original circle) and base $2 \cdot \pi \cdot r$ (the circumference of the original circle). The area of the "parallelogram" is base times height.

$$\text{Area of "parallelogram"} = \underset{b}{\underline{\quad}} \cdot h$$

$$\text{Area of "parallelogram"} = \overbrace{2 \cdot \pi \cdot r} \cdot r$$

$$\text{Area of "parallelogram"} = 2 \cdot \pi \cdot r^2 \leftarrow \text{Recall that } r \cdot r \text{ is } r^2$$

Because the "parallelogram" was formed from *two* circles, the area of *one* circle is half as much.

$$\frac{1}{\underset{1}{2}} \cdot \overset{1}{2} \cdot \pi \cdot r^2 = 1 \cdot \pi \cdot r^2 \quad \text{or simply} \quad \pi r^2$$

> **Finding the Area of a Circle**
>
> Area of a circle $= \pi \cdot \text{radius} \cdot \text{radius}$
>
> $$A = \pi r^2$$
>
> Remember to use **square units** when measuring **area**.

EXAMPLE 3   **Finding the Area of Circles**

Find the area of each circle. Use 3.14 for π. Round your answers to the nearest tenth.

(a) A circle with a radius of 8.2 cm
   Use the formula $A = \pi r^2$, which means $A = \pi \cdot r \cdot r$

$$A = \pi \cdot r \cdot r$$

$$A \approx 3.14 \cdot 8.2\text{ cm} \cdot 8.2\text{ cm}$$

$$A \approx 211.1\text{ cm}^2 \leftarrow \text{Rounded}$$

> This is **area**, so write **cm²** in the answer.

(b)

10 ft

To use the area formula $A = \pi r^2$, you need to know the radius (r). In this circle, the *diameter* is 10 ft. First find the radius.

$$r = \frac{d}{2}$$

$$r = \frac{10\text{ ft}}{2} = 5\text{ ft}$$

> You *cannot* use the diameter in the area formula, so **find the radius first.**

Now find the area.

$$A \approx 3.14 \cdot 5\text{ ft} \cdot 5\text{ ft}$$

$$A \approx 78.5\text{ ft}^2 \leftarrow \text{Square units for area}$$

> **❗ CAUTION**
>
> When finding *circumference*, you can start with either the radius or the diameter. When finding *area*, you must use the *radius*. If you are given the diameter, divide it by 2 to find the radius. Then find the area.

◀ **Work Problem ❸ at the Side.**

▦ **Calculator Tip**

You can find the area of the circle in **Example 3(a)** on the previous page using your calculator. The first method works on all types of calculators.

$$3.14 \; \boxed{\times} \; 8.2 \; \boxed{\times} \; 8.2 \; \boxed{=} \; \textbf{211.1336}$$

You round the answer to 211.1 (nearest tenth).

On a *scientific* calculator you can also use the $\boxed{x^2}$ key, which squares the number you enter automatically (that is, multiplies the number times itself).

$$3.14 \; \boxed{\times} \; 8.2 \; \boxed{x^2} \; \boxed{=} \; \textbf{211.1336}$$

In the next example we will find the area of a *semicircle,* which is half the area of a circle.

EXAMPLE 4 **Finding the Area of a Semicircle**

Find the area of the semicircle below. Use 3.14 for π. Round your answer to the nearest tenth.

12 ft

First, find the area of a whole circle with a radius of 12 ft.

$$A = \pi \cdot r \cdot r$$
$$A \approx 3.14 \cdot 12 \text{ ft} \cdot 12 \text{ ft}$$
$$A \approx 452.16 \text{ ft}^2 \quad \text{Do not round yet.}$$

Divide the area of the whole circle by 2 to find the area of the semicircle.

$$\frac{452.16 \text{ ft}^2}{2} = 226.08 \text{ ft}^2$$

The *last* step is rounding 226.08 to the nearest tenth.

Area of semicircle ≈ 226.1 ft²   Rounded

> This is **area**, so write **ft²** in your answer.

─── **Work Problem ④ at the Side.** ▶

EXAMPLE 5 **Applying the Concept of Circumference**

A circular rug is 8 feet in diameter. The cost of fringe for the edge is $2.25 per foot. What will it cost to add fringe to the rug? Use 3.14 for π.

$$\text{Circumference} = \pi \cdot d$$
$$C \approx 3.14 \cdot 8 \text{ ft}$$
$$C \approx 25.12 \text{ ft}$$

$$\text{cost} = \text{cost per foot} \cdot \text{Circumference}$$

$$\text{cost} = \frac{\$2.25}{1 \text{ ft}} \cdot \frac{25.12 \text{ ft}}{1}$$

> The common units divide out.

$$\text{cost} = \$56.52$$

The **cost of adding fringe to the rug is $56.52**

─── **Work Problem ⑤ at the Side.** ▶

VOCABULARY TIP

Semicircle **Semi** means "half."

④ Find the area of each semicircle. ▦ Use 3.14 for π. Round your answers to the nearest tenth.

(a)

24 m

(b)

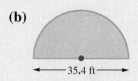

←——— 35.4 ft ———→

(c)

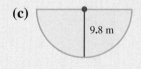

9.8 m

⑤ Find the cost of binding around the edge of a circular rug that is 3 meters in diameter. The binder charges $4.50 per meter. Use 3.14 for π.

Answers

4. All answers are rounded.
   (a) $A \approx 904.3 \text{ m}^2$ (b) $A \approx 491.9 \text{ ft}^2$
   (c) $A \approx 150.8 \text{ m}^2$

5. $42.39

6 Find the cost of covering the underside of the rug in **Margin Problem 5** on the previous page with a nonslip rubber backing. The rubber backing costs $3.89 per square meter.

EXAMPLE 6 **Applying the Concept of Area**

Find the cost of covering the underside of the rug in **Example 5** (on the previous page) with a nonslip rubber backing. The rubber backing costs $1.50 per square foot. Use 3.14 for π.

First find the radius. $r = \dfrac{d}{2} = \dfrac{8 \text{ ft}}{2} = 4 \text{ ft}$

Then find the area. $A = \pi \cdot r^2$

$A \approx 3.14 \cdot 4 \text{ ft} \cdot 4 \text{ ft}$

$A \approx 50.24 \text{ ft}^2$

Write a $ in your answer.

$\text{cost} = \dfrac{\$1.50}{1 \text{ ft}^2} \cdot \dfrac{50.24 \text{ ft}^2}{1} = \mathbf{\$75.36}$

◀ **Work Problem** **6** **at the Side.**

---

7 Find the volume of each cylinder. Use 3.14 for π. Round your answers to the nearest tenth.

GS (a)

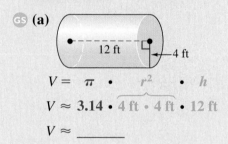

$V = \pi \cdot r^2 \cdot h$

$V \approx 3.14 \cdot \overbrace{4 \text{ ft} \cdot 4 \text{ ft}} \cdot 12 \text{ ft}$

$V \approx \underline{\hspace{1.5cm}}$

(b)

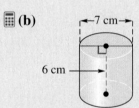

Hint: Find the **radius** first.

(c) radius 14.5 yd, height 3.2 yd

OBJECTIVE **4** **Find the volume of a cylinder.** Three *cylinders* are shown.

The height must be perpendicular to the circular top and bottom of the cylinder.

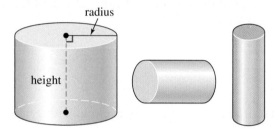

These are called *right circular cylinders* because the top and bottom are circles, and the side makes a right angle with the top and bottom. Examples of cylinders are a soup can, a home water heater, and a piece of pipe.

Use the following formula to find the *volume* of a *cylinder*. Notice that the first part of the formula, $\pi \cdot r \cdot r$, is the *area* of the circular base.

Finding the Volume of a Cylinder

Volume of a cylinder $= \pi \cdot r \cdot r \cdot h$

$V = \pi r^2 h$

Remember to use **cubic units** when measuring **volume**.

EXAMPLE 7 **Finding the Volume of Cylinders**

Find the volume of each cylinder. Use 3.14 as the approximate value of π. Round your answers to the nearest tenth if necessary.

(a)

The diameter is 20 m, so the radius is 20 m ÷ 2 = 10 m. The height is 9 m.

$V = \pi \cdot r \cdot r \cdot h$

$V \approx 3.14 \cdot 10 \text{ m} \cdot 10 \text{ m} \cdot 9 \text{ m}$

$V \approx \mathbf{2826 \text{ m}^3}$ ← Cubic units for **volume**

(b)

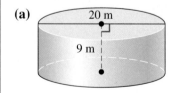

$V \approx 3.14 \cdot 6.2 \text{ cm} \cdot 6.2 \text{ cm} \cdot 38.4 \text{ cm}$

$V \approx 4634.94144$ Now round to tenths.

$V \approx \mathbf{4634.9 \text{ cm}^3}$ ← Cubic units for **volume**

◀ **Work Problem** **7** **at the Side.**

Answers

6. $27.48

7. **(a)** $V \approx 602.9 \text{ ft}^3$ (rounded)
(b) $V \approx 230.8 \text{ cm}^3$ (rounded)
(c) $V \approx 2112.6 \text{ yd}^3$ (rounded)

OBJECTIVE ▶ **5** **Find the surface area of a rectangular solid.** You have just learned how to find the *volume* of a cylinder. Earlier you found the *volume* of a rectangular solid. For example, the volume of the cereal box shown below is $V = lwh = (7 \text{ in.})(2 \text{ in.})(10 \text{ in.}) = 140 \text{ in.}^3$ But if your company makes cereal boxes, you also need to know how much cardboard is needed for each box. You need to find the *surface area* of the box.

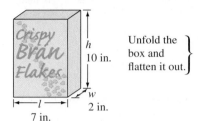

Unfold the box and flatten it out. }

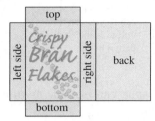

The unfolded box is made up of six rectangles: front, back, top, bottom, left side, right side.

Surface area is the area on the surface of a three-dimensional object (a solid). For a rectangular solid like the cereal box, the surface area is the sum of the areas of the six rectangular sides. Notice that the top and bottom have the same area, the front and back have the same area, and the left and right sides have the same area.

Surface area $=$ top $w +$ bottom $w +$ front $h +$ back $h +$ left side $h +$ right side h

$$S = l \cdot w \quad + \ l \cdot w \quad + l \cdot h \ + \ l \cdot h \ + w \cdot h + w \cdot h$$

$$S = \quad 2lw \quad\quad + \quad\quad 2lh \quad\quad + \quad\quad 2wh$$

Finding the Surface Area of a Rectangular Solid

Surface area $= (2 \cdot l \cdot w) + (2 \cdot l \cdot h) + (2 \cdot w \cdot h)$

$$S = \quad 2lw \quad + \quad 2lh \quad + \quad 2wh$$

Use **square units** when measuring surface **area**.

EXAMPLE 8   **Finding the Volume and Surface Area of a Rectangular Solid**

Find the volume and surface area of this shipping carton.

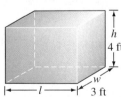

First find the volume.

$V = lwh$ （ft · ft · ft is ft³）

$V = 5 \text{ ft} \cdot 3 \text{ ft} \cdot 4 \text{ ft}$

$V = 60 \text{ ft}^3$ ← Cubic units for **volume**

Now find the surface area.

$$S = \quad 2lw \quad\quad + \quad\quad 2lh \quad\quad + \quad\quad 2wh$$

$S = (2 \cdot 5 \text{ ft} \cdot 3 \text{ ft}) + (2 \cdot 5 \text{ ft} \cdot 4 \text{ ft}) + (2 \cdot 3 \text{ ft} \cdot 4 \text{ ft})$

$S = \quad 30 \text{ ft}^2 \quad + \quad 40 \text{ ft}^2 \quad + \quad 24 \text{ ft}^2$ （Write ft² for area.）

$S = 94 \text{ ft}^2$ ← Square units for **area**

— **Work Problem** **8** **at the Side.** ▶

---

8 Find the volume and surface area of each rectangular solid.

GS **(a)**

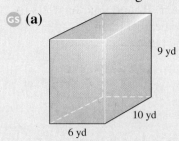

$V = lwh$

$V = \underline{\quad} \cdot \underline{\quad} \cdot \underline{\quad}$

$V = \underline{\quad\quad}$

$S = 2lw + 2lh + 2wh$

(b)

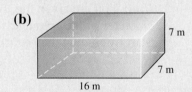

Answers

8. **(a)** $V = 10 \text{ yd} \cdot 6 \text{ yd} \cdot 9 \text{ yd};$
   $V = 540 \text{ yd}^3$ (*cubic* yd for *volume*)
   $S = 408 \text{ yd}^2$ (*square* yd for *area*)
   (b) $V = 784 \text{ m}^3$
   $S = 546 \text{ m}^2$

9 Find the volume and surface area of each cylinder. Use 3.14 for π. Round your answers to the nearest tenth.

(a)

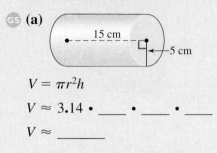

$V = \pi r^2 h$

$V \approx 3.14 \cdot \underline{\quad} \cdot \underline{\quad} \cdot \underline{\quad}$

$V \approx \underline{\qquad}$

$S = 2\pi rh + 2\pi r^2$

(b)

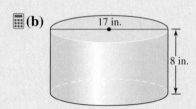

Hint: You are given the *diameter* of the cylinder. Start by finding the *radius*.

OBJECTIVE 6 **Find the surface area of a cylinder.** You can use the same idea of "unfolding" a shape to find the surface area of a cylinder, such as the soup can shown below. Finding the surface area will tell you how much aluminum you need to make the can.

The unfolded soup can is made up of a rectangular side, a circular top, and a circular bottom.

Remember that the formula for the area of a circle is πr^2.

Surface area =

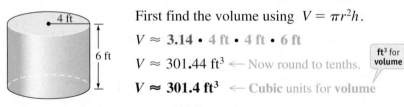

$S = \quad 2\pi r \cdot h \quad + \quad \pi r^2 \quad + \quad \pi r^2$

$S = \quad 2\pi rh \quad + \quad 2\pi r^2$

Finding the Surface Area of a Right Circular Cylinder

Surface area = $(2 \cdot \pi \cdot r \cdot h) + (2 \cdot \pi \cdot r \cdot r)$

$S = \quad 2\pi rh \quad + \quad 2\pi r^2$

Remember that area is measured in square units, so use **square units** when measuring surface **area**.

EXAMPLE 9 **Finding the Volume and Surface Area of a Right Circular Cylinder**

Find the volume and surface area of this water tank. Use 3.14 as the approximate value for π. Round your answers to the nearest tenth when necessary.

First find the volume using $V = \pi r^2 h$.

$V \approx 3.14 \cdot 4 \text{ ft} \cdot 4 \text{ ft} \cdot 6 \text{ ft}$

$V \approx 301.44 \text{ ft}^3$ ← Now round to tenths. [ft³ for volume]

$V \approx 301.4 \text{ ft}^3$ ← Cubic units for **volume**

Now find the surface area.

$S = \quad 2\pi rh \quad + \quad 2\pi r^2$

$S \approx (2 \cdot 3.14 \cdot 4 \text{ ft} \cdot 6 \text{ ft}) + (2 \cdot 3.14 \cdot 4 \text{ ft} \cdot 4 \text{ ft})$

$S \approx \quad 150.72 \text{ ft}^2 \quad + \quad 100.48 \text{ ft}^2$

$S \approx 251.2 \text{ ft}^2$ ← Square units for area [ft² for area]

Answers

9. **(a)** $V \approx 3.14 \cdot 5 \text{ cm} \cdot 5 \text{ cm} \cdot 15 \text{ cm}$
   $V \approx 1177.5 \text{ cm}^3$ (*cubic* units for *volume*)
   $S \approx 628 \text{ cm}^2$ (*square* units for *area*)
   (b) $V \approx 1814.9 \text{ in.}^3$ (rounded)
   $S \approx 880.8 \text{ in.}^2$ (rounded)

◀ **Work Problem 9** at the Side.

5.10 Exercises

FOR EXTRA HELP *Go to* MyMathLab *for worked-out, step-by-step solutions to exercises enclosed in a square* ▢ *and video solutions to* ▶ *exercises.*

1. **CONCEPT CHECK** Fill in the blanks. Choose from height, perimeter, area, base, ft, and ft².

   Finding the circumference of a circle is like finding the _____ of a rectangle. If the circle's radius and diameter are measured in feet, then the units for the circumference will be _____ .

2. **CONCEPT CHECK** Fill in the blanks. Choose from radius, diameter, circumference, divide by 2, and multiply by 2.

   To use the formula for finding the area of a circle, you need to know the length of the _____ . If you are given the diameter of a circle, what should you do to it to find the radius? _____

Find the unknown length in each circle. ***See Example 1.***

3.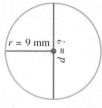

$d = 2 \cdot r$

4.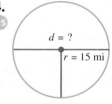

$d = 2 \cdot r$

5.

6.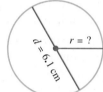

Find the circumference and area of each circle. Use 3.14 as the approximate value for π. Round your final answers to the nearest tenth. ***See Examples 2 and 3.***

7.

11 ft

8.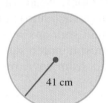

41 cm

9.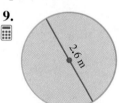

2.6 m

10.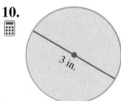

3 in.

Find the circumference and area of circles having the following diameters. Use 3.14 for π. Round your final answers to the nearest tenth. ***See Examples 2 and 3.***

11. $d = 15$ cm

$C = \pi \cdot d$

$C \approx 3.14 \cdot 15$ cm

$C \approx$ _____

To find the area, first find the radius.

12. $d = 39$ ft

$C = \pi \cdot d$

$C \approx 3.14 \cdot 39$ ft

$C \approx$ _____

To find the area, first find the radius.

13. $d = 7\frac{1}{2}$ ft

14. $d = 4\frac{1}{2}$ yd

Find each shaded area. Note that Exercises 15–18 all contain semicircles. Use 3.14 as the approximate value of π. Round your final answers to the nearest tenth. ***See Example 4.***

15.

7 in.

16.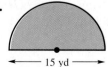

15 yd

17.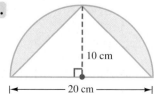

10 cm

20 cm

18.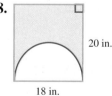

20 in.

18 in.

Solve each application problem. Use 3.14 as the approximate value of π. Round your final answers to the nearest tenth. ***See Examples 5 and 6.***

19. An irrigation system moves around a center point to water a circular area for crops. If the irrigation system is 50 yd long, how large is the watered area?

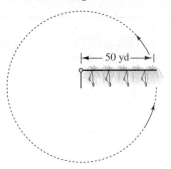

20. If you swing a ball held at the end of a string 20 cm long, how far will the ball travel on each turn?

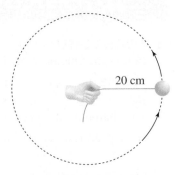

21. A tire has an overall diameter of 29.10 inches. How far will a point on the tire tread move in one complete turn?

Bonus question: How many revolutions does the tire make per mile? Round to the nearest whole number.

22. In October 2012, Hurricane Sandy slammed into the northeast coast of the United States. Sandy was one of the largest Atlantic tropical storms ever recorded. The huge circular storm had a diameter of 1000 miles. By contrast, Hurricane Katrina, which flooded New Orleans in 2005, had a diameter of 210 miles. What was the area of each hurricane? Round your answers to the nearest hundred square miles. (Data from www.nws.noaa.gov)

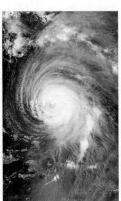

Circular hurricane as seen from a weather satellite.

For Exercises 23–28, first draw a circle and label the radius or diameter. Then solve the problem. For Exercises 23–34, use 3.14 for π and round final answers to the nearest tenth.

23. A radio station can be heard 150 miles in all directions during evening hours. How many square miles are in the station's broadcast area?

The radius is 150 miles.

24. An earthquake was felt by people 900 km away in all directions from the epicenter (the source of the earthquake). How much area was affected by the quake?

Hint: The affected area is a circle with a radius of 900 km. Make a drawing of the circle.

25. The diameter of Diana Hestwood's wristwatch is 1 in. and the radius of the clock face on her kitchen wall is 3 in. Find the circumference and the area of each clock face.

26. A farmer claims to have the world's largest ball of twine, with a diameter of 12 ft 9 in. The sign posted near the ball says it has a circumference of 40 ft. Is the sign correct? (*Hint:* First change 9 in. to feet and add it to 12 ft.)

27. Blaine Fenstad wants to buy a pair of two-way radios. Some models have a range of 16 miles under ideal conditions. More expensive models have a range of 35 miles. What is the difference in the area covered by the 16-mile and 35-mile models? (Data from Best Buy.)

28. The National Audubon Society holds an end-of-year bird count. Volunteers count all the birds they see in a circular area during a 24-hour period. Each circle has a diameter of 15 miles. About 1900 circular areas are counted across the United States each December. What is the total area covered by the count, to the nearest hundred square miles?

29. To find the age of a living tree, a forestry specialist measures the circumference of the tree. (See photo.)

  (a) If the circumference of a tree is 144 cm, what is the diameter?

  (b) Explain how you solved part (a).

30. In Atlanta, Interstate 285 circles the city and is known as the "perimeter." If the circumference of the circle made by the highway is 62.8 miles, find:

  (a) the diameter of the circle

  (b) the area inside the circle.

31. The Mormons traveled west to Utah by covered wagon in 1847. They tied a rag to a wagon wheel to keep track of the distance they traveled. The radius of the wheel was 2.33 ft. How far did the rag travel each time the wheel made a complete revolution? (Data from *Trail of Hope*.)

32. First work **Exercise 31.** Then find how many wheel revolutions equaled one mile. There are 5280 ft in one mile.

33. Find the shaded area.

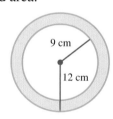

9 cm

12 cm

34. Find the area of this skating rink.

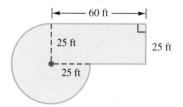

60 ft

25 ft

25 ft

25 ft

⊞ *Find the volume and surface area of each cylinder or rectangular solid. Use 3.14 as the approximate value of π. Round your final answers to the nearest tenth when necessary.* **See Examples 7–9.**

35.

5 ft
6 ft

36.
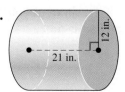
12 in.
21 in.

37.
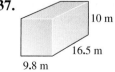
10 m
16.5 m
9.8 m

38.
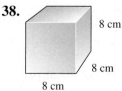
8 cm
8 cm
8 cm

39.

40.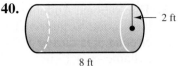

41.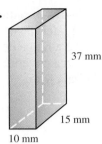

42.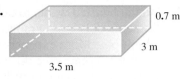

43. A box for graham crackers measures 5.5 in. by 2.8 in. by 8 in. high. Find the amount of cardboard needed to make the box.

44. A soda can is 12.5 cm tall and the round top has a diameter of 6.7 cm. How much aluminum is needed to make the can? (Assume the can has a flat top and bottom.)

45. CONCEPT CHECK   The work shown below has **two** mistakes in it.

What Went Wrong?   First write a sentence explaining what each mistake is. Then fix the mistakes and find the correct solution.

Bronte found the volume of a cylinder with a diameter of 7 cm and a height of 5 cm. Here is her work.

$$V \approx 3.14 \cdot 7 \cdot 7 \cdot 5$$
$$V \approx 769.3 \text{ cm}^2$$

46. CONCEPT CHECK   The work shown below has a mistake in it.

What Went Wrong?   First write a sentence explaining what the mistake is. Then fix the mistake and find the correct solution.

Olivia was working with Bronte on the same cylinder as in **Exercise 45.** Olivia calculated the surface area of the cylinder this way:

$$S \approx (2 \cdot 3.14 \cdot 3.5 \cdot 5) + (2 \cdot 3.14 \cdot 3.5 \cdot 3.5)$$
$$S \approx 186.83 \text{ cm}^3$$

Relating Concepts (Exercises 47–48) For Individual or Group Work

Work Exercises 47 and 48 in order.

47. Look again at **Exercise 38** on the previous page. The figure is a *cube*. What is special about the measurements of a cube?

48. Find a shortcut you can use to calculate the surface area of a cube.

Chapter 5 Summary

Key Terms

5.1

decimals Decimals, like fractions, are used to show parts of a whole.

decimal point A decimal point is the dot that is used to separate the whole number part from the fractional part of a decimal number.

place value Place value is the value assigned to each place to the right or left of the decimal point. Whole numbers, such as ones and tens, are to the *left* of the decimal point. Fractional parts, such as tenths and hundredths, are to the *right* of the decimal point.

5.2

round To round is to "cut off" a number after a certain place, such as to round to the nearest hundredth. The rounded number is less accurate than the original number. You can use the symbol " ≈ " to mean "is approximately equal to."

decimal places Decimal places are the number of digits to the *right* of the decimal point; for example, 6.37 has two decimal places, and 4.706 has three decimal places.

5.5

repeating decimal A repeating decimal is a decimal number with one or more digits that repeat forever; it never ends. For example, in 0.1555 . . . , the digit 5 continues to repeat. Use three dots to indicate that it is a repeating decimal. Or write a bar above the repeating digits, as in 0.1$\overline{5}$. (Use the dots or the bar, but not both.)

5.7

mean The mean is the sum of all the values divided by the number of values. It is often called the *average*.

weighted mean The weighted mean is a mean calculated so that each value is multiplied by its frequency.

median The median is the middle number in a group of values that are listed from least to greatest. It divides a group of values in half. If there is an even number of values, the median is the mean (average) of the two middle values.

mode The mode is the value that occurs most often in a group of values. A group of values that has two modes is called *bimodal*.

5.8

square root A positive square root of a positive number is one of two identical positive factors of the number.

hypotenuse The hypotenuse is the side of a right triangle opposite the 90° angle; it is the longest side.
Example: See the red side in the triangle at the right.

5.10

circle A circle is a two-dimensional (flat) figure with all points the same distance from a fixed center point.
Example: See the figure at the right.

radius Radius is the distance from the center of a circle to any point on the circle.
Example: See the red radius in the circle at the right.

diameter Diameter is the distance across a circle, passing through the center.
Example: See the blue diameter in the circle at the right.

circumference Circumference is the distance around a circle.

π (pi) π is the ratio of the circumference to the diameter of any circle. It is approximately equal to 3.14.

surface area Surface area is the area on the surface of a three-dimensional object (a solid). Surface area is measured in square units.

New Symbols

3.85 ⟵ Bar above repeating digit(s) in a decimal

3.8555 . . . Three dots indicate a repeating decimal

≈   is approximately equal to

√   square root

π   Greek letter pi (pronounced PIE); ratio of the circumference of any circle to its diameter

New Formulas

$$\text{mean} = \frac{\text{sum of all values}}{\text{number of values}}$$

$$\text{hypotenuse} = \sqrt{(\text{leg})^2 + (\text{leg})^2}$$

$$\text{leg} = \sqrt{(\text{hypotenuse})^2 - (\text{leg})^2}$$

diameter of a circle: $d = 2r$

radius of a circle: $r = \dfrac{d}{2}$

Circumference of a circle: $C = \pi d$

$$\text{or} \quad C = 2\pi r$$

Area of a circle: $A = \pi r^2$

Volume of a cylinder: $V = \pi r^2 h$

Surface area of a rectangular solid: $S = 2lw + 2lh + 2wh$

Surface area of a cylinder: $S = 2\pi rh + 2\pi r^2$

Test Your Word Power

See how well you have learned the vocabulary in this chapter.

1 **Decimal numbers** are like fractions in that they both
   A. must be written in lowest terms
   B. have decimal points
   C. represent parts of a whole.

2 **Decimal places** refer to
   A. digits to the left of the decimal point
   B. digits to the right of the decimal point
   C. the number of zeros in a decimal number.

3 The **hypotenuse** is
   A. the long base in a rectangle
   B. the height in a parallelogram
   C. the longest side in a right triangle.

4 The **decimal point**
   A. separates the whole number part from the fractional part
   B. is always moved when finding a quotient
   C. is at the far left side of a whole number.

5 The number $0.\overline{3}$ is an example of
   A. a repeating decimal
   B. a rounded number
   C. a truncated number.

6 π is the ratio of
   A. the diameter to the radius of a circle
   B. the circumference to the diameter of a circle
   C. the diameter to the circumference of a circle.

7 The **median** for a set of values is
   A. the mathematical average
   B. the value that occurs most often
   C. the middle value when the values are listed from least to greatest.

8 The **circumference** of a circle is
   A. the perimeter of the circle
   B. found using the expression πr^2
   C. the distance across the circle.

9 The **radius** of a circle is
   A. twice the circumference
   B. measured in square units
   C. half the diameter.

Answers to Test Your Word Power

1. C; *Example:* For 0.7, the whole is cut into ten parts, and you are interested in 7 of the parts.

2. B; *Examples:* The number 6.87 has two decimal places; 0.309 has three decimal places.

3. C; *Example:* In triangle *ABC*, side *AC* (the red side) is the hypotenuse.

4. A; *Example:* In 5.42, the decimal point separates the whole number part, 5 ones, from the fractional part, 42 hundredths.

5. A; *Example:* The bar above the 3 in $0.\overline{3}$ indicates that the 3 repeats forever.

6. B; *Example:* The ratio of a circumference of 12.57 cm to a diameter of 4 cm is $\dfrac{12.57}{4} \approx 3.14$ (rounded).

7. C; *Example:* For this set of prices—$4, $3, $7, $2, $5, $6, $7—the median is $5.

8. A; *Example:* If the radius of a circle is 4 ft, then $C \approx 2 \cdot 3.14 \cdot 4 \approx 25.1$ ft. So the distance around the circle is about 25.1 ft.

9. C; *Example:* If the diameter of a circle is 10 yd, the radius is $\dfrac{10 \text{ yd}}{2} = 5$ yd.

Quick Review

Concepts	Examples

5.1 Reading and Writing Decimals

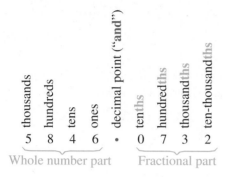

Write each decimal in words.

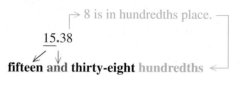

→ 8 is in hundredths place.

15.38

fifteen and thirty-eight hundredths ←

→ 3 is in ten-thousandths place.

0.0103

one hundred three ten-thousandths ←

5.1 Writing Decimals as Fractions

The digits to the right of the decimal point are the numerator. The place value of the rightmost digit determines the denominator.

Always write the fraction in lowest terms.

Write 0.45 as a fraction in lowest terms.

The numerator is 45. The rightmost digit, 5, is in the hundredths place, so the denominator is 100. Then write the fraction in lowest terms.

$$\frac{45}{100} = \frac{45 \div 5}{100 \div 5} = \frac{9}{20} \leftarrow \text{Lowest terms}$$

5.2 Rounding Decimals

Find the place to which you are rounding. Draw a cut-off line to the right of that place; the rest of the digits will be dropped.

Look *only* at the first digit being cut. If it is *4 or less,* the part you are keeping stays the same. If it is *5 or more,* the part you are keeping rounds up.

Do not move the decimal point when rounding. Use the sign "≈" to mean "is approximately equal to."

Round 0.17952 to the nearest thousandth.

⎯ First digit cut is *5 or more* so round up.

0.179 | 52

0.179 ←⎯ Keep this part.

+ 0.001 ←⎯ To round up, add 1 thousandth.

0.180

0.17952 rounds to 0.180
Write **0.17952 ≈ 0.180**

5.3 Adding Positive and Negative Decimals

To find the sum, line up the decimal points. If needed, write in zeros as placeholders. Add or subtract the absolute values as if they were whole numbers. Line up the decimal point in the answer.

If the numbers have the same sign, use the common sign as the sign of the sum. If the numbers have different signs, the sign of the sum is the sign of the number with the greater absolute value.

Add. $-5.23 + (-0.792)$

Both addends are negative, so the sum will be negative.

5.23**0**   Use zero as a
+ 0.792   placeholder.

6.022

↑⎯⎯ Line up decimal points.

$-5.23 + (-0.792) = \mathbf{-6.022}$

Concepts	**Examples**

5.3 **Subtracting Positive and Negative Decimals**

Rewrite subtracting as adding the opposite of the second number. Then follow the rules for adding positive and negative decimals.

Subtract.
$$4.2 - \quad 12.91$$
$$\downarrow \qquad \downarrow$$
$$4.2 + (-12.91)$$

$|4.2|$ is 4.2 and $|-12.91|$ is 12.91

Subtract $12.91 - 4.2$ to get 8.71
Because -12.91 has the greater absolute value and is negative, the answer will be negative.

$$4.2 + (-12.91) = -8.71$$

5.4 **Multiplying Positive and Negative Decimals**

Step 1 Multiply as you would for whole numbers.

Step 2 Count the total number of decimal places in both factors.

Step 3 Write the decimal point in the answer so it has the same number of decimal places as the total from *Step 2*. You may need to write extra zeros on the left side of the product in order to get enough decimal places in the answer.

Step 4 If two factors have the *same sign*, the product is *positive*. If two factors have *different signs*, the product is negative.

Multiply $(0.169)(-0.21)$

$$
\begin{array}{r}
0.169 \leftarrow 3 \text{ decimal places} \\
\times 0.21 \leftarrow 2 \text{ decimal places} \\
\hline
169 \qquad 5 \text{ total decimal places} \\
338 \\
\hline
.03549 \leftarrow 5 \text{ decimal places in product}
\end{array}
$$

Write in a 0 so you can count over 5 decimal places.
    The factors have *different* signs, so the product is *negative*.

$$(0.169)(-0.21) = -0.03549$$

5.5 **Dividing by a Decimal**

Step 1 Change the divisor to a whole number by moving the decimal point to the right.

Step 2 Move the decimal point in the dividend the same number of places to the right.

Step 3 Write the decimal point in the quotient directly above the decimal point in the dividend. Then divide as with whole numbers.

Step 4 If the numbers have the *same sign*, the quotient is *positive*. If they have *different signs*, the quotient is *negative*.

Divide -52.8 by -0.75
First consider $52.8 \div 0.75$

$$
\begin{array}{r}
7\,0.4 \\
0.7\,5\,)\overline{5\,2.8\,0\,0} \\
5\,2\,5 \\
\hline
3\,0\,0 \\
3\,0\,0 \\
\hline
0
\end{array}
$$

Move decimal point two places to the right in divisor and dividend.

Write zeros in the dividend so you can move the decimal point and continue dividing until the remainder is 0.

The **quotient is 70.4** and is positive because both the divisor and dividend were negative (same signs means positive quotient).

5.6 **Writing Fractions as Decimals**

Divide the numerator by the denominator. If necessary, round to the place indicated.

Write $\frac{1}{8}$ as a decimal.

$\frac{1}{8}$ means $1 \div 8$. Write it as $8)\overline{1}$
The decimal point is on the right side of 1

$$
\begin{array}{r}
0.125 \\
8)\overline{1.000} \leftarrow \\
8 \\
\hline
20 \\
16 \\
\hline
40 \\
40 \\
\hline
0
\end{array}
$$

Write the decimal point and three zeros so you can continue dividing.

Therefore, $\frac{1}{8}$ **is equivalent to 0.125**

Concepts	Examples

5.6 Comparing the Size of Fractions and Decimals

Step 1 Write any fractions as decimals.

Step 2 Write zeros so that all the numbers being compared have the same number of decimal places.

Step 3 Use $<$ to mean "is less than," $>$ to mean "is greater than," or list the numbers from least to greatest.

Arrange in order from least to greatest.

$$0.505 \quad \frac{1}{2} \quad 0.55$$

$0.505 = 505$ thousandths $\leftarrow$ 505 is in the middle.

$\frac{1}{2} = 0.5 = 0.500 = 500$ thousandths $\leftarrow$ 500 is least.

$0.55 = 0.550 = 550$ thousandths $\leftarrow$ 550 is greatest.

(least) $\quad \frac{1}{2} \quad$ **0.505** **0.55** $\quad$ (greatest)

5.7 Finding the Mean (Average) of a Set of Values

Step 1 Add all values to obtain a total.

Step 2 Divide the total by the number of values.

Here are Heather Hall's test scores in her math course.

$$93 \quad 76 \quad 83 \quad 93$$
$$78 \quad 82 \quad 87 \quad 85$$

Find Heather's mean score to the nearest tenth.

$$\text{mean} = \frac{93 + 76 + 83 + 93 + 78 + 82 + 87 + 85}{8}$$

$$\text{mean} = \frac{677}{8} \approx \mathbf{84.6} \leftarrow \text{Mean test score (rounded)}$$

5.7 Finding the Median of a Set of Values

Step 1 Arrange the values from least to greatest.

Step 2 If there is an odd number of values, select the middle value. If there is an even number of values, find the average of the two middle values.

Find the median for Heather Hall's scores from the previous example.

List the scores from least to greatest.

$$76 \quad 78 \quad 82 \quad \underline{83 \quad 85} \quad 87 \quad 93 \quad 93$$

Least $\qquad$ Middle values $\qquad$ Greatest

The middle two values are 83 and 85. Find the average of these two values.

$$\frac{83 + 85}{2} = \mathbf{84} \leftarrow \text{Median test score}$$

5.7 Finding the Mode of a Set of Values

Find the value that appears most often in the list of values. This is the mode.

If no value appears more than once, there is no mode. If two different values appear the same number of times (and no other number appears more often), the list is bimodal.

Find the mode for Heather's scores in the previous example.

The most frequently occurring score is 93 (it occurs twice). Therefore, the **mode is 93**

Concepts	Examples

5.8 Finding the Square Root of a Number

Use the square root key on a calculator, or . Round to the nearest thousandth when necessary.

$$\sqrt{64} = 8$$

$$\sqrt{43} \approx 6.557 \quad \text{6.557438524 is rounded to the nearest thousandth.}$$

5.8 Finding the Unknown Length in a Right Triangle

To find the hypotenuse, use:

$$\text{hypotenuse} = \sqrt{(\text{leg})^2 + (\text{leg})^2}$$

The hypotenuse is the side opposite the right angle. It is the longest side in a right triangle.

Find the unknown length in this right triangle. Round to the nearest tenth.

6 m

5 m

$$\text{hypotenuse} = \sqrt{(6)^2 + (5)^2}$$

$$\text{hypotenuse} = \sqrt{36 + 25}$$

$$\text{hypotenuse} = \sqrt{61} \approx 7.8$$

The **hypotenuse is about 7.8 m** long.

To find a leg, use:

$$\text{leg} = \sqrt{(\text{hypotenuse})^2 - (\text{leg})^2}$$

The legs are the sides that form the right angle.

Find the unknown length in this right triangle. Round to the nearest tenth.

25 cm

16 cm

$$\text{leg} = \sqrt{(25)^2 - (16)^2}$$

$$\text{leg} = \sqrt{625 - 256}$$

$$\text{leg} = \sqrt{369} \approx 19.2$$

The **leg is about 19.2 cm** long.

5.9 Solving Equations Containing Decimals

Step 1 Use the addition property of equality to get the variable term by itself on one side of the equal sign.

Solve the equation and check the solution.

$$4.5x + 0.7 = -5.15$$

$$\underline{-0.7 -0.7} \quad \text{Add } -0.7 \text{ to both sides.}$$

$$4.5x + 0 -5.85$$

Step 2 Divide both sides by the coefficient of the variable term to find the solution.

$$\frac{\overset{1}{4.5x}}{\underset{1}{4.5}} = \frac{-5.85}{4.5} \quad \text{Divide both sides by 4.5}$$

$$x = -1.3$$

The solution is -1.3
To check the solution, go back to the *original* equation.

Step 3 Check the solution by going back to the *original* equation. Replace the variable with the solution. If the equation balances, the solution is correct.

CHECK $\quad 4.5x + 0.7 = -5.15 \quad$ Original equation

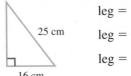

$$4.5\,(-1.3) + 0.7 = -5.15 \quad \text{Replace } x \text{ with } -1.3$$

$$-5.85 + 0.7 = -5.15$$

$$-5.15 = -5.15 \checkmark \text{ Balances}$$

When x is replaced with -1.3, the equation balances, so -1.3 is the correct solution (*not* -5.15).

Concepts	Examples

5.10 Circles

Use this formula to find the *diameter* of a circle when you are given the radius.

$$\text{diameter} = 2 \cdot \text{radius} \quad \text{or} \quad d = 2r$$

Find the diameter of a circle if the radius is 3 ft.

$$d = 2 \cdot r$$
$$d = 2 \cdot 3\text{ ft} = \mathbf{6\ ft}$$

Use this formula to find the *radius* of a circle when you are given the diameter.

$$\text{radius} = \frac{\text{diameter}}{2} \quad \text{or} \quad r = \frac{d}{2}$$

Find the radius of a circle if the diameter is 5 cm.

$$r = \frac{d}{2}$$
$$r = \frac{5\text{ cm}}{2} = \mathbf{2.5\ cm}$$

Use these formulas to find the *circumference* of a circle.

When you know the *radius,* use this formula.

$$C = 2 \cdot \pi \cdot \text{radius}$$
$$\text{or} \quad C = 2\pi r$$

When you know the *diameter,* use this formula.

$$C = \pi \cdot \text{diameter}$$
$$\text{or} \quad C = \pi d$$

Use 3.14 as the approximate value for π.

Find the circumference of a circle with a radius of 7 yd. Round your final answer to the nearest tenth.

$$\text{Circumference} = 2 \cdot \pi \cdot r$$
$$C \approx 2 \cdot 3.14 \cdot 7\text{ yd}$$
$$C \approx \mathbf{44.0\ yd} \leftarrow \text{Rounded}$$

Use this formula to find the *area* of a circle.

$$A = \pi \cdot r \cdot r$$
$$\text{or} \quad A = \pi r^2$$

Use 3.14 as the approximate value of π.
Area is measured in square units.

Find the area of the circle. Round your final answer to the nearest tenth.

$$\text{Area} = \pi \cdot r \cdot r$$
$$A \approx 3.14 \cdot 3\text{ cm} \cdot 3\text{ cm}$$
$$A \approx \mathbf{28.3\ cm^2} \leftarrow \text{Rounded;}$$
square units for area

5.10 Volume of a Cylinder

Use this formula to find the volume of a cylinder.

$$\text{Volume} = \pi \cdot r \cdot r \cdot h$$
$$\text{or} \quad V = \pi r^2 h$$

where r is the radius of the circular base and h is the height of the cylinder.

Volume is measured in cubic units.

Find the volume of this cylinder.

First, find the radius. $\quad r = \dfrac{8\text{ m}}{2} = 4\text{ m}$

$$V = \pi \cdot r \cdot r \cdot h$$
$$V \approx 3.14 \cdot 4\text{ m} \cdot 4\text{ m} \cdot 10\text{ m}$$
$$V \approx \mathbf{502.4\ m^3} \leftarrow \text{Cubic units for volume}$$

Concepts	Examples

5.10 Surface Area of a Rectangular Solid

Use this formula to find the surface area of a rectangular solid.

Surface area = $(2 \cdot l \cdot w) + (2 \cdot l \cdot h) + (2 \cdot w \cdot h)$

or $S = 2lw + 2lh + 2wh$

where l is the length, w is the width, and h is the height of the solid.

Surface area is measured in square units.

Find the surface area of this packing crate.

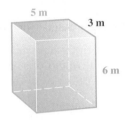

5 m 3 m 6 m

$S = 2lw + 2lh + 2wh$

$S = (2 \cdot 5\ m \cdot 3\ m) + (2 \cdot 5\ m \cdot 6\ m) + (2 \cdot 3\ m \cdot 6\ m)$

$S = 30\ m^2 + 60\ m^2 + 36\ m^2$

$S = 126\ m^2$ ← Square units for surface area

5.10 Surface Area of a Cylinder

Use this formula to find the surface area of a cylinder.

Surface area = $(2 \cdot \pi \cdot r \cdot h) + (2 \cdot \pi \cdot r \cdot r)$

or $S = 2\pi rh + 2\pi r^2$

where r is the radius of the circular base and h is the height of the cylinder. Use 3.14 as the approximate value of π.

Surface area is measured in square units.

Find the surface area of a cylindrical hot water tank with a height of 4.5 ft and a diameter of 1.8 ft. Round your answer to the nearest tenth.

First, find the radius. $r = \dfrac{1.8\ ft}{2} = 0.9\ ft$

$S = 2\pi rh + 2\pi r^2$

$S \approx (2 \cdot 3.14 \cdot 0.9\ ft \cdot 4.5\ ft) + (2 \cdot 3.14 \cdot 0.9\ ft \cdot 0.9\ ft)$

$S \approx 25.434\ ft^2 + 5.0868\ ft^2$

$S \approx 30.5208\ ft^2$ Now round to tenths.

$S \approx 30.5\ ft^2$ ← Square units for surface area

Chapter 5 Review Exercises

5.1 *Name the digit that has the given place value.*

1. 243.059

tenths

hundredths

2. 0.6817

ones

tenths

3. $5824.39

hundreds

hundredths

4. 896.503

tenths

tens

5. 20.73861

tenths

ten-thousandths

Write each decimal as a fraction or mixed number in lowest terms.

6. 0.5

7. 0.75

8. 4.05

9. 0.875

10. 0.027

11. 27.8

Write each decimal in words.

12. 0.8

13. 400.29

14. 12.007

15. 0.0306

Write each decimal in numbers.

16. Eight and three tenths

17. Two hundred five thousandths

18. Seventy and sixty-six ten-thousandths

19. Thirty hundredths

5.2 *Round to the place indicated.*

20. 275.635 to the nearest tenth

21. 72.789 to the nearest hundredth

22. 0.1604 to the nearest thousandth

23. 0.0905 to the nearest thousandth

24. 0.98 to the nearest tenth

Round each money amount to the nearest cent.

25. $15.8333

26. $0.698

27. $17,625.7906

Round each income or expense item to the nearest dollar.

28. The income from the pancake breakfast was $350.48.

29. Each member paid $129.50 in dues.

30. The refreshments cost $99.61.

31. The bank charges were $29.37.

5.3 *Find each sum or difference.*

32. $0.4 - 6.07$

33. $-20 + 19.97$

34. $-1.35 + 7.229$

35. $0.005 + (3 - 9.44)$

First, use front end rounding to estimate each answer. Then find the exact answer.

36. American's favorite outdoor activity is walking. About 151.6 million adults walk for 10 minutes or more on a regular basis. Jogging is second with about 38.6 million adults, and biking is fourth with 31.3 million adults. How many more people go walking than biking? (Data from Outdoor Foundation, CDC.)

Estimate:

Exact:

37. Today, Jasmin had $406 in her bank account. She paid $315.53 to the day care center and $74.67 for groceries. What is the new balance in her account?

Estimate:

Exact:

38. Joey spent $1.59 for toothpaste, $5.33 for vitamins, and $18.94 for a toaster. He gave the clerk three $10 bills. How much change did he get?

Estimate:

Exact:

39. Roseanne is training for a wheelchair race. She raced 2.3 kilometers on Monday, 4 kilometers on Wednesday, and 5.25 kilometers on Friday. How far did she race altogether?

Estimate:

Exact:

5.4 *First, use front end rounding to estimate each answer. Then find the exact answer.*

40. *Estimate:*   *Exact:*

$$\begin{array}{r} 6.138 \\ \times\ \ 3.7 \\ \hline \end{array}$$

$\times\ \underline{\hspace{1cm}}$

41. *Estimate:*   *Exact:*

$$\begin{array}{r} 42.9 \\ \times\ 3.3 \\ \hline \end{array}$$

$\times\ \underline{\hspace{1cm}}$

Find each product.

42. $(-5.6)(-0.002)$

43. $(0.071)(-0.005)$

5.5 *Decide if each answer is* reasonable *or* unreasonable *by rounding the numbers and estimating the answer. If the exact answer is* unreasonable, *find and correct the error.*

44. $706.2 \div 12 = 58.85$

Estimate:

45. $26.6 \div 2.8 = 0.95$

Estimate:

Divide. Round quotients to the nearest thousandth when necessary.

46. $3\overline{)43.4}$

47. $\dfrac{-72}{-0.06}$

48. $-0.00048 \div 0.0012$

5.4–5.5 *Solve each application problem.*

49. Adrienne worked 46.5 hours this week. Her hourly wage is $14.24 for the first 40 hours and 1.5 times that rate over 40 hours. Find her total earnings. Round your final answer to the nearest dollar.

50. A book of 30 tickets costs $59.75 at the State Fair midway. What is the cost per ticket? Round your final answer to the nearest cent.

51. Stock in Math Tronic sells for $3.75 per share. Kenneth is thinking of investing $500. How many whole shares could he buy?

52. Grapes are on sale at $0.99 per pound. How much will Ms. Lee pay for 3.5 pounds of grapes? Round your final answer to the nearest cent.

Simplify.

53. $3.5^2 + 8.7(-1.95)$

54. $11 - 3.06 \div (3.95 - 0.35)$

5.6 *Write each fraction as a decimal. Round to the nearest thousandth when necessary.*

55. $3\dfrac{4}{5}$

56. $\dfrac{16}{25}$

57. $1\dfrac{7}{8}$

58. $\dfrac{1}{9}$

Arrange each group of numbers in order from least to greatest.

59. 3.68, 3.806, 3.6008

60. 0.215, 0.22, 0.209, 0.2102

61. $0.17, \dfrac{3}{20}, \dfrac{1}{8}, 0.159$

5.7 *Find the mean and the median for each set of data.*

62. Smartphones sold each day:
18, 12, 15, 24, 9, 42, 54, 87, 21, 3

63. Number of student loans processed:
54, 28, 35, 43, 17, 37, 68, 75, 39

64. Find the weighted mean.

Dollar Value	Frequency
$42	3
$47	7
$53	2
$55	3
$59	5

65. Find the mode or modes for each set of data.

Hiking boots at store J priced at
$107, $69, $139, $107, $160, $84, $160

Hiking boots at store K priced at
$119, $136, $99, $119, $139, $119, $95

5.8 *Find the unknown length in each right triangle. Use a calculator to find square roots. Round your answers to the nearest tenth when necessary.*

66.
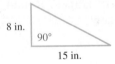
8 in.
90°
15 in.

67.

24 cm
25 cm

68.

15 cm
90°
11 cm

69.
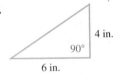
4 in.
90°
6 in.

70.

2.2 m
1.3 m

71.

12 km
8.5 km

5.9 *Solve each equation. Show your work.*

72. $-0.1 = b - 0.35$

73. $-3.8x = 0$

74. $6.8 + 0.4n = 1.6$

75. $-0.375 + 1.75a = 2a$

76. $0.3y - 5.4 = 2.7 + 0.8y$

5.10 *Find the unknown radius or diameter.*

77. The radius of a circular irrigation field is 68.9 m. What is the diameter of the field?

78. The diameter of a juice can is 3 in. What is the radius of the can?

Find the circumference and area of each circle. Use 3.14 as the approximate value for π. Round your final answers to the nearest tenth.

79.

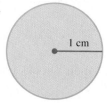

1 cm

80.

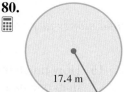

17.4 m

81.

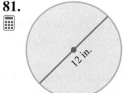

12 in.

Find the volume and surface area of each solid. Use 3.14 as the approximate value for π. Round your final answers to the nearest tenth when necessary.

82.

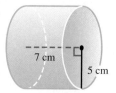

7 cm

5 cm

83.

24 m   4 m

84. A rectangular cooler that measures 3.5 ft by 1.5 ft and is 1.5 ft high

Chapter 5   Mixed Review Exercises

Practicing exercises in mixed-up order helps you prepare for tests.

Simplify.

1. $89.19 + 0.075 + 310.6 + 5$

2. $72.8\,(-3.5)$

3. $1648.3 \div 0.46$
▦ Round to thousandths.

4. $30 - 0.9102$

5. $(4.38)\,(0.007)$

6. $0.005\overline{)0.047}$

7. $72.105 + 8.2 - 95.37$

8. $\dfrac{-81.36}{9}$

9. $(0.6 - 1.22) + 4.8\,(-3.15)$

10. $0.455\,(18)$

11. $(-1.6)\,(-0.58)$

12. $0.218\overline{)7.63}$
▦

13. $-21.059 - 20.8$

14. $18.3 - 3^2 \div 0.5$

Use the information in the ad to solve Exercises 15–17. Round money answers to the nearest cent. (Disregard any sales tax.)

15. How much more would one pair of men's socks cost than one pair of children's socks?

16. How much would Akiko pay for five pairs of teen jeans and four pairs of women's jeans?

17. What is the difference between the cheapest sale price for athletic shoes and the highest regular price?

Solve each equation. Show your work.

18. $4.62 = -6.6y$

19. $1.05x - 2.5 = 0.8x + 5$

Solve each application problem. Round final answers to the nearest tenth when necessary.

20. A circular table has a diameter of 5 ft. What length ▦ of rubber strip is needed to go around the edge of the table? What is the area of the tabletop?

21. Jerry missed one math test, so his test scores are 82, 0, 78, 93, 85. Find his mean and median scores.

22. LaRae drove 16 miles south, then made a 90° right ▦ turn and drove 12 miles west. How far is she, in a straight line, from her starting point?

23. A juice can that is 7 in. tall has a diameter of 3 in. ▦ What is the volume of the can?

Chapter 5 *Test*

The Chapter Test Prep Videos with step-by-step solutions are available in MyMathLab or on YouTube at **https://goo.gl/c3befo**

Write each decimal as a fraction or mixed number in lowest terms.

1. 18.4

2. 0.075

Write each decimal in words.

3. 60.007

4. 0.0208

First, use front end rounding to round each number and estimate the answer. Then find the exact answer.

5. $7.6 + 82.0128 + 39.59$

Estimate:

Exact:

6. $-5.79\,(1.2)$

Estimate:

Exact:

7. $-79.1 - 3.602$

Estimate:

Exact:

8. $-20.04 \div (-4.8)$

Estimate:

Exact:

Find the exact answer.

9. $670 - 0.996$

10. $0.15\overline{)72}$

11. $(-0.006)\,(-0.007)$

Solve each application problem.

12. Pat bought 6.5 ft of decorative gold chain to hang a light over her dining table. She paid \$24.64. What was the cost per foot, to the nearest cent?

13. Davida ran a race in 3.059 minutes. Angela ran the race in 3.5 minutes. Who won? By how much?

14. Mr. Yamamoto bought 1.85 pounds of cheese at \$2.89 per pound. What was the total amount he paid, to the nearest cent?

Solve each equation. Show your work.

15. $-5.9 = y + 0.25$

16. $-4.2x = 1.47$

17. $3a - 22.7 = 10$

18. $-0.8n + 1.88 = 2n - 6.1$

19. Arrange in order from least to greatest:

$$0.44, 0.451, \frac{9}{20}, 0.4506$$

20. Simplify: $6.3^2 - 5.9 + 3.4(-0.5)$

21. Find the mean number of books loaned:
52, 61, 68, 69, 73, 75, 79, 84, 91, 98

22. Find the mode for hot tub temperatures (Fahrenheit) of 96°, 104°, 103°, 104°, 103°, 104°, 91°, 74°, 103°

23. Find the weighted mean.

Cost	Frequency
$ 6	7
$10	3
$11	4
$14	2
$19	3
$24	1

24. Find the median cost of an online textbook:
$54.50, $48, $39.75, $89, $56.25, $49.30, $46.90, $51.80

Find the unknown lengths. Round your answers to the nearest tenth when necessary.

25.
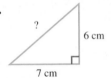
? 6 cm 7 cm

26.

11 ft ? 20 ft

Use 3.14 as the approximate value for π. Round your final answers to the nearest tenth when necessary.

27. Find the radius.

25 in.

28. Find the circumference.

0.9 km

29. Find the area.

16.2 cm

30. Find the volume.

18 ft
5 ft

31. Find the surface area of the solid in **Problem 30.**

Chapters 1–5   *Cumulative Review Exercises*

1. Write these numbers in words.

   (a) 45.0203

   (b) 30,000,650,008

2. Write these numbers using digits.

   (a) One hundred sixty million, five hundred

   (b) Seventy-five thousandths

Round each number as indicated.

3. 46,908 to the nearest hundred

4. 6.197 to the nearest hundredth

5. 0.66148 to the nearest thousandth

6. 9951 to the nearest hundred

Simplify.

7. $-5 - 8$

8. $-0.003\,(0.02)$

9. $\dfrac{-7}{0}$

10. $8 + 4\,(2 - 5)^2$

11. $-\dfrac{3}{8}(-48)$

12. $|4| - |-10|$

13. $0.721 + 55.9$

14. $3\dfrac{1}{3} - 1\dfrac{5}{6}$

15. $\left(\dfrac{3}{b^2}\right)\left(\dfrac{b}{8}\right)$

16. $12 - 0.853$

17. $\dfrac{3}{10} - \dfrac{3}{4}$

18. $\dfrac{\frac{5}{16}}{-10}$

19. $-3.75 \div (-2.9)$
Round your answer to the nearest tenth.

20. $\dfrac{x}{2} + \dfrac{3}{5}$

21. $5 \div \left(-\dfrac{5}{8}\right)$

22. $2^5 - 4^3$

23. $\dfrac{-36 \div (-2)}{-6 - 4\,(0 - 6)}$

24. $(0.8)^2 - 3.2 + 4\,(-0.8)$

25. $\dfrac{3}{4} \div \dfrac{3}{10}\left(\dfrac{1}{4} + \dfrac{2}{3}\right)$

The ages (in years) of the students in a math class are: 19, 23, 24, 19, 20, 29, 26, 35, 20, 22, 26, 23, 25, 26, 20, 30. *Use this data for Exercises 26–28.*

26. Find the mean age, to the nearest tenth.

27. Find the median age.

28. Find the mode.

Solve each equation. Show your work.

29. $3h - 4h = 16 - 12$

30. $-2x = x - 15$

31. $20 = 6r - 45.4$

32. $-3(y + 4) = 7y + 8$

33. $-0.8 + 1.4n = 2 + 0.7n$

Find the unknown length, perimeter, area, or volume. Use 3.14 as the approximate value of π and round your final answers to the nearest tenth.

34. Find the perimeter and the area.

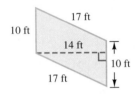

35. Find the circumference and the area.

36. Find the length of the third side and the area.

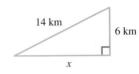

37. Find the volume and surface area.

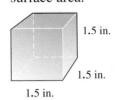

First, use front end rounding to estimate each answer. Then find the exact answer.

38. Lameck had three $20 bills. He spent $47.96 on gasoline and $0.87 for a candy bar at the convenience store. How much money does he have left?

Estimate:

Exact:

39. Toshihiro bought $2\frac{1}{3}$ yards of cotton fabric and $3\frac{7}{8}$ yards of wool fabric. How many yards of fabric did he buy in all? Write your answer as a mixed number.

Estimate:

Exact:

40. Paulette bought 2.7 pounds of strawberries for $6.18. What was the cost per pound, to the nearest cent?

Estimate:

Exact:

41. Carter Community College received a $78,000 grant from a local hospital to help students pay tuition for nursing classes. How much money could be given to each of 107 students? Round to the nearest dollar.

Estimate:

Exact:

Solve each application problem using the six problem-solving steps. Show your work.

42. A $30,000 scholarship is being divided between two students so that one student receives three times as much as the other. How much will each student receive?

43. The perimeter of a rectangular game board is 48 in. The length is 4 in. more than the width. Find the length and width of the game board.

6

Ratio, Proportion, and Line/Angle/ Triangle Relationships

In the United States, 8 out of 10 adults use unit prices to find the best buy when shopping. In this chapter you will learn how to use rates, ratios, and proportions to compare prices and save money on everything from data plans to dinner menus. (Data from *Consumer Reports.*)

6.1 Ratios

A **ratio** compares two quantities. You can compare two numbers, such as 8 and 4, or two measurements that have the *same* type of units, such as 3 days and 12 days. (*Rates* compare measurements with different types of units and are covered in the next section.)

Ratios can help you see important relationships. For example, if the ratio of your monthly expenses to your monthly income is 10 to 9, then you are spending $10 for every $9 you earn and going deeper into debt.

OBJECTIVE ▶ 1 Write ratios as fractions. A ratio can be written in three ways.

Writing a Ratio

The ratio of $7 to $3 can be written as follows.

$$7 \text{ to } 3 \quad \text{or} \quad 7{:}3 \quad \text{or} \quad \frac{7}{3} \leftarrow \text{Fraction bar indicates "to"}$$

"**:**" indicates "**to**"

Writing a ratio as a fraction is the most common method, and the one we will use here. All three ways are read, "the ratio of 7 to 3." The word to separates the quantities being compared.

Writing a Ratio as a Fraction

Order is important when you're writing a ratio. The quantity mentioned **first** is the **numerator**. The quantity mentioned second is the denominator. For example:

The ratio of **5** to **12** is written $\dfrac{5}{12}$

EXAMPLE 1   Writing Ratios

Ancestors of the Pueblo Indians built multistory apartment towns in New Mexico about 1100 years ago. A room might measure 14 feet long, 11 feet wide, and 15 feet high.

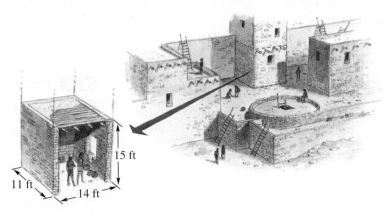

Continued on Next Page

Write each ratio as a fraction, using the room measurements.

(a) Ratio of length to width

> Do **not** rewrite the ratio as $1\frac{3}{11}$

The ratio of **length** to **width** is $\dfrac{14 \text{ feet}}{11 \text{ feet}} = \dfrac{14}{11}$

↗ Numerator (mentioned first)   ↖ Denominator (mentioned second)

You can divide out common *units* just as you divided out common *factors* when writing fractions in lowest terms. However, do *not* rewrite the fraction as a mixed number. Keep it as the ratio of 14 to 11.

(b) The ratio of width to height is $\dfrac{11 \text{ feet}}{15 \text{ feet}} = \dfrac{11}{15}$

> Divide out the common units (ft).

❗ CAUTION

Remember, the *order* of the numbers is important in a ratio. Look for the words "ratio of *a* to *b*." Write the ratio as $\dfrac{a}{b}$, **not** $\dfrac{b}{a}$. The quantity mentioned first is the numerator.

—————— **Work Problem ❶ at the Side.** ▶

You can write a ratio in *lowest terms,* just as you do with any fraction.

EXAMPLE 2   Writing Ratios in Lowest Terms

Write each ratio as a fraction in lowest terms.

(a) 60 days of sun to 20 days of rain

The ratio is $\frac{60 \text{ days}}{20 \text{ days}}$. Divide out the common units. Then write this ratio in lowest terms by dividing the numerator and the denominator by 20.

$$\frac{60 \text{ days}}{20 \text{ days}} = \frac{60}{20} = \frac{60 \div 20}{20 \div 20} = \frac{3}{1} \quad \left\{ \begin{array}{l} \text{Ratio in} \\ \text{lowest terms} \end{array} \right.$$

So, the ratio of 60 days to 20 days is 3 to 1, or, written as a fraction, $\frac{3}{1}$. In other words, for every 3 days of sun, there is 1 day of rain.

❗ CAUTION

With fractions, you would rewrite $\frac{3}{1}$ as 3. But a *ratio* compares **two** quantities, so you need to keep both parts of the ratio and write it as $\frac{3}{1}$.

(b) 50 ounces of medicine to 120 ounces of water

The ratio is $\frac{50 \text{ ounces}}{120 \text{ ounces}}$. Divide out the common units. Then divide the numerator and the denominator by 10.

$$\frac{50 \text{ ounces}}{120 \text{ ounces}} = \frac{50}{120} = \frac{50 \div 10}{120 \div 10} = \frac{5}{12} \quad \left\{ \begin{array}{l} \text{Ratio in} \\ \text{lowest terms} \end{array} \right.$$

So, the ratio of 50 ounces to 120 ounces is $\frac{5}{12}$. In other words, for every 5 ounces of medicine, there is 12 ounces of water.

—————— **Continued on Next Page**

---

❶ Shane spent $14 on meat, $5 on milk, and $7 on fresh fruit. Write the following ratios as fractions.

ɢꜱ (a) The ratio of amount spent on fruit to amount spent on milk

Spent on fruit ⟶ ☐
Spent on milk ⟶ 5

(b) The ratio of amount spent on milk to amount spent on meat

(c) The ratio of amount spent on meat to amount spent on milk

Answers

1. (a) $\dfrac{7}{5}$ **(b)** $\dfrac{5}{14}$ **(c)** $\dfrac{14}{5}$

2 Write each ratio as a fraction in lowest terms.

(a) 9 hours to 12 hours

$$\frac{9 \text{ hours}}{12 \text{ hours}} = \frac{9}{12} = \frac{9 \div 3}{12 \div 3} = \frac{\square}{\square}$$

(b) 100 meters to 50 meters

(c) The ratio of width to length for this rectangle

Length
48 ft

Width
24 ft

3 Write each ratio as a ratio of whole numbers in lowest terms.

(a) The price of Tamar's favorite brand of lipstick increased from $5.50 to $9.00. Find the ratio of the increase in price to the original price.

(b) Last week, Lance worked 4.5 hours each day. This week he cut back to 3 hours each day. Find the ratio of the decrease in hours to the original number of hours.

(c) 15 people in a large van to 6 people in a small van

$$\text{The ratio is } \frac{15}{6} = \frac{15 \div 3}{6 \div 3} = \frac{5}{2} \longleftarrow \left\{ \begin{array}{l} \text{Ratio in} \\ \text{lowest terms} \end{array} \right.$$

Note

Although $\frac{5}{2} = 2\frac{1}{2}$, ratios are *not* written as mixed numbers. Nevertheless, in **Example 2(c)** above, the ratio $\frac{5}{2}$ does mean the large van holds $2\frac{1}{2}$ times as many people as the small van.

◀ **Work Problem 2** at the Side.

OBJECTIVE ▶ **2** **Solve ratio problems involving decimals or mixed numbers.** Sometimes a ratio compares two decimal numbers or two fractions. It is easier to understand if we rewrite the ratio as a ratio of two whole numbers.

EXAMPLE 3 **Using Decimal Numbers in a Ratio**

The price of a Sunday newspaper increased from $1.50 to $1.75. Find the ratio of the <u>increase in price</u> to the <u>original price</u>.

The words <u>increase in price</u> are mentioned first, so the increase will be the numerator. How much did the price go up? Use subtraction.

new price − original price = increase

$1.75 − $1.50 = $0.25

Subtract first to find how much the price went up.

The words <u>original price</u> are mentioned second, so the original price of $1.50 is the denominator.

The ratio of <u>increase in price</u> to <u>original price</u> is shown below.

$$\frac{0.25}{1.50} \begin{array}{l} \leftarrow \text{increase in price} \\ \leftarrow \text{original price} \end{array}$$

Now rewrite the ratio as a ratio of whole numbers. Recall that if you multiply both the numerator and denominator of a fraction by the same number, you get an equivalent fraction. The decimals in this example are hundredths, so multiply by 100 to get whole numbers. (If the decimals are tenths, multiply by 10. If thousandths, multiply by 1000.) Then write the ratio in lowest terms.

$$\frac{0.25}{1.50} = \frac{(0.25)(100)}{(1.50)(100)} = \frac{25}{150} = \frac{25 \div 25}{150 \div 25} = \frac{1}{6} \longleftarrow \left\{ \begin{array}{l} \text{Ratio in} \\ \text{lowest terms} \end{array} \right.$$

Ratio as two whole numbers

◀ **Work Problem 3** at the Side.

EXAMPLE 4 **Using Mixed Numbers in Ratios**

Write each ratio as a comparison of whole numbers in lowest terms.

(a) 2 days to $2\frac{1}{4}$ days

Write the ratio as follows. Divide out the common units.

$$\frac{2 \text{ days}}{2\frac{1}{4} \text{ days}} = \frac{2}{2\frac{1}{4}}$$

Answers

2. (a) $\frac{3}{4}$ **(b)** $\frac{2}{1}$ **(c)** $\frac{1}{2}$

3. (a) $\frac{(3.50)(100)}{(5.50)(100)} = \frac{350 \div 50}{550 \div 50} = \frac{7}{11}$

(b) $\frac{(1.5)(10)}{(4.5)(10)} = \frac{15 \div 15}{45 \div 15} = \frac{1}{3}$

Continued on Next Page

Next, write 2 as $\frac{2}{1}$ and $2\frac{1}{4}$ as the improper fraction $\frac{9}{4}$.

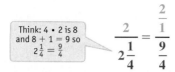

Think: 4 · 2 is 8
and 8 + 1 = 9 so
$2\frac{1}{4} = \frac{9}{4}$

$$\frac{2}{2\frac{1}{4}} = \frac{\frac{2}{1}}{\frac{9}{4}}$$

Now rewrite the problem in horizontal format, using the "÷" symbol for division. Finally, multiply by the reciprocal of the divisor.

$$\frac{\frac{2}{1}}{\frac{9}{4}} = \frac{2}{1} \div \frac{9}{4} = \frac{2}{1} \cdot \frac{4}{9} = \frac{8}{9} \quad \left\{ \begin{array}{l} \text{Ratio in} \\ \text{lowest terms} \end{array} \right.$$

Reciprocals

The ratio, in lowest terms, is $\frac{8}{9}$

(b) $3\frac{1}{4}$ to $1\frac{1}{2}$

Write the ratio as $\dfrac{3\frac{1}{4}}{1\frac{1}{2}}$. Then write $3\frac{1}{4}$ and $1\frac{1}{2}$ as improper fractions.

Think: 4 · 3 = 12
and 12 + 1 = 13
so $3\frac{1}{4} = \frac{13}{4}$

$$3\frac{1}{4} = \frac{13}{4} \quad \text{and} \quad 1\frac{1}{2} = \frac{3}{2}$$

Think: 2 · 1 = 2
and 2 + 1 = 3
so $1\frac{1}{2} = \frac{3}{2}$

The ratio is shown below.

$$\frac{3\frac{1}{4}}{1\frac{1}{2}} = \frac{\frac{13}{4}}{\frac{3}{2}}$$

Rewrite as a division problem in horizontal format, using the "÷" symbol. Then multiply by the reciprocal of the divisor.

$$\frac{13}{4} \div \frac{3}{2} = \frac{13}{4} \cdot \frac{2}{3} = \frac{13 \cdot \overset{1}{2}}{\underset{1}{2} \cdot 2 \cdot 3} = \frac{13}{6} \quad \left\{ \begin{array}{l} \text{Ratio in} \\ \text{lowest terms} \end{array} \right.$$

Reciprocals

> **Note**
>
> We can also work **Examples 4(a) and 4(b)** above by using decimals.
>
> **(a)** $2\frac{1}{4}$ is equivalent to 2.25, so we have the ratio shown below.
>
> $$\frac{2}{2\frac{1}{4}} = \frac{2}{2.25} = \frac{(2)(100)}{(2.25)(100)} = \frac{200}{225} = \frac{200 \div 25}{225 \div 25} = \frac{8}{9} \quad \leftarrow \text{Same result}$$
>
> **(b)** $3\frac{1}{4}$ is equivalent to 3.25 and $1\frac{1}{2}$ is equivalent to 1.5.
>
> $$\frac{3\frac{1}{4}}{1\frac{1}{2}} = \frac{3.25}{1.5} = \frac{(3.25)(100)}{(1.5)(100)} = \frac{325}{150} = \frac{325 \div 25}{150 \div 25} = \frac{13}{6} \quad \leftarrow \text{Same result}$$
>
> This method would *not* work for fractions that are repeating decimals, such as $\frac{1}{3}$ or $\frac{5}{6}$.

— **Work Problem ④ at the Side.** ▶

---

④ Write each ratio as a ratio of whole numbers in lowest terms.

ⒼⓈ **(a)** $3\frac{1}{2}$ to 4

$$\frac{3\frac{1}{2}}{4} = \frac{\frac{7}{2}}{\frac{4}{1}} =$$

$$\frac{7}{2} \div \frac{4}{1} = \frac{7}{2} \cdot \frac{1}{4} = \frac{\Box}{\Box}$$

Reciprocals

(b) $5\frac{5}{8}$ pounds to $3\frac{3}{4}$ pounds

(c) $3\frac{1}{3}$ inches to $\frac{5}{6}$ inch

Answers

4. (a) $\dfrac{7}{8}$ (b) $\dfrac{3}{2}$ (c) $\dfrac{4}{1}$

5 Write each ratio as a fraction in lowest terms. (*Hint:* Recall that it is usually easier to write the ratio using the smaller measurement unit.)

(a) 9 inches to 6 feet

Change 6 ft to inches

$6 \cdot 12 \text{ in.} = \underline{\hspace{2cm}} \text{ in.}$

$$\frac{9 \text{ in.}}{6 \text{ ft}} = \frac{9 \text{ in.}}{\boxed{} \text{ in.}} = \frac{9 \div \boxed{}}{\boxed{} \div \boxed{}}$$

$$= \frac{\boxed{}}{\boxed{}}$$

(b) 2 days to 8 hours

(c) 7 yards to 14 feet

(d) 3 quarts to 3 gallons

(e) 25 minutes to 2 hours

(f) 4 pounds to 12 ounces

OBJECTIVE **3** **Solve ratio problems after converting units.** When a ratio compares measurements, both measurements must be in the *same* units. For example, *feet* must be compared to *feet*, *hours* to *hours*, and so on.

EXAMPLE 5 Ratio Applications Using Measurement

(a) Write the ratio of the length of the shorter board on the left to the length of the longer board on the right. Compare in inches.

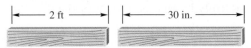

First, express 2 feet in inches. Because 1 foot has 12 inches, 2 feet is

$$2 \cdot 12 \text{ inches} = 24 \text{ inches}$$

The length of the board on the left is 24 inches, so the ratio of the lengths is shown below. The common units divide out.

Once the units match, you can divide them out.

$$\frac{2 \text{ ft}}{30 \text{ in.}} = \frac{24 \text{ inches}}{30 \text{ inches}} = \frac{24}{30}$$

Write the ratio in lowest terms.

$$\frac{24}{30} = \frac{24 \div 6}{30 \div 6} = \frac{4}{5} \quad \left\{ \begin{array}{l} \text{Ratio in} \\ \text{lowest terms} \end{array} \right.$$

The **shorter board on the left is $\frac{4}{5}$ the length of the longer board** on the right.

Note

Notice in the example above that we wrote the ratio using the smaller unit (inches are smaller than feet). Using the smaller unit will help you avoid working with fractions.

(b) Write the ratio of 28 days to 3 weeks.

It is usually easier to write the ratio using the smaller measurement so compare in *days* (because days are shorter than weeks).

First express 3 weeks in days. Because 1 week has 7 days, 3 weeks is

$$3 \cdot 7 \text{ days} = 21 \text{ days}$$

The ratio in days is shown below.

$$\frac{28 \text{ days}}{3 \text{ weeks}} = \frac{28 \text{ days}}{21 \text{ days}} = \frac{28}{21} = \frac{28 \div 7}{21 \div 7} = \frac{4}{3} \quad \left\{ \begin{array}{l} \text{Ratio in} \\ \text{lowest terms} \end{array} \right.$$

Use the table below to help set up ratios that compare measurements.

Measurement Comparisons

Length	Weight
12 inches = 1 foot	16 ounces = 1 pound
3 feet = 1 yard	2000 pounds = 1 ton
5280 feet = 1 mile	
	Time
Capacity (Volume)	60 seconds = 1 minute
2 cups = 1 pint	60 minutes = 1 hour
2 pints = 1 quart	24 hours = 1 day
4 quarts = 1 gallon	7 days = 1 week

◀ **Work Problem** **5** **at the Side.**

6.1 Exercises

FOR EXTRA HELP Go to MyMathLab *for worked-out, step-by-step solutions to exercises enclosed in a square* ■ *and video solutions to* ▶ *exercises.*

1. **CONCEPT CHECK** What is a ratio? Write an explanation. Then make up an example of a ratio.

2. **CONCEPT CHECK** Both quantities in a ratio have the same units. That means you can do what to the units?

3. **CONCEPT CHECK** To rewrite the ratio $\frac{80}{20}$ in lowest terms, divide both the numerator and the denominator by _____ . What is the ratio in lowest terms?

4. **CONCEPT CHECK** To rewrite the ratio $\frac{20}{75}$ in lowest terms, divide both the numerator and the denominator by _____ . What is the ratio in lowest terms?

Write each ratio as a fraction in lowest terms. ***See Examples 1 and 2.***

5. 8 days to 9 days

6. $11 to $15

7. $100 to $50

8. 35¢ to 7¢

9. 30 minutes to 90 minutes
▶

10. 9 pounds to 36 pounds

Write each ratio as a ratio of whole numbers in lowest terms. ***See Examples 3 and 4.***

11. $4.50 to $3.50

12. $0.08 to $0.06

13. $1\frac{1}{4}$ to $1\frac{1}{2}$
▶

14. $2\frac{1}{3}$ to $2\frac{2}{3}$

15. **CONCEPT CHECK** The work shown below has a mistake in it.

    What Went Wrong? First write a sentence explaining what the mistake is. Then fix the mistake and find the correct solution.

    Chander simplified the ratio of 15 inches to 30 inches this way:

    $$\frac{15 \cancel{\text{ in.}}}{30 \cancel{\text{ in.}}} = \frac{15 \div 15}{30 \div 15} = 2$$

16. **CONCEPT CHECK** The work shown below has a mistake in it.

    What Went Wrong? First write a sentence explaining what the mistake is. Then fix the mistake and find the correct solution.

    Reena simplified the ratio of $5\frac{1}{3}$ to $1\frac{1}{3}$ this way:

    $$\frac{5\frac{\cancel{1}}{\cancel{3}}}{1\frac{\cancel{1}}{\cancel{3}}} = \frac{5}{1}$$

Write each ratio as a fraction in lowest terms. For help, use the table of measurement comparisons on the previous page. ***See Example 5.***

17. 4 feet to 30 inches
GS First convert 4 ft to inches.
    4 • 12 in. = _____ in.

18. 8 feet to 4 yards
GS First convert 4 yd to feet.
    4 • 3 ft = _____ ft

19. 5 minutes to 1 hour

20. 3 pounds to 6 ounces

21. 5 gallons to 5 quarts
▶

22. 3 cups to 3 pints

The table shows the number of greeting cards that Americans buy for various occasions. Use the information to answer Exercises 23–26. Write each ratio as a fraction in lowest terms.

Holiday/Event	Cards Sold
Valentine's Day	145 million
Mother's Day	130 million
Father's Day	90 million
Graduation	70 million
Halloween	20 million
St. Patrick's Day	10 million

Data from Hallmark Cards.

23. Find the ratio of St. Patrick's Day cards to graduation cards.

24. Find the ratio of Halloween cards to Mother's Day cards.

25. Find the ratio of Mother's Day cards to St. Patrick's Day cards.

26. Find the ratio of Valentine's Day cards to Father's Day cards.

For each figure in Exercises 27–30, find the ratio of the length of the longest side to the length of the shortest side. Write each ratio as a fraction in lowest terms.
See Examples 2–4.

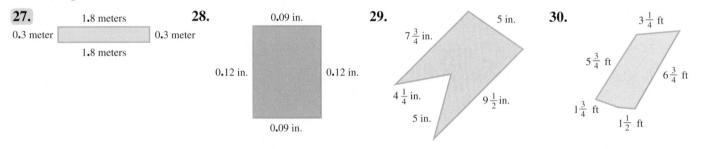

27.

1.8 meters

0.3 meter _____ 0.3 meter

1.8 meters

28.

0.09 in.

0.12 in.   0.12 in.

0.09 in.

29.

$7\frac{3}{4}$ in.

5 in.

$4\frac{1}{4}$ in.

$9\frac{1}{2}$ in.

5 in.

30.

$3\frac{1}{4}$ ft

$5\frac{3}{4}$ ft

$6\frac{3}{4}$ ft

$1\frac{3}{4}$ ft

$1\frac{1}{2}$ ft

Write each ratio as a fraction in lowest terms.

31. An auto mechanic used to pay $10 for the 5 quarts of oil he needs to do an oil change. Now he pays $12.50. Find the ratio of the increase in price to the original price.

32. The price of an antibiotic decreased from $8.80 to $5.60 for 10 tablets. Find the ratio of the decrease in price to the original price.

33. The first time a movie was made in Minnesota, the cast and crew spent $59\frac{1}{2}$ days filming winter scenes. The next year, another movie was filmed in $8\frac{3}{4}$ weeks. Find the ratio of the first movie's filming time to the second movie's time. Compare in weeks.

34. The percheron, a large draft horse, measures about $5\frac{3}{4}$ feet at the shoulder. The prehistoric ancestor of the horse measured only $15\frac{3}{4}$ inches at the shoulder. Find the ratio of the percheron's height to its prehistoric ancestor's height. Compare in inches. (Data from *Eyewitness Books: Horse.*)

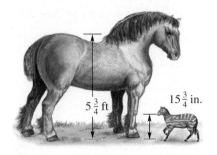

$5\frac{3}{4}$ ft

$15\frac{3}{4}$ in.

6.2 | Rates

A *ratio* compares two measurements with the **same** units, such as 9 feet to 12 feet (both length measurements). But many of the comparisons we make use measurements with **different** units, such as shown here.

160 dollars for 8 hours (money to time)

450 miles on 15 gallons (distance to capacity)

This type of comparison is called a **rate.**

OBJECTIVE ▶ **1** **Write rates as fractions.** Suppose that yesterday you hiked 14 miles in 4 hours. The *rate* at which you hiked can be written as a fraction in lowest terms.

$$\frac{14\ \text{miles}}{4\ \text{hours}} = \frac{14\ \text{miles} \div 2}{4\ \text{hours} \div 2} = \frac{7\ \text{miles}}{2\ \text{hours}} \Bigg\} \text{Rate in lowest terms}$$

In a rate, you often find one of these words separating the quantities you are comparing.

in   for   on   per   from

> ❗ **CAUTION**
>
> When writing a rate, always include the units, such as miles, hours, dollars, and so on. Because the units in a rate are different, the units do *not* divide out.

EXAMPLE 1 | **Writing Rates in Lowest Terms**

Write each rate as a fraction in lowest terms.

(a) 5 gallons of chemical for $60

$$\frac{5\ \text{gallons} \div 5}{60\ \text{dollars} \div 5} = \frac{\textbf{1 gallon}}{\textbf{12 dollars}}$$

> The units are different; you **cannot** divide them out.

(b) $4500 wages in 10 weeks

$$\frac{4500\ \text{dollars} \div 10}{10\ \text{weeks} \quad \div 10} = \frac{\textbf{450 dollars}}{\textbf{1 week}}$$

> Be sure to write the units in a *rate:* dollars, miles, gallons, etc.

(c) 2225 miles on 75 gallons of gas

$$\frac{2225\ \text{miles} \div 25}{75\ \text{gallons} \div 25} = \frac{\textbf{89 miles}}{\textbf{3 gallons}}$$

────── **Work Problem 1 at the Side.** ▶

OBJECTIVE ▶ **2** **Find unit rates.** When the *denominator* of a rate is 1, it is called a **unit rate.** We use unit rates frequently. For example, you earn $12.75 for *1 hour* of work. This unit rate is written:

$12.75 **per** hour   or $12.75 / hour

Use **per** or a / mark when writing unit rates.

---

1 Write rates as fractions.

2 Find unit rates.

3 Find the best buy based on cost per unit.

VOCABULARY TIP

Ratio vs. Rate

Ratios compare two quantities that have the **same** units, so the units divide out. But **rates** compare two quantities with **different** units, so you must write the units in a rate.

1 Write each rate as a fraction in lowest terms.

(a) $6 for 30 bars

$$\frac{\$6}{30\ \text{bars}} = \frac{\$6 \div \square}{30\ \text{bars} \div \square} = \frac{\$\square}{\square\ \text{bars}}$$

(b) 500 miles in 10 hours

(c) 4 teachers for 90 students

Answers

1. (a) $\dfrac{\$6}{30\ \text{bars}} = \dfrac{\$6 \div 6}{30\ \text{bars} \div 6} = \dfrac{\$1}{5\ \text{bars}}$

(b) $\dfrac{50\ \text{miles}}{1\ \text{hour}}$ **(c)** $\dfrac{2\ \text{teachers}}{45\ \text{students}}$

VOCABULARY TIP

Per means "for each." For example, if your car gets 30 miles **per** gallon, it gets 30 miles *for each gallon* of gas. Use **per** or a slash mark, like this: 30 miles/gallon.

2 Find each unit rate.

(a) $4.35 for 3 pounds of cheese

$$\frac{\$4.35}{3 \text{ pounds}} \leftarrow \begin{array}{l}\text{Fraction bar} \\ \text{indicates} \\ \text{division.}\end{array}$$

$$3\overline{)4.35} \quad \begin{array}{l}\text{Finish the} \\ \text{division.}\end{array}$$

(b) 304 miles on 9.5 gallons of gas

(c) $850 in 5 days

(d) 24-pound turkey for 15 people

Answers

2. **(a)** $1.45/pound **(b)** 32 miles/gallon
   (c) $170/day **(d)** 1.6 pounds/person

EXAMPLE 2 Finding Unit Rates

Find each unit rate.

(a) 337.5 miles on 13.5 gallons of gas

Write the rate as a fraction.

$$\frac{337.5 \text{ miles}}{13.5 \text{ gallons}} \leftarrow \text{The fraction bar indicates division.}$$

Divide 337.5 by 13.5 to find the unit rate.

$$13.5\overline{)337.5} \quad \to 2\,5.$$

$$\frac{337.5 \text{ miles} \div 13.5}{13.5 \text{ gallons} \div 13.5} = \frac{25 \text{ miles}}{1 \text{ gallon}} \quad \begin{array}{l}\text{Be sure to write} \\ \textbf{miles} \text{ and } \textbf{gallons} \\ \text{in your answer.}\end{array}$$

The unit rate is **25 miles** per **gallon**, or **25 miles/gallon**.

(b) 549 miles in 18 hours

$$\frac{549 \text{ miles}}{18 \text{ hours}} \quad \text{Divide.} \quad 18\overline{)549.0} \to 30.5$$

The unit rate is **30.5 miles** per **hour**, or **30.5 miles/hour**.

(c) $810 in 6 days

$$\frac{810 \text{ dollars}}{6 \text{ days}} \quad \text{Divide.} \quad 6\overline{)810} \to 135$$

Use *per* or a slash mark to write unit rates.

The unit rate is **$135** per **day**, or **$135/day**.

◀ **Work Problem 2 at the Side.**

OBJECTIVE 3 Find the best buy based on cost per unit. When shopping for groceries, household supplies, and health and beauty items, you will find many different brands and package sizes. You can save money by finding the lowest *cost per unit*.

Cost per Unit

Cost per unit is a rate that tells how much you pay for *one* item or *one* unit. Examples are $3.95 per gallon, $47 per shirt, and $2.98 per pound.

EXAMPLE 3 Finding the Best Buy

The local store charges the following prices for pancake syrup. Find the best buy.

12 fl oz	24 fl oz	36 fl oz
$1.28	$1.81	$2.73

Continued on Next Page

The best buy is the container with the *lowest* cost per unit. All the containers are measured in *ounces* (oz), so you first need to find the *cost per ounce* for each one. Divide the price of the container by the number of ounces in it. Round to the nearest thousandth if necessary.

Let the *order* of the *words* help you set up the rate.

cost ⟶ **$1.28**
per (means divide) → ⎯⎯⎯⎯⎯
ounce ⟶ **12 ounces**

Size	Cost per Unit (Rounded)
12 ounces	$\frac{\$1.28}{12 \text{ ounces}} \approx \$0.107$ per ounce (highest)
24 ounces	$\frac{\$1.81}{24 \text{ ounces}} \approx \$0.075$ per ounce (lowest)
36 ounces	$\frac{\$2.73}{36 \text{ ounces}} \approx \$0.076$ per ounce

The lowest cost per ounce is $0.075, so the **24-ounce container is the best buy.**

Note

Earlier we rounded money amounts to the nearest hundredth (nearest cent). But when comparing unit costs, you may need more decimal places to see the difference between very similar unit costs. Notice that the 24-ounce and 36-ounce syrup containers above both round to $0.08 per ounce if we round to hundredths.

⎯⎯⎯⎯⎯⎯⎯ **Work Problem ③ at the Side.** ▶

▦ Calculator Tip

When using a calculator to find unit prices, remember that division is *not* commutative. In **Example 3** above you wanted to find cost per ounce. Let the *order* of the *words* help you enter the numbers in the correct order.

cost   **per**   **ounce**
  ↓   ↓   ↓
Enter   Per means   Enter
the cost.   divide.   number
                             of ounces.
  ↓   ↓   ↓
$2.73   ÷   36   =   0.076 (rounded)

0.076 is the *cost* per *ounce*. This means it will cost $0.076 to buy **one** ounce. If you enter 36 ÷ 2.73 =, you will get the number of *ounces* per *dollar*. *Ounces per dollar* tells you how many ounces you can buy with **one** dollar.

Finding the best buy is sometimes a complicated process. Things that affect the cost per unit can include "cents off" coupons and differences in how much use you'll get out of each unit.

③ Find the best buy (lowest cost per unit) for each purchase.

(a) 2 quarts for $3.25
3 quarts for $4.95
4 quarts for $6.48

Divide to find each unit cost.

cost →
per → $\frac{\$3.25}{2 \text{ quarts}} = \$$____ per quart
quart →

cost →
per → $\frac{}{3 \text{ quarts}} = \$$____ per quart
quart →

Now find the unit cost for 4 quarts. Then compare. The *lowest* unit cost is the best buy. Which one is the best buy?

(b) 6 cans of cola for $1.99
12 cans of cola for $3.49
24 cans of cola for $7

Answers

3. (a) $1.625 per quart; $1.65 per quart; $1.62 per quart. Lowest unit cost is 4 quarts for $6.48, or $1.62 per quart
(b) 12 cans for $3.49, about $0.291 per can

4 Solve each problem.

(a) Some batteries claim to last longer than others. If you believe these claims, which brand is the best buy?

$1.19

Lasts twice as long!

AA-Size

4-Pack $2.79
Just $2.99

AA-Size

(b) Which tube of toothpaste is the best buy? You have a coupon for 85¢ off Brand C and a coupon for 50¢ off Brand D.

Brand C is $3.89 for 6 ounces.

Brand D is $1.89 for 2.5 ounces.

EXAMPLE 4	Solving Best Buy Applications

Solve each application problem.

(a) There are many brands of liquid laundry detergent. If you think they all do a good job of cleaning your clothes, you can pick one based on cost per load of clothes. Two bottles of detergent are shown below. Find the best buy.

60 50 loads

Sudzy

50% Fresher!
4× concentrated
45 ounces

125 loads

WHITE-O

Now washes 60% more loads!
Twice as clean!
188 ounces

$7.89

$14.49

Find Sudzy's unit cost. You need to know the price of the bottle and how many loads it washes. Find the same information for White-O. Be careful! The labels have extra information you do not need.

$$\text{Sudzy} \quad \frac{\$7.89}{60 \text{ loads}} \approx \$0.132 \text{ per load}$$

$$\text{White-O} \quad \frac{\$14.49}{125 \text{ loads}} \approx \$0.116 \text{ per load}$$

White-O has the lower cost per load and is the best buy. (However, if you try it and it really doesn't do a good job of cleaning your clothes, Sudzy may be worth the extra cost.)

(b) "Cents-off" coupons also affect the best buy. Suppose you are looking at these choices for "extra-strength" pain reliever. Both brands have the same amount of pain reliever in each tablet.

Brand X is $2.29 for 50 tablets.

Brand Y is $10.75 for 200 tablets.

You have a 40¢ coupon for Brand X and a 75¢ coupon for Brand Y. Which choice is the best buy?

To find the best buy, first subtract the coupon amounts, then divide to find the lowest cost per ounce.

Brand X costs $2.29 − $0.40 = $1.89

$$\frac{\$1.89}{50 \text{ tablets}} \approx \$0.038 \text{ per tablet}$$

Brand Y costs $10.75 − $0.75 = $10.00

$$\frac{\$10.00}{200 \text{ tablets}} = \$0.05 \text{ per tablet}$$

> Look for the *lower* cost per tablet.

Brand X has the lower cost per tablet and is the best buy.

◀ **Work Problem 4** at the Side.

Answers

4. (a) One battery that lasts twice as long (like getting two) is the best buy. The cost per unit is $0.595 per battery. The four-pack is about $0.698 per battery.
(b) Brand C with the 85¢ coupon is the best buy at about $0.507 per ounce. Brand D with the 50¢ coupon is $0.556 per ounce.

6.2 Exercises

FOR EXTRA HELP

Go to MyMathLab *for worked-out, step-by-step solutions to exercises enclosed in a square* [ ] *and video solutions to* ▶ *exercises.*

Write each rate as a fraction in lowest terms. **See Example 1.**

1. 15 feet in 35 seconds
▶

2. 100 miles in 30 hours

3. 25 letters in 5 minutes
▶

4. 68 pills for 17 people

5. 72 miles on 4 gallons
▶

6. 132 miles on 8 gallons

7. CONCEPT CHECK To find a unit rate, do you add, subtract, multiply, or divide? _____ Show the correct set-up to find this unit rate: $5.85 for 3 boxes.

8. CONCEPT CHECK Circle the correct answer. When finding this unit rate, 15 hospital rooms for 3 nurses, the final answer will be:

5 to 1   5 rooms for 1 nurse   5 nurses for 1 room

Find each unit rate. **See Example 2.**

9. $60 in 5 hours
▶

10. $2500 in 20 days

11. 7.5 pounds for 6 people
▶

12. 44 bushels from 8 trees

13. $413.20 for 4 days

14. $74.25 for 9 hours

▦ *Earl kept the following record of the gas he bought for his car. For each entry, find the number of miles he traveled and the unit rate. Round your answers to the nearest tenth.*

	Date	Odometer at Start	Odometer at End	Miles Traveled	Gallons Purchased	Miles per Gallon
15.	2/4	27,432.3	27,758.2		15.5	
16.	2/9	27,758.2	28,058.1		13.4	
17.	2/16	28,058.1	28,396.7		16.2	
18.	2/20	28,396.7	28,704.5		13.3	

Data from author's car records.

*Find the best buy (based on the cost per unit) for each item. Round to the nearest thousandth, if necessary. **See Example 3.** (Data from Cub Foods, Target, Rainbow Foods.)*

19. Black pepper

20. Shampoo

21. Cereal

12 ounces for $2.49

14 ounces for $2.89

18 ounces for $3.96

22. Soup (same size cans)

2 cans for $2.18

3 cans for $3.57

5 cans for $5.29

23. Chunky peanut butter

12 ounces for $1.29

18 ounces for $1.79

28 ounces for $3.39

40 ounces for $4.39

24. Baked beans

8 ounces for $0.59

16 ounces for $0.99

21 ounces for $1.29

28 ounces for $1.89

25. You are choosing between two brands of chicken noodle soup. Brand A is $1.88 per can and Brand B is $1.98 per can. The cans are the same size, but Brand B is organic and lower in salt. Which soup is the best buy? Explain your choice.

26. A small bag of potatoes costs $0.19 per pound. A large bag costs $0.15 per pound. But there are only two people in your family, so half the large bag would probably spoil before you used it up. Which bag is the best buy? Explain.

*Solve each application problem. **See Examples 2–4.***

27. Makesha lost 10.5 pounds in six weeks. What was her rate of loss in pounds per week?

The rate is $\dfrac{10.5 \text{ pounds}}{6 \text{ weeks}}$ Divide: $6\overline{)10.5}$

28. Enrique's taco recipe uses three pounds of meat to feed 10 people. Give the rate in pounds per person.

The rate is $\dfrac{3 \text{ pounds}}{10 \text{ people}}$ Divide: $10\overline{)3.0}$

29. Russ works 7 hours to earn $85.82. What is his pay rate per hour?

30. Find the cost of 1 gallon of Hawaiian Punch beverage if 18 gallons for a graduation party cost $55.62.

The table lists information about three transportation options. The initial fare is charged no matter how long the ride is, even if it is less than one mile. Use the table to answer Exercises 31–34. Round answers to the nearest cent when necessary.

Company	Initial Fare	Cost per Mile
Uber	$1.25	$2.89
Taxi	$4.75	$2.55
Lyft	$2.60	$2.76

31. (a) Find the actual total cost, including the initial fare, for a 5-mile ride using each company.

(b) For the 5-mile ride, find the cost per mile using each company, and select the best buy.

32. (a) Find the actual total cost, including the initial fare, for a 20-mile ride using each company.

(b) For the 20-mile ride, find the cost per mile using each company, and select the best buy.

33. Find the actual total cost and the cost per mile for a 10-mile ride using each company. Explain what you see when you compare the unit rates for this ride.

34. All the companies charge an additional $6 fee if they are going to or from the airport.

(a) You live 38 miles away from the airport. Find the cost for a ride to the airport using each company.

(b) How much would you save by using a taxi instead of Uber?

35. If you believe the claims that some batteries last longer, which is the best buy?

36. Which is the best buy, assuming these laundry detergents both clean equally well?

37. Three brands of cornflakes are available. Brand G is priced at $2.39 for 10 ounces. Brand K is $3.99 for 20.3 ounces and Brand P is $3.39 for 16.5 ounces. You have a coupon for 50¢ off Brand P and a coupon for 60¢ off Brand G. Which cereal is the best buy based on cost per unit?

38. Two brands of facial tissue are available. Brand K is priced at $5 for three boxes of 175 tissues each. (You may buy just one box.) Brand S is priced at $1.29 per box of 125 tissues. You have a coupon for 20¢ off one box of Brand S and a coupon for 45¢ off one box of Brand K. How can you get the best buy on one box of tissue?

Relating Concepts (Exercises 39–44) For Individual or Group Work

On the first page of this chapter we said that unit rates can help you find the best buy. This includes the best buy on smartphone data plans. Use the information in the table to **work Exercises 39–44 in order.** *Round all money answers to the nearest cent.*

Plan	Data (in MegaBytes)	Monthly Charge†	Cost per MB (to nearest cent)	Additional Data*
A	1500 MB	$45		500 MB for $10
B	500 MB	$30		100 MB for $5
C	2500 MB	$60		$1.50 per MB

† All plans include unlimited talk and text; the prices do not include taxes or fees.
* Cost for data above your monthly plan amount (overage). You must pay for the entire amount listed; partial amounts cannot be purchased.
Data from www.virginmobileusa.com and www.att.com

39. Find the cost per MB for each plan. Round your answers to the nearest cent, and write them in the table above. Which plan is the best buy based on cost per MB?

40. In a 30-day month, what is the average amount of data you can use each day on each plan?
   (a) Plan A
   (b) Plan B
   (c) Plan C

41. You are on Plan B. Last month you used 130 MB of data above your monthly plan amount. Find the new cost per MB for last month. (Assume 30 days in last month).

42. You are on Plan C. Last month you used 90 MB of data above your monthly plan amount. Find the new cost per MB for last month. (Assume 30 days in last month).

43. You usually use 2250 MB of data each month. Find the monthly cost for each plan if you use 2250 MB of data. What is the cost per MB for each plan if you use 2250 MB of data?

44. One month you had a group class project, so you used more data than usual. You used 3000 MB of data that month. What was the cost of using 3000 MB of data on each plan?

6.3 | Proportions

OBJECTIVE 1 Write proportions. A **proportion** states that two ratios (or rates) are equal. Here is an example.

$$\frac{\$20}{4\ \text{hours}} = \frac{\$40}{8\ \text{hours}}$$

This is a proportion that says the rate $\frac{\$20}{4\ \text{hours}}$ is equal to the rate $\frac{\$40}{8\ \text{hours}}$. As the amount of money doubles, the number of hours also doubles. This proportion is read:

20 dollars **is to** 4 hours   **as**   40 dollars **is to** 8 hours

EXAMPLE 1 Writing Proportions

Write each proportion.

(a) 6 feet is to 11 feet **as** 18 feet is to 33 feet.

$$\frac{6\ \text{feet}}{11\ \text{feet}} = \frac{18\ \text{feet}}{33\ \text{feet}} \quad \text{so} \quad \frac{6}{11} = \frac{18}{33}$$

The common units (feet) divide out and are not written.

(b) $9 is to 6 liters **as** $3 is to 2 liters.

$$\frac{\$9}{6\ \text{liters}} = \frac{\$3}{2\ \text{liters}}$$

The units are **different**, so you must write them in the proportion.

— **Work Problem 1 at the Side.** ▶

OBJECTIVE 2 Determine whether proportions are true or false. There are two ways to see whether a proportion is true. One way is to *write both of the ratios in lowest terms.*

EXAMPLE 2 Writing Both Ratios in Lowest Terms

Determine whether each proportion is true or false by writing both ratios in lowest terms.

(a) $\dfrac{5}{9} = \dfrac{18}{27}$

Write each ratio in lowest terms.

$$\frac{5}{9} \leftarrow \text{Already in lowest terms} \qquad \frac{18 \div 9}{27 \div 9} = \frac{2}{3} \leftarrow \text{Lowest terms}$$

Because $\frac{5}{9}$ is *not* equal to $\frac{2}{3}$, the proportion is *false*. The ratios are *not* proportional.

(b) $\dfrac{16}{12} = \dfrac{28}{21}$

Write each ratio in lowest terms.

Both ratios are in lowest terms.

$$\frac{16 \div 4}{12 \div 4} = \frac{4}{3} \quad \text{and} \quad \frac{28 \div 7}{21 \div 7} = \frac{4}{3}$$

Both ratios are equal to $\frac{4}{3}$, so the proportion is *true*. The ratios are proportional.

— **Work Problem 2 at the Side.** ▶

OBJECTIVES

1. Write proportions.
2. Determine whether proportions are true or false.
3. Find the unknown number in a proportion.

1 Write each proportion.

GS (a) $7 is to 3 cans as $28 is to 12 cans.

$$\frac{\$7}{3\ \text{cans}} = \frac{\square}{\square}$$

Complete the proportion. Remember to write the units!

(b) 9 meters is to 16 meters as 18 meters is to 32 meters.

2 Determine whether each proportion is true or false by writing both ratios in lowest terms.

(a) $\dfrac{6}{12} = \dfrac{15}{30}$

(b) $\dfrac{20}{24} = \dfrac{3}{4}$

(c) $\dfrac{35}{45} = \dfrac{12}{18}$

Answers

1. **(a)** $\dfrac{\$7}{3\ \text{cans}} = \dfrac{\$28}{12\ \text{cans}}$ **(b)** $\dfrac{9}{16} = \dfrac{18}{32}$

2. **(a)** $\dfrac{1}{2} = \dfrac{1}{2}$; true **(b)** $\dfrac{5}{6} \neq \dfrac{3}{4}$; false

   (c) $\dfrac{7}{9} \neq \dfrac{2}{3}$; false

A second way to see whether a proportion is true is to find *cross products*.

Using Cross Products to Determine Whether a Proportion Is True

To see whether a proportion is true, first multiply along one diagonal, then multiply along the other diagonal, as shown here.

$$5 \cdot 4 = 20$$

$$\frac{2}{5} = \frac{4}{10}$$

Cross products are equal.

$$2 \cdot 10 = 20$$

In this case the **cross products** are both 20. When cross products are *equal*, the proportion is *true*. If the cross products are *unequal*, the proportion is *false*.

Note

Why does the cross products test work? It is based on rewriting both fractions with a common denominator of 5 · 10 or 50. (We do not search for the *lowest* common denominator. We simply use the product of the two given denominators.)

$$\frac{2 \cdot 10}{5 \cdot 10} = \frac{20}{50} \quad \text{and} \quad \frac{4 \cdot 5}{10 \cdot 5} = \frac{20}{50}$$

We see that $\frac{2}{5}$ and $\frac{4}{10}$ are equivalent because both can be rewritten as $\frac{20}{50}$. The cross products test takes a shortcut by comparing only the two numerators (20 = 20).

EXAMPLE 3   Using Cross Products

Use cross products to see whether each proportion is true or false.

(a) $\dfrac{3}{5} = \dfrac{12}{20}$ Multiply along one diagonal and then multiply along the other diagonal.

$$5 \cdot 12 = 60$$

$$\frac{3}{5} = \frac{12}{20}$$

Equal — When cross products are equal, the proportion is true.

$$3 \cdot 20 = 60$$

The cross products are *equal*, so the proportion is *true*.

---

❶ CAUTION

Use cross products *only* when working with *proportions*. Do **not** use cross products when multiplying fractions, adding fractions, or writing fractions in lowest terms.

———— **Continued on Next Page**

(b) $\dfrac{2\frac{1}{3}}{3\frac{1}{3}} = \dfrac{9}{16}$ Find the cross products.

> Write $3\frac{1}{3}$ as $\frac{10}{3}$ and write 9 as $\frac{9}{1}$

$$3\frac{1}{3} \cdot 9 = \frac{10}{\overset{1}{\cancel{3}}} \cdot \frac{\overset{3}{\cancel{9}}}{1} = \frac{30}{1} = 30$$

$$\dfrac{2\frac{1}{3}}{3\frac{1}{3}} = \dfrac{9}{16}$$

Unequal; proportion is false.

$$2\frac{1}{3} \cdot 16 = \frac{7}{3} \cdot \frac{16}{1} = \frac{112}{3} = 37\frac{1}{3}$$

The cross products are *unequal*, so the proportion is *false*.

Note

The numbers in a proportion do *not* have to be whole numbers. They may be fractions, mixed numbers, decimal numbers, and so on.

— **Work Problem ❸ at the Side.** ▶

OBJECTIVE ▶ ❸ Find the unknown number in a proportion. Four numbers are used in a proportion. If any three of these numbers are known, the fourth can be found. For example, find the unknown number that will make this proportion true.

$$\frac{3}{5} = \frac{x}{40}$$

The variable x represents the unknown number. First find the cross products.

$$\frac{3}{5} \overset{5 \cdot x}{\underset{3 \cdot 40}{=}} \frac{x}{40} \quad \text{Cross products}$$

To make the proportion true, the cross products must be equal. This gives us the following equation.

On the left side, $5 \cdot x$ is $5x$ $5 \cdot x = 3 \cdot 40$ On the right side, $3 \cdot 40$ is 120
$$5x = 120$$

Recall that we can solve an equation of this type by dividing both sides by the coefficient of the variable term. In this case, the coefficient of $5x$ is 5.

$$\frac{5x}{5} = \frac{120}{5} \quad \leftarrow \text{Divide both sides by 5}$$

On the left side, divide out the common factor of 5; slashes indicate the division.

$$\frac{\overset{1}{\cancel{5}} \cdot x}{\underset{1}{\cancel{5}}} = 24 \qquad \text{On the right side, divide 120 by 5 to get 24}$$

$$x = 24$$

---

❸ Find the cross products to see whether each proportion is true or false.

$(9)(10) = $ ___

(a) $\dfrac{5}{9} = \dfrac{10}{18}$

$(5)(18) = $ ___

(b) $\dfrac{32}{15} = \dfrac{16}{8}$

(c) $\dfrac{10}{17} = \dfrac{20}{34}$

$(6)(5) = $ ___

(d) $\dfrac{2.4}{6} = \dfrac{5}{12}$

$(2.4)(12) = $ ___

(e) $\dfrac{3}{4.25} = \dfrac{24}{34}$

(f) $\dfrac{1\frac{1}{6}}{2\frac{1}{3}} = \dfrac{4}{8}$

Answers

3. **(a)** $(9)(10) = 90$; $(5)(18) = 90$; true
   (b) $240 \neq 256$; false
   (c) $340 = 340$; true
   (d) $(6)(5) = 30$; $(2.4)(12) = 28.8$; false
   (e) $102 = 102$; true
   (f) $9\frac{1}{3} = 9\frac{1}{3}$; true

The unknown number in the proportion is 24. The complete proportion is shown below.

$$\frac{3}{5} = \frac{24}{40} \leftarrow x \text{ is } 24$$

Check by finding the cross products. If they are equal, you solved the problem correctly. If they are unequal, rework the problem.

$$5 \cdot 24 = 120 \leftarrow$$

$$\frac{3}{5} = \frac{24}{40}$$

Equal; proportion is true.

$$3 \cdot 40 = 120 \leftarrow$$

The solution is 24, **not** 120

The cross products are equal, so the solution, $x = 24$, is correct.

> **! CAUTION**
>
> **The solution is 24,** which is the unknown number in the proportion. 120 is **not** the solution; it is the cross product you get when *checking* the solution.

Solve a proportion for an unknown number using these steps.

> **Solving a Proportion to Find an Unknown Number**
>
> **Step 1** Find the cross products.
>
> **Step 2** Show that the cross products are equal.
>
> **Step 3** Divide both sides of the equation by the coefficient of the variable term.
>
> **Step 4** Check by writing the solution in the *original* proportion and finding the cross products.

EXAMPLE 4 **Solving Proportions**

Find the unknown number in each proportion. Round answers to the nearest hundredth when necessary.

(a) $\dfrac{16}{x} = \dfrac{32}{20}$ ← You can write $\frac{32}{20}$ in lowest terms as $\frac{8}{5}$

Recall that ratios can be rewritten in lowest terms. If desired, you can do that *before* finding the cross products. In this example, write $\frac{32}{20}$ in lowest terms as $\frac{8}{5}$, which gives this proportion: $\dfrac{16}{x} = \dfrac{8}{5}$

Step 1

$$\frac{16}{x} = \frac{8}{5}$$

$x \cdot 8 \leftarrow$

Find the cross products.

$16 \cdot 5 \leftarrow$

——— **Continued on Next Page**

Step 2 $x \cdot 8 = \underbrace{16 \cdot 5}$ Show that the cross products are equal.

$x \cdot 8 = \quad 80$

Step 3 $\dfrac{x \cdot \overset{1}{8}}{\underset{1}{8}} = \dfrac{80}{8}$ ← Divide both sides by 8

$x = 10$ ← Find x (no rounding is necessary).

Step 4   Write the solution in the *original* proportion and check by finding the cross products.

x is 10 → $\dfrac{16}{10} = \dfrac{32}{20}$

$10 \cdot 32 = 320$ ←

$16 \cdot 20 = 320$ ←

Equal; proportion is true.

The cross products are equal, so **10 is the correct solution.**

> The solution is 10, **not** 320

Note

It is not necessary to write the ratios in lowest terms before solving. However, if you do, you will have smaller numbers to work with.

(b) $\dfrac{7}{12} = \dfrac{15}{x}$

Step 1 $\dfrac{7}{12} = \dfrac{15}{x}$

$12 \cdot 15 = 180$ ←

$7 \cdot x$ ←

Find the cross products.

Step 2 $7 \cdot x = 180$ ← Show that cross products are equal.

Step 3 $\dfrac{\overset{1}{7} \cdot x}{\underset{1}{7}} = \dfrac{180}{7}$ ← Divide both sides by 7

> The rounded solution is 25.71

$x \approx 25.71$ ← Rounded to nearest hundredth

When the division does not come out even, check for directions on how to round your answer. Divide out one more place, then round.

$\dfrac{25.714}{7)\overline{180.000}}$ ← Divide out to thousandths so you can round to hundredths.

Step 4   Write the solution in the original proportion and check by finding the cross products.

$\dfrac{7}{12} \approx \dfrac{15}{25.71}$

$(12)(15) = 180$ ←

$(7)(25.71) = 179.97$ ←

Very close, but *not* equal due to rounding the solution

The cross products are slightly different because you rounded the value of x. However, they are close enough to see that the problem was done correctly and **25.71 is the approximate solution** (**not** 179.97).

— **Work Problem ④ at the Side.** ▶

---

④ Find the unknown numbers. Round to hundredths when necessary. Check your answers by finding the cross products.

(a) $\dfrac{1}{2} = \dfrac{x}{12}$

$2 \cdot x$ ←

$1 \cdot 12$ ←

Cross products

$2 \cdot x = 12$ Now find the solution. Divide both sides by 2

(b) $\dfrac{6}{10} = \dfrac{15}{x}$

(c) $\dfrac{28}{x} = \dfrac{21}{9}$

(d) $\dfrac{x}{8} = \dfrac{3}{5}$

(e) $\dfrac{14}{11} = \dfrac{x}{3}$

Answers

4. **(a)** $\dfrac{\overset{1}{2} \cdot x}{\underset{1}{2}} = \dfrac{12}{2}$; $x = 6$ **(b)** $x = 25$

(c) $x = 12$ **(d)** $x = 4.8$

(e) $x \approx 3.82$ (rounded to nearest hundredth)

The next example shows how to solve for the unknown number in a proportion with fractions or decimals.

EXAMPLE 5   **Solving Proportions with Mixed Numbers and Decimals**

Find the unknown number in each proportion. Write fraction answers in simplest form and, when possible, as mixed numbers.

(a) $\dfrac{2\frac{1}{5}}{6} = \dfrac{x}{10}$

Step 1

$$\dfrac{2\frac{1}{5}}{6} = \dfrac{x}{10}$$

$6 \cdot x$

$2\frac{1}{5} \cdot 10$

Find the cross products.

Find $2\frac{1}{5} \cdot 10$.

$$2\frac{1}{5} \cdot 10 = \frac{11}{5} \cdot \frac{10}{1} = \frac{11 \cdot 2 \cdot \cancel{5}^{1}}{\cancel{5} \cdot 1} = \frac{22}{1} = 22$$

Changed to improper fraction

Step 2

$6 \cdot x = 22$ ← Show that cross products are equal.

Step 3

$$\dfrac{\cancel{6}^{1} \cdot x}{\cancel{6}_{1}} = \dfrac{22}{6}$$ ← Divide both sides by 6

Write the solution in simplest form and as a mixed number.

$$x = \frac{22 \div 2}{6 \div 2} = \frac{11}{3} = 3\frac{2}{3}$$

$\frac{11}{3}$ is in simplest form. It can be rewritten as the mixed number $3\frac{2}{3}$

So, **the solution is $3\frac{2}{3}$**

Step 4   Write the solution in the original proportion and check by finding the cross products.

$$6 \cdot 3\frac{2}{3} = \frac{2 \cdot \cancel{3}^{1}}{1} \cdot \frac{11}{\cancel{3}_{1}} = \frac{22}{1} = 22$$

$$\dfrac{2\frac{1}{5}}{6} = \dfrac{3\frac{2}{3}}{10}$$

Equal

$$2\frac{1}{5} \cdot 10 = \frac{11}{\cancel{5}_{1}} \cdot \frac{2 \cdot \cancel{5}^{1}}{1} = \frac{22}{1} = 22$$

The cross products are equal, so **$3\frac{2}{3}$ is the correct solution.**

The solution is $3\frac{2}{3}$, **not** 22

Continued on Next Page

Note

You can use decimal numbers and your calculator to solve **Example 5(a)** on the previous page. $2\frac{1}{5}$ is equivalent to 2.2, so the cross products are

$$6 \cdot x = (2.2)(10)$$

$$\frac{\overset{1}{6} \cdot x}{\underset{1}{6}} = \frac{22}{6}$$

When you divide 22 by 6 on your calculator, it shows 3.666666667. Write the answer using a bar to show the repeating digit: $3.\overline{6}$. Or round the answer to 3.67 (nearest hundredth).

(b) $\dfrac{1.5}{0.6} = \dfrac{2}{x}$

$(1.5)(x) = (0.6)(2)$ ← Show that cross products are equal.

$(1.5)(x) = 1.2$ ← On the right side, multiply $(0.6)(2)$ to get 1.2

$\dfrac{(\overset{1}{1.5})(x)}{\underset{1}{1.5}} = \dfrac{1.2}{1.5}$ ← Divide both sides by 1.5

$x = \dfrac{1.2}{1.5}$ Complete the division. $1.5\overline{)1.20}$

$x = 0.8$

So the solution is 0.8

Write the solution in the original equation, and check by finding the cross products.

$(0.6)(2) = 1.2$

$\dfrac{1.5}{0.6} = \dfrac{2}{0.8}$ Equal

$(1.5)(0.8) = 1.2$

The cross products are equal, so **0.8 is the correct solution.** ← The solution is 0.8 (**not** 1.2)

─── **Work Problem ⑤ at the Side.** ▶

Study Skills Reminder

There are many new mathematical words and procedures in the last several sections. To help you learn and remember them, see the Study Skills activity "Using Study Cards." As you continue in this chapter, other types of study cards will also be helpful. See the Study Skills activity "Using Study Cards Revisited" for even more ideas on how to make and use different types of study cards.

⑤ Find the unknown numbers. Write fraction answers in simplest form. Round decimal answers to hundredths when necessary. Check your solutions by finding the cross products.

GS (a) $\dfrac{3\frac{1}{4}}{2} = \dfrac{x}{8}$ $\overset{2 \cdot x}{}$

$$3\frac{1}{4} \cdot 8 = \frac{13}{\underset{1}{4}} \cdot \frac{\overset{2}{8}}{1}$$

$$= \frac{26}{1}$$

$2 \cdot x = 26$ Find the solution.

(b) $\dfrac{x}{3} = \dfrac{1\frac{2}{3}}{5}$

(c) $\dfrac{0.06}{x} = \dfrac{0.3}{0.4}$

(d) $\dfrac{2.2}{5} = \dfrac{13}{x}$

(e) $\dfrac{x}{6} = \dfrac{0.5}{1.2}$

(f) $\dfrac{0}{2} = \dfrac{x}{7.092}$

Answers

5. (a) $x = 13$ **(b)** $x = 1$ **(c)** $x = 0.08$
  (d) $x \approx 29.55$ (rounded to nearest hundredth)
  (e) $x = 2.5$ **(f)** $x = 0$

6.3 Exercises

FOR EXTRA HELP — Go to MyMathLab for worked-out, step-by-step solutions to exercises enclosed in a square ▢ and video solutions to ▶ exercises.

CONCEPT CHECK *Write each proportion. **See Example 1.***

1. $9 is to 12 cans as $18 is to 24 cans.

2. 28 people is to 7 cars as 16 people is to 4 cars.

3. 200 adults is to 450 children as 4 adults is to 9 children.

4. 150 trees is to 1 acre as 1500 trees is to 10 acres.

5. 120 feet is to 150 feet as 8 feet is to 10 feet.

6. $6 is to $9 as $10 is to $15.

Determine whether each proportion is true *or* false *by writing the ratios in lowest terms. Show the simplified ratios and then write* true *or* false*. **See Example 2.***

7. $\dfrac{6}{10} = \dfrac{3}{5}$

8. $\dfrac{1}{4} = \dfrac{9}{36}$

9. $\dfrac{5}{8} = \dfrac{25}{40}$

10. $\dfrac{2}{3} = \dfrac{20}{27}$

11. $\dfrac{150}{200} = \dfrac{200}{300}$

12. $\dfrac{100}{120} = \dfrac{75}{100}$

13. **CONCEPT CHECK** The work shown below has a mistake in it.

    What Went Wrong? First write a sentence explaining what the mistake is. Then fix the mistake and find the correct solution.

    Gary said $\dfrac{7}{9} = \dfrac{49}{81}$ was a true proportion because

    $$\frac{49 \div 7}{81 \div 9} = \frac{7}{9}$$

14. **CONCEPT CHECK** The work shown below has a mistake in it.

    What Went Wrong? First write a sentence explaining what the mistake is. Then fix the mistake and find the correct solution.

    Margo said $\dfrac{5}{8} = \dfrac{30}{40}$ was a true proportion because

    $$5 \cdot 8 = 40$$

In Exercises 15–26, use cross products to determine whether each proportion is true *or* false*. Show the cross products and then circle* True *or* False*. **See Example 3.***

15. $\dfrac{2}{9} = \dfrac{6}{27}$

    True   False

16. $\dfrac{20}{25} = \dfrac{4}{5}$

    True   False

17. $\dfrac{20}{28} = \dfrac{12}{16}$

    True   False

18. $\dfrac{16}{40} = \dfrac{22}{55}$

True   False

19. $\dfrac{110}{18} = \dfrac{160}{27}$

True   False

20. $\dfrac{600}{420} = \dfrac{20}{14}$

True   False

21. $\dfrac{3.5}{4} = \dfrac{7}{8}$

True   False

22. $\dfrac{36}{23} = \dfrac{9}{5.75}$

True   False

23. $\dfrac{18}{15} = \dfrac{2\frac{5}{6}}{2\frac{1}{2}}$

True   False

24. $\dfrac{1\frac{3}{10}}{3\frac{1}{5}} = \dfrac{4}{9}$

True   False

25. $\dfrac{6}{3\frac{2}{3}} = \dfrac{18}{11}$

True   False

26. $\dfrac{16}{13} = \dfrac{2}{1\frac{5}{8}}$

True   False

27. CONCEPT CHECK   Freddie Freeman of the Atlanta Braves had 168 hits in 600 times at bat, and Kurt Suzuki of the Minnesota Twins was at bat 450 times and got 126 hits. Paula is trying to convince Jenny that the two men hit equally well. Show how to use a proportion and cross products to see whether Paula is correct. (Data from www.espn.go.com)

28. CONCEPT CHECK   Jay worked 3.5 hours and packed 91 cartons. Craig packed 126 cartons in 5.25 hours. To see if the men worked equally fast, Barry set up this proportion:

$$\dfrac{3.5}{91} = \dfrac{126}{5.25}$$

Explain what is wrong with Barry's proportion and write a correct one. Is the correct proportion true or false?

Solve each proportion to find the unknown number. Round decimal answers to the nearest hundredth when necessary. Check your answers by finding cross products.
See Examples 4 and 5.

29. $\dfrac{1}{3} = \dfrac{x}{12}$

$3 \cdot x$

$1 \cdot 12$

$3 \cdot x = 1 \cdot 12$

$3 \cdot x = \underline{\qquad}$

$\dfrac{\overset{1}{\cancel{3}} \cdot x}{\underset{1}{\cancel{3}}} = \dfrac{\square}{3}$

$x = \underline{\qquad}$

30. $\dfrac{x}{6} = \dfrac{15}{18}$

$6 \cdot 15$

$x \cdot 18$

$x \cdot 18 = 6 \cdot 15$

$x \cdot 18 = \underline{\qquad}$

$\dfrac{x \cdot \overset{1}{\cancel{18}}}{\underset{1}{\cancel{18}}} = \dfrac{\square}{18}$

$x = \underline{\qquad}$

31. $\dfrac{15}{10} = \dfrac{3}{x}$

32. $\dfrac{5}{x} = \dfrac{20}{8}$

33. $\dfrac{x}{11} = \dfrac{32}{4}$

34. $\dfrac{12}{9} = \dfrac{8}{x}$

35. $\dfrac{42}{x} = \dfrac{18}{39}$

36. $\dfrac{49}{x} = \dfrac{14}{18}$

37. $\dfrac{x}{25} = \dfrac{4}{20}$

38. $\dfrac{6}{x} = \dfrac{4}{8}$

39. $\dfrac{8}{x} = \dfrac{24}{30}$

40. $\dfrac{32}{5} = \dfrac{x}{10}$

41. $\dfrac{99}{55} = \dfrac{44}{x}$

42. $\dfrac{x}{12} = \dfrac{101}{147}$

43. $\dfrac{0.7}{9.8} = \dfrac{3.6}{x}$

44. $\dfrac{x}{3.6} = \dfrac{4.5}{6}$

45. $\dfrac{250}{24.8} = \dfrac{x}{1.75}$

46. $\dfrac{4.75}{17} = \dfrac{43}{x}$

Find the unknown number in each proportion. Write your answers in simplest form and, when possible, as mixed numbers. **See Example 5.**

47. $\dfrac{15}{1\frac{2}{3}} = \dfrac{9}{x}$

48. $\dfrac{x}{\frac{3}{10}} = \dfrac{2\frac{2}{9}}{1}$

49. $\dfrac{2\frac{1}{3}}{1\frac{1}{2}} = \dfrac{x}{2\frac{1}{4}}$

50. $\dfrac{1\frac{5}{6}}{x} = \dfrac{\frac{3}{14}}{\frac{6}{7}}$

Solve each proportion in two different ways. First change all the numbers to decimal form and solve. Then change all the numbers to fraction form and solve. Write fraction answers in simplest form and, when possible, as mixed numbers.

51. $\dfrac{\frac{1}{2}}{x} = \dfrac{2}{0.8}$

52. $\dfrac{\frac{3}{20}}{0.1} = \dfrac{0.03}{x}$

53. $\dfrac{x}{\frac{3}{50}} = \dfrac{0.15}{1\frac{4}{5}}$

54. $\dfrac{8\frac{4}{5}}{1\frac{1}{10}} = \dfrac{x}{0.4}$

Relating Concepts (Exercises 55–56) For Individual or Group Work

Work Exercises 55–56 in order. *In each exercise, first prove that the proportion is* **false.** *Then create a true proportion by changing just one of the numbers in the original proportion. Create three more* **true** *proportions. Change a different number from the original proportion each time.*

55. $\dfrac{10}{4} = \dfrac{5}{3}$

56. $\dfrac{6}{8} = \dfrac{24}{30}$

Summary Exercises *Ratios, Rates, and Proportions*

Use the bar graph of the countries with the most websites to complete Exercises 1–4. Write each ratio as a fraction in lowest terms.

1. Write the ratio of the number of websites in Japan to the number of websites in China.

2. What is the ratio of the number of websites in China to the number of websites in Brazil?

3. Find the ratio of the number of websites in Italy and China compared to the number of websites in Japan.

4. Write the ratio of the number of websites in the USA compared to the total number of websites in all the other countries shown in the graph.

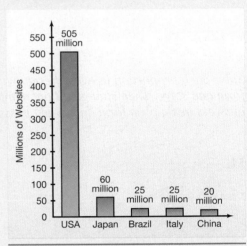

Number of Websites in Selected Countries

Data from *The World Factbook*.

The bar graph shows the number of Americans who play various instruments. Use the graph to complete Exercises 5 and 6. Write each ratio as a fraction in lowest terms.

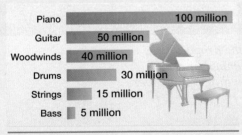

Americans Make Music by the Millions

How many people play each type of instrument?

Piano 100 million
Guitar 50 million
Woodwinds 40 million
Drums 30 million
Strings 15 million
Bass 5 million

Data from www.namm.org

5. Write five ratios that compare the least popular instrument type to each of the other instrument types.

6. Which two instrument types give each of these ratios? There may be more than one correct answer.

(a) $\frac{8}{1}$

(b) $\frac{2}{1}$

CONCEPT CHECK *The table lists data on the top three individual scoring NBA basketball games of all time. Use the data to answer Exercises 7 and 8. Round answers to the nearest tenth.*

7. What was Wilt Chamberlain's scoring rate in points per minute and in minutes per point?

Player	Date	Points	Min
Wilt Chamberlain	3/2/62	100	48
Kobe Bryant	1/22/06	81	48
David Thompson	9/4/78	73	43

Data from www.landofbasketball.com

8. Find Kobe Bryant's scoring rate in points per minute and minutes per point.

Wilt Chamberlain (#13)

9. Lucinda's paycheck showed gross pay of $652.80 for 40 hours of work and $195.84 for 8 hours of overtime work. Find her regular hourly pay rate and her overtime rate.

10. Satellite TV service is being offered to new subscribers at $29.99 per month for 150 channels, $34.99 per month for 210 channels, or $39.99 per month for 225 channels. What is the monthly cost per channel under each plan, to the nearest cent? (Data from DIRECTV.)

11. Find the best buy on gourmet coffee beans. Round to the nearest cent.

11 ounces for $6.79

12 ounces for $7.24

16 ounces for $10.99

(Data from Cub Foods.)

12. Which brand of cat food is the best buy? You have a coupon for $2 off on Brand P, and another for $1 off on Brand N.

Brand N is $3.75 for 3.5 pounds

Brand P is $5.99 for 7 pounds

Brand R is $10.79 for 18 pounds

Use either the method of writing in lowest terms or the method of finding cross products to decide whether each proportion is true *or* false. *Show your work and then write* true *or* false.

13. $\dfrac{28}{21} = \dfrac{44}{33}$

14. $\dfrac{2.3}{8.05} = \dfrac{0.25}{0.9}$

15. $\dfrac{2\frac{5}{8}}{3\frac{1}{4}} = \dfrac{21}{26}$

Solve each proportion to find the unknown number. Round decimal answers to the nearest hundredth when necessary. Write fraction answers in simplest form.

16. $\dfrac{7}{x} = \dfrac{25}{100}$

17. $\dfrac{15}{8} = \dfrac{6}{x}$

18. $\dfrac{x}{84} = \dfrac{78}{36}$

19. $\dfrac{10}{11} = \dfrac{x}{4}$

20. $\dfrac{x}{17} = \dfrac{3}{55}$

21. $\dfrac{2.6}{x} = \dfrac{13}{7.8}$

22. $\dfrac{0.14}{1.8} = \dfrac{x}{0.63}$

23. $\dfrac{\frac{1}{3}}{8} = \dfrac{x}{24}$

24. $\dfrac{6\frac{2}{3}}{4\frac{1}{6}} = \dfrac{\frac{6}{5}}{x}$

6.4 | Problem Solving with Proportions

OBJECTIVE **1** Use proportions to solve application problems. Proportions can be used to solve a wide variety of problems. Watch for problems in which you are given a ratio or rate and then asked to find part of a corresponding ratio or rate. Remember that a ratio or rate compares two quantities and often includes one of these indicator words.

in   for   on   per   from   to

Use the six problem-solving steps that you learned earlier. When setting up the proportion, use a variable to represent the unknown number. We have used the letter x, but you may use any letter you like.

EXAMPLE 1 | **Solving a Proportion Application**

Mike's car can travel 163 **miles** **on** 6.4 **gallons** of gas. How far can it travel on a full tank of 14 **gallons** of gas? Round to the nearest mile.

Step 1 **Read** the problem. It is about how far a car can travel on a certain amount of gas.

> Unknown: miles traveled on 14 gallons of gas
> Known: 163 miles traveled on 6.4 gallons of gas

Step 2 **Assign a variable.** There is only one unknown, so let x be the number of miles traveled on 14 gallons.

Step 3 **Write an equation using your variable.** The equation is in the form of a proportion. Decide what is being compared. This problem compares **miles** to **gallons**. Write the two rates described in the problem. Be sure that *both* rates compare miles to gallons in the same order. In other words, miles is in both numerators and gallons is in both denominators.

Matching units

This rate compares miles to **gallons**. $\quad \dfrac{163\ \text{miles}}{6.4\ \text{gallons}} = \dfrac{x\ \text{miles}}{14\ \text{gallons}} \quad$ This rate compares miles to **gallons**.

Matching units

Step 4 **Solve** the equation. Ignore the units while finding the cross products.

$$\frac{163\ \text{miles}}{6.4\ \text{gallons}} = \frac{x\ \text{miles}}{14\ \text{gallons}}$$

$(6.4)(x) = (163)(14)$ Show that cross products are equal.

$(6.4)(x) = 2282$

$$\frac{\overset{1}{(6.4)}(x)}{\underset{1}{6.4}} = \frac{2282}{6.4}$$ Divide both sides by 6.4

$x = 356.5625$ Round to 357   *Always check the problem for rounding directions.*

Step 5 **State the answer.** Mike's car can travel **357 miles,** rounded to the nearest mile, on a full tank of gas.

Continued on Next Page

Step 6   **Check** that the solution is reasonable and fits the facts given in the original statement of the problem. The car traveled 163 miles on 6.4 gallons of gas; 14 gallons is a little more than *twice as much* gas, so the car should travel a little more than *twice as far*.

$$(2)\,(163 \text{ miles}) = 326 \text{ miles} \leftarrow \text{Estimate}$$

The solution, 357 miles, is a little more than the estimate of 326 miles, so it is reasonable.

> ❗ **CAUTION**
>
> When setting up the proportion, do *not* mix up the units in the rates.
>
> compares miles to gallons $\left.\right\}$ $\dfrac{163 \text{ miles}}{6.4 \text{ gallons}} \neq \dfrac{14 \text{ gallons}}{x \text{ miles}}$ $\left\{\right.$ compares gallons to miles
>
> These rates do *not* compare things in the same order and *cannot* be set up as a proportion.

—————— **Work Problem ① at the Side.** ▶

| **EXAMPLE 2** | **Solving a Proportion Application** |

A newspaper report says that 7 out of 10 people surveyed watch the news on TV. At that rate, how many of the 3200 people in town would you expect to watch the news?

Step 1   **Read** the problem. It is about people watching the news on TV.

Unknown: how many people in town are expected to watch the news on TV
Known: 7 out of 10 people surveyed watched the news on TV.

Step 2   **Assign a variable.** There is only one unknown, so let x be the number of people in town who watch the news on TV.

Step 3   **Write an equation using the variable.** Set up the two rates as a proportion. You are comparing people who watch the news to people surveyed. Write the two rates described in the example. Be sure that both rates make the same comparison. "People who watch the news" is mentioned first, so it should be in the numerator of *both* rates.

People who watch the news $\rightarrow$ $\dfrac{7}{10} = \dfrac{x}{3200}$ $\leftarrow$ People who watch the news
Total group $\rightarrow$ (people surveyed) $\leftarrow$ Total group (people in town)

Step 4   **Solve** the equation.

$$\frac{7}{10} = \frac{x}{3200}$$

$$(10)\,(x) = (7)\,(3200) \quad \text{Show that cross products are equal.}$$

$$(10)\,(x) = 22{,}400$$

$$\frac{\overset{1}{\cancel{10}}\,(x)}{\underset{1}{\cancel{10}}} = \frac{22{,}400}{10} \quad \text{Divide both sides by 10}$$

$$x = 2240 \quad \boxed{\text{No rounding is needed here.}}$$

—————— **Continued on Next Page**

---

① Set up and solve a proportion for each problem using the six problem-solving steps.

(a) If 2 pounds of fertilizer will cover 50 square feet of garden, how many pounds are needed for 225 square feet?

$$\frac{2 \text{ pounds}}{50 \text{ square feet}} = \frac{\boxed{} \text{ pounds}}{\boxed{} \text{ square feet}}$$

(b) A U.S. map has a scale of 1 inch to 75 miles. Lake Superior is 4.75 inches long on the map. What is the lake's actual length, to the nearest whole mile?

(c) Cough syrup should be given at the rate of 30 milliliters for each 100 pounds of body weight. How much should be given to a 34-pound child? Round to the nearest whole milliliter.

Answers

1. **(a)** $\dfrac{2 \text{ pounds}}{50 \text{ sq. feet}} = \dfrac{x \text{ pounds}}{225 \text{ sq. feet}}$
   $x = 9$ pounds

(b) $\dfrac{1 \text{ inch}}{75 \text{ miles}} = \dfrac{4.75 \text{ inches}}{x \text{ miles}}$
   $x = 356.25$ miles, rounds to 356 miles

(c) $\dfrac{30 \text{ milliliters}}{100 \text{ pounds}} = \dfrac{x \text{ milliliters}}{34 \text{ pounds}}$
   $x \approx 10$ milliliters (rounded)

2 Solve each problem to find a reasonable answer. Then flip one side of your proportion to see what answer you get with an **incorrect** set-up. Explain why the second answer is **unreasonable**.

GS **(a)** A survey showed that 2 out of 3 people would like to lose weight. At this rate, how many people in a group of 150 want to lose weight?

Correct set-up

$$\text{Lose weight} \rightarrow \frac{2}{3} = \frac{x}{\square} \leftarrow \text{Lose weight}$$
People surveyed ↗   ↖ People in group

Incorrect Set-up

$$\text{Lose weight} \rightarrow \frac{2}{3} = \frac{\square}{x} \leftarrow \text{People in group}$$
People surveyed ↗   ↖ Lose weight

(b) In one state, 3 out of 5 college students receive financial aid. At this rate, how many of the 4500 students at Central Community College receive financial aid?

Step 5 **State the answer.** You would expect **2240 people** in town to watch the news on TV.

> Be sure to write **people** in your answer.

Step 6 **Check** that the solution is reasonable by putting it back into the original statement of the problem.

Notice that 7 out of 10 people is more than half the people, but less than all the people. Half of the 3200 people in town is $3200 \div 2 = 1600$, so between 1600 and 3200 people would be expected to watch the news on TV. The solution, 2240 people, is between 1600 and 3200, so it is reasonable.

❗ CAUTION

Always check that your answer is reasonable. If it isn't, look at the way your proportion is set up. Be sure you have matching units in the numerators and matching units in the denominators.

For example, suppose you set up the last proportion **incorrectly,** as shown below.

$$\frac{7}{10} = \frac{3200}{x} \leftarrow \text{Incorrect set-up}$$

$$(7)(x) = (10)(3200)$$

$$\frac{\overset{1}{\cancel{(7)}}(x)}{\underset{1}{\cancel{7}}} = \frac{32,000}{7}$$

$$x \approx 4571 \text{ people} \leftarrow \text{Unreasonable answer}$$

This answer is **unreasonable** because there are only 3200 people in the town; it is **not** possible for 4571 people to watch the news.

—————— ◀ **Work Problem 2** at the Side.

Study Skills Reminder

Mathematical skills and ideas always build from and connect to each other. A mind map is a visual way to see the connections between the different things you have learned. It also is a great tool to use when reviewing. See the Study Skills activity "Making a Mind Map" to learn how to make one.

Answers

2. **(a)** $\frac{2}{3} = \frac{x}{150}$; 100 people (reasonable);
   incorrect set-up of $\frac{2}{3} = \frac{150}{x}$; 225 people
   (unreasonable, only 150 people in the group).
   (b) 2700 students (reasonable);
   incorrect set-up gives 7500 students
   (only 4500 students at the college).

6.4 Exercises

FOR EXTRA HELP Go to *MyMathLab for worked-out, step-by-step solutions to exercises enclosed in a square* ▢ *and video solutions to* ▶ *exercises.*

CONCEPT CHECK *Complete the proportion for each situation in Exercises 1–2. Do **not** solve the proportion.*

1. When Linda makes potato salad, she uses 6 potatoes and 4 hard-boiled eggs. How many eggs will she need if she uses 12 potatoes? Write the rest of the units in the proportion below.

$$\frac{6 \text{ potatoes}}{4 \underline{\hspace{1cm}}} = \frac{12 \underline{\hspace{1cm}}}{x \underline{\hspace{1cm}}}$$

2. A potato chip bag says that 1 serving is 15 chips. Joe ate 70 chips. How many servings did he eat? Write the rest of the units in the proportion below.

$$\frac{1 \underline{\hspace{1cm}}}{15 \text{ chips}} = \frac{x \underline{\hspace{1cm}}}{70 \underline{\hspace{1cm}}}$$

Set up and solve a proportion for each application problem. ***See Example 1.***

3. Caroline can sketch four cartoon strips in five hours. How long will it take her to sketch 18 strips?

Compares cartoon strips to hours $\left.\right\}$ $\dfrac{4 \text{ strips}}{5 \text{ hours}} = \dfrac{\Box \text{ strips}}{\Box \text{ hours}}$

4. The Cosmic Toads recorded eight songs on their first album in 26 hours. At this same rate, how long will it take them to record 14 songs for their second album?

Compares songs to hours $\left.\right\}$ $\dfrac{8 \text{ songs}}{26 \text{ hours}} = \dfrac{\Box \text{ songs}}{\Box \text{ hours}}$

5. Sixty newspapers cost $27. Find the cost of 16 newspapers.

6. Twenty-two guitar lessons cost $528. Find the cost of 12 lessons.

7. If three pounds of fescue grass seed cover about 350 square feet of ground, how many pounds are needed for 4900 square feet?

8. Anna earns $1242.08 in 14 days. How much does she earn in 260 days?

9. Tom makes $672.80 in five days. How much does he make in three days?

10. If 5 ounces of a medicine must be mixed with 8 ounces of water, how many ounces of medicine would be mixed with 20 ounces of water?

11. The bag of rice noodles shown below makes 7 servings. At that rate, how many ounces of noodles do you need for 12 servings, to the nearest ounce?

12. This can of sweet potatoes is enough for 4 servings. How many ounces are needed for 9 servings, to the nearest ounce?

13. Three quarts of a latex enamel paint will cover about 270 square feet of wall surface. How many quarts will you need to cover 350 square feet of wall surface in your kitchen and 100 square feet of wall surface in your bathroom?

14. One gallon of clear gloss wood finish covers about 550 square feet of surface. If you need to apply three coats of finish to 400 square feet of surface, how many gallons do you need, to the nearest tenth?

Use the floor plan shown to answer Exercises 15–18. On the plan, one inch represents four feet.

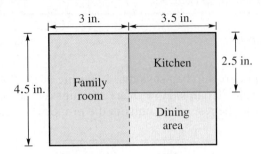

15. What is the actual length and width of the kitchen?

16. What is the actual length and width of the family room?

17. What is the actual length and width of the dining area?

18. What is the actual length and width of the entire floor plan?

The table below lists recommended amounts of food to order for 25 party guests. Use the table to answer Exercises 19 and 20. (Data from Cub Foods.)

FOOD FOR 25 GUESTS

Item	Amount
Fried chicken	40 pieces
Lasagna	14 pounds
Deli meats	4.5 pounds
Sliced cheese	$2\frac{1}{3}$ pounds
Bakery buns	3 dozen
Potato salad	6 pounds

19. How much of each food item should Nathan and Amanda order for a graduation party with 60 guests?

20. Taisha is having 20 neighbors over for a Fourth of July picnic. How much food should she buy?

21. CONCEPT CHECK 8 out of 10 students voted to take a fifteen-minute break in the middle of a two-hour class. There are 35 students in the class. How many students voted for the break? Carl set up the proportion below. Explain how to tell whether Carl's answer is reasonable or not.

$$\frac{8}{10} = \frac{35}{x}$$

$$8 \cdot x = 350$$

$$\frac{\overset{1}{\cancel{8}} \cdot x}{\underset{1}{\cancel{8}}} = \frac{350}{8}$$

$$x \approx 44 \text{ students (rounded)}$$

22. CONCEPT CHECK Many experts recommend 30 minutes of exercise each day. If Vera follows that advice for 60 days, how many minutes of exercise will she get? Vera set up the proportion below. Explain how to tell whether Vera's answer is reasonable or not.

$$\frac{1}{30} = \frac{x}{60}$$

$$30 \cdot x = 60$$

$$\frac{\overset{1}{\cancel{30}} \cdot x}{\underset{1}{\cancel{30}}} = \frac{60}{30}$$

$$x = 2 \text{ minutes}$$

Set up and solve a proportion for each problem.

23. The stock market report says that five stocks went up for every six stocks that went down. If 750 stocks went down yesterday, how many went up?

24. The human body contains 90 pounds of water for every 100 pounds of body weight. How many pounds of water are in a child who weighs 80 pounds?

25. The ratio of the length of an airplane wing to its width is 8 to 1. If the length of a wing is 32.5 meters, how wide must it be? Round to the nearest hundredth.

26. The Rosebud School District wants a student-to-teacher ratio of 19 to 1. How many teachers are needed for 1850 students? Round to the nearest whole number.

27. At 3 P.M., Coretta's shadow is 1.05 meters long. Her height is 1.68 meters. At the same time, a tree's shadow is 6.58 meters long. How tall is the tree? Round to the nearest hundredth.

28. (Complete **Exercise 27** first.) Later in the day, Coretta's shadow was 2.95 meters long. How long a shadow did the tree have at that time? Round to the nearest hundredth.

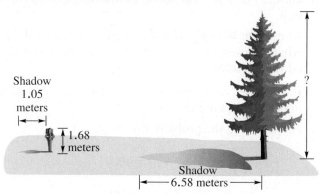

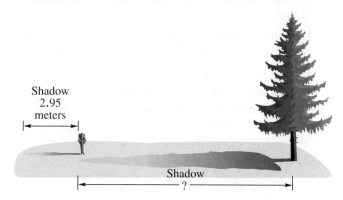

29. A survey of college students shows that 4 out of 5 drink coffee. Of the students who drink coffee, 1 out of 8 adds cream to it. How many of the 56,100 students at Ohio State University would be expected to use cream in their coffee?

30. About 9 out of 10 adults think it is a good idea to exercise regularly. But of the ones who think it is a good idea, only 1 in 6 actually exercises at least three times a week. At this rate, how many of the 300 employees in our company exercise regularly?

31. The nutrition label on a carton of diet ice cream says that a $\frac{1}{3}$ cup serving provides 80 calories and 8 grams of sugar. At that rate, how many calories and grams of sugar are in a $\frac{1}{2}$ cup serving?

32. A $\frac{3}{4}$ cup serving of whole grain penne pasta has 210 calories and 6 grams of dietary fiber. How many calories and grams of fiber would be in a 1 cup serving? (Data from Barilla Pasta.)

33. CONCEPT CHECK Can you set up a proportion to solve this problem? Explain why or why not. Jim is 25 years old and weighs 180 pounds. How much will he weigh when he is 50 years old?

34. CONCEPT CHECK Can you set up a proportion to solve this problem? Explain why or why not. Five pounds of chicken wings cost $11.75. How much will 3 pounds of chicken breasts cost?

CONCEPT CHECK *Set up a proportion to solve each problem. Check to see whether your answer is reasonable. Then flip one side of your proportion to see what answer you get with an incorrect set-up. Explain why the second answer is unreasonable. See Example 2.*

35. About 7 out of 10 people entering our community college need to take a refresher math course. If we have 2950 entering students, how many will probably need refresher math? (Data from Minneapolis Community and Technical College.)

36. In a survey, only 3 out of 100 people like their eggs poached. At that rate, how many of the 60 customers who ordered eggs at Soon-Won's restaurant this morning asked to have them poached? Round to the nearest whole person.

37. CONCEPT CHECK The work shown below has a mistake in it.

What Went Wrong? First write a sentence explaining what the mistake is. Then fix the mistake and find the correct solution.

A 150-pound person burns 210 calories during 30 minutes of tennis. How many calories would a 180-pound person burn?

Tali solved the problem this way:

$$\frac{150}{210} = \frac{30}{x}$$

$$150 \cdot x = 6300$$

$$\frac{\overset{1}{\cancel{150}} \cdot x}{\underset{1}{\cancel{150}}} = \frac{6300}{150}$$

$$x = 42 \text{ calories}$$

38. CONCEPT CHECK The work shown below has a mistake in it.

What Went Wrong? First write a sentence explaining what the mistake is. Then fix the mistake and find the correct solution.

If 2 cookies have 198 calories and 4 grams of fat, how many calories are in 5 cookies?

Amar solved the problem this way:

$$\frac{2}{198} = \frac{5}{x}$$

$$2 \cdot x = 990$$

$$\frac{\overset{1}{\cancel{2}} \cdot x}{\underset{1}{\cancel{2}}} = \frac{990}{2}$$

$$x = 495 \text{ grams of fat}$$

Relating Concepts (Exercises 39–40) For Individual or Group Work

*A box of instant mashed potatoes has the list of ingredients shown in the table. Use this information to **work Exercises 39–40 in order**. (Data from General Mills.)*

Ingredient	For 4 Servings
Water	$1\frac{1}{3}$ cups
Butter/Margarine	2 tablespoons
Milk	$\frac{2}{3}$ cup
Potato flakes	$1\frac{1}{2}$ cups

39. Find the amount of each ingredient needed for two servings. Show *two* different methods for finding the amounts. One method should use proportions.

40. Find the amount of each ingredient needed for six servings. Show *two* different methods for finding the amounts, one using proportions and one using your answers from **Exercise 39.**

6.5 Geometry: Lines and Angles

Geometry starts with the idea of a point. A **point** can be described as a location in space. It has no length or width. A point is represented by a dot and is named by writing a capital letter next to the dot.

$$\overset{\bullet}{}\,P$$

Point P

OBJECTIVE ➤ **1** **Identify and name lines, line segments, and rays.** A **line** is a straight row of points that goes on forever in both directions. A line is drawn by using arrowheads to show that it never ends. The line is named by using the letters of any two points on the line.

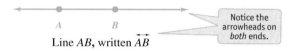

Line AB, written $\overleftrightarrow{AB}$

Notice the arrowheads on *both* ends.

A piece of a line that has two endpoints is called a **line segment.** A line segment is named for its endpoints. The segment with endpoints P and Q is shown below. It can be named $\overline{PQ}$ or $\overline{QP}$.

Line segment PQ, written $\overline{PQ}$

There are *no* arrowheads.

A **ray** is a part of a line that has only one endpoint and goes on forever in one direction. A ray is named by using the endpoint and some other point on the ray. The endpoint is always mentioned first.

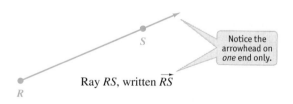

Ray RS, written $\overrightarrow{RS}$

Notice the arrowhead on *one* end only.

EXAMPLE 1   **Identifying and Naming Lines, Rays, and Line Segments**

Identify each figure below as a line, line segment, or ray and name it using the appropriate symbol.

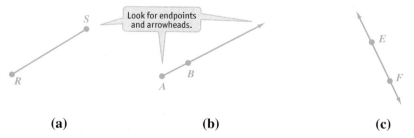

Look for endpoints and arrowheads.

(a)   (b)   (c)

Figure **(a)** has two endpoints, so it is a **line segment named $\overline{RS}$ or $\overline{SR}$.**
Figure **(b)** starts at point A and goes on forever in one direction, so it is a **ray named $\overrightarrow{AB}$.**
Figure **(c)** goes on forever in both directions, so it is a **line named $\overleftrightarrow{EF}$ or $\overleftrightarrow{FE}$.**

—————— **Work Problem ❶ at the Side.** ▶

OBJECTIVES

1 Identify and name lines, line segments, and rays.

2 Identify parallel and intersecting lines.

3 Identify and name angles.

4 Classify angles as right, acute, straight, or obtuse.

5 Identify perpendicular lines.

6 Identify complementary angles and supplementary angles, and find the measure of a complement or supplement of a given angle.

7 Identify congruent angles and vertical angles, and use this knowledge to find the measures of angles.

8 Identify corresponding angles and alternate interior angles, and use this knowledge to find the measures of angles.

---

❶ Identify each figure as a line, line segment, or ray, and name it.

(a)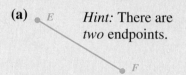
E *Hint:* There are *two* endpoints.

(b)

(c)

Answers

1. (a) line segment named $\overline{EF}$ or $\overline{FE}$
   (b) ray named $\overrightarrow{SR}$
   (c) line named $\overleftrightarrow{WX}$ or $\overleftrightarrow{XW}$

2 Label each pair of lines as appearing to be parallel or as intersecting.

GS (a)

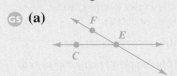

These lines cross, so they are _____ lines.

(b)

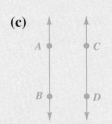

(c)

OBJECTIVE 2 Identify parallel and intersecting lines. A *plane* is an infinitely large, flat surface. A floor or a wall is part of a plane. Lines that are in the *same plane*, but that never intersect (never cross), are called **parallel lines,** while lines that cross are called **intersecting lines.** (Think of an intersection, where two streets cross each other.)

EXAMPLE 2   Identifying Parallel and Intersecting Lines

Label each pair of lines as appearing to be parallel or as intersecting.

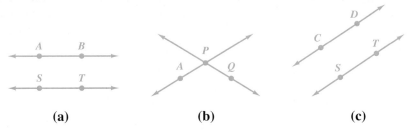

| (a) | (b) | (c) |

The lines in Figures **(a)** and **(c)** do not intersect; they appear to be **parallel lines.**

The lines in Figure **(b)** cross at *P*, so they are **intersecting lines.**

> **⊘ CAUTION**
>
> Appearances may be deceiving! Do not assume that lines are parallel unless it is stated that they are parallel.

◀ **Work Problem 2 at the Side.**

OBJECTIVE 3 Identify and name angles. An **angle** is made up of two rays that start at a common endpoint. This common endpoint is called the *vertex.*

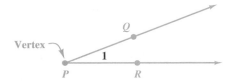

$\overrightarrow{PQ}$ and $\overrightarrow{PR}$ are called the *sides* of the angle. The angle can be named in four different ways, as shown below.

| ∠1 | ∠*P* | ∠*QPR* | ∠*RPQ* |

Vertex alone   Vertex in the middle

> **Naming an Angle**
>
> To name an angle, write the vertex alone or write the vertex in the middle of two other points, one from each side. If two or more angles have the *same vertex,* as in **Example 3** on the next page, do *not* use the vertex alone to name an angle.

Answers

2. (a) intersecting   (b) appear to be parallel
   (c) appear to be parallel

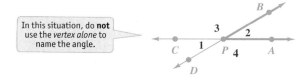

| EXAMPLE 3 | **Identifying and Naming an Angle** |

Name the highlighted angle in three different ways.

In this situation, do **not** use the *vertex alone* to name the angle.

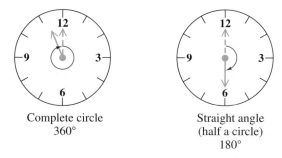

The angle can be named ∠**BPA**, ∠**APB**, or ∠**2**. It *cannot* be named ∠*P*, using the vertex alone, because four different angles have *P* as their vertex.

————————————— **Work Problem ③ at the Side.** ▶

| OBJECTIVE ▶ ④ | **Classify angles as right, acute, straight, or obtuse.** Angles can be measured in **degrees.** The symbol for degrees is a small, raised circle °. Think of the minute hand on a clock as a ray of an angle. Suppose it is at 12:00. During one hour of time, the minute hand moves around in a complete circle. It moves 360 *degrees*, or 360°. In half an hour, at 12:30, the minute hand has moved halfway around the circle, or 180°. An angle of 180° is called a **straight angle.** When two rays go in opposite directions and form a straight line, then the rays form a straight angle.

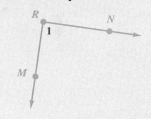

Complete circle
360°

Straight angle
(half a circle)
180°

In a quarter of an hour, at 12:15, the minute hand has moved $\frac{1}{4}$ of the way around the circle, or 90°. An angle of 90° is called a **right angle.** The rays of a right angle form one corner of a square. So, to show that an angle is a **right angle**, we draw a **small square** at the vertex.

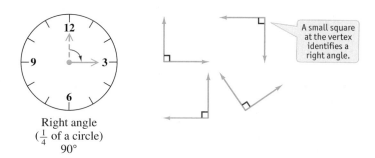

A small square at the vertex identifies a right angle.

Right angle
($\frac{1}{4}$ of a circle)
90°

An angle that measures 1° is shown below. You can see that an angle of 1° is very small.

➤ 1° angle

---

③ **(a)** Name the highlighted angle in three different ways.

(b) Darken the rays that make up ∠*ZTW*.

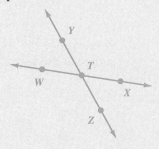

(c) Name this angle in four different ways.

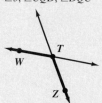

Answers

3. **(a)** ∠3, ∠*CQD*, ∠*DQC*
   (b)

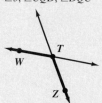

   (c) ∠1, ∠*R*, ∠*MRN*, ∠*NRM*

VOCABULARY TIP

Acute and Obtuse angles The word **acute** means "sharp." **Acute** angles make a sharper point than **obtuse** or right angles.

4 Label each angle as acute, right, obtuse, or straight. State the number of degrees in the right angle and in the straight angle.

(a)
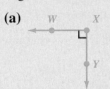

Hint: Notice the small red square at the vertex.

(b)
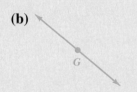

(c)
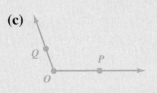

(d)

Some other terms used to describe angles are shown below.

Acute angles measure less than 90°.

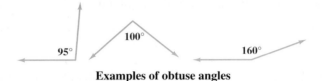

Examples of acute angles

Obtuse angles measure more than 90° but less than 180°.

Examples of obtuse angles

A tool called a *protractor* is used to measure the number of degrees in an angle.

> **Classifying Angles: Four Types of Angles**
>
> **Acute angles** measure less than 90°.
> **Right angles** measure *exactly* 90°.
> **Obtuse angles** measure more than 90° but less than 180°.
> **Straight angles** measure *exactly* 180°.

Note

Angles can also be measured in radians, which you will learn about in a later math course.

EXAMPLE 4 **Classifying an Angle**

Label each angle as acute, right, obtuse, or straight.

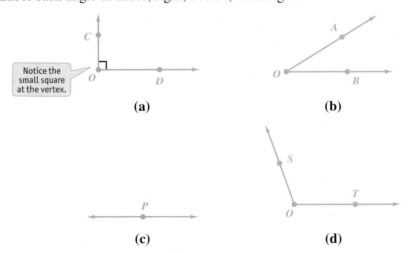

Notice the small square at the vertex.

(a) **(b)**

(c) **(d)**

Figure **(a)** shows a **right angle** (exactly 90° and identified by a small square at the vertex).

Figure **(b)** shows an **acute angle** (less than 90°).

Figure **(c)** shows a **straight angle** (exactly 180°).

Figure **(d)** shows an **obtuse angle** (more than 90° but less than 180°).

◀ **Work Problem** **4** at the Side.

Answers

4. (a) right angle; 90° **(b)** straight angle; 180°
   (c) obtuse angle **(d)** acute angle

OBJECTIVE ▶ ⑤ Identify perpendicular lines. Two lines are called **perpendicular lines** if they intersect to form a right angle.

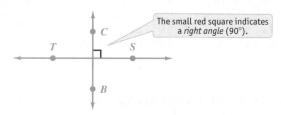

The small red square indicates a *right angle* (90°).

$\overleftrightarrow{CB}$ and $\overleftrightarrow{ST}$ are **perpendicular** lines because they intersect at right angles, as indicated by the small red square in the figure.

Perpendicular lines can be written in the following way: $\overleftrightarrow{CB} \perp \overleftrightarrow{ST}$.

EXAMPLE 5   Identifying Perpendicular Lines

Which pairs of lines are perpendicular?

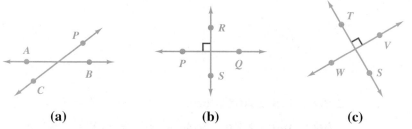

(a)   **(b)**   **(c)**

The lines in Figures **(b)** and **(c)** are **perpendicular** to each other, because they intersect at right angles.

The lines in Figure **(a)** are **intersecting lines,** but they are *not* perpendicular because they do *not* form a right angle.

— **Work Problem ⑤ at the Side.** ▶

OBJECTIVE ▶ ⑥ Identify complementary angles and supplementary angles, and find the measure of a complement or supplement of a given angle. Two angles are called **complementary angles** if the sum of their measures is 90°. If two angles are complementary, each angle is the *complement* of the other.

EXAMPLE 6   Identifying Complementary Angles

Identify each pair of complementary angles.

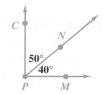

$\angle MPN$ **(40°)** and $\angle NPC$ **(50°)** are **complementary angles** because:

$$40° + 50° = 90°$$

$\angle CAB$ **(30°)** and $\angle FHG$ **(60°)** are **complementary angles** because:

$$30° + 60° = 90°$$

— **Work Problem ⑥ at the Side.** ▶

---

⑤ Which pair of lines is perpendicular? (*Hint:* Look for a right angle.) How can you describe the other pair of lines?

(a)

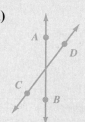

(b)

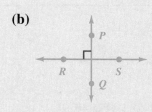

VOCABULARY TIP

Complementary angles Hint: The letter "**C**" in **c**omplementary can remind you of "**c**orner," like a 90° angle. **Complementary angles** add up to 90 degrees.

---

⑥ Identify each pair of complementary angles.

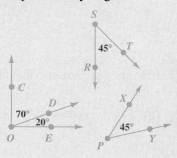

Answers

5. Figure **(b)** shows perpendicular lines; Figure **(a)** shows intersecting lines.

6. $\angle COD$ and $\angle DOE$; $\angle RST$ and $\angle XPY$

7 Find the complement of each angle.

GS **(a)** 35°   90° − 35° = _____

(b) 80°

VOCABULARY TIP

Supplementary angles *Hint:* The letter "**S**" in **s**upplementary can remind you of "**s**traight," like a 180° angle. **Supplementary angles** add up to 180 degrees.

---

8 Identify each pair of supplementary angles. (*Hint:* There are four pairs.)

9 Find the supplement of each angle.

GS **(a)** 175°   180° − 175° = _____

(b) 30°

Answers

7. **(a)** 55° **(b)** 10°
8. ∠CRF and ∠BRF; ∠CRE and ∠BRE; ∠FRB and ∠ERB; ∠ERC and ∠FRC
9. **(a)** 5° **(b)** 150°

> **EXAMPLE 7** **Finding the Complement of Angles**
>
> Find the complement of each angle.
>
>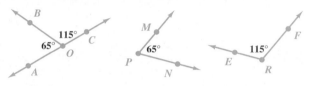
> Subtract from 90° to find the complement.
>
> **(a)** 30°
> Find the complement of 30° by subtracting. $90° − 30° = \mathbf{60°}$ ← Complement
>
> **(b)** 75°
> Find the complement of 75° by subtracting. $90° − 75° = \mathbf{15°}$ ← Complement

───────── ◀ **Work Problem 7 at the Side.**

Two angles are called **supplementary angles** if the sum of their measures is 180°. If two angles are supplementary, each angle is the *supplement* of the other.

> **EXAMPLE 8** **Identifying Supplementary Angles**
>
> Identify each pair of supplementary angles.
>
> ∠*BOA* **and** ∠*BOC,* because $65° + 115° = 180°$
> ∠*BOA* **and** ∠*ERF,* because $65° + 115° = 180°$
> ∠*BOC* **and** ∠*MPN,* because $115° + 65° = 180°$
> ∠*MPN* **and** ∠*ERF,* because $65° + 115° = 180°$

───────── ◀ **Work Problem 8 at the Side.**

> **EXAMPLE 9** **Finding the Supplement of Angles**
>
> Find the supplement of each angle.
>
> Subtract from 180° to find the supplement.
>
> **(a)** 70°
> Find the supplement of 70° by subtracting. $180° − 70° = \mathbf{110°}$ ← Supplement
>
> **(b)** 140°
> Find the supplement of 140° by subtracting. $180° − 140° = \mathbf{40°}$ ← Supplement

───────── ◀ **Work Problem 9 at the Side.**

OBJECTIVE 7 Identify congruent angles and vertical angles, and use this knowledge to find the measures of angles. Two angles are called **congruent angles** if they measure the same number of degrees. If two angles are congruent, this is written as ∠*A* ≅ ∠*B* and read as, "angle *A* is congruent to angle *B*." Here is an example.

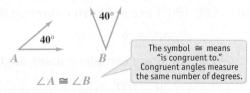

The symbol ≅ means "is congruent to." Congruent angles measure the same number of degrees.

Example of congruent angles

EXAMPLE 10 **Identifying Congruent Angles**

Identify the angles that are congruent.

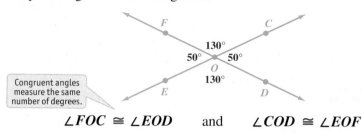

Congruent angles measure the same number of degrees.

∠FOC ≅ ∠EOD   and   ∠COD ≅ ∠EOF

— Work Problem ⑩ at the Side. ▶

Angles that share a common side and a common vertex are called *adjacent* angles, such as ∠FOC and ∠COD in **Example 10** above. Angles that do *not* share a common side are called *nonadjacent* angles. Two nonadjacent angles formed by two intersecting lines are called **vertical angles.**

EXAMPLE 11 **Identifying Vertical Angles**

Identify the vertical angles in this figure.

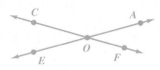

∠AOF and ∠COE are vertical angles because they do *not* share a common side and they are formed by two intersecting lines ($\overleftrightarrow{CF}$ and $\overleftrightarrow{EA}$).

∠COA and ∠EOF are also vertical angles.

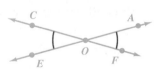

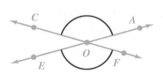

— Work Problem ⑪ at the Side. ▶

Look back at **Example 10** at the top of the page. Notice that the two *congruent* angles that measure 130° are also *vertical* angles. Also, the two congruent angles that measure 50° are vertical angles. This illustrates the following property.

Vertical Angles Are Congruent

If two angles are *vertical* angles, they are *congruent;* that is, they measure the same number of degrees.

⑩ Identify the angles that are congruent.

Hint: Congruent angles measure the same number of degrees.

⑪ Identify the vertical angles.

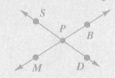

Answers

10. ∠BOC ≅ ∠AOD; ∠AOB ≅ ∠DOC
11. ∠SPB and ∠MPD; ∠BPD and ∠SPM.

12 In the figure below, find the measure of each unlabeled angle. Write the angle measures on the figure.

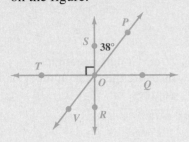

(a) ∠*TOS*

∠*TOS* has a small red square at the vertex, so it is a right angle and measures _____.

(b) ∠*QOR*

∠*QOR* and ∠*TOS* are vertical angles, so they are congruent. That means ∠*QOR* also measures _____.

(c) ∠*VOR*

(d) ∠*POQ*

(e) ∠*TOV*

Answers

12. (a) 90° **(b)** 90° **(c)** 38° **(d)** 52°
(e) 52°

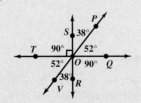

EXAMPLE 12   **Finding the Measures of Vertical Angles**

In the figure below, find the measure of each unlabeled angle.

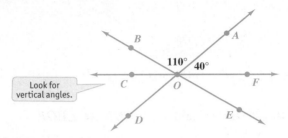

(a) ∠*COD*

∠*COD* and ∠*AOF* are vertical angles, so they are congruent. This means they measure the same number of degrees.

The measure of ∠*AOF* is 40°   so   the measure of ∠*COD* is **40°** also.

(b) ∠*DOE*

∠*DOE* and ∠*BOA* are vertical angles, so they are congruent.

The measure of ∠*BOA* is 110°   so   the measure of ∠*DOE* is **110°** also.

(c) ∠*COB*

Look at ∠*COB*, ∠*BOA*, and ∠*AOF*. Notice that $\overrightarrow{OC}$ and $\overrightarrow{OF}$ go in opposite directions. Therefore, ∠*COF* is a straight angle and measures 180°. To find the measure of ∠*COB*, subtract the sum of the other two angles from 180°.

$$180° - (110° + 40°) = 180° - (150°) = 30°$$

The measure of ∠**COB** is **30°**

(d) ∠*EOF*

∠*EOF* and ∠*COB* are vertical angles, so they are congruent. We know from part (c) above that the measure of ∠*COB* is 30° so the measure of ∠*EOF* is **30°** also.

◄ **Work Problem 12 at the Side.**

OBJECTIVE ▶ 8 **Identify corresponding angles and alternate interior angles, and use this knowledge to find the measures of angles.** We can also find congruent angles (angles with the same measure) when two *parallel lines* are crossed by a third line, called a *transversal*. When a transversal crosses two *parallel* lines, eight angles are formed, as shown below. There are special names for certain pairs of angles.

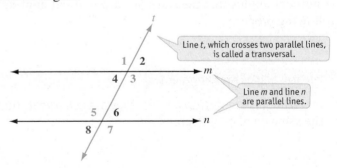

∠1 and ∠5 are called **corresponding angles.** Notice that they are both on the same side of the transversal (line *t*) and in the same relative position.

Corresponding angles are congruent, so ∠1 and ∠5 measure the same number of degrees. There are four pairs of corresponding angles.

∠1 and ∠5 are corresponding angles, so ∠1 ≅ ∠5

∠2 and ∠6 are corresponding angles, so ∠2 ≅ ∠6

∠3 and ∠7 are corresponding angles, so ∠3 ≅ ∠7

∠4 and ∠8 are corresponding angles, so ∠4 ≅ ∠8

When a transversal crosses two parallel lines, angles 3, 4, 5, and 6 are called *interior angles.* You can see that they are "inside" the *parallel* lines.

∠3 and ∠5 are alternate interior angles.

∠4 and ∠6 are alternate interior angles.

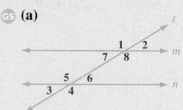

When two lines are *parallel,* then **alternate interior angles** *are congruent* (they have the same measure). Notice that alternate interior angles are on opposite (alternate) sides of the transversal.

$$\angle 3 \cong \angle 5 \quad \text{and} \quad \angle 4 \cong \angle 6$$

Angles Formed by Parallel Lines and a Transversal

When two parallel lines are crossed by a transversal:

1. Corresponding angles are congruent, and

2. Alternate interior angles are congruent.

EXAMPLE 13 **Identifying Corresponding Angles and Alternate Interior Angles**

In each figure, line *m* is parallel to line *n*. Identify all pairs of corresponding angles and all pairs of alternate interior angles.

(a)

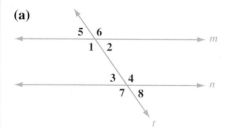

(b)

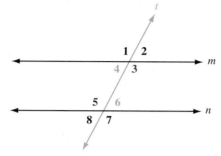

There are four pairs of **corresponding angles:**

∠5 and ∠3   ∠6 and ∠4
∠1 and ∠7   ∠2 and ∠8

Alternate interior angles:

∠1 and ∠4   ∠2 and ∠3

Corresponding angles:

∠1 and ∠5   ∠3 and ∠7
∠2 and ∠6   ∠4 and ∠8

Alternate interior angles:

∠3 and ∠6   ∠4 and ∠5

---

13 In each figure below, line *m* is parallel to line *n*. Identify all pairs of corresponding angles and all pairs of alternate interior angles.

GS (a)

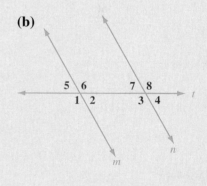

∠1 and ∠5 are *corresponding* angles. (Find three more pairs.)

∠7 and ∠6 are *alternate interior* angles. (Find one more pair.)

(b)

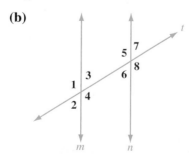

Answers

13. **(a)** corresponding angles: ∠1 and ∠5;
  ∠2 and ∠6; ∠7 and ∠3; ∠8 and ∠4
  alternate interior angles: ∠7 and ∠6;
  ∠8 and ∠5
  (b) corresponding angles: ∠5 and ∠7;
  ∠6 and ∠8; ∠1 and ∠3; ∠2 and ∠4
  alternate interior angles: ∠6 and ∠3;
  ∠2 and ∠7

Work Problem 13 at the Side. ▶

14 In each figure below, line *m* is parallel to line *n*.

GS **(a)** The measure of ∠6 is 150°. Find the measures of the other angles.

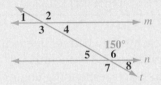

∠6 ≅ ∠2 (corresponding angles) so ∠2 is also _____

(b) The measure of ∠1 is 45°. Find the measures of the other angles.

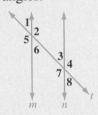

Recall that two angles are supplementary angles if the sum of their measures is 180°. Also remember that two rays that form a 180° angle form a straight line. Now you can combine your knowledge about supplementary angles with the information on parallel lines.

EXAMPLE 14 **Working with Parallel Lines**

In the figure at the right, line *m* is parallel to line *n* and the measure of ∠4 is 70°. Find the measures of the other angles.

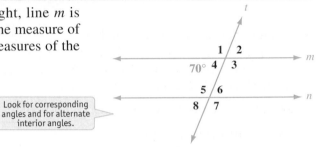

Look for corresponding angles and for alternate interior angles.

As you find the measure of each angle, write it on the figure.

∠4 ≅ ∠8 (corresponding angles), so the measure of **∠8 is also 70°**.

∠4 ≅ ∠6 (alternate interior angles), so the measure of **∠6 is also 70°**.

∠6 ≅ ∠2 (corresponding angles), so the measure of **∠2 is also 70°**.

Notice that the exterior sides of ∠4 and ∠3 form a straight line, that is, a straight angle of 180°. Therefore, ∠4 and ∠3 are supplementary angles and the sum of their measures is 180°. If ∠4 is 70° then ∠3 must be 110° because 180° − 70° = 110°. So the measure of **∠3 is 110°**.

∠3 ≅ ∠7 (corresponding angles), so the measure of **∠7 is also 110°**.

∠3 ≅ ∠5 (alternate interior angles), so the measure of **∠5 is also 110°**.

∠5 ≅ ∠1 (corresponding angles), so the measure of **∠1 is also 110°**.

With the measures of all the angles labeled, you can double-check that each pair of angles that forms a straight angle also adds up to 180°.

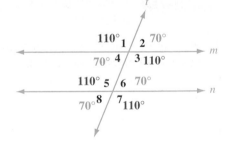

◀ **Work Problem 14 at the Side.**

Answers

14. (a) ∠2 is also 150°.

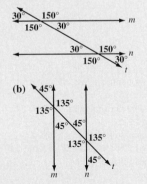

Study Skills Reminder

You are almost to the end of this chapter. You may have a test coming up soon. Would you like to improve your test scores? Do you feel anxious before or during tests? See the Study Skills activity "Tips for Taking Math Tests." You'll find helpful ideas on what to do during a test to reduce anxiety and get your best possible score.

6.5 Exercises

FOR EXTRA HELP Go to MyMathLab *for worked-out, step-by-step solutions to exercises enclosed in a square* ☐ *and video solutions to* ▶ *exercises.*

1. **CONCEPT CHECK** Explain the difference between a line, a line segment, and a ray. Draw a picture of each one.

2. **CONCEPT CHECK** Explain the difference between acute, obtuse, straight, and right angles. Draw a picture of each type of angle.

Identify each figure as a line, line segment, *or* ray *and name it using the appropriate symbol.* ***See Example 1.***

3.

4.

5.

6.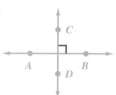

7.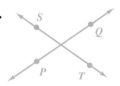

8.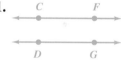

Label each pair of lines as appearing to be parallel, *as* perpendicular, *or as* intersecting. ***See Examples 2 and 5.***

9.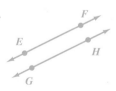

10.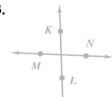

11.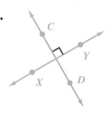

12.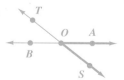

13.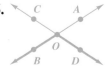

14.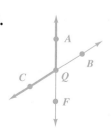

Name each highlighted angle by using the three-letter form of identification. ***See Example 3.***

15. ▶

16.

17.

18.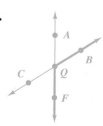

Label each angle as acute, right, obtuse, *or* straight. *For* right angles *and* straight angles, *indicate the number of degrees in the angle.* **See Example 4.**

19.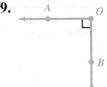

20.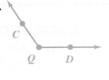

21.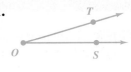

22.

23.

24.

Identify each pair of complementary angles. **See Example 6.**

25.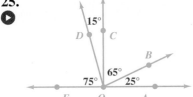

26.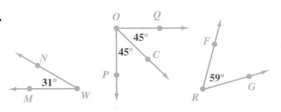

Identify each pair of supplementary angles. **See Example 8.**

27.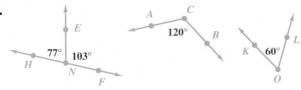

28.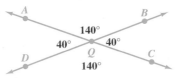

Find the complement of each angle. **See Example 7.**

29. 40°

GS   90° − 40° = _____

30. 35°

GS   90° − 35° = _____

31. 86°

32. 59°

Find the supplement of each angle. **See Example 9.**

33. 130°

GS   180° − 130° = _____

34. 75°

GS   180° − 75° = _____

35. 90°

36. 5°

*In Exercises 37 and 38, identify the angles that are congruent. **See Examples 10 and 11.***

37.
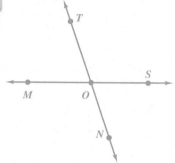

38.

*In Exercises 39 and 40, find the measure of each of the angles. **See Example 12.***

39. In the figure below, $\angle AOH$ measures 37° and $\angle COE$ measures 63°.

Hint: $\angle AOC$ completes a straight line (180° angle). Subtract $180° - (63° + 37°)$ to find $\angle AOC$.

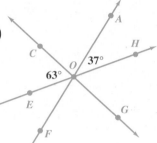

40. In the figure below, $\angle POU$ measures 105° and $\angle UOT$ measures 40°.

Hint: $\angle TOS$ completes a straight line (180° angle). Subtract $180° - (105° + 40°)$ to find $\angle TOS$.

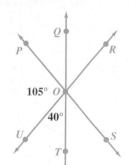

CONCEPT CHECK   *Use the figure to work Exercises 41–46. Decide whether each statement is* true *or* false. *If it is* true, *explain why. If it is* false, *rewrite it to make a true statement.*

41. $\angle UST$ is 90°.

42. $\overleftrightarrow{SQ}$ and $\overleftrightarrow{PQ}$ are perpendicular.

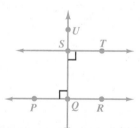

43. The measure of $\angle USQ$ is less than the measure of $\angle PQR$.

44. $\overleftrightarrow{ST}$ and $\overleftrightarrow{PR}$ are intersecting.

45. $\overleftrightarrow{QU}$ and $\overleftrightarrow{TS}$ are parallel.

46. $\angle UST$ and $\angle UQR$ measure the same number of degrees.

*In each figure, line m is parallel to line n. Identify all pairs of corresponding angles and all pairs of alternate interior angles. **See Example 13.***

47.

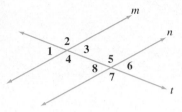

48.

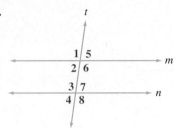

In each figure, line m is parallel to line n. Find the measure of each angle.
See Example 14.

49. ∠8 measures 130°

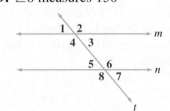

50. ∠2 measures 80°

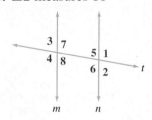

51. ∠6 measures 47°

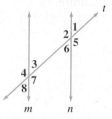

52. ∠2 measures 108°

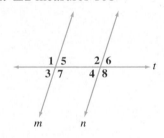

53. ∠6 measures 114°

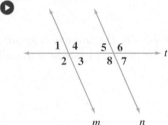

54. ∠3 measures 59°

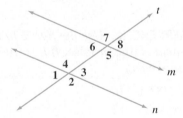

In each figure, $\overrightarrow{BA}$ is parallel to $\overrightarrow{CD}$. Find the measure of each numbered angle.

55.

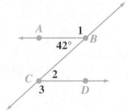

56.

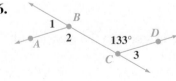

6.6 Geometry Applications: Congruent and Similar Triangles

Two useful concepts in geometry are *congruence* and *similarity*. If two figures are *identical*, both in *shape* and in *size*, we say the figures are **congruent,** or perfect duplicates of each other. This is like getting two exact copies of a photo. If two figures have the *same shape* but are *different sizes,* we say the figures are **similar,** like getting a photo and then enlarging it. We'll explore the ideas of congruence and similarity using triangles.

OBJECTIVES

1 Identify corresponding parts of congruent triangles.

2 Prove that triangles are congruent using ASA, SSS, or SAS.

3 Identify corresponding parts of similar triangles.

4 Find the unknown lengths of sides in similar triangles.

5 Solve application problems involving similar triangles.

OBJECTIVE ▶ **1** **Identify corresponding parts of congruent triangles.** The two triangles below are *congruent* because they are the *same shape* and the *same size*.

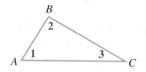

One way to name this triangle is $\triangle ABC$.

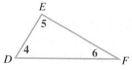

One way to name this triangle is $\triangle DEF$.

If you pick up $\triangle ABC$ and slide it over on top of $\triangle DEF$, the triangles will be a perfect match. $\angle 1$ will be on top of $\angle 4$, so they are *corresponding angles*. Similarly, $\angle 2$ and $\angle 5$ are corresponding angles, and $\angle 3$ and $\angle 6$ are corresponding angles. Corresponding angles have the same measure, as indicated below.

$$m \angle 1 = m \angle 4 \qquad m \angle 2 = m \angle 5 \qquad m \angle 3 = m \angle 6$$

The abbreviation for measure is m, so $m \angle 1$ is read, "the measure of angle 1."

When you put $\triangle ABC$ on top of $\triangle DEF$, you will also see that side AB is on top of side DE. We say that $\overline{AB}$ and $\overline{DE}$ are *corresponding sides*. Similarly, $\overline{BC}$ and $\overline{EF}$ are corresponding sides, and $\overline{AC}$ and $\overline{DF}$ are corresponding sides. You will see that corresponding sides have the same length.

$$AB = DE \qquad BC = EF \qquad AC = DF$$

Because corresponding angles have the same measure, and corresponding sides have the same length, we know that $\triangle ABC$ **is congruent to** $\triangle DEF$. We can write this as $\triangle ABC \cong \triangle DEF$.

Congruent Triangles

If two triangles are congruent, then

1. Corresponding angles have the same measure, and

2. Corresponding sides have the same length.

EXAMPLE 1 **Identifying Corresponding Parts in Congruent Triangles**

Each pair of triangles is congruent. List the corresponding angles and corresponding sides.

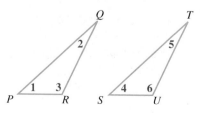

(a) If you slide $\triangle PQR$ on top of $\triangle STU$, the two triangles will match.

The **corresponding parts are:**

$\angle 1$ and $\angle 4$	$\overline{PQ}$ and $\overline{ST}$
$\angle 2$ and $\angle 5$	$\overline{PR}$ and $\overline{SU}$
$\angle 3$ and $\angle 6$	$\overline{QR}$ and $\overline{TU}$

—— **Continued on Next Page**

1 Each pair of triangles is congruent. List the corresponding angles and the corresponding sides.

(a)

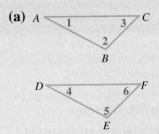

(b)

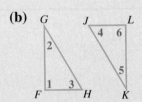

Hint: Rotate △ *FGH*, then slide it on top of △ *JLK*.

(c)

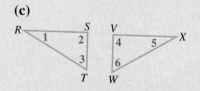

Hint: Flip △ *RST* over, then slide it on top of △ *VWX*.

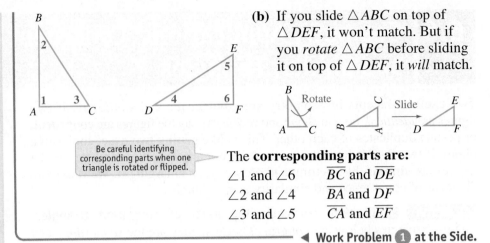

(b) If you slide △*ABC* on top of △*DEF*, it won't match. But if you *rotate* △*ABC* before sliding it on top of △*DEF*, it *will* match.

Be careful identifying corresponding parts when one triangle is rotated or flipped.

The **corresponding parts are:**

∠1 and ∠6 $\overline{BC}$ and $\overline{DE}$
∠2 and ∠4 $\overline{BA}$ and $\overline{DF}$
∠3 and ∠5 $\overline{CA}$ and $\overline{EF}$

◀ Work Problem **1** at the Side.

OBJECTIVE **2** **Prove that triangles are congruent using ASA, SSS, or SAS.**
One way to prove that two triangles are congruent would be to measure all the angles and all the sides. If the measures of the corresponding angles and sides are equal, then the triangles are congruent. But here are three quicker methods to prove that two triangles are congruent.

Proving That Two Triangles Are Congruent

1. Angle–Side–Angle (ASA) Method
If two angles and the side between them on one triangle measure the same as the corresponding parts on another triangle, the triangles are congruent.

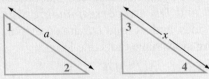

If $m\angle 1 = m\angle 3$ and $m\angle 2 = m\angle 4$ and $a = x$ then the two triangles are congruent.

2. Side–Side–Side (SSS) Method
If three sides of one triangle measure the same as the corresponding sides of another triangle, the triangles are congruent.

If $a = x$ and $b = y$ and $c = z$ then the two triangles are congruent.

3. Side–Angle–Side (SAS) Method
If two sides and the angle between them on one triangle measure the same as the corresponding parts on another triangle, the triangles are congruent.

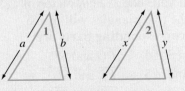

If $a = x$ and $b = y$ and $m\angle 1 = m\angle 2$ then the two triangles are congruent.

Answers

1. **(a)** ∠1 and ∠4, ∠2 and ∠5, ∠3 and ∠6;
$\overline{AC}$ and $\overline{DF}$, $\overline{AB}$ and $\overline{DE}$, $\overline{BC}$ and $\overline{EF}$

(b) ∠1 and ∠6, ∠2 and ∠5, ∠3 and ∠4;
$\overline{GF}$ and $\overline{KL}$, $\overline{FH}$ and $\overline{LJ}$, $\overline{GH}$ and $\overline{KJ}$

(c) ∠1 and ∠5, ∠2 and ∠4, ∠3 and ∠6;
$\overline{RS}$ and $\overline{XV}$, $\overline{RT}$ and $\overline{XW}$, $\overline{ST}$ and $\overline{VW}$

EXAMPLE 2 **Proving That Two Triangles Are Congruent**

Explain which method can be used to prove that each pair of triangles is congruent. Choose from ASA, SSS, and SAS.

(a)

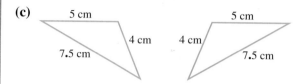

16 ft   16 ft
110°   110°
9 ft   9 ft

(b)

50°   50°
6 m   6 m

(c)

5 cm   5 cm
4 cm   4 cm
7.5 cm   7.5 cm

(a) On both triangles, two corresponding sides and the angle between them measure the same, so the **Side–Angle–Side (SAS) method** can be used to prove that the triangles are congruent.

(b) On both triangles, two corresponding angles and the side between them measure the same, so the **Angle–Side–Angle (ASA) method** can be used to prove that the triangles are congruent.

(c) Each pair of corresponding sides has the same length, so the **Side–Side–Side (SSS) method** can be used to prove that the triangles are congruent.

—————————————— **Work Problem ❷ at the Side.** ▶

OBJECTIVE ❸ Identify corresponding parts of similar triangles. Now that you've worked with *congruent* triangles, let's look at *similar* triangles. Remember that congruent triangles match exactly, both in shape and in size. Similar triangles, on the other hand, have the *same shape* but are *different sizes*. Three pairs of similar triangles are shown here.

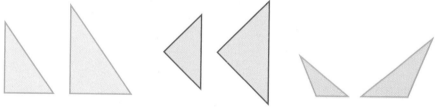

Each pair of triangles has the same shape because the corresponding angles have the same measure. But the corresponding sides are *not* the same length, so the triangles are of *different sizes*.

Two similar triangles are shown to the right. Notice that corresponding angles have the same measure, but corresponding sides have different lengths.
$\overline{RQ}$ corresponds to $\overline{CB}$. Also, $\overline{RP}$ corresponds to $\overline{CA}$, and $\overline{QP}$ corresponds to $\overline{BA}$. Notice that each side in the smaller triangle is *half* the length of the corresponding side in the larger triangle.

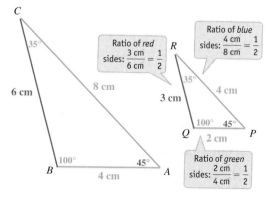

Ratio of *red* sides: $\frac{3 \text{ cm}}{6 \text{ cm}} = \frac{1}{2}$

Ratio of *blue* sides: $\frac{4 \text{ cm}}{8 \text{ cm}} = \frac{1}{2}$

Ratio of *green* sides: $\frac{2 \text{ cm}}{4 \text{ cm}} = \frac{1}{2}$

Work Problem ❸ at the Side. ▶

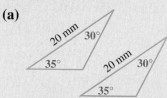
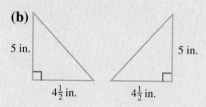
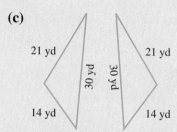
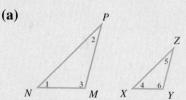
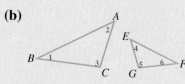

4 Find the length of $\overline{EF}$ in
GS **Example 3** by setting up and
solving a proportion. Let x
represent the unknown length.

$$\begin{array}{cc} EF \rightarrow & x \\ CB \rightarrow & \overline{33} \end{array} = \underline{\qquad} \begin{array}{c} \leftarrow ED \\ \leftarrow CA \end{array}$$

OBJECTIVE ▶ **4** **Find the unknown lengths of sides in similar triangles.**
Similar triangles are useful because of the following definition.

Similar Triangles

If two triangles are similar, then

1. Corresponding angles have the same measure, and

2. The *ratios* of the lengths of corresponding sides are equal.

EXAMPLE 3 Finding the Unknown Lengths of Sides in Similar Triangles

Find the length of $\overline{DF}$ in the smaller triangle. Assume the triangles are similar.

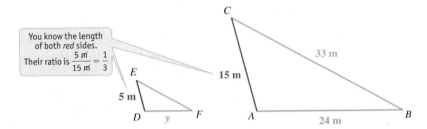

The length you want to find in the smaller triangle is $\overline{DF}$, and it corresponds
to $\overline{AB}$ in the *larger* triangle. Then, notice that $\overline{ED}$ in the smaller triangle
corresponds to $\overline{CA}$ in the larger triangle, and you know both of their lengths.
Since the *ratios* of the lengths of corresponding sides are equal, you can set
up a proportion. (Recall that a proportion states that two ratios are equal.)

Corresponding sides
Smaller triangle in numerator
Larger triangle in denominator
$\left\{ \begin{array}{cc} DF \rightarrow & y \\ AB \rightarrow & \overline{24} \end{array} = \begin{array}{c} 5 \\ \overline{15} \end{array} \begin{array}{c} \leftarrow ED \\ \leftarrow CA \end{array} \right\}$
Corresponding sides
Smaller triangle in numerator
Larger triangle in denominator

$$\frac{y}{24} = \frac{1}{3}$$ Write $\frac{5}{15}$ in lowest terms as $\frac{1}{3}$

Find the cross products.

$$24 \cdot 1 = 24$$

$$\frac{y}{24} \diagdown \frac{1}{3}$$

$$y \cdot 3$$

$$y \cdot 3 = 24$$ Show that the cross products are equal.

$$\frac{y \cdot \overset{1}{3}}{\underset{1}{3}} = \frac{24}{3}$$ Divide both sides by 3

$$y = 8$$

Write **m** in the answer.

$\overline{DF}$ has a length of 8 m

◀ **Work Problem** **4** **at the Side.**

Answer

4. $\dfrac{x}{33} = \dfrac{5}{15}; \quad x = 11$ m

| EXAMPLE 4 | Finding an Unknown Length and the Perimeter |

Find the perimeter of the smaller triangle. Assume the triangles are similar.

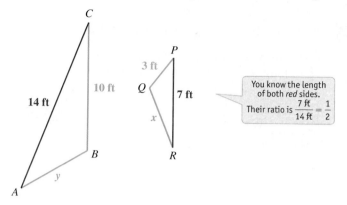

You know the length of both *red* sides.
Their ratio is $\dfrac{7\text{ ft}}{14\text{ ft}} = \dfrac{1}{2}$

First find *x*, the length of $\overline{QR}$ in the smaller triangle, then add the lengths of all three sides to find the perimeter.

The smaller triangle is turned "upside down" compared to the larger triangle, so be careful when identifying corresponding sides.

$\overline{PR}$ is the longest side in the smaller triangle, and $\overline{AC}$ is the longest side in the larger triangle. So $\overline{PR}$ and $\overline{AC}$ are corresponding sides and you know both of their lengths. $\overline{QR}$, the length you want to find in the smaller triangle, corresponds to $\overline{BC}$ in the larger triangle. The ratios of the lengths of corresponding sides are equal, so you can set up a proportion.

$$\begin{array}{c} QR \rightarrow \\ BC \rightarrow \end{array} \dfrac{x}{10} = \dfrac{7}{14} \begin{array}{c} \leftarrow PR \\ \leftarrow AC \end{array}$$

$$\dfrac{x}{10} = \dfrac{1}{2} \qquad \text{Write } \dfrac{7}{14} \text{ in lowest terms as } \dfrac{1}{2}$$

Find the cross products.

$$10 \cdot 1 = 10$$

$$\dfrac{x}{10} \bowtie \dfrac{1}{2}$$

$$x \cdot 2$$

$$x \cdot 2 = 10 \qquad \text{Show that the cross products are equal.}$$

$$\dfrac{x \cdot \overset{1}{2}}{\underset{1}{2}} = \dfrac{10}{2} \qquad \text{Divide both sides by 2}$$

Write **ft** for the length.

$$x = 5$$

$\overline{QR}$ has a length of 5 ft

Now add the lengths of all three sides to find the perimeter of the smaller triangle.

$$\text{Perimeter} = 5\text{ ft} + 3\text{ ft} + 7\text{ ft}$$

$$\textbf{Perimeter} = \textbf{15 ft}$$

────── **Work Problem ⑤ at the Side.** ▶

⑤ (a) Find the perimeter of ⑤ triangle *ABC* in **Example 4.**

Let *y* represent the unknown length of side *AB*.

$$\begin{array}{c} PQ \rightarrow \\ AB \rightarrow \end{array} \dfrac{3}{y} = \underline{\hspace{1cm}} \begin{array}{c} \leftarrow PR \\ \leftarrow AC \end{array}$$

(b) Find the perimeter of each triangle. Assume the triangles are similar.

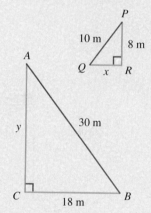

Answers

5. (a) $\dfrac{3}{y} = \dfrac{7}{14}$; $y = 6$ ft

Perimeter $= 14$ ft $+ 10$ ft $+ 6$ ft $= 30$ ft

(b) $x = 6$ m, Perimeter $= 24$ m
$y = 24$ m, Perimeter $= 72$ m

6 Find the height of each flagpole.

GS **(a)**

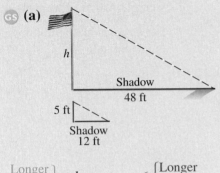

Shadow
48 ft

Shadow
12 ft

$$\begin{matrix}\text{Longer} \\ \text{height}\end{matrix}\Big\} \rightarrow \frac{h}{5} = \underline{\quad} \begin{matrix}\leftarrow \begin{cases}\text{Longer} \\ \text{shadow}\end{cases} \\ \leftarrow \begin{cases}\text{Shorter} \\ \text{shadow}\end{cases}\end{matrix}$$

(b)

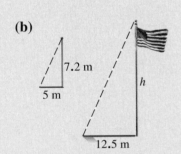

7.2 m

5 m

h

12.5 m

OBJECTIVE ▶ **5** **Solve application problems involving similar triangles.**
The next example shows an application of similar triangles.

EXAMPLE 5 **Using Similar Triangles in an Application**

A flagpole casts a shadow 99 m long at the same time that a pole 10 m tall casts a shadow 18 m long. Find the height of the flagpole.

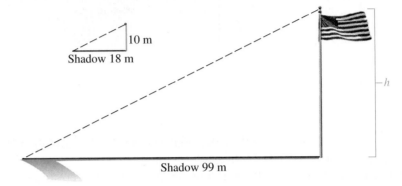

10 m

Shadow 18 m

Shadow 99 m

The triangles shown are similar, so write a proportion to find h.

$$\text{Height in \textbf{larger} triangle} \rightarrow \frac{h}{10} = \frac{99}{18} \leftarrow \text{\textbf{Shadow} in \textbf{larger} triangle}$$
$$\text{Height in \textbf{smaller} triangle} \rightarrow \qquad\qquad \leftarrow \text{\textbf{Shadow} in \textbf{smaller} triangle}$$

Find the cross products and show that they are equal.

$$h \cdot 18 = 10 \cdot 99$$
$$h \cdot 18 = 990$$
$$\frac{h \cdot \overset{1}{\cancel{18}}}{\underset{1}{\cancel{18}}} = \frac{990}{18} \qquad \text{Divide both sides by 18}$$
$$h = 55$$

The **flagpole is 55 m high.**

Note

There are several other correct ways to set up the proportion in **Example 5** above. One way is to simply flip the ratios on *both* sides of the equal sign.

$$\frac{10}{h} = \frac{18}{99}$$

But there is another option, shown below.

$$\text{Height in larger triangle} \rightarrow \frac{h}{99} = \frac{10}{18} \leftarrow \text{Height in smaller triangle}$$
$$\text{Shadow in larger triangle} \rightarrow \qquad\qquad \leftarrow \text{Shadow in smaller triangle}$$

Notice that both ratios compare *height* to *shadow* in the same order. The ratio on the left describes the larger triangle, and the ratio on the right describes the smaller triangle.

◀ **Work Problem** **6** **at the Side.**

Answers

6. **(a)** $\frac{h}{5} = \frac{48}{12}$; $h = 20$ ft **(b)** $h = 18$ m

6.6 Exercises

FOR EXTRA HELP Go to MyMathLab for worked-out, step-by-step solutions to exercises enclosed in a square ☐ and video solutions to ▶ exercises.

1. **CONCEPT CHECK** Look up the word *congruent* in a dictionary. What is the nonmathematical definition of this word? Describe two examples of congruent objects at home, school, or work.

2. **CONCEPT CHECK** Look up the word *similar* in a dictionary. What is the nonmathematical definition of this word? Describe two examples of similar objects at home, school, or work.

Each pair of triangles is congruent. List the corresponding angles and the corresponding sides. **See Example 1.**

3.
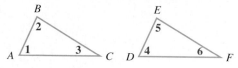

4.
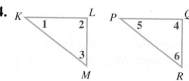

5.
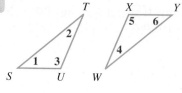

6.
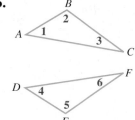

7.
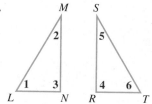

8.
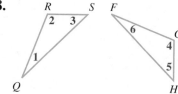

State which of these methods can be used to prove that each pair of triangles is congruent; Angle–Side–Angle (ASA), Side–Side–Side (SSS), or Side–Angle–Side (SAS). See Example 2.

9.

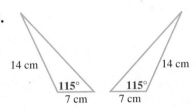

10.

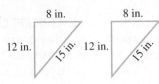

11.

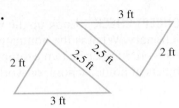

12.

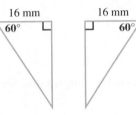

13.

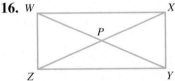

14.

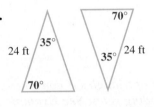

Given the information in Exercises 15–18, show how you can prove that the indicated triangles are congruent. Note: A midpoint divides a segment into two congruent parts.

15.

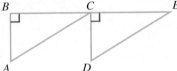

C is the midpoint of $\overline{BE}$, and $CD = BA$.
Prove that $\triangle ABC \cong \triangle DCE$.

16.

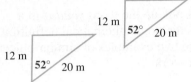

P is the midpoint of both $\overline{WY}$ and $\overline{XZ}$, and $WZ = XY$.
Prove that $\triangle WPZ \cong \triangle YPX$.

17.

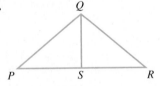

$\overline{QS} \perp \overline{PR}$, and S is the midpoint of $\overline{PR}$.
Prove that $\triangle PQS \cong \triangle RQS$.

18.

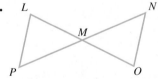

M is the midpoint of both $\overline{LO}$ and $\overline{PN}$.
Prove that $\triangle PLM \cong \triangle NOM$.

In Exercises 19–24, find the unknown lengths in each pair of similar triangles.
See Example 3.

19.
GS

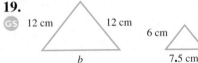

Solve two proportions.

$$\frac{a}{12} = \frac{6}{12} \qquad \frac{7.5}{b} = \frac{\square}{\square}$$

20.
GS

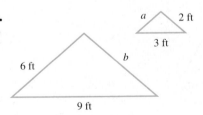

Solve two proportions.

$$\frac{a}{6} = \frac{3}{\square} \qquad \frac{2}{b} = \frac{\square}{\square}$$

21.

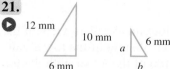

22.

23.

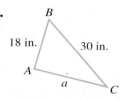

24.

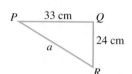

In Exercises 25 and 26, find the perimeter of each triangle. Assume the triangles are similar. See Example 4.

25.

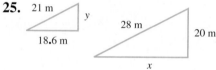

26.

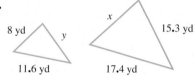

27. Triangles *CDE* and *FGH* are similar. Find the perimeter and area of triangle *FGH*. *Note:* The heights of similar triangles have the same ratio as corresponding sides. Round the height to the nearest tenth if necessary.

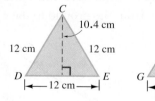

28. Triangles *JKL* and *MNO* are similar. Find the perimeter and area of triangle *MNO*.

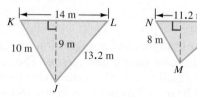

Solve each application problem. See Example 5.

29. The height of the house shown here can be found by comparing its shadow to the shadow cast by a 3-foot stick. Find the height of the house by writing a proportion and solving it.

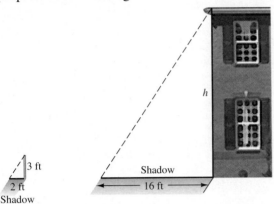

30. A fire lookout tower provides an excellent view of the surrounding countryside. The height of the tower can be found by lining up the top of the tower with the top of a 2-meter stick. Use similar triangles to find the height of the tower.

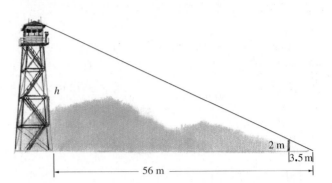

Find the unknown length in Exercises 31–34. Round your answers to the nearest tenth. Note: When a line is drawn parallel to one side of a triangle, the smaller triangle that is formed will be similar to the original triangle. In Exercises 31–32, the red segments are parallel.

31.

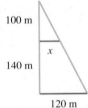

Hint: Redraw the two triangles and label the sides.

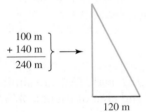

32.

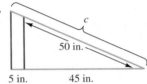

Hint: Redraw the two triangles and label the sides.

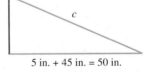 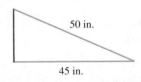

33. Use similar triangles and a proportion to find the length of the lake shown here. (*Hint:* The side 100 m long in the smaller triangle corresponds to the side of 100 m + 120 m = 220 m in the larger triangle.)

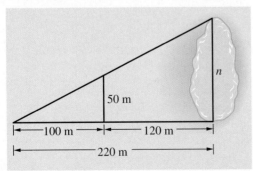

34. To find the height of the tree, find *y* and then add $5\frac{1}{2}$ ft for the distance from the ground to the eye level of the person.

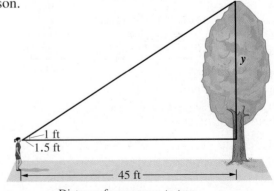

Distance from person to tree.

Chapter 6 **Summary**

Key Terms

6.1

ratio A ratio compares two quantities that have the same units. For example, the ratio of 6 apples to 11 apples is written in fraction form as $\frac{6}{11}$. The common units (apples) divide out.

6.2

rate A rate compares two measurements with different units. Examples are 96 dollars for 8 hours, or 450 miles on 18 gallons.

unit rate A unit rate has 1 in the denominator.

cost per unit Cost per unit is a rate that tells how much you pay for one item or one unit. The lowest cost per unit is the best buy.

6.3

proportion A proportion states that two ratios or rates are equal.

cross products Multiply along one diagonal and then along the other diagonal to find the cross products of a proportion. If the cross products are equal, the proportion is true.

6.5

point A point is a location in space. *Example:* Point *P* at the right.

line A line is a straight row of points that goes on forever in both directions. *Example:* Line *AB*, written $\overleftrightarrow{AB}$, at the right.

line segment A line segment is a piece of a line with two endpoints. *Example:* Line segment *PQ*, written $\overline{PQ}$, at the right.

ray A ray is a part of a line that has one endpoint and extends forever in one direction. *Example:* Ray *RS*, written $\overrightarrow{RS}$, at the right.

parallel lines Parallel lines are two lines in the same plane that never intersect (never cross). *Example:* $\overleftrightarrow{AB}$ is parallel to $\overleftrightarrow{ST}$ at the right.

intersecting lines Intersecting lines cross. *Example:* $\overleftrightarrow{RQ}$ intersects $\overleftrightarrow{AB}$ at point *P* at the right.

angle An angle is made up of two rays that have a common endpoint called the vertex. *Example:* Angle 1 at the right.

degrees Degrees are used to measure angles; a complete circle is 360 degrees, written 360°.

straight angle A straight angle is an angle that measures *exactly* 180°; its sides form a straight line. *Example:* Angle *G* at the right.

right angle A right angle is an angle that measures *exactly* 90°, identified by a small square at the vertex. *Example:* Angle *AOB* at the right.

acute angle An acute angle is an angle that measures less than 90°. *Example*: Angle *E* at the right.

obtuse angle An obtuse angle is an angle that measures more than 90° but less than 180°. *Example:* Angle *F* at the right.

perpendicular lines Perpendicular lines are two lines that intersect to form a right angle. *Example:* $\overleftrightarrow{PQ}$ is perpendicular to $\overleftrightarrow{RS}$ at the right.

complementary angles Complementary angles are two angles whose measures add up to 90°.

supplementary angles Supplementary angles are two angles whose measures add up to 180°.

congruent angles Congruent angles are angles that measure the same number of degrees.

vertical angles Vertical angles are two nonadjacent congruent angles formed by two intersecting lines. *Example*: ∠*COA* and ∠*EOF* are vertical angles at the right.

corresponding angles Corresponding angles are formed when two parallel lines are crossed by a transversal; corresponding angles are congruent and are on the same side of the transversal and in the same relative position. *Example:* In the figure at the right, line *m* is parallel to line *n*. The pairs of corresponding angles are ∠1 and ∠5, ∠2 and ∠6, ∠3 and ∠7, ∠4 and ∠8.

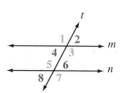

Key Terms

alternate interior angles When two parallel lines are crossed by a transversal, there are two pairs of alternate interior angles and each pair is congruent. They are on opposite sides of the transversal. *Example*: In the figure at the right, line *m* is parallel to line *n*. The pairs of alternate interior angles are $\angle 3$ and $\angle 5$, $\angle 4$ and $\angle 6$.

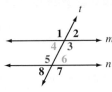

6.6

congruent figures Congruent figures are identical both in shape and in size.

similar figures Similar figures have the same shape but are different sizes.

congruent triangles Congruent triangles are triangles with the same shape and the same size; corresponding angles measure the same number of degrees and corresponding sides have the same length.

similar triangles Similar triangles are triangles with the same shape but not necessarily the same size; corresponding angles measure the same number of degrees and the *ratios* of the lengths of corresponding sides are equal.

New Symbols

$\overleftrightarrow{AB}$	line *AB*	$\perp$	is perpendicular to	$\cong$	is congruent to
$\overline{EF}$	line segment *EF*	$\angle MRN$	angle *MRN*	$m\angle 2$	the measure of angle 2
$\overrightarrow{RS}$	ray *RS*	$1°$	one degree	$\triangle ABC$	triangle *ABC*

Test Your Word Power

See how well you have learned the vocabulary in this chapter.

1 A **rate**
 A. can be written only as a decimal
 B. compares two quantities that have the same units
 C. compares two quantities that have different units.

2 **Cost per unit** is
 A. the best buy
 B. a ratio written in lowest terms
 C. the price of one item or one unit.

3 A **proportion**
 A. shows that two ratios or rates are equal
 B. contains only whole numbers or decimals
 C. always has one unknown number.

4 An **obtuse angle**
 A. is formed by perpendicular lines
 B. measures more than 90° but less than 180°
 C. measures less than 90°.

5 Two angles that are **complementary**
 A. have measures that add up to 180°
 B. form a straight angle
 C. have measures that add up to 90°.

6 **Perpendicular lines**
 A. intersect to form a right angle
 B. intersect to form an acute angle
 C. never intersect.

Answers to Test Your Word Power

1. C; *Example:* $\$4.50$ for 3 pounds is a rate comparing dollars to pounds.

2. C; *Example:* $\$1.95$ per pound tells the price of one pound (one unit).

3. A; *Example:* The proportion $\frac{5}{6} = \frac{25}{30}$ says that $\frac{5}{6}$ is equal to $\frac{25}{30}$.

4. B; *Examples:* Angles that measure 91°, 120°, and 175° are all obtuse angles.

5. C; *Example:* If $\angle 1$ measures 35° and $\angle 2$ measures 55°, the angles are complementary because $35° + 55° = 90°$.

6. A; *Example:* $\overleftrightarrow{EF}$ is perpendicular to $\overleftrightarrow{GH}$.

Quick Review

Concepts	Examples

6.1 Writing a Ratio

A ratio compares two quantities that have the same units. A ratio is usually written as a fraction with the number that is mentioned first in the numerator. The common units divide out and are not written in the answer. Check that the fraction is in lowest terms.

Write this ratio as a fraction in lowest terms.

60 ounces of medicine to 160 ounces of water

$$\frac{60 \text{ ounces}}{160 \text{ ounces}} = \frac{60 \div 20}{160 \div 20} = \frac{3}{8} \quad \left\{ \begin{array}{l} \text{Ratio in} \\ \text{lowest terms} \end{array} \right.$$

Divide out common units.

6.1 Using Mixed Numbers in a Ratio

If a ratio has mixed numbers, change the mixed numbers to improper fractions. Rewrite the problem in horizontal form, using the "÷" symbol for division. Finally, multiply by the reciprocal of the divisor.

Write as a ratio of whole numbers in lowest terms.

$$2\frac{1}{2} \quad \text{to} \quad 3\frac{3}{4}$$

$$\frac{2\frac{1}{2}}{3\frac{3}{4}} \quad \text{Ratio with mixed numbers}$$

$$\frac{\frac{5}{2}}{\frac{15}{4}} \quad \text{Ratio with improper fractions}$$

$$\frac{5}{2} \div \frac{15}{4} = \frac{5}{2} \cdot \frac{4}{15} = \frac{\overset{1}{5} \cdot \overset{1}{2} \cdot 2}{\underset{1}{2} \cdot 3 \cdot \underset{1}{5}} = \frac{2}{3} \quad \left\{ \begin{array}{l} \text{Ratio in} \\ \text{lowest} \\ \text{terms} \end{array} \right.$$

Reciprocals

6.1 Using Measurements in Ratios

When a ratio compares measurements, both measurements must be in the *same* units. It is usually easier to compare the measurements using the smaller unit, for example, inches instead of feet.

Write 8 inches to 6 feet as a ratio in lowest terms.

Compare using the smaller unit, inches. Because 1 foot has 12 inches, 6 feet is

$$6 \cdot 12 \text{ inches} = 72 \text{ inches}$$

The ratio is shown below.

$$\frac{8 \text{ inches}}{6 \text{ feet}} = \frac{8 \text{ inches}}{72 \text{ inches}} = \frac{8 \div 8}{72 \div 8} = \frac{1}{9} \quad \left\{ \begin{array}{l} \text{Ratio in} \\ \text{lowest terms} \end{array} \right.$$

6.2 Writing Rates

A rate compares two measurements with different units. The units do *not* divide out, so you must write them as part of the rate.

Write the rate as a fraction in lowest terms.

475 miles in 10 hours

$$\frac{475 \text{ miles} \div 5}{10 \text{ hours} \div 5} = \frac{\textbf{95 miles}}{\textbf{2 hours}} \quad \left] \begin{array}{l} \text{Must write units:} \\ \text{miles and hours} \end{array} \right.$$

Concepts	Examples

6.2 Finding a Unit Rate

A unit rate has 1 in the denominator. To find the unit rate, divide the numerator by the denominator. Write unit rates using the word **per** or a **/** mark.

Write as a unit rate: $1278 in 9 days.

$$\frac{\$1278}{9 \text{ days}} \leftarrow \text{Fraction bar indicates division.}$$

$$\begin{array}{r} 142 \\ 9)\overline{1278} \end{array} \quad \text{so} \quad \frac{\$1278 \div 9}{9 \text{ days} \div 9} = \frac{\$142}{1 \text{ day}}$$

Write the answer as **$142 per day** or **$142/day**.

6.2 Finding the Best Buy

The best buy is the item with the lowest cost per unit. Divide the price by the number of units. You may need to divide past the hundredths place to see a difference. Then compare to find the lowest cost per unit.

Find the best buy on apples.

2 pounds for $2.25

3 pounds for $3.40

Find cost per unit (cost per pound).

$$\frac{\$2.25}{2} = \$1.125 \text{ per pound}$$

$$\frac{\$3.40}{3} \approx \$1.133 \text{ per pound}$$

The lower cost per pound is $1.125, so **2 pounds of apples is the best buy.**

6.3 Writing Proportions

A proportion states that two ratios or rates are equal. The proportion "5 is to 6 as 25 is to 30" is written as shown below.

$$\frac{5}{6} = \frac{25}{30}$$

To see whether a proportion is true or false, multiply along one diagonal, then multiply along the other diagonal. If the two cross products are equal, the proportion is true. If the two cross products are unequal, the proportion is false.

Write as a proportion: 8 is to 40 as 32 is to 160

$$\frac{8}{40} = \frac{32}{160}$$

Is this proportion true or false?

$$\frac{6}{8\frac{1}{2}} = \frac{24}{34}$$

Find the cross products.

$$8\frac{1}{2} \cdot 24 = \frac{17}{2} \cdot \frac{\overset{1}{2} \cdot 12}{1} = 204$$

$$\frac{6}{8\frac{1}{2}} = \frac{24}{34}$$

$$6 \cdot 34 = 204 \qquad \text{Equal}$$

The **cross products are equal, so the proportion is true.**

Concepts	Examples

6.3 Solving Proportions

Solve for an unknown number in a proportion by using these steps.

Find the unknown number.

$$\frac{12}{x} = \frac{6}{8}$$

$$\left.\begin{array}{c}\\[-0.5em]\end{array}\right]\ \text{Write } \tfrac{6}{8} \text{ in lowest terms.}$$

$$\frac{12}{x} = \frac{3}{4}$$

Step 1 Find the cross products. (If desired, you can rewrite the ratios in lowest terms before finding the cross products.)

Step 1

$$\frac{12}{x} = \frac{3}{4}$$

$$x \cdot 3$$
$$12 \cdot 4$$
$$\left.\begin{array}{c}\\[-0.5em]\end{array}\right]\ \text{Find the cross products.}$$

Step 2 Show that the cross products are equal.

Step 2

$$x \cdot 3 = 12 \cdot 4 \quad \text{Show that cross products are equal.}$$

$$x \cdot 3 = \underbrace{}_{} 48$$

Step 3 Divide both sides of the equation by the coefficient of the variable term.

Step 3

$$\frac{x \cdot \overset{1}{3}}{\underset{1}{3}} = \frac{48}{3} \quad \text{Divide both sides by 3}$$

$$x = 16 \quad \text{The solution is 16}$$

Step 4 Check by writing the solution in the *original* proportion and finding the cross products.

Step 4

$$x \text{ is } 16 \rightarrow \quad \frac{12}{16} = \frac{6}{8}$$

$$16 \cdot 6 = 96$$
$$12 \cdot 8 = 96$$
$$\left.\begin{array}{c}\\[-0.5em]\end{array}\right]\ \text{Equal}$$

The cross products are equal, so **16 is the correct solution** (**not** 96).

6.4 Applications of Proportions

Use the six problem-solving steps.

If 3 pounds of grass seed cover 450 square feet of lawn, how much seed is needed for 1500 square feet of lawn?

Step 1 Read the problem.

Step 1 The problem is about the amount of grass seed needed for a lawn.

Unknown: pounds of seed needed for 1500 square feet of lawn

Known: 3 pounds cover 450 square feet of lawn

Step 2 Assign a variable.

Step 2 There is only one unknown, so let x be the pounds of seed needed for 1500 square feet of lawn.

Step 3 Write an equation using your variable.

Step 3 The equation is in the form of a proportion. Be sure that both rates in the proportion compare pounds to square feet in the same order.

Matching units

$$\frac{3 \text{ pounds}}{450 \text{ square feet}} = \frac{x \text{ pounds}}{1500 \text{ square feet}}$$

Matching units

(Continued)

Concepts	Examples

6.4 Applications of Proportions *(Continued)*

Step 4 **Solve** the equation.

Step 4 Ignore the units while finding the cross products and solving for x.

$$450 \cdot x = 3 \cdot 1500 \quad \text{Show that cross products are equal.}$$

$$450 \cdot x = 4500$$

$$\frac{\overset{1}{450} \cdot x}{\underset{1}{450}} = \frac{4500}{450} \quad \text{Divide both sides by 450}$$

$$x = 10$$

Step 5 **State the answer.**

Step 5 **10 pounds of grass seed** are needed for 1500 square feet of lawn.

Step 6 **Check** that the solution is reasonable by putting it back into the original statement of the problem.

Step 6 450 square feet of lawn needs 3 pounds of seed; 1500 square feet is about *three times as much* lawn, so about *three times as much* seed is needed.

$$(3)(3 \text{ pounds}) = 9 \text{ pounds} \leftarrow \text{Estimate}$$

The solution, 10 pounds, is close to the estimate of 9 pounds, so it is reasonable.

6.5 Lines

A *line* is a straight row of points that goes on forever in both directions.

If part of a line has one endpoint, it is a *ray*.

If part of a line has two endpoints, it is a *line segment*.

Identify each figure as a line, line segment, or ray and name it using the appropriate symbol.

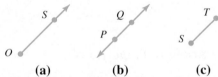

| (a) | (b) | (c) |

Figure **(a)** shows a **ray named** $\overrightarrow{OS}$.

Figure **(b)** shows a **line named** $\overleftrightarrow{PQ}$ or $\overleftrightarrow{QP}$.

Figure **(c)** shows a **line segment named** $\overline{ST}$ or $\overline{TS}$.

If two lines intersect at right angles, they are *perpendicular*.

If two lines in the same plane never intersect, they are *parallel*.

Label each pair of lines below as appearing to be parallel or as perpendicular.

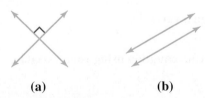

| (a) | (b) |

Figure **(a)** shows **perpendicular lines** (they intersect at 90°).

Figure **(b)** shows **lines that appear to be parallel** (they never intersect).

Concepts	Examples

6.5 Angles

If the sum of the measures of two angles is 90°, they are *complementary*.

If the sum of the measures of two angles is 180°, they are *supplementary*.

If two angles measure the same number of degrees, the angles are *congruent*. The symbol for congruent is ≅.

Two nonadjacent angles formed by two intersecting lines are called *vertical angles*. Vertical angles are congruent.

Find the complement and supplement of a 35° angle.

$$90° - 35° = \mathbf{55°} \text{ (the complement)}$$
$$180° - 35° = \mathbf{145°} \text{ (the supplement)}$$

Identify the vertical angles in this figure. Which angles are congruent?

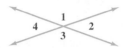

∠1 and ∠3 are vertical angles.

∠2 and ∠4 are vertical angles.

Vertical angles are congruent, so **∠1 ≅ ∠3** and **∠2 ≅ ∠4**.

6.5 Parallel Lines

When two parallel lines are crossed by a transversal, corresponding angles are congruent, and alternate interior angles are congruent. Use this information to find the measures of the other angles.

Read $m \angle 1$ as "the measure of angle 1."

Line m is parallel to line n and the measure of ∠4 is 125°. Find the measures of the other angles.

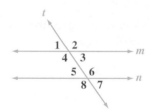

∠4 ≅ ∠8 (corresponding angles), so $\boldsymbol{m \angle 8 = 125°}$

∠4 ≅ ∠6 (alternate interior angles), so $\boldsymbol{m \angle 6 = 125°}$

∠6 ≅ ∠2 (corresponding angles), so $\boldsymbol{m \angle 2 = 125°}$

∠4 and ∠3 are supplements, so $\boldsymbol{m \angle 3 = 180° - 125° = 55°}$

∠3 ≅ ∠7 (corresponding angles), so $\boldsymbol{m \angle 7 = 55°}$

∠3 ≅ ∠5 (alternate interior angles), so $\boldsymbol{m \angle 5 = 55°}$

∠5 ≅ ∠1 (corresponding angles), so $\boldsymbol{m \angle 1 = 55°}$

Concepts	Examples

6.6 Proving That Two Triangles Are Congruent

Congruent triangles are identical both in shape and in size. This means that corresponding angles have the same measure and corresponding sides have the same length.

Here are three ways to prove that two triangles are congruent.

1. Angle–Side–Angle (ASA) method: If two angles and the side between them on one triangle measure the same as the corresponding parts on another triangle, the triangles are congruent.

2. Side–Side–Side (SSS) method: If three sides of one triangle measure the same as the corresponding sides of another triangle, the triangles are congruent.

3. Side–Angle–Side (SAS) method: If two sides and the angle between them on one triangle measure the same as the corresponding parts on another triangle, the triangles are congruent.

State which method can be used to prove that each pair of triangles is congruent.

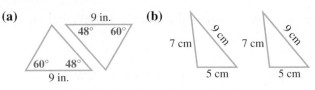

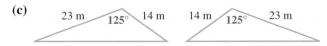

(a) On both triangles, two corresponding angles and the side between them measure the same, so use **ASA**.

(b) Each pair of corresponding sides has the same length, so use **SSS**.

(c) On both triangles, two corresponding sides and the angle between them measure the same, so use **SAS**.

6.6 Finding the Unknown Lengths in Similar Triangles

Use the fact that in similar triangles, the *ratios* of the lengths of corresponding sides are equal.

Find the unknown lengths in this pair of similar triangles.

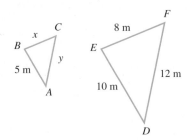

Write a proportion. Then find the cross products and show that they are equal. Finish solving for the unknown length.

$$\frac{x}{8} = \frac{5}{10}$$

$$x \cdot 10 = 8 \cdot 5$$

$$\frac{x \cdot \overset{1}{10}}{\underset{1}{10}} = \frac{40}{10}$$

$$x = 4 \text{ m}$$

$$\frac{y}{12} = \frac{5}{10}$$

$$y \cdot 10 = 12 \cdot 5$$

$$\frac{y \cdot \overset{1}{10}}{\underset{1}{10}} = \frac{60}{10}$$

$$y = 6 \text{ m}$$

Chapter 6 Review Exercises

6.1 *Write each ratio as a fraction in lowest terms. Change to the same units when necessary, using the table of measurement comparisons in the first section of this chapter. Use the information in the graph to answer Exercises 1–3.*

Average Length of Sharks And Whales

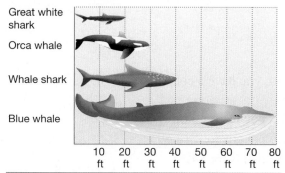

Great white shark

Orca whale

Whale shark

Blue whale

10 ft 20 ft 30 ft 40 ft 50 ft 60 ft 70 ft 80 ft

Data from *Grolier Multimedia Encyclopedia.*

1. Orca whale's length to whale shark's length

2. Blue whale's length to great white shark's length

3. Which two animals' lengths give a ratio of $\frac{1}{2}$? There is more than one correct answer.

4. $2.50 to $1.25

5. $0.30 to $0.45

6. $1\frac{2}{3}$ cups to $\frac{2}{3}$ cup

7. $2\frac{3}{4}$ miles to $16\frac{1}{2}$ miles

8. 5 hours to 100 minutes

9. 9 in. to 2 ft

10. 1 ton to 1500 pounds

11. 8 hours to 3 days

12. Jake sold $350 worth of his kachina figures. Ramona sold $500 worth of her pottery. What is the ratio of Ramona's sales to Jake's sales?

13. Ms. Wei's new car gets 35 miles per gallon. Her old car got 25 miles per gallon. Find the ratio of the new car's mileage to the old car's mileage.

14. This fall, 6000 students are taking math courses and 7200 students are taking English courses. Find the ratio of math students to English students.

6.2 *Write each rate as a fraction in lowest terms.*

15. $88 for 8 dozen

16. 96 children in 40 families

17. When entering data into his computer, Patrick can type four pages in 20 minutes. Give his rate in pages per minute and minutes per page.

18. Elena made $60 in three hours. Give her earnings in dollars per hour and hours per dollar.

Find the best buy.

19. Minced onion

  8 ounces for $4.98

  3 ounces for $2.49

  2 ounces for $1.89

20. Dog food; you have a coupon for $1 off on 8 pounds or more.

  35.2 pounds for $36.96

  17.6 pounds for $18.69

  3.5 pounds for $4.25

6.3 *Use either the method of writing in lowest terms or of finding cross products to decide whether each proportion is* true *or* false. *Show your work and then write* true *or* false.

21. $\dfrac{6}{10} = \dfrac{9}{15}$

22. $\dfrac{6}{48} = \dfrac{9}{36}$

23. $\dfrac{47}{10} = \dfrac{98}{20}$

24. $\dfrac{1.5}{2.4} = \dfrac{2}{3.2}$

25. $\dfrac{3\frac{1}{2}}{2\frac{1}{3}} = \dfrac{6}{4}$

Find the unknown number in each proportion. Round answers to the nearest hundredth when necessary.

26. $\dfrac{4}{42} = \dfrac{150}{x}$

27. $\dfrac{16}{x} = \dfrac{12}{15}$

28. $\dfrac{100}{14} = \dfrac{x}{56}$

29. $\dfrac{5}{8} = \dfrac{x}{20}$

30. $\dfrac{x}{24} = \dfrac{11}{18}$

31. $\dfrac{7}{x} = \dfrac{18}{21}$

32. $\dfrac{x}{3.6} = \dfrac{9.8}{0.7}$

33. $\dfrac{13.5}{1.7} = \dfrac{4.5}{x}$

34. $\dfrac{0.82}{1.89} = \dfrac{x}{5.7}$

6.4 *Set up and solve a proportion for each application problem.*

35. The ratio of cats to dogs at the animal shelter is 3 to 5. If there are 45 dogs, how many cats are there?

36. Danielle had 8 hits in 28 times at bat during last week's games. If she continues to hit at the same rate, how many hits will she get in 161 times at bat?

37. If 3.5 pounds of ground beef cost $14.84, what will 5.6 pounds cost? Round your answer to the nearest cent.

38. About 4 out of 10 students are expected to vote in campus elections. There are 8247 students. How many are expected to vote? Round to the nearest whole number.

39. The scale on Brian's model railroad is 1 inch to 16 feet. One of the scale model boxcars is 4.25 inches long. What is the length of a real boxcar in feet?

40. Marvette makes necklaces to sell at a local gift shop. She made 2 dozen necklaces in $16\frac{1}{2}$ hours. How long will it take her to make 40 necklaces?

41. A 180-pound person burns 272 calories playing basketball for 25 minutes. At this rate, how many calories would the person burn in 45 minutes? Round your answer to the nearest whole number. (Data from www.myfitnesspal.com)

42. In the hospital pharmacy, Amit sees that a medicine is given at the rate of 3.5 milligrams for every 50 pounds of body weight. How much medicine should be given to a patient who weighs 210 pounds?

6.5 *Identify each figure as a* line, line segment, *or* ray *and name it using the appropriate symbol.*

43.

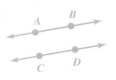

44.

45.

Label each pair of lines as appearing to be parallel, *as* perpendicular, *or as* intersecting.

46.

47.

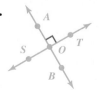

48.

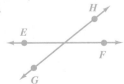

Label each angle as an acute, right, obtuse, *or straight angle. For* right *and* straight angles, *indicate the number of degrees in the angle.*

49.

50.

51.

52.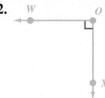

Find the complement or supplement of each angle.

53. Find each complement.

(a) 80°

(b) 45°

(c) 7°

54. Find each supplement.

(a) 155°

(b) 90°

(c) 33°

55. In the figure below, ∠2 measures 60°. Find the measure of each of the other angles.

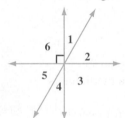

56. Line *m* is parallel to line *n* and ∠8 measures 160°. Find the measures of the other angles.

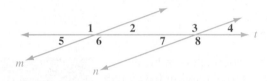

6.6 *State which method can be used to prove that each pair of triangles is congruent.*

57.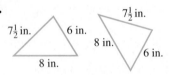

58.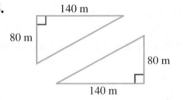

59.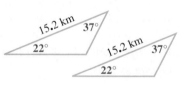

Find the unknown lengths in each pair of similar triangles. Then find the perimeter of the larger triangle in each pair.

60.

61.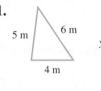

62.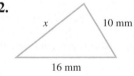

Chapter 6 Mixed Review Exercises

Find the unknown number in each proportion. Round to hundredths when necessary.

1. $\dfrac{x}{45} = \dfrac{70}{30}$

2. $\dfrac{x}{52} = \dfrac{0}{20}$

3. $\dfrac{64}{10} = \dfrac{x}{20}$

4. $\dfrac{15}{x} = \dfrac{65}{100}$

5. $\dfrac{7.8}{3.9} = \dfrac{13}{x}$

6. $\dfrac{34.1}{x} = \dfrac{0.77}{2.65}$

Write each ratio as a fraction in lowest terms. Change to the same units when necessary.

7. 4 dollars to 10 quarters

8. $4\dfrac{1}{8}$ inches to 10 inches

9. 10 yards to 8 feet

10. $3.60 to $0.90

11. 12 eggs to 15 eggs

12. 37 meters to 7 meters

13. 3 pints to 4 quarts

14. 15 minutes to 3 hours

15. $4\dfrac{1}{2}$ miles to $1\dfrac{3}{10}$ miles

Set up and solve a proportion for each application problem.

16. Nearly 7 out of 8 fans buy something to drink at rock concerts. How many of the 28,500 fans at today's concert would be expected to buy a beverage? Round to the nearest hundred fans.

17. Emily spent $150 on car repairs and $400 on car insurance. What is the ratio of the amount spent on insurance to the amount spent on repairs?

18. Antonio is choosing among three packages of plastic wrap. Is the best buy 25 feet for $0.78; 75 feet for $1.99; or 100 feet for $2.59? He has a coupon for 50¢ off that is good for either of the larger two packages.

19. On this scale drawing of a backyard patio, 0.5 inch represents 6 feet. If the patio measures 1.75 inches long and 1.25 inches wide on the drawing, what will be the actual length and width of the patio when it is built?

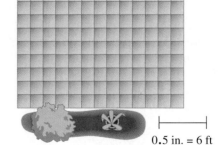

0.5 in. = 6 ft

20. A vitamin supplement for cats should be given at the rate of 1000 milligrams for a 5-pound cat. (Data from St. Jon Pet Care Products.)

(a) How much should be given to a 7-pound cat?

(b) How much should be given to an 8-ounce kitten?

21. Charles made 251 points during 169 minutes of playing time last year. At that same rate, how many points would you expect him to make if he plays 14 minutes in tonight's game? Round to the nearest whole number.

22. An antibiotic should be given at the rate of $1\dfrac{1}{2}$ teaspoons for every 24 pounds of body weight. How much should be given to an infant who weighs 8 pounds?

Label each figure. Choose from these labels: line segment, ray, parallel lines, perpendicular lines, intersecting lines, acute angle, right angle, straight angle, obtuse angle. *Indicate the number of degrees in the* right angle *and the* straight angle.

23.

24.

25.

26.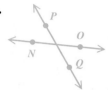

27.

28.

29.

30.

31.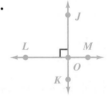

32. Explain what is happening in each sentence.

  (a) The road was so slippery that my car did a 360.

  (b) After the election, the governor's view on new taxes took a 180° turn.

33. (a) Can two obtuse angles be supplementary? Explain why or why not.

  (b) Can two acute angles be complementary? Explain why or why not.

34. In the figure below, ∠2 measures 45° and ∠7 measures 55°. Find the measure of each of the other angles.

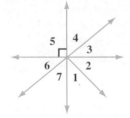

35. In the figure below, line *m* is parallel to line *n* and ∠5 measures 75°. Find the measures of the other angles.

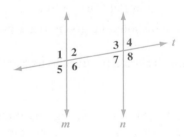

Chapter 6 *Test*

The Chapter Test Prep Videos with step-by-step solutions are available in MyMathLab or on YouTube at **https://goo.gl/c3befo**

Write each rate or ratio as a fraction in lowest terms. Change to the same units when necessary.

1. $15 for 75 minutes

2. 3 hours to 40 minutes

3. The little theater at our college has 320 seats. The auditorium has 1200 seats. Find the ratio of auditorium seats to theater seats.

4. Find the best buy on spaghetti sauce. You have a coupon for 75¢ off Brand X and a coupon for 50¢ off Brand Y.

26 ounces of Brand X for $3.89

16 ounces of Brand Y for $1.89

14 ounces of Brand Z for $1.29

Find the unknown number in each proportion. In Problems 5–7, round answers to the nearest hundredth when necessary.

5. $\dfrac{5}{9} = \dfrac{x}{45}$

6. $\dfrac{3}{1} = \dfrac{8}{x}$

7. $\dfrac{x}{20} = \dfrac{6.5}{0.4}$

8. $\dfrac{2\frac{1}{3}}{x} = \dfrac{\frac{8}{9}}{4}$

Set up and solve a proportion for each application problem.

9. Pedro entered 18 orders into his computer in 30 minutes at his job. At that rate, how many orders could he enter in forty minutes?

10. About 2 out of every 15 people are left-handed. How many of the 650 students in our school would you expect to be left-handed? Round to the nearest whole number.

11. A medication is given at the rate of 8.2 grams for every 50 pounds of body weight. How much should be given to a 145-pound person? Round to the nearest tenth of a gram.

12. On a scale model, 1 inch represents 8 feet. If a building in the model is 7.5 inches tall, what is the actual height of the building in feet?

Choose the figure that matches each label. For right and straight angles, indicate the number of degrees in the angle.

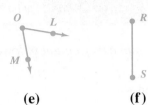

(a)　　　　　　　　(b)　　　　　　　(c)　　　　　　　(d)　　　　　(e)　　　　　(f)　　　　(g)

13. Acute angle is figure _____ .

14. Right angle is figure _____ and its measure is _____ .

15. Ray is figure _____ .

16. Straight angle is figure _____ and its measure is _____ .

17. Write a definition of parallel lines and a definition of perpendicular lines. Make a sketch to illustrate each definition.

18. Find the complement of an 81° angle.

19. Find the supplement of a 20° angle.

20. In the figure below, $\angle 4$ measures 50° and $\angle 6$ measures 95°. Find the measures of the other angles.

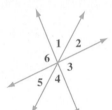

21. In the figure below, line *m* is parallel to line *n* and $\angle 3$ measures 65°. Find the measures of the other angles.

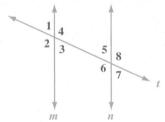

State which method can be used to prove that each pair of triangles is congruent.

22.

23.

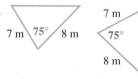

24. Find the unknown lengths in these similar triangles.

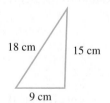

25. Find the perimeter of each of these similar triangles.

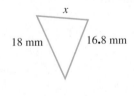

Chapters 1–6 *Cumulative Review Exercises*

1. Write these numbers in words.
   (a) 77,001,000,805
   (b) 0.02

2. Write these numbers using digits.
   (a) Three and forty thousandths
   (b) Five hundred million, thirty-seven thousand

Simplify.

3. $\dfrac{0}{-16}$

4. $|0| + |-6 - 8|$

5. $4\dfrac{3}{4} - 1\dfrac{5}{6}$

6. $\dfrac{h}{5} - \dfrac{3}{10}$

7. $100 - 0.0095$

8. $\dfrac{-4 + 7}{9 - 3^2}$

9. $-6 + 3(0 - 4)$

10. $\dfrac{-8}{\frac{4}{7}}$

11. $\dfrac{5n}{6m^3} \div \dfrac{10}{3m^2}$

12. $(0.06)(-0.007)$

13. $\dfrac{x}{14y} \cdot \dfrac{7}{xy}$

14. $\dfrac{9}{n} + \dfrac{2}{3}$

15. $-40 + 8(-5) + 2^4$

16. $5.8 - (-0.6)^2 \div 0.9$

17. $\left(-\dfrac{1}{2}\right)^3 \left(\dfrac{2}{3}\right)^2$

Solve each equation. Show your work.

18. $7y + 5 = -3 - y$

19. $-2 + \dfrac{3}{5}x = 7$

20. $\dfrac{4}{x} = \dfrac{14}{35}$

Solve each application problem using the six problem-solving steps.

21. Tuyen heard on the radio that the temperature had risen 15 degrees by noon, then dropped 23 degrees due to a storm, then risen 5 degrees and was now at 71 degrees. What was the starting temperature?

22. The width of a rectangular swimming pool is 14 ft less than the length. The perimeter of the pool is 100 ft. Find the length and the width.

Find the unknown length, perimeter, area, or volume. When necessary, use 3.14 as the approximate value of π and round your final answers to the nearest tenth.

23. Find the perimeter and the area.

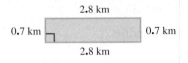

2.8 km

0.7 km 0.7 km

2.8 km

24. Find the diameter, circumference, and area.

8.5 m

25. Find the volume and surface area.

3 ft

12 ft

Solve each application problem. Show your work.

26. The honor society has a goal of collecting 1500 pounds of food to fill Thanksgiving baskets. So far they've collected $\frac{5}{6}$ of their goal. How many more pounds do they need?

27. The directions on a can of plant food call for $\frac{1}{2}$ teaspoon in two quarts of water. How much plant food is needed for five quarts?

28. The distance around Dunning Pond is $1\frac{1}{10}$ miles. Norma ran around the pond four times in the morning and $2\frac{1}{2}$ times in the afternoon. How far did she run in all?

29. Rodney used 40.8 gallons of gas for his car while driving 896.5 miles on a vacation. How many miles per gallon did he get, rounded to the nearest tenth?

Use the information in the table on international long-distance calling card rates to answer Exercises 30–34. Round answers to the nearest whole minute when necessary.

$20 International Phone Cards No Connection Fee!	
Place a call to	**Cost per minute**
Mexico (to a land line)	$0.005
Philippines (to a cell phone)	$0.084
France (to a cell phone)	$0.018
Argentina (to a cell phone)	$0.02
Bolivia (to a cell phone)	$0.099
Sudan (to a land line)	$0.106
Yemen (to a cell phone)	$0.10

Data from wwww.noblecom.com; www.sciencekids.co.nz/flags

30. List the per-minute rates from least to greatest.

31. If you buy a $20 calling card, how long a call can you make to a cell phone in Bolivia?

32. How many $20 cards would you have to buy to make 6 hours' worth of calls to a land line in Sudan?

33. Find the ratio of the minutes you can call Argentina for $20, to the minutes you can call Yemen for $20.

34. You made a 90-minute phone call to a land line in Mexico every Sunday for 24 weeks. With the amount left on one $20 phone card, how long can you talk if you call a cell phone in the Philippines?

Percent

Percents are used in many day-to-day activities. In this chapter you will learn how to calculate discounts and sales tax, estimate the amount to tip at a restaurant, figure out your grade on an exam, and calculate the interest on loans.

7.1 The Basics of Percent

OBJECTIVES

1. Learn the meaning of percent.
2. Write percents as decimals.
3. Write decimals as percents.
4. Write percents as fractions.
5. Write fractions as percents.
6. Use 100% and 50%.

OBJECTIVE ▶ 1 Learn the meaning of percent. You have probably seen percents frequently in daily life. The symbol for percent is %. For example, during one day you may leave a 15% tip for the server at dinner, pay 7% sales tax on a smartphone, and buy shoes at 25% off the regular price. The next day your score on a math test may be 89% correct.

The Meaning of Percent

A **percent** is a ratio with a denominator of 100. So percent means "per 100" or "how many out of 100." The symbol for percent is %. Read 15% as "fifteen percent."

EXAMPLE 1 Understanding Percent

Write a percent to describe each situation.

(a) If you left a $15 tip when the restaurant bill was $100, then you left $15 per $100 or $\frac{15}{100}$ or **15%**.

(b) If you pay $7 in tax on a $100 smartphone, then the tax rate is $7 per $100 or $\frac{7}{100}$ or **7%**.

(c) If you earn 89 points on a 100-point math test, then your score is 89 out of 100 or $\frac{89}{100}$ or **89%**.

◀ **Work Problem 1 at the Side.**

1. Write a percent to describe each situation.

(a) You leave a $20 tip for a restaurant bill of $100. What percent tip did you leave?

The tip is $20 per $100

or $\dfrac{\boxed{}}{100}$ or _____ %.

(b) The tax on a $100 graphing calculator is $5. What is the tax rate?

(c) You earn 94 points on a 100-point test. What percent of the points did you earn?

OBJECTIVE ▶ 2 Write percents as decimals. In order to work with percents, you will need to write them as decimals or as fractions. We'll start by writing percents as equivalent decimal numbers. Twenty-five percent, or 25%, means 25 parts out of 100 parts, or $\frac{25}{100}$. Remember that the fraction bar indicates division. So we can write $\frac{25}{100}$ as $25 \div 100$. When you do the division, $25 \div 100$ is 0.25.

Indicates division ⟶ $\dfrac{25}{100}$ can be written as $25 \div 100 = 0.25$
↑
Indicates division

Another way to remember how to write a percent as a decimal is to use the meaning of the word *percent*. The first part of the word, *per*, is an indicator word for *division*. The last part of the word, *cent*, comes from the Latin word for *hundred*. (Recall that there are 100 *cent*s in a dollar, and 100 years in a *cent*ury.)

25% is 25 percent

$25 \div 100 = 0.25$

Writing a Percent as an Equivalent Decimal

To write a percent as a decimal, drop the % symbol and divide by 100.

Answers

1. (a) $\frac{20}{100}$; 20%   (b) 5%   (c) 94%

| EXAMPLE 2 | Writing Percents as Decimals |

Write each percent as a decimal.

(a) 47%   47% =   47   ÷ 100 = **0.47**   Decimal form

(b) 3%   3% =   3   ÷ 100 = **0.03**   Decimal form

(c) 28.2%   28.2% = 28.2 ÷ 100 = **0.282**   Decimal form

(d) 100%   100% = 100   ÷ 100 = **1.00**   Decimal form

(e) 135%   135% = 135   ÷ 100 = **1.35**   Decimal form

> ❗ **CAUTION**
>
> In **Example 2(d)** above, notice that 100% is 1.00, or 1, which is a whole number. Whenever you have a percent that is *100% or greater*, the equivalent decimal number will be *1 or greater*. Notice in **Example 2(e)** above that 135% is 1.35 (greater than 1).

──── Work Problem ② at the Side. ▶

In the exercise set on dividing decimal numbers, you discovered a shortcut for *dividing* by 100: move the decimal point *two* places to the *left*. You can use this shortcut when writing percents as decimals.

| EXAMPLE 3 | Changing Percents to Decimals by Moving the Decimal Point |

Write each percent as a decimal by dropping the percent symbol and moving the decimal point two places to the left.

(a) 17%

┌─── Decimal point starts at far right side.
17% = 17.%

.17 ← Percent symbol is dropped.
┌─── Decimal point is moved
    two places to the *left*.

> Moving the decimal point *two places* to the *left* is a quick way to *divide by 100*.

17% = **0.17**

(b) 160%   ┌─── Decimal point starts at far right side.
160% = 160.% = **1.60 or 1.6**   1.60 is equivalent to 1.6

(c) 4.9%

04.9%   0 is attached so the decimal point can be moved
             two places to the *left*.

4.9% = **0.049** ← Drop the % symbol.

(d) 0.6%   ┌── Drop the % symbol. ──┐

00.6% = **0.006**   0 is attached so the decimal point can be
                     moved *two places* to the *left*.

──── Continued on Next Page

② Write each percent as a decimal.

GS **(a)** 68%

68% = 68 ÷ _____ = _____

(b) 5%

(c) 40.6%

GS **(d)** 200%

200% = 200 ÷ _____ = _____

(e) 350%

Answers

2. **(a)** 68 ÷ 100 = 0.68 **(b)** 0.05 **(c)** 0.406
      (d) 200 ÷ 100 = 2.00 or 2 **(e)** 3.50 or 3.5

3 Write each percent as a decimal by moving the decimal point.

(a) 90%

$$90\% = 90.\% = \underline{\hspace{1cm}}$$

Move the decimal point **two** places to the **left**.

Drop the % symbol.

(b) 9%

(c) 900%

(d) 9.9

$$9.9\% = 09.9\% = \underline{\hspace{1cm}}$$

Drop the % symbol.

(e) 0.9%

Note

In **Example 3 (d)** on the previous page, notice that 0.6% is less than 1%. Because 1% is equivalent to 0.01 or $\frac{1}{100}$, any fraction of a percent smaller than 1% is less than 0.01. The decimal equivalent of 0.6% is 0.006, which is less than 0.01.

◀ **Work Problem 3 at the Side.**

OBJECTIVE 3 Write decimals as percents. You can write a decimal as a percent. For example, the decimal 0.25 is the same as the fraction $\frac{25}{100}$.

This fraction means 25 out of 100 parts, or 25%. Notice that multiplying 0.25 by 100 gives the same result.

$$(0.25)(100) = 25 \quad \text{so} \quad 0.25 = 25\%$$

This result makes sense because we are doing the reverse of what we did to change a percent to a decimal.

To change a percent to a decimal, we *drop* the % symbol and *divide* by 100. So, to *reverse* the process and change a decimal to a percent, we *multiply* by 100 and *attach* a % symbol.

Writing a Decimal as a Percent

To write a decimal as a percent, multiply by 100 and attach a % symbol.

Note

A quick way to *multiply* a number by 100 is to move the decimal point *two* places to the *right*. Notice that this is the reverse of the shortcut for *dividing* by 100.

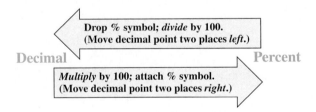

Drop % symbol; *divide* by 100.
(Move decimal point two places *left*.)

Decimal   **Percent**

Multiply by 100; attach % symbol.
(Move decimal point two places *right*.)

EXAMPLE 4   **Changing Decimals to Percents by Moving the Decimal Point**

Write each decimal as a percent.

(a) 0.21

0.21   Moving the decimal point *two places to the right* is a quick way to *multiply by 100.*

$0.21 = \mathbf{21}\%$ ◀── Percent symbol is attached after decimal point is moved.

⬆── Decimal point is *not* written with whole number percents.

(b) $0.529 = \mathbf{52.9}\%$ ◀── Percent symbol is attached after decimal point is moved.

(c) $1.92 = \mathbf{192}\%$ ◀── Percent symbol is attached after decimal point is moved.

─── **Continued on Next Page**

Answers

3. **(a)** 0.90 or 0.9   **(b)** 0.09   **(c)** 9.00 or 9
   (d) 0.099   **(e)** 0.009

(d) 2.5

> 2.5̲0̲   0 is attached so the decimal point can be moved
> two places to the right.

2.5 = **250%**

(e) 3

> Hint:
> 1 = 100%
> 2 = 200%
> 3 = 300%
> and so on.

3. = 3.00   so   3.0̲0̲ = **300%**

⊘ CAUTION

In **Examples 4(c), 4(d), and 4(e)** above, notice that 1.92, 2.5, and 3 are greater than 1. Because the number 1 is equivalent to 100%, all *numbers greater than 1* will be equivalent to *percents greater than 100%*.

———— Work Problem **4** at the Side. ▶

OBJECTIVE ▶ 4 **Write percents as fractions.** Percents can also be written as fractions. Recall that a percent is a ratio with a denominator of 100. For example, 89% is $\frac{89}{100}$. Because the fraction bar indicates division, we are dividing by 100, just as we did when writing a percent as a decimal.

Writing a Percent as a Fraction

To write a percent as a fraction, drop the % symbol and write the number over 100. Then write the fraction in lowest terms.

EXAMPLE 5   **Writing Percents as Fractions**

Write each percent as a fraction or mixed number in lowest terms or as a whole number.

(a) 25%   Drop the % symbol and write 25 over 100.

$$25\% = \frac{25}{100} \} \to 25 \text{ per } 100$$

$$= \frac{25 \div 25}{100 \div 25} = \frac{1}{4} \quad \text{Lowest terms}$$

As a check, write 25% as a decimal.

$$25\% = 25 \div 100 = 0.25 \leftarrow \text{Percent sign dropped}$$

Recall that 0.25 means 25 hundredths.

$$0.25 = \frac{25}{100} = \frac{25 \div 25}{100 \div 25} = \frac{1}{4} \leftarrow \text{Same result as above}$$

(b) 76%   Drop the % symbol and write 76 over 100.

This number becomes the numerator.

$$76\% = \frac{76}{100} \leftarrow \text{The } denominator \text{ is always 100 because percent means } parts \text{ per } 100.$$

Write $\frac{76}{100}$ in lowest terms. $$\frac{76 \div 4}{100 \div 4} = \frac{19}{25} \quad \text{Lowest terms}$$

———— **Continued on Next Page**

4 Write each number as a percent.

(a) 0.95

0.9̲5̲ = _____ ↑

Move the decimal point **two** places to the **right.**   Write the percent symbol.

(b) 0.16

(c) 0.09

(d) 0.617

(e) 0.4

(f) 5.34

(g) 2.8

2.8̲0̲ = _____ ↑

Write the % symbol.

(h) 4

Answers

4. (a) 95% (b) 16% (c) 9% (d) 61.7%
   (e) 40% (f) 534% (g) 280%
   (h) 400%

5 Write each percent as a fraction or mixed number in lowest terms or as a whole number.

(a) 50%

$$50\% = \frac{50}{100} = \frac{50 \div \square}{100 \div \square}$$

$$50\% = \frac{\square}{\square} \leftarrow \text{Lowest terms}$$

(b) 19%

(c) 80%

(d) 6%

(e) 125% Write your final answer as a mixed number.

(f) 300% Write your final answer as a whole number.

(c) 150%

$$150\% = \frac{150}{100} = \frac{150 \div 50}{100 \div 50} = \frac{3}{2} = 1\frac{1}{2} \leftarrow \text{Mixed number}$$

(d) $100\% = \dfrac{100}{100} = 1 \leftarrow$ Whole number

Always simplify fraction answers.

Note

Remember that percent means **per 100.**

◀ **Work Problem 5 at the Side.**

Example 6 below shows how you can write decimal percents and fraction percents as fractions.

EXAMPLE 6   **Writing Decimal Percents or Fraction Percents as Fractions**

Write each percent as a fraction in lowest terms.

(a) 15.5%

Drop the % symbol and write 15.5 over 100.

$$15.5\% = \frac{15.5}{100}$$

To get a whole number in the numerator, multiply the numerator and denominator by 10. (Multiplying by $\frac{10}{10}$ is the same as multiplying by 1.)

$$\frac{15.5}{100} = \frac{(15.5)(10)}{(100)(10)} = \frac{155}{1000}$$

Write the fraction in lowest terms.

$$\frac{155 \div 5}{1000 \div 5} = \frac{31}{200} \quad \text{Lowest terms}$$

(b) $33\frac{1}{3}\%$

Drop the % symbol and write $33\frac{1}{3}$ over 100.

$$33\frac{1}{3}\% = \frac{33\frac{1}{3}}{100}$$

When there is a mixed number in the numerator, write it as an improper fraction. So $33\frac{1}{3}$ is $\frac{100}{3}$.

*Think:
3 • 33 is 99
and 99 + 1 is 100
so $33\frac{1}{3}$ is $\frac{100}{3}$*

$$\frac{33\frac{1}{3}}{100} = \frac{\frac{100}{3}}{100}$$

Continued on Next Page

Answers

5. **(a)** $\dfrac{50 \div 50}{100 \div 50} = \dfrac{1}{2}$ **(b)** $\dfrac{19}{100}$ **(c)** $\dfrac{4}{5}$

(d) $\dfrac{3}{50}$ **(e)** $\dfrac{5}{4} = 1\dfrac{1}{4}$ **(f)** $\dfrac{300}{100} = 3$

Now you have a complex fraction. Rewrite the complex fraction using the ÷ symbol for division. Then follow the steps for dividing fractions.

$$\frac{\frac{100}{3}}{100} = \frac{100}{3} \div 100 = \frac{100}{3} \div \frac{100}{1} = \frac{100}{3} \cdot \frac{1}{100} = \frac{\overset{1}{100} \cdot 1}{3 \cdot \underset{1}{100}} = \frac{1}{3}$$

Reciprocals

Note

In **Example 6(a)** on the previous page, we could have changed 15.5% to $15\frac{1}{2}\%$ and then written it as the improper fraction $\frac{31}{2}$ over 100. But it is usually easier to work with decimal percents as they are.

──────── **Work Problem ⑥ at the Side.** ▶

OBJECTIVE ▶ ⑤ Write fractions as percents. Recall that to write a percent as a fraction, you *drop* the percent symbol and *divide* by 100. So, to reverse the process and change a fraction to a percent, you *multiply* by 100 and *attach* a percent symbol.

Writing a Fraction as a Percent

To write a fraction as a percent, multiply by 100 and attach a % symbol. This is the same as multiplying by 100%.

Note

Look back at **Example 5(d)** near the top of the previous page to see that 100% = 1. Recall that multiplying a number by 1 does *not* change the value of the number. So multiplying by 100% does not change the value of a number; it just gives us an *equivalent percent*.

EXAMPLE 7 Writing Fractions as Percents

Write each fraction as a percent. Round to the nearest tenth of a percent when necessary.

(a) $\frac{2}{5}$ Multiply $\frac{2}{5}$ by 100%.

$$\frac{2}{5} = \left(\frac{2}{5}\right)(100\%) = \left(\frac{2}{5}\right)\left(\frac{100}{1}\%\right) = \left(\frac{2}{5}\right)\left(\frac{5 \cdot 20}{1}\%\right) = \frac{2 \cdot \overset{1}{5} \cdot 20}{\underset{1}{5} \cdot 1}\%$$

$$= \frac{40}{1}\% = \mathbf{40\%}$$

To check the result, write 40% as $\frac{40}{100}$ and simplify the fraction.

$$40\% = \frac{40}{100} = \frac{40 \div 20}{100 \div 20} = \frac{2}{5} \quad \leftarrow \text{Original fraction}$$

──────── **Continued on Next Page**

---

⑥ Write each percent as a fraction in lowest terms.

GS (a) 18.5%

$$18.5\% = \frac{18.5}{100} = \frac{(18.5)(10)}{(100)(10)} =$$

$$\frac{185}{1000} = \frac{185 \div \square}{1000 \div \square} =$$

(b) 87.5%

(c) 6.5%

GS (d) $66\frac{2}{3}\%$

$$\frac{66\frac{2}{3}}{100} = \frac{\frac{200}{3}}{100} = \frac{200}{3} \div \frac{100}{1} =$$

$$\frac{\overset{2}{200}}{3} \cdot \frac{1}{\underset{1}{100}} =$$

(e) $12\frac{1}{3}\%$

(f) $62\frac{1}{2}\%$

Answers

6. (a) $\frac{185 \div 5}{1000 \div 5} = \frac{37}{200}$ **(b)** $\frac{7}{8}$ **(c)** $\frac{13}{200}$

(d) $\frac{2}{3}$ **(e)** $\frac{37}{300}$ **(f)** $\frac{5}{8}$

7 Write each fraction as a percent. If you're using a calculator, first work each one by hand. Then use your calculator and round to the nearest tenth of a percent when necessary.

GS **(a)** $\dfrac{1}{2}$

$$\dfrac{1}{2}(100\%) = \left(\dfrac{1}{2}\right)\left(\dfrac{100}{1}\%\right) =$$

$$\left(\dfrac{1}{\underset{1}{2}}\right)\left(\dfrac{\overset{}{2} \cdot 50}{1}\%\right) =$$

(b) $\dfrac{3}{4}$

(c) $\dfrac{1}{10}$

(d) $\dfrac{7}{8}$

(e) $\dfrac{5}{6}$

(f) $\dfrac{2}{3}$

Answers

7. **(a)** $\dfrac{50}{1}\% = 50\%$ **(b)** 75% **(c)** 10%

(d) $87\dfrac{1}{2}\%$ or 87.5% (Both are exact answers.)

(e) exactly $83\dfrac{1}{3}\%$, or 83.3% (rounded)

(f) exactly $66\dfrac{2}{3}\%$, or 66.7% (rounded)

(b) $\dfrac{5}{8}$ Multiply $\frac{5}{8}$ by 100%.

$$\dfrac{5}{8} = \left(\dfrac{5}{8}\right)(100\%) = \left(\dfrac{5}{8}\right)\left(\dfrac{100}{1}\%\right) = \left(\dfrac{5}{2 \cdot 4}\right)\left(\dfrac{4 \cdot 25}{1}\%\right) = \dfrac{5 \cdot \overset{1}{4} \cdot 25}{2 \cdot 4 \cdot 1}\%$$

$$= \dfrac{125}{2}\% = 62\dfrac{1}{2}\%$$

You can also do the last step of simplifying $\frac{125}{2}$ on your calculator. Enter $\frac{125}{2}$ as 125 ÷ 2 =. The result is 62.5, so $\frac{5}{8} = 62\frac{1}{2}\%$ or $\frac{5}{8} = 62.5\%$

(c) $\dfrac{1}{6}$ Multiply $\frac{1}{6}$ by 100%.

$$\dfrac{1}{6} = \left(\dfrac{1}{6}\right)(100\%) = \left(\dfrac{1}{6}\right)\left(\dfrac{100}{1}\%\right) = \left(\dfrac{1}{2 \cdot 3}\right)\left(\dfrac{2 \cdot 50}{1}\%\right) = \dfrac{1 \cdot \overset{1}{2} \cdot 50}{2 \cdot 3 \cdot 1}\%$$

$$= \dfrac{50}{3}\% = 16\dfrac{2}{3}\%$$

To simplify $\frac{50}{3}$ on your calculator, enter 50 ÷ 3 =. The result is 16.66666666, with the 6 continuing to repeat. The directions say to round to the nearest *tenth* of a percent.

Next digit is *5 or more.*

16.66666666%   rounds to   16.7%

When the next digit is *5 or more*, round up.

Tenths place

So, $\frac{1}{6} = 16\frac{2}{3}\%$ **(exact answer)**   or $\frac{1}{6} \approx 16.7\%$ **(rounded answer).**

🖩 **Calculator Tip**

In **Example 7 (a)** on the previous page, you can use your calculator to write $\frac{2}{5}$ as a percent.

Step 1   Enter $\frac{2}{5}$ as 2 ÷ 5 =. Your calculator shows   **0.4**

Decimal equivalent of $\dfrac{2}{5}$

Step 2   Change the decimal number 0.4 to a percent by moving the decimal point two places to the right (multiply by 100%).

0.40 = 40% ← Attach % symbol.

Try this technique on **Examples 7 (b) and 7 (c)** on this page.

For $\frac{5}{8}$, enter 5 ÷ 8 =. Your calculator shows   **0.625**

Move the decimal point in 0.625 two places to the right.

0.625 = 62.5% ← Attach % symbol.

Continued on Next Page

For $\frac{1}{6}$, enter 1 ⊙ 6 ⊜. Your calculator shows | **0.166666666**

Many calculators show 7 in the last place. ⟶

Move the decimal point two places to the right. Then round to the nearest tenth.

┌─ Next digit is *5 or more*

$$0.166666666 = 16.\underset{\uparrow}{6}6666666\% \approx 16.7\% \leftarrow \text{Attach \% symbol.}$$

Tenths place

──────── **Work Problem 7 on the Previous Page at the Side.**

OBJECTIVE ▶ 6 Use 100% and 50%. When working with percents, it is helpful to have several reference points. 100% and 50% are two helpful reference points.

100% means 100 parts out of 100 parts. That's *all* of the parts. If you pay 100% of a $245 dentist bill, you pay $245 (*all* of it).

EXAMPLE 8   Finding 100% of a Number

Fill in the blanks.

(a) 100% of $42 is _____.

100% is *all* of the money.

So 100% of $42 is ___**$42**___

(b) 100% of 9 miles is _____.

100% is *all* of the miles.

So 100% of 9 miles is ___**9 miles**___

──────── **Work Problem 8 at the Side. ▶**

50% means 50 parts out of 100 parts, which is *half* of the parts because $\frac{50}{100} = \frac{1}{2}$. So 50% of $12 is $6 (*half* of the money).

EXAMPLE 9   Finding 50% of a Number

Fill in the blanks.

(a) 50% of $42 is _____.

50% is *half* of the money.

So 50% of $42 is ___**$21**___

(b) 50% of 9 miles is _____.

50% is *half* of the miles.

So 50% of 9 miles is ___$4\frac{1}{2}$ **miles**___

──────── **Work Problem 9 at the Side. ▶**

Study Skills Reminder

What do you do when your instructor hands back a test? Do you spend time looking at it, or do you just file it away? Taking time to review your test after it is graded is an important step in improving your test-taking abilities. See the Study Skills activity "Analyzing Your Test Results" to learn how to review a test to find common errors and what steps you can take to prevent making the same errors on the next test.

---

8 Fill in the blanks.

GS (a) 100% of $4.60 is _____.

100% is *all* of the money.

(b) 100% of 3000 students is _____

(c) 100% of 7 pages is _____

(d) 100% of 272 miles is _____

(e) 100% of $10\frac{1}{2}$ hours is _____

9 Fill in the blanks.

GS (a) 50% of $4.60 is _____.

50% is *half* of the money.

(b) 50% of 3000 students is _____

(c) 50% of 7 pages is _____

(d) 50% of 272 miles is _____

(e) 50% of $10\frac{1}{2}$ hours is _____

Answers

8. (a) $4.60   **(b)** 3000 students   **(c)** 7 pages
(d) 272 miles   **(e)** $10\frac{1}{2}$ hours

9. (a) $2.30   **(b)** 1500 students
(c) $3\frac{1}{2}$ pages   **(d)** 136 miles
(e) $5\frac{1}{4}$ hours

7.1 Exercises

FOR EXTRA HELP　Go to MyMathLab *for worked-out, step-by-step solutions to exercises enclosed in a square* ▢ *and video solutions to* ▶ *exercises.*

1. **CONCEPT CHECK**
   (a) A shortcut for dividing by 100 is to move the decimal point _____.
   (b) Draw arrows to show how to use the shortcut when changing these percents to decimals.

   48 % =　　　120 % =

   3 5% =　　　6 % =

2. **CONCEPT CHECK**
   (a) A shortcut for multiplying by 100 is to move the decimal point _____.
   (b) Draw arrows to show how to use the shortcut when changing these decimals to percents.

   0.88 =　　　2.7 =

   0.105 =　　　0.3 =

Write each percent as a decimal. ***See Examples 2 and 3.***

3. 25%

4. 35%

5. 30%

6. 20%

7. 6%
   GS 0̮6.% = 0.___ ___

8. 3%
   GS 0̮3.% = 0.___ ___

9. 140%

10. 250%

11. 7.8%

12. 6.7%

13. 100%

14. 600%

15. 0.5%

16. 0.2%

17. 0.35%

18. 0.076%

Write each decimal as a percent. ***See Example 4.***

19. 0.5

20. 0.6

21. 0.62

22. 0.18

23. 0.03

24. 0.07

25. 0.125

26. 0.875

27. 2

28. 5

29. 2.6

30. 1.8

31. 0.0312

32. 0.0625

Write each percent as a fraction or mixed number in lowest terms. ***See Examples 5 and 6.***

33. 20%
    GS $\frac{20}{100} = \frac{\square}{\square}$ ← Lowest terms

34. 40%
    $\frac{40}{100} = \frac{\square}{\square}$ ← Lowest terms

35. 50%

36. 75%

37. 55%

38. 35%

39. 37.5%

40. 87.5%

41. 6.25%

42. 43.75%

43. $16\frac{2}{3}\%$

44. $83\frac{1}{3}\%$

45. 130%

46. 175%

47. 250%

48. 325%

Write each fraction as a percent. If you're using a calculator, first work each one by hand. Then use your calculator and round your final answers to the nearest tenth of a percent when necessary. ***See Example 7.***

49. $\dfrac{1}{4}$

$$\left(\dfrac{1}{4}\right)(100\%) = \left(\dfrac{1}{4}\right)\left(\dfrac{100}{1}\%\right) =$$

$$\left(\dfrac{1}{\overset{1}{\cancel{4}}}\right)\left(\dfrac{\overset{1}{\cancel{4}} \cdot 25}{1}\%\right) =$$

50. $\dfrac{1}{5}$

$$\left(\dfrac{1}{5}\right)(100\%) = \left(\dfrac{1}{5}\right)\left(\dfrac{100}{1}\%\right) =$$

$$\left(\dfrac{1}{\overset{1}{\cancel{5}}}\right)\left(\dfrac{\overset{1}{\cancel{5}} \cdot 20}{1}\%\right) =$$

51. $\dfrac{3}{10}$

52. $\dfrac{9}{10}$

53. $\dfrac{3}{5}$

54. $\dfrac{3}{4}$

55. $\dfrac{37}{100}$

56. $\dfrac{63}{100}$

57. $\dfrac{3}{8}$

58. $\dfrac{1}{8}$

59. $\dfrac{1}{20}$

60. $\dfrac{1}{50}$

61. $\dfrac{5}{9}$

62. $\dfrac{7}{9}$

63. $\dfrac{1}{7}$

64. $\dfrac{5}{7}$

In each statement, write percents as decimals and decimals as percents. ***See Examples 2–4.***

65. In 1900, only 8% of U.S. homes had a telephone. (Data from *Harper's Index.*)

66. In 1900, only 14% of homes in the United States had a bathtub. (Data from *Harper's Index.*)

67. Tornadoes can occur on any day of the year, but 54% of them appeared in May and June of 2014. (Data from www.ncdc.noaa.gov)

68. Only 2.8% of tornadoes occurred in March 2014, which was a lower percent than normal. (Data from www.ncdc.noaa.gov)

69. The property tax rate in Alpine County is 0.035.

70. The average human brain uses 0.2 of the body's oxygen and energy.

71. The number of people taking CPR training this session is 2 times that of the last session.

72. Attendance at this year's company picnic is 3 times last year's attendance.

Write a fraction and a percent for the shaded part of each figure. Then write a fraction and a percent for the unshaded part of each figure.

73. 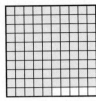   **74.**   **75.** 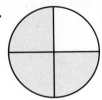   **76.**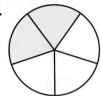

CONCEPT CHECK   *Find three items that are equivalent in each list and circle them.*

77. 50% $\dfrac{1}{4}$ 1.00   0.5 $\dfrac{1}{2}$ 5%   1.5

78. $\dfrac{3}{4}$ 100%   0.25   4% $\dfrac{1}{4}$ 0.3   25%

Complete this table. Write fractions and mixed numbers in lowest terms.

	Fraction	Decimal	Percent
79.	$\dfrac{1}{100}$	_____	_____
80.	$\dfrac{1}{10}$	_____	_____
81.	_____	0.2	_____
82.	_____	0.25	_____
83.	_____	_____	30%
84.	_____	_____	40%
85.	$\dfrac{1}{2}$	_____	_____
86.	$\dfrac{3}{4}$	_____	_____
87.	_____	_____	90%
88.	_____	_____	100%
89.	_____	1.5	_____
90.	_____	2.25	_____

The road signs in Exercises 91–92 tell drivers that they are approaching a steep, downward hill. For example, the 8% sign means that for every 100 ft of roadway length, the road will drop a total of 8 ft in elevation.

91. Write the percent on the sign as a decimal and as a fraction in lowest terms.

92. Write the percent on the sign as a decimal and as a fraction in lowest terms.

93. Suppose the roadway dropped 5 ft in elevation for every 100 ft of length. Write a percent, a decimal, and a fraction in lowest terms to describe this situation.

94. Suppose the roadway dropped 7 ft in elevation for every 100 ft of length. Write a percent, a decimal, and a fraction in lowest terms to describe this situation.

The diagram shows the different types of teeth in an adult's mouth. Use the diagram to answer Exercises 95–100. Write each answer in Exercises 95–98 as a fraction in lowest terms, as a decimal, and as a percent.

95. What portion of an adult's teeth are incisors, designed to bite and cut?

96. The pointy canine teeth tear and rip food. They are what portion of an adult's teeth?

97. The molars, which grind up food, are what portion of an adult's teeth?

98. What portion of an adult's teeth are premolars?

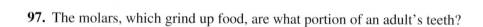

Incisors

Canines

Premolars

Molars

Upper teeth

Lower teeth

Molars

Premolars

Canines

Incisors

Data from Time-Life *Human Body.*

99. Some people have four fewer molars than shown in the tooth diagram. For these people, the canines are what portion of their teeth? Write your answer as a fraction in lowest terms and as a percent.

100. For the adults who have four fewer molars than shown, the incisors are what portion of their teeth? Write your answer as a fraction in lowest terms and as a percent.

101. CONCEPT CHECK The work shown below has **two** mistakes in it.

What Went Wrong? First write a sentence explaining what each mistake is. Then fix the mistakes and find the correct solutions.

Omaria used her calculator to do two problems.

For the first problem she had to write $\frac{7}{20}$ as a percent.

She entered 7 $\div$ 20 $=$, and the result was 0.35, so she said $\frac{7}{20} = 0.35\%$

For the second problem she had to write $\frac{16}{25}$ as a percent. She entered 25 $\div$ 16 $=$, and the result was 1.5625, so she said $\frac{16}{25} = 156.25\%$

102. CONCEPT CHECK The work shown below has **two** mistakes in it.

What Went Wrong? First write a sentence explaining what each mistake is. Then fix the mistakes and find the correct solutions.

Ezra did two problems by moving the decimal point.

For the first problem he had to write 3.2 as a percent. This was his work: $03.2 = 0.032$ so $3.2 = 0.032\%$

For the second problem he had to write 60% as a decimal. This was his work:

$$00.60 = 0.0060 \quad \text{so} \quad 60\% = 0.0060$$

Fill in the blanks in Exercises 103–118. Remember that 100% *is* all *of something and* 50% *is* half *of it.* **See Examples 8 and 9.**

103. (a) 100% of $78 is _____

  (b) 50% of $78 is _____

104. (a) 100% of 5 hours is _____

  (b) 50% of 5 hours is _____

105. (a) 100% of 15 inches is _____

  (b) 50% of 15 inches is _____

106. (a) 100% of $6000 is _____

  (b) 50% of $6000 is _____

107. (a) 100% of 2.8 miles is _____

  (b) 50% of 2.8 miles is _____

108. (a) 100% of $2.50 is _____

  (b) 50% of $2.50 is _____

109. There are 20 children in the preschool class. 100% of the children are served breakfast and lunch. How many children are served both meals?

_____

110. The Speedy Delivery company owns 345 vans. 100% of the vans are painted white with blue lettering. How many company vans are painted white with blue lettering?

_____

111. Alyssa needs 120 credits to graduate. She has earned 50% of the credits. How many credits has she earned?

_____

112. Ian has a limit of $1500 on his credit card. He has used 50% of the limit. How many dollars of credit has he used?

_____

113. **(a)** Joan owes $5019 for tuition. Financial aid will pay 50% of the cost. Financial aid will pay

_____

(b) What *percent* of the tuition will Joan have to pay?

_____

(c) How much *money* will Joan have to pay?

_____

114. **(a)** The Animal Humane Society took in 20,000 animals last year. About 50% of them were dogs. The number of dogs they took in was about how many?

_____

(b) What *percent* of the animals were *not* dogs?

_____

(c) How many *animals* were *not* dogs?

_____

115. **(a)** About 50% of the 8200 students at our college work more than 20 hours per week. How many *students* is this?

_____

(b) What *percent* of the students work 20 hours or less per week?

_____

116. **(a)** Shalayna's smartphone plan includes 1500 megabytes of data per month. Last month she used 50% of her megabytes. How many *megabytes* did she use?

_____

(b) What *percent* of her megabytes were *not* used?

_____

117. **(a)** Dylan's test score was 100%. How many of the 35 problems did he work correctly?

_____

(b) How many problems did Dylan miss?

_____

118. **(a)** Latrell made 100% of his 12 free throws during a practice game today. How many free throws did he make?

_____

(b) How many free throws did Latrell miss?

_____

119. **CONCEPT CHECK** Describe a shortcut way to find 50% of a number. Include two examples to illustrate the shortcut.

120. **CONCEPT CHECK** Describe a shortcut way to find 100% of a number. Include two examples to illustrate the shortcut.

7.2 | The Percent Proportion

OBJECTIVES

1 Identify the percent, whole, and part.

2 Solve percent problems using the percent proportion.

We will show you two ways to solve percent problems. One is the proportion method, which is discussed in this section. The other is the percent equation method, which is explained in the next section.

OBJECTIVE ▸ 1 Identify the percent, whole, and part. You have learned that a statement of two equal ratios is called a proportion. For example, the fraction $\frac{3}{5}$ is the same as the ratio 3 to 5, and 60% is the ratio 60 to 100. As the figure below shows, these two ratios are equal and make a proportion.

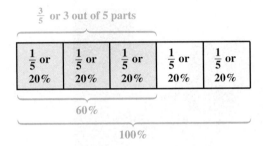

The **percent proportion** can be used to solve percent problems.

The Percent Proportion

Percent is to *100*   **as**   *part* is to *whole*.

$$\frac{\text{percent}}{100} = \frac{\text{part}}{\text{whole}}$$

Always 100 ⟶ because *percent* means "per 100"

In some textbooks the percent proportion is written using the words *amount* and *base*.

$$\frac{\text{percent}}{100} = \frac{\text{amount}}{\text{base}}$$

Here is the proportion for the figure at the top of the page.

60% means 60 parts out of 100 parts. $\Big\}$ $\dfrac{60}{100} = \dfrac{3}{5}$ ← Shaded (3 parts)
                                                                  ← Whole (5 parts)

If we write $\frac{60}{100}$ in lowest terms, it is equal to $\frac{3}{5}$, so the proportion is true.

$$\frac{60}{100} = \frac{60 \div 20}{100 \div 20} = \frac{3}{5} \quad \left\{ \begin{array}{l} \text{Matches ratio above on} \\ \text{right side of proportion} \end{array} \right.$$

When using the percent proportion, you must be able to pick out the *percent,* the *whole,* and the *part.* Look for the percent first; it is the easiest to identify.

Identifying the Percent

The **percent** is a ratio of a part to a whole, with 100 as the denominator. In a problem, the percent appears with the word *percent* or with the symbol % after it.

EXAMPLE 1   Identifying the Percent in Percent Problems

Identify the percent in each problem.

(a) 32% of the 900 women were retired. How many were retired?

Percent

The percent is 32. The number 32 appears with the symbol %.

(b) $150 is 25 percent of what number?

Percent

The percent is 25 because 25 appears with the word *percent*.

(c) If 7 students failed, what percent of the 350 students failed?

Percent (unknown)

The word *percent* has no number with it, so **the percent is the unknown** in the problem.

─────── **Work Problem 1 at the Side.** ▶

The second thing to look for is the *whole* (sometimes called the *base*).

Identifying the Whole

The **whole** is the entire quantity. In a percent problem, the *whole* often appears after the word of.

EXAMPLE 2   Identifying the Whole in Percent Problems

These problems are the same as those in **Example 1** above. Now identify the *whole*.

(a) 32% of the 900 women were retired. How many were retired?

Percent   Whole

The whole is 900 women. The number 900 appears after the word *of*.

(b) $150 is 25 percent of what number?

Percent   Whole (unknown; follows *of*)

The whole is unknown.

(c) If 7 students failed, what percent of the 350 students failed?

Percent (unknown)   Whole (follows *of*)

The whole is 350 students. The number 350 appears after the word *of*.

─────── **Work Problem 2 at the Side.** ▶

The final thing to identify is the *part* (sometimes called the *amount*).

Identifying the Part

The **part** is the number being compared to the whole.

---

1 Identify the percent.

(a) 15% of the $2000 was spent on a washing machine.

(b) 60 employees is what percent of 750 employees?

(c) The state sales tax is $6\frac{1}{2}$ percent of the $590 price.

(d) $30 is 48% of what amount of money?

(e) 75 of the 110 rental cars were rented today. What percent were rented?

2 Identify the whole.

(a) 15% of the $2000 was spent on a washing machine.

(b) 60 employees is what percent of 750 employees?

(c) The state sales tax is $6\frac{1}{2}$ percent of the $590 price.

(d) $30 is 48% of what amount of money?

(e) 75 of the 110 rental cars were rented today. What percent were rented?

Answers

1. (a) 15%   **(b)** unknown   **(c)** $6\frac{1}{2}$ percent
(d) 48%   **(e)** unknown

2. (a) $2000   **(b)** 750 employees   **(c)** $590
(d) unknown   **(e)** 110 rental cars

3 Identify the part.

(a) 15% of the $2000 was spent on a washing machine.

(b) 60 employees is what percent of 750 employees?

(c) The state sales tax is $6\frac{1}{2}$ percent of the $590 price.

(d) $30 is 48% of what amount of money?

(e) 75 of the 110 rental cars were rented today. What percent were rented?

Note

If you have trouble identifying the *part,* find the *whole* and the *percent* first. The remaining number is the *part.*

EXAMPLE 3 **Identifying the Part in Percent Problems**

These problems are the same as those in **Examples 1 and 2** on the previous page. Now identify the *part.*

(a) 32% of the 900 women were retired. How many were retired?

Percent   Whole   Part (unknown)

The part of the women who were retired **is unknown.** In other words, some part of 900 women were retired.

(b) $150 is 25 percent of what number?

Part   Percent   Whole (unknown)

$150 is the remaining number, so **$150 is the part.**

(c) If 7 students failed, what percent of 350 students failed?

Part   Percent (unknown) Whole

The part of the students who failed is **7 students.**

◀ **Work Problem** **3** **at the Side.**

OBJECTIVE ▶ **2** **Solve percent problems using the percent proportion.**

EXAMPLE 4 **Using the Percent Proportion to Find the Part**

Use the percent proportion to answer this question.

15% of $165 is how much money?

Recall that the percent proportion is: $\dfrac{\text{percent}}{100} = \dfrac{\text{part}}{\text{whole}}$ First identify the percent by looking for the % symbol or the word *percent.* Then look for the *whole* (usually follows the word *of*). Finally, identify the *part.*

15% of $165 is how much money?

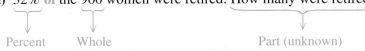

Percent   Whole   Part (unknown)
        (follows *of*)

Set up the percent proportion. Here we use *n* as the variable representing the unknown part. You may use any letter you like.

Percent → $\dfrac{15}{100} = \dfrac{n}{165}$ ← Part (unknown)
Always 100 →   ← Whole

Recall that the first step in solving a proportion is to find the cross products.

Step 1 $\dfrac{15}{100} \bowtie \dfrac{n}{165}$ $\begin{array}{l} 100 \cdot n \\[4pt] 15 \cdot 165 \end{array}$ Find the cross products.

──────── **Continued on Next Page**

Step 2

$100 \cdot n = \underbrace{15 \cdot 165}$ Show that the cross products are equal.

$100 \cdot n = \quad 2475$

Step 3

$\dfrac{\overset{1}{\cancel{100}} \cdot n}{\underset{1}{\cancel{100}}} = \dfrac{2475}{100}$ Divide both sides by 100, the coefficient of the variable term. On the left side, divide out the common factor of 100.

$n = 24.75$ On the right side, $2475 \div 100$ is 24.75

The part is $\$24.75$, so 15% of $165 is **$24.75** ◄ Write a $ in your answer.

> **⚠ CAUTION**
>
> When you use the percent proportion, do **not** move the decimal point in the percent or in the answer.

——— **Work Problem ④ at the Side.** ►

EXAMPLE 5 **Using the Percent Proportion to Find the Percent**

Use the percent proportion to answer this question.

8 pounds is what percent of 160 pounds?

Part   Percent (unknown)   Whole (follows of)

The percent proportion is: $\dfrac{percent}{100} = \dfrac{part}{whole}$ Set up the proportion

using p as the variable representing the unknown percent. Then find the cross products.

Percent (unknown) → $\dfrac{p}{100} = \dfrac{8}{160}$ ← Part
Always 100 → $\qquad\qquad$ ← Whole

$\dfrac{p}{100} \diagup \dfrac{8}{160}$ $100 \cdot 8 = 800$
$\qquad\qquad\qquad p \cdot 160$ Cross products

$p \cdot 160 = 800$ Show that the cross products are equal.

$\dfrac{p \cdot \overset{1}{\cancel{160}}}{\underset{1}{\cancel{160}}} = \dfrac{800}{160}$ Divide both sides by 160

$p = 5$ The *percent* was unknown. Write % in your answer.

The percent is 5%, so 8 pounds is **5%** of 160 pounds.

> **⚠ CAUTION**
>
> When you're finding an unknown percent, as in **Example 5** above, be careful to label your answer with the % symbol. Do **not** add a decimal point or move the decimal point in your answer.

——— **Work Problem ⑤ at the Side.** ►

---

④ Use the percent proportion to answer these questions.

(a) 9% of 3250 miles is how many miles?

Percent → $\dfrac{9}{100} = \dfrac{n}{3250}$ ← Part
Always 100 →   ← Whole

(b) 78% of $5.50 is how much?

(c) What is $12\frac{1}{2}\%$ of 400 homes? (*Hint:* Write $12\frac{1}{2}\%$ as 12.5%.)

⑤ Use the percent proportion to answer these questions.

(a) 1200 books is what percent of 5000 books?

Percent → $\dfrac{p}{100} = \dfrac{1200}{5000}$ ← Part
Always 100 →   ← Whole

(b) What percent of $6.50 is $0.52?

(c) 20 athletes is what percent of 32 athletes?

Answers

4. **(a)** The part is 292.5 miles.
   (b) $\dfrac{78}{100} = \dfrac{n}{5.50}$ The part is $4.29
   (c) $\dfrac{12.5}{100} = \dfrac{n}{400}$ The part is 50 homes.
5. **(a)** 24%
   (b) $\dfrac{p}{100} = \dfrac{0.52}{6.50}$;   8%
   (c) $\dfrac{p}{100} = \dfrac{20}{32}$;   62.5% or $62\frac{1}{2}\%$

6 Use the percent proportion to answer these questions.

GS **(a)** 37 cars is 74% **of** how many cars?

$$\text{Percent} \rightarrow \frac{74}{100} = \frac{37}{n} \begin{array}{l} \leftarrow \text{Part} \\ \leftarrow \text{Whole} \end{array}$$
Always 100 →

(b) 45% **of** how much money is $139.59?

(c) 1.2 tons is $2\frac{1}{2}$% **of** how many tons?

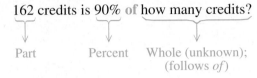
EXAMPLE 6 **Using the Percent Proportion to Find the Whole**

Use the percent proportion to answer this question.

162 credits is 90% of how many credits?

Part	Percent	Whole (unknown); (follows *of*)

$$\text{Percent} \rightarrow \frac{90}{100} = \frac{162}{n} \begin{array}{l} \leftarrow \text{Part} \\ \leftarrow \text{Whole (unknown)} \end{array}$$
Always 100 →

$$\frac{90}{100} = \frac{162}{n} \qquad \begin{array}{l} 100 \cdot 162 = 16{,}200 \\ 90 \cdot n \end{array} \text{Cross products}$$

$$90 \cdot n = 16{,}200 \qquad \text{Show that the cross products are equal.}$$

$$\frac{\overset{1}{\cancel{90}} \cdot n}{\underset{1}{\cancel{90}}} = \frac{16{,}200}{90} \qquad \text{Divide both sides by 90}$$

$$n = 180 \qquad \boxed{\text{Write } \textbf{credits} \text{ in your answer.}}$$

The whole is **180 credits**. So 162 credits is 90% of **180 credits**.

◀ **Work Problem 6 at the Side.**

So far in all the examples, the part has been *less* than the whole. This is because all the percents have been less than 100%. Recall that 100% of something is *all* of it. When the percent is *less* than 100%, you have *less* than all of it.

Now let's look at percents *greater* than 100%. For example,

100% of $20 is all of the money, or $20.

150% of $20 is *greater* than $20.

100%	+	50%	=	150%
of the money is		of the money is		of the money is
$20	+	$10	=	$30

So 150% of $20 is $30.

When the percent is *greater* than 100%, the part is *greater* than the whole, as you'll see in **Example 7** on the next page.

Answers

6. **(a)** 50 cars

(b) $\dfrac{45}{100} = \dfrac{139.59}{n}$; $310.20

(c) $\dfrac{2.5}{100} = \dfrac{1.2}{n}$; 48 tons

EXAMPLE 7   **Working with Percents Greater Than 100%**

Use the percent proportion to answer each question.

(a) How many students is 210% of 40 students?

Part (unknown)   Percent   Whole (follows *of*)

$$\text{Percent} \to \frac{210}{100} = \frac{n}{40} \begin{array}{l} \leftarrow \text{Part (unknown)} \\ \leftarrow \text{Whole} \end{array}$$
Always 100 →

$$\frac{210}{100} = \frac{n}{40}$$

$$\begin{array}{l} 100 \cdot n \leftarrow \\ \\ 210 \cdot 40 = 8400 \leftarrow \end{array} \text{Cross products}$$

$$100 \cdot n = 8400 \qquad \text{Show that the cross products are equal.}$$

$$\frac{\overset{1}{\cancel{100}} \cdot n}{\underset{1}{\cancel{100}}} = \frac{8400}{100} \qquad \text{Divide both sides by 100}$$

$$n = 84$$

The part is **84 students**, which is *greater* than the whole of 40 students. This result makes sense because the percent is 210%. If it was exactly 200%, we would have *2 times the whole,* and 2 times 40 students is 80 students. So 210% should be even a little more than 80 students. Our answer of 84 students is reasonable.

(b) What percent of $50 is $68?

Percent   Whole   Part
(unknown)   (follows *of*)

$$\text{Percent (unknown)} \to \frac{p}{100} = \frac{68}{50} \begin{array}{l} \leftarrow \text{Part} \\ \leftarrow \text{Whole} \end{array}$$
Always 100 →

$$\frac{p}{100} = \frac{68}{50}$$

$$\begin{array}{l} 100 \cdot 68 = 6800 \leftarrow \\ \\ p \cdot 50 \leftarrow \end{array} \text{Cross products}$$

$$p \cdot 50 = 6800 \qquad \text{Show that the cross products are equal.}$$

$$\frac{p \cdot \overset{1}{\cancel{50}}}{\underset{1}{\cancel{50}}} = \frac{6800}{50} \qquad \text{Divide both sides by 50}$$

$$p = 136$$

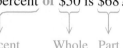
Write **%** in your answer.

The percent is **136%**. This result makes sense because $68 is *greater* than $50, so $68 has to be *greater than 100%* of $50.

─────── **Work Problem** ⑦ **at the Side.** ▶

---

⑦ Use the percent proportion to answer each question.

㏿ **(a)** 350% **of** $6 is how much?

$$\begin{array}{l} \text{Percent} \to \\ \text{Always 100} \to \end{array} \frac{350}{100} = \frac{n}{6} \begin{array}{l} \leftarrow \text{Part} \\ \leftarrow \text{Whole} \end{array}$$

(b) 23 hours is what percent of 20 hours?

🔢 **(c)** 225% of what amount is $106.47?

Answers

7. **(a)** $21 **(b)** $\dfrac{p}{100} = \dfrac{23}{20}$;   115%

   (c) $\dfrac{225}{100} = \dfrac{106.47}{n}$; $47.32

7.2 Exercises

FOR EXTRA HELP

Go to MyMathLab for worked-out, step-by-step solutions to exercises enclosed in a square ▢ and video solutions to ▶ exercises.

CONCEPT CHECK   *In Exercises 1–6, **(a)** identify the percent; **(b)** identify the whole; **(c)** identify the part; **(d)** write a percent proportion. Do **not** solve the proportion. See Examples 1–6.*

1. What is 10% of 3000 runners?

2. What is 35% of 2340 volunteers?

3. 16 pepperoni pizzas is what percent of 32 pizzas?

4. 35 hours is what percent of 140 hours?

5. 495 successful students is 90% of what number of students?

6. 84 e-mails is 28% of what number of e-mails?

Write a percent proportion and solve it to answer Exercises 7–24. If necessary, round money answers to the nearest cent and percent answers to the nearest tenth of a percent. See Examples 1–7.

7. 4% of 120 feet is how many feet?

GS ▶

$$\text{Percent} \rightarrow \frac{4}{100} = \frac{n}{120} \begin{array}{l} \leftarrow \text{Part} \\ \leftarrow \text{Whole} \end{array}$$
Always 100 →

8. 9% of \$150 is how much money?

GS

$$\text{Percent} \rightarrow \frac{9}{100} = \frac{n}{150} \begin{array}{l} \leftarrow \text{Part} \\ \leftarrow \text{Whole} \end{array}$$
Always 100 →

9. What percent of 200 calories is 16 calories?

10. What percent of 350 parking spaces is 7 accessible parking spaces?

11. $12\frac{1}{2}\%$ of what amount is \$3.50?

GS

$$\text{Percent} \rightarrow \frac{12.5}{100} = \frac{3.50}{n} \begin{array}{l} \leftarrow \text{Part} \\ \leftarrow \text{Whole} \end{array}$$
Always 100 →

12. $5\frac{1}{2}\%$ of what amount is \$17.60?

GS

$$\text{Percent} \rightarrow \frac{5.5}{100} = \frac{17.60}{n} \begin{array}{l} \leftarrow \text{Part} \\ \leftarrow \text{Whole} \end{array}$$
Always 100 →

13. 250% of 7 hours is how long?

14. What is 130% of 60 trees?

15. What percent of \$172 is \$32?

16. \$14 is what percent of \$398?

17. 748 e-books is 110% of what number of e-books?

18. 145% of what number of inches is 11.6 inches?

19. What is 14.7% of $274?

20. 8.3% of $43 is how much?

21. 105 employees is what percent of 54 employees?

22. What percent of 46 websites is 100 websites?

23. $0.33 is 5% of what amount?

24. 6% of what amount is $0.03?

25. CONCEPT CHECK   The work shown below has **two** mistakes in it.

What Went Wrong?   First write a sentence explaining what each mistake is. Then fix the mistakes and find the correct solution.

$14 is what percent of $8?
Jourden solved the problem this way:

$$\frac{p}{100} = \frac{8}{14} \qquad p \cdot 14 = 100 \cdot 8$$

$$\frac{p \cdot \overset{1}{\cancel{14}}}{\underset{1}{\cancel{14}}} = \frac{800}{14}$$

$$p \approx 57.1$$

Jourden said the answer was 57.1 (rounded).

26. CONCEPT CHECK   The work shown below has **two** mistakes in it.

What Went Wrong?   First write a sentence explaining what each mistake is. Then fix the mistakes and find the correct solution.

9 children is 30% of what number of children?
Edith solved the problem this way:

$$\frac{30}{100} = \frac{n}{9} \qquad 100 \cdot n = 30 \cdot 9$$

$$\frac{\overset{1}{\cancel{100}} \cdot n}{\underset{1}{\cancel{100}}} = \frac{270}{100}$$

$$n = 2.7$$

Edith said the answer is 2.7%

Relating Concepts (Exercises 27–28)   For Individual or Group Work

*Use your knowledge of percents as you work **Exercises 27 and 28 in order.***

27. A student turned in the following answers on a test. You can see that *two* of the answers are incorrect *without working the problems*. Find the incorrect answers and explain how you identified them (without actually solving the problems).

50% of $84 is ___$42___   25% of $16 is ___$32___

150% of $30 is ___$20___   100% of $217 is ___$217___

28. Name the three parts in a percent problem. For each of these three parts, write a sentence telling how you identify it.

7.3 The Percent Equation

OBJECTIVES

1 Estimate answers to percent problems involving 25%.

2 Find 10% and 1% of a number by moving the decimal point.

3 Solve basic percent problems using the percent equation.

OBJECTIVE 1 Estimate answers to percent problems involving 25%.
Before showing you the percent equation, we need to do some more estimation. Estimating answers helps you catch mistakes. Also, you can use estimating to figure out sales tax, discounted prices, and restaurant tips.

Earlier we used shortcuts for 100% of a number (all of the number) and 50% of a number (divide the number by 2). Now we look at a quick way to work with 25%.

25% means 25 parts out of 100 parts, or $\frac{25}{100}$, which is the same as $\frac{1}{4}$

25% of $40 would be $\frac{1}{4}$ of $40, or $10

A quick way to find $\frac{1}{4}$ of a number is to *divide it by 4*. The denominator, 4, tells you that the whole is divided into 4 equal parts.

EXAMPLE 1 Estimating 25% of a Number

Estimate the answer to each question.

(a) What is 25% of $817?

Use front end rounding to round $817 to $800. Then divide $800 by 4. The estimate is **$200**.

> 25% is $\frac{1}{4}$, so to find 25% of a number, divide it by 4.

(b) Find 25% of 19.7 miles.

Use front end rounding to round 19.7 miles to 20 miles. Then divide 20 miles by 4. The estimate is **5 miles**.

(c) 25% of 49 days is how long?

You could round 49 days to 50 days, using front end rounding. Then divide 50 by 4 to get an estimate of **12.5 days**.

However, the division step is simpler if you notice that 48 is a multiple of 4. You can round 49 days to 48 days and divide by 4 to get an estimate of **12 days**. Either way gives you a fairly good idea of the correct answer.

◀ **Work Problem 1 at the Side.**

---

1 *Estimate* the answer to each question.

(a) What is 25% of $110.38?

Round $110.38 to $100. 25% is $\frac{1}{4}$ so divide $100 by 4 to get _____
Estimate —↑

(b) Find 25% of 7.6 hours.

(c) 25% of 34 pounds is how many pounds?

OBJECTIVE 2 Find 10% and 1% of a number by moving the decimal point. There are also helpful shortcuts for finding 10% or 1% of a number.

Ten percent, or 10%, means 10 parts out of 100 parts or $\frac{10}{100}$, which is the same as $\frac{1}{10}$. A quick way to find $\frac{1}{10}$ of a number is to *divide it by 10*. The denominator, 10, tells you that the whole is divided into 10 equal parts. The shortcut for dividing by 10 is to move the decimal point *one* place to the *left*.

EXAMPLE 2 Finding 10% of a Number by Moving the Decimal Point

Find the *exact* answer to each question by moving the decimal point.

(a) What is 10% of $817?

To find 10% of $817, divide $817 by 10. Do the division by moving the decimal point *one* place to the *left*. The decimal point starts at the right side of $817.

10% of $81**7.** = $81.**70** ← Exact answer

> Do **not** leave it as $81.7

↑ Write this 0 because it's money.

So 10% of $817 is **$81.70**

Answers

1. (a) $25
(b) 8 hours ÷ 4 gives an estimate of 2 hours.
(c) 30 pounds ÷ 4 gives an estimate of 7.5 pounds, or 32 pounds ÷ 4 gives an estimate of 8 pounds.

Continued on Next Page

(b) Find 10% of 19.7 miles.

To find 10% of 19.7 miles, divide 19.7 by 10. Move the decimal point *one* place to the *left*.

$$10\% \text{ of } 19.7 \text{ miles} = 1.97 \text{ miles} \longleftarrow \text{Exact answer}$$

So 10% of 19.7 miles is 1.97 miles.

——————— **Work Problem ② at the Side.** ▶

One percent, or 1%, is 1 part out of 100 parts or $\frac{1}{100}$. This time the denominator of 100 tells you that the whole is divided into 100 parts. Recall that a quick way to divide by 100 is to move the decimal point *two* places to the *left*.

EXAMPLE 3   **Finding 1% of a Number by Moving the Decimal Point**

Find the *exact* answer to each question by moving the decimal point.

(a) What is 1% of $817?

To find 1% of $817, divide $817 by 100. Do the division by moving the decimal point *two* places to the *left*.

$$1\% \text{ of } \$817. = \$8.17 \longleftarrow \text{Exact answer}$$

So 1% of $817 is $8.17

(b) Find 1% of 19.7 miles.

To find 1% of 19.7 miles, divide 19.7 by 100. Move the decimal point *two* places to the *left*.

$$1\% \text{ of } 19.7 \text{ miles} = 0.197 \text{ mile} \longleftarrow \text{Exact answer}$$

So 1% of 19.7 miles is 0.197 mile.

——————— **Work Problem ③ at the Side.** ▶

Here is a summary of some of the shortcuts you can use with percents.

Percent Shortcuts

200% of a number is 2 times the number; **300% of a number** is 3 times the number; and so on.

100% of a number is the entire number.

To find **50% of a number,** divide the number by 2.

To find **25% of a number,** divide the number by 4.

To find **10% of a number,** divide the number by 10. To do the division, move the decimal point in the number *one* place to the *left*.

To find **1% of a number,** divide the number by 100. To do the division, move the decimal point in the number *two* places to the *left*.

---

② Find the *exact* answer to each question by moving the decimal point.

GS (a) What is 10% of $110.38?

$$10\% \text{ of } \$110.38 = \underline{\hspace{2cm}}$$
$$\text{Exact answer}$$

(b) Find 10% of 7.6 hours.

(c) 10% of 34 pounds is how many pounds?

③ Find the *exact* answer to each question by moving the decimal point.

GS (a) What is 1% of $110.38?

$$1\% \text{ of } \$110.38 = \underline{\hspace{2cm}}$$
$$\text{Exact answer}$$

(b) Find 1% of 7.6 hours.

(c) 1% of 34 pounds is how many pounds?

Answers

2. (a) $110.38 = $11.038 or $11.04 to the nearest cent
  (b) 7.6 hours = 0.76 hour
  (c) 34. pounds = 3.4 pounds
3. (a) $110.38 = $1.1038 or $1.10 to the nearest cent
  (b) 07.6 hours = 0.076 hour
  (c) 34. pounds = 0.34 pound

OBJECTIVE ▸ ③ Solve basic percent problems using the percent equation.
In the previous section you used a proportion to solve percent problems.
Now you will learn how to solve these problems using the percent equation.

Percent Equation

$$\text{percent of whole} = \text{part}$$

The word **of** indicates multiplication, so the **percent equation** becomes

$$\text{percent} \cdot \text{whole} = \text{part}$$

Be sure to write the percent as a decimal or fraction before using the equation.

The percent equation is just a rearrangement of the percent proportion. Recall that in the proportion you wrote the percent over 100. Because there is no 100 in the equation, you have to change the percent to a decimal or fraction by dividing by 100 *before* using the equation.

Note

Once you have set up a percent equation, we encourage you to use your calculator to do the multiplying or dividing needed to solve the equation. For this reason, we will always write the percent as a decimal. If you're doing the problems by hand, changing the percent to a fraction may be easier at times. Either method will work.

Examples 4, 5, and 6 below are the same percent questions that were in the examples in the previous section. There we used a proportion to answer each question. Now we will use an equation to answer them. You can then compare the equation method with the proportion method.

EXAMPLE 4   Using the Percent Equation to Find the Part

Write and solve a percent equation to answer each question.

(a) 15% of $165 is how much money?

Translate the sentence into an equation. Recall that *of* indicates multiplication and *is* translates to the equal sign. The percent must be written in decimal form. Use any letter you like to represent the unknown quantity. (We will use n for an unknown number and p for an unknown percent.)

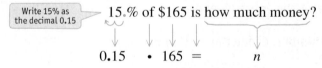

Write 15% as the decimal 0.15   15.% of $165 is how much money?

$$0.15 \quad \cdot \quad 165 \quad = \quad n$$

To solve the equation, simplify the left side, multiplying 0.15 by 165.

$$\underline{(0.15)(165)} = n$$
$$24.75 \quad = n$$

So 15% of $165 is $24.75, which matches the answer obtained using a proportion (see **Example 4** in the previous section).

──── Continued on Next Page

CHECK   Use estimation to check that the solution is reasonable. First find 10% of $165 by moving the decimal point.

$$10\% \text{ of } 165. \text{ is } \$16.50 \quad \text{and}$$

5% of $165 would be half as much, that is, half of $16.50 or about $8.

So the *estimate* for 15% of $165 is $16.50 + $8 = $24.50. The exact answer of $24.75 is very close to this estimate, so it is reasonable.

(b) How many students is 210% of 40 students?

Translate the sentence into an equation. Write the percent in decimal form.

How many students is 210.% of 40 students?

$$n \qquad = 2.10 \quad \cdot 40$$

> Write 210% as the decimal 2.10

This time the two sides of the percent equation are reversed, so

$$part = percent \cdot whole$$

Recall that the variable may be on either side of the equal sign. To solve the equation, simplify the right side, multiplying 2.10 by 40.

$$n = \underbrace{(2.10)\,(40)}$$
$$n = \qquad 84$$

So **84 students** is 210% of 40 students. This matches the answer obtained by using a proportion (see **Example 7(a)** in the previous section).

CHECK   Use estimation to check that the solution is reasonable. 210% is close to 200%.

$$200\% \text{ of } 40 \text{ students is } 2 \text{ times } 40 \text{ students} = 80 \text{ students} \longleftarrow \text{Estimate}$$

The exact answer of 84 students is close to the estimate, so it is reasonable.

──────── **Work Problem ④ at the Side.** ▶

EXAMPLE 5   **Using the Percent Equation to Find the Percent**

Write and solve a percent equation to answer each question.

(a) 8 pounds is what percent of 160 pounds?

Translate the sentence into an equation. This time the percent is unknown. Do **not** move the decimal point in the other numbers.

8 pounds is what percent of 160 pounds?

$$8 \quad = \quad p \quad \cdot \quad 160$$

To solve the equation, divide both sides by 160.

On the left side, divide 8 by 160 $\dfrac{8}{160} = \dfrac{p \cdot \overset{1}{\cancel{160}}}{\underset{1}{\cancel{160}}}$ On the right side, divide out the common factor of 160

Solution in *decimal* form } $0.05 = p$

> Multiply the solution by 100% to change it from a *decimal* to a *percent*.

$$0.05 = 5\%$$

So 8 pounds is **5%** of 160 pounds. This matches the answer obtained by using a proportion (see **Example 5** in the previous section).

──────── **Continued on Next Page**

---

④ Write and solve an equation to answer each question.

GS **(a)** 9% of 3250 miles is how many miles?

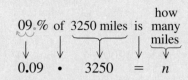

09.% of 3250 miles is   how many miles

$$0.09 \quad \cdot \quad 3250 \quad = \quad n$$

(b) 78% of $5.50 is how much?

(c) What is $12\frac{1}{2}\%$ of 400 homes?
(*Hint:* Write $12\frac{1}{2}\%$ as 12.5%, then move the decimal point two places to the left.)

(d) How much is 350% of $6?

Answers

4. (a) 292.5 miles
  (b) $(0.78)\,(5.50) = n;$ $4.29
  (c) $n = (0.125)\,(400);$ 50 homes
  (d) $n = (3.5)\,(6);$ $21

5 Write and solve an equation to answer each question.

(a) 1200 books is what percent of 5000 books?

$$\underbrace{1200}_{\substack{1200 \\ \text{books}}} \quad \underbrace{\text{is}}_{\downarrow} \quad \underbrace{\text{what}}_{\substack{\text{percent}}} \quad \underbrace{\text{of}}_{\downarrow} \quad \underbrace{5000}_{\substack{5000 \\ \text{books}}}$$

$$1200 = p \cdot 5000$$

CHECK The solution makes sense because 10% of 160 pounds would be 16 pounds.

$$\text{10\% of 160 pounds is 16 pounds} \quad \text{so}$$

5% of 160 pounds is half as much, that is, half of 16 pounds, or 8 pounds.

8 pounds matches the number given in the original problem, so 5% is the correct solution.

(b) What percent of \$50 is \$68?

Translate the sentence into an equation and solve it.

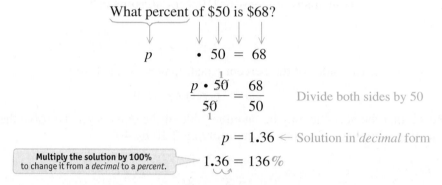

$$\text{What percent of \$50 is \$68?}$$

$$p \quad \cdot \quad 50 = 68$$

$$\frac{p \cdot \overset{1}{\cancel{50}}}{\underset{1}{\cancel{50}}} = \frac{68}{50} \qquad \text{Divide both sides by 50}$$

$$p = 1.36 \leftarrow \text{Solution in } \textit{decimal} \text{ form}$$

Multiply the solution by 100% to change it from a *decimal* to a *percent*. $\quad 1.36 = 136\%$

So **136%** of \$50 is \$68. This matches the answer obtained by using a proportion (see **Example 7(b)** in the previous section).

CHECK The solution makes sense because 100% of \$50 would be \$50 (all of it), and 200% of \$50 would be 2 times \$50, or \$100. So \$68 has to be between 100% and 200%.

(b) 23 hours is what percent of 20 hours?

$$\left.\begin{array}{l}136\% \text{ is between} \\ 100\% \text{ and } 200\%.\end{array}\right\} \quad \begin{array}{l}100\% \text{ of \$50} = \$50 \\ 200\% \text{ of \$50} = \$100\end{array} \quad \left\{\begin{array}{l}\$68 \text{ is between} \\ \$50 \text{ and } \$100\end{array}\right.$$

The solution of 136% fits the conditions.

❗ CAUTION

When you use an equation to solve for an unknown percent, *the solution will be in decimal form.* Remember to ***multiply the solution by 100%*** to change it from decimal form to a percent. The shortcut is to move the decimal point in the solution *two* places to the *right* and attach the % symbol.

Recall that 100% = 1, and multiplying a number by 1 does *not* change the value of the number; it just gives us an *equivalent percent.*

(c) What percent of \$6.50 is \$0.52?

◀ **Work Problem ⑤ at the Side.**

Answers

5. **(a)** $0.24 = 24\%$

(b) $23 = p \cdot 20; \quad 1.15 = 115\%$

(c) $p \cdot 6.50 = 0.52; \quad 0.08 = 8\%$

EXAMPLE 6   **Using the Percent Equation to Find the Whole**

Write and solve a percent equation to answer each question.

(a) 162 credits is 90% of how many credits?

Translate the sentence into an equation. Write the percent in decimal form.

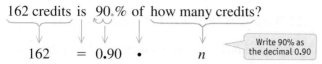

162 credits is 90.% of how many credits?

$$162 = 0.90 \cdot n$$

> Write 90% as the decimal 0.90

Recall that 0.90 is equivalent to 0.9, so use 0.9 in the equation.

$$\frac{162}{0.9} = \frac{\overset{1}{(0.9)}(n)}{\underset{1}{0.9}}$$ Divide both sides by 0.9

$$180 = n$$

> Write **credits** in your answer.

So 162 credits is 90% of **180 credits**. This matches the answer we got using a proportion (see **Example 6** in the previous section.)

CHECK   The solution makes sense because 90% of 180 credits should be 10% less than 100% of the credits, and 10% of 180. credits is 18 credits.

100%	−	10%	=	90%
of 180 credits		of 180. credits		of 180 credits
↓		↓		↓
180 credits	−	18 credits	=	162 credits

← Matches the number given in the original problem

(b) 250% of what amount is $75?

Translate the sentence into an equation. Write the percent in decimal form.

250.% of what amount is $75?

> Write 250% as the decimal 2.5

$$2.5 \cdot n = 75$$

$$\frac{\overset{1}{(2.5)}(n)}{\underset{1}{2.5}} = \frac{75}{2.5}$$ Divide both sides by 2.5

> Write a $ in your answer.

$$n = 30$$

So 250% of $30 is $75

CHECK   The solution makes sense because 200% of $30 is 2 times $30 = $60, and 50% of $30 is $30 ÷ 2 = $15.

200%	+	50%	=	250%
of $30		of $30		of $30
↓		↓		
(2)($30)		$30 ÷ 2		
↓		↓		
$60	+	$15	=	$75

← Matches the number given in the original problem

— **Work Problem 6 at the Side.** ▶

---

6 Write and solve an equation to answer each question.

GS **(a)** 74% of how many cars is 37 cars?

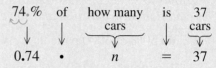

74.% of how many cars is 37 cars

$$0.74 \cdot n = 37$$

(b) 1.2 tons is $2\frac{1}{2}$% of how many tons?

(c) 216 calculators is 160% of how many calculators?

7.3 Exercises

Go to MyMathLab *for worked-out, step-by-step solutions to exercises enclosed in a square* ▢ *and video solutions to* ▶ *exercises.*

CONCEPT CHECK *In Exercises 1–14, use your estimation skills and the percent shortcuts to select the most reasonable answers. Circle your choices. Do **not** write an equation or proportion and solve it. See Examples 1–3.*

1. Find 50% of 3000 patients.

   150 patients   1500 patients   300 patients

2. What is 50% of 192 Facebook updates?

   48 updates   384 updates   96 updates

3. 25% of $60 is how much?

   $15 $6 $30

4. Find 25% of $2840.

   $28.40 $710 $284

5. What is 10% of 45 pounds?

   0.45 pound   22.5 pounds   4.5 pounds

6. 10% of 7 feet is how many feet?

   0.7 foot   3.5 feet   14 feet

7. What is 200% of $3.50?

   $0.35 $1.75 $7.00

8. Find 300% of $12.

   $4 $36 $1.20

9. 1% of 5200 students is how many students?

   520 students   52 students   2600 students

10. Find 1% of 460 miles.

   0.46 mile   46 miles   4.6 miles

11. Find 10% of 8700 smartphones.

   8700 phones   4350 phones   870 phones

12. 25% of 28 video games is how many games?

   14 games   7 games   112 games

13. What is 25% of 19 hours?

   4.75 hours   1.9 hours   2.5 hours

14. What is 1% of $37?

   $370 $3.70 $0.37

15. (a) Describe a shortcut for finding 10% of a number and explain *why* your shortcut works.

16. (a) Describe a shortcut for finding 1% of a number and explain *why* your shortcut works.

(b) Once you know 10% of a certain number, explain how you could use that information to find 20% and 30% of the same number.

(b) Once you know 1% of a certain number, explain how you could use that information to find 2% and 3% of the same number.

Write and solve an equation to answer each question in Exercises 17–48.
See Examples 4–6.

17. 35% of 660 programs is how many programs?

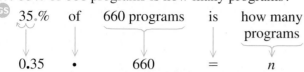

35.%	of	660 programs	is	how many programs
0.35	•	660	=	n

18. 55% of 740 students is how many students?

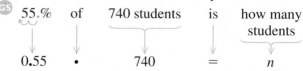

55.%	of	740 students	is	how many students
0.55	•	740	=	n

19. 70 coffee mugs is what percent of 140 mugs?

20. 30 fire extinguishers is what percent of 75 extinguishers?

21. 476 circuits is 70% of what number of circuits?

22. 621 tons is 45% of what number of tons?

23. $12\frac{1}{2}\%$ of what number of people is 135 people?

24. $6\frac{1}{2}\%$ of what number of iPads is 130 iPads?

25. What is 65% of 1300 blog posts?

26. What is 75% of 360 dosages?

27. 4% of $520 is how much?

28. 7% of $480 is how much?

29. 38 styles is what percent of 50 styles?

30. 75 offices is what percent of 125 offices?

31. What percent of $264 is $330?

32. What percent of $480 is $696?

33. 141 employees is 3% of what number of employees?

34. 16 e-readers is 8% of what number of e-readers?

35. 32% of 260 quarts is how many quarts?

36. 44% of 430 liters is how many liters?

37. $1.48 is what percent of $74?

38. $0.51 is what percent of $8.50?

39. How many tablets is 140% of 500 tablets?

40. How many patients is 175% of 540 patients?

41. 40% of what number of toddlers is 130 toddlers?

42. 75% of what number of pumpkins is 675 pumpkins?

43. What percent of 160 liters is 2.4 liters?

44. What percent of 600 miles is 7.5 miles?

45. 225% of what number of gallons is 11.25 gallons?

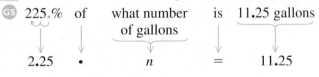

46. 180% of what number of ounces is 6.3 ounces?

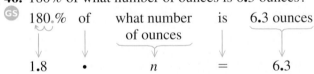

47. What is 12.4% of 8300 meters?

48. What is 13.2% of 9400 acres?

49. CONCEPT CHECK   The work shown below has **two** mistakes in it.

What Went Wrong?   First write a sentence explaining what each mistake is. Then fix the mistakes and find the correct solutions.

Chea made errors in two test questions shown below. The first test question was:

3 hours is what percent of 15 hours?

He solved it this way:

$$3 = p \cdot 15$$

$$\frac{3}{15} = \frac{p \cdot \overset{1}{\cancel{15}}}{\underset{1}{\cancel{15}}}$$

$$0.2 = p$$

He said the answer was 0.2%

The second test question was:

$50 is what percent of $20?

He solved it this way:

$$50 \cdot p = 20$$

$$\frac{\overset{1}{\cancel{50}} \cdot p}{\underset{1}{\cancel{50}}} = \frac{20}{50}$$

$$p = 0.40 = 40\%$$

He said the answer was 40%.

50. CONCEPT CHECK   The work shown below has **two** mistakes in it.

What Went Wrong?   First write a sentence explaining what each mistake is. Then fix the mistakes and find the correct solutions.

Amarianna made errors in two test questions shown below. The first test question was:

12 inches is 5% of what number of inches?

She solved it this way:

$$(12)(0.05) = n$$

$$0.6 = n$$

She said the answer was 0.6 inch.

The second test question was:

What is 4% of 30 pounds?

She solved it this way:

$$n = (4)(30)$$

$$n = 120$$

She said the answer was 120 pounds.

Relating Concepts (Exercises 51–54) For Individual or Group Work

Use your knowledge of percents as you **work Exercises 51–54 in order.**

51. 25% of the 80 children at the day care center are boys. How many boys are there? Show how to use the shortcut for 25% to solve this problem.

52. What percent of the children at the day care center are girls? Explain how you solved this problem.

53. How many girls are at the day care center? Explain **two** different ways to solve this problem *without* setting up an equation or proportion.

54. 10% of the boys at the day care center are left handed. There are no left-handed girls. How many children at the center are left handed? Explain how you can solve this problem *without* setting up an equation or proportion.

Summary Exercises *Percent Computation*

1. Complete this table. Write fractions in lowest terms and as whole or mixed numbers when possible.

	Fraction	Decimal	Percent
(a)	$\frac{3}{100}$		
(b)			30%
(c)		0.375	
(d)			160%
(e)	$\frac{1}{16}$		
(f)			5%
(g)		2.0	
(h)	$\frac{4}{5}$		
(i)		0.072	

2. Use percent shortcuts to answer these questions.

   (a) 10% of 35 ft is _____

   (b) 100% of 19 miles is _____

   (c) 50% of 210 cows is _____

   (d) 1% of $8 is _____

   (e) 25% of 2000 women is _____

   (f) 300% of $15 is _____

   (g) 10% of $875 is _____

   (h) 25% of 48 pounds is _____

   (i) 1% of 9500 students is _____

Use the percent proportion or percent equation to answer each question. If necessary, round money answers to the nearest cent and percent answers to the nearest tenth of a percent.

3. 9 websites is what percent of 72 websites?

4. 30 text messages is 40% of what number of text messages?

5. 6% of $8.79 is how much?

6. 945 students is what percent of 540 students?

7. $3\frac{1}{2}$% of 168 pounds is how much, to the nearest tenth of a pound?

8. 1.25% of what number of hours is 7.5 hours?

9. What percent of 80,000 deer is 40,000 deer?

10. 465 camp sites is 93% of what number of camp sites?

11. What number of golf balls is 280% of 35 golf balls?

12. What percent of $66 is $1.80?

13. 9% of what number of tweets is 207 tweets?

14. What weight is 84% of 0.75 ounce?

15. $1160 is what percent of $800?

16. Find 3.75% of 6500 voters, to the nearest whole number.

17. What is 300% of 0.007 inch?

18. 24 minutes is what percent of 6 minutes?

19. What percent of 60 yards is 4.8 yards?

20. $0.17 is 25% of what amount?

The circle graph shows the average costs for various wedding expenses. Use the graph to answer Exercises 21–25. Round money answers to the nearest cent, if necessary.

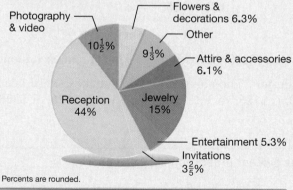

What It Costs to Ring the Wedding Bells
The average wedding has 136 guests and costs $30,200. Here is how couples are spending the money.

Photography & video
Flowers & decorations 6.3%
Other
$10\frac{1}{2}$%
$9\frac{1}{3}$%
Attire & accessories 6.1%
Reception 44%
Jewelry 15%
Entertainment 5.3%
Invitations $3\frac{2}{5}$%

Percents are rounded.

Data from www.costofawedding.com; www.theknot.com

21. Which item in the graph is least expensive? On average, how much is spent on that item?

22. What is the average amount spent on photography and video?

23. What amount is spent on jewelry?

24. Find the cost of flowers and decorations.

25. **(a)** How much is spent on the reception?

(b) On average, what is the cost for each guest at the reception?

7.4 Problem Solving with Percent

OBJECTIVES

1. Solve percent application problems.

2. Solve problems involving percent of increase or decrease.

OBJECTIVE **1** **Solve percent application problems.** Solving percent problems involves identifying three items: the *percent*, the *whole*, and the *part*. Then you can write a percent equation or percent proportion and solve it to answer the question in the problem. Use the six problem-solving steps.

EXAMPLE 1   Finding the Part

A new low-income housing project charges 30% of a family's income as rent. The Smiths' family income is $1260 per month. How much will the Smiths pay for rent each month?

Step 1   **Read the problem.** It is about a family paying part of its income for rent.

　　Unknown: amount of rent

　　Known: 30% of income paid for rent; $1260 monthly income

Step 2   **Assign a variable.** There is only one unknown, so let n be the amount paid for rent each month.

Step 3   **Write an equation.** Use the percent equation.

$$\text{percent} \cdot \text{whole} = \text{part}$$

Recall that the *whole* often follows the word *of*.

　　The percent is given in the problem: 30%. The key word *of* appears *right after* 30%, which means that you can use the phrase "30% of a family's income" to help you write one side of the equation.

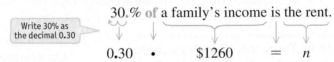

30.% of a family's income is the rent.

Write 30% as the decimal 0.30

$$0.30 \quad \cdot \quad \$1260 \quad = \quad n$$

Step 4   **Solve the equation.** Simplify the left side, multiplying 0.30 by 1260.

$$\underbrace{(0.30)(1260)}_{378} = n$$
$$378 = n \quad \text{Write a \$ in your answer.}$$

Step 5   **State the answer.** The Smiths will pay $378 for rent each month.

Step 6   **Check** the solution. The solution of $378 makes sense because 10% of $1260 is $126, so 30% would be 3 times $126, or $378.

> **Note**
>
> You could also use the percent proportion in **Step 3** above.
>
> Percent → $\dfrac{30}{100} = \dfrac{n}{1260}$ ← Part (unknown)
> Always 100 →   ← Whole
>
> The answer will be the same, $378.
>
> Throughout the rest of this chapter, we will use the percent equation. You may, if you wish, use the percent proportion instead. The final answers will be the same.

---

1 About 65% of the students at
GS City Center College receive some form of financial aid. How many of the 9280 students enrolled this year are receiving aid?

Let n be the number of students receiving aid.

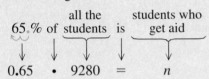

　　all the　　students who
65.% of students　is　get aid

$$0.65 \quad \cdot \quad 9280 \quad = \quad n$$

Answer

1. 6032 students receive aid.
   Check. 10% of 9280 students is 928, which rounds to 900 students. So 60% would be 6 times 900 students = 5400 students, and 70% would be 7 • 900 = 6300. The solution falls between 5400 and 6300 students, so it is reasonable.

◀ **Work Problem** **1** at the Side.

EXAMPLE 2 Finding the Part

When Britta received her first $180 paycheck as a math tutor, $12\frac{1}{2}\%$ was withheld for federal income tax. How much was withheld?

Step 1 **Read** the problem. It is about part of Britta's pay being withheld for taxes.

> Unknown: amount withheld for taxes
>
> Known: $12\frac{1}{2}\%$ of earnings withheld; $180 in pay

Step 2 **Assign a variable.** There is only one unknown, so let n be the amount withheld for taxes.

Step 3 **Write an equation.** Use the percent equation. The *percent* is given: $12\frac{1}{2}\%$. Write $12\frac{1}{2}\%$ as 12.5% and then move the decimal point two places to the left. So 12.5% becomes the decimal 0.125.

The key word *of* doesn't appear after $12\frac{1}{2}\%$. Instead, think about whether you know the *whole* or the *part*. You know Britta's *whole* paycheck is $180, but you do *not* know what *part* of it was withheld.

> Be careful!
> $12\frac{1}{2}\%$ is 12.5%, but remember to *move the decimal point two places to the left* to get 0.125

$$\text{percent} \cdot \text{whole} = \text{part}$$
$$12.5\% \cdot \$180 = n$$

Step 4 **Solve.**
$$(0.125)(180) = n$$
$$22.5 = n$$

Step 5 **State the answer.** $22.50 was withheld from Britta's paycheck.

Step 6 **Check the solution.** The solution of $22.50 makes sense because 10% of $180 is $18, so a little more than $18 should be withheld.

—— Work Problem ② at the Side. ▶

EXAMPLE 3 Finding the Percent

On a 15-point quiz, Zenitia earned 13 points.

What percent correct did she earn, *to the nearest whole percent*?

> This means round your answer to the nearest whole percent.

Step 1 **Read** the problem. It is about points earned on a quiz.

> Unknown: percent correct
>
> Known: earned 13 out of 15 points

Step 2 **Assign a variable.** Let p be the unknown percent.

Step 3 **Write an equation.** Use the percent equation. There is no number with a % symbol in the problem. The question "What percent correct did she earn?" tells you that the *percent* is unknown. The *whole* is all the points on the quiz (15 points), and 13 points is the *part* of the quiz that Zenitia did correctly.

$$\text{percent} \cdot \text{whole} = \text{part}$$
$$p \cdot 15 \text{ points} = 13 \text{ points}$$

—— Continued on Next Page

② There were 50 points on the first math test. Hue's score was 83% correct. How many points did Hue earn?

Let n be the points Hue earned. The whole is all the points on the test.

$$\text{percent} \cdot \text{whole} = \text{part}$$
$$83.\% \cdot 50 \text{ points} = n$$

Answer

2. $(0.83)(50) = n$
Hue earned 41.5 points.
Check. 10% of 50 points is 5 points, so 80% would be 8 times 5 points = 40 points. Hue earned a little more than 80%, so 41.5 points is reasonable.

③ The Los Angeles Lakers made 47 of 80 field goal attempts in one game. What percent is this, to the nearest whole percent?

Let p be the unknown percent. The whole is all the field goal attempts.

$$\underset{\underset{p}{\downarrow}}{\text{percent}} \cdot \underset{\underset{80 \text{ attempts}}{\downarrow}}{\text{whole}} = \underset{\underset{47 \text{ made}}{\downarrow}}{\text{part}}$$

④ Valley College predicted that 1200 new students would enroll in the fall. It actually had 1620 new students enroll. The actual enrollment is what percent of the predicted number?

Let p be the unknown percent.

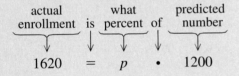

$$\underset{\underset{1620}{\downarrow}}{\overbrace{\substack{\text{actual} \\ \text{enrollment}}}} \underset{=}{\overset{\text{is}}{}} \underset{\underset{p}{\downarrow}}{\overbrace{\substack{\text{what} \\ \text{percent}}}} \text{ of } \underset{\underset{1200}{\downarrow}}{\overbrace{\substack{\text{predicted} \\ \text{number}}}}$$

Answers

3. $p = 0.5875$
$0.5875 = 58.75\% \approx 59\%$
The Lakers made 59% of their field goals, to the nearest whole percent.
Check. The Lakers made a little more than half of their field goals. $\frac{1}{2} = 50\%$, so the solution of 59% is reasonable.

4. $1.35 = p$
$1.35 = 135\%$
Enrollment is 135% of the predicted number.
Check. More than 1200 students enrolled (more than 100%), so 1620 students must be more than 100% of the predicted number. The solution of 135% is reasonable.

Step 4 **Solve** the equation.

$$\frac{p \cdot \overset{1}{\cancel{15}}}{\underset{1}{\cancel{15}}} = \frac{13}{15} \qquad \text{Divide both sides by 15}$$

> CAUTION! **Multiply the solution by 100%** to change it from a *decimal* to a *percent.*

$p = 0.8\overline{6} \leftarrow$ The solution is a repeating decimal.

$$0.866666667 = 86.6666667\% \approx 87\% \leftarrow \text{Rounded}$$

Step 5 **State the answer.** Zenitia earned **87%** correct, rounded to the nearest whole percent.

Step 6 **Check the solution.** The solution of 87% makes sense because she earned most of the possible points, so the percent should be fairly close to 100%.

◀ **Work Problem ③ at the Side.**

EXAMPLE 4 **Finding the Percent**

The rainfall in the Red River Valley was 33 inches this year. The average rainfall is 30 inches. This year's rainfall is what percent of the average rainfall?

Step 1 **Read the problem.** It is about comparing this year's rainfall to the average rainfall.

> **Unknown:** This year's rain is what percent of the average?
>
> **Known:** 33 inches this year; 30 inches is average

Step 2 **Assign a variable.** Let p be the unknown percent.

Step 3 **Write an equation.** The percent is unknown. The key word **of** appears *right after* the word *percent*, so you can use that sentence to help you write the equation.

$$\underset{\underset{33 \text{ inches}}{\downarrow}}{\text{This year's rainfall}} \underset{=}{\overset{\text{is}}{}} \underset{\underset{p}{\downarrow}}{\text{what percent}} \text{ of } \underset{\underset{30 \text{ inches}}{\downarrow}}{\text{the average rainfall?}}$$

Step 4 **Solve** the equation.

$$33 = p \cdot 30$$

$$\frac{33}{30} = \frac{p \cdot \overset{1}{\cancel{30}}}{\underset{1}{\cancel{30}}} \qquad \text{Divide both sides by 30}$$

Solution in *decimal* form → $1.1 = p$

> **Multiply the solution by 100%** to change it from a *decimal* to a *percent.*

$$1.10 = 110\%$$

Step 5 **State the answer.** This year's rainfall is **110%** of the average rainfall.

Step 6 **Check the solution.** The solution of 110% makes sense because 33 inches is *more* than 30 inches (more than 100% of the average rainfall), so 33 inches must be *more* than 100% of 30 inches.

◀ **Work Problem ④ at the Side.**

EXAMPLE 5 **Finding the Whole**

A newspaper article stated that 648 pints of blood were donated at the blood bank last month, which was only 72% of the number of pints needed. How many pints of blood were needed?

Step 1 **Read the problem.** It is about blood donations.

> Unknown: number of pints needed
>
> Known: 648 pints were donated;
>   648 pints is 72% of the number needed.

Step 2 **Assign a variable.** Let n be the number of pints needed.

Step 3 **Write an equation.** The percent is given in the problem: 72%. The key word **of** appears *right after* 72%, so you can use the phrase "72% of the number of pints needed" to help you write one side of the equation.

72.% of the number of pints needed is 648 pints.

$$0.72 \quad \cdot \quad\quad\quad n \quad\quad\quad = 648$$

Step 4 **Solve.** $0.72 \cdot n = 648$

$$\frac{(0.72)(n)}{0.72} = \frac{648}{0.72} \quad \text{Divide both sides by 0.72}$$

$$n = 900$$

> Write **pints** in your answer.

Step 5 **State the answer.** 900 pints of blood were needed.

Step 6 **Check the solution.** The solution of 900 pints makes sense because 10% of 900 pints is 90 pints, so 70% would be 7 times 90 pints, or 630 pints, which is close to the number given in the problem (648 pints).

——— **Work Problem 5 at the Side.** ▶

OBJECTIVE ▶ 2 Solve problems involving percent of increase or decrease. We are often interested in looking at increases or decreases in prices, earnings, population, and many other numbers. This type of problem involves finding the percent of change. Use the following steps to find the **percent of increase.**

Finding the Percent of Increase

Step 1 Use subtraction to find the *amount* of increase.

Step 2 Use a form of the percent equation to find the *percent* of increase.

percent of   whole   =   part

percent of original value = amount of increase

---

5 **(a)** Ezra did 15 problems correctly on a test, giving him a score of $62\frac{1}{2}$%. How many problems were on the test?

Let n be number of problems on the test. This is the *whole* test.

The *part* of the problems that Ezra did correctly is 15.

percent • whole = part

$$62.5\% \quad \cdot \quad n \quad = \quad 15$$

(b) A frozen dinner advertises that only 18% of its calories are from fat. If the dinner contains 55 calories from fat, what is the total number of calories in the dinner? Round to the nearest whole number.

Answers

5. (a) $(0.625)(n) = 15$
There were 24 problems on the test.
Check. 50% of 24 problems is
$24 \div 2 = 12$ problems correct, so
it is reasonable that $62\frac{1}{2}$% would be
15 problems correct.
(b) Let n be total number of calories.
$(0.18)(n) = 55$
There are 306 calories (rounded) in the
dinner. **Check.** 10% of 306 calories
is about 30 calories, so 20% would be
$2 \cdot 30 = 60$ calories, which is close to
the number given (55 calories).

6 **(a)** Over the last two years, Duyen's rent has increased from $650 per month to $767. What is the percent increase?

Let p be the percent of increase.

$$\$767 - 650 = \$117 \begin{cases} \text{amount of} \\ \text{increase} \end{cases}$$

$$\underset{\downarrow}{\text{percent}} \;\; \underset{\downarrow}{\text{of}} \;\; \underset{\downarrow}{\overbrace{\text{original}}_{\text{rent}}} \;\; = \;\; \overbrace{\text{amount of}}_{\text{increase}}$$

$$p \quad \bullet \quad 650 \quad = \quad 117$$

(b) A medical clinic increased the number of accessible parking spaces from 8 to 20. What is the percent increase?

EXAMPLE 6 **Finding the Percent of Increase**

Brad's hourly wage as assistant manager of a fast-food restaurant was raised from $9.40 to $9.87. What was the percent of increase?

Step 1 **Read the problem.** It is about an increase in wages.

> Unknown: percent of increase
>
> Known: original hourly wage was $9.40;
> new hourly wage is $9.87

Step 2 **Assign a variable.** Let p be the percent of increase.

Step 3 **Write an equation.** First subtract $9.87 - $9.40 to find how much Brad's wage went up. That is the *amount* of increase.

$$\$9.87 - 9.40 = \$0.47 \;\; \leftarrow \textit{Amount of increase}$$

Now write an equation to find the unknown *percent* of increase. Be sure to use his *original* wage ($9.40) in the equation because we are looking for the change from his *original* wage. Do **not** use the new wage of $9.87 in the equation.

$$\underset{\downarrow}{\text{percent}} \;\; \underset{\downarrow}{\text{of}} \;\; \overbrace{\text{original wage}} \;\; = \;\; \overbrace{\text{amount of increase}}$$

$$p \quad \bullet \quad \$9.40 \quad = \quad \$0.47$$

> Do **not** use $9.87 in the equation. Use the **original** wage of $9.40

Step 4 **Solve.**

$$\frac{(p)\,(\overset{1}{9.40})}{\underset{1}{9.40}} = \frac{0.47}{9.40} \qquad \text{Divide both sides by } 9.40$$

$$p = 0.05 \;\; \leftarrow \text{Solution in } \textit{decimal} \text{ form}$$

> Multiply the solution by **100%** to change it from a *decimal* to a *percent*.

$$0.05 = 5\%$$

Step 5 **State the answer.** Brad's hourly wage increased 5%.

Step 6 **Check the solution.** 10% of $9.40 would be a $0.94 raise. A raise of $0.47 is half as much, and half of 10% is 5%, so the solution checks.

> **❗ CAUTION**
>
> When writing the percent equation in **Step 3** above, be sure to multiply the percent (p) by the **original** wage. Do **not** use the *new* wage in the percent equation.

◀ **Work Problem** **6** **at the Side.**

Use a similar procedure to find the **percent of decrease.**

Finding the Percent of Decrease

Step 1 Use subtraction to find the *amount* of decrease.

Step 2 Use a form of the percent equation to find the *percent* of decrease.

$$\underset{\downarrow}{\text{percent}} \;\; \underset{\downarrow}{\text{of}} \;\; \underset{\downarrow}{\text{whole}} \;\; = \;\; \underset{\downarrow}{\text{part}}$$

$$\underbrace{\text{percent of original value}} = \underbrace{\text{amount of decrease}}$$

Answers

6. **(a)** $p = 0.18 = 18\%$

Duyen's rent increased 18%.

Check. 10% increase would be $650. = $65; 20% increase would be 2 • $65 = $130; so an 18% increase is reasonable.

(b) $20 - 8 = 12$ space increase

Let p be percent increase.

$$p \bullet 8 = 12$$

$$p = 1.50 = 150\%$$

The number of accessible parking spaces increased 150%.

Check. 150% is $1\frac{1}{2}$, and $1\frac{1}{2} \bullet 8 = 12$ space increase, so it checks.

EXAMPLE 7 Finding the Percent of Decrease

Rozenia trained for six months to run in a marathon. Her weight dropped from 137 pounds to 122 pounds. Find the percent of decrease, to the nearest whole percent.

> This means round your answer to the nearest whole percent.

Step 1 **Read the problem.** It is about a decrease in weight.

> Unknown: percent of decrease
>
> Known: original weight was 137 pounds; new weight is 122 pounds.

Step 2 **Assign a variable.** Let p be the percent of decrease.

Step 3 **Write an equation.** First subtract 137 pounds − 122 pounds to find how much Rozenia's weight went down. That is the *amount* of decrease.

$$137 \text{ pounds} - 122 \text{ pounds} = 15 \text{ pounds} \leftarrow \textit{Amount of decrease}$$

Now write an equation to find the unknown *percent* of decrease. Be sure to use her *original* weight (137 pounds) in the equation because we are looking for the change from her *original* weight. Do **not** use the new weight of 122 pounds in the equation.

percent of original weight = amount of decrease

$$p \quad \cdot \quad 137 \text{ pounds} \quad = \quad 15 \text{ pounds}$$

> Do **not** use 122 pounds in the equation. Use the **original** weight of 137 pounds.

Step 4 **Solve.**
$$\frac{p \cdot \cancel{137}^{1}}{\cancel{137}_{1}} = \frac{15}{137} \quad \text{Divide both sides by 137}$$

$$p \approx 0.109489051 \leftarrow \text{Solution in } \textit{decimal} \text{ form}$$

Multiply the solution by 100% to change it from a *decimal* to a *percent*.

$$0.109489051 = 10.9489051\% \approx 11\%$$

Rounded to nearest whole percent

Step 5 **State the answer.** Rozenia's weight decreased approximately 11%.

Step 6 **Check the solution.** A 10% decrease would be $137. = 13.7$ pounds, so an 11% decrease is a reasonable solution.

> **⚠ CAUTION**
> When writing the percent equation in **Step 3** above, be sure to multiply the percent (p) by the **original** weight. Do **not** use the *new* weight in the percent equation.

—————— Work Problem **7** at the Side. ▶

Study Skills Reminder

Preparing for and taking tests can cause both mental and physical stress. That is why it is important to prepare your body as well as your brain for a math test. See the Study Skills activity "Tips for Improving Test Scores" to learn how to get your body and your brain ready for a math test.

---

7 Find each percent of decrease. Round answers to the nearest whole percent.

GS **(a)** During a severe winter storm, average daily attendance at an elementary school fell from 425 students to 200 students. What was the percent decrease?

Let p be the percent decrease.

$$425 \text{ students} - 200 = \underline{225 \text{ students}}$$

Amount of decrease ⟵

percent of original attendance = amount of decrease

$$p \quad \cdot \quad 425 \quad = \quad 225$$

(b) The makers of a brand of spaghetti sauce claim that the number of calories from fat in each serving has been reduced by 20%. Is the claim correct if the number of calories from fat dropped from 70 calories to 60 calories per serving? Explain your answer.

Answers

7. **(a)** $p \approx 0.529 \approx 53\%$ (rounded)
Daily attendance decreased 53%.
Check. A 50% decrease would be $425 \div 2 \approx 212$, so a 53% decrease is reasonable.

(b) $70 - 60 = 10$ calorie decrease
Let p be percent of decrease.
$$p \cdot 70 = 10$$
$$p \approx 0.143 \approx 14\% \text{ (rounded)}$$
The claim of a 20% decrease is not true; the decrease in calories is about 14%.
Check. A 10% decrease would be $70. = 7$ calories; a 20% decrease would be $2 \cdot 7 = 14$ calories, so a 14% decrease is reasonable.

7.4 Exercises

FOR EXTRA HELP

Go to MyMathLab for worked-out, step-by-step solutions to exercises enclosed in a square ☐ and video solutions to ▶ exercises.

CONCEPT CHECK *Find each percent using the shortcuts for 50%, 25%, 10%, and 1%.*

1. An ATM machine charges $2 for any size cash withdrawal. The $2 fee is what percent of a
   (a) $20 withdrawal
   (b) $40 withdrawal
   (c) $200 withdrawal.

2. An online company charges $8 for shipping and handling on any order of $100 or less. The $8 shipping charge is what percent of a
   (a) $16 order
   (b) $32 order
   (c) $80 order

Use the six problem-solving steps in Exercises 3–22. Round percent answers to the nearest tenth of a percent when necessary. **See Examples 1–5.**

3. Robert Garrett, who works part-time, earns $210 per week and has 18% of this amount withheld for taxes, Social Security, and Medicare. Find the amount withheld.

   Let *n* be the amount withheld.

18.%	of	pay	is	withheld
↓	↓	↓	↓	↓
0.18	•	$210	=	*n*

4. Most shampoos contain 75% to 90% water. If a 16-ounce bottle of shampoo contains 78% water, find the number of ounces of water in the bottle, to the nearest tenth. (Data from *Consumer Reports*.)

   Let *n* be the ounces of water.

78.%	of	shampoo	is	water
↓	↓	↓	↓	↓
0.78	•	16 ounces	=	*n*

The circle graph below shows, on average, what portion of the human body is made up of water, protein, fat, and so on. Use the graph to answer Exercises 5 and 6. Round answers to the nearest tenth of a pound. (Data from www.fao.org)

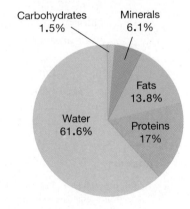

Composition of the Human Body

Carbohydrates 1.5%
Minerals 6.1%
Fats 13.8%
Water 61.6%
Proteins 17%

5. If an adult weighs 165 pounds, how much of that weight is
   (a) water.
   (b) minerals.
   (c) proteins.

6. If a teenager weighs 92 pounds, how much of that weight is
   (a) fat.
   (b) carbohydrates.
   (c) water.

7. When 480 people were asked what they planned to do with their tax refunds, 283 said they would pay off debt; 317 said they would save or invest the money; and 110 said they would buy something. (People could give more than one answer.) What percent of the people gave each answer? (Data from Synovate.)

8. Airline passengers generate 880 million tons of waste each year. Only 176 million tons are being recycled now. It is estimated that 660 million tons could be recycled if the airlines made an effort to do so. What percent of the waste is being recycled now? What percent could be recycled? (Data from *Washington Post*.)

9. The U.S. Census Bureau reported that Americans who are 65 years of age or older made up 14.5% of the total population in 2014. It said that there were 46.4 million Americans in this group. Find the total U.S. population for that year. (Data from www.census.gov)

10. Julie Ward has 8.5% of her earnings deposited into the credit union. If this amounts to $263.50 per month, find her monthly and annual earnings.

11. The campus honor society hoped to raise $50,000 in donations from businesses for scholarships. It actually raised $69,000. This amount was what percent of the goal?

12. Doug had budgeted $220 for textbooks but ended up spending $316.80. The amount he spent was what percent of his budget?

13. Alfonso earned a score of 85% on his test. He did 34 problems correctly. How many problems were on the test?

14. In an online survey, 182 adults said they order food delivered to their homes on Fridays. This was 36% of the adults in the survey. How many adults were surveyed, to the nearest whole number? (Data from Technomic Inc.)

15. During the 2014–2015 NBA season, Golden State Warrior Stephen Curry made 646 field goals, which was 48.9% of the shots he tried. How many shots did he try? Round your answer to the nearest whole number. (Data from www.NBA.com)

16. Maya Moore, who plays basketball in the WNBA, attempted 181 free throws during the 2014 season. She made 88.4% of her shots. How many free throws did Moore make? Round your answer to the nearest whole number. (Data from www.WNBA.com)

17. An ad for steel-belted radial tires promises 15% better mileage. If Sheera's SUV has gotten 20.6 miles per gallon in the past, what mileage can she expect after the new tires are installed? Round your answer to the nearest tenth of a mile.

18. There are 73,465,000 women age 18 or older who play video games in the United States. This represents 36% of all the people who play video games in the United States. (Data from Entertainment Software Association.)

   (a) How many people play video games in the United States? Round your answer to the nearest whole number.

   (b) There are only 34,692,000 boys age 18 or younger who play video games in the United States. Use your answer from part (a) to find the percent of the people playing video games in the United States who are boys age 18 or younger. Round your answer to the nearest whole percent.

⊞ *The bar graph shows the median weekly earnings for adults with various levels of education. Use the graph to answer Exercises 19–22. Round percent answers to the nearest whole percent.*

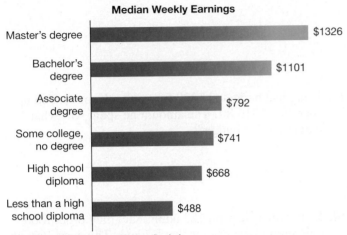

Median Weekly Earnings

Master's degree	$1326
Bachelor's degree	$1101
Associate degree	$792
Some college, no degree	$741
High school diploma	$668
Less than a high school diploma	$488

Data from U.S. Bureau of Labor Statistics.

19. The weekly earnings of a person with less than a high school diploma are what percent of the weekly earnings of a person with an associate degree? How much more does the person with the associate degree earn in a year? (Assume the people receive pay for 52 weeks each year.)

20. The weekly earnings of someone with a high school diploma are what percent of the weekly earnings of someone with an associate degree? How much more does the person with the associate degree earn in ten years? (Assume the people receive pay for 52 weeks each year.)

21. What percent are the weekly earnings of a person with a bachelor's degree compared to the weekly earnings of a person with an associate degree?

22. What percent are the weekly earnings of a person with a master's degree compared to the weekly earnings of a person with some college but no degree?

Use the six problem-solving steps to find the percent increase or decrease. Round your answers to the nearest tenth of a percent. **See Examples 6 and 7.**

23. Henry Ford started the Ford Motor Company. The Model T first appeared in 1908 and cost $825. In 1913, Ford started using an assembly line and the price of the Model T dropped to $290. What was the percent of decrease? (Data from Kenneth C. Davis.)

First subtract: $825 − $290 = _____

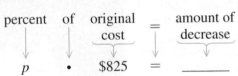

$$ p \cdot \$825 = \underline{\hspace{1cm}} $$

24. In 1914, Henry Ford became a hero to his workers because he doubled the daily minimum wage. He also cut the workday from nine hours to eight hours. Find the percent of decrease in the number of hours in the workday. (Data from Kenneth C. Davis.)

First subtract: 9 hours − 8 hours = _____

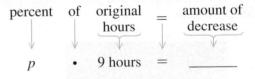

$$ p \cdot 9 \text{ hours} = \underline{\hspace{1cm}} $$

25. Students at Lane College were charged $1449 as tuition this semester. If the tuition was $1328 last semester, find the percent of increase.

26. Americans are eating more chicken. In 2015, the average American ate 90.1 pounds compared to only 38.7 pounds per year in 1975. Find the percent of increase. (Data from National Chicken Council.)

27. Jordan's part-time work schedule was reduced to 18 hours per week. He had been working 30 hours per week. What is the percent decrease?

28. Janis works as a hair stylist. During January, she cut her price on haircuts from $35 to $31.50 to try to get more customers. By what percent did she decrease the price?

29. In 1967 there were 78 animal species listed as threatened with extinction in the United States. In 2015 there were 683 animals on the list. What was the percent increase, to the nearest whole percent? (Data from U.S. Fish and Wildlife Service.)

Florida panther

Condor

Bighorn sheep

30. The world population was estimated at 6,980,445,600 people in 2011. It is projected to reach 8,002,978,434 people by 2025. By what percent will the world's population increase in those 14 years? (Data from United Nations.)

31. CONCEPT CHECK You can have an *increase* of 150% in the price of something. Could there be a 150% *decrease* in its price? Explain why or why not.

32. CONCEPT CHECK Show how to use a shortcut to find 25% of $80. Then explain how to use the result to find 75% of $80 and 125% of $80 *without* solving a proportion or equation.

Relating Concepts (Exercises 33–36) For Individual or Group Work

*As you **work Exercises 33–36 in order,** explain why each solution does **not** make sense. Then find and correct the error.*

33. George's doctor recommended that he eat only 65 grams of fat per day. George ate 78 grams of fat today. He ate what percent of the recommended amount?

$$p \cdot 78 = 65$$

$$\frac{p \cdot \overset{1}{\cancel{78}}}{\underset{1}{\cancel{78}}} = \frac{65}{78}$$

$$p = 0.833 = 83.3\%$$

34. The Goblers soccer team won 18 of its 25 games this season. What percent did the team win?

$$p \cdot 25 = 18$$

$$\frac{p \cdot \overset{1}{\cancel{25}}}{\underset{1}{\cancel{25}}} = \frac{18}{25}$$

$$p = 0.72\%$$

35. The human brain is $2\frac{1}{2}\%$ of total body weight. How much would the brain of a 150-pound person weigh?

$$2\frac{1}{2}\% \text{ of } 150 = n$$

$$(2.5)(150) = n$$

$$375 = n$$

36. Yesterday, because of an ice storm, 80% of the students were absent. How many of the 800 students made it to class?

$$80\% \text{ of } 800 = n$$

$$(0.80)(800) = n$$

$$640 = n$$

The answer is 640 students.

7.5 | Consumer Applications: Sales Tax, Tips, Discounts, and Simple Interest

OBJECTIVES

1. Find sales tax and total cost.
2. Estimate and calculate restaurant tips.
3. Find the discount and sale price.
4. Calculate simple interest and the total amount due on a loan.

Four of the more common uses of percent for consumers in their daily lives are sales taxes, tips, discounts, and simple interest on loans.

OBJECTIVE ▸ 1 Find sales tax and total cost. Most states collect **sales taxes** on the purchases you make in stores. Your county or city may also add on a small amount of sales tax. For example, your state may charge $6\frac{1}{2}\%$ on purchases and your city may add on another $\frac{1}{2}\%$ for a total of 7%. The exact percent varies from place to place but is usually from 4% to 9%. The stores collect the tax and send it to the city or state government where it is used to pay for things like road repair, public schools, parks, police and fire protection, and so on.

You can use a form of the percent equation to calculate sales tax. The **tax rate** is the *percent*. The cost of the item(s) you are buying is the *whole*. The amount of tax you pay is the *part*.

VOCABULARY TIP

Consumers are the buyers or users of products. You are a consumer too.

Finding Sales Tax and Total Cost

Use a form of the percent equation to find sales tax.

$$\text{percent} \cdot \text{whole} = \text{part}$$
$$\text{tax rate} \cdot \text{cost of item} = \text{amount you pay in sales tax}$$

Then add to find how much you will pay in all.

$$\text{cost of item} + \text{sales tax} = \text{total cost paid by you}$$

EXAMPLE 1   Finding Sales Tax and Total Cost

Suppose that you buy a tablet for $289 from A-1 Electronics. The sales tax rate in your state is $6\frac{1}{2}\%$. How much is the tax? What is the total cost of the tablet?

Step 1 **Read the problem.** It asks for the sales tax on a tablet and the total cost.

Step 2 **Assign a variable.** Let n be the amount of tax.

Step 3 **Write an equation.** Use the sales tax equation. Write $6\frac{1}{2}\%$ as 6.5% and then move the decimal point two places to the left.

$$\text{tax rate} \cdot \text{cost of item} = \text{sales tax}$$

Write $6\frac{1}{2}\%$ as 6.5%; then **move the decimal point to get 0.065**

$$06.5\% \cdot \$289 = n$$

Step 4 **Solve.**
$$(0.065)(289) = n$$
$$18.785 = n$$

The store rounds the tax to the nearest cent, so $18.785 rounds to $18.79. Add the sales tax to the cost of the tablet to find your total cost.

$$\text{cost of item} + \text{sales tax} = \text{total cost}$$
$$\$289 + \$18.79 = \$307.79$$

Continued on Next Page

Step 5 **State the answer.** The tax is $18.79 and the total cost of the tablet, including tax, is $307.79.

Step 6 **Check:** Use estimation to check that the amount of sales tax is reasonable. Round $289 to $300. Then 1% of $300. is $3. Round $6\frac{1}{2}\%$ to 7%. Then 7% would be 7 times $3 or $21 for sales tax. Our solution of $18.79 is close to $21, so it is reasonable.

——— **Work Problem ❶ at the Side.** ▶

EXAMPLE 2 **Finding the Sales Tax Rate**

Ms. Ortiz bought a $28,950 hybrid car. She paid an additional $2171.25 in sales tax. What was the sales tax rate?

Step 1 **Read** the problem. It asks for the sales tax rate on a car purchase.

Step 2 **Assign a variable.** Let p be the tax rate (the percent).

Step 3 **Write an equation.** Use the sales tax equation.

tax rate • cost of item = sales tax

$$p \cdot \$28{,}950 = \$2171.25$$

Step 4 **Solve.**
$$\frac{(p)\,(28{,}950)}{28{,}950} = \frac{2171.25}{28{,}950}$$
Divide both sides by 28,950

Multiply the solution by 100% $p = 0.075$ ◀— Solution in *decimal* form

Change the solution from a decimal to a percent: $0.075 = 7.5\%$.

Step 5 **State the answer.** The sales tax rate is 7.5% (or $7\frac{1}{2}\%$).

Step 6 **Check.** Use estimation to check that the solution is reasonable. If the tax rate was 1%, then 1% of $28950. = $289.50, or about $300.

Round 7.5% to 8%. Then 8% would be 8 times $300, or $2400.

The tax amount given in the original problem, $2171.25, is close to the estimate, so our solution of 7.5% is reasonable.

——— **Work Problem ❷ at the Side.** ▶

OBJECTIVE ❷ Estimate and calculate restaurant tips. Waiters and waitresses rely on tips as a major part of their income. The general rule of thumb is to leave at least 15% of your bill for food and beverages as a tip for the server. If you receive exceptional service or are eating in an upscale restaurant, consider leaving a 20% tip.

EXAMPLE 3 **Estimating 15% and 20% Tips**

First estimate each tip. Then calculate the exact amount.

(a) Kirby and his wife went out to dinner. The bill came to $77.85. How much should they leave for a 20% tip?

Estimate: Round $77.85 to $80. Then 10% of $80. is $8.

20% would be 2 times $8, or $16. ◀— Estimate

——— **Continued on Next Page**

❶ Find the sales tax and the total cost. Round the sales tax to the nearest cent when necessary. Check your answer by estimating the sales tax.

(a) $495 video game console; $5\frac{1}{2}\%$ sales tax

Let n be the amount of tax.

tax rate • cost of item = sales tax

$$05.5\% \cdot \$495 = n$$

(b) $1.19 candy bar; 4% sales tax

❷ Find the sales tax rate on each purchase. Then use estimation to check your solution.

(a) The tax on a $97 textbook is $5.82.

Let p be the tax rate.

tax rate • cost of item = sales tax

$$p \cdot \$97 = \$5.82$$

(b) The tax on a $4 notebook is $0.18.

Answers

1. **(a)** sales tax = $27.23 (rounded)
   total cost = $522.23
   Check. 1% of $500. is $5;
   6% is 6 times $5 = $30 (estimated tax).
   (b) Sales tax = $0.05 (rounded)
   total cost = $1.24
   Check. 1% of $1 is $0.01;
   4% is 4 times $0.01 = $0.04 (estimated tax).
2. **(a)** 6% **Check.** 1% of $100. is $1, so 6% would be 6 times $1, or $6 (which is close to $5.82 given in the problem).
   (b) 4.5% **Check.** 1% of $4 is $0.04; round 4.5% to 5%; 5% of $4 is 5 times $0.04, or $0.20 (which is close to $0.18 given in the problem).

3 First estimate each tip. Then calculate the exact tip.

(a) 20% tip on a bill of $58.37

Estimate: Round $58.37 to $60. Then 10% of $60. is $6.

20% is 2 times $6, or _____

Exact:

$$\text{percent} \cdot \text{whole} = \text{part}$$
$$20.\% \cdot \$58.37 = n$$

(b) 15% tip on a bill of $11.93

(c) A bill of $89.02 plus a 15% tip shared equally by four friends. How much will each friend pay?

Answers

3. **(a)** *Estimate:* $12
Exact: $11.67 (rounded to nearest cent)
(b) *Estimate:* 10% of $12 is $1.20; 5% is half of $1.20, or $0.60; $1.20 + 0.60 = $1.80
Exact: $1.79 (rounded to nearest cent)
(c) *Estimate* of tip: 10% of $90 is $9;
5% is half of $9, or $4.50;
$9 + $4.50 = $13.50.
Exact tip: $13.35 (rounded to nearest cent). Each friend pays:
($89.02 + $13.35) ÷ 4 = $25.59
(rounded).

Exact: Use the percent equation. Write 20% as a decimal. The bill for food and beverages is the *whole* and the tip is the *part*.

$$\text{percent} \cdot \text{whole} = \text{part}$$
$$20.\% \cdot \$77.85 = n$$

$20.\% = 0.20$

$$(0.20)(77.85) = n$$
$$15.57 = n$$

20% of $77.85 is $15.57, which is close to the estimate of $16.

A tip is usually rounded off to a convenient amount, such as the nearest quarter or nearest dollar, so Kirby and his wife left $16.

(b) Linda, Peggy, and Mary ordered similarly priced sandwiches and agreed to split the bill plus a 15% tip. How much should each woman pay if the bill is $21.63?

Estimate: Round $21.63 to $20. Then 10% of $20. is $2.

5% of $20 would be half as much, that is, half of $2, or $1.

So an estimate of the 15% tip is $2 + $1 = $3. ← Estimates

An estimate of the amount each woman should pay is ($20 + $3) ÷ 3 ≈ $8.

Exact: Use the percent equation to calculate the 15% tip. Add the tip to the bill. Then divide the total by 3 to find the amount each woman should pay.

$$\text{percent} \cdot \text{whole} = \text{part}$$
$$15.\% \cdot \$21.63 = n$$

$15.\% = 0.15$

$$(0.15)(21.63) = n$$
$$3.2445 = n$$

Round $3.2445 to $3.24 (nearest cent), which is close to the estimate of $3 for the tip.

Add: $21.63 + $3.24 = $24.87 ← Total cost of sandwiches and tip

Divide: $24.87 ÷ 3 = $8.29 ← Amount paid by each woman

◀ **Work Problem 3 at the Side.**

OBJECTIVE **3** **Find the discount and sale price.** Most people like to buy things when they are on sale. A store will reduce prices, or **discount,** to attract additional customers. You can use a form of the percent equation to calculate the discount. The *rate* of discount is the *percent*. The original price is the *whole*. The amount that will be discounted (subtracted from the original price) is the *part*.

Finding the Discount and Sale Price

Use a form of the percent equation to find the discount.

$$\text{percent} \cdot \text{whole} = \text{part}$$

rate of discount • original price = amount of discount

Then subtract to find the sale price.

original price − amount of discount = sale price

EXAMPLE 4	**Finding the Discount and Sale Price**

The Oak Mill Furniture Store has an oak dining room set with an original price of $840 on sale at 15% off. Find the sale price.

Step 1 **Read** the problem. It asks for the sale price on a dining room set.

Step 2 **Assign a variable.** Let n be the amount of discount.

Step 3 **Write an equation.** Use a form of the percent equation to find the discount. Write 15% as the decimal 0.15

$$\text{rate of discount} \cdot \text{original price} = \text{amount of discount}$$

Step 4 **Solve.**

$$(0.15)(840) = n$$
$$126 = n$$

The amount of discount is $126. Find the sale price by subtracting the amount of the discount ($126) from the original price.

$$\text{original price} - \text{amount of discount} = \text{sale price}$$

$$\$840 - \$126 = \$714$$

Remember to subtract the discount.

Step 5 **State the answer.** During the sale, you can buy the dining room set for $714.

Step 6 **Check.** Round $840 to $800. Then 10% of $800 is $80 and 5% is half as much, or $40. So 15% of $800 is $80 + $40 = $120. An estimate of the sale price is $800 − $120 = $680, so the exact answer of $714 is reasonable.

Work Problem ④ at the Side. ▶

▦ **Calculator Tip**

In **Example 4** above, you can use a *scientific* calculator to find the amount of discount and subtract the discount from the original price.

$$840 \ominus .15 \otimes 840 \oplus 714$$

Original price Amount of discount Sale price

Your *scientific* calculator observes the order of operations, so it will automatically do the multiplication before the subtraction. (Recall that simple, four-function calculators *may not* follow the order of operations; they might give an incorrect result.)

OBJECTIVE ▶ ④ **Calculate simple interest and the total amount due on a loan. Interest** is a fee paid, or a charge made, for lending or borrowing money. The amount of money borrowed is called the **principal**. The charge for interest is usually given as a percent, called the **interest rate**. The interest rate is assumed to be *per year* (for *one* year) unless stated otherwise.

When interest is calculated on the original principal, it is called **simple interest**. Use the following **interest formula** to find simple interest.

---

④ Find the amount of the discount and the sale price for each item.

GS **(a)** A dining table and four chairs originally priced at $950 are offered at a 35% discount.

Let n be the amount of discount.

$$\text{rate of discount} \cdot \text{original price} = \text{amount of discount}$$

$$35.\% \cdot \$950 = n$$

(b) In August, Eastside Boutique has women's swimsuits on sale at 40% off. One swimsuit was originally priced at $68. Another suit was originally priced at $97.

5 Find the simple interest and total amount due on each loan.

GS **(a)** $500 at 4% for 1 year

$$I = \quad p \quad \bullet \quad r \quad \bullet \quad t$$

$$I = (500)(0.04)(1)$$

Formula for Simple Interest

Interest = principal • rate • time

The formula is usually written using variables.

> In the formula, *p* is the principal, **not** the percent.

$$I = p \bullet r \bullet t \quad \text{or} \quad I = prt$$

When you repay a loan, the interest is added to the original principal to find the total amount due.

Finding the Total Amount Due

amount due = principal + interest

EXAMPLE 5 **Finding Simple Interest and Total Amount Due for 1 Year**

Find the simple interest and total amount due on a $2000 loan at 6% for 1 year.

The amount borrowed, or principal (p), is $2000. The interest rate (r) is 6%, and the time of the loan (t) is 1 year. Use the interest formula.

> The interest rate (r) is the percent. Write 6% as the decimal 0.06

$$I = \quad p \quad \bullet \quad r \quad \bullet \quad t$$

$$I = (2000)(0.06)(1)$$

$$I = 120$$

> Don't stop here. Add the interest to the principal.

The interest is $120

Now add the principal and the interest to find the total amount due.

$$\text{amount due} = \text{principal} + \text{interest}$$

$$\text{amount due} = \quad \$2000 \quad + \quad \$120$$

$$\text{amount due} = \quad \$2120$$

The total amount due is **$2120**

◀ **Work Problem** **5** **at the Side.**

(b) $1850 at $9\frac{1}{2}$% for 1 year

(*Hint:* Write $9\frac{1}{2}$% as 9.5%, then *move the decimal point* two places to the left to change 9.5% to a decimal.)

EXAMPLE 6 **Finding Simple Interest and Total Amount Due for More Than 1 Year**

Find the simple interest and total amount due on a $4200 loan at $8\frac{1}{2}$% for $3\frac{1}{2}$ years.

The principal (p) is $4200. The rate ($r$) is $8\frac{1}{2}$%, which is the same as 8.5%. Move the decimal point two places to the left to change 8.5% to a decimal.

$$8\frac{1}{2}\% = 8.5\% = 08.5\% = 0.085$$

> Remember to **move the decimal point** in 8.5% to get 0.085

The time (t) is $3\frac{1}{2}$ years or 3.5 years. Use the formula.

$$I = \quad p \quad \bullet \quad r \quad \bullet \quad t$$

$$I = (4200)(0.085)(3.5)$$

$$I = 1249.50$$

The interest is $1249.50

Continued on Next Page

Answers

5. **(a)** $20; $520   **(b)** $175.75; $2025.75

Now add the principal and the interest to find the total amount due.

$$\text{\textbf{amount due}} = \text{principal} + \text{interest}$$

$$\text{amount due} = \$4200 + \$1249.50$$

$$\text{amount due} = \$5449.50$$

The total amount due is **$5449.50**

> **⊘ CAUTION**
>
> Be careful when changing a mixed number percent, like $8\frac{1}{2}\%$, to a decimal. Writing $8\frac{1}{2}\%$ as 8.5% is only the first step. There is a decimal point in 8.5% but there is still a % sign. You must divide by 100 before dropping the % sign. **Remember to move the decimal point two places to the left,** as shown in **Example 6** on the previous page. So $8\frac{1}{2}\% = 8.5\% = 0.085$

——————— **Work Problem ⑥ at the Side.** ▶

Interest rates are given *per year*. For loan periods of less than one year, be careful to express the time as a fraction of a year.

If the time is given in months, use a denominator of 12, because there are 12 months in a year. A loan of 9 months would be for $\frac{9}{12}$ of a year, a loan of 7 months would be for $\frac{7}{12}$ of a year, and so on.

EXAMPLE 7 **Finding Simple Interest and Total Amount Due for Less Than 1 Year**

Find the simple interest and total amount due on $840 at $9\frac{3}{4}\%$ for 7 months.

The principal is $840. The rate is $9\frac{3}{4}\%$ or 0.0975.

$$9\frac{3}{4}\% = 9.75\% = 09.75\% = 0.0975$$

The time is $\frac{7}{12}$ of a year. Use the formula $I = prt$.

$$I = (840)(0.0975)\left(\frac{7}{12}\right) \quad \text{← 7 months is } \frac{7}{12} \text{ of a year.}$$

$$I = (81.9)\left(\frac{7}{12}\right)$$

$$I = \left(\frac{81.9}{1}\right)\left(\frac{7}{12}\right) \qquad \text{Multiply numerators.} \\ \text{Multiply denominators.}$$

$$I = \frac{573.3}{12} = 47.775 \qquad \text{Divide 573.3 by 12}$$

The interest is **$47.78**, rounded to the nearest cent. ← Money is usually rounded to the nearest cent.

The total amount due is $840 + **$47.78** = **$887.78**

▦ Calculator Tip

The calculator solution for finding the interest in **Example 7** above uses chain calculations.

840 ⊗ .0975 ⊗ 7 ⊘ 12 ⊜ 47.775 ← Round to $47.78

——————— **Work Problem ⑦ at the Side.** ▶

---

⑥ Find the simple interest and total amount due for each loan.

(a) $340 at 5% for $3\frac{1}{2}$ years

$$I = p \bullet r \bullet t$$

$$I = (340)(0.05)(3.5)$$

(b) $2450 at 8% for $3\frac{1}{4}$ years (*Hint:* Write $3\frac{1}{4}$ years as 3.25 years.)

(c) $14,200 at $1\frac{1}{2}\%$ for $2\frac{3}{4}$ years

---

⑦ Find the simple interest and total amount due for each loan.

(a) $1600 at 7% for 4 months

$$I = \underbrace{(1600)(0.07)}\left(\frac{4}{12}\right) =$$

$$\left(\frac{112}{1}\right)\left(\frac{4}{12}\right) =$$

$$\frac{448}{12} \approx \underline{\quad\quad} \begin{array}{l} \text{to} \\ \text{nearest} \\ \text{cent} \end{array}$$

Total amount due =

$$\$1600 + \underline{\quad} = \underline{\quad}$$

(b) $25,000 at $3\frac{3}{4}\%$ for 9 months

Answers

6. (a) $59.50; $399.50
   (b) $637; $3087
   (c) $585.75; $14,785.75
7. (a) $37.33 (rounded);
   $1600 + $37.33 = $1637.33
   (b) $703.13 (rounded); $25,703.13

7.5 Exercises

FOR EXTRA HELP

Go to MyMathLab for worked-out, step-by-step solutions to exercises enclosed in a square ⬜ and video solutions to ▶ exercises.

1. **CONCEPT CHECK** Change each of these percents to decimals.

   (a) $6\frac{1}{2}\%$

   (b) $1\frac{1}{4}\%$

   (c) $10\frac{1}{4}\%$

   (d) $5\frac{3}{4}\%$

2. **CONCEPT CHECK** Change each of these fractions or mixed numbers to decimals.

   (a) $\frac{3}{4}$ year

   (b) $11\frac{1}{2}$ years

   (c) $4\frac{1}{4}$ years

   (d) $8\frac{3}{4}$ years

Find the amount of the sales tax or the tax rate and the total cost. Round money answers to the nearest cent. **See Examples 1 and 2.**

	Cost of Item	Tax Rate	Amount of Tax	Total Cost
3. GS	$365.98	8%	_____	_____

tax rate • cost of item = sales tax

0.08 • $\$365.98$ = n

Remember to add the tax to the cost of the item.

4. GS	$28.49	7%	_____	_____

tax rate • cost of item = sales tax

0.07 • $\$28.49$ = n

Remember to add the tax to the cost of the item.

5.	$68	_____	$2.04	_____
6.	$185	_____	$9.25	_____
7. ▶	$2.10	$5\frac{1}{2}\%$	_____	_____
8.	$7.00	$7\frac{1}{2}\%$	_____	_____
9.	$12,600	_____	$567	_____
10.	$21,800	_____	$1417	_____

For each restaurant bill, estimate a 15% tip and a 20% tip. Then find the exact amounts for a 15% tip and a 20% tip. Round exact amounts to the nearest cent when necessary. See Example 3.

Bill	Estimate of 15% Tip	Exact 15% Tip	Estimate of 20% Tip	Exact 20% Tip
11. $32.17	_____ Round $32.17 to $30; 10% of $30. is $3 so 5% is half as much	_____ percent • whole = part 0.15 • $32.17 = n	_____ Round $32.17 to $30; 10% of $30. is $3 so 20% is 2 times $3	_____ percent • whole = part 0.20 • $32.17 = n
12. $21.94	_____ Round $21.94 to $20; 10% of $20. is $2 so 5% is half as much	_____ percent • whole = part 0.15 • $21.94 = n	_____ Round $21.94 to $20; 10% of $20. is $2 so 20% is 2 times $2	_____ percent • whole = part 0.20 • $21.94 = n
13. $78.33	_____	_____	_____	_____
14. $67.85	_____	_____	_____	_____
15. $9.55	_____	_____	_____	_____
16. $52.61	_____	_____	_____	_____

Find the amount or rate of discount and the sale price. Round money answers to the nearest cent when necessary. See Example 4.

	Original Price	Rate of Discount	Amount of Discount	Sale Price
17.	$100	15%	_____	_____
	rate of discount • original price = amount of discount 0.15 • $100 = n			Remember to subtract the discount to get the sale price.
18.	$200	20%	_____	_____
	rate of discount • original price = amount of discount 0.20 • $200 = n			Remember to subtract the discount to get the sale price.
19.	$180	_____	$54	_____
20.	$38	_____	$9.50	_____
21.	$17.50	25%	_____	_____

	Original Price	Rate of Discount	Amount of Discount	Sale Price
22.	$76	60%	_____	_____
23.	$37.88	10%	_____	_____
24.	$59.99	40%	_____	_____

Find the simple interest and total amount due on each loan. **See Examples 5–7.**
Round answers to the nearest cent when necessary.

	Principal	Rate	Time	Interest	Total Amount Due
25. GS	$300	14%	1 year	_____	_____

$I = prt$
$I = (300)(0.14)(1)$

Remember to add the interest to the principal.

26. GS	$600	11%	6 months	_____	_____

$I = prt$
$I = (600)(0.11)\left(\dfrac{6}{12}\right)$

Remember to add the interest to the principal.

27.	$740	6%	9 months	_____	_____
28.	$1180	4%	2 years	_____	_____
29.	$1500	$2\frac{1}{2}\%$	$1\frac{1}{2}$ years	_____	_____
30.	$3000	$3\frac{1}{2}\%$	$2\frac{1}{2}$ years	_____	_____
31.	$17,800	$7\frac{3}{4}\%$	8 months	_____	_____
32.	$20,500	$8\frac{1}{4}\%$	5 months	_____	_____

Solve each application problem. Round money answers to the nearest cent when necessary.

33. Diamonds at Discounts sells diamond engagement rings at 40% off the regular price. Find the sale price of a $\frac{1}{2}$-carat diamond ring normally priced at $1950.

34. A 32 GB iPod classic originally priced at $249 is marked down 8%. Find the price of the iPod after the markdown.

35. Evelina Jones lent $7500 to her son Rick, the owner of Rick's Limousine Service. He repaid the loan at the end of 9 months at $8\frac{1}{2}$% simple interest. What total amount did Rick pay his mother?

36. The owners of Delta Trucking purchased four diesel-powered tractors for cross-country hauling at a cost of $87,500 per tractor. If they borrowed the purchase price for $1\frac{1}{2}$ years at 11% simple interest, find the total amount due.

37. A bluetooth headset is sale priced at $24.99. The sales tax rate is $6\frac{1}{2}$%. Find the total cost of the headset.

38. A weekday "golf/breakfast special" includes breakfast, 18 holes of golf, and use of a cart for $47.95 plus tax per person. If the sales tax rate is $7\frac{1}{2}$%, find the total cost per person. How much will three friends pay to play golf?

39. A Texas Instruments TI-84 Plus graphing calculator is priced at $99 plus sales tax of $7.92. Find the sales tax rate.

40. Textbooks for two classes cost $185. Sales tax of $11.10 is added to the cost. What is the sales tax rate?

41. A "super 45% off sale" begins today. What is the sale price of a ski parka normally priced at $135?

42. A discontinued Whirlpool model side-by-side refrigerator with in-door icemaker originally sold for $1197. What is the sale price with a 35% discount?

43. Ricia and Seitu split a $43.70 dinner bill plus 15% tip. How much did each person pay?

44. Marvette took her brother out to dinner for his birthday. The bill was $58.36 for food and $15.44 for wine. How much was her 20% tip, rounded to the nearest dollar?

45. A 50″ flat screen TV normally priced at $1199 is on sale for 33% off. Find the discount and the sale price.

46. This week, Ford Fusion sedans are offered at 15% off the manufacturer's suggested price. Find the discount and the sale price of a Fusion originally priced at $28,700.

47. Ms. Henderson owes $1900 in taxes. She is charged a penalty of $12\frac{1}{4}\%$ annual simple interest and pays the taxes and penalty after 6 months. Find the total amount she must pay.

48. Norell Di Loreto started a technology services business in her home. She borrowed $27,000 to set up her office and advertise her business. The loan is for 24 months at $7\frac{3}{4}\%$ simple interest. What is the total amount due on the loan?

49. Vincente and Samuel ordered a large deep-dish pizza for $17.98, How much did they give the delivery person to pay for the pizza and a 15% tip, rounded to the nearest dollar?

50. Cher, Maya, and Adara shared a $28.50 bill for a buffet lunch. Because the server brought only their beverages, they left a 10% tip instead of the usual 15%. How much did each person pay?

Use the information in the store ad to answer Exercises 51–54. Round sale prices and sales tax to the nearest cent when necessary.

STORE CLOSE-OUT!

All clothing is now 45% off!

All jewelry is now 30% off!

All electronics are now 65% off!

6% sales tax added to jewelry and electronics purchases.

CASH ONLY! ALL SALES ARE FINAL!

51. Danika bought a wireless router originally priced at $129 and a $60 pair of earrings. What was her bill for the two items?

52. Find David's total bill for a $189 jacket and a $75 graphing calculator.

53. Sergei purchased a camcorder originally priced at $287.95, two pairs of $48 jeans, and a $95 ring. Find his total bill.

54. Richard picked out three pairs of $15 running shorts, two $28 shirts, and a smartphone originally priced at $99.99. How much did he pay in all?

Relating Concepts (Exercises 55–57) For Individual or Group Work

Learn about compound interest *as you* **work Exercises 55–57 in order.**

Simple interest is paid only on the original principal. But savings accounts and most investments earn **compound interest.** In that case, interest is paid on the principal *and* the interest earned. Let's see the difference made by compounding the interest. It's the way to make your investments grow.

If you invest $10,000 at 8% simple interest for 3 years, here is what you get.

$$I = (\$10,000)(0.08)(3) = \$2400 \leftarrow \text{Interest earned}$$

Total amount you have = $10,000 + $2400 = **$12,400**

But suppose you invest $10,000 in an account that earns 8% *compounded annually,* and you leave the initial investment and the interest in the account for 3 years. The diagram below shows the compounded amount in your account at the end of each year, to the nearest dollar.

Deposit today $10,000

($10,000)(1.08) = **$10,800**
($10,800)(1.08) = **$11,664** } Compound Amount
($11,664)(1.08) ≈ **$12,597**

0   1   2   3
    Year

You made an extra $197 with compounding ($12,597 − $12,400 = $197).

However, compound interest *really* makes a tremendous difference when you invest money over 10 or 20 or more years. The table at the right shows how a $10,000 investment, plus the interest it earns, grows over the years. Amounts are rounded to the nearest dollar.

COMPOUND AMOUNT FOR $10,000 INVESTMENT*

Years	2%	5%	8%
10	$12,190	$16,289	$21,589
20	14,859	26,533	46,610
30	18,114	43,219	100,627
40	22,080	70,400	217,245
50	26,916	114,674	469,016

*Interest compounded annually
Data from www.bankrate.com

Each investment in Exercises 55–57 is $10,000.

55. (a) Calculate the amount you would have after 20 years with a 2% *simple* interest investment.

   (b) In the table, find the amount you would have with *compound* interest. Then calculate how much more you would have with *compound* interest than with *simple* interest.

56. (a) Calculate the amount you would have after 30 years with a 5% simple interest investment.

   (b) In the table, find the amount you would have with *compound* interest. Then calculate how much more you would have with compound interest than with *simple* interest.

57. (a) Repeat the calculations you did in **Exercises 55 and 56** for an 8% simple interest account after 50 years.

   (b) The compounded amount is about how many *times* greater than the simple interest amount?

Study Skills

PREPARING FOR YOUR FINAL EXAM

OBJECTIVES

1 Create a final exam week plan.

2 Break studying into chunks and study over several days.

3 Practice all types of problems.

Your math final exam is likely to be a **comprehensive exam.** This means that it will cover material from the **entire term.** The end of the term will be less stressful if you **make a plan** for how you will prepare for each of your exams.

First, figure out the **score you need to earn on the final exam** to get the course grade you want. Check your course syllabus for grading policies, or ask your instructor if you are not sure of them. This allows you to set a goal for yourself.

> How many points do you need to earn on your mathematics final exam to get the grade you want? _____
>
> _____

Create a Plan

Second, create a **final exam week plan for your work and personal life.** If you need to make an adjustment in your work schedule, do it in advance, so you aren't scrambling at the last minute. If you have family members to care for, you might want to enlist some help from others so you can spend extra time studying. Try to plan in advance so you don't create additional stress for yourself. You will have to set some priorities, and studying has to be at the top of the list! Although life doesn't stop for finals, some things can be ignored for a short time. You don't want to "burn out" during final exam week; **get enough sleep and healthy food so you can perform your best.**

> What adjustments in your personal life do you need to make for final exam week? _____
>
> _____

Study and Review

Third, use the following suggestions to guide your studying and reviewing.

▶ **Know exactly which chapters and sections will be on the final exam.**

▶ **Divide up the chapters,** and decide how much you will review each day.

▶ Begin your reviewing **several days** before the exam.

▶ **Use returned quizzes and tests** to review earlier material (if you have them).

▶ **Practice all types of problems,** but emphasize the types that are most difficult for you. Use the **Cumulative Reviews** that are at the end of each chapter in your text.

▶ **Rewrite your notes or make mind maps** to create summaries.

▶ **Make study cards for all types of problems.** Be sure to use the same **direction words** (such as *simplify, solve, estimate*) that your exam will use. Carry the cards with you and review them whenever you have a few spare minutes.

Managing Stress

Of course, a week of final exams produces stress. **Students who develop skills for reducing and managing stress do better on their final exams and are less likely to "bomb" an exam.** You already know the damaging effect of adrenaline on your ability to think clearly. But several days (or weeks) of elevated stress is also harmful to your brain and your body. You will feel better if you make a conscious effort to reduce your stress level. Even if it takes you away from studying for a little while each day, the time will be well spent.

Reducing Physical Stress

Examples of ways to reduce **physical stress** are listed below. Can you add any of your own ideas to the list?

▶ *Laugh until your eyes water.* Watching your favorite funny movie, exchanging a joke with a friend, or viewing a comedy bit on the Internet are all ways to generate a healthy laugh. Laughing raises the level of *calming* chemicals (endorphins) in your brain.

▶ *Exercise for 20 to 30 minutes.* If you normally exercise regularly, do NOT stop during final exam week! Exercising helps relax muscles, diffuses adrenaline, and raises the level of endorphins in your body. If you don't exercise much, get some gentle exercise, such as a daily walk, to help you relax.

▶ *Practice deep breathing.* Several minutes of deep, smooth breathing will calm you. Close your eyes too.

▶ *Visualize a relaxing scene.* Choose something that you find peaceful and picture it. Imagine what it feels like and sounds like. Try to put yourself in the picture.

▶ If you feel stress in your muscles, such as in your shoulders or back, *slowly squeeze the muscles as much as you can, and then release them.* Sometimes we don't realize we are clenching our teeth or holding tension in our shoulders until we consciously work with them. Try to notice what it feels like when they are relaxed and loose. Squeezing and then releasing muscles is also something you can do during an exam if you feel yourself tightening up.

> **Which techniques will you try for reducing physical stress?**
>
> _____
>
> _____
>
> _____

Reducing Mental Stress

Mental stress reduction is also a powerful tool both before and during an exam. In addition to these suggestions, do you have any of your own techniques?

▶ *Talk positively to yourself.* Tell yourself you will get through it.

▶ *Reward yourself.* Give yourself small breaks, a little treat—something that makes you happy—every day of final exam week.

▶ *Make a list of things to do* and feel the sense of accomplishment when you cross each item off.

▶ When you take time to relax or exercise, *make sure you are relaxing your mind too.* Use your mind for something *completely* different from the kind of thinking you do when you study. Plan your garden, play your favorite music, walk your dog, read a good book.

▶ *Visualize.* Picture yourself completing exams and projects successfully. Picture yourself taking the test calmly and confidently.

> **Which techniques will you try for reducing mental stress?**
>
> _____
>
> _____
>
> _____

Chapter 7 *Summary*

Key Terms

7.1

percent Percent means "per one hundred." A percent is a ratio with a denominator of 100.

7.2

percent proportion The proportion used to solve percent problems is: $\dfrac{\text{percent}}{100} = \dfrac{\text{part}}{\text{whole}}$

whole The *whole* in a percent problem is the entire quantity or the total.

part The *part* in a percent problem is the number being compared to the *whole*.

7.3

percent equation The percent equation is: percent • whole = part. The equation can be used instead of the percent proportion to solve percent problems.

7.4

percent of increase or decrease Percent of increase or decrease is the amount of change (increase or decrease) expressed as a percent of the original value.

7.5

sales tax Sales tax is a percent of the total sales charged as a tax.

tax rate The tax rate is the percent used when calculating the amount of tax.

discount Discount is often expressed as a percent of the original price; it is then deducted from the original price, resulting in the sale price.

interest Interest is a fee paid or a charge made for lending or borrowing money.

principal Principal is the amount of money on which interest is charged.

interest rate Often referred to as *rate,* it is the charge for interest and is given as a percent.

simple interest When interest is calculated on the original principal, it is called simple interest.

interest formula The interest formula is used to calculate simple interest. It is: Interest = principal • rate • time or $I = prt$.

New Formulas

Finding simple interest: $I = prt$

New Symbols

% percent symbol (meaning "per 100")

Test Your Word Power

See how well you have learned the vocabulary in this chapter.

1 **Percent** means
A. per one thousand
B. a ratio with 10 in the denominator
C. per one hundred.

2 The **whole** in a percent problem is
A. the entire quantity or total
B. a ratio with a denominator of 100
C. always 100.

3 When calculating sales tax, the **rate** is
A. the whole
B. the total cost
C. the percent.

④ The **interest formula** is
A. $I = prt$
B. percent • whole = part
C. $\dfrac{\text{percent}}{100} = \dfrac{\text{part}}{\text{whole}}$.

⑤ When calculating interest on a loan, the **principal** is the
A. amount of money borrowed
B. total amount due
C. rate charged for borrowing money.

⑥ The **percent of increase or decrease** compares the amount of change to
A. 100
B. the original value before the change
C. the new value after the change.

⑦ The **percent equation** is
A. $I = prt$
B. percent • whole = part
C. $\dfrac{\text{percent}}{100} = \dfrac{\text{part}}{\text{whole}}$.

⑧ A **discount** is
A. added to the original price
B. divided by 100
C. subtracted from the original price.

⑨ **Sales tax** is
A. subtracted from the original price
B. added to the original price
C. multiplied by 100.

Answers to Test Your Word Power

1. C; *Example:* 7% means 7 per 100, or 7 out of 100.

2. A; *Example:* In the problem "15 computers is what percent of 75 computers," the whole is 75 computers, which is the total group of computers.

3. C; *Example:* In 2016, Houston had a sales tax rate of 8.25%.

4. A; *Example:* The interest formula is: Interest = principal • rate • time. If you borrow $4000 for 2 years at a rate of 9%, then $I = (4000)(0.09)(2) = \$720$.

5. A; *Example:* If you borrow $4000 for 2 years at a rate of 9%, the principal is $4000.

6. B; *Example:* If your rent increased from $800 to $850, the *percent of increase* compares the amount of change ($50) to the *original* rent ($800).

7. B; *Example:* To answer, "What percent of $40 is $12," let p be the unknown percent and write the equation as: $p \cdot 40 = 12$.

8. C; *Example:* If sunglasses are on sale at 10% off, then a pair of sunglasses regularly priced at $25 will have a discount of $2.50 subtracted from the price; you will pay $22.50. To find the amount of discount, multiply $(0.10)(\$25)$ to get $2.50.

9. B; *Example:* If sales tax is 6%, the amount of tax on a $70 video game is $(0.06)(\$70) = \$4.20$ and you will pay $70 + $4.20 = $74.20.

Quick Review

Concepts	Examples

7.1 Basics of Percent

Writing a Percent as a Decimal
To write a percent as a decimal, move the decimal point *two* places to the *left* and drop the % sign.

$50\% = 50.\% = 0.50$ or **0.5**

$3\% = 03.\% = \mathbf{0.03}$

Writing a Decimal as a Percent
To write a decimal as a percent, move the decimal point *two* places to the *right* and attach a % sign.

$0.75 = 0.75 = \mathbf{75}\%$

$3.6 = 3.60 = \mathbf{360}\%$

Writing a Percent as a Fraction
To write a percent as a fraction, drop the % symbol and write the number over 100. Then write the fraction in lowest terms.

$35\% = \dfrac{35}{100} = \dfrac{35 \div 5}{100 \div 5} = \dfrac{\mathbf{7}}{\mathbf{20}} \leftarrow$ Lowest terms

$125\% = \dfrac{125}{100} = \dfrac{125 \div 25}{100 \div 25} = \dfrac{5}{4} = \mathbf{1\dfrac{1}{4}}$

Writing a Fraction as a Percent
To write a fraction as a percent, multiply by 100 and attach a % symbol. This is the same as multiplying by 100%.

$\dfrac{3}{5} = \left(\dfrac{3}{5}\right)\left(\dfrac{100}{1}\%\right) = \dfrac{3 \cdot \overset{1}{5} \cdot 20}{\underset{1}{5} \cdot 1}\% = \dfrac{60}{1}\% = \mathbf{60}\%$

$\dfrac{7}{8} = \left(\dfrac{7}{8}\right)\left(\dfrac{100}{1}\%\right) = \dfrac{7 \cdot \overset{1}{4} \cdot 25}{2 \cdot \underset{1}{4} \cdot 1}\% = \dfrac{175}{2}\% =$

$\mathbf{87\dfrac{1}{2}\%}$ or   **87.5%**

Concepts	Examples

7.2 Using the Percent Proportion

The percent proportion is shown below.

$$\text{Always } 100 \rightarrow \frac{\text{percent}}{100} = \frac{\text{part}}{\text{whole}}$$

Identify the percent first. It appears with the word *percent* or the % symbol.

The *whole* is the entire quantity or total. It often appears after the word **of**.

The *part* is the number being compared to the whole.

---

30 children is what percent **of** 75 children?

$$\text{Percent (unknown)} \rightarrow \frac{p}{100} = \frac{30}{75} \leftarrow \text{Part}$$
$$\text{Always } 100 \rightarrow \qquad\qquad \leftarrow \text{Whole (follows } of)$$

To solve the proportion, find the cross products.

$$\frac{p}{100} \times \frac{30}{75} \qquad \begin{matrix} 100 \cdot 30 = 3000 \leftarrow \\ p \cdot 75 \leftarrow \end{matrix} \quad \text{Cross products}$$

$$p \cdot 75 = 3000 \qquad \text{Show that the cross products are equal.}$$

$$\frac{p \cdot \overset{1}{75}}{\underset{1}{75}} = \frac{3000}{75} \qquad \text{Divide both sides by 75}$$

$$p = 40$$

30 children is **40**% of 75 children.

7.3 Percent Shortcuts

200% of a number is 2 times the number.

100% of a number is the entire number.

To find 50% of a number, divide the number by 2.

To find 25% of a number, divide the number by 4.

To find 10% of a number, move the decimal point *one* place to the *left*.

To find 1% of a number, move the decimal point *two* places to the *left*.

200% of $35 is 2 times $35 = **$70**

100% of 600 women is *all* the women (**600 women**).

50% of $8000 is $8000 ÷ 2 = **$4000**

25% of 40 pens is 40 ÷ 4 = **10 pens**

10% of $92.40 is **$9.24**

1% of 62. miles is **0.62 mile**.

7.3–7.4 Using the Percent Equation

Use the six problem-solving steps.

Step 1 **Read** the problem.

Step 2 **Assign a variable.**

Step 3 **Write an equation.**

Use the percent equation:

$$\text{percent} \cdot \text{whole} = \text{part}$$

Be sure to write the percent as a decimal or fraction before using the equation.

Todd's regular pay is $540 per week but $43.20 is taken out of each paycheck for medical insurance. What percent is that?

Step 1 The problem asks for the percent of Todd's pay that is taken out for insurance.

Unknown: percent taken out

Known: Whole paycheck is $540; $43.20 is taken out.

Step 2 Let p be the unknown percent.

Step 3 Use the percent equation. The *whole* is Todd's entire pay of $540 and the *part* is $43.20 (the part taken out of his paycheck).

$$\text{percent} \cdot \text{whole} = \text{part}$$
$$p \quad \cdot \quad 540 \ = 43.20$$

(Continued)

Concepts

7.3–7.4   Using the Percent Equation *(Continued)*

Step 4   **Solve the equation.**

Step 5   **State the answer.**

Step 6   **Check** the solution.

7.4   Finding Percent of Increase or Decrease

Use subtraction to find the *amount* of increase or decrease. When writing the equation, be careful to use the *original* value as the whole.

7.5   Consumer Applications

Finding Sales Tax

To find sales tax, use this equation.

$$\text{tax rate} \cdot \text{cost of item} = \text{sales tax}$$

Write the tax rate (the percent) as a decimal.

Examples

Step 4
$$\frac{p \cdot \overset{1}{540}}{\underset{1}{540}} = \frac{43.20}{540} \quad \text{Divide both sides by 540}$$

$$p = 0.08 \leftarrow \text{Decimal form}$$

Multiply 0.08 by 100 so that $0.08 = 8\%$

Step 5   **8% of Todd's pay** is taken out for medical insurance.

Step 6   Use estimation. If 10% of Todd's pay were withheld, then 10% of $540. = $54, which is a little more than the $43.20 actually withheld. So 8% is a reasonable solution.

Enrollment rose from 3820 students to 5157 students. Find the percent of increase.

$$5157 \text{ students} - 3820 \text{ students} = 1337 \text{ students}$$

Amount of increase

percent of original value = amount of increase

$$p \quad \cdot \quad 3820 \quad = \quad 1337$$

$$p \cdot 3820 = 1337$$

$$\frac{p \cdot \overset{1}{3820}}{\underset{1}{3820}} = \frac{1337}{3820} \quad \text{Divide both sides by 3820}$$

$$p = 0.35 \leftarrow \text{Decimal form}$$

$$0.35 = 35\% \leftarrow \text{Percent increase}$$

The enrollment increased 35%.

Find the sales tax on an $89 pair of binoculars if the sales tax rate is 6%.

$$\text{tax rate} \cdot \text{cost of item} = \text{sales tax}$$

$$06.\% \cdot \$89 = n$$

$$(0.06)(89) = n$$

$$5.34 = n$$

The sales tax is $5.34

Concepts	**Examples**

Estimating and Calculating Restaurant Tips

To estimate a 15% tip, first round the food bill up to a convenient whole dollar amount. Then find 10% of the rounded amount by moving the decimal point *one* place to the *left*. Then add half of that amount for the other 5%.

To estimate a 20% tip, first find 10% by moving the decimal point *one* place to the *left*. Then double the amount.

To find the exact tip, write the percent as a decimal. The bill for food and beverages is the *whole* and the tip is the *part*.

Estimate a 15% tip on a restaurant bill of $38.72. Then find the exact tip.

Estimate: Round $38.72 to $40

   10% of $40 is $4

   Half of $4 is $2

So an **estimate of the 15% tip** is $4 + $2 = **$6**

Exact:

$$\text{percent} \cdot \text{whole} = \text{part}$$

$$15\% \cdot 38.72 = n$$

$$(0.15)(38.72) = n$$

$$5.808 = n$$

The **exact 15% tip is** $5.81 (rounded to nearest cent).

Finding a Discount

To find a discount, use this formula.

   rate of discount • original price = amount of discount

Then subtract to find the sale price.

   original price − amount of discount = sale price

All calculators are on sale at 20% off. Find the sale price of a calculator originally marked $35.

$$20\% \cdot \$35 = n$$

$$(0.20)(35) = n$$

$$7 = n$$

The amount of discount is $7

Then $35 − $7 = $28 ← Sale price

Finding Simple Interest and Total Amount Due

To find the simple interest on a loan, use this formula.

$$I = prt$$

$$\text{Interest} = \text{principal} \cdot \text{rate} \cdot \text{time}$$

Time (t) is in years. When the time is given in months, use a fraction with 12 in the denominator because there are 12 months in a year.

Write the rate (the percent) as a decimal.

To find the total amount due on the loan, add the interest to the original principal.

$2800 is borrowed at 8% for 5 months. Find the amount of interest and the total amount due.

$$I = p \cdot r \cdot t$$

$$I = (2800)(0.08)\left(\frac{5}{12}\right)$$

$$I = (224)\left(\frac{5}{12}\right) = \frac{(224)(5)}{12} \approx \$93.33$$

Total amount due = $2800 + $93.33 = **$2893.33**

Chapter 7 *Review Exercises*

7.1 *Write each percent as a decimal and each decimal as a percent.*

1. 25%   **2.** 180%   **3.** 12.5%   **4.** 7%

5. 2.65   **6.** 0.02   **7.** 0.3   **8.** 0.002

Write each percent as a fraction or mixed number in lowest terms. Write each fraction as a percent.

9. 12%   **10.** 37.5%   **11.** 250%   **12.** 5%

13. $\dfrac{3}{4}$ **14.** $\dfrac{5}{8}$ **15.** $3\dfrac{1}{4}$ **16.** $\dfrac{3}{50}$

Complete this table.

Fraction	Decimal	Percent
$\dfrac{1}{8}$	**17.** _____	**18.** _____
19. _____	0.15	**20.** _____
21. _____	**22.** _____	180%

Use percent shortcuts to fill in the blanks.

23. 100% of $46 is _____

24. 50% of $46 is _____

25. 100% of 9 hours is _____

26. 50% of 9 hours is _____

7.2 *Use a percent proportion to answer each question. If necessary, round percent answers to the nearest tenth of a percent.*

27. 338.8 meters is 140% of what number of meters?

28. 2.5% of what number of cases is 425 cases?

29. What is 6% of 450 smartphones?

30. 60% of 1450 math books is how many books?

31. What percent of 380 pairs is 36 pairs?

32. 1440 cans is what percent of 640 cans?

7.3 *Use the percent equation to answer each question. Round money answers to the nearest cent, if necessary.*

33. 11% of $23.60 is how much?

34. What is 125% of 64 days?

35. 1.28 ounces is what percent of 32 ounces?

36. $46 is 8% of what number of dollars?

37. 8 people is 40% of what number of people?

38. What percent of 174 ft is 304.5 ft?

7.4 *Use the six problem-solving steps to answer each question. If necessary, round percent answers to the nearest tenth of a percent.*

39. (a) A medical clinic found that 16.8% of the patients were late for their appointments in January. The number of patients who were late was 504. Find the total number of patients in January.

(b) In February, only 345 patients were late. Find the percent of decrease in late patients from January to February.

40. (a) Coreen budgeted $280 for food on her vacation. She actually spent 130% of that amount. How much did she spend on food?

(b) Coreen's vacation budget included $50 for gifts and souvenirs, but she spent $112. What percent of the budgeted amount did she spend?

41. (a) In the first part of a community tree-planting project, 64 of the 80 trees planted were still living one year later. What percent of the trees planted were still living?

(b) In the second part of the project, 85 trees were planted. What was the percent of increase in trees planted?

42. New types of radar help weather forecasters warn people sooner about tornadoes. The average lead time for tornado warnings was 7 minutes in 1975, and 13 minutes in 2015. What is the percent of increase in the lead time? (Data from National Weather Service.)

7.5 *Find the amount of sales tax or the tax rate and the total cost.*
Round to the nearest cent, if necessary.

	Amount of Sale	Tax Rate	Amount of Tax	Total Cost
43.	$2.79	4%	_____	_____
44.	$780	_____	$58.50	_____

For each restaurant bill, estimate a 15% tip and a 20% tip. Then find the exact
amount for a 15% tip and a 20% tip. Round to the nearest cent, if necessary.

	Bill	Estimated 15%	Exact 15%	Estimated 20%	Exact 20%
45.	$42.73	_____	_____	_____	_____
46.	$8.05	_____	_____	_____	_____

Find the amount or rate of discount and the sale price.

	Original Price	Rate of Discount	Amount of Discount	Sale Price
47.	$37.50	10%	_____	_____
48.	$252	_____	$63	_____

▦ *Find the simple interest and total amount due on each loan.*

	Principal	Rate	Time	Interest	Total Amount Due
49.	$350	$3\frac{1}{4}\%$	3 years	_____	_____
50.	$1530	16%	9 months	_____	_____

Chapter 7 Mixed Review Exercises

Practicing exercises in mixed-up order helps you prepare for tests.

The bar graph shows the types of video games Americans like to play on their mobile devices. Use the information in the graph to answer Exercises 1–4.

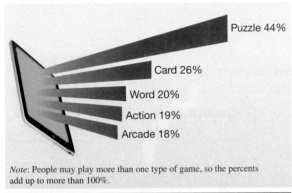

Care for a Game?

One out of three Americans play video games on a mobile device.* The most popular types of games are:

Puzzle 44%

Card 26%

Word 20%

Action 19%

Arcade 18%

Note: People may play more than one type of game, so the percents add up to more than 100%.

*Mobile devices include smartphones, tablets, and portable gaming devices.

Data from www.theesa.com

1. Write the portion of Americans who play video games on a mobile device as a fraction and as a percent.

2. What type of game is most popular? Write the portion of players who played this type as a percent, a decimal, and a fraction in lowest terms.

3. If 830 Americans were surveyed, how many of them play video games on a mobile device? Round your answer to the nearest whole number.

4. Using your answer from **Exercise 3,** find how many of the survey respondents choose:

   (a) the most popular type of game.

   (b) the least popular type of game.

The table shows the number of animals received during the first six months of 2015 by the Minneapolis/St. Paul Animal Humane Society, and the number placed in new homes or returned to their owners. Use the table to answer Exercises 5–10. Round percent answers to the nearest tenth of a percent.

Animals Received	Placed in New Homes or Returned to Owner
5725 dogs	4902 dogs
5283 cats	3774 cats
873 other*	719 other*

*"Other" includes birds, rabbits, guinea pigs, gerbils, etc.

Data from Animalhumanesociety.org

5. What percent of the dogs were placed in new homes or returned to their owners?

6. The number of dogs received during the first six months was 57% of the total number expected for the year. How many dogs were expected, to the nearest whole number?

7. What percent of the cats were placed in new homes or returned to their owners?

8. During the first six months of the previous year, 5862 cats were received. What is the percent decrease in cats received?

9. What percent of all the animals received were placed in new homes or returned to their owners?

10. The Society's goal was to place 85% of all animals received into new homes or return them to their owners. So far they have missed their goal by how many animals? Round your answer to the nearest whole number.

Write each percent as a decimal and each decimal as a percent.

1. 75%

2. 0.6

3. 1.8

4. 0.075

5. 300%

6. 2%

Write each percent as a fraction or mixed number in lowest terms.

7. 62.5%

8. 240%

Write each fraction or mixed number as a percent.

9. $\dfrac{1}{20}$

10. $\dfrac{7}{8}$

11. $1\dfrac{3}{4}$

Write and solve a proportion or an equation to answer Problems 12–17.
Show your work.

12. 16 laptops is 5% of what number of laptops?

13. $192 is what percent of $48?

14. Erica Green has saved 75% of the amount needed for a down payment on a condominium. If she has saved $14,625, find the total down payment needed.

15. The price of a used car is $7950 plus sales tax of $6\frac{1}{2}$%. Find the total cost of the car including sales tax.

16. Enrollment in mathematics courses increased from 1440 students last semester to 1925 students this semester. Find the percent of increase, to the nearest whole percent.

17. Explain a shortcut for finding 50% of a number and a shortcut for finding 25% of a number. Show an example of how to use each shortcut.

18. Explain how you would *estimate* a 15% tip on a restaurant bill of $31.94. Then explain how you would *estimate* a 20% tip on the same bill.

19. Find the exact 15% tip, to the nearest cent, for the restaurant bill in **Problem 18.** If you and two friends are sharing the bill and the exact tip, how much will each person pay?

Find the amount of discount and the sale price of each item in Problems 20 and 21. Round answers to the nearest cent when necessary. Show your work.

20. Jeremy plans to use his 8% employee discount to buy a $48 clock radio so he can get to work on time.

21. A Blu-ray player regularly priced at $229.95 is on sale at 18% off.

22. Jamal found a $1089 computer on sale at 30% off because it was a "discontinued" model. The store will let him pay for it over 6 months with no interest charge. How much will each monthly payment need to be to cover the discounted price plus 7% sales tax?

23. What is the simple interest and total amount due on a four-year loan of $5000 at $8\frac{1}{4}\%$?

24. Kendra borrowed $860 to pay medical expenses. The loan is for 6 months at 12% simple interest. Find the simple interest and total amount due on the loan.

1. Write these numbers in words.
   (a) 90.105

   (b) 125,000,670

2. Write these numbers in digits.
   (a) Thirty billion, five million

   (b) Seventy-eight ten-thousandths

3. Find the mean and median for these rent prices:
   $810, $880, $750, $885, $1225, $795, $840, $785

4. Find the best buy on cereal.
   17 ounces of Brand A for $2.89
   21 ounces of Brand B for $3.59
   15 ounces of Brand C for $2.79

Simplify.

5. $50 - 1.099$

6. $(-3)^2 + 2^3$

7. $\dfrac{3b}{10a} \cdot \dfrac{15ab}{4}$

8. $3\dfrac{3}{10} - 2\dfrac{4}{5}$

9. $3 + 2(-6 + 1)$

10. $\dfrac{4}{5} + \dfrac{x}{4}$

11. $(-0.5)(0.002)$

12. $\dfrac{-16 + 2^4}{-3 - 2}$

13. $\dfrac{7}{8} - \dfrac{2}{m}$

14. $\dfrac{4.8}{-0.16}$

15. $\dfrac{8}{9} \div 2n$

16. $1\dfrac{5}{6} + 1\dfrac{2}{3}$

Evaluate each expression when w is -4, *x is* -2, *and y is* 3.

17. $-6w - 5$

18. $5x + 3y$

19. $x^3 y$

20. $-5w^2 x$

Solve each equation. Show your work.

21. $12 - h = -3h$

22. $5a - 0.6 = 10.4$

23. $2(x + 3) = -4x + 18 + 2x$

24. Translate into an equation and solve. The sum of a number and 31 is three times the number plus 1. Find the number.

25. Use the six problem-solving steps. Susanna and Neoka are splitting $1620 for painting a house. Neoka worked twice as many hours, so she should earn twice as much. How much should each woman receive?

Find the unknown length, perimeter, circumference, or area. When necessary, use 3.14 *as the approximate value of* π *and round answers to the nearest tenth.*

26. Find the perimeter and the area.

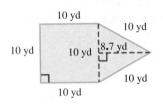

27. Find the circumference and the area.

28. Find the unknown length and the area of the figure.

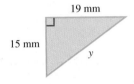

Solve each application problem. Show your work.

29. A survey of the 5600 students on our campus found that $\frac{3}{8}$ of the students work 20 hours or more per week. How many students is that?

30. The Americans With Disabilities Act provides the car parking space design below. Find the perimeter and area of this parking space, including the accessible aisle.

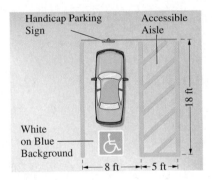

31. The Jackson family is making three kinds of holiday cookies that require brown sugar. The recipes call for $2\frac{1}{4}$ cups, $1\frac{1}{2}$ cups, and $\frac{3}{4}$ cup, respectively. They bought two packages of brown sugar, each holding $2\frac{1}{3}$ cups. The amount bought is how much more or less than the amount needed?

32. Leather jackets are on sale at 30% off the regular price. Tracy likes a jacket with a regular price of $189. Find the amount of discount and the sale price she will pay.

33. On the Illinois map, one centimeter represents 12 kilometers. The center of Springfield is 7.8 cm from the center of Bloomington on the map. What is the actual distance in kilometers?

34. On a 35-problem math test, Juana solved 31 problems correctly. What percent of the problems were correct? Round to the nearest tenth of a percent.

35. Daniel uses an online company to make prints from his digital camera. He ordered 18 prints at $0.12 per print plus $1.97 for shipping. What is the cost per print, including shipping, to the nearest cent? (Data from snapfish.com)

8

Measurement

In this chapter, you'll learn about measuring things. We need a way to measure very large things, like the coast redwood tree named Hyperion in California, which is 380 feet tall and about 24 feet wide at its base. And we need to measure tiny things, like a computer microchip that is only 5 mm long.

8.1 Problem Solving with U.S. Measurement Units

OBJECTIVES

1. Learn the basic U.S. measurement units.
2. Convert among U.S. measurement units using multiplication or division.
3. Convert among U.S. measurement units using unit fractions.
4. Solve application problems using U.S. measurement units.

We measure things all the time, from the distance traveled on vacation, to the weight of the bananas we buy at the store. In the United States, we still use the **United States (U.S.) measurement system** for many everyday tasks. Examples of U.S. measurement units are inches, feet, quarts, ounces, and pounds. However, science, medicine, sports, and manufacturing use the **metric system** (meters, liters, and grams). Because the rest of the world uses *only* the metric system, U.S. businesses have been changing to the metric system in order to compete internationally.

OBJECTIVE ▶ **1** **Learn the basic U.S. measurement units.** Until the switch to the metric system is complete, we still need to know how to use U.S. measurement units. The table below lists the relationships you should memorize. The time relationships are used in both the U.S. and metric systems.

U.S. Measurement Unit Relationships	
U.S. Length Units	**U.S. Weight Units**
12 inches (in.) = 1 foot (ft) 3 feet (ft) = 1 yard (yd) 5280 feet (ft) = 1 mile (mi)	16 ounces (oz) = 1 pound (lb) 2000 pounds (lb) = 1 ton
U.S. Capacity Units	**Time (U.S. and Metric)**
8 fluid ounces (fl oz) = 1 cup (c) 2 cups (c) = 1 pint (pt) 2 pints (pt) = 1 quart (qt) 4 quarts (qt) = 1 gallon (gal)	60 seconds (sec) = 1 minute (min) 60 minutes (min) = 1 hour (hr) 24 hours (hr) = 1 day 7 days = 1 week (wk)

1 After memorizing the U.S. measurement conversions, fill in the blanks.

(a) 1 c = _____ fl oz

(b) _____ qt = 1 gal

(c) 1 wk = _____ days

(d) _____ ft = 1 yd

(e) 1 ft = _____ in.

(f) _____ oz = 1 lb

(g) 1 ton = _____ lb

(h) _____ min = 1 hr

(i) 1 pt = _____ c

(j) _____ hr = 1 day

(k) 1 min = _____ sec

(l) 1 qt = _____ pt

(m) _____ ft = 1 mi

As you can see, there is no simple way to convert among these various measures. The units evolved over hundreds of years and were based on a variety of "standards." For example, one yard was the distance from the tip of a king's nose to his thumb when his arm was outstretched. An inch was three dried barleycorns laid end to end.

EXAMPLE 1 Knowing U.S. Measurement Units

Memorize the U.S. measurement conversions in the table above. Then fill in the blanks.

(a) 24 hr = _____ day   Answer: **1 day**

(b) 1 yd = _____ ft   Answer: **3 ft**

◀ Work Problem **1** at the Side.

OBJECTIVE ▶ **2** **Convert among U.S. measurement units using multiplication or division.** You often need to convert from one unit of measure to another. Two methods of converting measurements are shown here. Study each way and use the method you prefer. The first method involves deciding whether to multiply or divide.

Converting among Measurement Units

1. *Multiply* when converting from a larger unit to a smaller unit.
2. *Divide* when converting from a smaller unit to a larger unit.

Answers

1. (a) 8 (b) 4 (c) 7 (d) 3 (e) 12
   (f) 16 (g) 2000 (h) 60 (i) 2 (j) 24
   (k) 60 (l) 2 (m) 5280

EXAMPLE 2 Converting from One U.S. Unit of Measure to Another

Convert each measurement.

(a) 7 ft to inches

You are converting from a *larger* unit to a *smaller* unit (a *foot* is longer than an *inch*), so multiply.

Because *1 ft = 12 in.*, multiply by 12.

$$7 \text{ ft} = 7 \cdot 12 = \textbf{84 in.}$$

(b) $3\frac{1}{2}$ lb to ounces

You are converting from a *larger* unit to a *smaller* unit (a *pound* is heavier than an *ounce*), so multiply.

Because *1 lb = 16 oz*, multiply by 16.

> Divide 16 and 2 by their common factor of 2.
> 16 ÷ 2 is 8
> 2 ÷ 2 is 1

$$3\frac{1}{2} \text{ lb} = 3\frac{1}{2} \cdot 16 = \frac{7}{\underset{1}{2}} \cdot \frac{\overset{8}{16}}{1} = \frac{56}{1} = \textbf{56 oz}$$

(c) 20 qt to gallons

You are converting from a *smaller* unit to a *larger* unit (a *quart* is smaller than a *gallon*) so divide.

Because *4 qt = 1 gal*, divide by 4.

$$20 \text{ qt} = \frac{20}{4} = \textbf{5 gal}$$

Divide by 4 ⤴

> Remember to write **gal** in your answer.

(d) 45 min to hours

You are converting from a *smaller* unit to a *larger* unit (a *minute* is less than an *hour*), so divide.

Because *60 min = 1 hr*, divide by 60 and write the fraction in lowest terms.

$$45 \text{ min} = \frac{45}{60} = \frac{45 \div 15}{60 \div 15} = \frac{3}{4} \textbf{ hr} \leftarrow \text{Lowest terms}$$

Divide by 60 ⤴

─── **Work Problem 2 at the Side.** ▶

OBJECTIVE ▶ 3 Convert among U.S. measurement units using unit fractions. If you have trouble deciding whether to multiply or divide when converting measurements, use *unit fractions* to solve the problem. You'll also use this method in science courses. A **unit fraction** is equivalent to 1. Here is an example.

$$\frac{12 \text{ in.}}{12 \text{ in.}} = \frac{\overset{1}{12 \text{ in.}}}{\underset{1}{12 \text{ in.}}} = 1$$

Use the table of measurement relationships on the previous page to find that 12 in. is the same as 1 ft. So you can substitute 1 ft for 12 in. in the numerator, or you can substitute 1 ft for 12 in. in the denominator. This makes two useful unit fractions.

$$\frac{1 \text{ ft}}{12 \text{ in.}} = 1 \qquad \text{or} \qquad \frac{12 \text{ in.}}{1 \text{ ft}} = 1$$

---

2 Convert each measurement using multiplication or division.

GS (a) $5\frac{1}{2}$ ft to inches

Because you are converting from a *larger* to a *smaller* unit, do you multiply or divide? _____

GS (b) 64 oz to pounds

Because you are converting from a *smaller* to a *larger* unit, you _____.

(c) 6 yd to feet

(d) 2 tons to pounds

(e) 35 pt to quarts

(f) 20 min to hours

Answers

2. **(a)** multiply; 66 in. **(b)** divide; 4 lb
(c) 18 ft **(d)** 4000 lb **(e)** $17\frac{1}{2}$ qt **(f)** $\frac{1}{3}$ hr

VOCABULARY TIP

Uni- is a prefix that means "one," so a **unit fraction** is equivalent to 1.

3 First write the unit fraction needed to make each conversion. Then complete the conversion.

GS **(a)** 36 in. to feet

$$\left.\begin{array}{c}\text{unit}\\\text{fraction}\end{array}\right\} \dfrac{1\text{ ft}}{12\text{ in.}}$$

$$\dfrac{\overset{3}{\cancel{36}}\text{ in.}}{1} \cdot \dfrac{1\text{ ft}}{\underset{1}{\cancel{12}}\text{ in.}} =$$

(b) 14 ft to inches

$$\left.\begin{array}{c}\text{unit}\\\text{fraction}\end{array}\right\} \dfrac{\text{in.}}{\text{ft}}$$

(c) 60 in. to feet

$$\left.\begin{array}{c}\text{unit}\\\text{fraction}\end{array}\right\} \underline{\hspace{2cm}}$$

(d) 4 yd to feet

$$\left.\begin{array}{c}\text{unit}\\\text{fraction}\end{array}\right\} \underline{\hspace{2cm}}$$

(e) 39 ft to yards

$$\left.\begin{array}{c}\text{unit}\\\text{fraction}\end{array}\right\} \underline{\hspace{2cm}}$$

(f) 2 mi to feet

$$\left.\begin{array}{c}\text{unit}\\\text{fraction}\end{array}\right\} \underline{\hspace{2cm}}$$

Answers

3. (a) $\dfrac{3 \cdot 1\text{ ft}}{1} = 3\text{ ft}$ **(b)** $\dfrac{12\text{ in.}}{1\text{ ft}}$; 168 in.

(c) $\dfrac{1\text{ ft}}{12\text{ in.}}$; 5 ft **(d)** $\dfrac{3\text{ ft}}{1\text{ yd}}$; 12 ft

(e) $\dfrac{1\text{ yd}}{3\text{ ft}}$; 13 yd **(f)** $\dfrac{5280\text{ ft}}{1\text{ mi}}$; 10,560 ft

---

To convert from one measurement unit to another, just multiply by the appropriate unit fraction. Remember, a unit fraction is equivalent to 1. Multiplying something by 1 does *not* change its value.

Use these guidelines to choose the correct unit fraction.

Choosing a Unit Fraction

The **numerator** should use the measurement unit you want in the **answer**.
The **denominator** should use the measurement unit you want to **change**.

EXAMPLE 3 **Using Unit Fractions with U.S. Length Measurements**

(a) Convert 60 in. to feet.

Use a unit fraction with feet (the unit for your answer) in the numerator, and inches (the unit being changed) in the denominator. Because *1 ft = 12 in.*, the necessary unit fraction is

$$\dfrac{1\text{ ft}}{12\text{ in.}} \begin{array}{l} \leftarrow \text{Unit for your \textbf{answer} is \textbf{feet}.}\\ \leftarrow \text{Unit being \textbf{changed} is \textbf{inches}.}\end{array}$$

Next, multiply 60 in. times this unit fraction. Write 60 in. as the fraction $\dfrac{60\text{ in.}}{1}$ and divide out common units and factors wherever possible.

$$60\text{ in.} \cdot \dfrac{1\text{ ft}}{12\text{ in.}} = \dfrac{\overset{5}{\cancel{60}}\text{ in.}}{1} \cdot \dfrac{1\text{ ft}}{\underset{1}{\cancel{12}}\text{ in.}} = \dfrac{5 \cdot 1\text{ ft}}{1} = \textbf{5 ft}$$

Remember to write **ft** in your answer.

These units should match.
— Divide out inches.
— Divide **60** and **12** by 12

(b) Convert 9 ft to inches.

Select the correct unit fraction to change 9 ft to inches.

$$\dfrac{12\text{ in.}}{1\text{ ft}} \begin{array}{l} \leftarrow \text{Unit for your \textbf{answer} is \textbf{inches}.}\\ \leftarrow \text{Unit being \textbf{changed} is \textbf{feet}.}\end{array}$$

Multiply 9 ft times the unit fraction.

$$9\text{ ft} \cdot \dfrac{12\text{ in.}}{1\text{ ft}} = \dfrac{9\text{ ft}}{1} \cdot \dfrac{12\text{ in.}}{1\text{ ft}} = \dfrac{9 \cdot 12\text{ in.}}{1} = \textbf{108 in.}$$

These units should match.
— Divide out feet.

⚠ CAUTION
If no units will divide out, you used the wrong unit fraction.

◀ **Work Problem 3 at the Side.**

EXAMPLE 4 **Using Unit Fractions with U.S. Capacity and Weight Measurements**

(a) Convert 9 pt to quarts.

First select the correct unit fraction.

$$\dfrac{1\text{ qt}}{2\text{ pt}} \begin{array}{l} \leftarrow \text{Unit for your \textbf{answer} is \textbf{quarts}.}\\ \leftarrow \text{Unit being \textbf{changed} is \textbf{pints}.}\end{array}$$

Continued on Next Page

Next multiply.

Write as a mixed number.

$$9 \text{ pt} \cdot \frac{1 \text{ qt}}{2 \text{ pt}} = \frac{9 \text{ pt}}{1} \cdot \frac{1 \text{ qt}}{2 \text{ pt}} = \frac{9}{2} \text{ qt} = 4\frac{1}{2} \text{ qt}$$

These units should match.   Divide out pints.

(b) Convert $7\frac{1}{2}$ gal to quarts.

Write as an improper fraction.

$$\frac{7\frac{1}{2} \text{ gal}}{1} \cdot \frac{4 \text{ qt}}{1 \text{ gal}} = \frac{15}{2} \cdot \frac{4}{1} \text{ qt}$$

Divide out gallons.

> Divide 4 and 2 by their common factor of 2.
> $4 \div 2$ is 2
> $2 \div 2$ is 1

$$= \frac{15}{\underset{1}{2}} \cdot \frac{\overset{2}{4}}{1} \text{ qt}$$

$$= 30 \text{ qt}$$

(c) Convert 36 oz to pounds.

> Notice that oz divides out, leaving **lb**, the unit you want for the answer.

$$\frac{\overset{9}{36} \text{ oz}}{1} \cdot \frac{1 \text{ lb}}{\underset{4}{16} \text{ oz}} = \frac{9}{4} \text{ lb} = 2\frac{1}{4} \text{ lb}$$

Note

In **Example 4(c)** above you get $\frac{9}{4}$ lb. Recall that $\frac{9}{4}$ means $9 \div 4$. If you do $9 \div 4$ on your calculator, you get 2.25 lb. U.S. measurements usually use fractions or mixed numbers, like $2\frac{1}{4}$ lb. However, 2.25 lb is also correct and is the way grocery stores often show weights of produce, meat, and cheese.

———— **Work Problem 4 at the Side.** ▶

EXAMPLE 5   Using Several Unit Fractions

Sometimes you may need to use two or three unit fractions to complete a conversion.

(a) Convert 63 in. to yards.

Use the unit fraction $\frac{1 \text{ ft}}{12 \text{ in.}}$ to change inches to feet and the unit fraction $\frac{1 \text{ yd}}{3 \text{ ft}}$ to change feet to yards. Notice how all the units divide out except yards, which is the unit you want in the answer.

$$\frac{63 \text{ in.}}{1} \cdot \frac{1 \text{ ft}}{12 \text{ in.}} \cdot \frac{1 \text{ yd}}{3 \text{ ft}} = \frac{63}{36} \text{ yd} = \frac{63 \div 9}{36 \div 9} \text{ yd} = \frac{7}{4} \text{ yd} = 1\frac{3}{4} \text{ yd}$$

———— **Continued on Next Page**

---

4 Convert using unit fractions.

(a) 16 qt to gallons

unit fraction} $\frac{1 \text{ gal}}{4 \text{ qt}}$

$$\frac{16 \text{ qt}}{1} \cdot \frac{1 \text{ gal}}{4 \text{ qt}} = \frac{16}{4} \text{ gal} = \underline{\qquad}$$

(b) 3 c to pints

unit fraction} $\frac{\text{pt}}{\text{c}}$

(c) $3\frac{1}{2}$ tons to pounds

(d) $1\frac{3}{4}$ lb to ounces

(e) 4 oz to pounds

Answers

4. **(a)** 4 gal   **(b)** $\frac{1 \text{ pt}}{2 \text{ c}}$; $1\frac{1}{2}$ pt or 1.5 pt

(c) 7000 lb   **(d)** 28 oz   **(e)** $\frac{1}{4}$ lb or 0.25 lb

5 Convert using two or three unit fractions.

GS **(a)** 4 tons to ounces

$$\frac{4 \text{ tons}}{1} \cdot \frac{2000 \text{ lb}}{1 \text{ ton}} \cdot \frac{16 \text{ oz}}{1 \text{ lb}} =$$

(b) 3 mi to inches

(c) 36 pt to gallons

(d) 2 wk to minutes

You can also divide out common factors in the numbers.

$$\frac{63}{1} \cdot \frac{1}{12} \cdot \frac{1}{3} = \frac{\overset{1}{\cancel{3}} \cdot \overset{1}{\cancel{3}} \cdot 7}{1} \cdot \frac{1}{\underset{4}{\cancel{12}}} \cdot \frac{1}{\underset{1}{\cancel{3}}} = \frac{7}{4} = 1\frac{3}{4} \text{ yd}$$

Instead of changing $\frac{7}{4}$ to $1\frac{3}{4}$, you can enter $7 \div 4$ on your calculator to get 1.75 yd. Both answers are correct because 1.75 is equivalent to $1\frac{3}{4}$.

(b) Convert 2 days to seconds.

Use three unit fractions. The first one changes days to hours, the next one changes hours to minutes, and the last one changes minutes to seconds. All the units divide out except seconds, which is the unit you want in your answer.

$$\frac{2 \text{ days}}{1} \cdot \frac{24 \text{ hr}}{1 \text{ day}} \cdot \frac{60 \text{ min}}{1 \text{ hr}} \cdot \frac{60 \text{ sec}}{1 \text{ min}} = 172{,}800 \text{ sec}$$

Divide out **days.**
Divide out **hr**
Divide out **min**

Remember to write **sec** in your answer.

◀ **Work Problem 5 at the Side.**

OBJECTIVE **4** **Solve application problems using U.S. measurement units.** To solve application problems, we will use the steps summarized here.

Step 1 **Read** the problem carefully.

Step 2 **Work out a plan.**

Step 3 **Estimate** a reasonable answer.

Step 4 **Solve** the problem.

Step 5 **State the answer.**

Step 6 **Check** your work.

EXAMPLE 6 **Solving U.S. Measurement Applications**

(a) A 36 oz can of coffee is on sale at Jerry's Foods for $14.39. What is the cost per pound, to the nearest cent? (Data from Jerry's Foods.)

Step 1 **Read** the problem. The problem asks for the cost per *pound* of coffee.

Step 2 **Work out a plan.** The weight of the coffee is given in *ounces,* but the answer must be cost *per pound.* Convert ounces to pounds. The word *per* indicates division. You need to divide the cost by the number of pounds.

Step 3 **Estimate** a reasonable answer. To estimate, round $14.39 to $14. Then, there are 16 oz in a pound, so 36 oz are a little more than 2 pounds. Finally, $14 ÷ 2 = $7 per pound is our estimate.

─── **Continued on Next Page**

Answers

5. **(a)** 128,000 oz **(b)** 190,080 in.

(c) $4\frac{1}{2}$ gal or 4.5 gal **(d)** 20,160 min

Step 4 **Solve** the problem. Use a unit fraction to convert 36 oz to pounds.

> On your calculator,
> $9 \div 4 = 2.25$

> Notice that **oz** divides out, leaving **lb** for the answer.

$$\frac{36 \overset{9}{\cancel{oz}}}{1} \cdot \frac{1 \text{ lb}}{\underset{4}{16} \cancel{oz}} = \frac{9}{4} \text{ lb} = 2.25 \text{ lb}$$

Then divide to find the *cost* per *pound*.

> "To the nearest cent" means round to the hundredths place.

$$\begin{array}{l} \text{Cost} \rightarrow \\ \text{per} \rightarrow \\ \text{pound} \rightarrow \end{array} \frac{\$14.39}{2.25 \text{ lb}} = 6.39\overline{5} \approx 6.40 \quad \text{(rounded)}$$

Step 5 **State the answer.** The coffee costs **$6.40 per pound** (to the nearest cent).

Step 6 **Check** your work. The exact answer of $6.40 is close to our estimate of $7.

(b) Bilal's favorite cake recipe uses $1\frac{2}{3}$ cups of milk for one cake. If he makes six cakes for a bake sale at his son's school, how many quarts of milk will he need?

Step 1 **Read** the problem. The problem asks for the number of *quarts* of milk needed for six cakes.

Step 2 **Work out a plan.** Multiply to find the number of *cups* of milk for six cakes. Then convert *cups* to *quarts* (the unit required in the answer).

Step 3 **Estimate** a reasonable answer. To estimate, round $1\frac{2}{3}$ cups to 2 cups. Then, 2 cups times 6 is 12 cups. There are 4 cups in a quart, so 12 cups $\div$ 4 = 3 quarts is our estimate.

Step 4 **Solve** the problem. First multiply. Then use unit fractions to convert.

$$1\frac{2}{3} \cdot 6 = \frac{5}{\underset{1}{3}} \cdot \frac{\overset{2}{6}}{1} = \frac{10}{1} = 10 \text{ cups} \quad \left\} \begin{array}{l} \text{Milk needed for} \\ \text{six cakes} \end{array} \right.$$

$$\frac{\overset{5}{10} \cancel{cups}}{1} \cdot \frac{1 \cancel{pt}}{\underset{1}{2} \cancel{cups}} \cdot \frac{1 \text{ qt}}{2 \cancel{pt}} = \frac{5}{2} \text{ qt} = 2\frac{1}{2} \text{ qt}$$

> Both **cups** and **pt** divide out, leaving **qt**, the unit you want for the answer.

Step 5 **State the answer.** Bilal needs **$2\frac{1}{2}$ qt (or 2.5 qt) of milk**.

Step 6 **Check** your work. The exact answer of $2\frac{1}{2}$ qt is close to our estimate of 3 qt.

Note

In Step 2 above, we *first multiplied* $1\frac{2}{3}$ cups by 6 to find the number of cups needed, then *converted* 10 cups to $2\frac{1}{2}$ quarts. It would also work to *first convert* $1\frac{2}{3}$ cups to $\frac{5}{12}$ qt, then *multiply* $\frac{5}{12}$ qt by 6 to get $2\frac{1}{2}$ qt.

Work Problem ⑥ **at the Side.** ▶

---

⑥ Solve each application problem. Show your work for all the steps.

(a) Kristin paid $3.29 for 12 oz of extra sharp cheddar cheese. What is the price per pound, to the nearest cent?

Step 1 Problem asks for price per _____.

Step 2 So change 12 oz to what unit? _____

(b) A moving company estimates 11,000 lb of furnishings for an average 3-bedroom house. If the company made five such moves last week, how many tons of furnishings did it move? (Data from North American Van Lines.)

Answers

6. (a) pound; pounds; $4.39 per pound (rounded)

(b) 27.5 tons or $27\frac{1}{2}$ tons

8.1 Exercises

FOR EXTRA HELP   *Go to* MyMathLab *for worked-out, step-by-step solutions to exercises enclosed in a square* ⬚ *and video solutions to* ▶ *exercises.*

CONCEPT CHECK   *Rewrite each group of measurement units in order from smallest to largest.*

1. (a) foot, mile, inch, yard

(b) pint, fluid ounce, gallon, quart, cup

2. (a) ton, ounce, pound

(b) day, hour, week, second, minute

Fill in the blanks with the measurement relationships you have memorized.
See Example 1.

3. (a) 1 yd = _____ ft   **(b)** _____ in. = 1 ft

4. (a) 1 ft = _____ in.   **(b)** _____ ft = 1 mi

5. (a) _____ fl oz = 1 c   **(b)** 1 qt = _____ pt

6. (a) _____ qt = 1 gal   **(b)** 1 pt = _____ c

7. (a) _____ lb = 1 ton   **(b)** 1 lb = _____ oz

8. (a) _____ oz = 1 lb   **(b)** 1 ton = _____ lb

9. (a) 1 min = _____ sec   **(b)** _____ min = 1 hr

10. (a) 1 day = _____ hr   **(b)** _____ sec = 1 min

Convert each measurement by multiplying or dividing. ***See Example 2.***

11. (a) 120 sec = _____ min
▶ **(b)** 4 hr = _____ min

12. (a) 180 min = _____ hr
(b) 5 min = _____ sec

13. (a) 2 qt = _____ gal

(b) $6\frac{1}{2}$ ft = _____ in.

14. (a) $4\frac{1}{2}$ gal = _____ qt

(b) 12 oz = _____ lb

15. An adult African elephant could weigh 7 to 8 tons. How many pounds could it weigh? (Data from www.animals.nationalgeographic.com)

16. A reticulated python snake is the world's longest snake. It grows to a length of 18 to 33 feet. How many yards long can the snake be? (Data from www.animals.nationalgeographic.com)

Convert each measurement in Exercises 17–38 using unit fractions.
See Examples 3 and 4.

17. 9 yd = _____ ft
GS   unit fraction } $\dfrac{\text{ft}}{\text{yd}}$

18. 20,000 lb = _____ tons
GS   unit fraction } $\dfrac{\text{ton}}{\text{lb}}$

19. 7 lb = _____ oz

20. 96 oz = _____ lb

21. 5 qt = _____ pt

22. 26 pt = _____ qt

23. 90 min = _____ hr

24. 45 sec = _____ min

25. 3 in. = _____ ft

26. 30 in. = _____ ft

27. 24 oz = _____ lb

28. 36 oz = _____ lb

29. 5 c = _____ pt

30. 15 qt = _____ gal

Use the information in the bar graph below to answer Exercises 31–32.

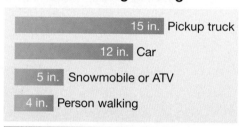

Thickness of Lake Ice Needed for Safe Walking/Driving

15 in. Pickup truck
12 in. Car
5 in. Snowmobile or ATV
4 in. Person walking

Data from Wisconsin DNR.

31. If the ice on a lake is $\frac{1}{2}$ ft thick, what will it safely support?

32. How many feet of ice are needed to safely drive a pickup truck on a lake?

33. $2\frac{1}{2}$ tons = _____ lb

34. $4\frac{1}{2}$ pt = _____ c

35. $4\frac{1}{4}$ gal = _____ qt

36. $2\frac{1}{4}$ hr = _____ min

37. After 15 years, a saguaro cactus is still only one-third to two-thirds of a foot tall, depending upon rainfall. How tall could the cactus be in inches? (Data from *Ecology of the Saguaro III.*)

38. Kevin Durant, an NBA basketball player, is $6\frac{3}{4}$ ft tall. What is his height in inches? (Data from www.NBA.com)

Use two or three unit fractions to make each conversion. ***See Example 5.***

39. 6 yd = _____ in.

⏵

40. 2 tons = _____ oz

41. 112 c = _____ qt

⏵

42. 336 hr = _____ wk

43. 6 days = _____ sec

44. 5 gal = _____ c

45. $1\frac{1}{2}$ tons = _____ oz

46. $3\frac{1}{3}$ yd = _____ in.

47. CONCEPT CHECK The statement 8 = 2 is *not* true. But with appropriate measurement units, it *is* true.

$$8 \text{ } quarts = 2 \text{ } gallons$$

Attach measurement units to these numbers to make the statement true.

(a) 1 _____ = 16 _____

(b) 10 _____ = 20 _____

(c) 120 _____ = 2 _____

(d) 2 _____ = 24 _____

(e) 6000 _____ = 3 _____

(f) 35 _____ = 5 _____

48. CONCEPT CHECK Explain in your own words why you can add 2 feet + 12 inches to get 3 feet, but you cannot add 2 feet + 12 pounds.

▦ *Convert each measurement.* ***See Example 5.***

49. $2\frac{3}{4}$ mi = _____ in.

50. $5\frac{3}{4}$ tons = _____ oz

51. $6\frac{1}{4}$ gal = _____ fl oz

52. $3\frac{1}{2}$ days = _____ sec

53. 24,000 oz = _____ ton

54. 57,024 in. = _____ mi

Solve each application problem. ***See Example 6.***

55. Geralyn bought 20 oz of strawberries for $2.79. What was the price per pound for the strawberries? Round your answer to the nearest cent.

56. Zach paid $1.35 for a 1.44 oz candy bar. What was the cost per pound?

57. Dan orders supplies for the science labs. Each of the
⏵ 24 stations in the chemistry lab needs 2 ft of rubber tubing. If rubber tubing sells for $8.75 per yard, how much will it cost to equip all the stations?

58. In 2014, Erie, Pennsylvania, had 105.0 inches of snowfall, while Omaha, Nebraska, had 14.5 inches. What was the difference in snowfall between the two cities, in feet? Round your answer to the nearest tenth of a foot. (Data from www.ncdc.noaa.gov)

59. The Australian tiger beetle is the fastest land insect. It can run about 8 feet per second. At this rate, how long would it take the beetle to travel one mile? (Data from www.bbc.com)

Give your answer

(a) in seconds.

(b) in minutes.

60. A snail moves at an average speed of 5 feet every 2 minutes. At this rate, how long would it take the snail to travel one mile? Give your answer in hours and in days. Round your answers to the nearest tenth. (*Hint:* First find the number of minutes to travel one mile.) (Data from www.snailworld.com)

61. At the day care center, each of the 15 toddlers drinks about $\frac{2}{3}$ cup of milk with lunch. The center is open 5 days a week.

(a) How many quarts of milk will the center need for one week of lunches?

(b) If the center buys milk in gallon containers, how many containers should be ordered for one week?

62. A chocolate factory uses about 3,500,000 pounds of whole milk every day to make chocolate. (Data from www.thechocolatestore.com)

(a) How many tons of whole milk are used to make chocolate during a 5-day workweek?

(b) The chocolate factory operates 24 hours per day. How many tons of whole milk are used to make chocolate each hour? Round your answer to the nearest tenth.

Relating Concepts (Exercises 63–66) For Individual or Group Work

On the first page of this chapter, we said that the coast redwood tree named Hyperion in California is 380 feet tall and about 24 feet across at its base. Use this information as you **work Exercises 63–66 in order.** *(Data from www.livescience.com)*

63. (a) How many inches tall is the tree?

(b) The height of the tree is how many yards? Round your answer to the nearest whole number.

(c) A park ranger tells you the tree is about as tall as a forty-story building. She is assuming each story is how many feet tall? Round your answer to the nearest tenth of a foot.

64. (a) The base of the tree is about 24 feet across. How many yards is that?

(b) The base of the tree is nearly circular. In a circle, the distance around the outside edge is about 3.14 times the distance across the circle. How far is it to walk around the tree, as measured in feet? Round your answer to the nearest tenth of a foot.

(c) How far is it to walk around the tree in yards? Round your answer to the nearest yard.

65. Hyperion is estimated to be about 700 years old. A young coast redwood near Hyperion is currently 30 inches tall.

(a) How tall is the young redwood tree in feet?

(b) If the young redwood tree grows at the same rate that Hyperion did, how old will it be when it reaches 100 feet? Round your answer to the nearest year.

66. Hyperion is the tallest tree, but it is not the widest or the heaviest tree. The widest tree is circular and its base is 38.1 ft wide. The heaviest tree is also circular and its base is 36.5 ft wide.

(a) Using the information about circles in **Exercise 64,** find the approximate distance you would walk around the base of the widest tree. Round your answer to the nearest foot.

(b) What is the distance you would walk around the base of the heaviest tree? Round your answer to the nearest foot.

8.2 The Metric System—Length

The metric system of measurement is an organized system based on multiples of 10, like our number system and our money. After you are familiar with metric units, you will see that they are easier to use than the U.S. measurement relationships in the last section.

> **Note**
>
> The metric system information in this text is consistent with usage guidelines from the National Institute of Standards and Technology, www.nist.gov/metric.

OBJECTIVE ▶ 1 Learn the basic metric units of length. The basic unit of length in the metric system is the **meter** (also spelled *metre*). If you put five pieces of printer paper side by side, they would measure a little more than 1 meter. Or, look at a yardstick. It is 36 inches long; a meter is just a little longer, about 39 inches long.

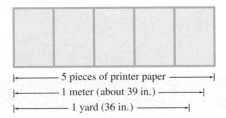

|← 5 pieces of printer paper →|
|← 1 meter (about 39 in.) →|
|← 1 yard (36 in.) →|

In the metric system, you use meters for things like measuring the length of your living room, talking about heights of buildings, or describing track and field athletic events.

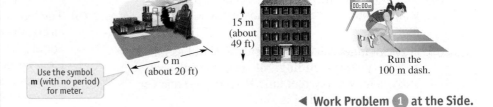

Use the symbol **m** (with no period) for meter.

6 m (about 20 ft)

15 m (about 49 ft)

Run the 100 m dash.

◀ **Work Problem 1 at the Side.**

To make longer or shorter length units in the metric system, **prefixes** are written in front of the word *meter*. For example, the prefix *kilo* means 1000, so a *kilo*meter is 1000 meters. The table below shows how to use the prefixes for length measurements. It is helpful to memorize the prefixes because they are also used with weight and capacity measurements. The arrows point to the units you will use most often in daily life.

Prefix	kilo-meter	hecto-meter	deka-meter	meter	deci-meter	centi-meter	milli-meter
Meaning	1000 meters	100 meters	10 meters	1 meter	$\frac{1}{10}$ of a meter	$\frac{1}{100}$ of a meter	$\frac{1}{1000}$ of a meter
Symbol	km	hm	dam	m	dm	cm	mm

Length units that are used most often

1. Circle the items that measure about 1 meter.

Length of a pencil

Length of a baseball bat

Height of doorknob from the floor

Height of a house

Basketball player's arm length

Length of a paper clip

Answer

1. baseball bat, height of doorknob, basketball player's arm length

Here are some comparisons to help you get acquainted with the commonly used length units: km, m, cm, mm.

*Kilo*meters are used instead of miles. A kilometer is **1000** meters. It is about 0.6 mile (a little more than half a mile). If you participate in a 10 km run, you'll run about 6 miles. (A 10 km run is often called a 10K.)

A meter is divided into 100 smaller pieces called *centi*meters. Each centimeter is $\frac{1}{100}$ of a meter. Centimeters are used instead of inches. A centimeter is a little shorter than $\frac{1}{2}$ inch. A nickel is about 2 cm across. A piece of printer paper is about 22 cm wide. Measure the width and length of your little finger on this centimeter ruler. The tip of your little finger is probably between 1 cm and 2 cm wide.

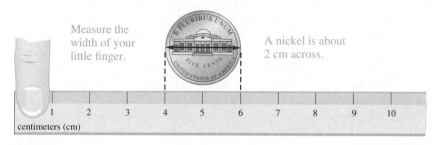

A meter is divided into 1000 smaller pieces called *milli*meters. Each millimeter is $\frac{1}{1000}$ of a meter. It takes 10 mm to equal 1 cm, so it is a very small length. The thickness of a dime is about 1 mm. Measure the width of your pen or pencil and the width of your little finger on this millimeter ruler.

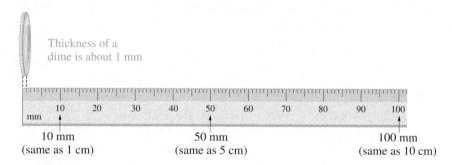

EXAMPLE 1　**Using Metric Length Units**

Write the most reasonable metric unit in each blank.
Choose from km, m, cm, and mm.

(a) The distance from home to work is 20 _____.

　20 **km** because kilometers are used instead of miles.
　20 km is about 12 miles.

(b) My wedding ring is 4 _____ wide.

　4 **mm** because the width of a ring is very small.

(c) The newborn baby is 50 _____ long.

　50 **cm**, which is half of a meter; a meter is about 39 inches,
　so half a meter is around 20 inches.

───── **Work Problem ② at the Side.** ▶

② Write the most reasonable metric unit in each blank. Choose from km, m, cm, and mm.

(a) The woman's height is 168 _____.

(b) The man's waist is 90 _____ around.

(c) Louise ran the 100 _____ dash in the track meet.

(d) A postage stamp is 22 _____ wide.

(e) Michael paddled his canoe 2 _____ down the river.

(f) The pencil lead is 1 _____ thick.

(g) A stick of gum is 7 _____ long.

(h) The highway speed limit is 90 _____ per hour.

(i) The classroom was 12 _____ long.

(j) A penny is about 18 _____ across.

Answers

2. (a) cm　**(b)** cm　**(c)** m　**(d)** mm　**(e)** km
　(f) mm　**(g)** cm　**(h)** km　**(i)** m　**(j)** mm

3 First write the unit fraction needed to make each conversion. Then complete the conversion.

GS **(a)** 3.67 m to cm

$$\left.\text{unit}\atop\text{fraction}\right\} \frac{100 \text{ cm}}{1 \text{ m}}$$

$$\frac{3.67 \text{ m}}{1} \cdot \frac{100 \text{ cm}}{1 \text{ m}}$$

$$= \frac{3.67 \cdot 100 \text{ cm}}{1} =$$

(b) 92 cm to m

$$\left.\text{unit}\atop\text{fraction}\right\} \frac{\text{m}}{\text{cm}}$$

(c) 432.7 cm to m

$$\left.\text{unit}\atop\text{fraction}\right\} \underline{}$$

(d) 65 mm to cm

$$\left.\text{unit}\atop\text{fraction}\right\} \underline{}$$

(e) 0.9 m to mm

$$\left.\text{unit}\atop\text{fraction}\right\} \underline{}$$

(f) 2.5 cm to mm

$$\left.\text{unit}\atop\text{fraction}\right\} \underline{}$$

Answers

3. **(a)** 367 cm **(b)** $\frac{1 \text{ m}}{100 \text{ cm}}$; 0.92 m

(c) $\frac{1 \text{ m}}{100 \text{ cm}}$; 4.327 m

(d) $\frac{1 \text{ cm}}{10 \text{ mm}}$; 6.5 cm

(e) $\frac{1000 \text{ mm}}{1 \text{ m}}$; 900 mm

(f) $\frac{10 \text{ mm}}{1 \text{ cm}}$; 25 mm

OBJECTIVE **2** **Use unit fractions to convert among metric units.** You can convert among metric length units using unit fractions. Keep these relationships in mind when setting up the unit fractions.

Metric Length Relationships

1 km = 1000 m so the unit fractions are:	1 m = 1000 mm so the unit fractions are:
$\frac{1 \text{ km}}{1000 \text{ m}}$ or $\frac{1000 \text{ m}}{1 \text{ km}}$	$\frac{1 \text{ m}}{1000 \text{ mm}}$ or $\frac{1000 \text{ mm}}{1 \text{ m}}$
1 m = 100 cm so the unit fractions are:	1 cm = 10 mm so the unit fractions are:
$\frac{1 \text{ m}}{100 \text{ cm}}$ or $\frac{100 \text{ cm}}{1 \text{ m}}$	$\frac{1 \text{ cm}}{10 \text{ mm}}$ or $\frac{10 \text{ mm}}{1 \text{ cm}}$

EXAMPLE 2 **Using Unit Fractions to Convert Metric Length Measurements**

Convert each measurement using unit fractions.

(a) 5 km to m

Put the unit for the answer (meters) in the numerator of the unit fraction; put the unit you want to change (km) in the denominator.

$$\text{Unit fraction}\atop\text{equivalent to 1}\left\{\frac{1000 \text{ m}}{1 \text{ km}}\right.\begin{array}{l}\leftarrow \text{Unit for your \textbf{answer} is \textbf{m}}\\ \leftarrow \text{Unit being \textbf{changed} is \textbf{km}}\end{array}$$

Multiply. Divide out common units where possible.

$$5 \text{ km} \cdot \frac{1000 \text{ m}}{1 \text{ km}} = \frac{5 \text{ km}}{1} \cdot \frac{1000 \text{ m}}{1 \text{ km}} = \frac{5 \cdot 1000 \text{ m}}{1} = 5000 \text{ m}$$

These units should match.

Here, **km** divides out leaving **m**, the unit you want for your answer.

Do **not** write a period here.

5 km = **5000 m**

The answer makes sense because a kilometer is much longer than a meter, so 5 km will contain many meters.

(b) 18.6 cm to m

Multiply by a unit fraction that allows you to divide out centimeters.

Unit fraction

$$\frac{18.6 \text{ cm}}{1} \cdot \frac{1 \text{ m}}{100 \text{ cm}} = \frac{18.6}{100} \text{ m} = 0.186 \text{ m}$$

Do **not** write a period here.

18.6 cm = **0.186 m**

There are 100 cm in a meter, so 18.6 cm will be a small part of a meter. The answer makes sense.

◀ **Work Problem** **3** **at the Side.**

OBJECTIVE ▶ **3** **Move the decimal point to convert among metric units.**
By now you have probably noticed that conversions among metric units are made by multiplying or dividing by 10, by 100, or by 1000. A quick way to *multiply* by 10 is to move the decimal point one place to the *right*. Move it two places to the right to multiply by 100, three places to multiply by 1000. *Dividing* is done by moving the decimal point to the *left* in the same manner.

Work Problem 4 at the Side. ▶

An alternate conversion method to unit fractions is moving the decimal point using this **metric conversion line.**

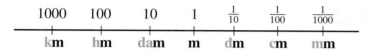

1000	100	10	1	$\frac{1}{10}$	$\frac{1}{100}$	$\frac{1}{1000}$
km	hm	dam	m	dm	cm	mm

Here are the steps for using the conversion line.

Using the Metric Conversion Line

Step 1   Find the unit you are given on the metric conversion line.

Step 2   Count the number of places to get from the unit you are given to the unit you want in the answer.

Step 3   Move the decimal point the **same number of places** and in the **same direction** as you did on the conversion line.

EXAMPLE 3   **Using the Metric Conversion Line**

Use the metric conversion line to make the following conversions.

(a) 5.702 km to m

Find **km** on the metric conversion line. To get to **m**, you move *three places* to the *right*. So move the decimal point in 5.702 *three places* to the *right*.

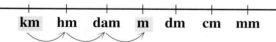

km   hm   dam   m   dm   cm   mm   5.702
         Three places to the right   Move decimal point three
                                           places to the right.

5.702 km = **5702 m**

(b) 69.5 cm to m

Find **cm** on the conversion line. To get to **m**, move *two places* to the *left*.

km   hm   dam   m   dm   cm   mm   69.5
              Two places   Move decimal point
              to the left   two places to the left.

69.5 cm = **0.695 m**

Continued on Next Page

4 Do each multiplication or division by hand or on a calculator. Compare your answer to the one you get by moving the decimal point.

GS **(a)** $(43.5)(10) = $ _____

43.5 gives 435

(b) $43.5 \div 10 = $ _____

43.5 gives _____

(c) $(28)(100) = $ _____

28.00 gives _____

(d) $28 \div 100 = $ _____

28. gives _____

(e) $(0.7)(1000) = $ _____

0.700 gives _____

(f) $0.7 \div 1000 = $ _____

000.7 gives _____

Answers

4. (a) 435   **(b)** 4.35; 4.35   **(c)** 2800; 2800
(d) 0.28; 0.28   **(e)** 700; 700
(f) 0.0007; 0.0007

5 Convert using the metric conversion line.

(GS) **(a)** 12.008 km to m

km to m is three places to the right.

12.008 km = _____ m

(GS) **(b)** 561.4 m to km

m to km is _____ places to the _____.

561.4 m = _____ km

(c) 20.7 cm to m

(d) 20.7 cm to mm

(e) 4.66 m to cm

6 Convert using the metric conversion line.

(GS) **(a)** 9 m to mm   9.000

m to mm is _____ places to the _____.

9 m = _____ mm

(GS) **(b)** 3 cm to m

cm to m is _____ places to the _____.

3 cm = _____ m

(c) 5 mm to cm

(d) 70 m to km

(e) 0.8 m to cm

Answers

5. **(a)** 12,008 m
   (b) three places to the left; 0.5614 km
   (c) 0.207 m   **(d)** 207 mm   **(e)** 466 cm
6. **(a)** three places to the right; 9000 mm
   (b) two places to the left; 0.03 m
   (c) 0.5 cm   **(d)** 0.07 km   **(e)** 80 cm

(c) 8.1 cm to mm

From **cm** to **mm** is *one place* to the *right*.

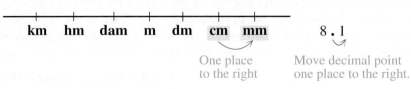

One place to the right

Move decimal point one place to the right.

8.1 cm = **81 mm**

◀ **Work Problem 5** at the Side.

EXAMPLE 4   **Practicing Length Conversions**

Convert using the metric conversion line.

(a) 1.28 m to mm

Moving from **m** to **mm** is going *three places to the right*. In order to move the decimal point in 1.28 three places to the right, you must write a 0 as a placeholder.

1.28**0**   Zero is written in as a placeholder.

Move decimal point three places to the right.

1.28 m = **1280 mm**

The answer is a *whole number,* so you do not need to write the decimal point.

(b) 60 cm to m

From **cm** to **m** is two places to the left. The decimal point in 60 starts at the *far right side* because 60 is a whole number. Then move it two places to the left.

60.   60.

Decimal point starts here.   Move decimal point two places to the left.

60 cm = 0.60 m, and 0.60 m is equivalent to **0.6 m**

Think:
$\frac{60}{100}$ in lowest terms is $\frac{6}{10}$

(c) 8 m to km

From **m** to **km** is three places to the left. The decimal point in 8 starts at the far right side. In order to move it three places to the left, you must write two zeros as placeholders.

Two zeros are written in as placeholders.

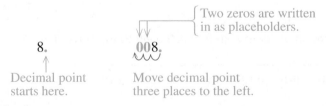

8.   008.

Decimal point starts here.   Move decimal point three places to the left.

8 m = **0.008 km**

◀ **Work Problem 6** at the Side.

8.2 Exercises

FOR EXTRA HELP

Go to MyMathLab *for worked-out, step-by-step solutions to exercises enclosed in a square* ▢ *and video solutions to* ▶ *exercises.*

CONCEPT CHECK *Use your knowledge of the meaning of metric prefixes to fill in the blanks.*

1. *kilo* means ____1000____ so

 1 km = _____ m

2. *deka* means ____10____ so

 1 dam = _____ m

3. *milli* means _____ so

 1 mm = _____ m

4. *deci* means _____ so

 1 dm = _____ m

5. *centi* means _____ so

 1 cm = _____ m

6. *hecto* means _____ so

 1 hm = _____ m

7. Circle the shorter length unit in each pair.

 (a) cm   m   **(b)** m   mm   **(c)** km   cm

8. Circle the longer length unit in each pair.

 (a) m   km   **(b)** mm   cm   **(c)** km   mm

Use this ruler to measure the width of your thumb and hand for Exercises 9–12.

9. The width of your hand in centimeters

10. The width of your hand in millimeters

11. The width of your thumb in millimeters

12. The width of your thumb in centimeters

Write the most reasonable metric length unit in each blank. Choose from km, m, cm, and mm. **See Example 1.**

13. The child was 91 _____ tall.

14. The cardboard was 3 _____ thick.

15. Ming-Na swam in the 200 _____ backstroke race.

16. The bookcase is 75 _____ wide.

17. Adriana drove 400 _____ on her vacation.

18. The door is 2 _____ high.

19. An aspirin tablet is 10 _____ across.

20. Lamard jogs 4 _____ every morning.

21. A paper clip is about 3 _____ long.

22. My pen is 145 _____ long.

23. Dave's truck is 5 _____ long.

24. Wheelchairs need doorways that are at least 81 _____ wide.

Convert each measurement. Use unit fractions or the metric conversion line.
See Examples 2–4.

25. 7 m to cm

$$\frac{7 \, \cancel{m}}{1} \cdot \frac{100 \text{ cm}}{1 \, \cancel{m}} = \frac{7 \cdot 100 \text{ cm}}{1} =$$

26. 18 m to cm

$$\frac{18 \, \cancel{m}}{1} \cdot \frac{100 \text{ cm}}{1 \, \cancel{m}} =$$

27. 40 mm to m

28. 6 mm to m

29. 9.4 km to m

30. 0.7 km to m

31. 509 cm to m

32. 30 cm to m

33. 400 mm to cm

34. 25 mm to cm

35. 0.91 m to mm

36. 4 m to mm

37. CONCEPT CHECK Is 82 cm greater than or less than 1 m? What is the difference in the lengths?

38. CONCEPT CHECK Is 1022 m greater than or less than 1 km? What is the difference in the lengths?

39. Computer microchips may be only 5 mm long and 1 mm wide. Using the ruler on the previous page, draw a rectangle that measures 5 mm by 1 mm. Then convert each measurement to centimeters.

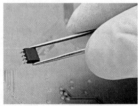

Using tweezers to hold a computer microchip.

40. The world's smallest butterfly has a wingspan of 13 mm. The smallest mouse is 50 mm long. Using the ruler on the previous page, draw a line that is 13 mm long and a line 50 mm long. Then convert each measurement to centimeters. (Data from www.entomon.net and www.mnn.com)

41. The Roe River near Great Falls, Montana, is the shortest river in the world, just 61 m long. How many kilometers long is the Roe River? (Data from *Guinness World Records*.)

42. There are over 100,000 km of blood vessels in the human body. How many meters of blood vessels are in the body? (Data from *The Human Body*.)

43. The median height for U.S. females who are 20 to 29 years old is about 1.63 m. Convert this height to centimeters and to millimeters. (Data from U.S. National Center for Health Statistics.)

44. The median height for 20- to 29-year-old males in the United States is about 177 cm. Convert this height to meters and to millimeters. (Data from U.S. National Center for Health Statistics.)

45. Use two unit fractions to convert 5.6 mm to km.

46. Use two unit fractions to convert 16.5 km to mm.

8.3 The Metric System—Capacity and Weight (Mass)

We use capacity units to measure liquids, such as the amount of gasoline in a car's gas tank. Recall that the capacity units in the U.S. system are cups, pints, quarts, and gallons. The basic metric unit for capacity is the **liter** (also spelled *litre*). The capital letter **L** is the symbol for liter.

OBJECTIVE ▶ 1 Learn the basic metric units of capacity. The liter is related to metric length in this way: A box that measures 10 cm on every side holds exactly one liter. A liter is just a little more than 1 quart.

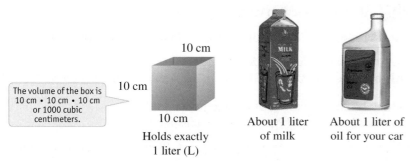

The volume of the box is 10 cm • 10 cm • 10 cm or 1000 cubic centimeters.

10 cm

10 cm

10 cm

Holds exactly 1 liter (L)

About 1 liter of milk

About 1 liter of oil for your car

A liter is a little more than one quart (just $\frac{1}{4}$ cup more).

In the metric system you use liters for things like buying milk and soda at the store, filling a pail with water, and describing the size of your home aquarium.

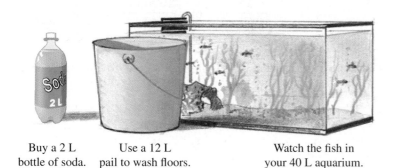

Buy a 2 L bottle of soda.

Use a 12 L pail to wash floors.

Watch the fish in your 40 L aquarium.

Work Problem ❶ at the Side. ▶

To make larger or smaller capacity units, we use the same **prefixes** as we did with length units. For example, the prefix *kilo* means 1000 so a *kilo*meter is 1000 meters. In the same way, a *kilo*liter is 1000 liters.

Prefix	kilo-liter	hecto-liter	deka-liter	liter	deci-liter	centi-liter	milli-liter
Meaning	1000 liters	100 liters	10 liters	1 liter	$\frac{1}{10}$ of a liter	$\frac{1}{100}$ of a liter	$\frac{1}{1000}$ of a liter
Symbol	kL	hL	daL	L	dL	cL	mL

Capacity units used most often

OBJECTIVES

❶ Learn the basic metric units of capacity.

❷ Convert among metric capacity units.

❸ Learn the basic metric units of weight (mass)

❹ Convert among metric weight (mass) units.

❺ Distinguish among basic metric units of length, capacity, and weight (mass).

VOCABULARY TIP

Liter Connect **liter** to a 2-liter bottle of soda. Use **liters** to measure **liquids** (things that pour). Write capital **L** for liter.

❶ Circle the things that can be measured in liters.

Amount of water in the bathtub

Length of the bathtub

Width of your car

Amount of gasoline you buy for your car

Weight of your car

Height of a pail

Amount of water in a pail

Answer

1. water in bathtub, gasoline, water in a pail

VOCABULARY TIP

L; mL Think: liter and quart; milliliter and "a drop."

2 Write the most reasonable metric unit in each blank. Choose from L and mL.

(a) I bought 8 _____ of soda at the store.

(b) The nurse gave me 10 _____ of cough syrup.

(c) This is a 100 _____ garbage can.

(d) It took 10 _____ of paint to cover the bedroom walls.

(e) My car's gas tank holds 50 _____.

(f) I added 15 _____ of oil to the pancake mix.

(g) The can of orange soda holds 350 _____.

(h) My friend gave me a 30 _____ bottle of expensive perfume.

Answers

2. **(a)** L **(b)** mL **(c)** L **(d)** L **(e)** L
   (f) mL **(g)** mL **(h)** mL

The capacity units you will use most often in daily life are liters (L) and *milli*liters (mL). A tiny box that measures 1 cm on every side holds exactly one milliliter. (In medicine, this small amount is also called 1 cubic centimeter, or 1 cc for short.) It takes 1000 mL to make 1 L. Here are some comparisons.

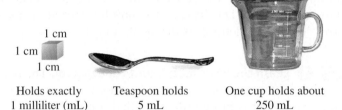

| Holds exactly 1 milliliter (mL) | Teaspoon holds 5 mL | One cup holds about 250 mL |

EXAMPLE 1 **Using Metric Capacity Units**

Write the most reasonable metric unit in each blank. Choose from L and mL.

(a) The bottle of shampoo held 500 _____.
500 **mL** because 500 L would be about 500 quarts, which is too much.

(b) I bought a 2 _____ carton of orange juice.
2 **L** because 2 mL would be less than a teaspoon.

◀ **Work Problem 2 at the Side.**

OBJECTIVE **2** **Convert among metric capacity units.** Just as with length units, you can convert between milliliters and liters using unit fractions.

Metric Capacity Relationships

1 L = 1000 mL, so the unit fractions are:

$$\frac{1\ L}{1000\ mL} \quad \text{or} \quad \frac{1000\ mL}{1\ L}$$

Or you can use a metric conversion line to decide how to move the decimal point.

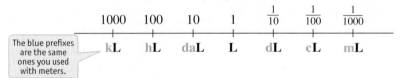

The blue prefixes are the same ones you used with meters.

| 1000 | 100 | 10 | 1 | $\frac{1}{10}$ | $\frac{1}{100}$ | $\frac{1}{1000}$ |
| **kL** | **hL** | **daL** | **L** | **dL** | **cL** | **mL** |

EXAMPLE 2 **Converting among Metric Capacity Units**

Convert using the metric conversion line or unit fractions.

(a) 2.5 L to mL

Using the metric conversion line:

From **L** to **mL** is *three places* to the *right*.

2.5̲0̲0̲ ← Write two zeros as placeholders.

2.5 L = **2500 mL**

Using unit fractions:

Multiply by a unit fraction that allows you to divide out liters.

$$\frac{2.5\ \cancel{L}}{1} \cdot \frac{1000\ mL}{1\ \cancel{L}} = \mathbf{2500\ mL}$$

L divides out, leaving mL for your answer.

Do **not** write a period here.

Continued on Next Page

(b) 80 mL to L

Using the metric conversion line:

From **mL** to **L** is *three places* to the *left*.

80.   080.

↑

Decimal point   Move decimal
starts here.   point three places
                to the left.

80 mL = **0.080 L** **or** **0.08 L**

Using unit fractions:

Multiply by a unit fraction that allows you to divide out mL.

$$\frac{80 \; \cancel{mL}}{1} \cdot \frac{1 \; L}{1000 \; \cancel{mL}}$$

mL divides out, leaving L for your answer.

$$= \frac{80}{1000} L = \textbf{0.08 L}$$

*Do **not** write a period here.*

— **Work Problem 3 at the Side.** ▶

OBJECTIVE 3 Learn the basic metric units of weight (mass).
The **gram** is the basic metric unit for *mass*. Although we often call it "weight," there is a difference that is important in science classes. But for everyday purposes, we will use the word *weight*.

The gram is related to metric length in this way: the weight of the water in a box measuring 1 cm on every side is 1 gram. This is a very tiny amount of water (1 mL) and a very small weight. One gram is also the weight of a dollar bill. See more examples below.

The 1 mL of water in this tiny box weighs 1 gram.

A dollar bill weighs 1 gram.

A nickel weighs 5 grams.

A plain hamburger weighs 175 to 200 grams.

Work Problem 4 at the Side. ▶

---

3 Convert using the metric conversion line or unit fractions.

GS (a) 9 L to mL

On the conversion line, L to mL is <u>three</u> places to the right **or** use the unit fraction

$$\frac{1000 \; mL}{1 \; L}$$

9 L = _____ mL

(b) 0.75 L to mL

(c) 500 mL to L

(d) 5 mL to L

(e) 2.07 L to mL

(f) 3275 mL to L

VOCABULARY TIP

gram The weight of a nickel (**5** cents) is **5** grams.

4 Circle the things that weigh about 1 gram.

A small paper clip

A pair of scissors

One playing card from a deck of cards

A calculator

An average-sized apple

A five-dollar bill

Answers

3. **(a)** 9000 mL **(b)** 750 mL **(c)** 0.5 L
   (d) 0.005 L **(e)** 2070 mL **(f)** 3.275 L

4. paper clip, playing card, five-dollar bill

5 Write the most reasonable metric unit in each blank. Choose from kg, g, and mg.

(a) A thumbtack weighs 800 ____ .

(b) A teenager weighs 50 ____ .

(c) This large cast-iron frying pan weighs 1 ____ .

(d) Jerry's basketball weighed 600 ____ .

(e) Tamlyn takes a 500 ____ calcium tablet every morning.

(f) On his diet, Greg can eat 90 ____ of meat for lunch.

(g) One strand of hair weighs 2 ____ .

(h) One banana might weigh 150 ____ .

To make larger or smaller weight units, we use the same **prefixes** as we did with length and capacity units. For example, the prefix *kilo-* means 1000 so a *kilo*meter is 1000 meters, a *kilo*liter is 1000 liters, and a *kilo*gram is 1000 grams.

Prefix	*kilo-*gram	*hecto-*gram	*deka-*gram	gram	*deci-*gram	*centi-*gram	*milli-*gram
Meaning	1000 grams	100 grams	10 grams	1 gram	$\frac{1}{10}$ of a gram	$\frac{1}{100}$ of a gram	$\frac{1}{1000}$ of a gram
Symbol	kg	hg	dag	g	dg	cg	mg

Weight (mass) units that are used most often

The units you will use most often in daily life are kilograms (kg), grams (g), and milligrams (mg). *Kilo*grams are used instead of pounds. A kilogram is 1000 grams. It is about 2.2 pounds. Two packages of butter plus one stick of butter weigh about 1 kg. An average newborn baby weighs 3 to 4 kg; a college football player might weigh 100 to 130 kg.

1 kilogram is about 2.2 pounds   100 to 130 kg   3 to 4 kg

Extremely small weights are measured in *milli*grams. It takes 1000 mg to make 1 g. Recall that a dollar bill weighs about 1 g. Imagine cutting it into 1000 pieces; the weight of one tiny piece would be 1 mg. Dosages of medicine and vitamins are given in milligrams. You will also use milligrams in science classes.

Cut a dollar bill into 1000 pieces. One tiny piece weighs 1 milligram.

EXAMPLE 3   **Using Metric Weight Units**

Write the most reasonable metric unit in each blank. Choose from kg, g, and mg.

(a) Ramon's suitcase weighed 20 _____ .

Write 20 **kg** because kilograms are used instead of pounds; 20 kg is about 44 pounds.

(b) LeTia took a 350 _____ aspirin tablet.

Write 350 **mg** because 350 g would be more than the weight of a hamburger, which is too much.

(c) Jenny mailed a letter that weighed 30 _____ .

Write 30 **g** because 30 kg would be much too heavy and 30 mg is less than the weight of a dollar bill.

◀ **Work Problem 5 at the Side.**

OBJECTIVE ▶ 4 Convert among metric weight (mass) units. As with length and capacity, you can convert among metric weight units by using unit fractions. The unit fractions you need are shown here.

Metric Weight (Mass) Relationships

1 kg = 1000 g, so the unit fractions are:

$$\frac{1 \text{ kg}}{1000 \text{ g}} \quad \text{or} \quad \frac{1000 \text{ g}}{1 \text{ kg}}$$

1 g = 1000 mg, so the unit fractions are:

$$\frac{1 \text{ g}}{1000 \text{ mg}} \quad \text{or} \quad \frac{1000 \text{ mg}}{1 \text{ g}}$$

Or you can use a metric conversion line to decide how to move the decimal point.

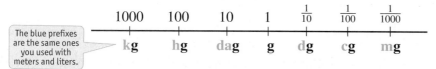

The blue prefixes are the same ones you used with meters and liters.

| 1000 | 100 | 10 | 1 | $\frac{1}{10}$ | $\frac{1}{100}$ | $\frac{1}{1000}$ |
| kg | hg | dag | g | dg | cg | mg |

EXAMPLE 4 Converting among Metric Weight Units

Convert using the metric conversion line or unit fractions.

(a) 7 mg to g

Using the conversion line:
From **mg** to **g** is *three places* to the *left*.

7.   007.

↑
Decimal point starts here.   Move decimal point three places to the left.

7 mg = **0.007 g**

Do **not** write a period here.

Using unit fractions:

Multiply by a unit fraction that allows you to divide out mg.

$$\frac{7 \text{ mg}}{1} \cdot \frac{1 \text{ g}}{1000 \text{ mg}} = \frac{7}{1000} \text{ g}$$

$$= \textbf{0.007 g}$$

Three decimal places for thousandths.

(b) 13.72 kg to g

Using the conversion line:
From **kg** to **g** is *three places* to the *right*.

13.720   Decimal point moves three places to the right.

13.72 kg = **13,720 g**

This is a comma (**not** a decimal point).

Using unit fractions:

Multiply by a unit fraction that allows you to divide out kg.

$$\frac{13.72 \text{ kg}}{1} \cdot \frac{1000 \text{ g}}{1 \text{ kg}} = \textbf{13,720 g}$$

This is a comma (**not** a decimal point).

Work Problem **6** at the Side. ▶

---

6 Convert using the metric conversion line or unit fractions.

GS (a) 10 kg to g

On the conversion line, kg to g is _____ places to the _____ **or** use

the unit fraction $\dfrac{1000 \text{ g}}{1 \text{ kg}}$

10 kg = _____ g

(b) 45 mg to g

(c) 6.3 kg to g

(d) 0.077 g to mg

(e) 5630 g to kg

(f) 90 g to kg

Answers

6. (a) three places to the right; 10,000 g
(b) 0.045 g **(c)** 6300 g **(d)** 77 mg
(e) 5.63 kg **(f)** 0.09 kg

7 First decide which type of units
are needed: length, capacity, or
weight. Then write the most
appropriate unit in the blank.
Choose from km, m, cm, mm,
L, mL, kg, g, and mg.

(a) Gail bought a 4 _____
can of paint.

Use _____ units.

(b) The bag of chips weighed
450 _____.

Use _____ units.

(c) Give the child 5 _____ of
cough syrup.

Use _____ units.

(d) The width of the window is
55 _____.

Use _____ units.

(e) Akbar drives 18 _____ to
work.

Use _____ units.

(f) The laptop computer
weighs 2 _____.

Use _____ units.

(g) A credit card is 55 _____
wide.

Use _____ units.

Answers

7. (a) L; capacity (b) g; weight
   (c) mL; capacity (d) cm; length
   (e) km; length (f) kg; weight
   (g) mm; length

OBJECTIVE **5** **Distinguish among basic metric units of length, capacity, and weight (mass).** As you encounter things to be measured at home, on the job, or in your classes at school, be careful to use the correct type of measurement unit.

Use *length units* (kilometers, meters, centimeters, millimeters) to measure:

how long	how high	how far away
how wide	how tall	how far around (perimeter)
how deep	distance	

Use *capacity units* (liters, milliliters) to measure liquids (things that can be poured) such as:

water	shampoo	gasoline
milk	perfume	oil
soft drinks	cough syrup	paint

Also use liters and milliliters to describe how much liquid something can hold, such as an eyedropper, measuring cup, pail, or bathtub.

Use *weight units* (kilograms, grams, milligrams) to measure:

the weight of something   how heavy something is

EXAMPLE 5   **Using a Variety of Metric Units**

First decide which type of units are needed: length, capacity, or weight. Then write the most appropriate metric unit in the blank. Choose from km, m, cm, mm, L, mL, kg, g, and mg.

(a) The letter needs another stamp because it weighs 40 _____.
Use _____ units.

Use **weight** units because of the word "weighs."

The letter weighs 40 **g** because 40 mg is less than the weight of a dollar bill and 40 kg would be about 88 pounds.

(b) The swimming pool is 3 _____ deep at the deep end.
Use _____ units.

Use **length** units because of the word "deep."

The pool is 3 **m** deep because 3 cm is only about an inch and 3 km is more than a mile.

(c) This is a 340 _____ can of juice.
Use _____ units.

Use **capacity** units because juice is a liquid.

It is a 340-**mL** can because 340 liters would be more than 340 quarts.

◄ Work Problem **7** at the Side.

8.3 Exercises

FOR EXTRA HELP

Go to MyMathLab for worked-out, step-by-step solutions to exercises enclosed in a square ▢ and video solutions to ▶ exercises.

CONCEPT CHECK *Fill in the blank with the best metric unit to measure each item. Choose from liters, milliliters, kilograms, grams, and milligrams.*

1. **(a)** For a dose of cough syrup, use _____.

    (b) For a large carton of milk, use _____.

    (c) For a vitamin pill, use _____.

    (d) For a heavyweight wrestler, use _____.

2. **(a)** For a serving of vegetables, use _____.

    (b) For a large bottle of soda, use _____.

    (c) For a travel size bottle of shampoo, use _____.

    (d) For a pain reliever tablet, use _____.

Write the most reasonable metric unit in each blank. Choose from L, mL, kg, g, and mg.
See Examples 1 and 3.

3. The glass held 250 _____ of water.

4. Hiromi used 12 _____ of water to wash the kitchen floor.

5. Dolores can make 10 _____ of soup in that pot.

6. Jay gave 2 _____ of vitamin drops to the baby.

7. ▶ Our yellow Labrador dog grew up to weigh 40 _____.

8. A small safety pin weighs 750 _____.

9. Lori caught a small sunfish weighing 150 _____.

10. One dime weighs 2 _____.

11. ▶ Andre donated 500 _____ of blood today.

12. Barbara bought the 2 _____ bottle of cola.

13. ▶ The patient received a 250 _____ tablet of medication each hour.

14. The 8 people on the elevator weighed a total of 500 _____.

15. The gas can for the lawn mower holds 4 _____.

16. Kevin poured 10 _____ of vanilla into the mixing bowl.

17. Kingston's backpack weighs 5 _____ when it is full of books.

18. One grain of salt weighs 2 _____.

CONCEPT CHECK *Today, medical prescriptions are usually given in the metric system. But sometimes a mistake is made. Indicate whether each dose is* reasonable *or* unreasonable. *If a dose is unreasonable, indicate whether it is* too much *or* too little.

19. Drink 4.1 L of Kaopectate after each meal.

20. Drop 1 mL of solution into the eye twice a day.

21. Soak your feet in 5 kg of Epsom salts per liter of water.

22. Inject 0.5 L of insulin each morning.

23. Take 15 mL of cough syrup every four hours.

24. Take 200 mg of vitamin C each day.

25. Take 350 mg of aspirin three times a day.

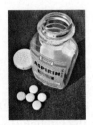

26. Buy a tube of ointment weighing 0.002 g.

27. CONCEPT CHECK The work shown below has a mistake in it.

What Went Wrong? First write a sentence explaining what the mistake is. Then fix the mistake and find the correct solution.

Marta converted 6.5 kg to grams this way:

$$\frac{6.5 \text{ kg}}{1} \cdot \frac{1 \text{ kg}}{1000 \text{ g}} = 0.0065 \text{ g}$$

28. CONCEPT CHECK The work shown below has a mistake in it.

What Went Wrong? First write a sentence explaining what the mistake is. Then fix the mistake and find the correct solution.

Maxwell converted 20 mg to grams this way:

$$20 \text{ mg} = 000.20 = 0.0002 \text{ g}$$

Convert each measurement. Use unit fractions or the metric conversion line. **See Examples 2 and 4.**

29. 15 L to mL

(GS)

unit fraction} $\dfrac{1000 \text{ mL}}{1 \text{ L}}$ **or** On conversion line, L to mL is _____ places to the _____.

15 L = _____ mL

30. 6 L to mL

(GS)

unit fraction} $\dfrac{1000 \text{ mL}}{1 \text{ L}}$ **or** On the conversion line, L to mL is _____ places to the _____.

6 L = _____ mL

31. 3000 mL to L

32. 18,000 mL to L

33. 925 mL to L

34. 200 mL to L

35. 8 mL to L

36. 25 mL to L

37. 4.15 L to mL

38. 11.7 L to mL

39. 8000 g to kg

40. 25,000 g to kg

41. 5.2 kg to g

42. 12.42 kg to g

43. 0.85 g to mg

44. 0.2 g to mg

45. 30,000 mg to g

46. 7500 mg to g

47. 598 mg to g

48. 900 mg to g

49. 60 mL to L

50. 6.007 kg to g

51. 3 g to kg

52. 12 mg to g

53. 0.99 L to mL

54. 13,700 mL to L

*Write the most appropriate metric unit in each blank. Choose from km, m, cm, mm, L, mL, kg, g, and mg. **See Example 5.***

55. The masking tape is 19 _____ wide.

56. The roll has 55 _____ of tape on it.

57. Buy a 60 _____ jar of acrylic
   ⏵ paint for art class.

58. One onion weighs
   200 _____ .

59. My waist measurement is 65 _____ .

60. Add 2 _____ of windshield washer fluid to your car.

61. A single postage stamp weighs 90 _____ .
   ⏵

62. The hallway is 10 _____ long.

Solve each application problem. (Data from The Human Body.*)*

63. Human skin has about 3 million sweat glands, which release an average of 300 mL of sweat per day. How many liters of sweat are released each day?

64. On a hot day, the sweat glands in a person's skin may release up to 1.2 L of sweat in one day. How many milliliters is that?

65. The average weight of an adult human brain is 1.34 kg. An adult human's skin weighs about 9 kg. How many grams does each body part weigh?

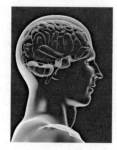

66. A healthy human heart pumps between 70 mL and 230 mL of blood per beat depending on the person's activity level. How many liters of blood does the heart pump per beat?

67. On average, we breathe in and out roughly 500 mL of air every 4 seconds. How many liters of air is that?

68. In the Victorian era, people believed that heavier brains meant greater intelligence. They were impressed that Otto von Bismarck's brain weighed 1907 g, which is how many kilograms?

69. A small adult cat weighs from 3000 g to 4000 g. How many kilograms is that? (Data from Lyndale Animal Hospital.)

70. If the letter you are mailing weighs 29 g, you must put additional postage on it. How many kilograms does the letter weigh? (Data from U.S. Postal Service.)

71. CONCEPT CHECK Is 1005 mg greater than or less than 1 g? What is the difference in the weights?

72. CONCEPT CHECK Is 990 mL greater than or less than 1 L? What is the difference in the amounts?

73. One nickel weighs 5 g. How many nickels are in 1 kg of nickels?

74. The ratio of the total length of all the fish to the amount of water in an aquarium can be 3 cm of fish for every 4 L of water. What is the total length of all the fish you can put in a 40 L aquarium? (Data from *Tropical Aquarium Fish*.)

Relating Concepts (Exercises 75–78) For Individual or Group Work

*Recall that the prefix **kilo** means 1000, so a **kilo**meter is 1000 meters. You'll learn about other prefixes for numbers greater than 1000 as you **work Exercises 75–78 in order**.*

75. (a) The prefix *mega* means one million. Use the symbol M (capitalized) for *mega*. So a *mega*meter (Mm) is how many meters?

1 Mm = _____ m

(b) Figure out a unit fraction that you can use to convert megameters to meters. Then use it to convert 3.5 Mm to meters.

76. (a) The prefix *giga* means one billion. Use the symbol G (capitalized) for *giga*. So a *giga*meter (Gm) is how many meters?

1 Gm = _____ m

(b) Figure out a unit fraction you can use to convert meters to gigameters. Then use it to convert 2500 m to gigameters.

77. (a) The prefix *tera* means one trillion. Use the symbol T (capitalized) for *tera*. So a *tera*meter (Tm) is how many meters?

1 Tm = _____ m

(b) Think carefully before you fill in the blanks:

1 Tm = _____ Gm

1 Tm = _____ Mm

78. A computer's memory is measured in *bytes*. A byte can represent a single letter, digit, or punctuation mark. The memory may be measured in megabytes (MB), gigabytes (GB), or terabytes (TB). Using the meanings of *mega*, *giga*, and *tera*, you'd think that

1 MB = _____ bytes

1 GB = _____ bytes

1 TB = _____ bytes

However, because computers use a base 2 or binary system, 1 MB is actually 2^{20}, 1 GB is 2^{30}, and 1 TB is 2^{40}. Use your calculator to find the actual values.

2^{20} = _____ 2^{30} = _____

2^{40} = _____   If your calculator does not show enough digits, ask your instructor for help.

Summary Exercises *U.S. and Metric Measurement Units*

CONCEPT CHECK *The most commonly used U.S. and metric measurement units are listed in mixed-up order. For* **Exercises 1 and 2** *write each unit in the correct box in the tables below. Then write the abbreviation or symbol next to each unit.*

pound	yard	liter	kilogram	ton	gallon	pint
millimeter	gram	quart	meter	inch	cup	centimeter
milliliter	mile	foot	milligram	kilometer	ounce	fluid ounce

1.

U.S. Measurement Units and Abbreviations		
Length	**Weight**	**Capacity**

2.

Metric Measurement Units and Symbols		
Length	**Weight**	**Capacity**

Fill in the blanks with the measurement relationships that you memorized.

3. (a) 1 _____ = 12 in.

  (b) 3 ft = 1 _____

  (c) 1 mi = _____ ft

4. (a) 60 sec = 1 _____

  (b) 1 hr = _____ min

  (c) _____ hr = 1 day

5. (a) 1 cup = _____ fl oz

  (b) 4 qt = 1 _____

  (c) _____ pt = 1 qt

6. (a) 16 oz = 1 _____

  (b) _____ lb = 1 ton

  (c) 1 lb = _____ oz

Write the most reasonable metric unit in each blank. Choose from km, m, cm, mm, L, mL, kg, g, and mg.

7. My water bottle holds 450 _____ .

8. Michael won the 200 _____ race today.

9. The child weighed 23 _____ .

10. Jifar took a 375 _____ aspirin tablet.

11. The red pen is 14 _____ long.

12. A dime is about 18 _____ across.

13. Merlene made 12 _____ of fruit punch for her daughter's birthday party.

14. This cereal has 4 _____ of protein in each serving.

Convert each measurement using unit fractions or the metric conversion line. Show your work.

15. 45 cm to meters

16. $\frac{3}{4}$ min to seconds

17. 0.6 L to milliliters

18. 8 g to milligrams

19. 300 mm to centimeters

20. 45 in. to feet

21. 50 mL to liters

22. 18 qt to gallons

23. 7.28 kg to grams

24. $2\frac{1}{4}$ lb to ounces

25. 9 g to kilograms

26. 5 yd to inches

Solve Exercises 27–30 using the data in the table at the right on some of the world's tallest people. (Data from Guinness World Records.)

27. What is the height of the tallest person in centimeters? In millimeters?

Robert Wadlow was 2.72 m tall.

WORLD'S TALLEST PEOPLE

Rank	Name/Dates/Country	Height
1st	Robert Wadlow (1918–1940) USA	2.72 m
2nd	John Rogan (1868–1905) USA	268 cm
3rd	John Carroll (1932–1969) USA	264 cm
4th	Sultan Kösen (1982–) Turkey	2.5 m
10th	Jeng Jinlian (tallest female) (1964–1982) China	248 cm

28. Find the 3rd tallest person's height in meters and in kilometers.

29. How much taller is the tallest person than the second tallest, in meters? Convert this difference to centimeters and millimeters.

30. What is the difference in height between the tallest person and the 10th tallest person, in centimeters? Convert the height difference to meters and millimeters.

Solve each application problem.

31. Vernice bought a 12 oz bag of chips for $3.89 today. What was the price per pound to the nearest cent?

32. Miami, Florida, gets an average of 62 in. of rain every year. At that rate, how many feet of rain does Miami get during two years? Round your answer to the nearest tenth.

8.4 | Problem Solving with Metric Measurement

OBJECTIVE ▶ ❶ Solve application problems involving metric measurements.
One advantage of the metric system is the ease of comparing measurements in application situations. Just be sure that you are comparing similar units: mg to mg, km to km, and so on.

The example below shows you the steps to use when solving application problems.

OBJECTIVE

❶ Solve application problems involving metric measurements.

EXAMPLE 1 | **Solving a Metric Application**

Ground turkey is on sale at $8.99 per kilogram. Jake bought 350 g of the turkey. How much did he pay, to the nearest cent?

Step 1 **Read** the problem. The problem asks for the cost of 350 g of turkey.

Step 2 **Work out a plan.** The price is $8.99 per *kilogram*, but the amount Jake bought is given in *grams*. Convert grams to kilograms (the unit in the price). Then multiply the weight by the cost per kilogram.

Step 3 **Estimate** a reasonable answer. Round the cost of 1 kg from $8.99 to $9. There are 1000 g in a kilogram, so 350 g is about $\frac{1}{3}$ of a kilogram. Jake is buying about $\frac{1}{3}$ of a kilogram, and $\frac{1}{3}$ of $9 is $3. So our estimate is $3.

Step 4 **Solve** the problem. Use a unit fraction to convert 350 g to kilograms.

> **g** divides out, leaving **kg** for your answer.

$$\frac{350 \cancel{g}}{1} \cdot \frac{1 \text{ kg}}{1000 \cancel{g}} = \frac{350}{1000} \text{ kg} = 0.35 \text{ kg}$$

Now multiply 0.35 kg times the cost per kilogram.

$$\frac{0.35 \cancel{kg}}{1} \cdot \frac{\$8.99}{1 \cancel{kg}} = \$3.1465 \approx \$3.15 \text{ (rounded)}$$

> Nearest cent is the nearest hundredth

Step 5 **State the answer.** Jake paid **$3.15** (rounded to the nearest cent).

Step 6 **Check** your work. The exact answer of $3.15 is close to our estimate of $3.

—————————— Work Problem ❶ at the Side. ▶

EXAMPLE 2 | **Solving a Metric Application**

Olivia has 2.6 m of lace. How many centimeters of lace can she use to trim each of six hair ornaments? Round to the nearest tenth of a centimeter.

Step 1 **Read** the problem. The problem asks for the number of centimeters of lace for each of six hair ornaments.

Step 2 **Work out a plan.** The given amount of lace is in *meters*, but the answer must be in *centimeters*. Convert meters to centimeters, then divide by 6 (the number of hair ornaments).

—————————— Continued on Next Page

---

❶ Solve this application problem.

GS Satin ribbon is on sale at $0.89 per meter. How much will 75 cm cost, to the nearest cent?

Step 1 The problem asks for the cost of 75 _____ of ribbon.

Step 2 The price is $0.89 per _____. So, convert cm to _____ (the unit in the price).

Step 3 There are 100 cm in a meter, so 75 cm will cost a little less than $0.89 (the price of a meter of ribbon).

Step 4 Finish the solution.

$$\frac{75 \cancel{cm}}{1} \cdot \frac{1 \text{ m}}{100 \cancel{cm}} = \frac{75}{100} \text{ m} = \_\_\_\_$$

Multiply the number of meters of ribbon by the cost per meter.

$$\frac{\boxed{} \text{ m}}{1} \cdot \frac{\$\boxed{}}{1 \text{ m}} = \$\_\_\_\_$$

Step 5 **State the answer.** 75 cm of ribbon costs _____

Step 6 **Check** your work.

Answer

1. *Step 1* cm;   *Step 2* m; m

Step 4 $\frac{75}{100}$ m = 0.75 m;

$\frac{0.75 \cancel{m}}{1} \cdot \frac{\$0.89}{1 \cancel{m}} = \$0.6675 \approx \$0.67$ (rounded)

Step 5 $0.67

Step 6 The exact answer is a little less than $0.89, which fits the estimate.

2 Lucinda's doctor wants her to take 1.2 g of medication each day in three equal doses. How many milligrams should be in each dose?

Step 3 **Estimate** a reasonable answer. To estimate, round 2.6 m of lace to 3 m. Then, 3 m = 300 cm, and 300 cm ÷ 6 = 50 cm is our estimate.

Step 4 **Solve** the problem. On the metric conversion line, moving from **m** to **cm** is two places to the right, so move the decimal point in 2.6 m two places to the right. Then divide by 6.

$$2.60 \text{ m} = 260 \text{ cm} \qquad \frac{260 \text{ cm}}{6 \text{ ornaments}} \approx 43.3 \text{ cm per ornament}$$

Step 5 **State the answer.** Olivia can use **43.3 cm** (rounded) of lace on each ornament.

> Be sure to write cm in your answer.

Step 6 **Check** your work. The exact answer of 43.3 cm is close to our estimate of 50 cm.

──────── ◄ **Work Problem 2 at the Side.**

Note

In **Example 1** we used a unit fraction to convert the measurement; in **Example 2** we moved the decimal point. Use whichever method you prefer.

Also, there is more than one way to solve an application problem. Another way to solve **Example 2** is to divide 2.6 m by 6 to get 0.4333 m of lace for each ornament. Then convert 0.4333 m to 43.3 cm (rounded to the nearest tenth).

3 Andrea has two pieces of fabric. One measures 2 m 35 cm and the other measures 1 m 85 cm. How many meters of fabric does she have in all?

EXAMPLE 3 **Solving a Metric Application**

Rubin measured a board and found that the length was 3 m plus an additional 5 cm. He cut off a piece measuring 1 m 40 cm for a shelf. Find the length in meters of the remaining piece of board.

Step 1 **Read** the problem. Part of a board is cut off. The problem asks what length of board, in meters, is left over. It may help to make a drawing of the board and label the lengths given in the problem.

Step 2 **Work out a plan.** The lengths involve two units, m and cm. Rewrite both lengths in meters (the unit called for in the answer), and then subtract.

Step 3 **Estimate** a reasonable answer. First, 3 m 5 cm can be rounded to 3 m, because 5 cm is less than half of a meter (less than 50 cm). Round 1 m 40 cm down to 1 m. Then, 3 m − 1 m = 2 m is our estimate.

Step 4 **Solve** the problem. Rewrite the lengths in meters. Then subtract.

$$
\begin{array}{ll}
3 \text{ m} \rightarrow & 3.00 \text{ m} \\
\text{plus } 5 \text{ cm} \rightarrow & + 0.05 \text{ m} \\
\hline
05. & 3.05 \text{ m}
\end{array}
\qquad
\begin{array}{ll}
40. \quad 1 \text{ m} \rightarrow & 1.0 \text{ m} \\
\text{plus } 40 \text{ cm} \rightarrow & + 0.4 \text{ m} \\
\hline
& 1.4 \text{ m}
\end{array}
$$

$$
\begin{array}{ll}
& 3.05 \text{ m} \leftarrow \text{Board} \\
\text{Subtract to find leftover length.} & - 1.40 \text{ m} \leftarrow \text{Shelf} \\
\hline
& 1.65 \text{ m} \leftarrow \text{Leftover piece}
\end{array}
$$

Step 5 **State the answer.** The remaining piece is **1.65 m** long.

Step 6 **Check** your work. The exact answer of 1.65 m is close to our estimate of 2 m.

──────── ◄ **Work Problem 3 at the Side.**

Answers

2. 400 mg per dose

3. 4.2 m

8.4 Exercises

FOR EXTRA HELP   Go to MyMathLab *for worked-out, step-by-step solutions to exercises enclosed in a square* and *video solutions to* ▶ *exercises.*

1. **CONCEPT CHECK** *Step 5* in the steps for solving application problems is **State the answer.** Which of these is an important thing to include in the answer?

   (a) Your estimated answer.

   (b) How many steps you followed.

   (c) The units, such as cm, kg, $, and so on.

2. **CONCEPT CHECK** *Step 6* in the steps for solving application problems is **Check** your work. How can you tell if your answer is reasonable?

   (a) Take a guess.

   (b) Compare your exact answer to the estimated answer.

   (c) Use your calculator.

Solve each application problem. Round money answers to the nearest cent.
See Examples 1–3.

3. Bulk rice at the food co-op is on special at $0.98 per kilogram. Pam scooped some rice into a bag and put it on the scale. How much will she pay for 850 g of rice?

   The price is $0.98 per _____, so convert 850 g to _____.

4. Lanh is buying a piece of plastic tubing measuring 315 cm for the science lab. The price is $5.50 per meter. How much will Lanh pay?

   The price is $5.50 per _____, so convert 315 cm to _____.

5. A miniature Yorkshire terrier, one of the smallest dogs, may weigh only 500 g. But a St. Bernard, the heaviest dog, could easily weigh 90 kg. What is the difference in the weights of the two dogs, in kilograms? (Data from *Big Book of Knowledge.*)

6. The world's longest insect is the giant stick insect of Borneo, measuring 57 cm. The fairy fly, the smallest insect, is just 0.2 mm long. How much longer is the giant stick insect, in millimeters? (Data from www.worldwildlife.org)

7. An adult human body contains about 5 L of blood. If each beat of the heart pumps 70 mL of blood, how many times must the heart beat to pass all the blood through the heart? Round to the nearest whole number of beats. (Data from *The Human Body.*)

8. A floor tile measures 30 cm by 30 cm and weighs 185 g. How many kilograms would a stack of 24 tiles weigh? How much would five stacks of tiles weigh? (Data from The Tile Shop.)

9. Each piece of lead for a mechanical pencil has a thickness of 0.5 mm and is 60 mm long. Find the total length in centimeters of the lead in a package with 30 pieces. If the price of the package is $3.29, find the cost per centimeter for the lead. (Data from Pentel.)

10. An apartment building caretaker puts 750 mL of chlorine into the swimming pool every day. How many liters should he order to have a one-month (30-day) supply on hand? If chlorine is sold in containers that hold 4 L, how many containers should be ordered for one month? How much chlorine will be left over at the end of the month?

11. Rosa is building a bookcase. She has one board that is 2 m 8 cm long and another that is 2 m 95 cm long. What is the total length of the two boards in meters?

12. Janet has a piece of fabric that is 10 m 30 cm in length. She wants to make curtains for three windows that are all the same size. What length of fabric can she use for each window, to the nearest tenth of a meter?

13. In a chemistry lab, each of the 45 students needs
⏵ 85 mL of acid. How many 1 L bottles of acid need
to be ordered?

14. James needs two 1.3 m pieces and two 85 cm pieces
of wood molding to frame a picture. The price is
$5.89 per meter plus 7% sales tax. How much will
James pay?

Use the bar graph below to answer Exercises 15 and 16.

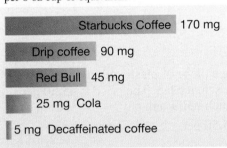

Caffeine Amounts
Average milligrams of caffeine
per 8 oz cup or equivalent

Starbucks Coffee 170 mg

Drip coffee 90 mg

Red Bull 45 mg

25 mg Cola

5 mg Decaffeinated coffee

Data from www.cspinet.org

15. If Agnete usually drinks three 8 oz cups of drip
coffee each day, how many grams of caffeine will
she consume in one week?

16. Lorenzo's doctor suggested that he cut down on
caffeine. So Lorenzo switched from drinking four
8 oz cups of cola every day to drinking two 8 oz cups
of decaffeinated coffee. How many fewer grams of
caffeine is he consuming each week?

17. The average winter snowfall in Boston is 111 cm.
Lander, Wyoming, usually gets 2.6 m of snow each
winter. Compare the yearly snowfall in the two cities.

(a) What is the difference for one year in meters?

(b) What is the difference over two years in
centimeters?
(Data from www.currentresults.com)

18. Many football stadiums have Field Turf instead of
grass. Use the drawing below to find the total thickness
in centimeters of the top two layers of Field Turf.

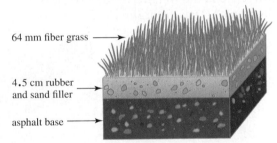

64 mm fiber grass →

4.5 cm rubber
and sand filler →

asphalt base →

Data from Sports Facilities Commission.

Relating Concepts (Exercises 19–22) For Individual or Group Work

*It is difficult to weigh very light objects, such as a single sheet of paper or a single
staple (unless you have an expensive scientific scale). But you can weigh a large
number of the items and then divide to find the weight of one item. Before dividing,
subtract the weight of the box or wrapper that the items are packaged in to find the
net weight.* **Work Exercises 19–22 in order,** *to complete the table.*

Item	Total Weight in Grams	Weight of Packaging	Net Weight	Weight of One Item in Grams	Weight of One Item in Milligrams
19. Box of 50 envelopes	255 g	40 g	_____	_____	_____
20. Box of 1000 staples	350 g	20 g	_____	_____	_____
21. Package of paper (500 sheets)	_____	50 g	_____	_____	3000 mg
22. Box of 100 small paper clips	_____	5 g	_____	_____	500 mg

8.5 Metric–U.S. Measurement Conversions and Temperature

OBJECTIVE ▶ ① Use unit fractions to convert between metric and U.S. measurement units. Until the United States has switched completely to the metric system, it will be necessary to make conversions between U.S. and metric units. *Approximate* conversions can be made with the help of the table below, in which the values have been rounded to the nearest hundredth or thousandth. (The only value that is exact, not rounded, is 1 inch = 2.54 cm.)

Metric to U.S. Units	U.S. to Metric Units
1 kilometer ≈ 0.62 mile	1 mile ≈ 1.61 kilometers
1 meter ≈ 1.09 yards	1 yard ≈ 0.91 meter
1 meter ≈ 3.28 feet	1 foot ≈ 0.30 meter
1 centimeter ≈ 0.39 inch	1 inch = 2.54 centimeters
1 liter ≈ 0.26 gallon	1 gallon ≈ 3.79 liters
1 liter ≈ 1.06 quarts	1 quart ≈ 0.95 liter
1 kilogram ≈ 2.20 pounds	1 pound ≈ 0.45 kilogram
1 gram ≈ 0.035 ounce	1 ounce ≈ 28.35 grams

EXAMPLE 1 Converting between Metric and U.S. Units of Length

Convert 10 m to yards using unit fractions. Round your final answer to the nearest tenth if necessary.

We're changing from a *metric length* unit to a *U.S. length* unit. In the "Metric to U.S. Units" side of the table, you see that 1 meter ≈ 1.09 yards. Two unit fractions can be written using that information.

$$\frac{1 \text{ m}}{1.09 \text{ yd}} \quad \text{or} \quad \frac{1.09 \text{ yd}}{1 \text{ m}}$$

Multiply by the unit fraction that allows you to divide out meters (that is, meters is in the denominator).

$$10 \text{ m} \cdot \frac{1.09 \text{ yd}}{1 \text{ m}} = \frac{10 \text{ m}}{1} \cdot \frac{1.09 \text{ yd}}{1 \text{ m}} = \frac{(10)(1.09 \text{ yd})}{1} = 10.9 \text{ yd}$$

These units should match.

Meters (**m**) divide out leaving **yd**, the unit you want for the answer.

10 m ≈ **10.9 yd**

Note

In **Example 1** above, you could also use the numbers from the "U.S. to Metric Units" side of the table that involve meters and yards:

$$\frac{10 \text{ m}}{1} \cdot \frac{1 \text{ yd}}{0.91 \text{ m}} = \frac{10}{0.91} \text{ yd} \approx 10.99 \text{ yd}$$

The answer is slightly different because the values in the table are rounded. Also, you have to divide instead of multiply, which is usually more difficult to do without a calculator. We will use the first method in this chapter.

Work Problem ① at the Side. ▶

OBJECTIVES

① Use unit fractions to convert between metric and U.S. measurement units.

② Learn common temperatures on the Celsius scale.

③ Use formulas to convert between Celsius and Fahrenheit temperatures.

① Convert using unit fractions. Round your final answers to the nearest tenth.

(a) 23 m to yards (Look at the "Metric to U.S. Units" side of the table.)

(b) 40 cm to inches

(c) 5 mi to kilometers (Look at the "U.S. to Metric Units" side of the table.)

(d) 12 in. to centimeters

Answers

1. **(a)** 23 m ≈ 25.1 yd
(b) 40 cm ≈ 15.6 in.
(c) 5 mi ≈ 8.1 km
(d) 12 in. ≈ 30.5 cm

2 Convert. Use the values from the table on the previous page to make unit fractions. Round your final answers to the nearest tenth.

GS **(a)** 17 kg to pounds. From the table, 1 kg ≈ 2.20 lb

$$\frac{17 \cancel{kg}}{1} \cdot \frac{2.20 \text{ lb}}{1 \cancel{kg}} =$$

(b) 5 L to quarts

(c) 90 g to ounces

(d) 3.5 gal to liters

(e) 145 lb to kilograms

(f) 8 oz to grams

| EXAMPLE 2 | Converting between Metric and U.S. Units of Weight and Capacity |

Convert using unit fractions. Round your final answers to the nearest tenth.

(a) 3.5 kg to pounds

Look in the "Metric to U.S. Units" side of the table on the previous page to see that 1 kilogram ≈ 2.20 pounds. Use this information to write a unit fraction that allows you to divide out kilograms.

$$\frac{3.5 \cancel{kg}}{1} \cdot \frac{2.20 \text{ lb}}{1 \cancel{kg}} = \frac{(3.5)(2.20 \text{ lb})}{1} = 7.7 \text{ lb}$$

3.5 kg ≈ **7.7 lb** — We use the ≈ symbol because the conversion value is approximate.

(b) 18 gal to liters

Look in the "U.S. to Metric Units" side of the table to see that 1 gallon ≈ 3.79 liters. Write a unit fraction that allows you to divide out gallons.

gal divides out, leaving **L** for your answer. →
$$\frac{18 \cancel{gal}}{1} \cdot \frac{3.79 \text{ L}}{1 \cancel{gal}} = \frac{(18)(3.79 \text{ L})}{1} = 68.22 \text{ L}$$

68.22 rounded to the nearest tenth is 68.2
18 gal ≈ **68.2 L**

(c) 300 g to ounces

In the "Metric to U.S. Units" side of the table, 1 gram ≈ 0.035 ounce.

$$\frac{300 \cancel{g}}{1} \cdot \frac{0.035 \text{ oz}}{1 \cancel{g}} = \frac{(300)(0.035 \text{ oz})}{1} = 10.5 \text{ oz}$$ — **g** divides out, leaving **oz** for your answer.

300 g ≈ **10.5 oz**

> ❗ **CAUTION**
> Because the metric and U.S. measurement systems were developed independently, almost all comparisons are approximate. So we use the ≈ symbol to show that the conversions are approximate.

──────────── ◀ **Work Problem 2 at the Side.**

OBJECTIVE 2 Learn common temperatures on the Celsius scale. In the metric system, temperature is measured on the **Celsius** scale. On the Celsius scale, water freezes at 0 °C and boils at 100 °C. The small raised circle stands for "degrees" and the capital **C** is for Celsius. Read the temperatures like this:

Water freezes at 0 degrees Celsius (0 °C). — In the metric system, °C is the symbol for "degrees Celsius."
Water boils at 100 degrees Celsius (100 °C).

The U.S. temperature system, used only in the United States, is measured on the **Fahrenheit** scale. On this scale:

Water freezes at 32 degrees Fahrenheit (32 °F). — In the U.S. system, °F stands for "degrees Fahrenheit."
Water boils at 212 degrees Fahrenheit (212 °F).

Answers

2. (a) $\frac{(17)(2.20 \text{ lb})}{1} = 37.4 \text{ lb}$; 17 kg ≈ 37.4 lb
(b) 5 L ≈ 5.3 qt
(c) 90 g ≈ 3.2 oz (rounded)
(d) 3.5 gal ≈ 13.3 L (rounded)
(e) 145 lb ≈ 65.3 kg (rounded)
(f) 8 oz ≈ 226.8 g

The thermometer below shows some typical temperatures in both Celsius and Fahrenheit. For example, comfortable room temperature is about 20 °C or 68 °F, and normal body temperature is about 37 °C or 98.6 °F.

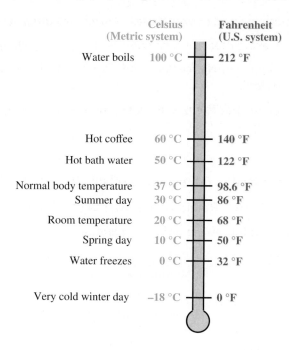

	Celsius (Metric system)	Fahrenheit (U.S. system)
Water boils	100 °C	212 °F
Hot coffee	60 °C	140 °F
Hot bath water	50 °C	122 °F
Normal body temperature	37 °C	98.6 °F
Summer day	30 °C	86 °F
Room temperature	20 °C	68 °F
Spring day	10 °C	50 °F
Water freezes	0 °C	32 °F
Very cold winter day	−18 °C	0 °F

Note

The freezing and boiling temperatures are exact. The other temperatures are approximate. Even normal body temperature varies slightly from person to person.

EXAMPLE 3 Using Celsius Temperatures

Circle the metric temperature that is most reasonable for each situation.

(a) Warm summer day   29 °C   64 °C   90 °C

29 °C is the reasonable temperature. 64 °C and 90 °C are too hot; they're both above the temperature of hot bath water (above 122 °F).

(b) Inside a freezer   −10 °C   3 °C   25 °C

−10 °C is the reasonable temperature because it is the only one below the freezing point of water (0 °C). Your frozen foods would start thawing at 3 °C or 25 °C.

─────── **Work Problem ❸ at the Side.** ▶

OBJECTIVE ❸ Use formulas to convert between Celsius and Fahrenheit temperatures. You can use these formulas to convert between Celsius and Fahrenheit temperatures.

Celsius–Fahrenheit Conversion Formulas

Converting from Fahrenheit (F) to Celsius (C)	Converting from Celsius (C) to Fahrenheit (F)
$C = \dfrac{5\,(F - 32)}{9}$	$F = \dfrac{9C}{5} + 32$

VOCABULARY TIP

Celsius (metric) temperatures used to be called **centi**grade temperatures, and **centi-** means hundred(th). Water boils at 100 degrees in the Celsius system.

❸ Circle the metric temperature that is *most* reasonable for each situation.

(a) Set the living room thermostat at:

11 °C   21 °C   71 °C

Hint: Look at the blue side of the thermometer in the text.

(b) The baby has a fever of:

29 °C   39 °C   49 °C

(c) Wear a sweater outside because it's:

15 °C   28 °C   50 °C

(d) My iced tea is:

−5 °C   5 °C   30 °C

(e) Let's go swimming! It's:

95 °C   65 °C   35 °C

(f) Inside a refrigerator (not the freezer) it's:

−15 °C   0 °C   3 °C

(g) There's a blizzard outside. It's:

12 °C   4 °C   −20 °C

(h) I need hot water to get these clothes clean. The water should be:

55 °C   105 °C   200 °C

Answers

3. (a) 21 °C   **(b)** 39 °C   **(c)** 15 °C
(d) 5 °C   **(e)** 35 °C   **(f)** 3 °C
(g) −20 °C   **(h)** 55 °C

4 Convert to Celsius. Round your answers to the nearest degree if necessary.

(a) 59 °F $\quad C = \dfrac{5(F - 32)}{9}$

$$C = \dfrac{5(59 - 32)}{9} \quad \text{Finish the work.}$$

(b) 20 °F

(c) 212 °F

(d) 98.6 °F

5 Convert to Fahrenheit. Round your answers to the nearest degree if necessary.

(a) 100 °C $\quad F = \dfrac{9C}{5} + 32$

$$F = \dfrac{9 \cdot 100}{5} + 32 \quad \text{Finish the work.}$$

(b) −25 °C

(c) 32 °C

(d) −18 °C

Answers

4. (a) $C = \dfrac{5(\overset{3}{\cancel{27}})}{\underset{1}{\cancel{9}}} = 15\,°C$ (b) −7 °C (rounded)

(c) 100 °C (d) 37 °C

5. (a) $F = \dfrac{9 \cdot \overset{20}{\cancel{100}}}{\underset{1}{\cancel{5}}} + 32 = 180 + 32 = 212\,°F$

(b) −13 °F (c) 90 °F (rounded)
(d) 0 °F (rounded)

As you use these formulas, be sure to follow the order of operations.

> **Order of Operations**
>
> 1. Do all operations inside **parentheses** or **other grouping symbols**.
> 2. Simplify any expressions with **exponents** and find any **square roots**.
> 3. Do the remaining **multiplications and divisions** as they occur from left to right.
> 4. Do the remaining **additions and subtractions** as they occur from left to right.

EXAMPLE 4 Converting Fahrenheit to Celsius (U.S. to Metric)

Convert 10 °F to Celsius. Round your answer to the nearest degree.

Use the formula and follow the order of operations.

$$C = \dfrac{5(F - 32)}{9} \quad \boxed{\text{Fahrenheit to Celsius formula}}$$

$$C = \dfrac{5(10 - 32)}{9} \quad \begin{array}{l}\text{Replace F with 10}\\ \text{Work inside parentheses first.}\\ 10 - 32 \text{ becomes } 10 + (-32)\end{array}$$

$$C = \dfrac{5(-22)}{9} \quad \begin{array}{l}\text{Multiply in numerator; positive times}\\ \text{negative gives a negative product.}\end{array}$$

$$C = \dfrac{-110}{9} \quad \begin{array}{l}\text{Divide; negative divided by positive gives}\\ \text{a negative quotient.}\end{array}$$

$$C = -12.\overline{2} \quad \text{Round to } -12 \text{ (nearest degree)}$$

Thus, 10 °F ≈ **−12 °C**

◀ **Work Problem 4 at the Side.**

EXAMPLE 5 Converting Celsius to Fahrenheit (Metric to U.S.)

Convert 15 °C to Fahrenheit.

Use the formula and follow the order of operations.

$$F = \dfrac{9C}{5} + 32 \quad \boxed{\text{Celsius to Fahrenheit formula}}$$

$$F = \dfrac{9 \cdot 15}{5} + 32$$

$$F = \dfrac{9 \cdot \overset{3}{\cancel{15}}}{\underset{1}{\cancel{5}}} + 32 \quad \begin{array}{l}\text{Divide 15 and 5 by their common factor:}\\ 15 \div 5 \text{ is 3, and } 5 \div 5 \text{ is 1}\\ \text{Multiply } 9 \cdot 3 \text{ in the numerator to get 27}\end{array}$$

$$F = 27 + 32 \quad \text{Add.}$$

$$F = 59$$

So, 15 °C = **59 °F**

◀ **Work Problem 5 at the Side.**

8.5 Exercises

FOR
EXTRA
HELP

Go to MyMathLab *for worked-out, step-by-step solutions to exercises enclosed in a square* ▢ *and video solutions to* ▶ *exercises.*

CONCEPT CHECK *Use the table on the first page of this section to answer Exercises 1 and 2.*

1. (a) Use only the "Metric to U.S." side of the table. Write **two** unit fractions using the information for converting between kilograms and pounds.

(b) When converting 5 kg to pounds, use the unit fraction from part (a) that allows you to divide out which unit? _____ Would you use the unit fraction with kg in the *numerator* or the *denominator*? _____

2. (a) Use only the "U.S. to Metric" side of the table. Write **two** unit fractions using the information for converting between quarts and liters.

(b) When converting 3 qt to liters, use the unit fraction from part (a) that allows you to divide out which unit? _____. Would you use the unit fraction with qt in the *numerator* or the *denominator*? _____

Use the table on the first page of this section and unit fractions to make approximate conversions from metric to U.S. units or U.S. to metric units. Round your final answers to the nearest tenth. **See Examples 1 and 2.**

3. 20 m to yards

$$\frac{20 \ \cancel{m}}{1} \cdot \frac{1.09 \text{ yd}}{1 \ \cancel{m}} =$$

4. 8 km to miles

$$\frac{8 \ \cancel{km}}{1} \cdot \frac{0.62 \text{ mi}}{1 \ \cancel{km}} =$$

5. 80 m to feet

6. 85 cm to inches

7. 16 ft to meters

8. 3.2 yd to meters

9. 150 g to ounces

10. 2.5 oz to grams

11. 248 lb to kilograms

12. 7.68 kg to pounds

13. 28.6 L to quarts

14. 15.75 L to gallons

15. For the 2000 Olympics, the 3M Company used 5 g of pure gold to coat Michael Johnson's track shoes. (Data from 3M Company.)

(a) How many ounces of gold were used, to the nearest tenth?

(b) Was this enough extra weight to slow him down?

16. The label on a Van Ness auto feeder for cats and dogs says it holds 1.4 kg of dry food. How many pounds of food does it hold, to the nearest tenth? (Data from Van Ness Plastics.)

17. The normal wash cycle in a dishwater uses 3.9 gal of water. How many liters does it use? Round your final answer to the nearest tenth. (Data from Kenmore.)

18. The quick-rinse cycle in a dishwasher uses 7.6 L of water. How many gallons does it use? Round your final answer to the nearest tenth. (Data from Kenmore.)

19. The smallest pet fish are dwarf gobies, which are half an inch long. How many centimeters long is a dwarf gobie, to the nearest tenth? (*Hint*: First write half an inch in decimal form.)

20. The fastest nerve signals in the human body travel 120 meters per second. How many feet per second do the signals travel? (Data from *The Human Body*.)

CONCEPT CHECK *Circle the most reasonable metric temperature for each situation.* *See Example 3.*

21. A snowy day

    12 °C   28 °C   −8 °C

22. Brewing coffee

    80 °C   180 °C   15 °C

23. A high fever

    21 °C   40 °C   103 °C

24. Swimming pool water

    90 °C   78 °C   25 °C

25. Oven temperature

    150 °C   50 °C   30 °C

26. Light jacket weather

    0 °C   10 °C   −10 °C

Use the conversion formulas from this section and the order of operations to convert Fahrenheit temperatures to Celsius, or Celsius temperatures to Fahrenheit. Round your final answers to the nearest degree if necessary. ***See Examples 4 and 5.***

27. 60 °F   Complete the conversion.

$$C = \frac{5(60-32)}{9} = \frac{5(28)}{9} =$$

28. 80 °F   Complete the conversion.

$$C = \frac{5(80-32)}{9} = \frac{5(48)}{9} =$$

29. −4 °F

30. 15 °F

31. 8 °C

32. 18 °C **33.** −5 °C **34.** 0 °C

35. CONCEPT CHECK The work shown below has **two** mistakes in it.

What Went Wrong? First write a sentence explaining what each mistake is. Then fix the mistakes and find the correct solution.

Floyd converted 42 °F to Celsius this way. He was told to round his answer to the nearest whole degree.

$$\frac{5(42-32)}{9} = \frac{210-32}{9} = \frac{178}{9} = 19.\overline{7} \approx 19\,°C$$

36. CONCEPT CHECK The work shown below has **two** mistakes in it.

What Went Wrong? First write a sentence explaining what each mistakes is. Then fix the mistakes and find the correct solution.

Lea converted −20 °C to Fahrenheit this way:

$$\frac{9(-20)}{5} + 32 = \frac{9(\overset{-4}{-20})}{\underset{1}{\cancel{5}}} + 32 = -4 + 32 = 28\,F$$

Solve each application problem. Round your final answers to the nearest degree if necessary.

37. (a) Here is the tag on a pair of boots. In what kind of weather would you wear these boots?

Comfort
range
24 °C to 4 °C

(b) For what Fahrenheit temperatures are the boots designed? Round your final answers to the nearest degree.

(c) What range of metric temperatures do you have in January where you live? Would you be comfortable in these boots?

38. Sleeping bags made by Eddie Bauer are sold around the world. Each type of sleeping bag is designed for outdoor camping in certain temperatures.

Junior bag	5 °C or warmer
Removable liner bag	0 °C to 15 °C
Conversion bag	−7 °C to 0 °C

Data from Eddie Bauer.

(a) At what Fahrenheit temperatures should you use the Junior bag?

(b) What Fahrenheit temperatures is the removable liner bag designed for?

39. The directions for a self-stick hook with adhesive on the back are as follows: "Apply to surfaces above 10 °C. Adhesive could soften and lose adhesion above 40 °C." What are these temperatures in the U.S. system?

40. The directions for the medication diazepam tell patients to store the tablets at temperatures between 60 °F and 80 °F. Rosita is going on a trip to Mexico, where temperatures are in the metric system. Convert the temperatures to metric for her.

Relating Concepts (Exercises 41–48) For Individual or Group Work

🖩 *The information below appeared in U.S. newspapers. However, both Newfoundland (part of Canada) and Ireland use the metric system. Their newspapers would have reported all the measurements in metric units. Complete the conversions to metric, rounding your final answers to the nearest tenth,* **as you work Exercises 41–48 in order.**

Q: **A recent news brief reported on some men who flew a model airplane from Newfoundland to Ireland. Can you provide some details of the flight?**

Newfoundland
Europe
Ireland
Africa

A: The model plane is 6 feet long and weighs 11 pounds. Made of balsa wood and Mylar, it crossed the Atlantic—the flight path took it 1888.3 miles—in 38 hours, 23 minutes. It soared at a cruising altitude of 1000 feet. The plane used a souped-up piston engine and carried less than a gallon of fuel, as mandated by rules of the Federation Aeronautique Internationale, the governing body of model airplane building. When it landed in County Galway, Ireland, it had less than 2 fluid ounces of fuel left. The plane was built by Maynard Hill of Silver Spring, Maryland.

(Data from www.barnardmicrosystems.com)

41. Length of model plane

42. Weight of plane

43. Length of flight path

44. Time of flight

45 Cruising altitude

46. Fuel at the start, in milliliters

47. Fuel left after landing, in milliliters
(*Hint:* First convert 2 fl oz to quarts.)

48. What *percent* of the fuel was left at the end of the flight?

Key Terms

8.1

U.S. measurement system The U.S. measurement system is used for many daily activities only in the United States. Commonly used units include quarts, pounds, feet, miles, and degrees Fahrenheit.

metric system The metric system of measurement is an international system used in manufacturing, science, medicine, sports, and other fields. Commonly used units in this system include meters, liters, grams, and degrees Celsius.

unit fraction A unit fraction involves measurement units and is equivalent to 1. Unit fractions are used to convert among different measurements.

8.2

meter The meter is the basic unit of length in the metric system. The symbol **m** is used for meter. One meter is a little longer than a yard.

prefixes Attaching a prefix such as *kilo-* or *milli-* to the words meter, liter, or gram gives names of larger or smaller units. For example, the prefix *kilo-* means 1000, so a *kilo*meter is 1000 meters.

metric conversion line The metric conversion line is a line showing the various metric measurement prefixes and their size relationship to each other.

8.3

liter The liter is the basic unit of capacity in the metric system. The symbol **L** is used for liter. One liter is a little more than one quart.

gram The gram is the basic unit of weight (mass) in the metric system. The symbol **g** is used for gram. One gram is the weight of 1 milliliter of water or one dollar bill.

8.5

Celsius The Celsius scale is used to measure temperature in the metric system. Water boils at 100 °C and freezes at 0 °C.

Fahrenheit The Fahrenheit scale is used to measure temperature in the U.S. system. Water boils at 212 °F and freezes at 32 °F.

New Symbols

Frequently used metric length units

- **km** kilometer
- **m** meter
- **cm** centimeter
- **mm** millimeter

Frequently used metric capacity units

- **L** liter
- **mL** milliliter

Frequently used metric weight (mass) units

- **kg** kilogram
- **g** gram
- **mg** milligram

°C degrees Celsius
(metric temperature unit)

°F degrees Fahrenheit
(U.S. temperature unit)

New Formulas

Converting from Celsius to Fahrenheit (metric to U.S.):

$$F = \frac{9C}{5} + 32$$

Converting from Fahrenheit to Celsius (U.S. to metric):

$$C = \frac{5(F - 32)}{9}$$

Test Your Word Power

See how well you have learned the vocabulary in this chapter.

① The **metric system**
   A. uses meters, liters, and degrees Fahrenheit
   B. is based on multiples of 10
   C. is used only in the United States.

② **U.S. measurement units**
   A. are used throughout the world
   B. are based on multiples of 12
   C. include feet, inches, quarts, and pounds.

③ A **unit fraction**
   A. has the unit you want to change in the numerator
   B. has a denominator of 1
   C. is equivalent to 1.

④ A **gram** is
   A. the weight of 1 mL of water
   B. equivalent to 1000 kg
   C. approximately equal to 2.2 pounds.

⑤ A **meter** is
   A. equivalent to 1000 cm
   B. approximately equal to $\frac{1}{2}$ inch
   C. abbreviated m with no period after it.

⑥ The **Celsius** temperature scale
   A. shows water freezing at 32°
   B. is used in the U.S. system of measurement
   C. shows water boiling at 100°.

Answers to Test Your Word Power

1. B; *Examples:* 10 meters = 1 dekameter; 100 meters = 1 hectometer; 1000 meters = 1 kilometer.

2. C; *Examples:* Feet and inches are used to measure length, quarts to measure capacity, pounds to measure weight.

3. C; *Example:* Because 12 in. = 1 ft, the unit fraction $\frac{12 \text{ in.}}{1 \text{ ft}}$ is equivalent to $\frac{12 \text{ in.}}{12 \text{ in.}} = 1$.

4. A; *Example:* A small box measuring 1 cm on every edge holds exactly 1 mL of water, and the water weighs 1 g.

5. C; *Example:* A measurement of 16 meters is written 16 m (without a period).

6. C; *Example:* In the metric system, water freezes at 0 °C and boils at 100 °C. The U.S. system uses the Fahrenheit temperature scale in which water freezes at 32 °F and boils at 212 °F.

Quick Review

Concepts

Examples

8.1 U.S. Measurement Units

Memorize the basic measurement relationships. Then, to convert units, multiply when changing from a larger unit to a smaller unit; divide when changing from a smaller unit to a larger unit.

Convert each measurement.
(a) 5 ft to inches (larger unit to smaller unit)
$$5 \text{ ft} = 5 \cdot 12 = \textbf{60 in.}$$

(b) 3 lb to ounces (larger unit to smaller unit)
$$3 \text{ lb} = 3 \cdot 16 = \textbf{48 oz}$$

(c) 15 qt to gallons (smaller unit to larger unit)
$$15 \text{ qt} = \frac{15}{4} = \mathbf{3\frac{3}{4} \text{ gal}}$$

8.1 Using Unit Fractions

Another, more useful, conversion method is multiplying by a unit fraction.
The unit you want in the answer should be in the numerator.
The unit you want to change should be in the denominator.

Convert 32 oz to pounds.

$$32 \text{ oz} \cdot \frac{1 \text{ lb}}{16 \text{ oz}} \Big\} \text{ Unit fraction}$$

These units should match.

$$\frac{\overset{2}{\cancel{32} \text{ oz}}}{1} \cdot \frac{1 \text{ lb}}{\underset{1}{\cancel{16} \text{ oz}}} \quad \begin{array}{l} \text{Divide out ounces.} \\ \text{Divide out common factors.} \end{array}$$

$$32 \text{ oz} = \textbf{2 lb}$$

Concepts	**Examples**
8.1 Solving U.S. Measurement Application Problems	Solve this application problem.
To solve application problems, use the steps summarized below.	Mr. Green has 10 yd of rope. He is cutting it into eight pieces so his sailing class can practice knot tying. How many feet of rope will each of his eight students get?
Step 1 **Read** the problem carefully.	*Step 1* The problem asks how many feet of rope can be given to each of eight students.
Step 2 **Work out a plan.**	*Step 2* Convert 10 yd to feet (the unit required in the answer). Then divide by eight students.
Step 3 **Estimate** a reasonable answer.	*Step 3* There are 3 ft in one yard, so there are 30 ft in 10 yd. Then 30 ft $\div$ 8 $\approx$ 4 ft is our estimate.
Step 4 **Solve** the problem.	*Step 4* Use a unit fraction to convert 10 yd to feet, then divide.

$$\frac{10 \text{ yd}}{1} \cdot \frac{3 \text{ ft}}{1 \text{ yd}} = 30 \text{ ft}$$

$$\frac{30 \text{ ft}}{8 \text{ students}} = 3\frac{3}{4} \text{ ft or } 3.75 \text{ ft per student}$$

Step 5 **State the answer.**	*Step 5* Each student gets $3\frac{3}{4}$ **ft or 3.75 ft of rope.**
Step 6 **Check** your work.	*Step 6* The exact answer of 3.75 ft is close to our estimate of 4 ft.

8.2 Basic Metric Length Units	
Use approximate comparisons to judge which length units are appropriate:	Write the most reasonable metric unit in each blank. Choose from km, m, cm, and mm.

1 mm is the thickness of a dime.	The room is 6 __m__ long.
1 cm is about $\frac{1}{2}$ inch.	A paper clip is 30 __mm__ long.
1 m is a little more than 1 yard.	He drove 20 __km__ to work.
1 km is about 0.6 mile.	

8.2–8.3 Converting within the Metric System	
Using Unit Fractions	Convert.
One conversion method is to multiply by a unit fraction. Use a fraction with the unit you want in the answer in the numerator and the unit you want to change in the denominator.	**(a)** 9 g to kg

$$\frac{9 \text{ g}}{1} \cdot \frac{1 \text{ kg}}{1000 \text{ g}} = \frac{9}{1000} \text{ kg} = 0.009 \text{ kg}$$

9 g = **0.009 kg**

(b) 3.6 m to cm

$$\frac{3.6 \text{ m}}{1} \cdot \frac{100 \text{ cm}}{1 \text{ m}} = 360 \text{ cm}$$

3.6 m = **360 cm**

(Continued)

Concepts	Examples

8.2–8.3 Converting within the Metric System (*Continued*)

Using the Metric Conversion Line
Another conversion method is to find the unit you are given on the metric conversion line. Count the number of places to get from the unit you are given to the unit you want. Move the decimal point the same number of places and in the same direction.

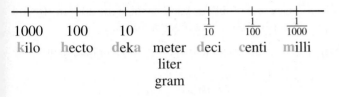

1000	100	10	1	$\frac{1}{10}$	$\frac{1}{100}$	$\frac{1}{1000}$
kilo	hecto	deka	meter liter gram	deci	centi	milli

Convert.
(a) 68.2 kg to g
From **kg** to **g** is three places to the right.

$$6\,8.2\,0\,0$$ Decimal point is moved three places to the right.

68.2 kg = **68,200 g**

(b) 300 mL to L
From **mL** to **L** is three places to the left.

$$3\,0\,0.$$ Decimal point is moved three places to the left.

300 mL = **0.3 L**

(c) 825 cm to m
From **cm** to **m** is two places to the left.

$$8\,2\,5.$$ Decimal point is moved two places to the left.

825 cm = **8.25 m**

8.3 Basic Metric Capacity Units

Use approximate comparisons to judge which capacity units are appropriate:

    1 L is a little more than 1 quart.

    1 mL is the amount of water in a cube that is 1 cm on each side.

    5 mL is about one teaspoon.

    250 mL is about one cup.

Write the most reasonable metric unit in each blank. Choose from L and mL.

The pail holds 12 __L__.

The milk carton from the vending machine holds 250 __mL__.

8.3 Basic Metric Weight (Mass) Units

Use approximate comparisons to judge which weight units are appropriate:

    1 kg is about 2.2 pounds.

    1 g is the weight of 1 mL of water or one dollar bill.

    1 mg is $\frac{1}{1000}$ of a gram; very tiny!

Write the most reasonable metric unit in each blank. Choose from kg, g, and mg.

The wrestler weighed 95 __kg__.

She took a 500 __mg__ aspirin tablet.

One banana weighs 150 __g__.

Concepts	Examples

8.4 **Solving Metric Measurement Application Problems**

Convert units so you are comparing kg to kg, cm to cm, and so on. When a measurement involves two units, such as 6 m 20 cm, write it in terms of the unit called for in the answer (6.2 m or 620 cm).

Use these steps for solving mesurement application problems.

Step 1 **Read** the problem carefully.

Step 2 **Work out a plan.**

Step 3 **Estimate** a reasonable answer.

Step 4 **Solve** the problem.

Step 5 **State the answer.**

Step 6 **Check** your work.

Solve this application problem.

George cut 1 m 35 cm off of a 3 m board. How long was the leftover piece, in meters?

Step 1 The problem asks for the length of the leftover piece in meters.

Step 2 Convert the cut-off measurement to meters (the unit required in the answer), and then subtract to find the "leftover."

Step 3 To estimate, round 1 m 35 cm to 1 m, because 35 cm is less than half a meter. Then, 3 m − 1 m = 2 m is our estimate.

Step 4 Convert the cut-off measurement to meters, then subtract.

$$
\begin{array}{r}
1\text{ m} \rightarrow \quad 1.00\text{ m} \\
\text{plus }35\text{ cm} \rightarrow \underline{+\ 0.35\text{ m}} \\
1.35\text{ m}
\end{array}
\qquad
\begin{array}{rl}
3.00\text{ m} & \leftarrow \text{Board} \\
\underline{-\ 1.35\text{ m}} & \leftarrow \text{Cut off} \\
1.65\text{ m} & \leftarrow \text{Left}
\end{array}
$$

Step 5 The leftover piece is **1.65 m long.**

Step 6 The exact answer of 1.65 m is close to our estimate of 2 m.

8.5 **Converting between Metric and U.S. Measurement Units**

Write a unit fraction using the values in the table of conversion factors in the last section of this chapter. Because the values in the table are rounded, your answers will be approximate.

Convert. Round your final answers to the nearest tenth.

(a) 23 m to yards
From the table, 1 meter ≈ 1.09 yards.

$$
\frac{23 \text{ m}}{1} \cdot \frac{1.09 \text{ yd}}{1 \text{ m}} = 25.07 \text{ yd}
$$

25.07 rounds to 25.1, so 23 m ≈ **25.1 yd**

(b) 4 oz to grams
From the table, 1 ounce ≈ 28.35 grams.

$$
\frac{4 \text{ oz}}{1} \cdot \frac{28.35 \text{ g}}{1 \text{ oz}} = 113.4 \text{ g}
$$

4 oz ≈ **113.4 g**

Concepts	**Examples**

8.5 Common Celsius (Metric) Temperatures

Use approximate and exact comparisons to judge which temperatures are appropriate:

Exact comparisons:

 0 °C is the freezing point of water

100 °C is the boiling point of water

Approximate comparisons:

10 °C for a spring day

20 °C for room temperature

30 °C for a summer day

37 °C for normal body temperature

Circle the metric temperature that is most reasonable.

(a) Hot summer day:

  $\boxed{35\,°C}$ 90 °C   110 °C

(b) The first snowy day in winter:

  −20 °C $\boxed{0\,°C}$ 15 °C

8.5 Converting between Fahrenheit and Celsius Temperatures

Use this formula to convert from Fahrenheit (F) to Celsius (C); that is, from U.S. to metric temperatures.

$$C = \frac{5\,(F - 32)}{9}$$

Convert 176 °F to Celsius. Round your answer to the nearest whole degree if necessary.

$$C = \frac{5\,(176 - 32)}{9}$$ Replace F with 176. Work inside parentheses: 176 − 32 is 144

$$C = \frac{5\,(\overset{16}{144})}{\underset{1}{9}}$$ Divide out common factors. Then multiply in the numerator.

$$C = 80$$

So 176 °F = **80 °C**

Use this formula to convert from Celsius (C) to Fahrenheit (F); that is, from metric to U.S. temperatures.

$$F = \frac{9C}{5} + 32$$

Convert −8 °C to Fahrenheit. Round your answer to the nearest whole degree if necessary.

$$F = \frac{9\,(-8)}{5} + 32 \quad \text{Replace C with } -8$$

$$F = \frac{-72}{5} + 32$$

$$F = -14.4 + 32$$

$$F = 17.6 \qquad \text{Round to 18}$$

So, −8 °C ≈ **18 °F**

Chapter 8 Review Exercises

8.1 *Fill in the blanks with the measurement relationships you have memorized.*

1. 1 lb = _____ oz

2. _____ ft = 1 yd

3. 1 ton = _____ lb

4. _____ qt = 1 gal

5. 1 hr = _____ min

6. 1 c = _____ fl oz

7. _____ sec = 1 min

8. _____ ft = 1 mi

9. _____ in. = 1 ft

Convert using unit fractions.

10. 4 ft = _____ in.

11. 6000 lb = _____ tons

12. 64 oz = _____ lb

13. 18 hr = _____ day

14. 150 min = _____ hr

15. $1\frac{3}{4}$ lb = _____ oz

16. $6\frac{1}{2}$ ft = _____ in.

17. 7 gal = _____ c

18. 4 days = _____ sec

19. The average depth of the world's oceans is 14,000 ft. (Data from www.noaa.gov)

   (a) What is the average depth in yards, to the nearest tenth?

   (b) What is the average depth in miles, to the nearest tenth?

20. During the first year of a program to recycle office paper, a company recycled 123,260 pounds of paper. The company received $40 per ton for the paper. How much money did the company make? (Data from I. C. System.)

8.2 *Write the most reasonable metric length unit in each blank. Choose from km, m, cm, and mm.*

21. My thumb is 20 _____ wide.

22. Her waist measurement is 66 _____.

23. The two towns are 40 _____ apart.

24. A basketball court is 30 _____ long.

25. The height of the picnic bench is 45 _____.

26. The eraser on the end of my pencil is 5 _____ long.

Convert using unit fractions or the metric conversion line.

27. 5 m to cm

28. 8.5 km to m

29. 85 mm to cm

30. 370 cm to m

31. 70 m to km

32. 0.93 m to mm

8.3 *Write the most reasonable metric unit in each blank.*
Choose from L, mL, kg, g, and mg.

33. The eyedropper
holds 1 _____ .

34. I can heat 3 _____
of water in this
saucepan.

35. Loretta's hammer weighed 650 _____ .

36. Yongshu's large suitcase weighed 20 _____
when it was packed.

37. My fish tank holds 80 _____ of water.

38. I'll buy the 500 _____ bottle of mouthwash.

39. Mara took a 200 _____ antibiotic pill.

40. This piece of chicken weighs 100 _____ .

Convert using unit fractions or the metric conversion line.

41. 5000 mL to L

42. 8 L to mL

43. 4.58 g to mg

44. 0.7 kg to g

45. 6 mg to g

46. 35 mL to L

8.4 *Solve each application problem. Show your work.*

47. Each serving of punch at the wedding reception will
be 180 mL. How many liters of punch are needed for
175 servings?

48. Jason is serving a 10 kg turkey to 28 people. How
many grams of meat is he allowing for each person?
Round to the nearest whole gram.

49. Chandni weighed 57 kg.
Then she lost 4 kg 750 g.
What is her weight now,
in kilograms?

50. Young-Mi bought 950 g of onions. The price was
$1.49 per kilogram. How much did she pay, to the
nearest cent?

8.5 *Use the table in the last section of this chapter and unit fractions to make*
approximate conversions. Round your final answers to the nearest tenth if necessary.

51. 6 m to yards

52. 30 cm to inches

53. 108 km to miles

54. 800 mi to kilometers

55. 23 qt to liters

56. 41.5 L to quarts

*Write the appropriate **metric** temperature in each blank.*

57. Water freezes at _____ .

58. Water boils at _____ .

59. Normal body temperature is about _____ .

60. Comfortable room temperature is about _____ .

Use the conversion formulas in the last section of this chapter to convert each
temperature to Fahrenheit or to Celsius. Round to the nearest degree if necessary.

61. 77 °F

62. 5 °F

63. −2 °C

64. 49 °C

Practicing exercises in mixed up order helps you prepare for tests.
Write the most reasonable metric unit in each blank. Choose from km, m, cm, mm,
L, mL, kg, g, and mg.

1. I added 1 _____ of oil to my car.

2. The box of books weighed 15 _____.

3. Larry's shoe is 30 _____ long.

4. Jan used 15 _____ of shampoo on her hair.

5. My fingernail is 10 _____ wide.

6. I walked 2 _____ to school.

7. The tiny bird weighed 15 _____.

8. The new library building is 18 _____ wide.

9. The cookie recipe uses 250 _____ of milk.

10. Renee's pet mouse weighs 30 _____.

11. One postage stamp weighs 90 _____.

12. I bought 30 _____ of gas for my car.

Convert the following using unit fractions, the metric conversion line, or the
temperature conversion formulas.

13. 10.5 cm to millimeters

14. 45 min to hours

15. 90 in. to feet

16. 1.3 m to centimeters

17. 25 °C to Fahrenheit

18. $3\frac{1}{2}$ gal to quarts

19. 700 mg to grams

20. 0.81 L to milliliters

21. 5 lb to ounces

22. 60 kg to grams

23. 1.8 L to milliliters

24. 86 °F to Celsius

25. 0.36 m to centimeters

26. 55 mL to liters

Solve each application problem. Show your work.

27. Peggy had a board measuring 2 m 4 cm. She cut off 78 cm. How long is the board now, in meters?

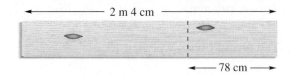

28. During the 12-day Minnesota State Fair, one of the biggest in the United States, Sweet Martha's booth sells an average of 3000 pounds of cookies per day. How many tons of cookies are sold in all? (Data from *Minneapolis Star Tribune.*)

29. Olivia is sending a recipe to her mother in Mexico. Among other things, the recipe calls for 4 oz of rice and a baking temperature of 350 °F. Convert these measurements to metric, rounding to the nearest gram and nearest degree.

30. While on vacation in Canada, Jalo became ill and went to a health clinic. They said he weighed 80.9 kg and was 1.83 m tall. Find his weight in pounds and height in feet. Round your final answers to the nearest tenth.

The largest two-axle trucks in the world are mining trucks used to haul huge loads of rock and iron ore. Some information about these $2 million trucks is given below. Fill in the blank spaces in the table. Then use the information to answer Exercises 37–38. Round your final answers to the nearest tenth.

WORLD'S LARGEST TWO-AXLE TRUCK

	Mining Truck	Measurements	
		Metric Units	U.S. Units
31.	Length of truck	_____ m	44 ft
32.	Height of truck	6.7 m	_____ ft
33.	Height of tire	_____ m	4 yd
34.	Width of tire tread	102 cm	_____ in.
35.	Weight of load truck can carry	220,000 kg	_____ lb
36.	Fuel needed to travel 1 mile	_____ to _____ L	5 to 6 gal

Data from Hull Rust Mahoning Mine.

37. Using U.S. measurements:

(a) The truck can carry a load weighing how many tons?

(b) How many inches high is each tire?

38. Using metric measurements:

(a) What is the width of the tire tread in meters?

(b) How tall is the truck in centimeters?

Chapter 8 *Test*

The Chapter Test Prep Videos with step-by-step solutions are available in MyMathLab or on YouTube at https://goo.gl/c3befo

Convert each measurement.

1. 9 gal = _____ qt

2. 45 ft = _____ yd

3. 135 min = _____ hr

4. 9 in. = _____ ft

5. $3\frac{1}{2}$ lb = _____ oz

6. 5 days = _____ min

Write the most reasonable metric unit in each blank. Choose from km, m, cm, mm, L, mL, kg, g, and mg.

7. My husband weighs 75 _____.

8. I hiked 5 _____ this morning.

9. She bought 125 _____ of cough syrup.

10. This apple weighs 180 _____.

11. A piece of printer paper is about 21 _____ wide.

12. My watch band is 10 _____ wide.

13. I bought 10 _____ of soda for the picnic.

14. The bracelet is 16 _____ long.

Convert the following measurements. Show your work.

15. 250 cm to meters

16. 4.6 km to meters

17. 5 mm to centimeters

18. 325 mg to grams

19. 16 L to milliliters

20. 0.4 kg to grams

21. 10.55 m to centimeters

22. 95 mL to liters

23. The rainiest place in the world is Meghalaya, India, which receives 467 inches of rain each year. What is the average rainfall per month, in feet? Round your answer to the nearest tenth. (Data from www.theatlantic.com)

24. A 6-inch Subway Veggie Delite sandwich has 280 mg of sodium. A 6-inch chicken and bacon ranch melt sandwich has 1.05 g of sodium. (Data from Subway.)

(a) How much more sodium is in the chicken and bacon ranch melt sandwich, in milligrams, than in the Veggie Delite?

(b) The recommended amount of sodium is no more than 1500 mg daily. How much more or less sodium does the chicken and bacon ranch melt sandwich have than the recommended daily amount?

Pick the metric temperature that is most reasonable in each situation.

25. The water is almost boiling.

210 °C   155 °C   95 °C

26. The tomato plants may freeze tonight.

30 °C   20 °C   0 °C

Use the U.S./metric conversion table from the last section of this chapter and unit fractions to convert each measurement. Round your final answers to the nearest tenth if necessary.

27. 6 ft to meters

28. 125 lb to kilograms

29. 50 L to gallons

30. 8.1 km to miles

In Problems 31 and 32, use the conversion formulas to convert each temperature. Round your answers to the nearest degree if necessary.

31. 74 °F to Celsius

32. −12 °C to Fahrenheit

33. Denise is making five matching pillows. She needs 120 cm of braid to trim each pillow. If the braid costs $3.98 per yard, how much will she spend to trim the pillows, to the nearest cent? (First find the number of meters of braid Denise needs.) Show your work.

34. Describe two benefits the United States would achieve by switching entirely to the metric system.

Chapters 1–8 *Cumulative Review Exercises*

Simplify.

1. $\dfrac{7}{6x^2} \cdot \dfrac{9x}{14y}$

2. $\dfrac{3}{c} - \dfrac{5}{6}$

3. $\dfrac{-3(-4)}{27 - 3^3}$

4. $3\dfrac{7}{12} - 4$

5. $\dfrac{9y^2}{8x} \div \dfrac{y}{6x^2}$

6. $\dfrac{2}{3} + \dfrac{n}{m}$

7. $1\dfrac{1}{6} + 1\dfrac{2}{3}$

8. $\dfrac{\frac{14}{15}}{-6}$

9. $\dfrac{12 \div (2 - 5) + 12(-1)}{2^3 - (-4)^2}$

10. $(-0.8)^2 \div (0.8 - 1)$

11. $\left(-\dfrac{1}{3}\right)^2 - \dfrac{1}{4}\left(\dfrac{4}{9}\right)$

Solve each equation. Show your work.

12. $-5n = n - 12$

13. $\dfrac{1.5}{45} = \dfrac{x}{12}$

14. $2 = \dfrac{1}{4}w - 3$

15. $4y - 3 = 7y + 12$

16. Translate this sentence into an equation and solve it. When twice a number is decreased by 8, the result is the number increased by 7. Find the number.

17. Use the six problem-solving steps. A 90 ft pipe is cut into two pieces so that one piece is 6 ft shorter than the other. Find the length of each piece.

Find the unknown length, perimeter, circumference, area, or volume. When necessary, use 3.14 as the approximate value of π and round your answers to the nearest tenth.

18. Find the perimeter and area.

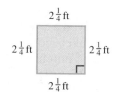

19. Find the circumference and the area.

20. Find the unknown length and the area of the figure.

21. Find the volume and the surface area of this rectangular solid.

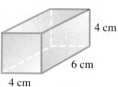

Convert the following using unit fractions or the metric conversion line.

22. $4\dfrac{1}{2}$ ft to inches

23. 72 hours to days

24. 3.7 kg to grams

25. 60 cm to meters

Solve each application problem. Show your work.

26. Mei Ling must earn 60 credits to receive an associate of arts degree. She has 35 credits. What percent of the necessary credits does she have? Round to the nearest whole percent.

27. Which bag of chips is the best buy? Brand T is $15\frac{1}{2}$ ounces for $2.99, Brand F is 14 ounces for $2.49, and Brand H is 18 ounces for $3.89. You have a coupon for 40¢ off Brand H and another for 30¢ off Brand T.

28. On the Illinois map, one centimeter represents 12 km. The center of Springfield is 7.8 cm from the center of Bloomington on the map.

(a) What is the actual distance in kilometers?

(b) What is the actual distance in miles, to the nearest tenth?

29. Bags of slivered almonds weigh 115 g each. They are packed in a carton that weighs 450 g when empty. How many kilograms would a carton containing 48 bags weigh?

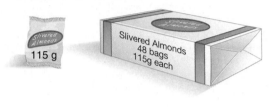

30. Steven bought a $4\frac{1}{2}$ yard length of canvas material to repair the tents used by the Scout troop. He used $1\frac{2}{3}$ yards on one tent and $1\frac{3}{4}$ yards on another. How much material is left?

31. Mark bought 650 grams of maple sugar candy on his vacation in Montreal. The candy is priced at $14.98 per kilogram. How much did Mark pay, to the nearest cent?

32. Calbert works at a Walmart store and used his employee discount to buy a $189.94 tablet at 10% off. Find the amount of his discount, to the nearest cent, and the sale price. (Data from Walmart.)

33. Akuba is knitting a scarf. Six rows of knitting result in 5 cm of scarf. At that rate, how many rows will she have to knit to make a 100 cm scarf?

34. Dimitri took out a $3\frac{1}{2}$ year car loan for $18,750 at 4.3% simple interest. Find the interest and the total amount Dimitri will pay on the loan, to the nearest cent.

35. A shelter for homeless people received a $400 donation to buy blankets. If the blankets cost $17 each, how many blankets can be purchased? How much money will be left over?

36. A spray-on bandage applies a transparent film over a wound that protects it from bacteria. The cost is $9.50 for about 50 applications. What is the cost per application? (Data from Walmart.)

37. The average hospital stay is now about 5 days, compared to about 8 days in 1970. What is the percent decrease in the length of hospital stays? (Data from Centers for Disease Control and Prevention.)

9

Graphs and Graphing

Information can be represented visually with tables and a variety of graphs.
In this chapter you'll learn how to interpret and make graphs.

9.1 Problem Solving with Tables and Pictographs

OBJECTIVES

1. Read and interpret data presented in a table.

2. Read and interpret data from a pictograph.

So far, you've used numbers, expressions, formulas, and equations to communicate information. Now you'll see how *tables, pictographs, circle graphs, bar graphs,* and *line graphs* are also used to communicate information.

OBJECTIVE 1 **Read and interpret data presented in a table.** A **table** presents data organized into rows and columns. The advantage of a table is that you can find very specific, exact values. The disadvantages are that you may have to spend some time searching through the table to find what you want, and it may not be easy to see trends or patterns.

The table below shows information about the performance of eight U.S. airlines during the year 2014 (January–December).

PERFORMANCE DATA FOR SELECTED U.S. AIRLINES
JANUARY–DECEMBER 2014

Airline	On-Time Performance	Luggage Handling*
Alaska	80%	2.7
Delta	89%	2.3
Frontier	68%	1.8
Hawaiian	88%	2.2
Jetblue	80%	2.0
Skywest	67%	4.7
United	72%	3.7
Virgin America	68%	1.0

*Luggage problems per 1000 passengers.
Data from www.Transportation.gov

For example, by reading from left to right along the row marked United, you first see that 72% of United's flights were on time during 2014. The next number is 3.7 and the heading at the top of that column is *Luggage Handling**. The little star (asterisk) tells you to look below the table for more information. Next to the asterisk below the table it says "Luggage problems per 1000 passengers." So, almost 4 passengers out of every 1000 passengers on United flights had some sort of problem with their luggage.

VOCABULARY TIP

Rows are read across from left to right.

Columns are read from top to bottom. Pay attention to the labels on rows and columns.

EXAMPLE 1 Reading and Interpreting Data from a Table

Use the table above on airline performance to answer these questions.

(a) What percent of Jetblue's flights were on time?

Look across the row labeled Jetblue. **80% of its flights were on time.**

(b) Which airline had the worst luggage handling record?

Look down the column headed Luggage Handling. To find the *worst* record, look for the *highest* number, which is 4.7. Then look to the left to find the airline, which is **Skywest.**

(c) What was the average percent of on-time flights for the four airlines with the best performance, to the *nearest whole percent*?

Look down the column headed On-Time Performance to find the four highest numbers: 89%, 88%, 80%, and 80%. To find the average, add the values and divide by 4.

——————— Continued on Next Page

$$\frac{89 + 88 + 80 + 80}{4} = \frac{337}{4} \approx 84$$

"To the nearest whole percent" means round your answer to the ones place.

The average on-time performance for the four best airlines was about 84%.

— **Work Problem ❶ at the Side.** ▶

EXAMPLE 2 | **Interpreting Data from a Table**

The table below shows the maximum cab fares in five different cities in 2015. The "flag drop" charge is made when the driver starts the meter. "Wait time" is the charge for having to wait in the middle of a ride.

MAXIMUM TAXICAB FARES IN SELECTED CITIES IN 2015

City	Flag Drop	Price per Mile	Wait Time (per Hour)
Chicago	$4	$2	$20
Houston	$2.75	$2.20	$24
Miami	$2.95	$5.10	$24
New York	$2.50	$2.50	$24
San Francisco	$3.50	$2.75	$32

Data from www.taxifarefinder.com

Use the table to answer these questions.

(a) What is the maximum fare for a 9-mile ride in Houston that includes having the cab wait 15 minutes while you pick up a package at a store?

The price per mile in Houston is $2.20, so the cost for 9 miles is $9 (\$2.20) = \$19.80$. Then, add the flag drop charge of $2.75. Finally, figure out the cost of the wait time. One way to do that is to set up a proportion. Recall that 1 hour is 60 minutes, so each side of the proportion compares the cost to the number of minutes of wait time.

Cost → $$\frac{\$24}{60 \text{ min}} = \frac{\$x}{15 \text{ min}}$$ ← Cost
Wait time →   ← Wait time

Show that cross products are equal. → $60 \cdot x = 24 \cdot 15$

$$\frac{\overset{1}{60}x}{\underset{1}{60}} = \frac{360}{60}$$ Divide both sides by 60

$x = \$6$ ——— Charge for waiting 15 minutes

Total fare = $19.80 + $2.75 + $6 = **$28.55**

(b) It is customary to give the cab driver a tip. Find the total cost of the cab ride in part (a) above if the passenger added a 15% tip, rounded to the nearest dollar.

Use the percent equation to find the exact tip.

$$\text{percent} \cdot \text{whole} = \text{part} \quad \text{Percent equation}$$
$$(15\,\%) (\$28.55) = n \quad \text{Write 15\% as a decimal.}$$

Write 15% as 0.15 → $(0.15) (\$28.55) = n$

$$\$4.2825 = n \quad \text{Round to \$4 (nearest dollar).}$$

The total cost of the cab ride is $28.55 + $4 tip = **$32.55**

— **Work Problem ❷ at the Side.** ▶

---

❶ Use the table of airline performance on the previous page to answer these questions.

(a) Which airline had the best on-time performance?

(b) Which airline had the best record for luggage handling? (*Hint*: Look for the *lowest* number of luggage problems.)

(c) Which airline(s) had 2 or fewer luggage handling problems per 1000 passengers?

(d) What was the average number of luggage problems for all eight airlines, to the nearest tenth?

❷ Use the table of cab fares on this page to answer these questions.

GS **(a)** What is the difference in the maximum fare for a cab ride of 6.5 miles in Chicago compared to San Francisco? Assume there is no wait time.

Chicago fare = 6.5 ($2) + $4 = _____

San Francisco fare = 6.5 ($2.75) + $3.50 = _____

(b) Find the total cost for a cab ride of 4.5 miles in Miami, including 30 minutes of wait time and a 15% tip. Round the tip to the nearest dollar.

Answers

1. (a) Delta   (b) Virgin America
    (c) Jetblue, Virgin America, Frontier
    (d) $20.4 \div 8 \approx 2.6$ (rounded)

2. (a) Chicago $17;
    San Francisco $21.38 (rounded);
    San Francisco fare is $4.38 higher.
    (b) $37.90 + $6 tip = $43.90

3 Use the pictograph in **Example 3** to answer these questions.

(a) What is the approximate population of Chicago?

How many whole symbols are shown for Chicago? ____

Each whole symbol represents _____ people.

Multiply to find that the approximate population of Chicago is _____

(b) What is the approximate population of Dallas?

(c) Approximately how much more is the population of New York than that of Chicago?

(d) The population of Atlanta is approximately how much less than that of Dallas?

Answers

3. **(a)** 5 symbols; 2 million, or, 2,000,000;
10 million people (or 10,000,000)
(b) 7 million people (or 7,000,000)
(c) 10 million people (or 10,000,000)
(d) 1 million people (or 1,000,000)

OBJECTIVE **2** **Read and interpret data from a pictograph.** Tables show numbers in rows and columns. Graphs, on the other hand, are a *visual* way to communicate data; that is, they show a *picture* of the information rather than a list of numbers. In this section you'll work with one type of graph, *pictographs,* and in the next two sections you'll learn about circle graphs, bar graphs, and line graphs.

The advantage of a graph is that you can easily make comparisons or see trends just by looking. The disadvantage is that the graph may not give you the specific, more exact numbers you need in some situations.

EXAMPLE 3 **Reading and Interpreting a Pictograph**

A **pictograph** uses symbols or pictures to represent various amounts. The pictograph below shows the population of five U.S. metropolitan areas (cities with their surrounding suburbs) in 2014. The *key* at the bottom of the graph tells you that each symbol of a person represents 2 million people. Half of a symbol represents half of 2 million people, or 1 million people.

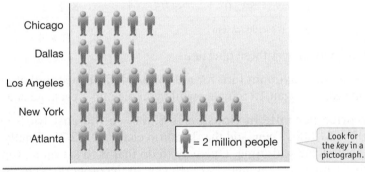

Population of U.S. Metropolitan Areas

Chicago
Dallas
Los Angeles
New York
Atlanta

= 2 million people

Look for the *key* in a pictograph.

Data from www.census.gov

Use the pictograph to answer these questions.

(a) What is the approximate population of Los Angeles?

The population of Los Angeles is represented by 6 whole symbols (6 • 2 million = 12 million) plus half of a symbol ($\frac{1}{2}$ of 2 million is 1 million) for a total of **13 million people.**

Because 2 million can be written as 2,000,000, another way to get the answer is to multiply 6.5 (the number of symbols) times 2,000,000 people for a total of **13,000,000 people.**

(b) How much greater is the population of New York than that of Atlanta?

New York shows 10 symbols and Atlanta shows 3 symbols. So New York has 7 more symbols than Atlanta, and 7 • 2 million = 14 million. Thus, **New York has about 14 million (or 14,000,000) more people than Atlanta.**

Note

One disadvantage of a pictograph is that you often have to round numbers a great deal; in this case, to the nearest *million* people. The actual population of the Atlanta area is 5,614,000, but the graph shows 3 symbols (6,000,000 people), so it is off by about 386,000 people.

◀ **Work Problem** **3** at the Side.

9.1 Exercises

FOR EXTRA HELP

Go to MyMathLab *for worked-out, step-by-step solutions to exercises enclosed in a square* ☐ *and video solutions to* ▶ *exercises.*

CONCEPT CHECK *Circle the letter of the correct answer to complete each statement.*

1. A table presents data organized into:
   (a) a circle graph   (b) rows and columns

2. An advantage of a table of data is that you can:
   (a) find exact values   (b) see trends and patterns

3. An advantage of a pictograph is that you can:
   (a) find exact values
   (b) easily make comparisons

4. In a pictograph, the *key* tells you:
   (a) what the graph is about
   (b) what each symbol represents

This table lists the basketball players in the NBA with the all-time highest number of points scored (at the time the 2014–2015 season ended). Use the table to answer Exercises 5–12. **See Examples 1 and 2.**

ALL-TIME NBA STATISTICAL LEADERS—POINTS SCORED (AT THE TIME THE 2014–2015 SEASON ENDED)

Player	Games	Points	Average Points Per Game
Kareem Abdul-Jabbar	1560	38,387	24.6
Karl Malone	1476	36,928	25.0
Kobe Bryant	1280	32,482	25.4
Michael Jordan	1072	32,292	30.1
Wilt Chamberlain	1045	31,419	
Shaquille O'Neal	1207	28,596	

Data from www.nba.com

5. (a) How many points did Wilt Chamberlain score during his NBA career?

   (b) Which player(s) scored more points than Chamberlain?

6. (a) How many games did Karl Malone play in during his career in the NBA?

   (b) Which player is closest to Malone in number of games?

7. (a) Which player has been in the greatest number of games?

   (b) Which player has been in the least number of games?

8. Find the total number of points scored by the first three players listed in the table.

9. What is the difference in points scored between the ▶ player with the greatest number of points and the player with the least number of points?

10. How many more games has Shaquille O'Neal played in than Michael Jordan?

11. Complete the table by finding the average number of points scored per game by Shaquille O'Neal and Wilt Chamberlain. Look at the other averages in the table to decide how to round your answers.

For Shaquille: $\dfrac{\text{Total points} \rightarrow\ 28{,}596}{\text{Total games} \rightarrow\ 1207}$

28,596 ÷ 1207 = _____   Show hundredths so you can round to tenths.

Round to the nearest tenth. _____ points per game.
Now find Wilt's average.

12. Find the overall scoring average (points per game) for all six players listed in the table. Use the numbers in the *Games* column and the *Points* column to calculate your answer.

Total points for all six players → _____

Total games for all six players → 7640

Divide. Show the digits in the tenths and hundredths places. Then round to the nearest tenth.

_____ ÷ 7640 ≈ _____ ≈ _____ points per game

This table shows the number of calories burned during 30 minutes of various types of exercise. The table also shows how the number of calories burned varies according to the weight of the person doing the exercise. Use the table to answer Exercises 13–22. See Examples 1 and 2.

CALORIES BURNED DURING 30 MINUTES OF EXERCISE BY PEOPLE OF DIFFERENT WEIGHTS

Activity	Calories Burned in 30 Minutes		
	110 Pounds	140 Pounds	170 Pounds
Moderate jogging	175	223	270
Moderate walking	69	88	107
Moderate bicycling	218	276	337
Low-impact aerobics	152	193	235
Racquetball	175	223	270
Tennis-singles	152	193	235

Data from www.healthstatus.com

13. A person weighing 140 pounds is looking at the table.

   (a) How many calories are burned during 30 minutes of low-impact aerobics?

   (b) Which activity burns the most calories?

14. A person weighing 170 pounds is looking at the table.

   (a) How many calories are burned during 30 minutes of singles tennis?

   (b) Which activity burns the fewest calories

15. (a) Which activities can a 110-pound person do to burn at least 200 calories in 30 minutes?

   (b) Which activities can a 170-pound person do to burn at least 200 calories in 30 minutes?

16. (a) Which activities burn fewer than 200 calories in 30 minutes for a 140-pound person?

   (b) Which activities burn fewer than 300 calories in 30 minutes for a 170-pound person?

17. How many total calories will a 140-pound person burn during 15 minutes of moderate bicycling and 60 minutes of moderate walking?

18. How many total calories will a 110-pound person burn during 90 minutes of singles tennis and 15 minutes of low-impact aerobics?

Set up and solve proportions to answer Exercises 19–22. Round your final answers to the nearest whole number.

19. How many calories would you expect a 125-pound person to burn

(a) during 30 minutes of moderate jogging?

From the table, a 110-pound person burns 175 calories during 30 minutes of moderate jogging.

Set up a proportion and find the cross products.

$$\frac{175 \text{ calories}}{110\text{-pound person}} = \frac{x \text{ calories}}{125\text{-pound person}}$$

$$(110)(x) = (175)(125) \quad \text{Now divide both sides by 110}$$

(b) during 30 minutes of singles tennis?

20. How many calories would you expect a 185-pound person to burn

(a) during 30 minutes of moderate walking?

From the table, a 170-pound person burns 107 calories during 30 minutes of moderate walking.

Set up a proportion and find the cross products.

$$\frac{107 \text{ calories}}{170\text{-pound person}} = \frac{x \text{ calories}}{185\text{-pound person}}$$

$$(170)(x) = (107)(185) \quad \text{Now divide both sides by 170}$$

(b) during 30 minutes of moderate bicycling?

21. How many more calories would a 158-pound person burn during 15 minutes of low-impact aerobics than during 20 minutes of moderate walking? Round to the nearest whole number.

22. How many fewer calories would a 196-pound person burn during 25 minutes of moderate walking than during 20 minutes of singles tennis?

This pictograph shows the approximate number of passenger arrivals and departures at selected U.S. airports in 2014. Use the pictograph to answer Exercises 23–30.
See Example 3.

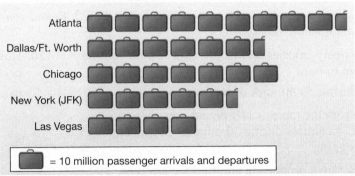

Data from www.aci-na.org

23. What is the approximate total number of arrivals and departures at the two busiest airports?

24. What is the difference in the number of arrivals and departures at the busiest airport and the least busy airport?

25. How many fewer arrivals and departures did the Dallas airport have compared to Chicago's airport?

26. Find the approximate total number of arrivals and departures for the three least busy airports.

27. What is the approximate total number of arrivals and departures for all five airports?

28. Find the average number of arrivals and departures for the five airports.

29. CONCEPT CHECK The work shown below has **two** mistakes in it.

What Went Wrong? First write a sentence explaining what each mistake is. Then fix the mistakes and find the correct solutions.

Britney used the pictograph above to find the approximate number of passenger arrivals and departures that took place at each of these airports.

At the Atlanta airport she said there were 95 arrivals and departures.

At the Chicago airport she said there were 70,000 arrivals and departures.

30. CONCEPT CHECK The work shown below has **two** mistakes in it.

What Went Wrong? First write a sentence explaining what each mistake is. Then fix the mistakes and find the correct solutions.

Lee used the pictograph above to find the approximate number of passenger arrivals and departures that took place at each of these airports.

At the Las Vegas airport he said there were 4 million arrivals and departures.

At the New York (JFK) airport he said there were 51 million arrivals and departures.

9.2 | Reading and Constructing Circle Graphs

A *circle graph* is another way to show a *picture* of a set of data. This picture can often be understood faster and more easily than a formula or a list of numbers.

OBJECTIVE ▸ **1** **Read a circle graph.** A **circle graph** is used to show how a total amount is divided into parts. The circle graph below shows you how 24 hours in the life of a college student are divided among different activities.

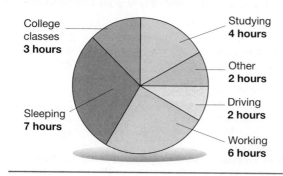

The Day of a College Student

College classes
3 hours

Studying
4 hours

Other
2 hours

Driving
2 hours

Working
6 hours

Sleeping
7 hours

Work Problem **1** at the Side. ▶

OBJECTIVE ▸ **2** **Use a circle graph.** The circle graph above uses pie-shaped pieces called *sectors* to show the amount of time spent on each activity (the total must be 24 hours). The circle graph can therefore be used to compare the time spent on one activity to the total number of hours in the day.

EXAMPLE 1 | **Using a Circle Graph**

Find the ratio of time spent in college classes to the total number of hours in a day. Write the ratio as a fraction in lowest terms.

The circle graph shows that 3 of the 24 hours in a day are spent in class. The ratio of class time to the hours in a day is shown below.

$$\frac{3 \text{ hours (college classes)}}{24 \text{ hours (whole day)}} = \frac{3 \text{ hours}}{24 \text{ hours}} = \frac{\overset{1}{\cancel{3}}}{\cancel{3} \cdot 8} = \frac{1}{8} \quad \left\{ \begin{array}{l} \text{Ratio in} \\ \text{lowest terms} \end{array} \right.$$

— Work Problem **2** at the Side. ▶

The circle graph above can also be used to find the ratio of the time spent on one activity to the time spent on any other activity. See the next example.

EXAMPLE 2 | **Finding a Ratio from a Circle Graph**

Find the ratio of working time to class time.

The circle graph shows 6 hours spent working and 3 hours spent in class.

$$\frac{6 \text{ hours (working)}}{3 \text{ hours (class)}} = \frac{6 \text{ hours}}{3 \text{ hours}} = \frac{\overset{1}{\cancel{3}} \cdot 2}{\cancel{3}} = \frac{2}{1} \quad \leftarrow \text{Ratio in lowest terms}$$

So the ratio is **2 hours spent working for every 1 hour in class.**

— Work Problem **3** at the Side. ▶

OBJECTIVES

1 Read a circle graph.

2 Use a circle graph.

3 Use a protractor to draw a circle graph.

VOCABULARY TIP

Data means "information." It may be pronounced "**day**-tuh" or "**dad**-uh."

1 Use the circle graph on this page to answer each question.

(a) The greatest number of hours is spent in which activity? How many hours is that?

(b) Find the total number of hours spent studying, working, and attending classes.

2 Use the circle graph to find each ratio. Write the ratios as fractions in lowest terms.

(a) Hours spent driving to whole day

(b) Hours spent studying to whole day

3 Use the circle graph to find each ratio. Write the ratios as fractions in lowest terms.

(a) Hours spent studying to hours spent working

(b) Hours spent studying to hours spent driving

Answers

1. (a) sleeping; 7 hours (b) 13 hours

2. (a) $\frac{1}{12}$ (b) $\frac{1}{6}$

3. (a) $\frac{2}{3}$ (b) $\frac{2}{1}$

4 Use the circle graph on the oil spill to find the amount of oil still at sea, skimmed, evaporated, or dispersed.

(a) Still at sea or on shores

$$\begin{array}{ccc} \text{percent} & \cdot & \text{whole} & = \text{part} \\ \downarrow & & \downarrow & \quad\downarrow \\ 26.\% & \cdot & 5{,}000{,}000 \text{ barrels} & = \quad n \end{array}$$

(b) Skimmed

(c) Evaporated or dissolved

(d) Chemically dispersed

A circle graph often shows data as percents. For example, the Deepwater Horizon oil well that exploded in the Gulf of Mexico in 2010 leaked a total of 5 million barrels of oil into the water. The circle graph shows where the oil ended up. The entire circle represents 5 million barrels of oil. Each sector represents the amount of oil that was removed from the Gulf, was dispersed in various ways, or still remains in the Gulf. The sector amounts are shown as a percent of the total oil spill. The total in a circle graph must be 100%, though it may be slightly more or less due to rounding the percent for each sector.

Where Did the Oil Go?

The Deepwater Horizon oil well leaked 5 million barrels of oil into the Gulf of Mexico. The percent of the oil that was removed, was dispersed in various ways, or still remains in the Gulf, is rounded to the nearest whole percent.

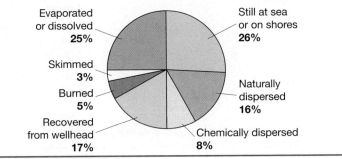

Evaporated or dissolved 25%
Still at sea or on shores 26%
Skimmed 3%
Burned 5%
Recovered from wellhead 17%
Naturally dispersed 16%
Chemically dispersed 8%

Data from www.noaa.gov

EXAMPLE 3 **Calculating an Amount Using a Circle Graph**

Use the circle graph above on the Deepwater Horizon oil spill to find the amount of oil that was burned.

Use the percent equation: percent • whole = part

The percent of oil burned was 5%. Rewrite 5% as the decimal 0.05. The *whole* is the total oil spill of 5 million barrels (the entire circle).

$$\begin{array}{ccc} \text{percent} \cdot & \text{whole} & = \text{part} \\ \downarrow & \downarrow & \downarrow \\ 05.\% & \cdot \ 5 \text{ million barrels} & = \quad n \\ \downarrow & \downarrow & \\ (0.05)(5{,}000{,}000 \text{ barrels}) & = \quad n \\ 250{,}000 \text{ barrels} & = \quad n \end{array}$$

> Write 5% as 0.05

> Write 5 million barrels as 5,000,000 barrels.

The **amount of oil burned was 250,000 barrels.**

◀ **Work Problem 4 at the Side.**

OBJECTIVE ▸ 3 Use a protractor to draw a circle graph. The coordinator of the Fair Oaks Youth Soccer League organizes teams in five age groups. She counts the number of registered players in each age group as shown in the table on the next page. Then she calculates what percent of the total each group represents. For example, there are 59 players in the "Under 8" group, out of 298 total players.

$$\text{percent} \cdot \text{whole} = \text{part}$$
$$p \quad \cdot \ 298 \ = \ 59$$
$$\frac{p \cdot 298}{298} = \frac{59}{298}$$

> Multiply the decimal answer by 100% to get the percent.

$$p \approx 0.198 \approx 19.8\% \quad \text{Rounds to 20\%}$$

Answers

4. **(a)** $(0.26)(5{,}000{,}000) = n$;
$1{,}300{,}000 = n$; 1,300,000 barrels
(b) 150,000 barrels
(c) 1,250,000 barrels
(d) 400,000 barrels

FAIR OAKS YOUTH SOCCER LEAGUE

Age Group	Number of Players	Percent of Total (rounded to nearest whole percent)
Under 8 years	59	20% $\leftarrow$ 59 players $\approx$ 20% of 298
Ages 8–9	46	15%
Ages 10–11	75	25%
Ages 12–13	74	25%
Ages 14–15	44	15%
Total	**298**	**100%**

You can show these percents in a circle graph. A circle has 360 degrees (written 360°). The 360° represents the entire league, or 100% of the players.

EXAMPLE 4   Drawing a Circle Graph

Using the data on *age groups,* find the number of degrees in the sector that would represent the "Under 8" group, and begin constructing a circle graph.

A complete circle has 360°. Because the "Under 8" group makes up 20% of the total number of players, the number of degrees needed for the "Under 8" sector of the circle graph is 20% of 360°.

$$20.\% \text{ of } 360° = n$$

Write 20% as 0.20 $\rightarrow$ $(0.20)(360°) = n$

$$72° = n$$

Use the circle drawn below and a tool called a **protractor** to make a circle graph. First, using a ruler or straightedge, draw a line from the center of the circle to the left edge. Place the hole in the protractor over the center of the circle, making sure that the 0 mark and the black line on the protractor are right over the line you drew. Find 72° and make a mark as shown in the illustration. Then remove the protractor and use the straightedge to draw a line from the 72° mark to the center of the circle. **This sector is 72° and represents the "Under 8" group.**

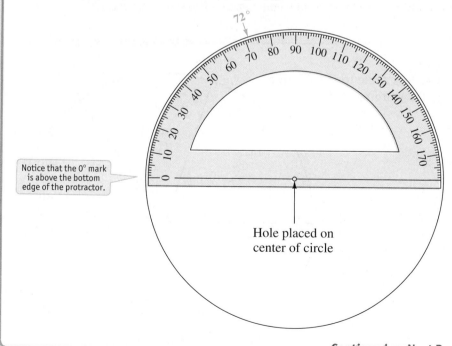

Notice that the 0° mark is above the bottom edge of the protractor.

Hole placed on center of circle

Continued on Next Page

5 Using the information on the soccer age groups in the table on the previous page, find the number of degrees needed for each sector. Then complete the circle graph at the bottom right on this page.

GS **(a)** "Ages 10–11" sector

"Ages 10–11" makes up 25% of the total number of players, so that sector needs 25% of the circle or 25% of 360°.

$$25\% \text{ of } 360° = n$$

$$(0.25)(360°) = n$$

$$\underline{} = n$$

(b) "Ages 12–13" sector

(c) "Ages 14–15" sector

To draw the "Ages 8–9" sector, begin by finding the number of degrees in the sector, which is 15% of the total circle, or 15% of 360°.

$$15\% \text{ of } 360° = n$$

Write 15% as 0.15 → $(0.15)(360°) = n$

$$54° = n$$

Again, place the hole of the protractor over the center of the circle, but this time align 0 with the previous 72° mark. Make a mark at 54° and draw a line from the mark to the center of the circle. **This sector represents the "Ages 8–9" group.**

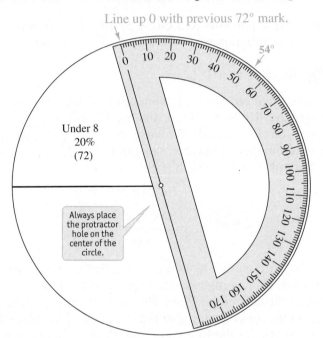

Line up 0 with previous 72° mark.

Under 8
20%
(72)

Always place the protractor hole on the center of the circle.

⚠ CAUTION

You must be certain that the hole in the protractor is placed on the exact center of the circle each time you measure the size of a sector.

◀ **Work Problem ⑤ at the Side.**

Answers

5. **(a)** 90° = n; 90° for "Ages 10–11" sector
   (b) 90°   **(c)** 54°

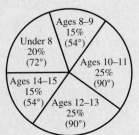

Use this circle for **Problem 5** in the margin.

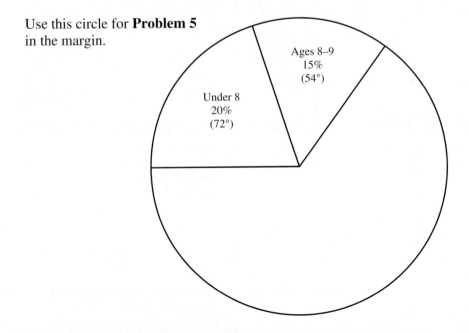

Ages 8–9
15%
(54°)

Under 8
20%
(72°)

9.2 Exercises

FOR EXTRA HELP   *Go to* MyMathLab *for worked-out, step-by-step solutions to exercises enclosed in a square* ▢ *and video solutions to* ▶ *exercises.*

CONCEPT CHECK   *Circle the letter of the correct answer to complete each statement.*

1. A circle graph shows:

(a) a trend or pattern

(b) how a total is divided into parts

(c) symbols to represent amounts

2. The pie-shaped pieces in a circle graph are called:

(a) bars

(b) lines

(c) sectors

3. The tool used to draw circle graphs is called a

(a) calculator

(b) compass

(c) protractor

4. When making a circle graph, remember that a circle has

(a) 360°

(b) 180°

(c) 90°

This circle graph shows the results of an online survey. Adults were asked on which day of the week they were most likely to order food delivered to their homes. Use the graph to answer Exercises 5–14. Write ratios as fractions in lowest terms. **See Examples 1 and 2.**

No Cooking Tonight!

Number of American adults who picked the day of the week they were most likely to order food for delivery:

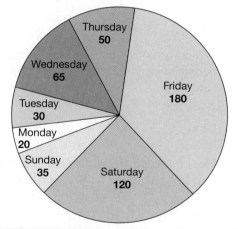

Thursday 50
Wednesday 65
Tuesday 30
Monday 20
Sunday 35
Friday 180
Saturday 120

Data from Technomic, Inc.

5. Find the total number of people in the survey.

6. Which day is the second most popular, and how many people picked it?

7. What is the most popular day of the week to order food, and how many people picked it?

8. Which day is the least popular day of the week to order food, and how many people picked it?

9. Find the ratio of the people who picked Thursday to
Ⓖ the total people in the survey.

$$\text{Picked Thursday} \rightarrow \frac{50}{500} = \frac{50 \div 50}{500 \div 50} = \frac{\square}{\square}$$
$$\text{Total people} \rightarrow$$

10. What is the ratio of the people who picked Friday to
Ⓖ the total people in the survey?

$$\text{Picked Friday} \rightarrow \frac{180}{500} = \frac{180 \div 20}{500 \div 20} = \frac{\square}{\square}$$
$$\text{Total people} \rightarrow$$

11. What is the ratio of the people who picked Friday to the people who picked Saturday?

12. Find the ratio of the people who picked Wednesday to the people who picked Tuesday.

13. What is the ratio of people who picked Friday or Saturday to the people who picked Monday?

14. What is the ratio of the people who picked Thursday to the people who picked Monday or Tuesday?

This circle graph shows the results of an online survey of 400 people who have a cat as a pet. They answered the question, "How does your cat react when you entertain?" The responses are expressed as percents of the 400 people in the survey. Use the graph to answer Exercises 15–20. **See Example 3.**

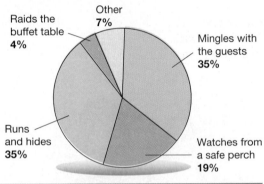

How Does Your Cat React When You Entertain?

Other 7%
Raids the buffet table 4%
Mingles with the guests 35%
Runs and hides 35%
Watches from a safe perch 19%

Data from www.catchannel.com

15. How many people said their cat mingles with the guests?

$$35\% \text{ of } 400 \text{ people} = n$$

16. How many cat owners said their cats watch from a safe perch?

$$19\% \text{ of } 400 \text{ people} = n$$

17. Which response was given least often and by how many people?

18. Which response category represented 7% of the people surveyed? How many people were in this category?

19. How many fewer people said "watches from a safe perch" than said "runs and hides"?

20. How many more people said "mingles with the guests" than said "raids the buffet table"?

This circle graph shows the results of a survey on favorite hot dog toppings in the United States. Each topping is expressed as a percent of the 3200 people in the survey. Use the graph to find the number of people favoring each of the toppings in Exercises 21–26.

21. Onions

22. Sauerkraut

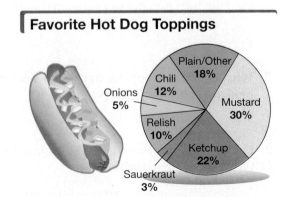

Favorite Hot Dog Toppings

Plain/Other 18%
Chili 12%
Onions 5%
Mustard 30%
Relish 10%
Ketchup 22%
Sauerkraut 3%

Data from National Hot Dog and Sausage Council.

23. The most popular topping

24. The second-most-popular topping

25. How many more people chose chili than chose relish?

26. How many fewer people chose onions than chose mustard?

27. Describe the steps for calculating the size of the sector for each item in a circle graph.

28. A protractor is the tool used to draw a circle graph. Give a brief explanation of what the protractor does and how you would use it to measure and draw each sector in the circle graph.

29. The grades in an algebra course are based on the number of points a student earns. The total number of points in the course is 800. The number of points that can be earned for various class activities is shown in the table below. Find all the missing numbers in the table. Then use the information to draw a circle graph. *See Example 4.*

Item	Number of Points	Percent of Total	Degrees of a Circle
(a) Chapter tests	400 points	50%	_____
(b) Final exam	160 points	_____	_____
(c) Homework	40 points	_____	_____
(d) Quizzes	120 points	_____	_____
(e) Project	80 points	_____	_____

(f) Draw a circle graph using the information from the table. Label each sector with the item and the percent of total for that item.

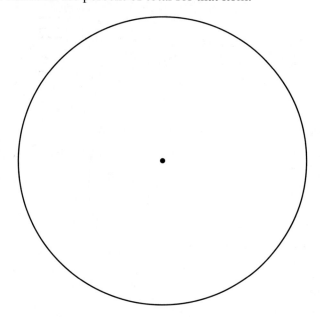

30. A survey of 4488 Americans asked how they fall asleep. The results are shown in the figure on the right.

Use this information to complete the table. Round to the nearest whole percent and to the nearest degree.

	Sleeping Position	Number of Americans	Percent of Total	Number of Degrees
(a)	Side			
(b)	Back			
(c)	Stomach			
(d)	Varies			
(e)	Not sure			

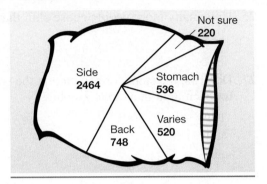

Set to Sleep

Number of Americans surveyed who fall asleep on their:

Not sure 220
Side 2464
Stomach 536
Varies 520
Back 748

(f) Add up the percents. Is the total 100%? Explain why or why not.

(g) Add up the degrees. Is the total 360°? Explain why or why not.

(h) Draw a circle graph using the information from the table above. Label each sector with the sleeping position and the percent of total for that position.

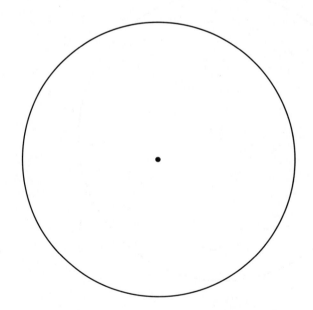

31. White Water Rafting Company divides its annual sales into five categories as shown in the table below.

Category	Annual Sales	Percent of Total
Adventure classes	$12,500	
Grocery/provision sales	$40,000	
Equipment rentals	$60,000	
Rafting tours	$50,000	
Equipment sales	$37,500	

(a) Find the total sales for the year.

(b) Find the percent of the total for each category and write it in the table.

(c) Find the number of degrees in a circle graph for each item.

(d) Make a circle graph showing the sales information. Label each sector with the category and the percent of total for that category.

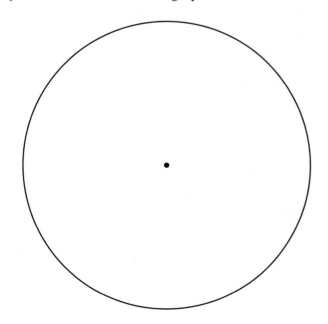

32. A book publisher had 25% of total sales in mysteries, 10% in biographies, 15% in cookbooks, 15% in romance novels, 20% in science, and the rest in travel books.

(a) Find the number of degrees in a circle graph for each type of book.

(b) Draw a circle graph, using the information given. Label each sector with the type of book and the percent of total for that type.

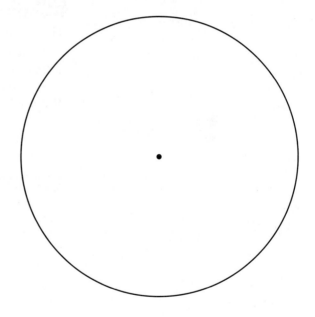

9.3 Bar Graphs and Line Graphs

OBJECTIVES

Read and understand

1 a bar graph;

2 a double-bar graph;

3 a line graph;

4 a comparison line graph.

1 Use the bar graph to find the number of K–12 students in the school district for each of these years.

(a) 2015

The bar for 2015 rises to 30. Multiply 30 by _____ to get _____ students.

(b) 2013

(c) 2012

(d) 2016

OBJECTIVE ▸ 1 **Read and understand a bar graph.** A **bar graph** is useful for showing comparisons. For example, the bar graph below compares the number of K–12 students in a city school district during each of five years.

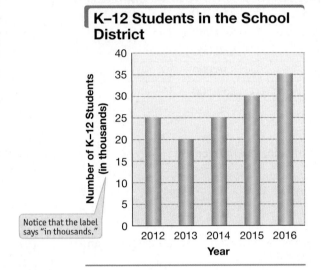

K–12 Students in the School District

Notice that the label says "in thousands."

EXAMPLE 1 **Using a Bar Graph**

How many K–12 students were in the district in 2014?

The bar for 2014 rises to 25. Notice the label along the left side of the graph that says "Number of K–12 Students (in thousands)." The phrase *in thousands* means you have to multiply 25 by 1000 to get 25,000. So **there were 25,000 students in the district in 2014.**

Careful! The answer is **not** 25.

◀ **Work Problem 1** at the Side.

OBJECTIVE ▸ 2 **Read and understand a double-bar graph.** A **double-bar graph** can be used to compare two sets of data. The graph below shows the number of new high-speed Internet connections in each quarter for two different years.

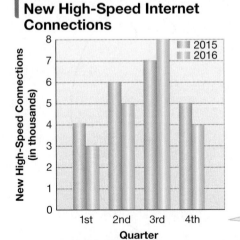

New High-Speed Internet Connections

1st quarter = Jan., Feb., Mar.
2nd quarter = Apr., May, June
3rd quarter = July, Aug., Sept.
4th quarter = Oct., Nov., Dec.

Answers

1. (a) 1000; 30,000 students
   (b) 20,000 students   **(c)** 25,000 students
   (d) 35,000 students

EXAMPLE 2   Reading a Double-Bar Graph

Use the double-bar graph on the previous page to find the following.

(a) The number of new high-speed connections in the second quarter of 2015.

There are two bars for the second quarter. The color code in the upper right-hand corner of the graph tells you that the **red bars** represent 2015. So the **red bar** on the *left* is for the 2nd quarter of 2015. It rises to 6.

Now multiply 6 by 1000 because the label on the left side of the graph says *in thousands.* **There were 6000 new high-speed connections in the second quarter of 2015.**

(b) The number of new high-speed connections in the second quarter of 2016.

The **green bar** for the second quarter rises to 5 and 5 times 1000 is 5000. **There were 5000 new high-speed connections in the second quarter of 2016.**

——————— **Work Problem ❷ at the Side.** ▶

OBJECTIVE ▶ ❸ Read and understand a line graph. A **line graph** is often useful for showing a trend. The line graph below shows the number of trout stocked in the Feather River during five months. Each dot indicates the number of trout stocked during the month directly below that dot.

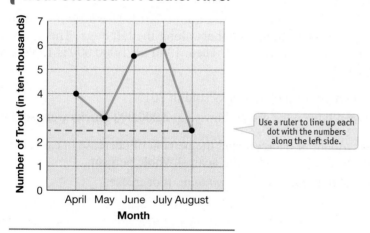

Trout Stocked in Feather River

EXAMPLE 3   Understanding a Line Graph

Use the line graph above to answer each question.

(a) In which month were the least number of trout stocked?

The lowest point on the graph is the dot directly above August, so **the least number of trout were stocked in August.**

(b) How many trout were stocked in August?

Use a ruler or straightedge to line up the August dot with the numbers along the left edge of the graph. (See the red dashed line on the graph.)

The August dot is halfway between the 2 and 3. Notice that the label on the left side says *in ten-thousands.* So the August dot is halfway between $(2 \cdot 10{,}000)$ and $(3 \cdot 10{,}000)$. It is halfway between 20,000 and 30,000. That means **25,000 trout were stocked in August.**

——————— **Work Problem ❸ at the Side.** ▶

❷ Use the double-bar graph on the previous page to find the number of new high-speed connections in 2015 and 2016 for each quarter.

GS (a) 1st quarter.

The red bar represents 2015. It rises to 4, and 4 times

———— is ———— new connections.

The green bar represents 2016. It rises to ————, and ———— times ———— is ———— new connections.

(b) 3rd quarter

(c) 4th quarter

VOCABULARY TIP
A **trend** is a pattern of change that happens over time.

❸ Use the line graph on trout to answer these questions.

(a) How many trout were stocked in June?

(b) How many fewer trout were stocked in May than in April?

(c) Find the total number of trout stocked during the five months.

(d) In which month were the most trout stocked? How many were stocked?

Answers

2. (a) 1000; 4000;
    3; 3 times 1000; 3000
  (b) 7000; 8000 connections
  (c) 5000; 4000 connections
3. (a) 55,000 trout   **(b)** 10,000 fewer trout
  (c) 210,000 trout   **(d)** July; 60,000 trout

4 Use the comparison line graph on car and truck sales to find the following.

(a) The number of cars sold in 2012, 2014, 2015, and 2016.

The dot on the blue line above 2012 lines up with 14 on the left edge. Then multiply 14 by _____ to get sales of _____ cars.

Cars sold in 2014 = _____

Cars sold in 2015 = _____

Cars sold in 2016 = _____

(b) The number of pickup trucks sold in 2012, 2013, 2014, and 2016

(c) In which year were the sales of pickup trucks unchanged from the previous year?

(d) The amount of increase and percent of increase in car sales from 2014 to 2015

Answers

4. **(a)** 1,000,000; 14,000,000;
  2014 is 10,000,000 cars;
  both 2015 and 2016 are 12,000,000 cars
  (b) 3,000,000 trucks in 2012;
  2,000,000 trucks in 2013;
  1,000,000 trucks in 2014;
  2,000,000 trucks in 2016
  (c) 2016
  (d) increase of 2,000,000 cars; 20% increase

OBJECTIVE ▶ **4** **Read and understand a comparison line graph.** Two sets of data can also be compared by drawing two line graphs together as a **comparison line graph.** For example, the line graph below compares the number of cars sold and the number of pickup trucks sold during each of five years.

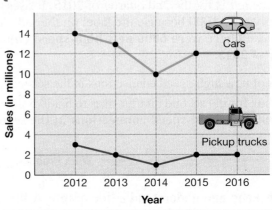

EXAMPLE 4 **Interpreting a Comparison Line Graph**

Use the comparison line graph above to find the following.

(a) The number of cars sold in 2013

Find the dot on the blue line above 2013. Use a ruler or straightedge to line up the dot with the numbers along the left edge. The dot is halfway between 12 and 14, which is 13. Then, 13 times 1,000,000 is **13,000,000 cars sold in 2013.**

(b) The number of pickup trucks sold in 2015

The **red line** on the graph shows that **2,000,000 trucks were sold in 2015.**

(c) The amount of decrease and the percent of decrease in car sales from 2013 to 2014. Round to the nearest whole percent.

Use subtraction to find the *amount* of decrease.

$$\begin{array}{c} \text{car} \\ \text{sales in 2013} \end{array} - \begin{array}{c} \text{car} \\ \text{sales in 2014} \end{array} = \begin{array}{c} \text{amount} \\ \text{of decrease} \end{array}$$

$$13{,}000{,}000 - 10{,}000{,}000 = 3{,}000{,}000$$

Now use the percent equation to find the *percent* of decrease.

percent of original sales = amount of decrease

$$p \cdot 13{,}000{,}000 = 3{,}000{,}000$$

$$\frac{p \cdot 13{,}000{,}000}{13{,}000{,}000} = \frac{3{,}000{,}000}{13{,}000{,}000} \quad \text{Divide both sides by 13,000,000}$$

Multiply by 100% to change the decimal to a percent.

$$p \approx 0.231 \leftarrow \text{Solution in } decimal \text{ form}$$

$$p \approx 0.231 = 23.1\% \approx 23\% \text{ to nearest whole percent}$$

The *amount* of decrease is **3,000,000 cars.**

The *percent* of decrease is **23%** (rounded).

◀ **Work Problem 4 at the Side.**

9.3 Exercises

FOR EXTRA HELP

Go to MyMathLab *for worked-out, step-by-step solutions to exercises enclosed in a square* ☐ *and video solutions to* ▶ *exercises.*

CONCEPT CHECK *Circle the letter of the correct answer to complete each statement.*

1. A comparison of two sets of data can be easily shown in a:

   (a) circle graph

   (b) pictograph

   (c) double-bar graph

2. A line graph is often useful for showing

   (a) trends or patterns

   (b) how a whole is divided into parts

   (c) specific, exact values

3. If a dot on a line graph lines up with "3" on the left side, and the label on the left side says, "in ten-thousands," then the dot represents

   (a) 3000

   (b) 30,000

   (c) 300,000

4. If the top of a bar on a bar graph lines up with "5" on the left side, and the label on the left side says, "in millions," then the bar represents

   (a) 50,000

   (b) 5,000,000

   (c) 5000

A survey of 15,800 U.S. adults who own smartphones asked, "How often do you look at, check, or use your phone?" The results are shown in the bar graph below.

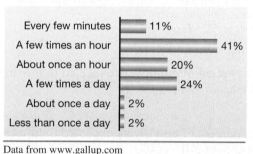

How Often Do Adults Look at Their Smartphones?

Every few minutes — 11%
A few times an hour — 41%
About once an hour — 20%
A few times a day — 24%
About once a day — 2%
Less than once a day — 2%

Data from www.gallup.com

5. The adults in the survey gave a variety of answers about how often they check or use their phones. Which answer was the most common? What percent of the people gave this answer?

6. What answer about smartphone use was given least often? What percent of adults gave this answer?

7. Of the 15,800 adults in the survey, how many said they check their phones "about once an hour."

$$20\% \cdot 15{,}800 = n$$

8. How many adults in the survey said they check their phones "a few times an hour"?

$$41\% \cdot 15{,}800 = n$$

9. Which answer was given by about $\frac{1}{4}$ of the adults?

10. Which answer was given by about $\frac{1}{10}$ of the adults?

This double-bar graph shows the number of workers who were unemployed in one city during the last six months of 2014 and 2015. Use this graph to answer Exercises 11–16. Round percent answers to the nearest whole percent. **See Example 2.**

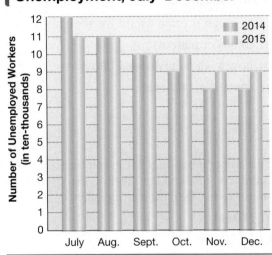

11. In the last half of 2014, which month had the greatest number of unemployed workers? What was the total number unemployed in that month?

12. How many workers were unemployed in October of 2015?

13. How many more workers were unemployed in December of 2015 than in December of 2014?

14. How many fewer workers were unemployed in July of 2015 than in July of 2014?

15. Find the amount of decrease and the percent of decrease in the number of unemployed workers from July 2014 to December 2014.

$120{,}000 -$ _____ $=$ _____ $\left\{\begin{array}{l}\text{amount of}\\\text{decrease}\end{array}\right.$

16. Find the amount of decrease and the percent of decrease in the number of unemployed workers from July 2015 to November 2015.

$110{,}000 -$ _____ $=$ _____ $\left\{\begin{array}{l}\text{amount of}\\\text{decrease}\end{array}\right.$

This double-bar graph shows the percent of employers in a survey who expected to hire more new college graduates, or to hire fewer new college graduates. The percents are given for four different academic years and are rounded to the nearest 5%. Use the graph to answer Exercises 17–22.

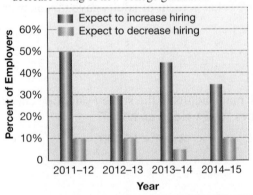

Data from National Association of Colleges and Employers.

17. What percent of employers expected to hire more graduates in 2012–13?

18. What percent of employers expected to hire fewer graduates in 2013–14?

19. Which year or years had the greatest difference between the percent of employers expecting to hire more graduates and the percent expecting to hire fewer graduates?

20. The percent of employers expecting to hire more graduates increased the most between which two consecutive years?

21. There were 172 employers in the survey in 2011–12. How many of them expected to hire fewer graduates? Round your answer to the nearest whole number.

$10.\% \cdot 172 \text{ employers} = n$

22. Of the 172 employers in the 2011–12 survey, how many expected to hire more graduates? Round your answer to the nearest whole number.

$50.\% \cdot 172 \text{ employers} = n$

This line graph shows how yearly sales of smartphones have increased from 2006 to 2015 in United States. Use the line graph to answer Exercises 23–28. ***See Example 3.***

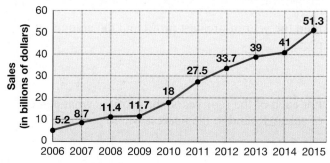

Yearly Smartphone Sales in the United States

Data from www.statista.com

23. (a) Smartphone sales were how much in 2010?

(b) In 2006?

24. (a) What were the dollar sales of smartphones in 2013?

(b) In 2015?

25. Sales of smartphones in 2014 were how much more than in 2010?

26. In which year did smartphone sales increase the least from the previous year? What was the amount of the increase?

27. Find the amount of increase in sales from 2010 to 2015. What was the percent of increase, to the nearest whole percent?

28. Find the amount of increase and the percent of increase in smartphone sales from 2006 to 2010.

This comparison line graph shows the number of DVDs sold by two different chain stores during each of five years. Use this graph to find the annual number of DVDs sold in each year listed in Exercises 29–32 and to answer Exercises 33–34. ***See Example 4.***

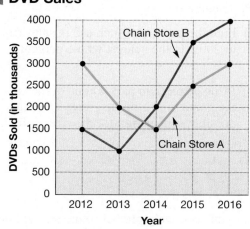

DVD Sales

29. (a) Chain Store A in 2012
(b) Chain Store B in 2012

30. (a) Chain Store A in 2013
(b) Chain Store B in 2013

31. (a) Chain Store A in 2015
(b) Chain Store A in 2016

32. (a) Chain Store B in 2014
(b) Chain Store B in 2015

33. Describe the pattern(s) or trend(s) you see in the graph.

34. Store B used to have lower sales than Store A. What might have happened to cause this change? Give four possible explanations.

This comparison line graph shows the sales and profits of Tacos-To-Go for each of four years. Use the graph to answer Exercises 35–40.

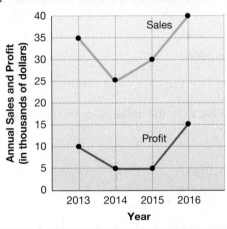

35. Find the year of lowest sales and the amount of sales.

36. Find the year of greatest profit and the amount of profit.

37. For each of the four years, the profit is what percent of sales? Round to the nearest whole percent.

38. Find the total sales for all four years and the total profit for all four years. Total profit is what percent of total sales? Round to the nearest whole percent.

39. Give two possible explanations for the decrease in sales from 2013 to 2014 and two possible explanations for the increase in sales from 2014 to 2016.

40. *Based on the graph,* what conclusion can you make about the relationship between sales and profits for Tacos-To-Go?

Relating Concepts (Exercises 41–44)   For Individual or Group Work

Use the line graph of yearly smartphone sales in Exercises 23–28 as you **work Exercises 41–44 in order.** *Round percent answers to the nearest whole percent.*

41. (a) What overall trend do you see in the line graph on smartphone sales?

   (b) Write two possible explanations for the trend you see in the line graph.

42. (a) Find the percent of increase in sales from 2008 to 2009, to the nearest whole percent.

   (b) Is the increase in part (a) one of the smallest on the graph or one of the greatest? Write one possible explanation for the size of the increase.

43. What future events might cause a decrease in smartphone sales?

44. Can you accurately predict future smartphone sales from the line graph? Explain why or why not.

9.4 The Rectangular Coordinate System

OBJECTIVE ❶ Plot a point, given the coordinates, and find the coordinates, given a point. A bar graph or line graph shows the relationship between two things. The line graph below is from **Example 3** in the previous section. It shows the relationship between the month of the year and the number of trout stocked in the Feather River.

OBJECTIVES

❶ Plot a point, given the coordinates, and find the coordinates, given a point.

❷ Identify the four quadrants and determine which points lie within each one.

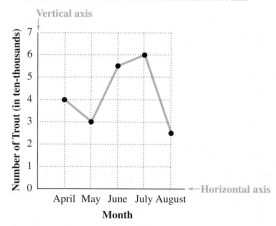

TROUT STOCKED IN FEATHER RIVER

Each black dot on the graph represents a particular month paired with a particular number of trout. This is an example of **paired data.** We write each pair inside parentheses, with a comma separating the two items. To be consistent, we always list the item on the *horizontal axis* first. In this case, the months are shown on the **horizontal axis** (the line that goes "left and right"), and the number of trout is shown along the **vertical axis** (the line that goes "up and down").

VOCABULARY TIP

Horizontal The "z" in **horizontal** zigs from side to side, reminding you that horizontal lines run from left to right.

Paired Data from Line Graph on Trout Stocked in Feather River

(Apr, 40,000) (May, 30,000) (June, 55,000) (July, 60,000) (Aug, 25,000)

Each data pair gives you the location of a particular spot on the graph, and that spot is marked with a dot. This idea of paired data can be used to locate particular places on any flat surface.

Think of a small town laid out in a grid of square blocks, as shown below. To tell a taxi driver where to go, you could say, "the corner of 4th Avenue and 2nd Street" or just "4th and 2nd." As an *ordered pair*, it would be (4, 2). Of course, both you and the taxi driver need to know that the avenue is mentioned first (the number on the horizontal axis) and that the street is mentioned second (the number on the vertical axis). If the driver goes to (2, 4) instead, you'll be at the wrong corner.

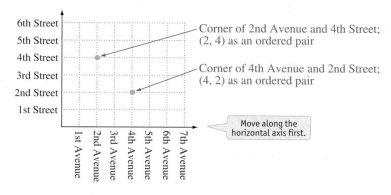

Corner of 2nd Avenue and 4th Street; (2, 4) as an ordered pair

Corner of 4th Avenue and 2nd Street; (4, 2) as an ordered pair

Move along the horizontal axis first.

1 Plot each point on the grid. Write the ordered pair next to each point.

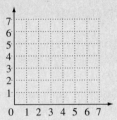

GS **(a)** $(1, 4)$

Start at 0.

Move to the right until you reach 1.

Then move up _____ units and make a dot.

Write (_____ , _____) next to the dot.

(b) $(5, 2)$

(c) $(4, 1)$

Careful! This is **not** the same point as $(1, 4)$.

(d) $(3, 3)$

EXAMPLE 1 **Plotting Points on a Grid**

Use the grid at the right to plot each point.

(a) $(3, 5)$

Start at 0. Move *to the right* along the horizontal axis until you reach 3. Then move *up* 5 units so that you are aligned with 5 on the vertical axis. Make a dot. This is the plot, or graph, of the point $(3, 5)$.

(b) $(5, 3)$

Start at 0. Move *to the right* along the horizontal axis until you reach 5. Then move *up* 3 units so that you are aligned with 3 on the vertical axis. Make a dot. This is the plot, or graph, of the point $(5, 3)$.

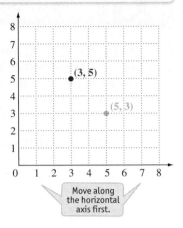

Move along the horizontal axis first.

! CAUTION

In **Example 1**, notice that the points $(3, 5)$ and $(5, 3)$ are *not* the same. The "address" of a point is called an **ordered pair** because the *order* within the pair is important. Always move along the horizontal axis first.

◄ **Work Problem 1** at the Side.

You have used both positive and negative numbers throughout this text. We can extend our grid system to include negative numbers, as shown below. The horizontal axis is now a number line with 0 at the center, positive numbers extending to the right and negative numbers to the left. This horizontal number line is called the **x-axis.**

The vertical axis is also a number line, with positive numbers extending upward from 0 and negative numbers extending downward from 0. The vertical axis is called the **y-axis.** Together, the x-axis and the y-axis form a rectangular **coordinate system.** The point $(0, 0)$ is where the x-axis crosses the y-axis; it is called the **origin.**

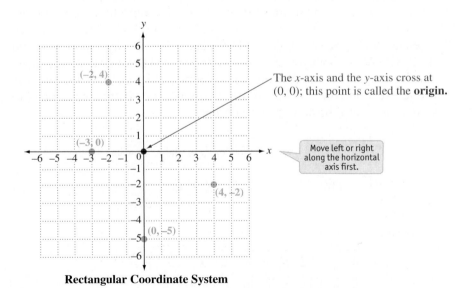

The x-axis and the y-axis cross at $(0, 0)$; this point is called the **origin.**

Move left or right along the horizontal axis first.

Rectangular Coordinate System

Answers

1. (a) move up 4 units; write $(1, 4)$ next to the dot.
(b)–(d) See points on the grid below.

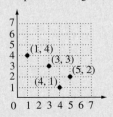

| EXAMPLE 2 | Plotting Points on a Rectangular Coordinate System |

Plot each point on the rectangular coordinate system shown **at the bottom of the previous page.**

(a) $(4, -2)$

Start at 0. Then move left or right along the horizontal x-axis first. Because 4 is *positive,* move *to the right* until you reach 4. Now, because the 2 is *negative,* move *down* 2 units so that you are aligned with -2 on the y-axis. Make a dot and label it $(4, -2)$.

(b) $(-2, 4)$

Starting at 0, move left or right along the horizontal x-axis first. In this case, move *to the left* until you reach -2. Then move *up* 4 units. Make a dot and label it $(-2, 4)$. Notice that $(-2, 4)$ is *not* the same as $(4, -2)$.

(c) $(0, -5)$

Start at 0. Move left or right along the horizontal x-axis first. However, because the first number is 0, do *not* move left or right. Then move *down* 5 units. Make a dot and label it $(0, -5)$.

(d) $(-3, 0)$

Starting at 0, move *to the left* along the horizontal x-axis to -3. Then, because the second number is 0, stay on -3. Do *not* move up or down. Make a dot and label it $(-3, 0)$.

> **Note**
>
> When the *first* number in an ordered pair is 0, the point is on the y-axis, as in **Example 2(c)** above. When the *second* number in an ordered pair is 0, the point is on the x-axis, as in **Example 2(d)** above.

—————— Work Problem **2** at the Side. ▶

A coordinate system and an ordered pair can show the location of any point. The numbers in the ordered pair are called the **coordinates** of the point.

| EXAMPLE 3 | Finding the Coordinates of Points |

Find the coordinates of points A, B, C, and D.

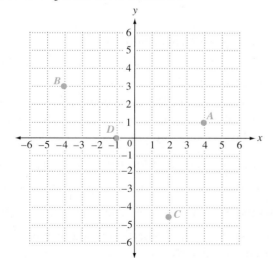

To reach **point A** from the origin, move 4 units *to the right;* then move *up* 1 unit. The coordinates are **(4, 1)**.

—————— **Continued on Next Page**

2 Plot each point on the coordinate system below. Write the ordered pair next to each point.

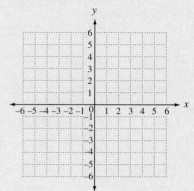

GS **(a)** $(5, -3)$ Start at 0.

Move to the _____ until you reach ____.

Then move _____ 3 units. Make a dot and label it (____, ____).

(b) $(-5, 3)$

GS **(c)** $(0, 3)$ Start at 0.
Do *not* move left or right.

Move _____ 3 units. Make a dot and label it (____, ____).

This point is on the ____-axis.

(d) $(-4, -4)$

GS **(e)** $(-2, 0)$ Start at 0.

Move to the _____ until you reach ____.
Do *not* move up or down.

This point is on the ____-axis.

Answers

2. (a) right, 5; down; $(5, -3)$
   (b) See the point labeled $(-5, 3)$ below.

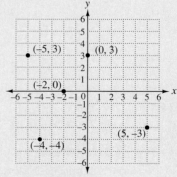

   (c) up; $(0, 3)$; y-axis
   (d) See the point labeled $(-4, -4)$ on the grid.
   (e) left, -2; x-axis

3 Find the coordinates of points A, B, C, D, and E.

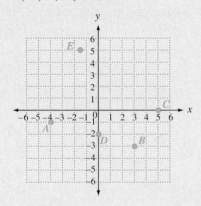

To reach **point B** from the origin, move 4 units *to the left;* then move *up* 3 units. The coordinates are $(-4, 3)$.

To reach **point C** from the origin, move 2 units *to the right;* then move *down* approximately $4\frac{1}{2}$ units. The approximate coordinates are $\left(2, -4\frac{1}{2}\right)$.

To reach **point D** from the origin, move 1 unit *to the left;* then do *not* move either up or down. The coordinates are $(-1, 0)$.

> **Note**
>
> If a point is between the lines on the coordinate system, you can use fractions to give the approximate coordinates. For example, the approximate coordinates of point C above are $\left(2, -4\frac{1}{2}\right)$.

◀ **Work Problem 3 at the Side.**

OBJECTIVE ▶ 2 Identify the four quadrants and determine which points lie within each one. The *x*-axis and *y*-axis divide the coordinate system into four regions, called **quadrants.** These quadrants are numbered with Roman numerals, as shown below. Points on the *x*-axis or *y*-axis are **not in any quadrant.**

4 **(a)** All points in the fourth quadrant are similar in what way? Give two examples of points in the fourth quadrant.

Hint: Look at the grid in the text and find the part that is labeled Quadrant IV. (In Roman numerals, IV represents "4.")

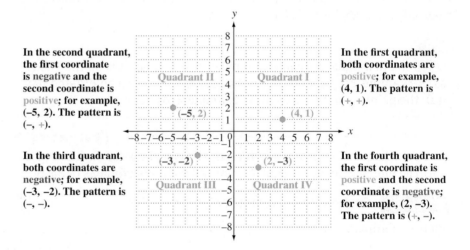

In the second quadrant, the first coordinate is negative and the second coordinate is positive; for example, (–5, 2). The pattern is (–, +).

In the third quadrant, both coordinates are negative; for example, (–3, –2). The pattern is (–, –).

In the first quadrant, both coordinates are positive; for example, (4, 1). The pattern is (+, +).

In the fourth quadrant, the first coordinate is positive and the second coordinate is negative; for example, (2, –3). The pattern is (+, –).

(b) In which quadrant is each point located: $(-2, -6)$; $(0, 5)$; $(-3, 1)$; $(4, -1)$?

EXAMPLE 4 **Working with Quadrants**

(a) All points in the third quadrant are similar in what way? Give two examples of points in the third quadrant.

  For all points in Quadrant III, both coordinates are negative. The pattern is $(-, -)$. There are many possible examples, such as $(-2, -5)$ and $(-4, -4)$. Just be sure that both numbers are negative.

(b) In which quadrant is each point located: $(3, 5)$; $(1, -6)$; $(-4, 0)$?

  For $(3, 5)$ the pattern is $(+, +)$, so the point is in **Quadrant I.**

  For $(1, -6)$ the pattern is $(+, -)$, so the point is in **Quadrant IV.**

  The point corresponding to $(-4, 0)$ is on the *x*-axis, so it **isn't in any quadrant.**

> When one of the numbers is 0, the point is on an axis line.

Answers

3. A is $(-4, -1)$; B is $(3, -3)$; C is $(5, 0)$;
   D is $(0, -2)$; E is approximately $\left(-1\frac{1}{2}, 5\right)$.

4. **(a)** The pattern for all points in Quadrant IV is $(+, -)$. Examples will vary; just be sure that they fit the $(+, -)$ pattern.
   (b) III; no quadrant; II; IV

◀ **Work Problem 4 at the Side.**

9.4 Exercises

FOR EXTRA HELP

Go to MyMathLab *for worked-out, step-by-step solutions to exercises enclosed in a square* ▢ *and video solutions to* ▶ *exercises.*

CONCEPT CHECK *A rectangular coordinate system is shown here.*

1. (a) Make a dot at the origin of the coordinate system.

(b) Label the *x*-axis and the *y*-axis on the coordinate system.

2. (a) Write the Roman numerals for 1, 2, 3, and 4.

(b) Use the Roman numerals to label the four quadrants on the coordinate system.

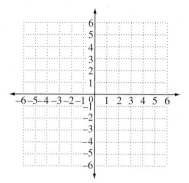

Plot each point on the rectangular coordinate system. Label each point with its coordinates. **See Examples 1 and 2.**

3. $(3, 5)$ $(-2, 2)$ $(-3, -5)$ $(2, -2)$
$(0, 6)$ $(6, 0)$ $(0, -4)$ $(-4, 0)$

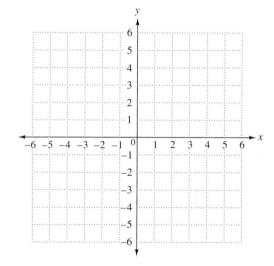

4. $(5, 2)$ $(-3, -3)$ $(4, -1)$ $(-4, 1)$
$(-1, 0)$ $(0, 3)$ $(2, 0)$ $(0, -5)$

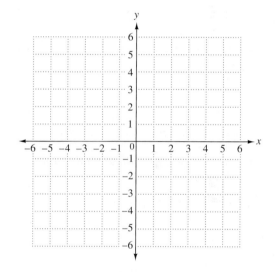

5. $(-5, 3)$ $(4, 4)$ $\left(-2\frac{1}{2}, 0\right)$ $(3, -5)$
$(0, 0)$ $\left(2, \frac{1}{2}\right)$ $(-3, -3)$ $(-1, -6)$

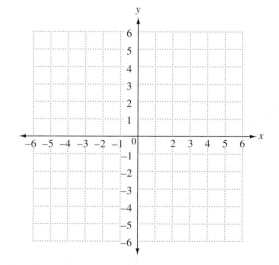

6. $(1, 5)$ $\left(0, 3\frac{1}{2}\right)$ $(-5, -1)$ $(6, -2)$
$(-2, 6)$ $(0, 0)$ $(-3, 3)$ $\left(-\frac{1}{2}, -2\right)$

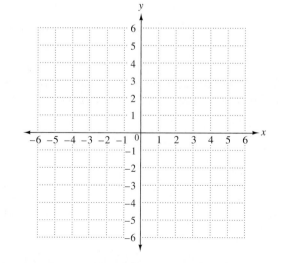

Give the coordinates of each point. See Example 3.

7.

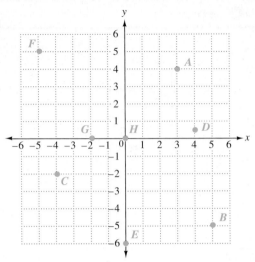

8.

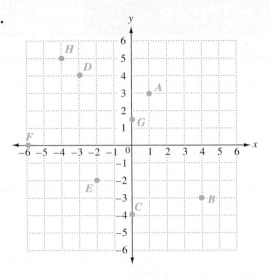

Identify the quadrant in which each point is located. See Example 4.

9. In which quadrant is each point located?

$(-3, -7)$ $(0, 4)$ $(10, -16)$ $(-9, 5)$

10. In which quadrant is each point located?

$(1, 12)$ $(20, -8)$ $(-5, 0)$ $(-14, 14)$

Complete each ordered pair with a number that will make the point fall in the specified quadrant.

11. (a) Quadrant II $(-4,$ ___ $)$

(b) Quadrant IV $(7,$ ___ $)$

(c) No quadrant $($ ___ $, -2)$

(d) Quadrant III $($ ___ $, -1\frac{1}{2})$

(e) Quadrant I $(3\frac{1}{4},$ ___ $)$

12. (a) Quadrant III $(-5,$ ___ $)$

(b) Quadrant I $($ ___ $, 3)$

(c) Quadrant IV $($ ___ $, -\frac{1}{2})$

(d) No quadrant $(6,$ ___ $)$

(e) Quadrant II $($ ___ $, 1\frac{3}{4})$

13. Explain how to graph the ordered pair (a, b), where a and b are positive or negative integers.

14. First explain how to graph the ordered pair (a, b) where a is 0 and b is an integer. Then explain how to graph (a, b) where a is an integer and b is 0.

9.5 | Introduction to Graphing Linear Equations

In earlier chapters you solved equations that had only one variable, such as $2n - 3 = 7$ or $\frac{1}{3}x = 10$. Each of these equations had exactly one solution; n is 5 in the first equation, and x is 30 in the second equation. In other words, there was only *one* number that could replace the variable and make the equation balance. As you take more algebra courses, you will work with equations that have two variables and many different numbers that will make the equation balance. This section will get you started.

OBJECTIVES

1. Graph linear equations in two variables.
2. Identify the slope of a line as positive or negative.

OBJECTIVE ▸ 1 Graph linear equations in two variables. Suppose that you have 6 hours of study time available during a weekend. You plan to study math and psychology. For example, you could spend 4 hours on math and 2 hours on psychology, for a total of 6 hours. Or you could spend $1\frac{1}{2}$ hours on math and $4\frac{1}{2}$ hours on psychology, for a total of 6 hours. Here is a list of *some* of the possible combinations.

Hours on Math	+	Hours on Psychology	=	Total Hours Studying
0	+	6	=	6
1	+	5	=	6
$1\frac{1}{2}$	+	$4\frac{1}{2}$	=	6
3	+	3	=	6
4	+	2	=	6
$5\frac{1}{2}$	+	$\frac{1}{2}$	=	6
6	+	0	=	6

We can write an equation to represent this situation.

$$\underset{m}{\underset{\text{math}}{\text{hours studying}}} + \underset{p}{\underset{\text{psychology}}{\text{hours studying}}} = \underset{6}{\underset{\text{6 hours}}{\text{total of}}}$$

This equation, $m + p = 6$, has *two* variables. The hours spent on math (m) can vary, and the hours spent on psychology (p) can vary.

As you can see, there is more than one solution for this equation. We can list possible solutions as *ordered pairs*. The first number in the pair is the value of m (hours studying math), and the second number in the pair is the corresponding value of p (hours studying psychology).

$(0, 6)$ $(1, 5)$ $\left(1\frac{1}{2}, 4\frac{1}{2}\right)$ $(3, 3)$ $(4, 2)$ $\left(5\frac{1}{2}, \frac{1}{2}\right)$ $(6, 0)$

Another way to show the solutions is to plot the ordered pairs, as you learned to do in the previous section. This method gives us a "picture" of the solutions that we listed on the previous page.

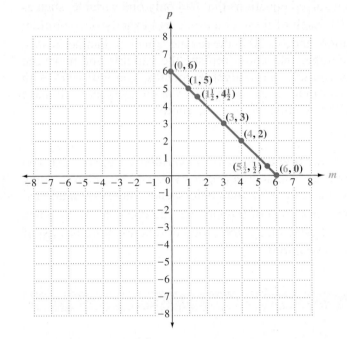

$m + p = 6$

The variables m and p represent hours, and hours can be 0 or positive numbers, but *not* negative numbers.

Notice that all the solutions (all the ordered pairs) lie on a straight line. When you draw a line connecting the ordered pairs, you have graphed the solutions. **Every point on the line is a solution.** You can use the line to find additional solutions besides the ones that we listed. For example, the point $(5, 1)$ is on the line. This point tells you that another solution is 5 hours on math and 1 hour on psychology. The fact that the line is a *straight* line tells you that $m + p = 6$ is a *linear equation.* (The word *line* is part of the word *line*ar.) Later on in algebra you will work with equations whose solutions form a curve rather than a straight line when you graph them.

To draw the line for $m + p = 6$, we really needed only two solutions (two ordered pairs). But it's a good idea to use a third ordered pair as a check. If the three ordered pairs are *not* in a straight line, there is an error in your work.

VOCABULARY TIP

Linear means "like a straight line."

> ### Graphing a Linear Equation
>
> To **graph a linear equation**, find at least three ordered pairs that satisfy the equation. Then plot the ordered pairs on a coordinate system and connect them with a straight line. *Every point on the line is a solution of the equation.*

EXAMPLE 1   Graphing a Linear Equation

Graph $x + y = 3$ by finding three solutions and plotting the ordered pairs. Then use the graph to find a fourth solution of the equation.

There are many possible solutions. Start by picking three different values for x. You can choose any numbers you like, but 0 and small numbers usually are easy to use. Then find the value of y that will make the sum equal to 3. Set up a table to organize the information.

— **Continued on Next Page**

Start by picking easy numbers for x.	x	y	Check that x + y = 3	Ordered Pair (x, y)
	0	3	0 + 3 = 3	(0, 3)
	1	2	1 + 2 = 3	(1, 2)
	2	1	2 + 1 = 3	(2, 1)

Plot the ordered pairs and draw a line through the points, extending it in both directions as shown below.

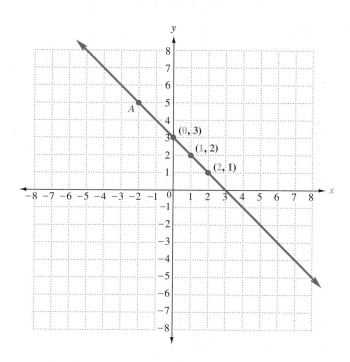

Now you can use the graph to find more solutions of $x + y = 3$. *Every point on the line is a solution.* Suppose that you pick **point A**. The coordinates are $(-2, 5)$.

To check that $(-2, 5)$ is a solution, substitute -2 for x and 5 for y in the original equation.

$$x + y = 3 \quad \text{Original equation}$$
$$-2 + 5 = 3$$

$(-2, 5)$ is a solution, **not** 3

$$3 = 3 \checkmark \quad \text{Balances}$$

The equation balances, so $(-2, 5)$ is another solution of $x + y = 3$.

Note

The line in **Example 1** above was extended in both directions because *every* point on the line is a solution of $x + y = 3$. However, when we graphed the line for the hours spent studying, $m + p = 6$, we did *not* extend the line. That is because the variables m and p represented hours, and hours can only be 0 or positive numbers; all the solutions had to be in the first quadrant.

Work Problem ① at the Side. ▶

① Graph $x + y = 5$ by finding
GS three solutions and plotting the ordered pairs. Then use the graph to find *two* other solutions of the equation.

x	y	Check that x + y = 5	Ordered Pair (x, y)
0	5	0 + 5 = ___	(0, ___)
1	4	1 + 4 = ___	(___, ___)
2	___	2 + ___ = ___	(___, ___)

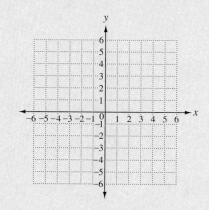

Two other solutions are (___, ___) and (___, ___).

Answers

1. First row of table: 0 + 5 = 5; (0, 5)
   Second row: 1 + 4 = 5; (1, 4)
   Third row: 3; 2 + 3 = 5; (2, 3)

   Plot (0, 5), (1, 4), and (2, 3)

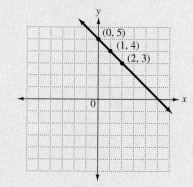

There are many other solutions. Some possibilities: $(-1, 6)$; $(3, 2)$; $(4, 1)$; $(5, 0)$; $(6, -1)$

2 Graph $y = 2x$ by finding three solutions and plotting the ordered pairs. Then use the graph to find *two* other solutions of the equation.

x	$y = 2 \cdot x$	Ordered Pair (x, y)
0	$2 \cdot 0$ is 0	$(0, \text{\_\_})$
1	$2 \cdot 1$ is __	$(1, \text{\_\_})$
2	$2 \cdot$ __ is __	$(\text{\_\_}, \text{\_\_})$

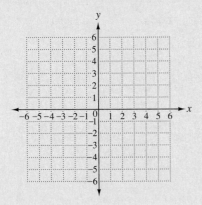

Two other solutions are $(\text{\_\_}, \text{\_\_})$ and $(\text{\_\_}, \text{\_\_})$.

Answers

2. First row of table: $(0, 0)$
Second row: $2 \cdot 1$ is 2; $(1, 2)$
Third row: $2 \cdot 2$ is 4; $(2, 4)$

Plot $(0, 0)$, $(1, 2)$, and $(2, 4)$

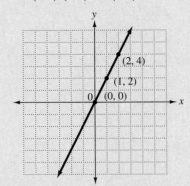

There are many other solutions. Some possibilities: $(3, 6)$; $(-1, -2)$; $(-2, -4)$; $(-3, -6)$

EXAMPLE 2 **Graphing a Linear Equation**

Graph $y = -3x$ by finding three solutions and plotting the ordered pairs. Then use the graph to find a fourth solution of the equation.

You can choose any three values for x, but small numbers such as 0, 1, and 2 are easy to use. Then $y = -3x$ tells you that y is -3 times the value of x.

$$y = -3x$$
$$y \text{ is } -3 \text{ times } x$$

First set up a table.

> Start by picking easy numbers for x.

x	$y = -3 \cdot x$	Ordered Pair (x, y)
0	$-3 \cdot 0$ is 0	$(0, 0)$
1	$-3 \cdot 1$ is -3	$(1, -3)$
2	$-3 \cdot 2$ is -6	$(2, -6)$

Plot the ordered pairs and draw a line through the points. Be sure to draw arrows on both ends of the line to show that it continues in both directions.

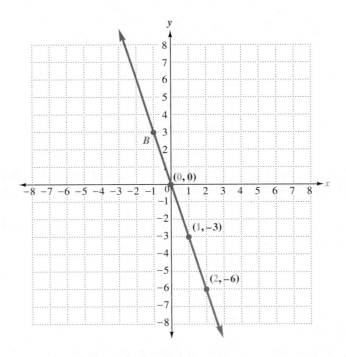

Now use the graph to find more solutions. *Every point on the line is a solution.*
Suppose that you pick **point B**. The coordinates are $(-1, 3)$. To check that $(-1, 3)$ is a solution, substitute -1 for x and 3 for y in the original equation.

$$y = -3x \qquad \text{Original equation}$$
$$3 = -3(-1)$$
$$3 = 3 \checkmark \qquad \text{Balances}$$

The equation balances, so $(-1, 3)$ is another solution of $y = -3x$.

◀ **Work Problem 2 at the Side.**

EXAMPLE 3 **Graphing a Linear Equation**

Graph $y = \dfrac{1}{2}x$ by finding three solutions and plotting the ordered pairs. Then use the graph to find a fourth solution of the equation.

Complete the table. The coefficient of x is $\frac{1}{2}$, so choose even numbers like 2, 4, and 6 as values for x because they are easy to divide in half. The equation $y = \frac{1}{2}x$ tells you that y is $\frac{1}{2}$ *times* the value of x.

Pick *even* numbers for x; they are easy to multiply by $\frac{1}{2}$

x	$y = \frac{1}{2} \cdot x$	Ordered Pair (x, y)
2	$\frac{1}{2} \cdot 2$ is **1**	**(2, 1)**
4	$\frac{1}{2} \cdot 4$ is **2**	**(4, 2)**
6	$\frac{1}{2} \cdot 6$ is **3**	**(6, 3)**

Plot the ordered pairs and draw a line through the points.

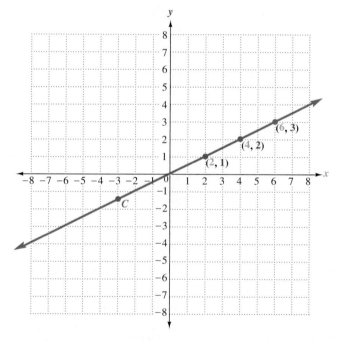

Now use the graph to find more solutions. ***Every point on the line is a solution.***

Suppose that you pick **point C**. The coordinates are $\left(-3, -1\frac{1}{2}\right)$. Check that $\left(-3, -1\frac{1}{2}\right)$ is a solution by substituting -3 for x and $-1\frac{1}{2}$ for y.

$$y = \frac{1}{2}x \quad \text{Original equation}$$

$$-1\frac{1}{2} = \frac{1}{2}(-3)$$

$-1\frac{1}{2}$ is equivalent to $-\frac{3}{2}$

$$-1\frac{1}{2} = -\frac{3}{2} \checkmark \quad \text{Balances}$$

The equation balances, so $\left(-3, -1\frac{1}{2}\right)$ is another solution of $y = \frac{1}{2}x$.

—— **Work Problem ③ at the Side.** ▶

③ Graph $y = -\frac{1}{2}x$ by finding three solutions and plotting the ordered pairs. Then use the graph to find *two* more solutions.

x	$y = -\frac{1}{2} \cdot x$	Ordered Pair (x, y)
2	$-\frac{1}{2} \cdot 2$ is -1	$(2, \underline{})$
4	$-\frac{1}{2} \cdot 4$ is $\underline{}$	$(4, \underline{})$
6	$-\frac{1}{2} \cdot \underline{}$ is $\underline{}$	$(\underline{}, \underline{})$

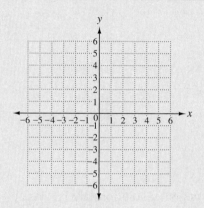

Two other solutions are $(\underline{}, \underline{})$ and $(\underline{}, \underline{})$.

Answers

3. First row of table: $(2, -1)$

Second row: $-\frac{1}{2} \cdot 4$ is -2; $(4, -2)$

Third row: $-\frac{1}{2} \cdot 6$ is -3; $(6, -3)$

Plot $(2, -1)$, $(4, -2)$, and $(6, -3)$

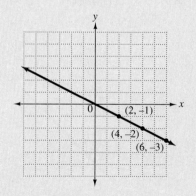

There are many other solutions. Some possibilities: $(0, 0)$; $(-2, 1)$; $(-4, 2)$; $(-6, 3)$; $\left(1, -\frac{1}{2}\right)$; $\left(3, -1\frac{1}{2}\right)$; $\left(5, -2\frac{1}{2}\right)$

④ **Graph the equation** $y = x - 5$
⑤ by finding three solutions and plotting the ordered pairs. Then use the graph to find *two* more solutions.

x	$y = x - 5$	Ordered Pair (x, y)
1	$1 - 5$ is -4	$(1, \underline{\quad})$
2	$2 - 5$ is $\underline{\quad}$	$(2, \underline{\quad})$
3	$\underline{\quad} - 5$ is $\underline{\quad}$	$(\underline{\quad}, \underline{\quad})$

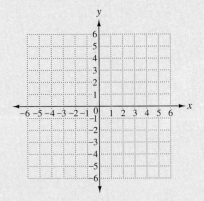

Two other solutions are $(\underline{\quad}, \underline{\quad})$ and $(\underline{\quad}, \underline{\quad})$.

Answers

4. First row of table: $(1, -4)$
Second row: $2 - 5$ is -3; $(2, -3)$
Third row: $3 - 5$ is -2; $(3, -2)$

Plot $(1, -4)$, $(2, -3)$, and $(3, -2)$

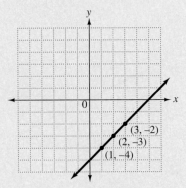

There are many other solutions. Some possibilities: $(-1, -6)$; $(0, -5)$; $\left(4\frac{1}{2}, -\frac{1}{2}\right)$; $(5, 0)$; $\left(5\frac{1}{2}, \frac{1}{2}\right)$

EXAMPLE 4 **Graphing a Linear Equation**

Graph the equation $y = x + 4$ by finding three solutions and plotting the ordered pairs. Then use the graph to find two more solutions of the equation.

First set up a table. The equation $y = x + 4$ tells you that y must be 4 more than the value of x.

x	$y = x + 4$	Ordered Pair (x, y)
0	$0 + 4$ is **4**	$(0, 4)$
1	$1 + 4$ is **5**	$(1, 5)$
2	$2 + 4$ is **6**	$(2, 6)$

Plot the ordered pairs and draw a line through the points.

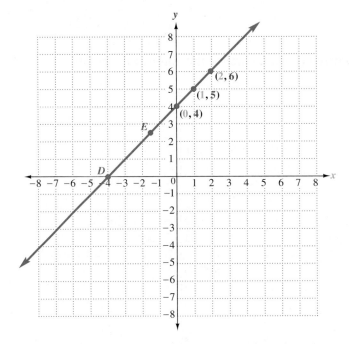

Now use the graph to find two more solutions. ***Every point on the line is a solution.*** Suppose that you pick **point D** at $(-4, 0)$ and **point E** at $\left(-1\frac{1}{2}, 2\frac{1}{2}\right)$. Check that both ordered pairs are solutions.

CHECK $(-4, 0)$

$$y = x + 4$$
$$\downarrow \qquad \downarrow$$
$$0 = -4 + 4$$

$(-4, 0)$ is a solution, **not** 0.

$$0 = 0 \checkmark \quad \text{Balances}$$

CHECK $\left(-1\frac{1}{2}, 2\frac{1}{2}\right)$

$$y = x + 4$$
$$\downarrow \qquad \downarrow$$
$$2\frac{1}{2} = -1\frac{1}{2} + 4$$
$$\downarrow \qquad \downarrow$$
$$\frac{5}{2} = -\frac{3}{2} + \frac{8}{2}$$
$$\frac{5}{2} = \frac{5}{2} \checkmark \quad \text{Balances}$$

$2\frac{1}{2}$ can be written as $\frac{5}{2}$ and $-1\frac{1}{2}$ can be written as $-\frac{3}{2}$

Both equations balance, so $(-4, 0)$ and $\left(-1\frac{1}{2}, 2\frac{1}{2}\right)$ are also solutions of $y = x + 4$.

◄ **Work Problem ④ at the Side.**

OBJECTIVE ▶ **2** **Identify the slope of a line as positive or negative.**
Let's look again at some of the lines that we graphed for various equations.
All are straight lines, but some are almost flat and some tilt steeply upward
or downward. The **slope** of a line is how much tilt it has.

Graph from **Example 1:** $x + y = 3$

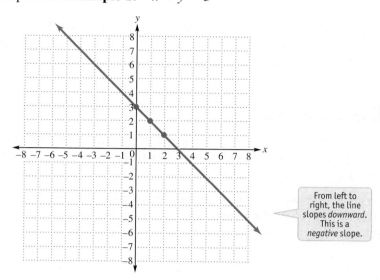

From left to
right, the line
slopes *downward*.
This is a
negative slope.

As you move from *left to right,* the line slopes downward, as if you were walking
down a hill. When a line tilts downward, we say that it has a **negative slope.**
Now look at the table of solutions we used to draw the line.

The value of
x is *increasing*
from 0 to 1 to 2

x	y
0	3
1	2
2	1

The value of
y is *decreasing*
from 3 to 2 to 1

As the value of *x increases,* the value of *y* does the *opposite*—it *decreases.*
Whenever one variable increases while the other variable decreases, the line
will have a negative slope.

Graph from **Example 3:** $y = \dfrac{1}{2}x$

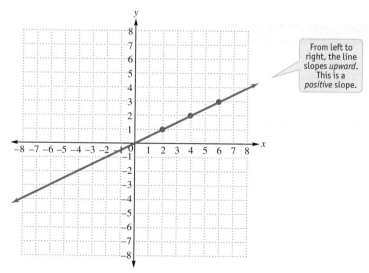

From left to
right, the line
slopes *upward*.
This is a
positive slope.

As you move from *left to right,* this line slopes upward, as if you were walking
up a hill. When a line tilts upward, we say that it has a **positive slope.**

5 Look back at the graphs in **Margin Problems 2 and 3.** Then complete these sentences.

(a) The graph of $y = 2x$ has a _____ slope. As the value of x increases, the value of y _____ .

Now look at the table of solutions we used to draw the line.

The value of x is *increasing* from 2 to 4 to 6

x	y
2	1
4	2
6	3

The value of y is *increasing* from 1 to 2 to 3

As the value of x *increases*, the value of y does the *same* thing—it also *increases*. Whenever both variables do the same thing (both increase or both decrease), the line will have a positive slope.

Positive and Negative Slopes

As you move from left to right, a line with a *positive* slope tilts *upward* or rises. As the value of one variable increases, the value of the other variable also increases (does the same).

As you move from left to right, a line with a *negative* slope tilts *downward* or falls. As the value of one variable increases, the value of the other variable decreases (does the opposite).

EXAMPLE 5   **Identifying Positive or Negative Slope in a Line**

Look back at the graph of $y = -3x$ in **Example 2.** Then complete these sentences.

The graph of $y = -3x$ has a _____ slope.

As the value of x increases, the value of y _____ .

The graph of $y = -3x$ has a **negative** slope (because it tilts downward).

As the value of x increases, the value of y **decreases** (does the opposite).

◀ **Work Problem** **5** **at the Side.**

(b) The graph of $y = -\frac{1}{2}x$ has a _____ slope. As the value of x increases, the value of y _____ .

Study Skills Reminder

Does your instructor give a final exam? If it covers everything taught in the course, it is called a cumulative exam or a comprehensive exam. In that case, you will need a plan to prepare for the final. See the Study Skills activity "Preparing for Your Final Exam." It has suggestions to help you prepare for a comprehensive final exam.

Answers

5. (a) positive; increases
   (b) negative; decreases

9.5 Exercises

FOR
EXTRA
HELP

Go to MyMathLab *for worked-out, step-by-step solutions to exercises enclosed in a square* [ ] *and video solutions to* ▶ *exercises.*

CONCEPT CHECK *Circle the letter of the correct answer to complete each statement.*

1. The graph of a linear equation is a
   (a) curved line
   (b) circle
   (c) straight line

2. When you draw the graph of a linear equation
   (a) only three points on the line are solutions
   (b) every point on the line is a solution
   (c) none of the solutions are on the line

Graph each equation by completing the table to find three solutions and plotting the ordered pairs. Then use the graph to find two *other solutions.* **See Example 1.**

3. $x + y = 4$

x	y	Ordered Pair (x, y)
0	4	(0, 4)
1	3	(1, ___)
2		(2, ___)

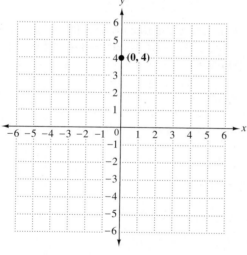

Two other solutions are (___, ___) and (___, ___).

4. $x + y = -4$

x	y	Ordered Pair (x, y)
0	-4	(0, -4)
1	-5	(1, ___)
2		(2, ___)

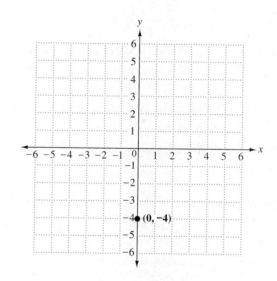

Two other solutions are (___, ___) and (___, ___).

5. $x + y = -1$

x	y	Ordered Pair (x, y)
0		
1		
2		

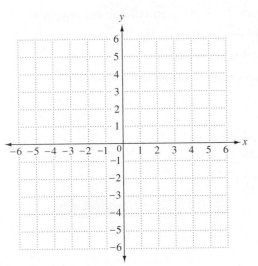

Two other solutions are (____, ____) and (____, ____).

6. $x + y = 1$

x	y	Ordered Pair (x, y)
0		
1		
2		

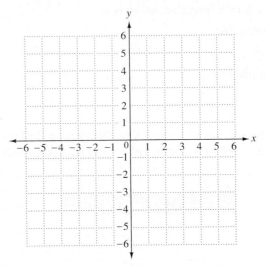

Two other solutions are (____, ____) and (____, ____).

7. Where does the line in **Exercise 3** cross the
y-axis? ____ Where does the line in **Exercise 5**
cross the y-axis? ____
Based on these examples, where would the graph of
$x + y = -6$ cross the y-axis? ____
Where would the graph of $x + y = 99$ cross the
y-axis? ____

8. Look at where the line crosses the *x*-axis and where it
crosses the *y*-axis in **Exercises 4 and 6.** What pattern
do you see?

Graph each equation. Make your own table using the listed values of x.
See Examples 2 and 4.

9. $y = x - 2$
Use 1, 2, and 3 as the values of *x*.

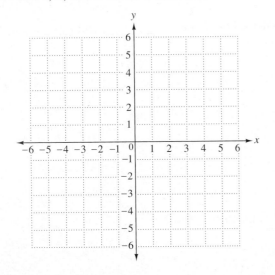

10. $y = x + 1$
Use 1, 2, and 3 as the values of *x*.

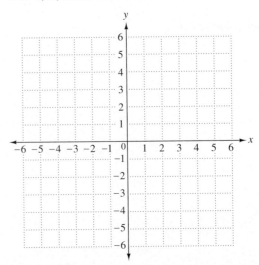

11. $y = x + 2$

Use $0, -1,$ and -2 as the values of x.

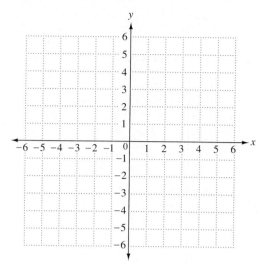

12. $y = x - 1$

Use $0, -1,$ and -2 as the values of x.

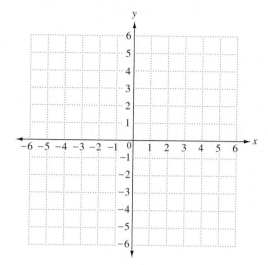

13. $y = -3x$

Use $0, 1,$ and 2 as the values of x.

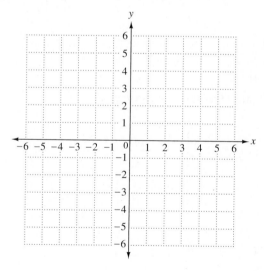

14. $y = -2x$

Use $0, 1,$ and 2 as the values of x.

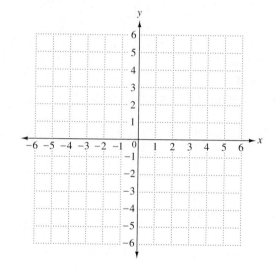

15. Look back at the graphs in **Exercises 3, 5, 9, 11, and 13.** Which lines have a positive slope? Which lines have a negative slope?

16. Look back at the graphs in **Exercises 4, 6, 10, 12, and 14.** Which lines have a positive slope? Which lines have a negative slope?

Graph each equation. Make your own table using the listed values of x.
See Examples 1–4.

17. $y = \dfrac{1}{3}x$

Use 0, 3, and 6 as the values of x.

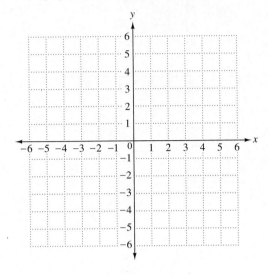

18. $y = \dfrac{1}{2}x$

Use 0, 2, and 4 as the values of x.

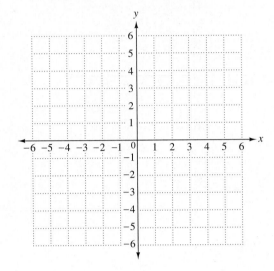

19. $y = x$

Use -1, -2, and -3 as the values of x.

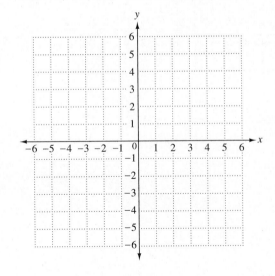

20. $x + y = 0$

Use 1, 2, and 3 as the values of x.

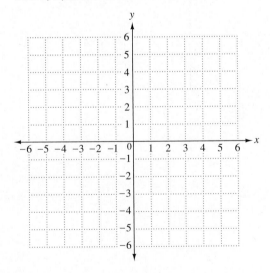

21. $y = -2x + 3$

Use 0, 1, and 2 as the values of x.

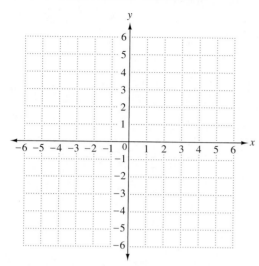

22. $y = 3x - 4$

Use 0, 1, and 2 as the values of x.

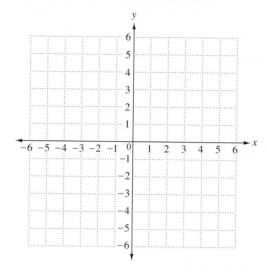

Graph each equation. Choose three values for x. Make a table showing your x values and the corresponding y values. After graphing the equation, state whether the line has a positive or negative slope. ***See Examples 1–5.***

23. $x + y = -3$

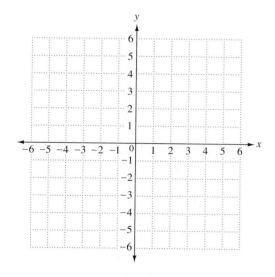

24. $x + y = 2$

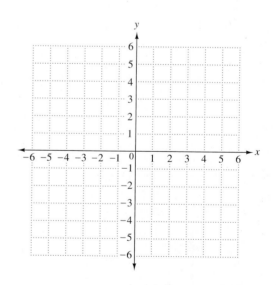

25. $y = \frac{1}{4}x$ (*Hint:* Try using 0 and multiples of 4 as the values of x.)

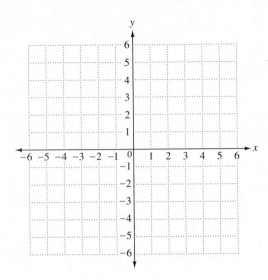

26. $y = -\frac{1}{3}x$ (*Hint:* Try using 0 and multiples of 3 as the values of x.)

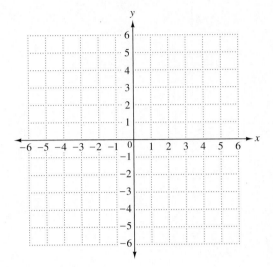

27. $y = x - 5$

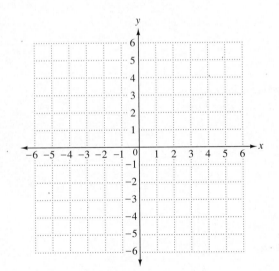

28. $y = x + 4$

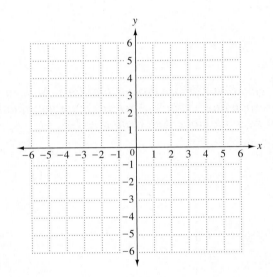

29. $y = -3x + 1$

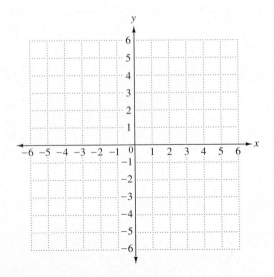

30. $y = 2x - 2$

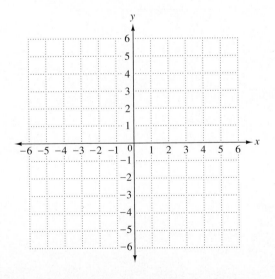

Chapter 9 *Summary*

Key Terms

9.1

table   A table presents data organized into rows and columns.

pictograph   A pictograph uses symbols or pictures to represent various amounts.

9.2

circle graph   A circle graph shows how a total amount is divided into parts or sectors. It is based on percents of 360°.

protractor   A protractor is a tool (usually in the shape of a half-circle) used to measure the number of degrees in an angle or part of a circle.

9.3

bar graph   A bar graph uses bars of various heights to show quantity or frequency.

double-bar graph   A double-bar graph compares two sets of data by showing two sets of bars.

line graph   A line graph uses dots connected by line segments to show trends.

comparison line graph   A comparison line graph shows how two or more sets of data relate to each other by showing a line graph for each set of data.

9.4

paired data   When each number in a set of data is matched with another number by some rule of association, we call it paired data.

horizontal axis   The horizontal axis is the number line in a coordinate system that goes "left and right."

vertical axis   The vertical axis is the number line in a coordinate system that goes "up and down."

ordered pair   An ordered pair is the "address" of a point in a coordinate system. The *order* of the numbers is important. The first number tells how far to move left or right from 0 along the horizontal axis. The second number tells how far to move up or down from 0 along the vertical axis.

x-axis   The horizontal axis is called the *x*-axis.

y-axis   The vertical axis is called the *y*-axis.

coordinate system   Together, the *x*-axis and the *y*-axis form a rectangular coordinate system. *Example:* See figure below.

origin   The *x*-axis and the *y*-axis in a rectangular coordinate system cross at $(0, 0)$; this point is called the origin. *Example:* See black dot marking $(0, 0)$ in figure below.

coordinates   Coordinates are the numbers in the ordered pair that specify the location of a point on a rectangular coordinate system. *Example:* See the point $(-4, 2)$ in the figure below.

quadrants   The *x*-axis and the *y*-axis divide the coordinate system into four regions called quadrants; they are designated with Roman numerals. *Example:* The point $(-4, 2)$ below is in Quadrant II.

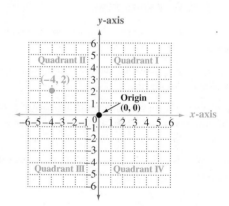

9.5

graph a linear equation   All the solutions of a linear equation (all the ordered pairs that satisfy the equation) lie along a straight line. When you draw the line, you have graphed the equation.

slope   The slope of a line is how much tilt the line has. As you move from left to right, an upward tilting line has a positive slope. A downward tilting line has a negative slope.

New Symbols

(x, y) ordered pair

Test Your Word Power

See how well you have learned the vocabulary in this chapter.

1 A **circle graph**
  A. uses symbols to represent various amounts
  B. is useful for showing trends
  C. shows how a total amount is divided into parts.

2 A **line graph**
  A. uses symbols to represent various amounts
  B. is useful for showing trends
  C. shows how a total amount is divided into parts.

3 A **protractor** is used to
  A. measure degrees in an angle
  B. draw circles of various sizes
  C. measure lengths.

4 Two sets of data can be compared using a
  A. circle graph
  B. pictograph
  C. double-bar graph.

5 The **x-axis** in a rectangular coordinate system is the
  A. number line that goes "up and down"
  B. number line that goes "left and right"
  C. center point of the grid.

6 **Quadrants** are the
  A. numbers in an ordered pair
  B. four regions in a coordinate system
  C. paired data shown on a graph.

7 **Coordinates** are
  A. points in a straight line on a coordinate system
  B. designated with Roman numerals
  C. numbers used to locate a point in a coordinate system.

8 A **rectangular coordinate system**
  A. is formed by the x-axis and y-axis
  B. is divided into eight quadrants
  C. has only positive numbers.

9 The **origin** is the point where
  A. the y-axis and x-axis cross
  B. the coordinates are $(1, -1)$
  C. there is no ordered pair.

Answers To Test Your Word Power

1. C; *Example:* A circle graph can show how a 24-hour day is divided among various activities.

2. B; *Example:* A line graph can show changes in the amount of sales over several months or several years.

3. A; *Example:* A protractor is a tool, usually in the shape of a half-circle, used to measure or draw angles of a certain number of degrees. A tool called a compass is used to draw circles of various sizes.

4. C; *Example:* Two sets of bars in different colors can compare monthly unemployment figures for two different years.

5. B; *Example:* The x-axis is the horizontal number line with 0 at the center, negative numbers extending to the left, and positive numbers extending to the right.

6. B; *Example:* The x-axis and the y-axis divide the coordinate system into four regions; each region is designated by a Roman numeral.

7. C; *Example:* To locate the point $(2, -3)$, start at the origin and move 2 units to the right along the x-axis, then move down 3 units.

8. A; *Example:* Together, a horizontal number line (x-axis) and a vertical number line (y-axis) form a rectangular coordinate system.

9. A; *Example*: The origin has coordinates of $(0, 0)$ because it is the point where the x-axis and y-axis cross.

Quick Review

Concepts

9.1 Reading a Table

The data in a table is organized into rows and columns. As you read from left to right along each row, check the heading at the top of each column.

Examples

Use the table below to answer the question.

PER CAPITA CONSUMPTION OF SELECTED BEVERAGES IN GALLONS

	2000	2005	2010	2015
Milk	23.9	22.5	21.0	20.6
Coffee	20.2	26.3	24.3	23.3
Bottled water	11.6	16.7	23.2	28.3
Soft drinks	50.6	53.2	52.3	46.0

In which year was the consumption of milk greater than the consumption of coffee?
  Read across the rows labeled "milk" and "coffee" from left to right, comparing the numbers in each column. **In 2000 the number for milk (23.9 gallons) is greater than for coffee (20.2 gallons).**

Concepts	Examples

9.1 Reading a Pictograph

A pictograph uses symbols or pictures to represent various amounts. The *key* tells you how much each symbol represents. A fractional part of a symbol represents a fractional part of the symbol's value.

Use the pictograph to answer these questions.

Sales of Food and Drink at U.S. Restaurants

2000 [symbol]

2005 [symbol] [symbol] [half symbol]

2010 [symbol] [symbol] [symbol] [symbol]

2015 [symbol] [symbol] [symbol] [symbol] [symbol] [symbol]

[symbol] = $100 billion

(a) How much was spent at U.S. restaurants in 2005?

Sales for 2005 are represented by 2 whole symbols (2 • $100 billion = $200 billion) plus half of a symbol ($\frac{1}{2}$ of $100 billion = $50 billion) for a total of **$250 billion.**

(b) How much did restaurant expenditures increase from 2010 to 2015?

The year 2015 shows two more symbols than 2010, so the increase is 2 • $100 billion = **$200 billion.**

9.2 Constructing a Circle Graph

Step 1 Determine the percent of the total for each item.

Step 2 Find the number of degrees out of 360° that each percent represents.

Step 3 Use a protractor to measure the number of degrees for each item in the circle.

Step 4 Label each sector in the circle with the item name and percent of total for that item.

Construct a circle graph for these expenses from a business trip.

Item	Amount
Transportation	$350
Lodging	$300
Food	$250
Other	$100
Total	$1000

(Continued)

Concepts	Examples

9.2 Constructing a Circle Graph (*Continued*)

Item	Amount	Percent of Total		Sector Size
Transportation	$350	$\dfrac{\$350}{\$1000} = \dfrac{7}{20} = 35\%$	so 35% of $360°$ $= (0.35)(360)$	$= \mathbf{126°}$
Lodging	$300	$\dfrac{\$300}{\$1000} = \dfrac{3}{10} = 30\%$	so 30% of $360°$ $= (0.30)(360)$	$= \mathbf{108°}$
Food	$250	$\dfrac{\$250}{\$1000} = \dfrac{1}{4} = 25\%$	so 25% of $360°$ $= (0.25)(360)$	$= \mathbf{90°}$
Other	$100	$\dfrac{\$100}{\$1000} = \dfrac{1}{10} = 10\%$	so 10% of $360°$ $= (0.10)(360)$	$= \mathbf{36°}$

Business Trip Expenses

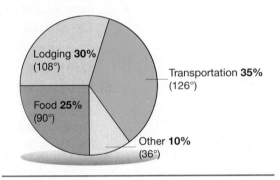

Lodging **30%** (108°)

Transportation **35%** (126°)

Food **25%** (90°)

Other **10%** (36°)

9.3 Reading a Bar Graph

The height of the bar is used to show the quantity or frequency (number) in a specific category. If necessary, use a ruler or straightedge to line up the top of each bar with the numbers on the left side of the graph.

Use the bar graph below to determine the number of students who earned each letter grade.

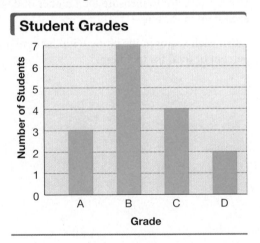

Student Grades

Write the number of students earning each letter grade.
A: 3 students; B: 7 students; C: 4 students; D: 2 students.

Concepts	**Examples**

9.3 Reading a Line Graph

A dot is used to show the number or quantity in a specific class. The dots are connected with line segments. This kind of graph is used to show a trend.

The line graph below shows the annual sales for the Fabric Supply Center for each of four years.

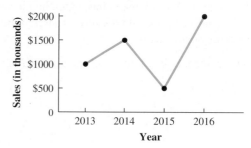

Which year had the lowest sales? What was the amount of sales that year?

2015; sales were $500 \cdot 1000 = $ **$500,000**

9.4 Plotting Points

Start at the center of the coordinate system (the origin). The first number in an ordered pair tells you how far to move *left* or *right* along the horizontal axis; *positive* numbers are to the *right*, and *negative* numbers are to the *left*. The second number in an ordered pair tells you how far to move *up* or *down*; *positive* numbers are *up*, and *negative* numbers are *down*.

To plot $(3, -2)$, start at 0, move to the *right* 3 units, and then move *down* 2 units.

To plot $(-2, 3)$, start at 0, move to the *left* 2 units, and then move *up* 3 units.

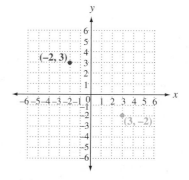

9.4 Identifying Quadrants

The *x*-axis and *y*-axis divide the coordinate system into four regions called quadrants. The quadrants are designated with the Roman numerals I, II, III, and IV.

Points in the first quadrant fit the pattern $(+, +)$.

Points in the second quadrant fit the pattern $(-, +)$.

Points in the third quadrant fit the pattern $(-, -)$.

Points in the fourth quadrant fit the pattern $(+, -)$.

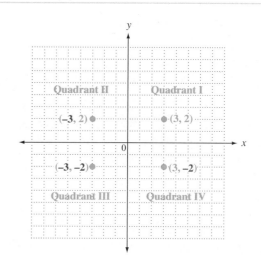

Concepts	Examples

9.5 Graphing Linear Equations

Choose any three values for *x*. Then find the corresponding values of *y*. Plot the three ordered pairs on a coordinate system. Draw a line through the points, extending it in both directions. If the three points do *not* lie on a straight line, there is an error in your work.

Every point on the line is a solution of the given equation.

Graph $y = -2x$. Then use the graph to find *two* other solutions.

x	$y = -2 \cdot x$	Ordered Pair (x, y)
1	$-2 \cdot 1$ is -2	$(1, -2)$
2	$-2 \cdot 2$ is -4	$(2, -4)$
3	$-2 \cdot 3$ is -6	$(3, -6)$

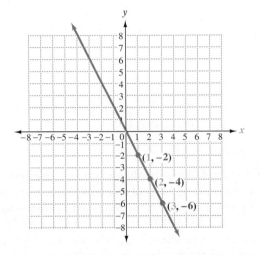

***Every point on the line* is a solution.** Some of the other solutions are $(0, 0)$; $(-1, 2)$; $(-2, 4)$; and $\left(-2\frac{1}{2}, 5\right)$.

Chapter 9 Review Exercises

9.1 *Use the table at the right to answer Exercises 1–4.*

1. Which sport had the
   (a) fewest men's teams?

   (b) second-greatest number of women's teams?

2. Which sport had
   (a) about 16,000 female athletes?

   (b) about 14,000 male athletes?

PARTICIPATION IN SELECTED NCAA SPORTS 2013–2014

Sport	Males			Females		
	Teams	Athletes	Average Squad	Teams	Athletes	Average Squad
Basketball	1081	18,320	16.9	1101	16,319	14.8
Soccer	818	23,602	28.9	1022	26,358	25.8
Gymnastics	17	370	21.8	82	1513	18.5
Cross country	982	14,218		1061	15,922	

Data from The National Collegiate Athletic Association (NCAA).

3. (a) How many more men participated in basketball than cross country?

   (b) How many fewer women participated in gymnastics than cross country?

4. Find the average squad size for men's and women's cross country teams and complete the table. Round your answers to the nearest tenth.

Use the pictograph on average yearly snowfall to answer Exercises 5–8.

5. What is the average yearly snowfall in
   (a) Juneau?

   (b) Washington, D.C.?

6. What is the average yearly snowfall in
   (a) Minneapolis?

   (b) Cleveland?

7. Find the difference in the average yearly snowfall between
   (a) Buffalo and Cleveland.

   (b) Memphis and Minneapolis.

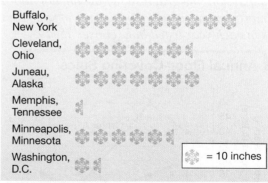

Average Yearly Snowfall in Selected U.S. Cities

Data from National Climatic Data Center.

8. Find the difference in the average yearly snowfall between the city with the greatest amount and the city with the least amount.

9.2 *Use this circle graph for Exercises 9–11.*

9. What was the largest single expense of the vacation? How much was that item?

Use the circle graph to find each ratio. Write the ratios as fractions in lowest terms.

10. Cost of lodging to the cost of food

11. Cost of gasoline to the total cost of the vacation

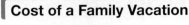
Cost of a Family Vacation

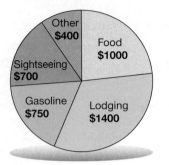

Other $400
Food $1000
Sightseeing $700
Gasoline $750
Lodging $1400

9.3 *This double-bar graph shows the number of acre-feet of water in the Lake Natoma reservoir for each of the first six months of 2015 and 2016. Use this graph to answer Exercises 12–14.*

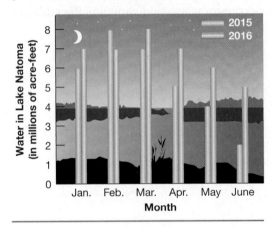

Water in Lake Natoma

Water in Lake Natoma (in millions of acre-feet)

2015
2016

Jan. Feb. Mar. Apr. May June
Month

12. From January through June 2016, which month had the greatest amount of water in the lake? How much was there?

13. Find the amount of decrease and the percent of decrease in the acre-feet of water in the lake from March 2016 to June 2016.

14. Find the amount of decrease and the percent of decrease in the acre-feet of water in the lake from April 2015 to June 2015.

This comparison line graph shows the annual floor-covering sales of two different home improvement centers during each of five years. Use this graph to find the amount of sales in each year listed in Exercises 15–19.

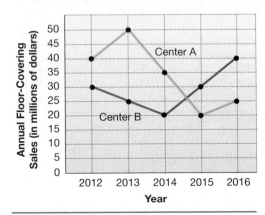

Annual Floor-Covering Sales

Annual Floor-Covering Sales (in millions of dollars)

Center A

Center B

2012 2013 2014 2015 2016
Year

15. Center A in 2013

16. Center A in 2015

17. Center B in 2014

18. Center B in 2016

19. What trend do you see in Center A's sales from 2013 to 2016? Why might this have happened?

9.2 *The Broadway Hair Salon spent a total of $22,400 to remodel their salon. The breakdown of the cost for various items is shown. Find all the missing numbers in Exercises 20–24.*

Item	Dollar Amount	Percent of Total	Degrees of Circle
20. Plumbing and electrical changes	$2240	10%	_____
21. Work stations	$7840	_____	_____
22. Small appliances	$4480	_____	_____
23. Interior decoration	$5600	_____	_____
24. Supplies	_____	_____	36°

25. Draw a circle graph using the information in **Exercises 20–24.** Label each sector with the item and the percent of total for that item.

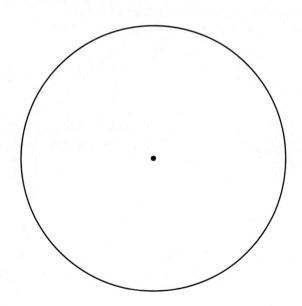

9.4 *In Exercise 26, plot each point on the rectangular coordinate system and label it with its coordinates. In Exercise 27, give the coordinates of each point.*

26. $(1, 7)$ $\left(-1\frac{1}{2}, 0\right)$ $(-4, -2)$ $(0, 3)$ $\left(2, -\frac{1}{2}\right)$ **27.**
$(-7, 1)$ $(5, -5)$ $(0, -4)$

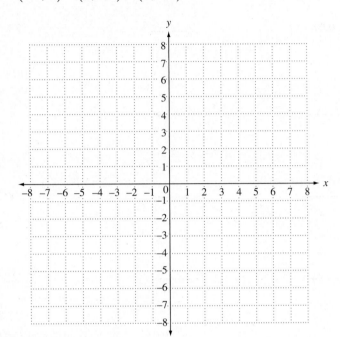

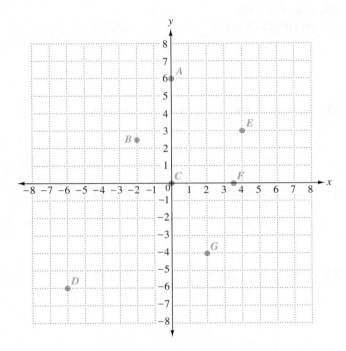

9.5 *Graph each equation. Complete the table using the listed values of x. State whether each line has a positive or negative slope. Use the graph to find two other solutions of the equation.*

28. $x + y = -2$

Use 0, 1, and 2 as the values of x.

x	y	(x, y)

The graph of $x + y = -2$ has a _____ slope.

Two other solutions of $x + y = -2$ are
(____, ____) and (____, ____)

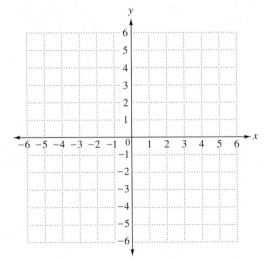

29. $y = x + 3$

Use 0, 1, and 2 as the values of x.

x	y	(x, y)

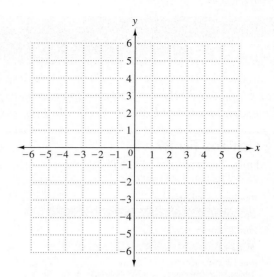

The graph of $y = x + 3$ has a _____ slope.

Two other solutions for $y = x + 3$ are
(____, ____)   and   (____, ____)

30. $y = -4x$

Use -1, 0, and 1 as the values of x.

x	y	(x, y)

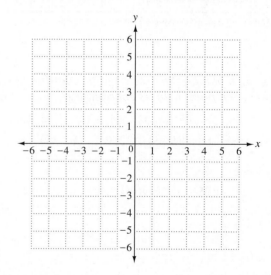

The graph of $y = -4x$ has a _____ slope.

Two other solutions of $y = -4x$ are
(____, ____)   and   (____, ____)

Chapter 9 Mixed Review Exercises

Practicing exercises in mixed-up order helps you prepare for a test.

This bar graph shows the top six home improvement projects that people do themselves.
Use the graph to answer Exercises 1–4. Multiple responses were allowed.

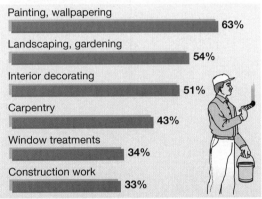

Can You Fix It?

Projects homeowners take on themselves according to a recent survey:

Painting, wallpapering — 63%
Landscaping, gardening — 54%
Interior decorating — 51%
Carpentry — 43%
Window treatments — 34%
Construction work — 33%

Data from American Express Home Improvement.

1. What is the most popular project? What percent of the homeowners in the survey gave that answer?

2. What percent of the homeowners in the survey selected carpentry projects? There were 341 people surveyed. How many of them selected carpentry, to the nearest whole number?

3. Of the 341 homeowners surveyed, how many selected landscaping or gardening projects? Round to the nearest whole number.

4. Name the project(s) selected by about
   (a) $\frac{1}{2}$ of the homeowners.

   (b) $\frac{1}{3}$ of the homeowners.

5. *Plot each point on the rectangular coordinate system. Label each point with its coordinates.*

   $(5, 2)$ $(-3, -3)$ $(4, -1)$ $(-4, 1)$
   $(-1, 0)$ $(0, 3)$ $(2, 0)$ $(0, -5)$

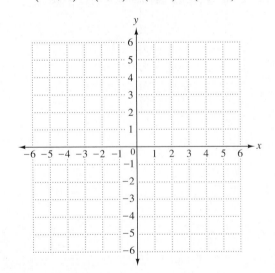

(a) Which point is in Quadrant I?
(b) Which point is in Quadrant II?
(c) Which point is in Quadrant III?
(d) Which point is in Quadrant IV?
(e) Which points are on the axes?

6. Graph $y = 3x$. Use $-1, 0,$ and 1 as the values of x.

x	y	(x, y)

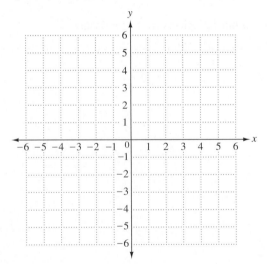

The graph of $y = 3x$ has a _____ slope.

Chapter 9 *Test*

The Chapter Test Prep Videos with step-by-step solutions are available in MyMathLab *or on* You Tube *at https://goo.gl/c3befo*

Use this table to answer Problems 1–3.

1. Which food has

   (a) the greatest amount of calcium in one serving?

   (b) the least amount of calcium in one serving?

AMOUNT OF CALCIUM IN SELECTED FOODS

Food	Serving size	Calories	Calcium (mg)
Swiss cheese	1 oz	95	219
Cream cheese	1 oz	100	23
Yogurt, fruit flavor	8 oz	230	345
Skim milk	1 cup	85	302
Sardines	3 oz	175	371

Data from U.S. Department of Agriculture.

2. A packet of Swiss cheese contains 1.75 oz. How many calories are in that amount of Swiss cheese? Round to the nearest whole number.

3. Find the amount of calcium in a 6 oz container of fruit-flavored yogurt. Round to the nearest whole number.

Use this pictograph to answer Problems 4–6.

4. **(a)** Which group has the greatest number of endangered species? How many species are in that group?

   (b) Which group has the fewest endangered species? How many species are in that group?

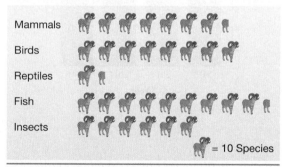

Number of U.S. Endangered Wildlife Species

Data from U.S. Fish and Wildlife Service.

5. How many more species of birds are endangered than species of mammals?

6. What is the total number of endangered species shown in the pictograph?

This circle graph shows the advertising budget for Lakeland Amusement Park. Find the dollar amount budgeted for each category listed in Problems 7–10. The total advertising budget is $2,800,000.

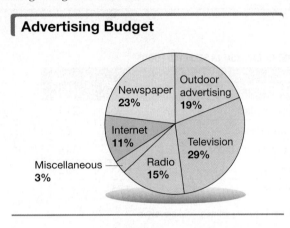

Advertising Budget

7. Which form of advertising has the largest budget? What amount is this?

8. Which form of advertising has the smallest budget? How much is this?

9. How much is budgeted for Internet advertising?

10. How much is budgeted for newspaper ads?

This graph shows one student's income and expenses for four years.

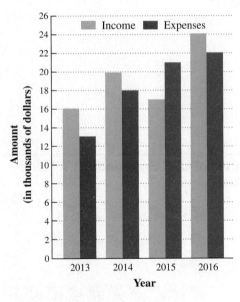

11. In what year did the student's expenses exceed income? By how much?

12. Find the amount of increase and the percent of increase in the student's expenses from 2013 to 2014. Round to the nearest whole percent.

13. In what year did the student's income decline? Give two possible explanations for the decline.

This graph shows enrollment at two community colleges.

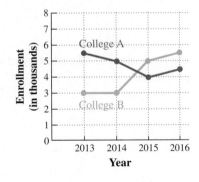

14. What was College A's enrollment in 2013? What was College B's enrollment in 2014?

15. Which college had higher enrollment in 2016? How much higher?

16. Give two possible explanations for the fact that College B's enrollment passed College A's.

During a one-year period, Oak Mill Furniture Sales had a total of $480,000 in expenses. Find all the numbers missing from the table.

Item	Dollar Amount	Percent of Total	Degrees of a Circle
17. Salaries	$168,000	_____	_____
18. Delivery expense	$24,000	_____	_____
19. Advertising	$96,000	_____	_____
20. Rent	$144,000	_____	_____
21. Other	_____	_____	36°

22. Draw a circle graph using a protractor and the information in **Problems 17–21.** Label each sector with the item and the percent of total for that item.

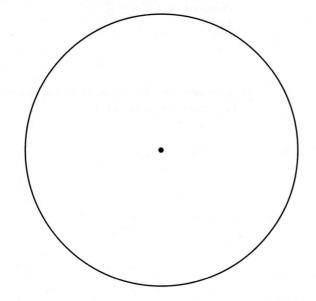

Plot each point on the coordinate system below. Label each point with its coordinates.

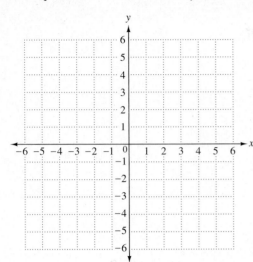

23. $(-5, 3)$

24. $(1, -4)$

25. $(0, 6)$

26. $(2, 0)$

Give the coordinates of each lettered point shown below, and state which quadrant the point is in.

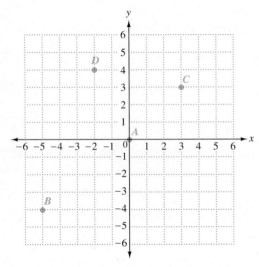

27. Point A

28. Point B

29. Point C

30. Point D

31. Graph $y = x - 4$ on the coordinate system below. Make your own table using 0, 1, and 2 as the values of x.

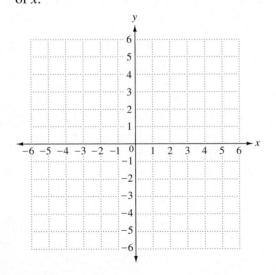

32. (a) Use the graph in **Problem 31** to find *two* other solutions of $y = x - 4$.

(b) State whether the graph of $y = x - 4$ has a positive or negative slope.

Chapters 1–9 Cumulative Review Exercises

Simplify.

1. $\dfrac{4}{5} + 2\dfrac{1}{3}$

2. $\dfrac{0.8}{-3.2}$

3. $5^2 + (-4)^3$

4. $(0.002)(-0.05)$

5. $\dfrac{4a}{9} \cdot \dfrac{6b}{2a^3}$

6. $1\dfrac{1}{4} - 3\dfrac{5}{6}$

7. $-13 + 2.993$

8. $\dfrac{2}{7} - \dfrac{8}{x}$

9. $-3 - 33$

10. $\dfrac{10m}{3n^2} \div \dfrac{2m^2}{5n}$

11. $\dfrac{-3}{-\dfrac{9}{10}}$

12. $\dfrac{3(-7)}{2^4 - 16}$

13. $10 - 2\dfrac{5}{8}$

14. $\dfrac{3}{w} + \dfrac{x}{6}$

15. $8 + 4(2 - 5)$

16. $\dfrac{7}{8}$ of 960

17. $\dfrac{(-4)^2 + 8(0 - 2)}{8 \div 2(-3 + 5) - 10}$

18. $0.5 - 0.25(3.2)^2$

19. $6\left(-\dfrac{1}{2}\right)^3 + \dfrac{2}{3}\left(\dfrac{3}{5}\right)$

Solve each equation. Show your work.

20. $-12 = 3(y + 2)$

21. $6x = 14 - x$

22. $-8 = \dfrac{2}{3}m + 2$

23. $3.4x - 6 = 8 + 1.4x$

24. $2(h - 1) = -3(h + 12) - 11$

25. Translate this sentence into an equation and solve it. If five times a number is subtracted from 12, the result is the number. Find the number.

26. Solve using the six problem-solving steps. An $1800 scholarship is to be split between two people so that one person gets $500 more than the other. How much will each person receive?

Find the unknown length, perimeter, circumference, area, or volume. Use 3.14 as the approximate value of π, and round answers to the nearest tenth when necessary.

27. Find the perimeter and area.

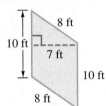

8 ft
10 ft
7 ft
10 ft
8 ft

28. A circular hot tub has a diameter of 6 ft. Find its circumference and the area of the tub's floor.

29. Find the unknown length, the perimeter, and the area.

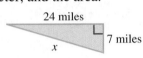

24 miles
7 miles
x

Solve each application problem. Write all answers in simplest form and, when possible, as mixed numbers.

30. A survey found that 4 out of 5 adults are nonsmokers. If Mathtronic has 732 employees, how many would you expect to be nonsmokers? Round your answer to the nearest whole number.

31. Esther bought a fitness tracker at a 15% discount. The regular price was $129. She also paid $6\frac{1}{2}$% sales tax. What was her total cost for the fitness tracker, to the nearest cent?

32. Abiola earned 167 points out of 180 on the prealgebra final exam. What percent of the points did she earn, to the nearest tenth?

33. Wayne bought $2\frac{1}{4}$ pounds of lunch meat. He used $\frac{1}{6}$ pound of meat on each of five sandwiches. How much meat is left?

34. Plot the following points and label each one with its coordinates. Which of the points are in Quadrant II? Which are in the Quadrant III?

$(3, -5)$ $(0, -4)$ $\left(-2, 3\frac{1}{2}\right)$ $(1, 0)$ $(4, 2)$

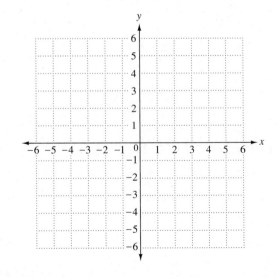

35. Graph $y = x + 6$. Use 0, -1, and -2 as the values of x. After graphing the equation, state whether the line has a positive or negative slope, and find two other solutions of $y = x + 6$.

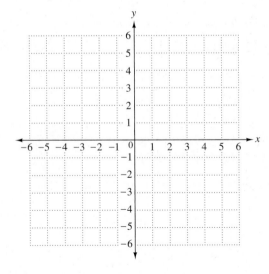

10

Exponents and Polynomials

The expression $100t - 13t^2$ gives the distance in feet that a car going approximately 68 mph will skid in t seconds. In this chapter you'll learn about expressions like this one, which are called *polynomials*. Accident investigators use polynomials to determine the length of a skid or the elapsed time during a skid.

10.1 The Product Rule and Power Rules for Exponents

OBJECTIVES

1. Review the use of exponents.
2. Use the product rule for exponents.
3. Use the rule $(a^m)^n = a^{mn}$
4. Use the rule $(ab)^m = a^m b^m$
5. Use the rule $\left(\dfrac{a}{b}\right)^m = \dfrac{a^m}{b^m}$

OBJECTIVE ▶ 1 **Review the use of exponents.** You have used exponents to write repeated products. Recall that in the expression 5^2, the number 5 is called the *base* and 2 is called the *exponent* or *power*. The expression 5^2 is written in *exponential form*. Usually we do not write an exponent of 1, but sometimes it is convenient to do so. In general, for any quantity x, we can write x as x^1.

EXAMPLE 1 **Review of Using Exponents**

Write $3 \cdot 3 \cdot 3 \cdot 3 \cdot 3$ in exponential form. Then evaluate the exponential expression.

Since 3 occurs as a factor five times, the base is 3 and the exponent is 5. The **exponential expression is 3^5**, read "3 to the fifth power" or simply "3 to the fifth." To evaluate 3^5, actually do the multiplication.

$$3^5 = 3 \cdot 3 \cdot 3 \cdot 3 \cdot 3 = \mathbf{243}$$

◀ **Work Problem 1 at the Side.**

1. Write $2 \cdot 2 \cdot 2 \cdot 2$ in exponential form. Then evaluate the exponential expression.

EXAMPLE 2 **Evaluating Exponential Expressions**

Evaluate each exponential expression. Name the base and the exponent.

2. Evaluate each exponential expression. Name the base and the exponent.

		Base	Exponent
(a) $5^4 = 5 \cdot 5 \cdot 5 \cdot 5 = 625$		5	4
(b) $-5^4 = -(5 \cdot 5 \cdot 5 \cdot 5) = -625$		5	4
(c) $(-5)^4 = (-5)(-5)(-5)(-5) = 625$		-5	4

GS **(a)** $-4^2 = -(4 \cdot 4) =$

GS **(b)** $(-4)^2 = (-4)(-4) =$

> **⚠ CAUTION**
>
> It is important to understand the difference between parts (b) and (c) of **Example 2** above. In -5^4 the exponent 4 applies only to the base 5. In $(-5)^4$ the parentheses show that the exponent 4 applies to the base -5. In summary, $-a^n$ *and* $(-a)^n$ *mean different things*. The exponent applies only to what is *immediately* to the left of it.
>
> | Expression | Base | Exponent | Example |
> |---|---|---|---|
> | $-a^n$ | a | n | $-3^2 = -(3 \cdot 3) = -9$ |
> | $(-a)^n$ | $-a$ | n | $(-3)^2 = (-3)(-3) = 9$ |

(c) $(-2)^5$

(d) -2^5

◀ **Work Problem 2 at the Side.**

OBJECTIVE ▶ 2 **Use the product rule for exponents.** To develop the product rule, we use the definition of an exponential expression.

$$2^4 \cdot 2^3 = \overbrace{(2 \cdot 2 \cdot 2 \cdot 2)}^{\text{4 factors}} \overbrace{(2 \cdot 2 \cdot 2)}^{\text{3 factors}}$$

$$= \underbrace{2 \cdot 2 \cdot 2 \cdot 2 \cdot 2 \cdot 2 \cdot 2}_{\text{4 factors + 3 factors = 7 factors}}$$

So, $2^4 \cdot 2^3 = \mathbf{2^7}$

Answers

1. $2^4 = 16$
2. **(a)** -16; base is 4; exponent is 2
   (b) 16; base is -4; exponent is 2
   (c) -32; base is -2; exponent is 5
   (d) -32; base is 2; exponent is 5

Here is another example.

$$6^2 \cdot 6^3 = (6 \cdot 6)(6 \cdot 6 \cdot 6)$$
$$= 6 \cdot 6 \cdot 6 \cdot 6 \cdot 6$$

So, $6^2 \cdot 6^3 = \mathbf{6^5} \leftarrow$ 2 factors + 3 factors = 5 factors

From these examples, $2^4 \cdot 2^3 = 2^{4+3} = 2^7$ and $6^2 \cdot 6^3 = 6^{2+3} = 6^5$, we can see that adding the exponents gives the exponent of the product, suggesting the **product rule for exponents.**

Product Rule for Exponents

If m and n are positive integers, then $a^m \cdot a^n = a^{m+n}$
(Keep the same base and add the exponents.)

Example: $6^2 \cdot 6^5 = 6^{2+5} = 6^7$

! CAUTION

Avoid the common error of multiplying the *bases* when using the product rule.

$6^2 \cdot 6^5 \neq 36^7 \leftarrow$ Error $\qquad 6^2 \cdot 6^5 = 6^7 \leftarrow$ Correct

Keep the *same base* and add the exponents.

EXAMPLE 3 **Using the Product Rule**

Use the product rule for exponents to find each product, if possible.

(a) $6^3 \cdot 6^5 = 6^{3+5} = \mathbf{6^8}$
   (Keep the same base.)

(b) $(-4)^7(-4)^2 = (-4)^{7+2} = \mathbf{(-4)^9}$

(c) $x^2 \cdot x = x^2 \cdot x^1 = x^{2+1} = \mathbf{x^3}$
   (x can be written as x^1)

(d) $m^4 \cdot m^3 \cdot m^4 = m^{4+3+4} = \mathbf{m^{11}}$

(e) $2^3 \cdot 3^2$

The product rule does *not* apply to the product $2^3 \cdot 3^2$ because the *bases are different.* We can still *evaluate* the expression:

$$2^3 \cdot 3^2 = (2 \cdot 2 \cdot 2)(3 \cdot 3) = (8)(9) = \mathbf{72}$$

! CAUTION

The *bases* must be the *same* before we can apply the product rule.

(f) $2^3 + 2^4$

The product rule does *not* apply to $2^3 + 2^4$ because it is a *sum*, not a *product.* We can still *evaluate* the expression:

$$2^3 + 2^4 = (2 \cdot 2 \cdot 2) + (2 \cdot 2 \cdot 2 \cdot 2) = (8) + (16) = \mathbf{24}$$

! CAUTION

Use the product rule **only when multiplying** (*not* when *adding*).

$2^3 \cdot 2^4 = 2^{3+4} = 2^7$ **BUT** $2^3 + 2^4 \neq 2^7 \leftarrow$ **Error**

Work Problem 3 at the Side. ▶

---

3 Find each product using the product rule, if possible. Write answers in exponential form.

(a) $8^2 \cdot 8^5$

$8^2 \cdot 8^5 = 8{-}{+}{-} = 8{-}$
   ↑ ↑ ↑
   same base

Keep the same base.
Add the exponents.

(b) $(-7)^5(-7)^3$

(c) $y^3 \cdot y \cdot y^3$

(d) $4^2 \cdot 3^5$ Careful!
   The bases are different.

(e) $6^4 + 6^2$ Careful!
   This is adding, **not** multiplying.

Answers

3. **(a)** $8^{2+5} = 8^7$ **(b)** $(-7)^8$ **(c)** y^7
   (d) Cannot use the product rule because the bases are different.
   (e) Cannot use the product rule because it is a sum, not a product.

4 Multiply.

(GS) **(a)** $5m^2 \cdot 2m^6$

$$(5 \cdot 2) \cdot (m^2 \cdot m^6)$$

$$10m \underset{\uparrow}{\overline{}} + \overline{} = 10m \overline{}$$

(b) $3p^5 \cdot 9p^4$

(c) $-7p^8 \cdot (3p^8)$

5 Simplify each expression. Write answers in exponential form.

(GS) **(a)** $(5^3)^4$

$$(5^3)^4 = 5 \underset{\underset{\text{Same base}}{\uparrow}}{\overline{}} \cdot \underset{\underset{\substack{\text{Multiply the} \\ \text{exponents.}}}{\uparrow}}{\overline{}} = 5 \overline{}$$

(b) $(6^2)^5$

(c) $(x^3)^3$

(d) $(n^6)^5$

EXAMPLE 4 Using the Product Rule

Multiply $8x^3$ and $5x^3$.

Recall that $8x^3$ means $8 \cdot x^3$ and $5x^3$ means $5 \cdot x^3$. Use the associative and commutative properties to regroup the factors. Then use the product rule.

$$8x^3 \cdot 5x^3 = (8 \cdot 5) \cdot (x^3 \cdot x^3) = 40x^{3+3} = \mathbf{40}x^6$$

> **⊘ CAUTION**
>
> Be sure you understand the difference between *adding* and *multiplying* exponential expressions. Here is a comparison.
>
> **Adding** expressions $\quad 8x^3 + 5x^3 = (8+5)x^3 \quad = 13x^3$
>
> **Multiplying** expressions $\quad (8x^3)(5x^3) = (8 \cdot 5)x^{3+3} = 40x^6$

◀ **Work Problem ④ at the Side.**

OBJECTIVE **3** **Use the rule** $(a^m)^n = a^{mn}$ We can simplify an expression such as $(8^3)^2$ with the product rule for exponents, as shown below.

$$(8^3)^2 = (8^3)(8^3) = 8^{3+3} = \mathbf{8}^6$$

Notice that the *product* of the original exponents, $3 \cdot 2$, gives 6, the exponent in 8^6. Here is another example.

$$(5^2)^4 = 5^2 \cdot 5^2 \cdot 5^2 \cdot 5^2 \qquad \text{Definition of exponent}$$

$$= 5^{2+2+2+2} \qquad\qquad \text{Product rule}$$

$$= \mathbf{5}^8 \qquad\qquad\qquad \begin{array}{l}\text{Add the exponents:} \\ 2+2+2+2 \text{ is } 8\end{array}$$

Notice that the *product* of the original exponents gives the exponent of the result: $2 \cdot 4 = 8$. These examples suggest **power rule (a) for exponents.**

> **Power Rule (a) for Exponents**
>
> If m and n are positive integers, then $\quad (a^m)^n = a^{mn}$
> (Raise a power to a power by multiplying exponents.)
>
> *Example:* $\quad (3^2)^4 = 3^{2 \cdot 4} = 3^8$

EXAMPLE 5 Using Power Rule (a)

Use power rule (a) for exponents to simplify each expression. Write answers in exponential form.

(a) $(2^5)^3 = 2^{5 \cdot 3} = 2^{15}$ ⟵ Multiply the exponents: 5 · 3 is 15

Keep the same base.

(b) $(5^7)^2 = 5^{7 \cdot 2} = \mathbf{5}^{14}$

(c) $(x^5)^5 = x^{5 \cdot 5} = \mathbf{x}^{25}$

(d) $(n^3)^2 = n^{3 \cdot 2} = \mathbf{n}^6$

◀ **Work Problem ⑤ at the Side.**

Answers

4. **(a)** $10m^{2+6} = 10m^8$ **(b)** $27p^9$ **(c)** $-21p^{16}$
5. **(a)** $5^{3 \cdot 4} = 5^{12}$ **(b)** 6^{10} **(c)** x^9 **(d)** n^{30}

OBJECTIVE ▶ **4** **Use the rule** $(ab)^m = a^m b^m$ We can use the definition of an exponential expression and the commutative and associative properties to develop two more rules for exponents. Here is an example.

$$(4x)^3 = (4x)(4x)(4x) \quad \text{Definition of exponent}$$

$$= 4 \cdot 4 \cdot 4 \cdot x \cdot x \cdot x \quad \text{Commutative and associative properties}$$

$$= 4^3 x^3 \quad \text{Definition of exponent}$$

This example suggests **power rule (b) for exponents.**

Power Rule (b) for Exponents

If m is a positive integer, then $(ab)^m = a^m b^m$
(Raise a product to a power by raising each factor to the power.)

Example: $(2p)^5 = 2^5 p^5$

EXAMPLE 6 **Using Power Rule (b)**

Use power rule (b) to simplify each expression and evaluate if possible.

(a) $(3xy)^2 = 3^2 x^2 y^2$ Power rule (b)

$$= 9x^2 y^2$$

$3^2 = 3 \cdot 3 = 9$

(b) $9(pq)^2 = 9(p^2 q^2)$ Power rule (b); notice that the 9 is *outside* the parentheses, so the exponent applies *only* to pq, which is *inside* the parentheses.

$$= 9p^2 q^2$$

(c) $5(2m^2 p^3)^4 = 5[2^4 (m^2)^4 (p^3)^4]$ Power rule (b); notice that the 5 is *outside* the parentheses so the exponent does *not* apply to the 5.

$$= 5(2^4 m^8 p^{12}) \quad \text{Power rule (a): } (m^2)^4 \text{ is } m^{2\cdot4} \text{ or } m^8 \\ \text{and } (p^3)^4 \text{ is } p^{3\cdot4} \text{ or } p^{12}$$

$$= 5 \cdot 2^4 m^8 p^{12}$$

$2^4 = 2 \cdot 2 \cdot 2 \cdot 2 = 16$
Then $5 \cdot 16 = 80$

$$= 80 m^8 p^{12}$$

──── **Work Problem 6 at the Side.** ▶

⊘ CAUTION

Power rule (b) *does not* apply to a *sum.*

$$(x + 4)^2 \neq x^2 + 4^2 \leftarrow \text{Error}$$

You will learn how to work with $(x + 4)^2$ later in this chapter.

---

6 Simplify.

GS **(a)** $5(mn)^3$

Because 5 is **outside** the parentheses, the exponent does **not** apply to 5.

$$5(mn)^3 = 5m\underline{\quad}n\underline{\quad}$$

GS **(b)** $(4ab)^2$

The 4 is **inside** the parentheses, so the exponent **does** apply to 4.

$$(4ab)^2 = 4^2 \, a\underline{\quad}b\underline{\quad}$$

$$= \underline{\quad}a\underline{\quad}b\underline{\quad}$$

$4 \cdot 4$
(not $4 \cdot 2$)

(c) $(3a^2 b^4)^5$

(d) $2(3xy^3)^4$

---

7 Simplify. Assume all variables represent nonzero numbers.

GS (a) $\left(\dfrac{5}{2}\right)^4 = \dfrac{5\underline{}}{2\underline{}} =$

$\dfrac{5 \cdot 5 \cdot 5 \cdot 5}{2 \cdot 2 \cdot 2 \cdot 2} = \dfrac{\Box}{16}$

(b) $\left(\dfrac{p}{q}\right)^2$

(c) $\left(\dfrac{r}{t}\right)^3$

OBJECTIVE ▶ **5** Use the rule $\left(\dfrac{a}{b}\right)^m = \dfrac{a^m}{b^m}$ Because the quotient $\dfrac{a}{b}$ can be written as $a \cdot \dfrac{1}{b}$, we can use power rule (b), together with some of the properties of real numbers, to get **power rule (c) for exponents.**

Power Rule (c) for Exponents

If m is a positive integer, then $\left(\dfrac{a}{b}\right)^m = \dfrac{a^m}{b^m}$ as long as $b \neq 0$.

(Raise a quotient to a power by raising both the numerator and the denominator to the power. The denominator cannot be 0.)

Example: $\left(\dfrac{5}{3}\right)^2 = \dfrac{5^2}{3^2}$

EXAMPLE 7 Using Power Rule (c)

Simplify each expression and evaluate if possible.

$2^5 = 2 \cdot 2 \cdot 2 \cdot 2 \cdot 2 = 32$

(a) $\left(\dfrac{2}{3}\right)^5 = \dfrac{2^5}{3^5} = \dfrac{32}{243}$

$3^5 = 3 \cdot 3 \cdot 3 \cdot 3 \cdot 3 = 243$

(b) $\left(\dfrac{m}{n}\right)^4 = \dfrac{m^4}{n^4}$ when $n \neq 0$ Recall that division by 0 is undefined, so the denominator cannot be equal to 0

◀ **Work Problem 7 at the Side.**

Here is a list of the rules for exponents discussed in this section. These rules are basic to the study of algebra. Be sure you understand them.

Rules for Exponents

If m and n are positive integers, then:	**Examples**
Product rule $a^m \cdot a^n = a^{m+n}$	$6^2 \cdot 6^5 = 6^{2+5} = 6^7$
Power rule (a) $(a^m)^n = a^{mn}$	$(3^2)^4 = 3^{2 \cdot 4} = 3^8$
Power rule (b) $(ab)^m = a^m b^m$	$(2p)^5 = 2^5 p^5$
Power rule (c) $\left(\dfrac{a}{b}\right)^m = \dfrac{a^m}{b^m}$ when $b \neq 0$	$\left(\dfrac{5}{3}\right)^2 = \dfrac{5^2}{3^2}$

In the next section, the rules will be expanded to include zero and negative integers as exponents.

Answers

7. (a) $\dfrac{5^4}{2^4} = \dfrac{625}{16}$ **(b)** $\dfrac{p^2}{q^2}$ **(c)** $\dfrac{r^3}{t^3}$

10.1 Exercises

FOR
EXTRA
HELP

Go to MyMathLab *for worked-out, step-by-step solutions to exercises enclosed in a square* ▢ *and video solutions to* ▶ *exercises.*

1. CONCEPT CHECK What is the understood exponent for x in the expression xy^2?

2. CONCEPT CHECK How are the expressions 3^2, 5^3, and 7^4 read?

CONCEPT CHECK *Decide whether each statement is* true *or* false. *If a statement is* false, *correct it.*

3. $3^3 = 9$

4. $2^4 + 2^3 = 2^7$

5. $(a^2)^3 = a^5$

6. $4^5 \cdot 4^3 = 16^8$

Write each expression using exponents. **See Example 1.**

7. $(-2)(-2)(-2)(-2)(-2)$

8. $w \cdot w \cdot w \cdot w \cdot w \cdot w$

9. $\left(\frac{1}{2}\right)\left(\frac{1}{2}\right)\left(\frac{1}{2}\right)\left(\frac{1}{2}\right)\left(\frac{1}{2}\right)\left(\frac{1}{2}\right)$

10. $\left(-\frac{1}{4}\right)\left(-\frac{1}{4}\right)\left(-\frac{1}{4}\right)\left(-\frac{1}{4}\right)\left(-\frac{1}{4}\right)$

11. $(-8p)(-8p)$

12. $(-7x)(-7x)(-7x)(-7x)$

13. CONCEPT CHECK The work shown below has a mistake in it.

What Went Wrong? First write a sentence explaining what the mistake is. Then fix the mistake and find the correct solution.

Betsy simplified the expression this way:
$(-3)^4 = -3 \cdot 3 \cdot 3 \cdot 3 = -81$

14. CONCEPT CHECK The work shown below has a mistake in it.

What Went Wrong? First write a sentence explaining what the mistake is. Then fix the mistake and find the correct solution.

Devin simplified the expression this way:
$(5x)^3 = 5x^3$

Identify the base and the exponent for each exponential expression. In Exercises 15–18, also evaluate the expression. **See Example 2.**

15. 3^5

16. 2^7

17. $(-3)^5$

18. $(-2)^7$

19. $(-6x)^4$

20. $(-8x)^4$

21. $-6x^4$

22. $-8x^4$

23. CONCEPT CHECK Explain why the product rule does *not* apply to the expression $5^2 + 5^3$. Then evaluate the expression by finding the individual powers and adding the results.

24. CONCEPT CHECK Explain why the product rule does *not* apply to the expression $3^2 \cdot 4^3$. Then evaluate the expression by finding the individual powers and multiplying the results.

Use the product rule to simplify each expression. Write each answer in exponential form. See Examples 3 and 4.

25. $5^2 \cdot 5^6 = 5 \underline{} + \underline{} = 5 \underline{}$

$\underset{\text{same base}}{\underbrace{\text{Keep the}}}$

26. $3^6 \cdot 3^7 = 3 \underline{} + \underline{} = 3 \underline{}$

$\underset{\text{same base}}{\underbrace{\text{Keep the}}}$

27. $4^2 \cdot 4^7 \cdot 4^3$

28. $5^3 \cdot 5^8 \cdot 5^2$

29. $(-7)^3 (-7)^6$

30. $(-9)^8 (-9)^5$

31. $t^3 \cdot t \cdot t^{13}$

32. $n \cdot n^6 \cdot n^9$

33. $(-8r^4)(7r^3)$

34. $(10a^7)(-4a^3)$

35. $(-6p^5)(-7p^5)$

36. $(-5w^8)(-9w^8)$

For each group of terms, first add the given terms. Then start over and multiply them.

37. $5x^4, 9x^4$

38. $8t^5, 3t^5$

39. $-7a^2, 2a^2, 10a^2$

40. $6x^3, 9x^3, -2x^3$

Use the power rules for exponents to simplify each expression. Write each answer in exponential form. See Examples 5–7.

41. $(4^3)^2$

42. $(8^3)^6$

43. $(t^4)^5$

44. $(y^6)^5$

45. $(7r)^3$

46. $(11x)^4$

47. $(5xy)^5$

48. $(9pq)^6$

49. $8(qr)^3$

50. $4(vw)^5$

51. $\left(\dfrac{1}{2}\right)^3$

52. $\left(\dfrac{1}{3}\right)^5$

53. $\left(\dfrac{a}{b}\right)^3$ when $b \neq 0$

54. $\left(\dfrac{r}{t}\right)^4$ when $t \neq 0$

55. $\left(\dfrac{9}{5}\right)^8$

56. $\left(\dfrac{12}{7}\right)^3$

57. $(-2x^2y)^3$

58. $(-5m^4p^2)^3$

59. $(3a^3b^2)^2$

60. $(4x^3y^5)^4$

Find the area of each figure. Write each answer in exponential form.

61.

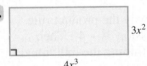

To find the area of a rectangle, multiply length times width.

62.

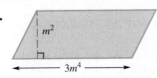

To find the area of a parallelogram, multiply the base times the height.

10.2 Integer Exponents and the Quotient Rule

In the last section we studied the product rule for exponents. In all our earlier work, exponents were positive integers. Now we want to develop meaning for exponents that are not positive integers.

Look at this list of exponential expressions.

Exponent is decreasing by 1 each time. ↓

$2^4 = 16$ ⎞
$2^3 = 8$ ⎠ $16 \div 2$ is 8
$2^2 = 4$ ⎠ $8 \div 2$ is 4

OBJECTIVES

1. Use 0 as an exponent.
2. Use negative numbers as exponents.
3. Use the quotient rule for exponents.
4. Use the product rule with negative exponents.

Do you see the pattern in the values? Each time we decrease the exponent by 1, the value is divided by 2. Using this pattern, we can continue the list to smaller and smaller integer exponents.

$2^1 = 2$ ⎞
⎠ $2 \div 2$ is 1
$2^0 = 1$ ⎞
⎠ $1 \div 2$ is $\frac{1}{2}$
$2^{-1} = \frac{1}{2}$ ⎞
⎠ $\frac{1}{2} \div 2$ is $\frac{1}{4}$
$2^{-2} = \frac{1}{4}$ ⎠

Work Problem 1 at the Side. ▶

From the list above and the answers to **Problem 1** at the side, it appears that we should define 2^0 as 1, and define negative exponents as reciprocals.

OBJECTIVE ▶ 1 Use 0 as an exponent. The definitions of 0 as an exponent and negative exponents need to satisfy the rules for exponents from the previous section. For example, if $6^0 = 1$, we can check the product rule as shown below.

$$6^0 \cdot 6^2 = 1 \cdot 6^2 = 6^2 \quad \text{and} \quad 6^0 \cdot 6^2 = 6^{0+2} = 6^2$$

The results match, so the product rule is satisfied. If you check the power rules, you will see that they are also valid for a base raised to an exponent of 0. Thus, we define an exponent of 0 as follows.

Zero Exponent

If a is any *nonzero* number, then, $\quad a^0 = 1$

Example: $\quad 17^0 = 1$

EXAMPLE 1 Using Zero Exponents

Evaluate each exponential expression.

(a) $60^0 = \mathbf{1}$

(b) $y^0 = \mathbf{1}$, if $y \neq 0$

(c) $6y^0 = 6(1) = \mathbf{6}$, if $y \neq 0$

(d) $(6y)^0 = \mathbf{1}$, if $y \neq 0$

(e) $(-60)^0 = \mathbf{1}$

(f) $-60^0 = -(1) = -\mathbf{1}$

⚠ **CAUTION**

Notice the difference between parts (e) and (f) above. In part (e), the base is **− 60** and the exponent is 0. But in part (f), the base is **60**.

---

1 Continue the list of exponentials using -3 and -4 as exponents.

$2^4 = 16$

$2^3 = 8$

$2^2 = 4$

$2^1 = 2$

$2^0 = 1$

$2^{-1} = \frac{1}{2}$

$2^{-2} = \frac{1}{4}$

$2^{-3} = $ _____

$2^{-4} = $ _____

2 Evaluate.

(a) 28^0 **(b)** $(-16)^0$

(c) -7^0 **(d)** m^0
                when $m \neq 0$

Answers

1. $2^{-3} = \frac{1}{8}$; $2^{-4} = \frac{1}{16}$

2. **(a)** 1   **(b)** 1   **(c)** −1   **(d)** 1

Work Problem 2 at the Side. ▶

③ Simplify by writing each expression with positive exponents. Then evaluate the expression, if possible.

GS **(a)** 4^{-3}

$$4^{-3} = \frac{1}{4^3} = \frac{1}{\boxed{}} \leftarrow 4 \cdot 4 \cdot 4$$

(b) 6^{-2}

GS **(c)** $2^{-1} + 5^{-1}$

$$\underbrace{\frac{1}{2}} + \underbrace{\frac{1}{5}} \quad \begin{array}{l}\text{The common} \\ \text{denominator} \\ \text{is 10}\end{array}$$

$$= \frac{\boxed{}}{10} + \frac{\boxed{}}{10} = \frac{\boxed{}}{10}$$

(d) m^{-5} when $m \neq 0$

Answers

3. **(a)** $\frac{1}{4^3} = \frac{1}{64}$ **(b)** $\frac{1}{6^2} = \frac{1}{36}$

(c) $\frac{1}{2} + \frac{1}{5} = \frac{5}{10} + \frac{2}{10} = \frac{7}{10}$ **(d)** $\frac{1}{m^5}$

OBJECTIVE ▶ ❷ Use negative numbers as exponents. In **Margin Problem 1** on the previous page, we saw that $2^{-2} = \frac{1}{4}$, $2^{-3} = \frac{1}{8}$ and $2^{-4} = \frac{1}{16}$. If we write the denominators as powers of 2, we see an interesting pattern in the exponents.

$$2^{-2} = \frac{1}{4} = \frac{1}{2^2} \qquad 2^{-3} = \frac{1}{8} = \frac{1}{2^3} \qquad 2^{-4} = \frac{1}{16} = \frac{1}{2^4}$$

So it seems that 2^{-n} should equal $\frac{1}{2^n}$. Is the product rule valid in such cases? For example, using the product rule to multiply 6^{-2} by 6^2 gives the following.

$$6^{-2} \cdot 6^2 = 6^{-2+2} = 6^0 = 1$$

The expression 6^{-2} behaves as if it were the reciprocal of 6^2, because their product is 1. The reciprocal of 6^2 may be written $\frac{1}{6^2}$, leading us to define 6^{-2} as $\frac{1}{6^2}$. This example illustrates the definition of negative exponents.

Negative Exponents

If a is any *nonzero* number and n is any integer, then $a^{-n} = \dfrac{1}{a^n}$

Example: $3^{-2} = \dfrac{1}{3^2}$

EXAMPLE 2   Using Negative Exponents

Simplify by writing each expression with positive exponents. Then evaluate the expression, if possible.

(a) $4^{-2} = \dfrac{1}{4^2} = \dfrac{1}{\mathbf{16}}$ ⟵ Think: 4 · 4 is 16

(b) $5^{-3} = \dfrac{1}{5^3} = \dfrac{1}{\mathbf{125}}$ ⟵ Think: 5 · 5 · 5 is 125

(c) $4^{-1} - 2^{-1} = \dfrac{1}{4} - \dfrac{1}{2}$ Apply the exponents first.

$$= \frac{1}{4} - \frac{2}{4} \quad \begin{array}{l}\text{The common denominator} \\ \text{is 4, so rewrite } \frac{1}{2} \text{ as } \frac{2}{4}\end{array}$$

$$= -\frac{1}{4}$$

(d) $p^{-2} = \dfrac{1}{p^2}$ when $p \neq 0$

❗ CAUTION

A negative exponent does **not** indicate a negative number; negative exponents lead to reciprocals.

Expression	Example
a^{-n}	$3^{-2} = \dfrac{1}{3^2} = \dfrac{1}{9}$ ⟵ Not a negative number

---

◀ **Work Problem ❸ at the Side.**

OBJECTIVE ❸ **Use the quotient rule for exponents.** What about the quotient of two exponential expressions with the same base? We know the following.

$$\frac{6^5}{6^3} = \frac{\overset{1}{6} \cdot \overset{1}{6} \cdot \overset{1}{6} \cdot 6 \cdot 6}{\underset{1}{6} \cdot \underset{1}{6} \cdot \underset{1}{6}} = \frac{6^2}{1} = 6^2$$

Notice that subtracting the exponents, $5 - 3$, gives 2, which is the exponent in the quotient. Here is another example.

$$\frac{6^2}{6^4} = \frac{\overset{1}{6} \cdot \overset{1}{6}}{\underset{1}{6} \cdot \underset{1}{6} \cdot 6 \cdot 6} = \frac{1}{6^2} = 6^{-2}$$

In this case, subtracting the exponents gives $2 - 4$, which is -2. These examples suggest the quotient rule for exponents.

Quotient Rule for Exponents

If a is any nonzero number, and m and n are integers, then

$$\frac{a^m}{a^n} = a^{m-n} \qquad \begin{array}{l}\text{Keep the same base and} \\ \text{subtract the exponents.}\end{array}$$

Example: $\dfrac{5^8}{5^4} = 5^{8-4} = 5^4$

❗ CAUTION

A common **error** is to write $\dfrac{5^8}{5^4} = 1^{8-4} = 1^4$. When using the quotient rule, the quotient should have the **same base.** The base here is 5.

$$\frac{5^8}{5^4} = 5^{8-4} = 5^4$$

If you are not sure, use the definition of an exponent to write out the factors.

$$\frac{5^8}{5^4} = \frac{\overset{1}{5} \cdot \overset{1}{5} \cdot \overset{1}{5} \cdot \overset{1}{5} \cdot 5 \cdot 5 \cdot 5 \cdot 5}{\underset{1}{5} \cdot \underset{1}{5} \cdot \underset{1}{5} \cdot \underset{1}{5}} = \frac{5^4}{1} = 5^4$$

EXAMPLE 3 Using the Quotient Rule for Exponents

Simplify, using the quotient rule for exponents. Write answers with positive exponents.

(a) $\dfrac{5^8}{5^6} = 5^{8-6} = \mathbf{5^2}$ *Keep the same base.*

(c) $\dfrac{5^{-3}}{5^{-7}} = 5^{-3-(-7)} = \mathbf{5^4}$ *Be careful with signs.*

(b) $\dfrac{4^2}{4^9} = 4^{2-9} = 4^{-7} = \dfrac{1}{4^7}$

(d) $\dfrac{x^5}{x^{-3}} = x^{5-(-3)} = \mathbf{x^8}$ when $x \neq 0$

—————— **Work Problem ❹ at the Side.** ▶

OBJECTIVE ❹ **Use the product rule with negative exponents.** As shown in the next example, we may use the product rule to simplify expressions with negative exponents.

---

❹ Simplify. Write answers with positive exponents.

GS (a) $\dfrac{5^{11}}{5^8} = 5^{11-8} = 5^{\underline{}}$

(b) $\dfrac{4^7}{4^{10}}$

GS (c) $\dfrac{6^{-5}}{6^{-3}} = 6^{-5-(-3)} = 6^{\underline{}}$

Be careful when subtracting the exponents. Change subtraction to adding the opposite.

$$\begin{array}{cc} -5 & - \; (-3) \\ \downarrow & \downarrow \\ -5 & + \; (+3) \end{array}$$

$$\underline{}$$

Then, $6^{-2} = \dfrac{1}{6^{\underline{}}}$

(d) $\dfrac{a^4}{a^{-2}}$ when $a \neq 0$

Answers

4. (a) 5^3 (b) $4^{-3} = \dfrac{1}{4^3}$ (c) $-2; 6^{-2} = \dfrac{1}{6^2}$
    (d) a^6

5 Simplify. Assume all variables represent nonzero numbers. Write answers with positive exponents.

(a) $10^{-7}(10^9) = 10^{-7+9} = 10^{\underline{\ \ }}$

(b) $(10^{-3})(10^{-3})$

(c) $\underbrace{x^{-3} \cdot x^{-2}}\ \cdot x^4$

$= \underbrace{x^{-3+(-2)}} \cdot x^4$

$= \quad x^{-5} \quad \cdot x^4$

$= \quad x^{-5+4}$

$= x^{\underline{\ \ }}$

Now rewrite x^{-1} so it has a positive exponent.

$x^{-1} = \dfrac{1}{x^{\underline{\ \ }}} = \dfrac{1}{\boxed{\ }}$

(d) $(a^{-2})(a^{12})(a^{-4})$

Answers

5. (a) 10^2 (b) $\dfrac{1}{10^6}$ (c) $x^{-1} = \dfrac{1}{x^1} = \dfrac{1}{x}$
    (d) a^6

EXAMPLE 4   **Using the Product Rule with Negative Exponents**

Simplify each expression. Assume all variables represent nonzero numbers. Write answers with positive exponents.

(a) $10^5(10^{-4}) = 10^{5+(-4)}$ Using the product rule, add the exponents.

     $= 10^1$ or   **10**

(b) $(10^{-5})(10^{-8}) = 10^{-5+(-8)}$ Product rule

     $= 10^{-13} = \dfrac{1}{10^{13}}$ Definition of negative exponent

(c) $y^{-2} \cdot y^5 \cdot y^{-8} = \underbrace{y^{-2+5}} \cdot y^{-8}$ Product rule

     $= y^3 \cdot y^{-8}$

     $= y^{3+(-8)}$ Product rule

     $= y^{-5}$

     $= \dfrac{1}{y^5}$ Definition of negative exponent

◀ **Work Problem 5 at the Side.**

The definitions and rules for exponents are summarized below.

Definitions and Rules for Exponents

If m and n are any integers,		**Examples**
Product rule	$a^m \cdot a^n = a^{m+n}$	$7^4 \cdot 7^5 = 7^{4+5} = 7^9$
Power rule (a)	$(a^m)^n = a^{mn}$	$(4^2)^3 = 4^{2 \cdot 3} = 4^6$
Power rule (b)	$(ab)^m = a^m b^m$	$(3k)^4 = 3^4 k^4$
Power rule (c)	$\left(\dfrac{a}{b}\right)^m = \dfrac{a^m}{b^m}$ when $b \neq 0$	$\left(\dfrac{2}{3}\right)^2 = \dfrac{2^2}{3^2}$
Zero exponent	$a^0 = 1$ when $a \neq 0$	$(-3)^0 = 1$
Negative exponent	$a^{-n} = \dfrac{1}{a^n}$ when $a \neq 0$	$5^{-3} = \dfrac{1}{5^3}$
Quotient rule	$\dfrac{a^m}{a^n} = a^{m-n}$ when $a \neq 0$	$\dfrac{2^2}{2^5} = 2^{2-5} = 2^{-3} = \dfrac{1}{2^3}$

⚠ CAUTION

Read directions carefully. In the examples above, we **simplified** each expression and wrote the answer using positive exponents.

If you are told to **evaluate** an expression, go one step further and do all possible calculations.

Example: $\dfrac{2^2}{2^5} \xrightarrow[\text{by using the quotient rule.}]{\text{Simplify}} 2^{2-5} = 2^{-3} = \dfrac{1}{2^3} \xrightarrow[\text{in the denominator.}]{\text{Evaluate}} \dfrac{1}{8}$ by multiplying $2 \cdot 2 \cdot 2$

10.2 Exercises

FOR EXTRA HELP

Go to MyMathLab for worked-out, step-by-step solutions to exercises enclosed in a square ▢ and video solutions to ▶ exercises.

CONCEPT CHECK *Decide whether each expression gives a positive or negative result.*

1. $(-2)^{-3}$

2. $(-3)^{-2}$

3. -2^4

4. -3^6

5. $(-2)^6$

6. $(-2)^3$

*Each expression is equal to either 0, 1, or −1. Decide which is correct. **See Example 1.***

7. $(-4)^0$

8. $(-10)^0$

9. -9^0

10. -5^0

11. $(-2)^0 - 2^0$

12. $(-8)^0 - 8^0$

13. $\dfrac{0^{10}}{10^0}$

14. $\dfrac{0^5}{5^0}$

*Evaluate each expression. **See Examples 1 and 2.***

15. $7^0 + 9^0$

16. $8^0 + 6^0$

17. 2^{-5}

18. 3^{-3}

19. $1 - 5^0$

20. $1 - 7^0$

21. 4^{-3} ▶

22. 5^{-4}

23. $5^{-1} + 3^{-1}$ ▶

24. $6^{-1} + 2^{-1}$

25. $8^{-1} - 4^{-1}$

26. $10^{-1} - 5^{-1}$

*Use the quotient rule to simplify each expression. Write each expression with positive exponents. Assume that all variables represent nonzero numbers. **See Example 3.***

27. $\dfrac{6^7}{6^2} = 6^{7-2} = 6\underline{}$ GS

28. $\dfrac{8^5}{8^3} = 8^{5-3} = 8\underline{}$ GS

29. $\dfrac{10^4}{10}$ ▶

30. $\dfrac{7^8}{7}$

31. $\dfrac{y^2}{y^6}$

32. $\dfrac{d^9}{d^5}$

33. $\dfrac{c^6}{c^5}$ ▶

34. $\dfrac{x^4}{x^3}$

35. $\dfrac{5^3}{5^7}$

36. $\dfrac{9^2}{9^8}$

37. $\dfrac{m^7}{m^8}$

38. $\dfrac{n^2}{n^3}$

39. $\dfrac{3^{-4}}{3^{-8}} = 3^{-4-(-8)} =$

$3^{-4+(+8)} = 3^{\underline{}}$

40. $\dfrac{5^{-3}}{5^{-10}} = 5^{-3-(-10)} =$

$5^{-3+(+10)} = 5^{\underline{}}$

41. $\dfrac{a^{-2}}{a^{-5}}$

42. $\dfrac{b^{-4}}{b^{-6}}$

43. $\dfrac{2^{-10}}{2^{-2}}$

44. $\dfrac{4^{-7}}{4^{-3}}$

45. $\dfrac{r^{-12}}{r^{-8}}$

46. $\dfrac{v^{-10}}{v^{-5}}$

47. $\dfrac{10^6}{10^{-4}}$

48. $\dfrac{10^4}{10^{-2}}$

49. $\dfrac{10^{-2}}{10^3}$

50. $\dfrac{10^{-1}}{10^7}$

51. $\dfrac{10^3}{10^{-8}}$

52. $\dfrac{10^5}{10^{-7}}$

53. $\dfrac{10^{-4}}{10^4}$

54. $\dfrac{10^{-6}}{10^6}$

Simplify each expression. Write answers with only positive exponents. Assume that all variables represent nonzero numbers. ***See Example 4.***

55. $10^6 \left(10^{-2}\right) =$

$10^{6+(-2)} = 10^{\underline{}}$

56. $10^{-7} \left(10^4\right) =$

$10^{-7+4} = 10^{-3} = \dfrac{1}{10^{\underline{}}}$

57. $a^{-4} \left(a^3\right)$

58. $w^9 \left(w^{-8}\right)$

59. $\left(2^{-4}\right) \left(2^{-4}\right)$

60. $\left(2^{-1}\right) \left(2^{-5}\right)$

61. $\left(x^{-10}\right) \left(x^{-1}\right)$

62. $\left(h^{-5}\right) \left(h^{-5}\right)$

63. $10^8 \cdot 10^{-2} \cdot 10^{-4}$

64. $10^{-2} \cdot 10^7 \cdot 10^{-2}$

65. $\left(y^{-3}\right) \left(y^5\right) \left(y^{-4}\right)$

66. $\left(n^{-8}\right) \left(n^{-2}\right) \left(n^5\right)$

67. $m^{-6} \cdot m^3 \cdot m^{12}$

68. $b^{-3} \cdot b^3 \cdot b^8$

69. $\left(10^{-1}\right) \left(10^{-2}\right) \left(10^{-3}\right)$

70. $\left(10^{-4}\right) \left(10^{-5}\right) \left(10^{-6}\right)$

Summary Exercises *Using Exponent Rules*

Evaluate each expression. Assume all variables represent nonzero numbers.

1. $(-3)^2$

2. $5^3 + 5^2$

3. 15^0

4. $3^{-2} - 5^{-1}$

5. p^0

6. 4^{-3}

7. -3^2

8. $2^3 \cdot 3^2$

9. -2^3

10. $-(8)^0$

11. $6^{-1} + 2^{-2}$

12. $9g^0$

13. 2^{-5}

14. $(-12)^0$

15. $(-2)^3$

16. $(-7n)^0$

Simplify. Write all answers in exponential form. Assume all variables represent nonzero numbers.

17. $(6^4)(6^4)$

18. x^{-7}

19. $10^2(10^{-10})$

20. $(2^3)^3$

21. $(y^{-12})(y^6)(y^6)$

22. $(2m^2n)^5$

23. $(-5)^2(-5)^8$

24. $\dfrac{w^{-4}}{w^{-7}}$

25. $(m^6)^4$

26. $\dfrac{n^{-5}}{n^{-3}}$

27. $(10^{-2})(10^{-2})(10^{-2})$

28. $d^5 \cdot d \cdot d^3$

29. $4x^4 \cdot 4x^4$

30. $\dfrac{c^4}{c^5}$

31. $(10^4)^2$

32. $\dfrac{4^{-1}}{4^{-6}}$

33. $\left(\dfrac{x}{y}\right)^6$

34. $3^3 \cdot 3^3 \cdot 3^5$

35. $\dfrac{7^{12}}{7^{11}}$

36. $h^{-6}\left(h^7\right)$

37. $-8a^3\left(2a\right)$

38. $\left(\dfrac{1}{2}\right)^4$

39. $(5xy)^3$

40. $\left(m^{13}\right)\left(m^{-2}\right)\left(m^{-6}\right)$

41. $\dfrac{10^{-3}}{10^4}$

42. $b^9\left(b^{-10}\right)$

43. $(b^3)^5$

44. $\dfrac{5^3}{5^{10}}$

45. $10^{-8}\left(10^5\right)$

46. $\left(-7y^4\right)\left(-3y^4\right)$

47. $\dfrac{10^5}{10^{-8}}$

48. $\left(k^{-4}\right)\left(k^{-4}\right)$

49. $10\left(ab\right)^4$

50. $x^{-5} \cdot x \cdot x^{-5}$

10.3 An Application of Exponents: Scientific Notation

OBJECTIVE ▸ 1 **Express numbers in scientific notation.** The numbers used in science are often extremely large (such as the distance from Earth to the sun, 93,000,000 miles) or extremely small (the wavelength of yellow-green light, approximately 0.0000006 meter). To simplify their work, scientists often express such numbers with exponents. Each number is written as $a \times 10^n$, where n is an integer and a is greater than or equal to 1 and less than 10 (that is, $1 \le |a| < 10$). This form is called **scientific notation.** There is always one nonzero digit before the decimal point. Here are some examples.

> **OBJECTIVES**
>
> **1** Express numbers in scientific notation.
>
> **2** Convert numbers in scientific notation to numbers without exponents.
>
> **3** Use scientific notation in calculations.
>
> **4** Solve application problems using scientific notation.

35 is written 3.5×10^1

because $35 = 3.5 \times 10 = 3.5 \times 10^1$

$56{,}200$ is written 5.62×10^4

because $56{,}200 = 5.62 \times 10{,}000 = 5.62 \times 10^4$

0.09 is written 9×10^{-2}

because $0.09 = \dfrac{9}{100} = 9 \times \dfrac{1}{100} = 9 \times \dfrac{1}{10^2} = 9 \times 10^{-2}$

To write a number in scientific notation, follow the steps listed below.

Writing a Number in Scientific Notation

Step 1   Move the decimal point to the right of the first nonzero digit.

Step 2   To figure out the exponent on 10, count the number of places you moved the decimal point. The number of places is the absolute value of the exponent on 10.

Step 3   To decide on the sign of the exponent, look at the *original* number. If the *original* number was "large" (10 or more), the exponent on 10 is positive. If the *original* number was "small" (between 0 and 1), the exponent is negative.

EXAMPLE 1   **Writing Numbers in Scientific Notation**

Write each number in scientific notation.

(a) 93,000,000

Move the decimal point to the right of the first nonzero digit (the 9). Count the number of places the decimal point was moved.

9 3 000 000.   9.3 000 000   ◁ Move the decimal point 7 places.

↑   ↑
Decimal point   Decimal point
moves here.   starts here.

We will write the number as 9.3×10^n. The *original* number (93,000,000) was "large" so the exponent n will be *positive 7*.

$$93{,}000{,}000 = \mathbf{9.3 \times 10^7}$$

(b) $302{,}100 = \mathbf{3.021 \times 10^5}$ Move the decimal point 5 places. *Original* number was "large" so the exponent is *positive 5*.

—————————————— **Continued on Next Page**

1 Write each number in scientific notation.

(a) 63,000

6 3000 Move the decimal point ____ places.

The original number is "large," so the exponent is positive.

$6.3 \times 10^{-}$

(b) 5,870,000

(c) 0.0571

(d) 0.000062

2 Write without exponents.

(a) 4.2×10^3 ← Positive exponent tells you to make the number larger; move the decimal point 3 places.

$4.200 = $ _____ Attach zeros as needed.

(b) 8.7×10^5

(c) 6.42×10^{-3}

3 Simplify. Write the answers both in scientific notation and as numbers without exponents.

(a) $(2.6 \times 10^4)(2 \times 10^{-6})$

$= (2.6 \times 2)(10^4 \times 10^{-6})$

$= 5.2 \times 10^{4+(-6)} = 5.2 \times 10^{-}$

$= $ _____

(b) $\dfrac{4.8 \times 10^2}{2.4 \times 10^{-3}}$

Answers

1. **(a)** 4 places; 6.3×10^4 **(b)** 5.87×10^6
   (c) 5.71×10^{-2} **(d)** 6.2×10^{-5}
2. **(a)** 4200 **(b)** 870,000 **(c)** 0.00642
3. **(a)** $5.2 \times 10^{-2} = 0.052$
   (b) $2 \times 10^5 = 200,000$

(c) 0.00462

Move the decimal point to the right of the first nonzero digit (the 4). Count the number of places it was moved.

0 004.62 ← Move the decimal point 3 places.

The *original* number (0.00462) was "small" so the exponent will be *negative 3*.

$$0.00462 = \mathbf{4.62 \times 10^{-3}}$$

(d) $0.0000762 = \mathbf{7.62 \times 10^{-5}}$ Move the decimal point 5 places; *original* number was "small" so the exponent is *negative 5*.

◀ **Work Problem ❶ at the Side.**

OBJECTIVE ❷ Convert numbers in scientific notation to numbers without exponents. To convert a number written in scientific notation to a number without exponents, remember that multiplying a number by a positive power of 10 will make the number larger; multiplying by a negative power of 10 will make the number smaller.

EXAMPLE 2 | **Writing Numbers without Exponents**

Write each number without exponents.

(a) 6.2×10^3

Because the exponent is positive, make 6.2 larger by moving the decimal point 3 places to the *right*. Attach zeros as needed.

$$6.2 \times 10^3 = 6.200 = \mathbf{6200}$$

(b) $4.283 \times 10^5 = 4.28300 = \mathbf{428,300}$ Move 5 places to the *right*. Attach zeros as needed.

(c) $9.73 \times 10^{-2} = 09.73 = \mathbf{0.0973}$ Move 2 places to the *left*; multiplying by a negative power of 10 makes the number *smaller*.

The exponent tells you the number of places that the decimal point is moved, and the direction it moves.

◀ **Work Problem ❷ at the Side.**

OBJECTIVE ❸ Use scientific notation in calculations. The next example uses scientific notation with products and quotients.

EXAMPLE 3 | **Multiplying and Dividing with Scientific Notation**

Simplify. Write the answers both in scientific notation and as numbers without exponents.

(a) $(2 \times 10^3)(4 \times 10^{-4})$

$= (2 \times 4)(10^3 \times 10^{-4})$ Commutative and associative properties

$= \mathbf{8 \times 10^{-1}}$ Product rule for exponents

$= 8. = \mathbf{0.8}$ Write without exponents.

(b) $\dfrac{6.4 \times 10^{-5}}{2 \times 10^3} = \dfrac{6.4}{2} \times \dfrac{10^{-5}}{10^3} = \mathbf{3.2 \times 10^{-8}} = \mathbf{0.000000032}$ ← Move the decimal point 8 places to the *left*.

◀ **Work Problem ❸ at the Side.**

OBJECTIVE ▶ **4** **Solve application problems using scientific notation.** Sometimes application problems may involve one or more numbers written in scientific notation and other numbers written without exponents. By writing all the numbers in scientific notation, we can use the product and quotient rules for exponents to help solve the problem.

EXAMPLE 4 **Solving an Application Problem**

There are approximately 3.03×10^{23} molecules in one gram of hydrogen gas. How many molecules are in 2000 g of hydrogen gas? Write your answer both in scientific notation and as a number without exponents.

First write 2000 in scientific notation.

2.000 Move 3 places to the left; *original* number was "large" so the exponent is *positive*.

$$2000 = 2.0 \times 10^3$$

Now multiply the number of molecules in one gram times the number of grams.

$$(3.03 \times 10^{23})(2.0 \times 10^3) =$$
$$(3.03 \times 2.0)(10^{23} \times 10^3) = \quad \text{Use the product rule for exponents.}$$
$$6.06 \times 10^{26}$$

Written in scientific notation, there are **6.06×10^{26} molecules** in 2000 g of hydrogen gas. Written without an exponent, the number is:

606,000,000,000,000,000,000,000,000 molecules

──────── **Work Problem** **4** **at the Side.** ▶

▦ Calculator Tip

Most scientific calculators have a key that allows you to enter numbers in scientific notation. Usually this key is labeled $\boxed{\text{EE}}$ or $\boxed{\text{EXP}}$. Then you could solve **Example 4,** above, by pressing:

3.03 $\boxed{\text{EXP}}$ 23 $\boxed{\times}$ 2000 $\boxed{=}$ $\boxed{6.06 \times 10^{26}}$

The exponent in the answer is 26. **You may have to write the $\times$ 10 part of the answer** to get 6.06×10^{26}. (Some calculators show the $\times$ 10 part.)

To enter a number in scientific notation that has a *negative* exponent, use the *change of sign* key. For example, to enter 8.45×10^{-6}, press:

8.45 $\boxed{\text{EXP}}$ 6 $\boxed{+/-}$ $\boxed{8.45 \times 10^{-06}}$

Your calculator may have a separate negative sign key to use: $\boxed{(-)}$

Also, on some calculators, you may have to use the 2nd function key in order to access EE or EXP. Check your calculator's instructions.

---

4 There are 4.356×10^4 square feet in one acre. How many square feet are in 1500 acres? Write your answer both in scientific notation and as a number without exponents.

Write 1500 in scientific notation.

$$1.500 = 1.5 \times 10 —$$

Now multiply.

$$(4.356 \times 10^4)(1.5 \times 10 —) =$$

$6.534 \times 10 —$ square feet

Now write the answer without exponents.

Answer

4. $1500 = 1.5 \times 10^3$; 6.534×10^7 square feet, or 65,340,000 square feet

⑤ In a laboratory experiment, a test group of 950 common house spiders weighed 2.09×10^{-1} pound. Find the weight of each spider. Write your answer both in scientific notation and as a number without exponents.

EXAMPLE 5 **Solving an Application Problem**

At a printing shop, a stack of 5×10^3 sheets of paper measures 0.38 meter high. What is the thickness of one sheet of paper? Write your answer both in scientific notation and as a number without exponents.

First write 0.38 in scientific notation.

$$0\;3.8$$

Move 1 place to the right; *original* number was "small" so exponent is negative.

$$0.38 = 3.8 \times 10^{-1}$$

Now divide the total height of the stack by the number of sheets in the stack.

$$\frac{3.8 \times 10^{-1}}{5 \times 10^3} = \frac{3.8}{5} \times \frac{10^{-1}}{10^3} = 0.76 \times 10^{-4}$$

Use the quotient rule for exponents.

> Don't stop! This number is *not* in scientific notation because 0.76 is not between 1 and 10.

The answer of 0.76×10^{-4} is *not* in proper scientific notation yet, because 0.76 is *not* between 1 and 10. The decimal point in 0.76 must be moved *1 place* to get 7.6. Then, because the original number (0.76) was "small," add *negative* 1 to the exponent.

$$\overset{(-4) + (-1)}{0.76 \times 10^{-4}} = 7.6 \times 10^{-5}$$

Move decimal point 1 place.

The thickness of one sheet of paper is $\mathbf{7.6 \times 10^{-5}}$ **meter**.

Written without an exponent, the thickness is **0.000076 meter**.

◀ **Work Problem ⑤ at the Side.**

10.3 Exercises

FOR EXTRA HELP

Go to MyMathLab for worked-out, step-by-step solutions to exercises enclosed in a square ▢ and video solutions to ▶ exercises.

CONCEPT CHECK   *Decide whether or not each number is written in scientific notation. Write* yes *or* no.

1. 4.56×10^3

2. 7.34×10^5

3. $5,600,000$

4. $34,000$

5. 20.32×10^4

6. 44.9×10^6

7. 0.8×10^2

8. 0.9×10^3

9. 9.007×10^{-5}

10. 6.611×10^{-4}

Write the numbers (other than dates) in these statements in scientific notation.
See Example 1.

11. The mass of Mercury, the smallest planet, is 0.0553 times that of Earth. The mass of Jupiter, the largest planet, is 317.83 times that of Earth. (Data from *The World Almanac and Book of Facts.*)

12. NASA budgeted $3,290,000,000 to operate the international space station in 2018. (Data from NASA.)

13. There are 1,240,000,000 pounds of pumpkins grown every year in the United States. Illinois leads the nation with 560,000,000 pounds. (Data from www.agr.state.il.us)

14. The human body contains trace amounts of several minerals. For example, a 150-pound body has about 0.00023 pound of copper in it.

*Write each number in scientific notation. **See Example 1.***

15. $5,876,000,000$

16. $9,994,000,000$

17. $82,350$ ▶

18. $78,330$

19. 0.000007

20. 0.0000004

21. 0.00203 ▶

22. 0.0000578

*Write each number without exponents. **See Example 2.***

23. 7.5×10^5 ▶

24. 8.8×10^6

25. 5.677×10^{12}

26. 8.766×10^9

27. 6.21×10^0

28. 8.56×10^0

29. 7.8×10^{-4} ▶

30. 8.9×10^{-5}

31. 5.134×10^{-9}

32. 7.123×10^{-10}

Simplify. Write the answers both in scientific notation and as numbers without exponents. **See Example 3.**

33. $(2 \times 10^8)(3 \times 10^3) =$

$(2 \times 3)(10^8 \times 10^3) =$

$6 \times 10^{8+3} = 6 \times 10$—

34. $(4 \times 10^7)(2 \times 10^3) =$

$(4 \times 2)(10^7 \times 10^3) =$

$8 \times 10^{7+3} = 8 \times 10$—

35. $(5 \times 10^4)(3 \times 10^2)$

36. $(8 \times 10^5)(2 \times 10^3)$

37. $(3.15 \times 10^{-4})(2.04 \times 10^8)$

38. $(4.92 \times 10^{-3})(2.25 \times 10^7)$

39. $\dfrac{9 \times 10^{-5}}{3 \times 10^{-1}} =$

$\dfrac{9}{3} \times \dfrac{10^{-5}}{10^{-1}} = 3 \times 10$—

40. $\dfrac{12 \times 10^{-4}}{4 \times 10^{-3}} =$

$\dfrac{12}{4} \times \dfrac{10^{-4}}{10^{-3}} = 3 \times 10$—

41. $\dfrac{8 \times 10^3}{2 \times 10^2}$

42. $\dfrac{5 \times 10^4}{1 \times 10^3}$

43. $\dfrac{(2.6 \times 10^{-3})(7.0 \times 10^{-1})}{(2 \times 10^2)(3.5 \times 10^{-3})}$

44. $\dfrac{(9.5 \times 10^{-1})(2.4 \times 10^4)}{(5 \times 10^3)(1.2 \times 10^{-2})}$

Solve each problem. Write the answers both in scientific notation and without exponents. **See Examples 4 and 5.**

45. The population of China in 2015 was about 1.376×10^9 people. The country's land area is 3,700,000 square miles. What is the average number of people per square mile in China? (Data from *United Nations World Population Prospects*.)

46. The mass of Earth is approximately 5.98×10^{24} kg, and the mass of the moon is approximately 7.36×10^{22} kg. Earth's mass is how many times greater than the moon's mass? (Data from Time Life *The Universe*.)

47. The average diameter of a hydrogen atom is 1×10^{-10} meter. (Data from *FactFinder*.)

(a) If one million hydrogen atoms are lined up in a row, how long would the row be?

(b) Find the length of a row of one billion hydrogen atoms.

48. The speed of light is about 1.86×10^5 miles per second. (Data from Time Life *The Universe*.)

(a) How far does light travel in one minute?

(b) How far does light travel in one hour?

49. During a recent year, the U.S. Postal Service handled 1.6×10^{11} pieces of mail. The U.S. population was about 3.2×10^8. On average, how many mailed items did each person receive that year? (Data from USPS.)

50. Americans send 8×10^{12} text messages each year. On average, how many text messages are sent each month? (Data from www.Bloomberg.com)

10.4 Adding and Subtracting Polynomials

Terms are the quantities being added in an expression. Here is an example.

$$4x^3 + 6x^2 + 5x + 8$$

In the term $4x^3$, the number 4 is called the *numerical coefficient,* or simply the *coefficient,* of x^3. In the same way, 6 is the coefficient of x^2 in the term $6x^2$, and 5 is the coefficient of x in the term $5x$. The 8 is the constant term.

OBJECTIVE 1 **Review combining like terms.** Remember that *like terms* are terms with exactly the same variables, and with the same exponents on the variables. Only the coefficients may differ. Like terms are combined, or *added,* by adding their coefficients using the distributive property.

EXAMPLE 1 **Review of Combining Like Terms**

Simplify each expression by combining (adding) like terms.

(a) $-4x^3 + 6x^3 = (-4 + 6)x^3 = \mathbf{2x^3}$ Distributive property

(b) $9x^6 - 14x^6 + x^6 = (9 - 14 + 1)x^6 = \mathbf{-4x^6}$

(c) $12m^2 + 5m + 4m^2 = (12 + 4)m^2 + 5m = \mathbf{16m^2 + 5m}$

(d) $3x^2y + 4x^2y - x^2y = (3 + 4 - 1)x^2y = \mathbf{6x^2y}$

> **! CAUTION**
> In **Example 1(c)** above, we *cannot* add $16m^2$ and $5m$. These two terms are *unlike* because the exponents on the variables are different. *Unlike terms* have different variables or different exponents on the same variables.

Work Problem **1** *at the Side.* ▶

OBJECTIVE 2 **Use the vocabulary for polynomials.** A **polynomial in x** is a term, or the sum of several terms, of the form ax^n, where a is any real number and n is any whole number. Here is an example.

$$16x^8 - 7x^6 + 5x^4 - 3x^2 + 4 \longleftarrow \begin{cases} \text{Example of} \\ \text{polynomial in } x \end{cases}$$

The polynomial above is written in **descending powers,** which means the exponents on x decrease from left to right. Look at the next example.

$$2x^3 - x^2 + \frac{4}{x} \longleftarrow \begin{cases} \textit{Not} \text{ a polynomial because there} \\ \text{is a variable in the denominator} \end{cases}$$

This is *not* a polynomial, because a variable appears in the denominator. Of course, we could define a *polynomial* using any variable, not just x, as in **Example 1(c)** above. In fact, polynomials may have terms with more than one variable, as in **Example 1(d)** above.

Work Problem **2** *at the Side.* ▶

---

1 Simplify by adding like terms.

GS (a) $5x^4 - 7x^4 =$

$(5 - 7)x^4 = $ ____

(b) $9pq + 3pq - 2pq$

(c) $r^2 + 3r + 5r^2$

(d) $8t + 6w$

2 Choose all descriptions that apply for each of the expressions in parts (a)–(d) below.

  A. Polynomial

  B. Polynomial written in descending powers

  C. Not a polynomial

(a) $3m^5 + 5m^2 - 2m + 1$

(b) $2p^4 + p^6$

(c) $\dfrac{1}{x} + 2x^2 + 3$

(d) $x - 3$

Answers

1. **(a)** $-2x^4$ **(b)** $10pq$ **(c)** $6r^2 + 3r$
   (d) cannot be added—they are unlike terms
2. **(a)** A and B **(b)** A **(c)** C **(d)** A and B

3 Simplify each polynomial, if possible. Then give the degree and tell whether the simplified polynomial is a monomial, binomial, trinomial, or none of these.

GS **(a)** $3x^2 + 2x - 4$

There are no like terms, so this polynomial cannot be simplified.

Look at the term $3x^2$ to find the degree of the polynomial.

It is degree ____

There are *three* terms, so this is a _____

(b) $x^3 + 4x^3$

(c) $x^8 - x^7 + 2x^8$

The **degree of a term** is the *sum* of the exponents on the variables. For example, $3x^4$ has degree 4, while $6x^{17}$ has degree 17. The term $5x$ has degree 1, and $2x^2y$ has degree $2 + 1 = 3$ (because y has an understood exponent of 1).

The **degree of a polynomial** is the highest degree of any nonzero term in the polynomial. For example, $3x^4 - 5x^2 + 6$ is a polynomial of degree 4, the polynomial $5x + 7$ is of degree 1, and $x^2y + xy - 5xy^2$ is of degree 3.

Three types of polynomials are very common and are given special names. A polynomial with only one term is called a **monomial** because *mon(o)*- means "one," as in *mono*rail. Some examples are shown below.

$$9m \qquad -6y^5 \qquad a^2 \qquad 6 \quad \boxed{\text{Monomials}}$$

A polynomial with exactly two terms is called a **binomial** because *bi*- means "two," as in *bi*cycle. Examples are shown below.

$$-9x^4 + 9x^3 \qquad 8m^2 + 6m \qquad 3m^5 - 9m^2 \quad \boxed{\text{Binomials}}$$

A polynomial with exactly three terms is called a **trinomial** because *tri*- means "three," as in *tri*angle. Examples are shown below.

$$9m^3 - 4m^2 + 6 \qquad -3m^5 - 9m^2 + 2 \quad \boxed{\text{Trinomials}}$$

EXAMPLE 2 Classifying Polynomials

Simplify each polynomial, if possible, by combining like terms. Then give the degree and tell whether the simplified polynomial is a monomial, a binomial, a trinomial, or none of these.

(a) $2x^3 + 5$

The polynomial cannot be simplified. The degree is 3. The polynomial is a *binomial* because it has exactly *two* terms.

(b) $4x - 5x + 2x$

Add like terms to simplify: $4x - 5x + 2x = x$. The degree is 1 (since $x = x^1$). The simplified polynomial is a *monomial* because it has only *one* term.

◀ **Work Problem 3** at the Side.

OBJECTIVE 3 Evaluate polynomials. A polynomial usually represents different numbers for different values of the variable, as shown in the next example.

EXAMPLE 3 Evaluating a Polynomial

Evaluate $5x^3 - 4x - 4$ when $x = -2$ and when $x = 3$.

First, replace x with -2. ⎫ Use parentheses to avoid errors.

$$5x^3 - 4x - 4 = 5(-2)^3 - 4(-2) - 4 \qquad \text{Apply the exponent.}$$
$$= 5(-8) - 4(-2) - 4 \qquad \text{Multiply.}$$
$$= -40 + 8 - 4 \qquad \text{Add and subtract.}$$
$$= \mathbf{-36}$$

Now replace x with 3.

$$5x^3 - 4x - 4 = 5(3)^3 - 4(3) - 4$$
$$= 5(27) - 4(3) - 4$$
$$= 135 - 12 - 4$$
$$= \mathbf{119}$$

─────── **Continued on Next Page**

Answers

3. (a) degree 2; trinomial
 (b) simplify to $5x^3$; degree 3; monomial
 (c) simplify to $3x^8 - x^7$; degree 8; binomial

CAUTION

Notice the use of parentheses around the numbers that are replacing the variable in **Example 3** on the previous page. This is particularly helpful when a negative number replaces a variable that is raised to a power, so the sign of the product will be correct.

─── **Work Problem 4 at the Side.** ▶

OBJECTIVE 4 Add polynomials. Polynomials may be added, subtracted, multiplied, and divided.

Adding Polynomials

To add two polynomials, add like terms.

EXAMPLE 4 Adding Polynomials Vertically

(a) Add $6x^3 - 4x^2 + 3$ and $-2x^3 + 7x^2 - 5$ vertically.

Write like terms in columns. Then add column by column.

$$\begin{array}{r} 6x^3 - 4x^2 + 3 \\ -2x^3 + 7x^2 - 5 \\ \hline 4x^3 + 3x^2 - 2 \end{array}$$

(b) Add $2x^2 - 4x + 3$ and $x^3 + 5x$ vertically.

Write like terms in columns and add column by column.

$$\begin{array}{r} 2x^2 - 4x + 3 \\ x^3 \qquad + 5x \\ \hline x^3 + 2x^2 + x + 3 \end{array}$$

Leave spaces for missing terms.

─── **Work Problem 5 at the Side.** ▶

The polynomials in **Example 4** above also could be added horizontally by combining like terms, as shown in the next example.

EXAMPLE 5 Adding Polynomials Horizontally

Find each sum by adding horizontally.

(a) Add $(6x^3 - 4x^2 + 3)$ and $(-2x^3 + 7x^2 - 5)$

$$(6x^3 - 4x^2 + 3) + (-2x^3 + 7x^2 - 5)$$ Combine like terms.

The sum is the same as in **Example 4(a)** above: $4x^3 + 3x^2 - 2$

(b) Add $(2x^2 - 4x + 3)$ and $(x^3 + 5x)$

$= (2x^2 - 4x + 3) + (x^3 + 5x)$ Combine like terms.

$= x^3 + 2x^2 + x + 3$

─── **Continued on Next Page**

4 Evaluate $2y^3 + 8y - 6$ when:

(a) $y = -1$

$$2y^3 + 8y - 6$$
$$= 2(-1)^3 + 8(-1) - 6$$
$$= 2(-1) + (-8) - 6$$
$$= -2 - 8 - 6$$
$$= \underline{\quad}$$

(b) $y = 4$

5 Add each pair of polynomials vertically.

(a) $4x^3 - 3x^2 + 2x$ and $6x^3 + 2x^2 - 3x$

$$\begin{array}{r} 4x^3 - 3x^2 + 2x \\ 6x^3 + 2x^2 - 3x \\ \hline \underline{\quad}x^3 - \underline{\quad} - \underline{\quad} \end{array}$$

(b) $x^2 - 2x + 5$ and $4x^2 - 2$

Answers
4. (a) -16 (b) 154
5. (a) $10x^3 - x^2 - x$ (b) $5x^2 - 2x + 3$

6 Find each sum by adding horizontally.

(a) $(2x^4 - 6x^2 + 7)$
$+ (-3x^4 + 5x^2 + 2)$

Combine like terms.

Blue   Green   **Red**
terms   terms   **terms**

$-1x^4 - 1x^2 + \underline{}$

or $-x^4 - x^2 + \underline{}$

(b) $(3x^2 + 4x + 2)$
$+ (6x^3 - 5x - 7)$

7 Subtract, and check your answers using addition.

(a) $(14y^3 - 6y^2 + 2y - 5)$
$- (2y^3 - 7y^2 - 4y + 6)$

Change   Change *every*
subtraction   sign in the
to addition.   second polynomial.

(b) $(2y^2 - 6y + 6)$ from
$3y^3 - 7y^2 - 11y + 6$

Answers

6. **(a)** $-1x^4 - 1x^2 + 9$ or $-x^4 - x^2 + 9$
   (b) $6x^3 + 3x^2 - x - 5$
7. **(a)** $12y^3 + y^2 + 6y - 11$
   (b) $3y^3 - 9y^2 - 5y$

Note

Notice the order of the terms in the final answer to **Example 5(b)** on the previous page. Mathematicians generally write polynomials in descending order of power. The answers in this text are shown that way.

◀ **Work Problem 6** at the Side.

OBJECTIVE 5 **Subtract polynomials.** Recall that we change subtraction problems to addition problems. To subtract two numbers, we *add* the first number to the *opposite* of the second number. Here are two examples as a quick review.

$$7 - 2 \qquad\qquad -8 - (-5)$$
$$\downarrow \quad \downarrow \qquad\qquad \downarrow \quad \downarrow$$
$$7 + (-2) = 5 \qquad -8 + (+5) = -3$$

A similar method is used to subtract polynomials.

Subtracting Polynomials

To subtract two polynomials, change *all* the signs of the second polynomial and add the result to the first polynomial.

EXAMPLE 6   **Subtracting Polynomials**

(a) Perform this subtraction: $(5x - 2) - (3x - 8)$

$$(5x - 2) - (3x - 8)$$

Change subtraction to addition.
Change *every* sign in the *second* polynomial to its opposite.

$$= (5x - 2) + (-3x + 8)$$

$$= \mathbf{2x + 6} \qquad \text{Combine like terms.}$$

(b) Subtract $6x^3 - 4x^2 + 2$ from $11x^3 + 2x^2 - 8$

Be careful to rewrite the problem in the correct order.

$$(11x^3 + 2x^2 - 8) - (6x^3 - 4x^2 + 2)$$

Change every sign in the *second* polynomial; then *add* the two polynomials.

$$(11x^3 + 2x^2 - 8) + (-6x^3 + 4x^2 - 2)$$

$$= \mathbf{5x^3 + 6x^2 - 10}$$

Change *every* sign in the second polynomial.

To check a subtraction problem, use the fact that if $a - b = c$, then $a = b + c$. For example, $6 - 2 = 4$, so check by writing $6 = 2 + 4$, which is correct.

Check the polynomial subtraction above by adding $6x^3 - 4x^2 + 2$ and $5x^3 + 6x^2 - 10$. Because the sum is $11x^3 + 2x^2 - 8$, the subtraction was done correctly.

◀ **Work Problem 7** at the Side.

10.4 Exercises

FOR EXTRA HELP   Go to MyMathLab for worked-out, step-by-step solutions to exercises enclosed in a square ☐ and video solutions to ▶ exercises.

CONCEPT CHECK   *For each polynomial, state the number of terms and identify the coefficient of each term. Then state whether the polynomial is a* monomial, binomial, *or* trinomial.

1. $6x^4$

2. $-9y^5$

3. t^4

4. s^7

5. $-19r^2 - r$

6. $2y^3 - y$

7. $x + 8x^2 - 8x^3$

8. $v - 2v^3 + 2v^4$

CONCEPT CHECK   *Circle the like terms in each list.*

9. $-3mn \qquad 3m^2n \qquad mn \qquad 9mn$

10. $4x^2 \qquad 4x^3 \qquad -x^2 \qquad 4x$

11. $8w^4 \qquad -8 \qquad 8x^4 \qquad 4w^8$

12. $-2ab^2 \qquad a^2b^2 \qquad ab^2 \qquad ab$

13. $xy^3 \qquad 16 \qquad -x^3y \qquad -1$

14. $7wxy \qquad -wy \qquad 7 \qquad 7w^2x^2y^2$

Simplify each polynomial, if possible. Write the result with descending powers.
See Example 1.

15. $-3m^5 + 5m^5$
GS $(-3 + 5)m^5 = $ ____

16. $-4y^3 + 3y^3$
GS $(-4 + 3)y^3 = -1y^3$ or   ____

17. $2r^5 + (-3r^5)$

18. $-19y^2 + 9y^2$

19. $0.2m^5 - 0.5m^2$

20. $-0.9y + 0.9y^2$

21. $-3x^5 + 2x^5 - 4x^5$

22. $6x^3 - 8x^3 + 9x^3$

23. $-4p^7 + 8p^7 + 5p^9$
▶

24. $-3a^8 + 4a^8 - 3a^2$

25. $-4y^2 + 3y^2 - 2y^2 + y^2$

26. $3r^5 - 8r^5 + r^5 + 2r^5$

Simplify each polynomial, if possible, and write it with descending powers. Then give the degree of the simplified polynomial, and tell whether it is a monomial, binomial, trinomial, *or* none of these. **See Examples 1 and 2.**

27. $6x^4 - 9x$

28. $7t^3 - 3t$

29. $5m^4 - 3m^2 + 6m^5 - 7m^3$

30. $6p^5 + 4p^3 - 8p^4 + 10p^2$

31. $\dfrac{5}{3}x^4 - \dfrac{2}{3}x^4 + \dfrac{1}{3}x^2 - 4$

32. $\dfrac{4}{5}r^6 + \dfrac{1}{5}r^6 - r^4 + \dfrac{2}{5}r$

33. $0.8x^4 - 0.3x^4 - 0.5x^4 + 7x$

34. $1.2t^3 - 0.9t^3 - 0.3t^3 + 9$

*Evaluate each polynomial (**a**) when $x = 2$ and (**b**) when $x = -1$. **See Example 3.***

35. $-2x + 3$

  (a) $\underbrace{-2\,(2) + 3}$

     $\underbrace{-4\ +3}$

     _____

  (b) $\underbrace{-2\,(-1) + 3}$

     $\underbrace{2\ \ +3}$

     _____

36. $5x - 4$

  (a) $\underbrace{5\,(2) - 4}$

     $\underbrace{10\ -4}$

     _____

  (b) $\underbrace{5\,(-1) - 4}$

     $\underbrace{-5\ \ -4}$

     _____

37. $2x^2 + 5x + 1$

38. $-3x^2 + 14x - 2$

39. $2x^5 - 4x^4 + 5x^3 - x^2$

40. $x^4 - 6x^3 + x^2 + 1$

41. $-4x^5 + x^2$

42. $2x^6 - 4x$

*Add each pair of polynomials vertically. **See Example 4.***

43. $3m^2 + 5m$ and $2m^2 - 2m$

44. $4a^3 - 4a^2$ and $6a^3 + 5a^2$

45. $-6x^3 - 4x + 1$ and $5x + 5$

46. $7n^2 + 2n - 4$
and $-6n^2 - 8$

47. $3w^3 - 2w^2 + 8w$
and $2w^2 - 8w + 5$

48. $-5r^3 - 4r^2 - 3r$
and $r^4 - 4r^2 + 3r$

Perform the indicated operations. See Examples 5 and 6.

49. $(12x^4 - x^2) - (8x^4 + 3x^2)$

50. $(13y^5 - y^3) - (7y^5 + 5y^3)$

51. $(2r^2 + 3r - 12) + (6r^2 + 2r)$

52. $(3r^2 + 5r - 6) + (2r - 5r^2)$

53. $(8m^2 - 7m) - (3m^2 + 7m - 6)$
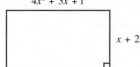

54. $(x^2 + x) - (3x^2 + 2x - 1)$

55. $(16x^3 - x^2 + 3x) + (-12x^3 + 3x^2 + 2x)$

56. $(-2b^6 + 3b^4 - b^2) + (b^6 + 2b^4 + 2b^2)$

57. Subtract $5m^2 - 4$ from $12m^3 - 8m^2 + 6m + 7$

58. Subtract $a^2 + a - 1$ from $-3a^3 + 2a^2 - a + 6$

59. Subtract $9x^2 - 3x + 7$ from $-2x^2 - 6x + 4$

60. Subtract $-5w^3 + 5w^2 - 7$ from $6w^3 + 8w + 5$

Find the perimeter of each rectangle or triangle.

61.
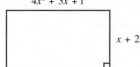
$4x^2 + 3x + 1$

$x + 2$

62.
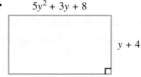
$5y^2 + 3y + 8$

$y + 4$

63.
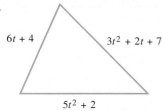
$6t + 4$

$3t^2 + 2t + 7$

$5t^2 + 2$

64.
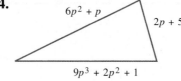
$6p^2 + p$

$2p + 5$

$9p^3 + 2p^2 + 1$

Relating Concepts (Exercises 65–68) For Individual or Group Work

At the start of this chapter, we gave a polynomial that models the distance in feet that a car going approximately 68 mph will skid in t seconds. If we let D represent the skidding distance, then we have this formula.

$$D = 100t - 13t^2$$

Use this formula as you **work Exercises 65–68 in order.**

65. To find the distance that the car will skid in 1 second, evaluate the polynomial when *t* is 1. Finish the work that is started below.

$$D = 100t - 13t^2 \qquad \text{Replace } t \text{ with 1}$$
$$D = 100\,(1) - 13\,(1)^2$$
$$D =$$
$$D =$$

66. From **Exercise 65,** you should have found that when *t* is 1 second, then *D* is 87 ft. Now evaluate the polynomial for the other values of *t* listed in the table at the right. Write the ordered pairs in the table. Round distance (*D*) to the nearest whole number.

t (sec)	*D* (ft)	(*t, D*)
0		
0.5		
1	87	(1, 87)
1.5		
2		
2.5		
3		
3.5		

67. Graph each of the ordered pairs by making a dot at the appropriate point on the grid at the right.

68. Connect the points on the grid with a smooth curve. Describe how this graph is different from the graphs of linear equations.

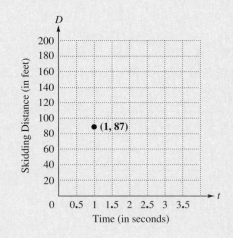

10.5 | Multiplying Polynomials: An Introduction

OBJECTIVE ▶ ① Multiply a monomial and a polynomial. Earlier in this chapter, you found the product of two monomials by using the rules for exponents and the commutative and associative properties. Here is an example.

$$(-8m^6)(-9n^6) = (-8)(-9)(m^6)(n^6) = 72m^6n^6$$

OBJECTIVES

① Multiply a monomial and a polynomial.

② Multiply two polynomials.

! CAUTION

Do not confuse *addition* of terms with *multiplication* of terms. Here is an example.

Adding Terms **Multiplying Terms**

$7x^5 + 2x^5 = (7 + 2)x^5 = 9x^5$ $(7x^5)(2x^5) = (7 \cdot 2)x^{5+5} = 14x^{10}$

To find the product of a monomial and a polynomial with more than one term, we use the distributive property and multiplication of monomials.

EXAMPLE 1 Multiplying a Monomial and a Polynomial

Use the distributive property to find each product.

(a) $4x^2(3x + 5)$

$4x^2(3x + 5) = 4x^2(3x) + 4x^2(5)$ Distributive property

$= 12x^3 + 20x^2$ Multiply monomials.

(b) $-8m^3(4m^3 + 3m^2 + 2m)$

$= (-8m^3)(4m^3) + (-8m^3)(3m^2) + (-8m^3)(2m)$ Distributive property

$= -32m^6 + (-24m^5) + (-16m^4)$ Multiply monomials.

$= -32m^6 - 24m^5 - 16m^4$

— **Work Problem ① at the Side.** ▶

OBJECTIVE ▶ ② Multiply two polynomials. We can use the distributive property repeatedly to find the product of any two polynomials. For example, to find the product of the polynomials $x^2 + 3x + 5$ and $x - 4$, think of $x - 4$ as a single quantity and use the distributive property as follows.

$$(x^2 + 3x + 5)(x - 4) = x^2(x - 4) + 3x(x - 4) + 5(x - 4)$$

Now use the distributive property three times to first find $x^2(x - 4)$, then $3x(x - 4)$, and finally $5(x - 4)$.

$x^2(x - 4) + 3x(x - 4) + 5(x - 4)$

$= x^2(x) + x^2(-4) + 3x(x) + 3x(-4) + 5(x) + 5(-4)$

$= (x^3) + (-4x^2) + (3x^2) + (-12x) + (5x) + (-20)$

$= x^3 - x^2 - 7x - 20$ Combine like terms.

This example suggests a rule for multiplying polynomials.

① Find each product.

(a) $5m^3(2m + 7) =$

$5m^3(2m) + 5m^3(7) =$

$10 \_\_\_\_ + 35 \_\_\_\_$

(b) $2x^4(3x^2 + 2x - 5)$

(c) $-4y^2(3y^3 + 2y^2 - 4y + 8)$

Answers

1. (a) $10m^4 + 35m^3$ **(b)** $6x^6 + 4x^5 - 10x^4$
(c) $-12y^5 - 8y^4 + 16y^3 - 32y^2$

2 Multiply.

(a) $(k + 6)(k + 7)$

$$\underbrace{k(k)} + \underbrace{k(7)} + \underbrace{6(k)} + \underbrace{6(7)}$$

$$\underline{} + \underline{} + \underline{} + \underline{}$$

$$\underline{} + \underline{} + \underline{}$$
Combine like terms.

(b) $(n - 2)(4n + 3)$

(c) $(2m + 1)(2m^2 + 4m - 3)$

(d) $(6p^2 - 2p - 4)(3p^2 + 5)$

Multiplying Polynomials

To multiply two polynomials, multiply each term of the second polynomial by each term of the first polynomial; then add the products.

EXAMPLE 2 **Multiplying Two Polynomials Horizontally**

(a) Multiply: $(x + 1)(x - 4)$

Multiply each term of the second polynomial by each term of the first. Then combine like terms.

$$(x + 1)(x - 4) = x(x) + x(-4) + 1(x) + 1(-4)$$
$$= x^2 + (-4x) + x + (-4) \quad \text{Combine like terms.}$$
$$= x^2 - 3x - 4$$

(b) Multiply: $(m^2 + 5)(4m^3 - 2m^2 + 4m)$

Multiply each term of the second polynomial by each term of the first.

$$(m^2 + 5)(4m^3 - 2m^2 + 4m)$$
$$= \underbrace{m^2(4m^3)} + \underbrace{m^2(-2m^2)} + \underbrace{m^2(4m)} + \underbrace{5(4m^3)} + \underbrace{5(-2m^2)} + \underbrace{5(4m)}$$
$$= (4m^5) + (-2m^4) + (4m^3) + (20m^3) + (-10m^2) + (20m)$$

Now combine like terms.

$$= 4m^5 - 2m^4 + 24m^3 - 10m^2 + 20m$$

◀ **Work Problem** **2** at the Side.

When multiplying polynomials with three or more terms, simplify the process by writing one polynomial above the other vertically.

EXAMPLE 3 **Multiplying Polynomials Vertically**

Multiply $(x^3 + 2x^2 + 4x + 1)(3x + 5)$ using the vertical method.

Write the polynomials vertically, as shown below.

$$\begin{array}{r} x^3 + 2x^2 + 4x + 1 \\ 3x + 5 \\ \hline \end{array}$$

Here are the steps.

Step 1 Multiply each term in the top row by 5.

Step 2 Multiply each term in the top row by $3x$. Be careful to place like terms in columns.

Step 3 Add like terms.

Place *like* terms in the same columns.

$$\begin{array}{r} x^3 + 2x^2 + 4x + 1 \\ 3x + 5 \\ \hline 5x^3 + 10x^2 + 20x + 5 \quad \leftarrow \text{Step 1} \quad 5(x^3 + 2x^2 + 4x + 1) \\ 3x^4 + 6x^3 + 12x^2 + 3x \quad \leftarrow \text{Step 2} \quad 3x(x^3 + 2x^2 + 4x + 1) \\ \hline 3x^4 + 11x^3 + 22x^2 + 23x + 5 \quad \leftarrow \text{Step 3} \quad \text{Add like terms.} \end{array}$$

The product is $3x^4 + 11x^3 + 22x^2 + 23x + 5$

◀ **Work Problem** **3** at the Side.

---

3 Find the product.

$$\begin{array}{r} 2x^3 + 3x^2 + 4x - 5 \\ x + 4 \\ \hline 8x^3 + 12x^2 + 16x - 20 \quad \textbf{\textit{Step 1}} \\ \hline \quad \textbf{\textit{Step 2}} \\ \hline \quad \textbf{\textit{Step 3}} \\ \text{Add like terms.} \end{array}$$

Answers

2. **(a)** $k^2 + 7k + 6k + 42 = k^2 + 13k + 42$
   (b) $4n^2 - 5n - 6$
   (c) $4m^3 + 10m^2 - 2m - 3$
   (d) $18p^4 - 6p^3 + 18p^2 - 10p - 20$
3. **Step 2:** $2x^4 + 3x^3 + 4x^2 - 5x$
   Step 3: $2x^4 + 11x^3 + 16x^2 + 11x - 20$

10.5 Exercises

Go to MyMathLab *for worked-out, step-by-step solutions to exercises enclosed in a square ▢ and video solutions to ▶ exercises.*

CONCEPT CHECK *Fill in each blank in Exercises 1–4.*

1. (a) In the term $7x$, the coefficient is _____ and the exponent is _____

   (b) In the term $-y^4$, the coefficient is _____ and the exponent is _____

2. (a) How many terms does $a^3b - 3b$ have?

   _____

   (b) How many terms does $-10w + x - 6$ have?

   _____

3. (a) A polynomial with exactly two terms is called a _____

   (b) A polynomial with just one term is called a _____

4. Write "addition" next to the example(s) below that show addition of two polynomials. Write "multiplication" next to the example(s) that show multiplication of two polynomials.

   $$(5n + 8)(n - 1) \underline{\hspace{2cm}}$$
   $$(2c + 3)(4c^2 + c + 7) \underline{\hspace{2cm}}$$
   $$(x^2 - 7x) + (9x + 10) \underline{\hspace{2cm}}$$

5. (a) Add these polynomials.
   $$9b^4 + 3b^4 =$$
   (b) Now multiply the same polynomials.
   $$(9b^4)(3b^4) =$$

6. (a) Add these polynomials.
   $$y^3 + 5y^3 =$$
   (b) Now multiply the same polynomials.
   $$(y^3)(5y^3) =$$

Find each product. **See Example 1.**

7. $5y(8y^2 + 3)$

   GS $\underbrace{5y(8y^2)} + \underbrace{5y(3)}$

   _____ + _____

8. $3a(4a^2 + 6a)$

   GS $\underbrace{3a(4a^2)} + \underbrace{3a(6a)}$

   _____ + _____

9. $-2m(3m + 2)$

10. $-6p(3p - 1)$

11. $4x^2(6x^2 - 3x + 2)$

12. $7n^2(5n^2 + n - 4)$

13. $-3k^3(2k^3 - 3k^2 - k + 1)$
    ▶

14. $-2b^4(-3b^3 - b^2 + 6b - 8)$

In Exercises 15–20, find each product horizontally. **See Example 2.**

15. $(n - 2)(n + 3) =$

    GS ▶ $\underbrace{n(n)} + \underbrace{n(3)} + \underbrace{(-2)n} + \underbrace{(-2)(3)}$

16. $(r - 6)(r + 8) =$

    GS $\underbrace{r(r)} + \underbrace{r(8)} + \underbrace{(-6)r} + \underbrace{(-6)(8)}$

17. $(4r + 1)(2r - 3)$

18. $(5x + 2)(2x - 7)$

19. $(6x + 1)(2x^2 + 4x + 1)$

20. $(9y - 2)(8y^2 - 6y + 1)$

In Exercises 21–28, find each product vertically. **See Example 3.**

21. $(4m + 3)(5m^3 - 4m^2 + m - 5)$

22. $(y + 4)(3y^3 - 2y^2 + y + 3)$

23. $(5x^2 + 2x + 1)(x^2 - 3x + 5)$

24. $(2m^2 + m - 3)(m^2 - 4m + 5)$

25. $(-x - 3)(-x - 4)$

26. $(-x + 5)(-x - 2)$

27. $(2x + 1)(x^2 + 3x)$

28. $(x + 4)(3x^2 + 2x)$

Match each product with the correct polynomial.

29. Product   **Polynomial**

$(a + 4)(a + 5)$ ___   **A.** $a^2 - a - 20$

$(a - 4)(a + 5)$ ___   **B.** $a^2 + a - 20$

$(a + 4)(a - 5)$ ___   **C.** $a^2 - 9a + 20$

$(a - 4)(a - 5)$ ___   **D.** $a^2 + 9a + 20$

30. Product   **Polynomial**

$(x - 5)(x + 3)$ ___   **A.** $x^2 + 8x + 15$

$(x + 5)(x + 3)$ ___   **B.** $x^2 - 8x + 15$

$(x - 5)(x - 3)$ ___   **C.** $x^2 - 2x - 15$

$(x + 5)(x - 3)$ ___   **D.** $x^2 + 2x - 15$

31. CONCEPT CHECK   The work shown below has a mistake in it.

What Went Wrong?   First write a sentence explaining what the mistake is. Then fix the mistake and find the correct solution.

Mohamed multiplied this way:

$$(x - 3)(x - 2) = x(x) + x(-2) + (-3)x + (-3)(-2)$$
$$= 2x^2 - 2x - 3x + 6$$
$$= 2x^2 - 5x + 6$$

32. CONCEPT CHECK   The work shown below has a mistake in it.

What Went Wrong?   First write a sentence explaining what the mistake is. Then fix the mistake and find the correct solution.

Fatima multiplied this way:

$$(2y + 3)(y + 5) = 2y(y) + 2y(5) + 3(y) + 3(5)$$
$$= 2y^2 + 10y + 3y + 15$$
$$= 15y^4 + 15$$

Key Terms

10.3

scientific notation Scientific notation is used to write very large or very small numbers using exponents. The numbers are written as $a \times 10^n$, where $1 \le |a| < 10$ and n is an integer.

10.4

polynomial A polynomial is a term, or the sum of terms, with whole number exponents.

descending powers A polynomial in x is written in descending powers if the exponents on x decrease from left to right.

degree of a term The degree of a term is the sum of the exponents on the variables in that term.

degree of a polynomial The degree of a polynomial is the highest degree of any term in the polynomial.

monomial A monomial is a polynomial with only one term.

binomial A binomial is a polynomial with exactly two terms.

trinomial A trinomial is a polynomial with exactly three terms.

New Symbols

a^{-n} a to the negative n power

a^0 a to the zero power

$a \times 10^n$ scientific notation

Test Your Word Power

See how well you have learned the vocabulary in this chapter.

1 A **polynomial** is an algebraic expression made up of
- **A.** a term or the product of several terms with whole number exponents
- **B.** a term or the sum of several terms with whole number exponents
- **C.** the product of two or more terms with negative exponents.

2 The **degree of a term** is
- **A.** the number of variables in the term
- **B.** the product of the exponents on the variables
- **C.** the sum of the exponents on the variables.

3 A **trinomial** is a polynomial with
- **A.** exactly two terms
- **B.** exactly three terms
- **C.** more than three terms.

4 A **binomial** is a polynomial with
- **A.** only one term
- **B.** exactly two terms
- **C.** more than three terms.

5 A **monomial** is a polynomial with
- **A.** only one term
- **B.** exactly two terms
- **C.** more than three terms.

6 An example of a polynomial written in **descending powers** is
- **A.** $7x + 6x^2 - 4x^3 - 2x^4$
- **B.** $x^6 - 2x^5 + 3x^2 - 6x$
- **C.** $16 + x^5 - x^3 + x^2$

7 An example of a number written in **scientific notation** is
- **A.** 0.25×10^5 **B.** 57.6×10^{-3}
- **C.** 3.61×10^{-4}

8 The **degree of a polynomial** is
- **A.** the number of terms in the polynomial
- **B.** the product of the exponents on the variables
- **C.** the highest degree of any term in the polynomial.

Answers to Test Your Word Power

1. B; *Example:* $5x^3 + 2x^2 - 7$

2. C; *Examples:* The term 6 has degree 0, the term $3x$ has degree 1, the term $-2x^8$ has degree 8, and the term $5x^2y^4$ has degree 6.

3. B; *Example:* $2a^2 - 3ab + b^2$

4. B; *Example:* $3t^3 + 5t$

5. A; *Examples:* -5 and $4xy^5$ are both monomials.

6. B; *Example:* $3n^4 + n^3 - 2n + 8$ is written in descending powers.

7. C; *Example:* 2.5×10^4 is written in scientific notation.

8. C; *Example:* $x^2y + xy - 5$ is a polynomial of degree 3, because the sum of the exponents in x^2y is $2 + 1 = 3$.

Quick Review

Concepts

Examples

10.1 The Product Rule and Power Rules for Exponents

For any integers m and n:

Product rule $a^m \cdot a^n = a^{m+n}$

Power rule (a) $(a^m)^n = a^{mn}$

Power rule (b) $(ab)^m = a^m b^m$

Power rule (c) $\left(\dfrac{a}{b}\right)^m = \dfrac{a^m}{b^m}$ when $b \neq 0$

$2^4 \cdot 2^5 = 2^{4+5} = \mathbf{2^9}$ Product rule

$(3^4)^2 = 3^{4 \cdot 2} = \mathbf{3^8}$ Power rule (a)

$(6a)^5 = \mathbf{6^5 a^5}$ Power rule (b)

$\left(\dfrac{2}{3}\right)^4 = \dfrac{\mathbf{2^4}}{\mathbf{3^4}}$ Power rule (c)

10.2 Integer Exponents and the Quotient Rule

If $a \neq 0$, for integers m and n:

Zero exponent $a^0 = 1$

Negative exponent $a^{-n} = \dfrac{1}{a^n}$

Quotient rule $\dfrac{a^m}{a^n} = a^{m-n}$

$15^0 = \mathbf{1}$ Zero exponent

$5^{-2} = \dfrac{1}{5^2} = \dfrac{\mathbf{1}}{\mathbf{25}}$ Negative exponent

$\dfrac{4^8}{4^3} = 4^{8-3} = \mathbf{4^5}$ Quotient rule

10.3 An Application of Exponents: Scientific Notation

To write a number in scientific notation ($a \times 10^n$), move the decimal point to the right of the first nonzero digit. The number of places moved is the absolute value of the exponent n. If the *original* number was "large" (10 or more), n is positive. If the *original* number was "small" (between 0 and 1), n is negative.

$247 = \mathbf{2.47 \times 10^2}$

$0.0051 = \mathbf{5.1 \times 10^{-3}}$

$3.25 \times 10^5 = \mathbf{325{,}000}$

$8.44 \times 10^{-6} = \mathbf{0.00000844}$

10.4 Adding and Subtracting Polynomials

Addition: Add like terms.

Subtraction: Change the signs of all the terms in the second polynomial and add it to the first polynomial.

Add: $\begin{array}{r} 2x^2 + 5x - 3 \\ 5x^2 - 2x + 7 \\ \hline \mathbf{7x^2 + 3x + 4} \end{array}$

Subtract: $(2x^2 + 5x - 3) - (5x^2 - 2x + 7)$

$(2x^2 + 5x - 3) + (-5x^2 + 2x - 7)$

$\mathbf{-3x^2 + 7x - 10}$

10.5 Multiplying Polynomials

Multiply each term of the first polynomial by each term of the second polynomial. Then add like terms.

Multiply: $\begin{array}{r} 3x^3 - 4x^2 + 2x - 7 \\ 4x + 3 \\ \hline 9x^3 - 12x^2 + 6x - 21 \\ 12x^4 - 16x^3 + 8x^2 - 28x \\ \hline \mathbf{12x^4 - 7x^3 - 4x^2 - 22x - 21} \end{array}$

Chapter 10 Review Exercises

10.1 *Use the product rule or power rules to simplify each expression. Write the answer in exponential form.*

1. $(-5)^6(-5)^5$

2. $(-8x^4)(9x^3)$

3. $(2x^2)(5x^3)(x^9)$

4. $(19x)^5$

5. $5(pt)^4$

6. $\left(\dfrac{7}{5}\right)^6$

7. $(3x^2y^3)^3$

8. Explain why the product rule for exponents does *not* apply to this expression: $7^2 + 7^4$

10.2 *Evaluate each expression.*

9. $5^0 + 8^0$

10. 2^{-5}

11. $10w^0$

12. $4^{-2} + 4^{-1}$

Simplify. Write answers in exponential form, using only positive exponents. Assume all variables are nonzero.

13. $\dfrac{p^{-8}}{p^4}$

14. $\dfrac{r^{-2}}{r^{-6}}$

15. $\dfrac{5^4}{5^5}$

16. $10^3(10^{-10})$

17. $n^{-2} \cdot n^3 \cdot n^{-4}$

18. $(10^4)(10^{-1})$

19. $(2^{-4})(2^{-4})$

20. $\dfrac{x^2}{x^{-4}}$

10.3 *Write each number in scientific notation or as a number without exponents.*

21. 48,000,000

22. 0.0000000824

23. 2.4×10^4

24. 9.95×10^{-12}

Simplify. Write the answers both in scientific notation and as numbers without exponents.

25. $(2 \times 10^{-3})(4 \times 10^5)$

26. $\dfrac{12 \times 10^{-8}}{4 \times 10^{-3}}$

27. $\dfrac{(2.5 \times 10^5)(4.8 \times 10^{-4})}{(7.5 \times 10^8)(1.6 \times 10^{-5})}$

*Solve each application problem. Write your answers both in scientific notation and
as numbers without exponents.*

28. The population of the United States in 2015 was about
3.22×10^8 people spread over 3.79×10^6 square miles
of land. What is the average number of people per
square mile? (Data from www.esa.un.org)

29. Light travels at a speed of 3.0×10^5 kilometers per
second. It takes about 360 seconds for light from the
sun to reach Venus. How far is Venus from the sun?
(Data from NASA.)

10.4 *Simplify each polynomial and write it with descending powers of the variable.
Give the degree of the simplified polynomial, and tell whether it is a* monomial,
binomial, trinomial, *or* none of these.

30. $9m^2 + 11m^2 + 2m^2$

31. $-7y^5 - 8y^4 - y^5 + y^4 + 9y$

Add or subtract as indicated.

32. $(-2a^3 + 5a^2) + (-3a^3 - a^2)$

33. Subtract $-5y^2 + 2y - 7$ from $6y^2 - 8y + 2$

34. $(-5y^2 + 3y + 11) + (4y^3 - 7y + 15)$

35. $(12r^4 - 7r^3 + 2r^2) - (5r^4 - 3r^3 + 2r^2 + 1)$

10.5 *Find each product.*

36. $5x(2x + 14)$

37. $-3p^3(2p^2 - 5p)$

38. $(3r - 2)(2r^2 + 4r - 3)$

39. $(3k - 6)(2k + 1)$

40. $(6p - 3q)(2p - 7q)$

41. $(m^2 + m - 9)(2m^2 + 3m - 1)$

Chapter 10 Mixed Review Exercises

Practicing exercises in mixed-up order helps you prepare for tests.

Simplify. Write answers in exponential form, using only positive exponents. Assume all variables are nonzero.

1. 7^{-2}

2. $(3p)^4$

3. $\dfrac{h^6}{h^5}$

4. $(-2c^2d^6)^3$

5. $(-3)^4(-3)^3$

6. $\dfrac{7^4}{7}$

7. -9^0

8. $(-6b^7)(-5b)$

9. $\dfrac{m^9}{m^{10}}$

10. $\dfrac{10^6}{10^{-6}}$

11. $4(xy)^5$

12. $(10)(10^{-3})(10^{-3})$

13. $\dfrac{0^8}{8^0}$

14. $\dfrac{10^{-10}}{10^{-5}}$

15. $\dfrac{n^{-5}}{n^9}$

16. $(b^{-3})(b^8)(b^3)$

17. $2^{-1} + 4^{-1}$

18. $19^0 - 3^0$

19. $-m^5(8m^2 + 10m + 6)$

20. $6^{-1} - 4^{-1}$

Perform the indicated operations. Write answers with positive exponents. Assume that all variables represent nonzero numbers.

21. $7m^3(-m^3 + 2m^2 - 6)$

22. $(5t^3 - 3t - 5) + (-5t^3 - 3t + 5)$

23. $(a + 2)(a^2 - 4a + 1)$

24. Subtract $-a^3 + 3$ from $-3a^3 + 2a^2 - 6$

25. Americans send 2×10^{11} tweets each year. How many tweets are sent each month, on average? (Data from www.internetlivestats.com)

26. In 2016, the United States spent about \$10,300 per person on health care. If the population was about 3.25×10^8 in 2016, how much did the United States spend on health care? (Data from www.pbs.org)

| Chapter 10 **Test** | The Chapter Test Prep Videos with step-by-step solutions are available in MyMathLab or on YouTube at **https://goo.gl/c3befo** |

Evaluate each expression.

1. 5^{-4}

2. $(-3)^0 + 4^0$

3. $4^{-1} + 3^{-1}$

Simplify, and write each answer using only positive exponents. Assume that variables represent nonzero numbers.

4. $6^{-3} \cdot 6^4$

5. $12 (xy)^3$

6. $\dfrac{10^5}{10^9}$

7. $r^{-4} \cdot r^{-4} \cdot r^3$

8. $(7x^2)(-3x^5)$

9. $\dfrac{m^{-5}}{m^{-8}}$

10. $(3a^4b)^2$

11. $3^4 + 3^2$

Write each number in scientific notation.

12. 300,000,000,000

13. 0.00000557

Write each number without exponents.

14. 2.9×10^7

15. 6.07×10^{-8}

16. The mass of the sun is about 330,000 times the mass of Earth. If Earth's mass is about 5.98×10^{24} kg, find the approximate mass of the sun. Write your answer in scientific notation.

Simplify each polynomial, when possible, and write it with descending powers of the variable. Give the degree of the simplified polynomial, and tell whether it is a monomial, binomial, trinomial, or none of these.

17. $5x^2 + 8x - 12x^2$

18. $13n^3 - n^2 + n^4 + 3n^4 - 9n^2$

Perform the indicated operations.

19. $(5t^4 - 3t^2 + 3) - (t^4 - t^2 + 3)$

20. $(2y^4 - 8y + 8) + (-3y^2 + 2y - 8)$

21. Subtract $9t^3 + 8t^2 - 6$ from $9t^3 - 4t^2 + 2$

22. $-4x^3(3x^2 - 5x)$

23. $(y + 2)(3y - 1)$

24. $(2r - 3)(r^2 + 2r - 5)$

25. Write a polynomial that represents the perimeter of this square, and a polynomial that represents the area.

$3x + 9$

Chapters 1-10 *Cumulative Review Exercises*

Simplify. Assume that all variables are nonzero.

1. $-12 + 7.829$

2. $7 + 5(3 - 8)$

3. $1\frac{2}{3} + \frac{3}{5}$

4. $\dfrac{(-6)^2 + 9(0 - 4)}{10 \div 5(-7 + 5) - 10}$

5. $\dfrac{8x^2}{9} \cdot \dfrac{12w}{2x^3}$

6. $\dfrac{-0.7}{5.6}$

7. $\dfrac{10^5}{10^6}$

8. $2\frac{1}{4} - 2\frac{5}{6}$

9. $(-6)^2 + (-2)^3$

10. $(-4y^3)(3y^4)$

11. $\dfrac{5b}{2a^2} \div \dfrac{3b^2}{10a}$

12. $\dfrac{r}{8} + \dfrac{6}{t}$

13. $\dfrac{-4(6)}{3^3 - 27}$

14. $\dfrac{5}{6}$ of 900

15. $(-0.003)(0.04)$

16. $(2a^3b^2)^4$

17. $\dfrac{4}{9} - \dfrac{6}{m}$

18. $t^5 \cdot t^{-2} \cdot t^{-4}$

19. $\dfrac{-\frac{15}{16}}{-6}$

20. $\left(-\dfrac{1}{2}\right)^4 + 6\left(\dfrac{2}{3}\right)$

21. $\dfrac{n^{-3}}{n^{-4}}$

22. $-8 - 88$

23. $8^0 + 2^{-1}$

24. $9 - 7\frac{4}{5}$

Solve each equation. Show your work.

25. $4 + h = 3h - 6$

26. $-1.65 = 0.5x + 2.3$

27. $3(a - 5) = -3 + a$

28. This month Dallas received 2.61 inches of rain. This is what percent of the normal rainfall of 1.8 inches?

29. Calbert bought a video game priced at $64.95. He paid $7\frac{1}{2}\%$ sales tax. Find the amount of tax, to the nearest cent, and the total cost of the game.

30. Find the perimeter and area of this shape.

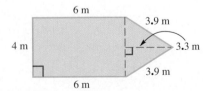

31. Find the circumference and area of this circle. Use 3.14 for π and round answers to the nearest tenth.

Whole Numbers Review

R.1 Adding Whole Numbers

R.2 Subtracting Whole Numbers

R.3 Multiplying Whole Numbers

R.4 Dividing Whole Numbers

R.5 Long Division

Note

Use the **Pretest: Whole Numbers Computation** to decide which sections you need to work in this chapter.

R.1 | Adding Whole Numbers

There are 4 triangles on the left and 2 triangles on the right. In all, there are 6 triangles.

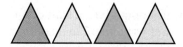 

The process of finding the total is called *addition*. Here 4 and 2 are added to get 6. Addition is written with a $+$ symbol, as shown below.

$$4 + 2 = 6$$

OBJECTIVE ① Add two or three single-digit numbers. In addition problems, the numbers being added are called **addends,** and the answer is called the **sum** or **total.**

$$
\begin{array}{r}
4 \leftarrow \text{Addend} \\
+\,2 \leftarrow \text{Addend} \\
\hline
6 \leftarrow \text{Sum (answer)}
\end{array}
$$

Addition problems can also be written horizontally as shown below.

$$
\underset{\text{Addend}}{4} \quad + \quad \underset{\text{Addend}}{2} \quad = \quad \underset{\text{Sum}}{6}
$$

OBJECTIVES

① Add two or three single-digit numbers.

② Add more than two numbers.

③ Add when regrouping is not required.

④ Add with regrouping.

⑤ Use addition to solve application problems.

⑥ Check the sum in addition.

1 Add. Then use the commutative property to write another addition problem and find the sum.

(a) $3 + 4$

(b) $9 + 9$

(c) $7 + 8$

(d) $6 + 9$

2 Find each sum.

(a)
$$\begin{array}{r} 5 \\ 4 \\ 6 \\ 9 \\ + 2 \\ \hline \end{array}$$

(b)
$$\begin{array}{r} 7 \\ 5 \\ 1 \\ 2 \\ + 6 \\ \hline \end{array}$$

(c)
$$\begin{array}{r} 9 \\ 2 \\ 1 \\ 3 \\ + 4 \\ \hline \end{array}$$

(d)
$$\begin{array}{r} 3 \\ 8 \\ 6 \\ 4 \\ + 8 \\ \hline \end{array}$$

Answers

1. (a) $3 + 4 = 7; 4 + 3 = 7$
   (b) $9 + 9 = 18$; no change
   (c) $7 + 8 = 15; 8 + 7 = 15$
   (d) $6 + 9 = 15; 9 + 6 = 15$
2. (a) 26 (b) 21 (c) 19 (d) 29

Commutative Property of Addition

The **commutative property of addition** states that changing the *order* of two addends in an addition problem does *not* change the sum.

For example, the sum of $4 + 2$ is the same as the sum of $2 + 4$. Both sums are 6. This allows the addition of the same numbers in a different order.

EXAMPLE 1   Adding Two Single-Digit Numbers

Add. Then use the commutative property to write another addition problem and find the sum.

(a) $6 + 2 = \mathbf{8}$ and $2 + 6 = \mathbf{8}$ (b) $5 + 9 = \mathbf{14}$ and $9 + 5 = \mathbf{14}$

(c) $8 + 8 = \mathbf{16}$ (No change occurs when commutative property is used.)

◀ Work Problem **1** at the Side.

OBJECTIVE **2** Add more than two numbers. Some additions involve more than two addends.

Associative Property of Addition

By the **associative property of addition**, changing the *grouping* of addends does *not* change the sum.

For example, find the sum of $3 + 5 + 6$.

$(3 + 5) + 6 = 8 + 6 = \mathbf{14}$ Parentheses tell you to add $3 + 5$ first.

$3 + (5 + 6) = 3 + 11 = \mathbf{14}$ Parentheses tell you to add $5 + 6$ first.

By the associative property, either grouping gives a sum of 14.

To add four or more numbers, first write them in a column. Add the first number to the second. Add this sum to the third number and so on.

EXAMPLE 2   Adding More Than Two Numbers

Find the sum of 2, 5, 6, 1, and 4.

$$\begin{array}{r} 2 \\ 5 \\ 6 \\ 1 \\ + 4 \\ \hline \mathbf{18} \end{array}$$
$2 + 5 = 7$
$7 + 6 = 13$
$13 + 1 = 14$
$14 + 4 = 18$
← Sum

Note

You may also add numbers by starting at the bottom of a column. Adding down from the top or adding up from the bottom will give the same sum.

◀ Work Problem **2** at the Side.

OBJECTIVE **3** Add when regrouping is not required. If numbers have two or more digits, arrange the numbers in columns so that the ones digits are in the same column, tens are in the same column, hundreds are in the same column, and so on. Then add column by column, starting with the ones column.

EXAMPLE 3 Adding without Regrouping

Add: 511 + 23 + 154 + 10

First line up the numbers in columns, with the ones column at the right.

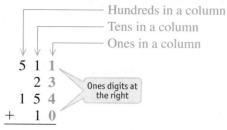

Hundreds in a column
Tens in a column
Ones in a column

$$\begin{array}{r} 5\ 1\ 1 \\ 2\ 3 \\ 1\ 5\ 4 \\ +\ \ \ 1\ 0 \end{array}$$

Ones digits at the right

Now start at the right and add the ones digits. Add the tens digits next, and finally, add the hundreds digits.

$$\begin{array}{r} 5\ 1\ 1 \\ 2\ 3 \\ 1\ 5\ 4 \\ +\ \ \ 1\ 0 \\ \hline 6\ 9\ 8 \end{array}$$

Sum of ones
Sum of tens
Sum of hundreds

The sum of the four numbers is **698**.

─────────── Work Problem **3** at the Side. ▶

OBJECTIVE ▶ **4** **Add with Regrouping.** If the sum of the digits in a column is more than 9, use **regrouping**.

EXAMPLE 4 Adding with Regrouping

Find the sum of 47 and 29.

Add the digits in the ones column.

$$\begin{array}{r} 47 \\ +\ 29 \end{array}$$

7 ones + 9 ones = 16 ones

Regroup 16 ones as 1 ten and 6 ones. Write 6 ones in the ones column and write 1 ten in the tens column.

Write 1 ten in the tens column.

$$\begin{array}{r} \overset{1}{4}7 \\ +\ 29 \\ \hline 6 \end{array}$$

7 ones + 9 ones = 16 ones

Regroup 16 ones as 1 ten and 6 ones.

Write 6 ones in the ones column.

Add the digits in the tens column, including the regrouped 1.

$$\begin{array}{r} \overset{1}{4}7 \\ +\ 29 \\ \hline 76 \end{array}$$

Remember to add the 1 regrouped ten.

1 ten + 4 tens + 2 tens = 7 tens

The **sum is 76**.

─────────── Work Problem **4** at the Side. ▶

3 Add.

(a) $\begin{array}{r} 25 \\ +\ 73 \end{array}$

(b) $\begin{array}{r} 364 \\ +\ 532 \end{array}$

(c) $\begin{array}{r} 42{,}305 \\ +\ 11{,}563 \end{array}$

4 Find each sum using regrouping.

(a) $\begin{array}{r} 69 \\ +\ 26 \end{array}$

(b) $\begin{array}{r} 76 \\ +\ 18 \end{array}$

(c) $\begin{array}{r} 56 \\ +\ 37 \end{array}$

(d) $\begin{array}{r} 34 \\ +\ 49 \end{array}$

Answers

3. (a) 98 (b) 896 (c) 53,868
4. (a) 95 (b) 94 (c) 93 (d) 83

5 Find each sum, regrouping when necessary.

(a)   481
        79
        38
      + 395

(b)   4271
        372
       8976
      + 162

(c)   57
         4
        392
        804
         51
      + 27

(d)   7821
        435
         72
        305
      + 1693

(e)   15,829
        765
         78
         15
          9
          7
      + 13,179

EXAMPLE 5   **Adding with Regrouping**

Add: 324 + 7855 + 23 + 7 + 86

Step 1   Add the digits in the ones column.

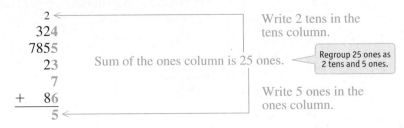

Step 2   Add the digits in the tens column, including the regrouped 2.

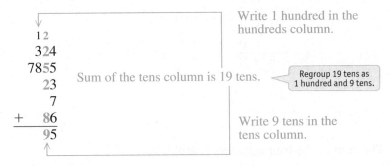

Step 3   Add the hundreds column, including the regrouped 1.

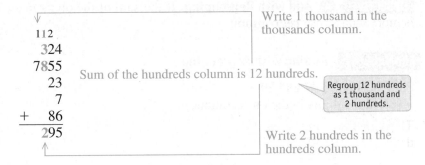

Step 4   Add the thousands column, including the regrouped 1.

```
 1 12
  324
 7855
   23      Sum of the thousands
    7      column is 8 thousands.
+  86
 8295
  ↑
```

Thus, 324 + 7855 + 23 + 7 + 86 = **8295**.

◄ **Work Problem 5 at the Side.**

OBJECTIVE ▶ 5 Use addition to solve application problems.

| EXAMPLE 6 | Applying Addition Skills |

On this map, the distance in miles from one town to another is written along-side the road. Find the shortest route from Altamonte Springs to Clear Lake.

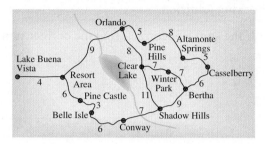

Approach   Add the mileage along various routes from Altamonte Springs to Clear Lake. Then select the shortest route.

Solution   One way from Altamonte Springs to Clear Lake is through Orlando. Add the mileage numbers along this route.

$$
\begin{array}{rl}
8 & \text{Altamonte Springs to Pine Hills} \\
5 & \text{Pine Hills to Orlando} \\
+\ 8 & \text{Orlando to Clear Lake} \\
\hline
21 & \longrightarrow \text{miles from Altamonte Springs to Clear Lake,} \\
& \text{going through Orlando}
\end{array}
$$

Another way is through Casselberry, Bertha, and Winter Park. Add the mileage numbers along this route.

$$
\begin{array}{rl}
5 & \text{Altamonte Springs to Casselberry} \\
6 & \text{Casselberry to Bertha} \\
7 & \text{Bertha to Winter Park} \\
+\ 7 & \text{Winter Park to Clear Lake} \\
\hline
25 & \longrightarrow \text{miles from Altamonte Springs to Clear Lake} \\
& \text{through Casselberry, Bertha, and Winter Park}
\end{array}
$$

The **shortest route** from Altamonte Springs to Clear Lake **is 21 miles** through Orlando.

─────────────── **Work Problem ⑥ at the Side.** ▶

| EXAMPLE 7 | Applying Addition Skills |

Using the map in **Example 6** above, find the total mileage from Shadow Hills to Casselberry to Orlando and back to Shadow Hills through Clear Lake.

Solution   Add the mileage numbers from the map.

$$
\begin{array}{rl}
9 & \text{Shadow Hills to Bertha} \\
6 & \text{Bertha to Casselberry} \\
5 & \text{Casselberry to Altamonte Springs} \\
8 & \text{Altamonte Springs to Pine Hills} \\
5 & \text{Pine Hills to Orlando} \\
8 & \text{Orlando to Clear Lake} \\
+\ 11 & \text{Clear Lake to Shadow Hills} \\
\hline
52 & \longrightarrow \text{miles from Shadow Hills to Casselberry to} \\
& \text{Orlando and back to Shadow Hills}
\end{array}
$$

─────────────── **Work Problem ⑦ at the Side.** ▶

⑥ Use the map on this page to find the shortest route from Conway to Pine Hills.

⑦ The road is closed between Orlando and Clear Lake, so this route cannot be used. Find the next shortest route from Orlando to Clear Lake.

Answers

6. 29 miles through Belle Isle

7.
Orlando to Pine Hills	5
Pine Hills to Altamonte Springs	8
Altamonte Springs to Casselberry	5
Casselberry to Bertha	6
Bertha to Winter Park	7
Winter Park to Clear Lake	+ 7

38 miles

8 Check each addition. If the sum is incorrect, find the correct sum.

(a)
```
    32
     8
     5
 + 14
    59
```

(b)
```
   872
   539
    46
 + 152
  1609
```

(c)
```
    79
   218
     7
 + 639
   953
```

(d)
```
  21,892
  11,746
+ 43,925
  79,563
```

Answers

8. **(a)** correct   **(b)** correct
   (c) incorrect; should be 943
   (d) incorrect; should be 77,563

OBJECTIVE ▶ 6 Check the sum in addition. Checking the answer is an important part of problem solving. A common method for checking addition is to re-add from the bottom to the top. This is an application of the commutative and associative properties of addition.

EXAMPLE 8 **Checking Addition**

Check each sum. If the sum is incorrect, find the correct sum.

(a)
```
         1428  ↑
Add down. 738
           63
          125
           17
        + 485  ── To check, add up.
         1428
```

Adding down and adding up should give the same sum. In this case, the answers agree, so the sum is probably correct.

(b)
```
        1033  ↑   Correct, because both
 785     785       answers are the same
  63      63
+ 185   + 185  ── To check, add up.
1033    1033
```

(c)
```
        2454  ↑   Error, because answers
 635     635       are different
  73      73
 831     831
+ 915   + 915  ── To check, add up.
2444    2444
```

Re-add to find that **the correct sum is 2454.**

◀ **Work Problem 8 at the Side.**

R.1 Exercises

FOR EXTRA HELP

Go to MyMathLab for worked-out, step-by-step solutions to exercises enclosed in a square and video solutions to ▶ exercises.

*Add. **See Examples 2 and 3.***

1. (a)
```
    5
    7
    6
  + 5
```
(b)
```
    9
    2
    1
    3
  + 4
```
2. (a)
```
    2
    9
    5
  + 1
```
(b)
```
    3
    8
    6
    4
  + 8
```

3. (a) $3213 + 5715$

(b) $38{,}204 + 21{,}020$

4. (a) $6344 + 1655$

(b) $63{,}251 + 36{,}305$

*Find each sum, regrouping when necessary. **See Examples 4 and 5.***

5.
```
   67
 + 83
```
6.
```
   78
 + 36
```
7.
```
  746
 + 905
```
8.
```
  621
 + 359
```
9.
```
  798
 + 206
```

10.
```
  172
 + 156
```
11.
```
  7968
 + 1285
```
12.
```
  1768
 + 8275
```
13.
```
  7896
 + 3728
```
14.
```
  9382
 + 7586
```

15.
```
  3705
  3916
 + 9037
```
16.
```
  6629
  6076
 + 8218
```
17.
```
    32
 + 4977
```
18.
```
   402
 + 9938
```
19.
```
  3077
     8
 + 421
```

20.
```
   56
 7721
 + 172
```
21.
```
  9056
    78
  6089
 + 731
```
22.
```
  4022
   709
  8621
 +  37
```
23.
```
    18
   708
  9286
 + 636
```
24.
```
  1708
   321
    61
 + 8926
```

*Check each sum by adding from bottom to top. If an answer is incorrect, find the correct sum. **See Example 8.***

25.
```
  179
  214
 + 376
  759
```
26.
```
   17
  296
  713
 + 94
 1220
```
27.
```
 4713
   28
  615
 + 64
 5420
```
28.
```
  6 215
   744
    36
 + 4 284
 11,279
```

Using the map below, find the shortest route between each pair of cities.
See Examples 6 and 7.

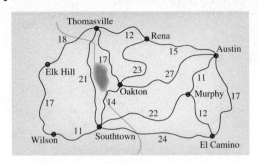

29. Southtown and Rena

30. Elk Hill and Oakton

31. Thomasville and Murphy

32. Austin and Wilson

Solve each application problem.

33. The sale price for a basic auto tune-up is $159, a tire rotation is $24, and an oil change is $29. Find the total cost for all the services.

34. Jane Lim bought an 11-piece set of golf clubs for $327, a dozen golf balls for $20, golf shoes for $109, and a golf glove for $25. How much did she spend in all? (Data from Golfsmith.)

35. There are 413 women and 286 men on the sales staff. How many people are on the sales staff?

36. One department in an office building has 283 employees while another department has 218 employees. How many employees are in the two departments?

37. According to the Census Bureau, the two states with the greatest populations are California with 38,802,500 people and Texas with 26,956,958 people. How many people live in those two states? (Data from U.S. Census Bureau.)

38. The two states with the lowest populations are Wyoming with 584,153 people and Vermont with 626,562 people. What is the total population of the two states? (Data from U.S. Census Bureau.)

Find the perimeter of each figure. Perimeter is the total distance around the outside edges of a flat shape.

39.

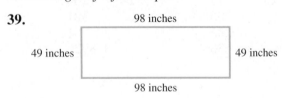

40.

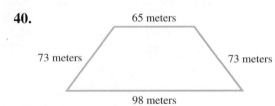

41.

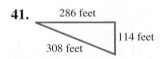

42.

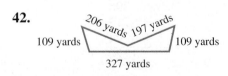

R.2 | Subtracting Whole Numbers

Suppose you have $18, and you spend $15 for food. You then have $3 left. There are two different ways of looking at these numbers.

As an addition problem:

$$\$15 \quad + \quad \$3 \quad = \quad \$18$$

    Amount   Amount   Original
    spent   left   amount

As a subtraction problem:

$$\$18 \quad - \quad \$15 \quad = \quad \$3$$

 Original Subtraction Amount Amount
 amount   symbol   spent   left

OBJECTIVES

1 Change addition problems to subtraction problems or the reverse.

2 Identify the minuend, subtrahend, and difference.

3 Subtract when no regrouping is needed.

4 Use addition to check subtraction answers.

5 Subtract with regrouping.

6 Use subtraction to solve application problems.

OBJECTIVE ▶ 1 Change addition problems to subtraction problems or the reverse. As this example shows, an addition problem can be changed to a subtraction problem, and a subtraction problem can be changed to an addition problem.

EXAMPLE 1 | **Changing Addition Problems to Subtraction**

Change each addition problem to a subtraction problem.

(a) $4 + 1 = 5$

Two subtraction problems are possible, as shown below.

$$5 - 1 = 4 \quad \text{or} \quad 5 - 4 = 1$$

These figures show each subtraction problem.

$$5 - 1 = 4 \qquad\qquad 5 - 4 = 1$$

(b) $10 + 17 = 27$

$$27 - 17 = 10 \quad \text{or} \quad 27 - 10 = 17$$

Work Problem 1 at the Side. ▶

EXAMPLE 2 | **Changing Subtraction Problems to Addition**

Change each subtraction problem to an addition problem.

(a) $8 - 3 = 5$

$$8 = 3 + 5$$

It is also correct to write $8 = 5 + 3$ (using the commutative property).

Continued on Next Page

1 Write two subtraction problems for each addition problem.

(a) $4 + 3 = 7$

(b) $6 + 5 = 11$

(c) $150 + 220 = 370$

(d) $623 + 55 = 678$

Answers

1. (a) $7 - 3 = 4$ or $7 - 4 = 3$
(b) $11 - 5 = 6$ or $11 - 6 = 5$
(c) $370 - 220 = 150$ or $370 - 150 = 220$
(d) $678 - 55 = 623$ or $678 - 623 = 55$

2 Write an addition problem for each subtraction problem.

(a) $5 - 3 = 2$

(b) $8 - 3 = 5$

(c) $21 - 15 = 6$

(d) $58 - 42 = 16$

(b) $19 - 14 = 5$
$\mathbf{19 = 14 + 5}$ $19 = 5 + 14$ is also correct.

(c) $290 - 130 = 160$
$\mathbf{290 = 130 + 160}$ $290 = 160 + 130$ is also correct.

◄ **Work Problem ❷ at the Side.**

OBJECTIVE ▶ ❷ Identify the minuend, subtrahend, and difference. In subtraction, as in addition, the numbers in a problem have names. For example, in the problem $8 - 5 = 3$, the number 8 is the **minuend,** 5 is the **subtrahend,** and 3 is the **difference** or answer.

$$8 \quad - \quad 5 \quad = 3 \leftarrow \text{Difference (answer)}$$
$$\uparrow \qquad\qquad \uparrow$$
$$\text{Minuend} \quad\;\; \text{Subtrahend}$$

$$\begin{array}{r} 8 \leftarrow \text{Minuend} \\ - \; 5 \leftarrow \text{Subtrahend} \\ \hline 3 \leftarrow \text{Difference} \end{array}$$

OBJECTIVE ▶ ❸ Subtract when no regrouping is needed. Subtract two numbers by lining up the numbers in columns so that the digits in the ones place are in the same column, tens are in the same column, and so on. Then, subtract column by column, starting at the right with the ones column.

3 Subtract.

(a) $\begin{array}{r} 56 \\ - 31 \\ \hline \end{array}$

(b) $\begin{array}{r} 38 \\ - 14 \\ \hline \end{array}$

(c) $\begin{array}{r} 378 \\ - 235 \\ \hline \end{array}$

(d) $\begin{array}{r} 3927 \\ - 2614 \\ \hline \end{array}$

(e) $\begin{array}{r} 5464 \\ - 324 \\ \hline \end{array}$

EXAMPLE 3   **Subtracting Two Numbers without Regrouping**

Ones digits are lined up in the same column.

(a) $\begin{array}{r} 5\mathbf{3} \\ - 2\mathbf{1} \\ \hline \mathbf{3}\mathbf{2} \end{array}$ Subtract in the ones column first.

3 ones − 1 one = 2 ones
5 tens − 2 tens = 3 tens

Ones digits are lined up.

(b) $\begin{array}{r} 385 \\ - 165 \\ \hline \mathbf{220} \end{array}$ ← 5 ones − 5 ones = 0 ones

8 tens − 6 tens = 2 tens
3 hundreds − 1 hundred = 2 hundreds

(c) $\begin{array}{r} 9837 \\ - \;\; 210 \\ \hline \mathbf{9627} \end{array}$ ← 7 ones − 0 ones = 7 ones

3 tens − 1 ten = 2 tens
8 hundreds − 2 hundreds = 6 hundreds
9 thousands − 0 thousands = 9 thousands

◄ **Work Problem ❸ at the Side.**

OBJECTIVE ▶ ❹ Use addition to check subtraction answers. You can check $8 - 3 = 5$ by *adding* 3 and 5.

$$3 + 5 = 8 \quad \text{so} \quad 8 - 3 = 5 \quad \text{is correct.}$$

Answers

2. (a) $5 = 3 + 2$ (b) $8 = 3 + 5$
    (c) $21 = 15 + 6$ (d) $58 = 42 + 16$
3. (a) 25   (b) 24   (c) 143   (d) 1313
    (e) 5140

EXAMPLE 4 Checking Subtraction

Use addition to check each answer. If incorrect, find the correct answer.

(a)
```
  89
- 47
-----
  42
```

Rewrite as an addition problem.

Subtraction problem
```
  89
- 47
-----
  42
-----
  89
```
Addition problem
```
  47
+ 42
-----
  89
```

Because $47 + 42 = 89$, the subtraction was done correctly.

(b) $72 - 41 = 21$

Rewrite as an addition problem.

Does $72 = 41 + 21$? $41 + 21 = 62$, *not* 72, so there is an error!

But, $41 + 21 = 62$, *not* 72, so the subtraction was done *incorrectly*. Rework the original subtraction to get the correct answer of 31. Then, $41 + 31 = 72$.

(c)
```
  374   ← Match
- 141
-----
  233
```
$141 + 233 = 374$ The answer checks.

─── **Work Problem 4 at the Side.** ▶

OBJECTIVE ▶ 5 Subtract with regrouping. If a digit in the minuend is *less* than the one directly below it, **regrouping** is necessary.

EXAMPLE 5 Subtracting with Regrouping

Subtract 19 from 57.

Watch the wording! Subtract 19 **from** 57 is
```
57        19
-19  not  -57
```

Rewrite the problem in vertical format.

```
 5 7
-1 9
```

In the ones column, 7 is *less* than 9, so, in order to subtract, you must regroup 1 ten as 10 ones.

5 tens − 1 ten = 4 tens ──→ 4 17 ← 1 ten = 10 ones,
```
 5 7
-1 9
```
and 10 ones + 7 ones = 17 ones

Now subtract 17 ones minus 9 ones in the ones column.
Then subtract 4 tens minus 1 ten in the tens column.

```
 4 17
 5 7
-1 9
-----
 3 8
```
4 tens − 1 ten = 3 tens → 3 8 ← 17 ones − 9 ones = 8 ones

Thus, **$57 - 19 = 38$**. Check by adding 19 and 38. You should get 57.

─── **Work Problem 5 at the Side.** ▶

4 Use addition to check each answer. If incorrect, find the correct answer.

(a)
```
  65     CHECK
- 23
-----
  42
```

(b)
```
  46     CHECK
- 32
-----
  24
```

(c)
```
  374    CHECK
- 251
-----
  113
```

(d)
```
  7531   CHECK
- 4301
------
  3230
```

5 Find each difference.

(a)
```
  67
- 38
```

(b)
```
  97
- 29
```

(c)
```
  31
- 17
```

(d)
```
  863
-  47
```

(e)
```
  762
- 157
```

Answers

4. (a) $23 + 42 = 65$; correct
(b) $32 + 24 \neq 46$; incorrect; should be 14
(c) $251 + 113 \neq 374$; incorrect; should be 123
(d) $4301 + 3230 = 7531$; correct
5. (a) 29 **(b)** 68 **(c)** 14 **(d)** 816 **(e)** 605

6 Subtract.

(a)   354
   − 82

(b)   457
   − 68

(c)   874
   − 486

(d)   1437
   − 988

(e)   8739
   − 3892

| EXAMPLE 6 | **Subtracting with Regrouping** |

Subtract, regrouping when necessary.

(a)   7856
   − 137

Regroup 1 ten ——————→ ↓ ↓ ←—— 10 ones + 6 ones = 16 ones
as 10 ones.

             4 16
      7 8 5̶ 6
    −   1 3 7
      7 7 1 9 ← Difference

(b)   635
   − 546

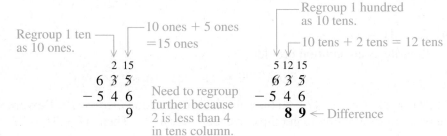

Regroup 1 ten ——┐   ┌— 10 ones + 5 ones
as 10 ones.   ↓ ↓   =15 ones

             2 15
      6 3̶ 5̶
    − 5 4 6   Need to regroup
          9   further because
                2 is less than 4
                in tens column.

Regroup 1 hundred
as 10 tens.
┌— 10 tens + 2 tens = 12 tens
↓ ↓
  5 12 15
  6̶ 3̶ 5̶
− 5 4 6
    8 9 ← Difference

(c)   412
   − 225

             0 12
      4 1̶ 2̶   Need to regroup
    − 2 2 5   further because 0
          7   is less than 2 in
                tens column.

  3 10 12
  4̶ 1̶ 2̶
− 2 2 5
  1 8 7 ← Difference

◀ **Work Problem 6 at the Side.**

Sometimes a minuend has zeros in some of the positions. In such cases, regrouping may be a little more complicated than what we have shown so far.

| EXAMPLE 7 | **Regrouping with Zeros** |

Subtract.

       4607
     − 3168

There are no tens that can be regrouped into ones. So you must first regroup 1 hundred as 10 tens.

Regroup 1 hundred ——————→ ↓ ↓ ←—— Write 10 tens.
as 10 tens

           5 10
      4 6̶ 0̶ 7
    − 3 1 6 8

Now regroup 1 ten as 10 ones.

                    Regroup 1 ten as 10 ones;
          9 ←——┌— 10 tens − 1 ten = 9 tens
      5 1̶0 17 ←—— 10 ones + 7 ones = 17 ones
      4 6̶ 0̶ 7
    − 3 1 6 8
              9

———— **Continued on Next Page**

Complete the problem.

$$
\begin{array}{r}
\overset{9}{} \\
\overset{5\ \ \overset{}{1}0\ \ 17}{} \\
4\ 6\ \cancel{0}\ \cancel{7} \\
-\ 3\ 1\ 6\ 8 \\
\hline
\mathbf{1\ 4\ 3\ 9} \leftarrow \text{Difference}
\end{array}
$$

Check by adding 1439 and 3168; you should get 4607.

──────── **Work Problem 7 at the Side.** ▶

EXAMPLE 8 **Regrouping with Zeros**

Find each difference.

(a)
$$
\begin{array}{r}
708 \\
-\ 149 \\
\hline
\end{array}
$$

Write 10 tens. ──────────────────┐ ┌── Regroup 1 ten as 10 ones.

$$
\begin{array}{r}
\overset{9}{} \\
6\ \ 10\ \ 18 \leftarrow \text{10 ones + 8 ones = 18 ones} \\
7\ \cancel{0}\ 8 \\
-\ 1\ 4\ 9 \\
\hline
\mathbf{5\ 5\ 9}
\end{array}
$$

Regroup 1 hundred as 10 tens.

(b)
$$
\begin{array}{r}
380 \\
-\ 276 \\
\hline
\end{array}
$$

Regroup 1 ten ──────┐ ┌── Write 10 ones.
as 10 ones.

$$
\begin{array}{r}
7\ \ 10 \\
3\ 8\ \cancel{0} \\
-\ 2\ 7\ 6 \\
\hline
\mathbf{1\ 0\ 4}
\end{array}
$$

(c)
$$
\begin{array}{r}
9000 \\
-\ 6999 \\
\hline
\end{array}
$$

$$
\begin{array}{r}
9\ \ \ 9 \\
8\ \ 10\ \ 10\ \ 10 \\
9\ \cancel{0}\ \cancel{0}\ \cancel{0} \\
-\ 6\ 9\ 9\ 9 \\
\hline
\mathbf{2\ 0\ 0\ 1}
\end{array}
$$

──────── **Work Problem 8 at the Side.** ▶

Recall that an answer to a subtraction problem can be checked by adding.

EXAMPLE 9 **Checking Subtraction**

Use addition to check each answer. If incorrect, find the correct answer.

CHECK

(a)
$$
\begin{array}{r}
613 \\
-\ 275 \\
\hline
338
\end{array}
\qquad
\begin{array}{r}
275 \\
+\ 338 \\
\hline
613
\end{array}
$$

Matches Correct

──────── **Continued on Next Page**

---

7 Find each difference.

(a)
$$
\begin{array}{r}
308 \\
-\ 285 \\
\hline
\end{array}
$$

(b)
$$
\begin{array}{r}
206 \\
-\ 148 \\
\hline
\end{array}
$$

(c)
$$
\begin{array}{r}
5073 \\
-\ 1632 \\
\hline
\end{array}
$$

8 Find each difference.

(a)
$$
\begin{array}{r}
405 \\
-\ 267 \\
\hline
\end{array}
$$

(b)
$$
\begin{array}{r}
370 \\
-\ 163 \\
\hline
\end{array}
$$

(c)
$$
\begin{array}{r}
1570 \\
-\ 983 \\
\hline
\end{array}
$$

(d)
$$
\begin{array}{r}
7001 \\
-\ 5193 \\
\hline
\end{array}
$$

(e)
$$
\begin{array}{r}
4000 \\
-\ 1782 \\
\hline
\end{array}
$$

Answers

7. (a) 23 (b) 58 (c) 3441
8. (a) 138 (b) 207 (c) 587 (d) 1808
 (e) 2218

9 Use addition to check each answer. If an answer is incorrect, find the correct answer.

(a)   425   **CHECK**
     − 368
     ———
       57

(b)   670   **CHECK**
     − 439
     ———
      241

(c) 14,726   **CHECK**
     − 8 839
     ———
      5 887

10 Use the table from **Example 10** at the right.

(a) How many fewer deliveries did Ms. Lopez make on Tuesday than on Friday?

(b) What was the difference in the number of deliveries on Tuesday from the number on Wednesday?

(c) How many more deliveries did she make on Thursday than on Saturday?

CHECK

(b)   1915 ← Matches   1635
    − 1635   Correct + 280
    ———   ———
     280   1915

CHECK

(c) 15,803 ← Does not match   7 325
    − 7 325   Error   + 8 578
    ———   ———
     8 578   15,903

Rework the original problem to get the correct answer, **8478**.
Then, 7325 + 8478 *does* equal 15,803.

◀ **Work Problem 9 at the Side.**

OBJECTIVE 6 Use subtraction to solve application problems.

EXAMPLE 10   Applying Subtraction Skills

Diana Lopez drives a delivery truck. Using the table below, decide how many more deliveries were made by Ms. Lopez on Monday than on Thursday.

PACKAGE DELIVERY (LOPEZ)

Day	Number of Deliveries
Monday	137
Tuesday	126
Wednesday	119
Thursday	89
Friday	147
Saturday	0

Ms. Lopez made 137 deliveries on Monday, but had only 89 deliveries on Thursday. Find how many more deliveries were made on Monday than on Thursday by subtracting 89 from 137.

    137 ← Deliveries on Monday
  − 89 ← Deliveries on Thursday
  ———
     48 ← More deliveries on Monday

Ms. Lopez made 48 more deliveries on Monday than she made on Thursday.

◀ **Work Problem 10 at the Side.**

Answers

9. (a) 368 + 57 = 425; correct
  (b) 439 + 241 ≠ 670; incorrect; should be 231
  (c) 8839 + 5887 = 14,726; correct
10. (a) 21 fewer deliveries
  (b) 7 fewer deliveries on Wednesday
  (c) 89 more deliveries

R.2 Exercises

FOR EXTRA HELP

Go to MyMathLab *for worked-out, step-by-step solutions to exercises enclosed in a square* ▢ *and video solutions to* ▶ *exercises.*

Use addition to check each subtraction. If an answer is incorrect, find the correct answer.
See Examples 3 and 4.

1. 89
    − 27
    ‾‾‾‾
    63

2. 47
    − 35
    ‾‾‾‾
    13

3. 382
    − 261
    ‾‾‾‾‾
    131

4. 838
    − 516
    ‾‾‾‾‾
    322

Find each difference, regrouping when necessary. ***See Examples 5–8.***

5. 36
    − 28

6. 97
    − 39

7. 83
    − 58

8. 65
    − 28

9. 45
    − 29

10. 93
    − 37

11. 719
    − 658

12. 916
    − 618

13. 771
    − 252

14. 973
    − 788

15. 9861
    − 684

16. 6171
    − 1182

17. 9988
    − 2399

18. 3576
    − 1658

19. 38,335
    − 29,476

20. 61,278
    − 3 559

21. 40
    − 37

22. 80
    − 73

23. 60
    − 37

24. 70
    − 27

25. 6020
    − 4078

26. 7050
    − 6045

27. 8503
    − 2816

28. 16,004
    − 5 087

29. 80,705
    − 61,667

30. 72,000
    − 44,234

31. 66,000
    − 444

32. 77,000
    − 308

33. 20,080
    − 96

34. 80,056
    − 69

Use addition to check each subtraction. If an answer is incorrect, find the
correct answer. ***See Example 9.***

35. 3070
    − 576
    ‾‾‾‾‾
    2596

36. 1439
    − 1169
    ‾‾‾‾‾
    270

37. 27,600
    − 807
    ‾‾‾‾‾‾
    26,793

38. 34,021
    − 33,708
    ‾‾‾‾‾‾
    727

*Solve each application problem. **See Example 10.***

39. Swimming laps burns 255 calories in 30 minutes. Hiking burns 185 calories in 30 minutes. How many fewer calories are burned in 30 minutes by hiking than by swimming?

40. Lynn Couch had $553 in her bank account. She paid $308 for school fees. How much is left in her account?

41. An airplane was carrying 254 passengers. When it landed in Atlanta, 183 passengers got off and 109 passengers got on. How many passengers were on the plane then?

42. On Tuesday, 5822 people went to a soccer game, and on Friday, 7994 people went to a soccer game. How many more people went to the game on Friday? What was the total attendance at the two games?

This table shows the yearly starting salaries for various occupations. Use the table to answer Exercises 43 and 44.

43. (a) Identify the occupations with the greatest and lowest salaries.

   (b) What is the difference in the yearly starting salaries for these two occupations?

44. (a) How much less is the starting salary of an elementary school teacher than a state police officer?

   (b) How much more is the starting salary of a registered nurse than a licensed practical nurse?

All Walks of Life

Average yearly starting salaries for selected occupations. Yearly earnings increase with experience.

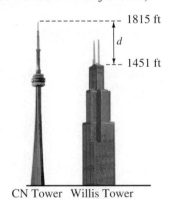

Occupation	Starting salary
Accountant (not a CPA)	$53,300
Auto mechanic	$35,935
Computer programmer	$50,796
Licensed practical nurse (LPN)	$36,000
Registered nurse (RN)	$52,000
State police officer	$42,570
Elementary school teacher	$36,669

Data from www.payscale.com

45. Downtown Toronto's skyline is dominated by the CN Tower, which rises 1815 feet. The Willis Tower in Chicago is 1451 feet high. Find the difference in height between the two structures. (Data from *The World Almanac and Book of Facts*.)

46. The fastest animal in the world, the peregrine falcon, dives at 242 miles per hour. A Boeing 747 cruises at 560 miles per hour. How much faster does the plane cruise than the falcon dives? (Data from *Audubon* magazine.)

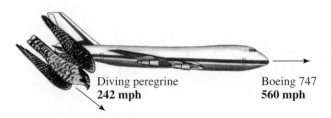

Diving peregrine
242 mph

Boeing 747
560 mph

1815 ft

d

1451 ft

CN Tower   Willis Tower

R.3 | Multiplying Whole Numbers

Suppose we want to know the total number of computers in a computer lab. The computers are arranged in three rows with four computers in each row. Adding the number 4 a total of 3 times gives 12.

$$4 + 4 + 4 = 12$$

This result is illustrated below. There are 4 computers in each row.

$$\begin{array}{r} 4 \\ 4 \\ + \ 4 \\ \hline 12 \end{array}$$

3 rows

OBJECTIVE ▶ 1 Identify the parts of a multiplication problem. Multiplication is a shortcut for repeated addition. The numbers being multiplied are called **factors.** The answer is called the **product.** The product of 3 and 4 can be written with the symbol ×, a raised dot, parentheses, or, in computer work, an asterisk.

$$\begin{array}{r} 3 \\ \times \ 4 \\ \hline 12 \end{array} \begin{array}{l} \leftarrow \text{Factor (also called } \textit{multiplicand)} \\ \leftarrow \text{Factor (also called } \textit{multiplier)} \\ \leftarrow \text{Product (answer)} \end{array}$$

$$12 = 3 \times 4 \qquad 3 \cdot 4 = 12 \qquad (3)(4) = 12 \qquad 3 * 4 = 12$$

Raised dot   In computer work

Work Problem 1 at the Side. ▶

Commutative Property of Multiplication

By the **commutative property of multiplication,** changing the *order* of two factors does *not* change the product.

For example: $3 \times 5 = 15$ and $5 \times 3 = 15$ Both products are 15.

---

❶ CAUTION

Remember, addition also has a commutative property. For example, $4 + 2$ has the same sum as $2 + 4$. Subtraction, however, is *not* commutative.

EXAMPLE 1   Multiplying Two Numbers

Multiply. (Remember, a raised dot means to multiply.) Do the work mentally.

(a) $3 \times 4 = \mathbf{12}$ By the commutative property, $4 \times 3 = 12$ also.

(b) $6 \cdot 0 = \mathbf{0}$ The product of any number and 0 is 0; if you give no money to each of 6 relatives, you give no money.

(c) $(4)(8) = \mathbf{32}$ By the commutative property, $(8)(4) = 32$ also.

Work Problem 2 at the Side. ▶

1. Identify the factors and the product in each multiplication problem.

   (a) $3 \times 6 = 18$

   (b) $35 = 5 \cdot 7$

   (c) $(3)(9) = 27$

2. Multiply. Do the work mentally. Then use the commutative property to write another multiplication problem and find the product.

   (a) 4×7 **(b)** 0×9

   (c) $8 \cdot 6$ **(d)** $(3)(8)$

Answers

1. **(a)** factors: 3, 6; product: 18
   (b) factors: 5, 7; product: 35
   (c) factors: 3, 9; product: 27
2. **(a)** 28; $7 \times 4 = 28$ **(b)** 0; $9 \times 0 = 0$
   (c) 48; $6 \cdot 8 = 48$ **(d)** 24; $(8)(3) = 24$

3 Find each product.

(a) $2 \times 3 \times 4$

(b) $6 \cdot 1 \cdot 5$

(c) $(8)(3)(0)$

(d) $3 \times 3 \times 7$

(e) $4 \cdot 2 \cdot 8$

(f) $(2)(2)(9)$

Answers

3. (a) 24 (b) 30 (c) 0 (d) 63
   (e) 64 (f) 36

OBJECTIVE **2** **Do chain multiplications.** Some multiplications involve more than two factors.

> **Associative Property of Multiplication**
>
> By the **associative property of multiplication**, changing the *grouping* of factors does *not* change the product.

EXAMPLE 2 **Multiplying Three Numbers**

Multiply $2 \times 3 \times 5$.

$(2 \times 3) \times 5 = 6 \times 5 = \mathbf{30}$ Parentheses tell you to multiply 2×3 first.

$2 \times (3 \times 5) = 2 \times 15 = \mathbf{30}$ Parentheses tell you to multiply 3×5 first.

By the associative property, either grouping results in the same product.

> **Calculator Tip**
>
> The calculator approach to **Example 2** above uses chain calculations. Notice that you can enter *all* the factors before pressing the ⊜ key.
>
> 2 ⓧ 3 ⓧ 5 ⊜ 30

◀ **Work Problem** **3** **at the Side.**

OBJECTIVE **3** **Multiply by single-digit numbers.** Regrouping may be needed in multiplication problems with larger factors.

EXAMPLE 3 **Regrouping with Multiplication**

Find each product.

(a) $\begin{array}{r} 53 \\ \times\ 4 \\ \hline \end{array}$

Start by multiplying in the ones column. Multiply 4 times 3 ones.

$\begin{array}{r} 1 \\ 5\mathbf{3} \\ \times\ 4 \\ \hline 2 \end{array}$ Write 1 ten in the tens column.

4×3 ones $= 12$ ones

Write 2 ones in the ones column.

Next, multiply 4 times 5 tens.

$\begin{array}{r} 1 \\ 53 \\ \times\ 4 \\ \hline 2 \end{array}$ 4×5 tens $= 20$ tens

$\begin{array}{r} 1 \\ 53 \\ \times\ 4 \\ \hline 212 \end{array}$ Add the 1 ten that was written at the top of the tens column.

20 tens $+$ 1 ten $= \mathbf{21}$ tens

The product is **212**.

Continued on Next Page

(b) 724
× 5

Work as shown below.

12
724
× 5
3620 ← 5 × 4 ones = 20 ones; write 0 ones;
write 2 tens in the tens column.

5 × 2 tens = 10 tens; add the 2 regrouped tens to get 12 tens;
write 2 tens; write 1 hundred in the hundreds column.

5 × 7 hundreds = 35 hundreds; add the
1 regrouped hundred to get 36 hundreds.

The product is **3620**.

Work Problem 4 at the Side. ▶

OBJECTIVE ▶ **4 Multiply quickly by numbers ending in zeros.** The product of two whole number factors is also called a **multiple** of either factor. For example, since 4 • 2 = 8, the number 8 is a multiple of 4, and 8 is also a multiple of 2. Multiples of 10 are very useful when multiplying. A *multiple of 10* is a whole number that ends in 0, such as 10, 20, or 30; 100, 200, or 300; 1000, 2000, or 3000; and so on. There is a quick way to multiply by multiples of 10. Look at the following examples.

$$26 \times 1 = 26$$
$$26 \times 10 = 260$$
$$26 \times 100 = 2600$$
$$26 \times 1000 = 26,000$$

Do you see a pattern in the multiplications? These examples suggest the following rule.

Multiplying by Multiples of 10

To multiply a whole number by 10, by 100, or by 1000, attach one, two, or three zeros to the right of the whole number.

EXAMPLE 4 Multiplying by Multiples of 10

Multiply.

(a) 59 × 10 = **590**
Attach 0

(b) 100 × 74 = **7400**
Attach 00

(c) 803 × 1000 = **803,000**
Attach 000

Work Problem 5 at the Side. ▶

You can also find the product of other multiples of ten by attaching zeros.

4 Find each product.

(a) 52
× 5

(b) 79
× 0

(c) 862
× 9

(d) 2831
× 7

(e) 4714
× 8

5 Multiply by attaching zeros.

(a) 45 × 10

(b) 102 × 100

(c) 1000 × 571

(d) 100 × 3625

(e) 69 × 1000

Answers

4. (a) 260 (b) 0 (c) 7758 (d) 19,817
(e) 37,712

5. (a) 450 (b) 10,200 (c) 571,000
(d) 362,500 (e) 69,000

6 Find each product by attaching zeros.

(a) 14×50

(b) $(68)(400)$

(c) $\begin{array}{r} 180 \\ \times \ 30 \\ \hline \end{array}$

(d) $\begin{array}{r} 6100 \\ \times \ \ 90 \\ \hline \end{array}$

(e) $\begin{array}{r} 800 \\ \times 200 \\ \hline \end{array}$

(f) $(5000)(700)$

(g) $(9)(20,000)$

EXAMPLE 5 Multiplying with Other Multiples of 10

Find each product.

(a) 75×3000

Multiply 75 by 3, and then attach three zeros.

$$\begin{array}{r} 75 \\ \times \ \ 3 \\ \hline 225 \end{array} \qquad 75 \times 3000 = \mathbf{225{,}000}$$

(b) 150×70

Multiply 15 by 7, and then attach two zeros.

$$\begin{array}{r} 15 \\ \times \ \ 7 \\ \hline 105 \end{array} \qquad 150 \times 70 = \mathbf{10{,}500}$$

◀ **Work Problem 6 at the Side.**

OBJECTIVE ▶ 5 Multiply by numbers having more than one digit. The next example shows multiplication when both factors have more than one digit.

EXAMPLE 6 Multiply by Numbers with More Than One Digit

Multiply.

(a) $(46)(23)$

Rewrite the problem in vertical format. Then start by multiplying 46 by 3.

$$\begin{array}{r} {\scriptstyle 1} \\ 46 \\ \times \ \ 3 \\ \hline 138 \end{array} \leftarrow 46 \times 3 = 138$$

Next, multiply 46 by 20.

$$\begin{array}{r} {\scriptstyle 1} \\ 46 \\ \times \ 20 \\ \hline 138 \\ 920 \end{array} \leftarrow 46 \times 20 = 920$$

Add the results.

$$\begin{array}{r} 46 \\ \times \ 23 \\ \hline 138 \leftarrow 46 \times 3 \\ + \ 920 \leftarrow 46 \times 20 \\ \hline 1058 \leftarrow \text{Add to find the product.} \end{array}$$

Both 138 and 920 are called *partial products*.
To save time, the 0 in 920 is usually not written.

$$\begin{array}{r} 46 \\ \times \ 23 \\ \hline 138 \\ 92 \\ \hline 1058 \end{array}$$

0 not written. Be very careful to place the 2 in the tens column.

Answers

6. (a) 700   (b) 27,200   (c) 5400
   (d) 549,000   (e) 160,000   (f) 3,500,000
   (g) 180,000

Continued on Next Page

(b)
```
      2 3 3
   ×  1 3 2
   ─────────
      4 6 6
```
      6 9 9   Tens lined up
    2 3 3   Hundreds lined up
```
   ─────────
   3 0,7 5 6  ← Product
```

(c)
```
      5 3 8
   ×    4 6
```

First multiply by 6.

```
                    2 4  ← Regrouping is
                    5 3 8   needed here.
                  ×   4 6
                  ───────
                    3 2 2 8
```

Now multiply by 4, being careful to line up the tens.

```
                  1 3
                  2 4
                5 3 8
              ×   4 6
              ───────
                3 2 2 8  ┐─ Add the partial products.
              2 1 5 2    ┘
              ─────────
              2 4,7 4 8
```

────────────── Work Problem **7** at the Side. ▶

When 0 appears in the multiplier, be sure to move the partial product to the left to account for the position held by the 0.

EXAMPLE 7 **Multiplication with Zeros**

Find each product.
(a)
```
        1 3 7
     ×  3 0 6
     ─────────
        8 2 2
```
      0 0 0   Tens lined up
    4 1 1   Hundreds lined up
```
   ─────────
   4 1,9 2 2
```

(b)
```
        1 4 0 6                    1 4 0 6
     ×  2 0 0 1                 ×  2 0 0 1
     ─────────                  ─────────
        1 4 0 6                    1 4 0 6
      0 0 0 0 ← 0 to line up tens
      0 0 0 0 ← 0 to line up hundreds      2 8 1 2 0 0
    2 8 1 2                    ─────────────
    ─────────                  2,8 1 3,4 0 6
    2,8 1 3,4 0 6
```

Zeros are written so that you start writing the partial product 2812 in the *thousands* column.

⚠ CAUTION

In **Example 7(b)** above, in the solution on the right, be careful to insert zeros so that thousands are lined up in the thousands column.

─────────── **Continued on Next Page**

7 Find each product.

(a)
```
      3 8
   ×  1 5
```

(b)
```
      3 1
   ×  4 3
```

(c)
```
      6 7
   ×  5 9
```

(d)
```
     2 3 4
   ×   7 3
```

(e)
```
     8 3 5
   ×  1 8 9
```

Answers

7. (a) 570   **(b)** 1333   **(c)** 3953
   (d) 17,082   **(e)** 157,815

8 Find each product.

(a) 28
  $\times$ 60

(b) 817
  $\times$ 30

(c) 481
  $\times$ 206

(d) 3526
  $\times$ 6002

◄ **Work Problem 8 at the Side.**

OBJECTIVE ▶ 6 Use multiplication to solve application problems.

EXAMPLE 8   Applying Multiplication Skills

Find the total cost of 24 months of Internet service at $59 per month.

Approach   To find the total cost, multiply the cost for one month of service ($59) by the number of months (24).

Solution   Multiply $59 by 24.

$$
\begin{array}{r}
59 \\
\times\ 24 \\
\hline
236 \\
118 \\
\hline
1416
\end{array}
$$

The **total cost of 24 months of Internet service is $1416.**

⌨ **Calculator Tip**

If you are using a calculator for **Example 8** above, press the following keys.

59 ⊗ 24 ⊜ **1416**

─────── ◄ **Work Problem 9 at the Side.**

9 Find the total cost of these items.

(a) 36 months of cable TV costing $79 each month

(b) 15 laptop computers priced at $1090 each

(c) 60 months of car payments at $389 per month

Answers

8. (a) 1680 **(b)** 24,510 **(c)** 99,086
   (d) 21,163,052
9. (a) $2844 **(b)** $16,350 **(c)** $23,340

R.3 Exercises

FOR EXTRA HELP Go to MyMathLab for worked-out, step-by-step solutions to exercises enclosed in a square ▢ and video solutions to ▶ exercises.

Find each product. Try to do the work mentally. **See Examples 1 and 2.**

1. $3 \times 1 \times 3$

2. $2 \times 8 \times 2$

3. $9 \times 1 \times 7$

4. $2 \times 4 \times 5$

5. $9 \cdot 5 \cdot 0$

6. $6 \cdot 0 \cdot 8$

7. $(4)(1)(6)$

8. $(1)(5)(7)$

9. $(2)(3)(6)$

10. $(4)(1)(9)$

Find each product. **See Examples 3–7.**

11. $\begin{array}{r} 35 \\ \times\ 7 \\ \hline \end{array}$

12. $\begin{array}{r} 76 \\ \times\ 9 \\ \hline \end{array}$

13. $\begin{array}{r} 28 \\ \times\ 6 \\ \hline \end{array}$

14. $\begin{array}{r} 83 \\ \times\ 5 \\ \hline \end{array}$

15. $\begin{array}{r} 3182 \\ \times\ \ \ 6 \\ \hline \end{array}$

16. $\begin{array}{r} 7326 \\ \times\ \ \ 5 \\ \hline \end{array}$

17. $\begin{array}{r} 36{,}921 \\ \times\ \ \ \ \ 7 \\ \hline \end{array}$

18. $\begin{array}{r} 28{,}116 \\ \times\ \ \ \ \ 4 \\ \hline \end{array}$

19. $\begin{array}{r} 125 \\ \times\ 100 \\ \hline \end{array}$

20. $\begin{array}{r} 246 \\ \times\ 100 \\ \hline \end{array}$

21. $\begin{array}{r} 1485 \\ \times\ \ \ 30 \\ \hline \end{array}$

22. $\begin{array}{r} 8522 \\ \times\ \ \ 50 \\ \hline \end{array}$

23. $\begin{array}{r} 900 \\ \times\ 300 \\ \hline \end{array}$

24. $\begin{array}{r} 400 \\ \times\ 700 \\ \hline \end{array}$

25. $\begin{array}{r} 43{,}000 \\ \times\ \ 2000 \\ \hline \end{array}$

26. $\begin{array}{r} 11{,}000 \\ \times\ \ 9000 \\ \hline \end{array}$

27. $\begin{array}{r} 68 \\ \times\ 22 \\ \hline \end{array}$

28. $\begin{array}{r} 82 \\ \times\ 32 \\ \hline \end{array}$

29. $(83)(45)$

30. $(43)(27)$

31. $(32)(475)$

32. $(67)(218)$

33. $(729)(45)$

34. $(681)(47)$

35. $\begin{array}{r} 538 \\ \times\ 342 \\ \hline \end{array}$

36. $\begin{array}{r} 3228 \\ \times\ \ \ 751 \\ \hline \end{array}$

37. $\begin{array}{r} 8162 \\ \times\ \ \ 407 \\ \hline \end{array}$

38. $\begin{array}{r} 528 \\ \times\ 106 \\ \hline \end{array}$

39. $\begin{array}{r} 6310 \\ \times\ 3008 \\ \hline \end{array}$

40. $\begin{array}{r} 3533 \\ \times\ 5001 \\ \hline \end{array}$

*Solve each application problem. **See Example 8.***

41. Giant kelp plants in the ocean can grow 18 inches each day. How much could kelp grow in two weeks? How much could it grow in a 30-day month? (Data from Natural Bridges State Park, CA.)

42. A hospital has 20 bottles of thyroid medication, with each bottle containing 2500 tablets. How many of these tablets does the hospital have in all?

43. There are 12 tomato plants to a flat. If a garden center has 48 flats, find the total number of tomato plants.

44. A hummingbird's wings beat about 65 times per second, a chickadee's wings about 27 times per second. How many times does each bird's wings beat in one minute? (Data from *Birder's Handbook*.)

45. A hybrid automobile gets 48 miles per gallon on the highway. How many highway miles can it travel on 11 gallons of gas?

Ruby-throated hummingbird
65 wing beats per second

46. Find the total cost of 16 gallons of paint at $36 per gallon.

Use addition, subtraction, or multiplication to solve each problem.

47. The distance from Reno, Nevada, to the Atlantic Ocean is 2695 miles, while the distance from Reno to the Pacific Ocean is 255 miles. How much farther is it to the Atlantic Ocean than it is to the Pacific Ocean? If you make three round trips from Reno to the Atlantic Ocean, how many frequent flier miles will you earn?

48. The largest living land mammal is the African elephant, and the largest mammal of all time is the blue whale. An African bull elephant may weigh 15,225 pounds and a blue whale may weigh 12 to 25 times that amount. Find the range of weights for the blue whale and the difference between the lightest and heaviest. (Data from *Big Book of Knowledge*.)

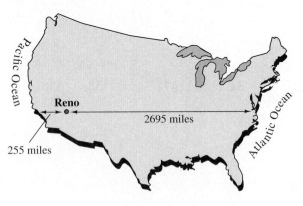

49. A high-fat meal contains 1406 calories, while a low-fat meal contains 348 calories. How many more calories are in seven high-fat meals than in seven low-fat meals?

50. Dannie Sanchez bought four tires at $110 each, two seat covers at $49 each, and a set of socket wrenches for $20. Sales tax was $43. Find the total amount that he spent.

R.4 | Dividing Whole Numbers

$12 is divided equally among 3 children. Each child gets $4, as shown here.

$12 total

3 equal parts

$4 in each part

OBJECTIVES

OBJECTIVES

1. Write division problems in three ways.
2. Identify the parts of a division problem.
3. Divide 0 by a number.
4. Recognize that division by 0 is undefined.
5. Divide a number by itself.
6. Use short division.
7. Use multiplication to check quotients.
8. Use tests for divisibility.

OBJECTIVE ❶ **Write division problems in three ways.** Recall that $3 \cdot 4$, 3×4, and $(3)(4)$ are different ways to write the *multiplication* of 3 and 4. There are several ways to write 12 *divided* by 3.

Being divided

$$12 \div 3 = 4$$

Divided by

Divided by

$$3\overline{)12}^{\,4}$$

Being divided

Being divided

$$\frac{12}{3} = 4$$

Divided by

Being divided

$$12 / 3 = 4$$

Divided by

We will use three division symbols: $\div$, $\overline{)}\,$, and —. In algebra the bar, —, is frequently used. In computer science a slash, $/$, is used.

EXAMPLE 1 | **Using Division Symbols**

Rewrite each division using two other symbols.

(a) $20 \div 4 = 5$ can also be written $4\overline{)20}^{\,5}$ or $\dfrac{20}{4} = 5$

(b) $\dfrac{18}{6} = 3$ can also be written $18 \div 6 = 3$ or $6\overline{)18}^{\,3}$

(c) $5\overline{)40}^{\,8}$ can also be written $40 \div 5 = 8$ or $\dfrac{40}{5} = 8$

— **Work Problem ❶ at the Side.** ▶

OBJECTIVE ❷ **Identify the parts of a division problem.** In division, the number being divided is the **dividend,** the number you are dividing by is the **divisor,** and the answer is the **quotient.**

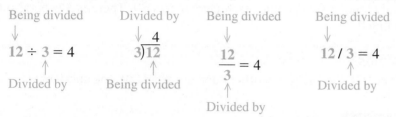

$$\text{dividend} \div \text{divisor} = \text{quotient}$$

$$\text{divisor}\overline{)\text{dividend}}^{\,\text{quotient}} \qquad \frac{\text{dividend}}{\text{divisor}} = \text{quotient}$$

EXAMPLE 2 | **Identifying the Parts in a Division Problem**

Identify the dividend, divisor, and quotient.

(a) $35 \div 7 = 5$

$$35 \div 7 = 5 \leftarrow \text{Quotient}$$

Dividend Divisor

— **Continued on Next Page**

❶ Rewrite each division using two other symbols.

(a) $48 \div 6 = 8$

(b) $4 = 24 \div 6$

(c) $9\overline{)36}^{\,4}$

(d) $\dfrac{42}{6} = 7$

Answers

1. (a) $6\overline{)48}^{\,8}$ and $\dfrac{48}{6} = 8$

(b) $6\overline{)24}^{\,4}$ and $4 = \dfrac{24}{6}$

(c) $36 \div 9 = 4$ and $\dfrac{36}{9} = 4$

(d) $6\overline{)42}^{\,7}$ and $42 \div 6 = 7$

2 Identify the dividend, divisor, and quotient.

(a) $10 \div 2 = 5$

(b) $6 = 30 \div 5$

(c) $\dfrac{28}{7} = 4$

(d) $2\overline{)36}^{\,18}$

3 Divide.

(a) $0 \div 9$

(b) $\dfrac{0}{36}$

(c) $57\overline{)0}$

4 Write each division problem as a multiplication problem.

(a) $6\overline{)18}^{\,3}$

(b) $\dfrac{28}{4} = 7$

(c) $48 \div 8 = 6$

$$\overset{\text{Dividend}}{\underset{}{\downarrow}}$$

(b) $\dfrac{100}{20} = 5$ 　　　 $\dfrac{\overset{\text{Dividend}}{100}}{\underset{\uparrow}{20}} = 5 \leftarrow \text{Quotient}$

　　　　　　　　　　　　　　　Divisor

(c) $12\overline{)72}^{\,6}$ 　　　 $12\overline{)72}^{\,6 \leftarrow \text{Quotient}} \leftarrow \text{Dividend}$

　　　　　　　　　　　　　　$\uparrow$
　　　　　　　　　　　　　Divisor

◀ **Work Problem ② at the Side.**

OBJECTIVE ③ Divide 0 by a number. If no money, or $0, is divided equally among five people, each person gets $0. There is a general rule for dividing 0.

> **Dividing 0**
>
> When 0 is divided by any other number (except 0), the quotient is 0.

EXAMPLE 3　Dividing 0 by a Number

Divide.

(a) $0 \div 12 = 0$

(b) $0 \div 1728 = 0$

(c) $\dfrac{0}{375} = 0$

(d) $129\overline{)0}^{\,0}$

◀ **Work Problem ③ at the Side.**

Recall that a subtraction such as $8 - 3 = 5$ can be written as the addition $8 = 3 + 5$. In a similar way, any division can be written as a multiplication. For example, $12 \div 3 = 4$ can be written as

$$3 \times 4 = 12 \quad \text{or, by the commutative property,} \quad 4 \times 3 = 12$$

EXAMPLE 4　Changing Division to Multiplication

Change each division to multiplication.

(a) $\dfrac{20}{4} = 5$　becomes　$4 \cdot 5 = 20$ ◁ $\begin{smallmatrix} 5 \cdot 4 = 20 \\ \text{is also correct.} \end{smallmatrix}$

(b) $8\overline{)48}^{\,6}$　becomes　$8 \cdot 6 = 48$ ◁ $\begin{smallmatrix} 6 \cdot 8 = 48 \\ \text{is also correct.} \end{smallmatrix}$

(c) $72 \div 9 = 8$　becomes　$9 \cdot 8 = 72$ ◁ $\begin{smallmatrix} 8 \cdot 9 = 72 \\ \text{is also correct.} \end{smallmatrix}$

◀ **Work Problem ④ at the Side.**

Answers

2. (a) dividend: 10; divisor: 2; quotient: 5
   (b) dividend: 30; divisor: 5; quotient: 6
   (c) dividend: 28; divisor: 7; quotient: 4
   (d) dividend: 36; divisor: 2; quotient: 18

3. all 0

4. (a) $6 \cdot 3 = 18$　(b) $4 \cdot 7 = 28$
   (c) $8 \cdot 6 = 48$

OBJECTIVE **4** **Recognize that division by 0 is undefined.** Division by 0 cannot be done. To see why, try to find the answer to this division.

$$9 \div 0 = ?$$

Remember, any division problem can be changed to a multiplication problem.

$$\text{divisor} \cdot \text{quotient} = \text{dividend}$$

If you convert the problem $9 \div 0 = ?$ to its multiplication counterpart, it reads as follows.

$$0 \cdot ? = 9$$

You already know that 0 times any number must always equal 0. Try any number you like to replace the ? and you'll always get 0 instead of 9. Therefore, the division problem $9 \div 0$ *cannot* be done. Mathematicians say it is *undefined* and have agreed never to divide by 0. However, $0 \div 9$ *can* be done. Check by rewriting it as a multiplication problem.

$$0 \div 9 = 0 \quad \text{because} \quad 9 \cdot 0 = 0 \text{ is true.}$$

Dividing by 0

Dividing by 0 cannot be done. We say that **division by 0 is undefined.** It is impossible to calculate an answer.

EXAMPLE 5 **Dividing by 0 Is Undefined**

All of these divisions are **undefined**.

(a) $\dfrac{6}{0}$ is **undefined**.

It is **not possible** to divide by 0. Write *undefined* as the answer (**not** 0).

(b) $0\overline{)8}$ is **undefined**.

(c) $18 \div 0$ is **undefined**.

Division Involving 0

$$\frac{0}{\text{nonzero number}} = 0 \quad \textbf{but} \quad \frac{\text{number}}{0} \text{ is } \textbf{\textit{undefined.}}$$

⊘ CAUTION

When 0 is the divisor in a problem, write **undefined.**
Never divide by 0.

──────── **Work Problem 5 at the Side.** ▶

▦ Calculator Tip

Try these two problems on your calculator. Jot down your answers.

$9 \div 0 =$ _____ $0 \div 9 =$ _____

When you try to divide by 0, the calculator cannot do it, so it shows the word "Error" in the display, or the letters "ERR" or "E" (for "error").

5 Find the quotient whenever possible.

(a) $\dfrac{8}{0}$

(b) $\dfrac{0}{8}$

(c) $0\overline{)32}$

(d) $32\overline{)0}$

(e) $100 \div 0$

(f) $0 \div 100$

Answers

5. **(a)** undefined **(b)** 0 **(c)** undefined
   (d) 0 **(e)** undefined **(f)** 0

6 Divide.

(a) $5 \div 5$

(b) $14\overline{)14}$

(c) $\dfrac{37}{37}$

7 Divide using short division.

(a) $2\overline{)18}$

(b) $3\overline{)39}$

(c) $4\overline{)88}$

(d) $2\overline{)462}$

OBJECTIVE 5 Divide a number by itself. What happens when a number is divided by itself? For example, $4 \div 4$ or $97 \div 97$?

Dividing a Number by Itself

When a nonzero number is divided by itself, the quotient is 1.

EXAMPLE 6   Dividing a Nonzero Number by Itself

Divide.

(a) $16 \div 16 = 1$

(b) $32\overline{)32}^{\,1}$

(c) $\dfrac{57}{57} = 1$

◀ **Work Problem 6 at the Side.**

OBJECTIVE 6 Use short division. **Short division** is a quick method of dividing a number by a one-digit divisor.

EXAMPLE 7   Using Short Division

Divide $2\overline{)86}$.

First, divide 8 by 2.

$2\overline{)86}^{\,4}$ $\dfrac{8}{2} = 4$

Next, divide 6 by 2.

$2\overline{)86}^{\,43}$ $\dfrac{6}{2} = 3$

The quotient is 43.

◀ **Work Problem 7 at the Side.**

When two numbers do *not* divide evenly, the leftover portion is called the **remainder.** The remainder is always *less* than the divisor.

EXAMPLE 8   Using Short Division with a Remainder

Divide 147 by 4.

Rewrite the problem. $4\overline{)147}$

Because 1 cannot be divided by 4, divide 14 by 4.

$4\overline{)14^27}^{\,3}$ $\dfrac{14}{4} = 3$ with 2 left over

Continued on Next Page

Next, divide 27 by 4. The final number left over is the remainder. Write the remainder to the side. "R" stands for remainder.

$$\overset{3\ 6\ \textbf{R3}}{4)14\,{}^{2}7} \qquad \frac{27}{4} = 6 \text{ with 3 left over}$$

The quotient is 36 R3.

— Work Problem ❽ at the Side. ▶

EXAMPLE 9	**Writing Zeros in the Quotient**

Divide 1439 by 7.

Rewrite the problem. Then, divide 14 by 7.

$$\overset{2}{7)1439} \qquad \frac{14}{7} = 2 \text{ with 0 left over}$$

No need to write 0

Next, divide 3 by 7. But 7 will not go into 3 even one time, so write a 0 in the quotient.

Be sure to write the 0 in the quotient.

$$\overset{20}{7)143\,{}^{3}9} \qquad \frac{3}{7} = 0 \text{ with 3 left over}$$

Finally, divide 39 by 7.

$$\overset{20\ 5\ \textbf{R4}}{7)143\,{}^{3}9} \qquad \frac{39}{7} = 5 \text{ with 4 left over}$$

The quotient is 205 R4.

— Work Problem ❾ at the Side. ▶

OBJECTIVE ▶ ❼ Use multiplication to check quotients.

Checking Division

$$(\text{divisor} \times \text{quotient}) + \text{remainder} = \text{dividend}$$

Parentheses tell you what to do first. In this case, multiply the divisor by the quotient first and then add the remainder.

EXAMPLE 10	**Using Multiplication to Check Division**

Check each quotient. If incorrect, find the correct quotient.

(a) $\overset{91\ \textbf{R3}}{5)458}$

$(\text{divisor} \times \text{quotient}) + \text{remainder} = \text{dividend}$

$(5 \times 91) + 3$

$455 + 3 = 458 \leftarrow$ Matches original dividend so the quotient is correct.

— **Continued on Next Page**

❽ Divide.

(a) $2)\overline{225}$

(b) $3)\overline{275}$

(c) $4)\overline{538}$

(d) $\dfrac{819}{5}$

❾ Divide.

(a) $4)\overline{837}$

(b) $\dfrac{747}{7}$

(c) $5)\overline{4538}$

(d) $8)\overline{2440}$

Answers

8. (a) 112 **R1** (b) 91 **R2** (c) 134 **R2**
   (d) 163 **R4**

9. (a) 209 **R1** (b) 106 **R5** (c) 907 **R3**
   (d) 305

10 Check each division. If a quotient is incorrect, find the correct quotient.

$$\begin{array}{r} 38 \; \textbf{R1} \\ \textbf{(a)} \; 3\overline{)115} \end{array}$$

$$\begin{array}{r} 92 \; \textbf{R2} \\ \textbf{(b)} \; 8\overline{)743} \end{array}$$

$$\begin{array}{r} 328 \\ \textbf{(c)} \; 4\overline{)1312} \end{array}$$

$$\begin{array}{r} 46 \; \textbf{R3} \\ \textbf{(d)} \; 5\overline{)2033} \end{array}$$

Answers

10. (a) correct
  (b) incorrect; should be 92 **R7**
  (c) correct
  (d) incorrect; should be 406 **R3**

$$\begin{array}{r} 29 \; \textbf{R4} \\ \textbf{(b)} \; 6\overline{)1258} \end{array}$$

$$(\text{divisor} \times \text{quotient}) + \text{remainder} = \text{dividend}$$

$$(6 \quad \times \quad 29) \quad + \quad 4$$

$$174 \quad + \quad 4 = \textbf{178}$$

Does **not** match original dividend of 1258

The quotient does *not* check. Rework the original problem to get the correct quotient, **209 R4**. Then, to check, $(6 \times 209) + 4 = 1254 + 4 = 1258$, the original dividend.

> **❶ CAUTION**
>
> A common error is forgetting to add the remainder. Be sure to add any remainder when checking a division problem.

◀ **Work Problem 10 at the Side.**

OBJECTIVE 8 Use tests for divisibility. It is often important to know whether a number is divisible by another number. You will find this useful when working with fractions.

> **Divisibility**
>
> A whole number is *divisible* by another whole number if the remainder is 0.

There are some quick tests you can use to decide whether one number is divisible by another.

Tests for Divisibility

A number is divisible by	
2	if it ends in 0, 2, 4, 6, or 8.
3	if the sum of its digits is divisible by 3.
4	if the last two digits make a number that is divisible by 4.
5	if it ends in 0 or 5.
6	if it is divisible by both 2 and 3.
8	if the last three digits make a number that is divisible by 8.
9	if the sum of its digits is divisible by 9.
10	if it ends in 0.

The most commonly used divisibility tests are those for 2, 3, 5, and 10.

Divisibility by 2

A number is divisible by **2** if the number ends in 0, 2, 4, 6, or 8.

EXAMPLE 11 Testing for Divisibility by 2

Which numbers are divisible by 2?

(a) 986

$\uparrow$
————— Ends in 6

Because the number ends in 6, which is in the list $(0, 2, 4, 6,$ or $8)$, the number **986 is divisible by 2**.

(b) 3255 is *not* **divisible by 2.**

$\uparrow$
————— Ends in 5, and *not* in 0, 2, 4, 6, or 8

———— **Work Problem 11 at the Side.** ▶

Divisibility by 3

A number is divisible by **3** if the sum of its digits is divisible by **3.**

EXAMPLE 12 Testing for Divisibility by 3

Which numbers are divisible by 3?

(a) 4251

Add the digits.

4 2 5 1

$4 + 2 + 5 + 1 = 12$

> Is 12 divisible by 3? Yes. So the *original* number of 4251 is *also* divisible by 3.

Because 12 is divisible by 3, the number **4251 is divisible by 3**.

(b) 29,806

Add the digits.

2 9 8 0 6

$2 + 9 + 8 + 0 + 6 = 25$

> Is 25 divisible by 3? No. So the original number of 29,806 is *not* divisible by 3.

Because 25 is *not* divisible by 3, the number **29,806 is** *not* **divisible by 3**.

❗ CAUTION

Be careful when testing for divisibility by *adding the digits*. This method works only when testing for divisibility by 3 or by 9.

———— **Work Problem 12 at the Side.** ▶

---

11 Which numbers are divisible by 2?

(a) 612

(b) 315

(c) 2714

(d) 36,000

12 Which numbers are divisible by 3?

(a) 836

(b) 7005

(c) 242,913

(d) 102,484

Answers

11. (a), (c), (d)
12. (b) and (c)

13 Which numbers are divisible by 5?

(a) 160

(b) 635

(c) 3381

(d) 108,605

14 Which numbers are divisible by 10?

(a) 290

(b) 218

(c) 2020

(d) 11,670

Divisibility by 5 and by 10

A number is divisible by **5** if it ends in 0 or 5.

A number is divisible by **10** if it ends in 0.

EXAMPLE 13 **Determining Divisibility by 5**

Which numbers are divisible by 5?

(a) 12,900   ends in 0, so **it is divisible by 5**.

(b) 4325   ends in 5, so **it is divisible by 5**.

(c) 592   ends in 2, so **it is *not* divisible by 5**.

◄ **Work Problem 13 at the Side.**

EXAMPLE 14 **Determining Divisibility by 10**

Which numbers are divisible by 10?

(a) 700 and 9140   end in 0, so **both numbers are divisible by 10**.

(b) 355 and 18,043   do *not* end in 0, so **these numbers are *not* divisible by 10**.

◄ **Work Problem 14 at the Side.**

Answers

13. (a), (b), (d)
14. (a), (c), (d)

R.4 Exercises

FOR EXTRA HELP Go to MyMathLab *for worked-out, step-by-step solutions to exercises enclosed in a* square ▢ *and video solutions to* ▶ *exercises.*

Divide. Then rewrite each division using two other division symbols.
See Examples 1, 3, 5, and 6.

1. $\dfrac{12}{12}$

2. $\dfrac{9}{0}$

3. $24 \div 0$

4. $4 \div 4$

5. $\dfrac{0}{4}$

6. $0 \div 8$

7. $0 \div 12$

8. $\dfrac{0}{7}$

9. $0\overline{)21}$

10. $2 \div 0$

Find each quotient using short division. **See Examples 7–9.** *Also, in Exercises 11–14, identify the dividend, the divisor, and the quotient.* **See Example 2.**

11. $4\overline{)84}$

12. $2\overline{)66}$

13. $9\overline{)324}$

14. $8\overline{)176}$

15. $6\overline{)9125}$

16. $9\overline{)8371}$

17. $6\overline{)1854}$

18. $8\overline{)856}$

19. $4024 \div 4$

20. $16,024 \div 8$

21. $15,019 \div 3$

22. $32,013 \div 8$

23. $\dfrac{26,684}{4}$

24. $\dfrac{16,398}{9}$

25. $\dfrac{74,751}{6}$

26. $\dfrac{72,543}{5}$

27. $\dfrac{71,776}{7}$

28. $\dfrac{77,621}{3}$

29. $\dfrac{128,645}{7}$

30. $\dfrac{172,255}{4}$

Check each quotient. If a quotient is incorrect, find the correct quotient. **See Example 10.**

31. $7\overline{)4692}$ 67 **R**2

32. $9\overline{)5974}$ 663 **R**5

33. $6\overline{)21,409}$ 3 568 **R**5

34. $4\overline{)103,516}$ 25,879

35. $6\overline{)18,023}$ 3 003 **R**5

36. $8\overline{)33,664}$ 4 208

37. $6\overline{)69,140}$ 11,523 **R**2

38. $3\overline{)82,510}$ 2 753 **R**1

Solve each application problem.

39. Kaci Salmon, a supervisor at Albany Electric, earns $184 for an 8-hour shift. The workers she supervises each earn $112 or $152 for an 8-hour shift. What are the hourly wages for Kaci and for her workers?

40. In a hospital weight loss program, Patient A lost 72 pounds in six months, Patient B lost 81 pounds in nine months, and Patient C lost 91 pounds in seven months. On average, how much did each patient lose each month?

41. Rosita's Bakery bought six identical delivery vans for a total of $124,320. Find the cost of each van.

42. Ted Slauson, the local coordinator of Books for Africa, has collected 2628 donated textbooks. If his group sends an equal number of books to four schools, how many books will each school receive?

A college theater group is looking at two different budgets to cover the cost of props, costumes, and programs for the spring play. They are not sure whether to charge $5, $7, or $9 per ticket. The play's director organized the information in the table at the right. Use the table for Exercises 43–46.

Budget Amount	Number of $5 Tickets	Number of $7 Tickets	Number of $9 Tickets
$1890			
$2205			

43. For the $1890 budget, find the number of tickets that would need to be sold at $5 each; at $7 each; at $9 each. Write your answers in the table above.

44. For the $2205 budget, find the number of tickets that would need to be sold at $5 each; at $7 each; at $9 each. Write your answers in the table above.

45. If the director thinks that 300 tickets can be sold, which ticket price ($5, $7, or $9) is the lowest price that will cover the $1890 budget? How much extra money will there be?

46. If the director thinks that 300 tickets can be sold, which ticket price will cover the $2205 budget? How much money will be left over?

47. Circle the numbers that are
   (a) divisible by 2
        358   2047   190   85
   (b) divisible by 3
        736   10,404   5603   78
   (c) divisible by 5.
        53   13,740   985   5506

48. Circle the numbers that are
   (a) divisible by 2
        443   3500   256   74
   (b) divisible by 3
        26,001   9316   92   840
   (c) divisible by 5.
        5051   95   30,652   710

*Put a ✓ mark in the blank if the number at the left is divisible by the number at the top of each column. Use the divisibility tests from **Examples 11–14**.*

	2	3	5	10
49. 30	——	——	——	——
51. 184	——	——	——	——
53. 445	——	——	——	——
55. 903	——	——	——	——
57. 5166	——	——	——	——
59. 21,763	——	——	——	——

	2	3	5	10
50. 25	——	——	——	——
52. 192	——	——	——	——
54. 897	——	——	——	——
56. 500	——	——	——	——
58. 8302	——	——	——	——
60. 32,472	——	——	——	——

R.5 | Long Division

Long division is used to divide by a number with more than one digit.

OBJECTIVE ▶ ❶ Use long division. In long division, estimate the various numbers by using a *trial divisor* to get a *trial quotient*.

EXAMPLE 1   Using a Trial Divisor and a Trial Quotient

Divide: $42\overline{)3066}$

Because 42 is closer to 40 than to 50, use 4 as a trial divisor.

$$42$$
$$\uparrow$$
$$\text{Trial divisor is 4}$$

Try to divide the first digit of the dividend by 4. Because 3 cannot be divided by 4, use the first *two* digits, 30.

$$\frac{30}{4} = \mathbf{7} \text{ with remainder 2}$$
$$\downarrow$$
$$\mathbf{7} \leftarrow \text{Trial quotient is 7}$$
$$42\overline{)3066}$$

Be careful to write the 7 over the 6, because $\frac{306}{42}$ is about 7

Multiply 7 times 42 to get 294; then subtract 294 from 306.

$$\begin{array}{r} 7 \\ 42\overline{)3066} \\ 294 \leftarrow 7 \times 42 = 294 \\ \hline 12 \leftarrow 306 - 294 = 12 \end{array}$$

Bring down the 6 at the right.

$$\begin{array}{r} 7 \\ 42\overline{)3066} \\ 294\downarrow \\ \hline 126 \leftarrow 6 \text{ brought down} \end{array}$$

Use the trial divisor, 4.

$$\begin{array}{r} 73 \leftarrow \\ 42\overline{)3066} \\ 294 \\ \hline 126 \leftarrow \text{First two digits of 126} \rightarrow \frac{12}{4} = 3 \\ 126 \leftarrow 3 \times 42 = 126 \\ \hline 0 \end{array}$$

The quotient is 73

Check the quotient by multiplying 42 and 73. The product should be 3066, which matches the original dividend.

❗ CAUTION

The first digit in the answer in long division must be placed in the proper position over the dividend.

— **Work Problem ❶ at the Side.** ▶

❶ Divide.

(a) $64\overline{)4608}$ *Hint:* 64 is closer to 60 than 70, so use 6 as the trial divisor.

(b) $32\overline{)1792}$ Use _____ as the trial divisor.

(c) $51\overline{)2295}$ Use _____ as the trial divisor.

(d) $\dfrac{6391}{83}$ Use _____ as the trial divisor.

Answers

1. **(a)** 72 **(b)** Use 3; 56 **(c)** Use 5; 45
   (d) Use 8; 77

2 Divide.

(a) $56\overline{)2352}$

(b) $38\overline{)1599}$

(c) $65\overline{)5416}$

(d) $89\overline{)6649}$

EXAMPLE 2 **Dividing to Find a Trial Quotient**

Divide: $58\overline{)2730}$

Use 6 as a trial divisor, because 58 is closer to 60 than to 50.

First two digits → $\dfrac{27}{6} = \mathbf{4}$ with remainder 3
of dividend

Be careful to write the 4 over the 3.

$$
\begin{array}{r}
4 \;\leftarrow \text{Trial quotient is 4} \\
58\overline{)2730} \\
232 \;\leftarrow 4 \times 58 = 232 \\
\hline
41 \;\leftarrow 273 - 232 = 41 \text{ (smaller than 58,} \\
\text{the divisor)}
\end{array}
$$

Bring down the 0.

$$
\begin{array}{r}
4 \\
58\overline{)2730} \\
232\downarrow \\
\hline
410 \;\leftarrow 0 \text{ brought down}
\end{array}
$$

$$
\begin{array}{r}
46 \\
58\overline{)2730} \\
232 \\
\hline
410 \\
348 \;\leftarrow 6 \times 58 = 348 \\
\hline
62 \;\leftarrow \textbf{Greater than 58}
\end{array}
$$

→First two digits of 410

$\dfrac{41}{6} = \mathbf{6}$ with remainder 5

The remainder, **62**, is *greater than the divisor*, 58, so **7** should be used in the quotient instead of **6**.

$$
\begin{array}{r}
\mathbf{47 \ R4} \;\leftarrow \\
58\overline{)2730} \\
232 \\
\hline
410 \\
406 \;\leftarrow 7 \times 58 = 406 \\
\hline
4 \;\leftarrow 410 - 406 = 4
\end{array}
$$

The quotient is 47 R4

◀ **Work Problem** **2** **at the Side.**

Sometimes it is necessary to write a 0 in the quotient.

EXAMPLE 3 **Writing Zeros in the Quotient**

Divide: $42\overline{)8734}$

Start as in **Example 2** above.

$$
\begin{array}{r}
2 \\
42\overline{)8734} \\
84 \;\leftarrow 2 \times 42 = 84 \\
\hline
3 \;\leftarrow 87 - 84 = 3
\end{array}
$$

Bring down the 3.

$$
\begin{array}{r}
2 \\
42\overline{)8734} \\
84\downarrow \\
\hline
33 \;\leftarrow 3 \text{ brought down}
\end{array}
$$

Answers

2. (a) 42 (b) 42 **R3** (c) 83 **R21**
(d) 74 **R63**

Continued on Next Page

Since 33 *cannot* be divided by 42, write a 0 in the quotient as a placeholder.

$$\begin{array}{r} 2\underline{0} \\ 42\overline{)8734} \\ 84 \\ \hline 33 \end{array}$$ Write a 0 in the quotient.

Bring down the final digit, the 4.

$$\begin{array}{r} 20 \\ 42\overline{)8734} \\ 84\downarrow \\ \hline 334 \end{array}$$ ← 4 brought down

Complete the problem.

$$\begin{array}{r} 207\ \textbf{R}40 \\ 42\overline{)8734} \\ 84 \\ \hline 334 \\ 294 \\ \hline 40 \end{array}$$
← 7 × 42 = 294
← 334 − 294 = 40

The quotient is 207 R40

> **❶ CAUTION**
>
> There must be a digit in the quotient (answer) above *every* digit in the dividend *once the answer has begun*. Notice that in **Example 3** above, a 0 was used to assure an answer digit above every digit in the dividend.

───── **Work Problem ❸ at the Side.** ▶

OBJECTIVE ❷ Divide by multiples of 10. When the divisor and dividend both contain zeros at the far right, recall that these numbers are multiples of 10. There is a short way to divide multiples of 10. Look at the following examples.

$$26{,}000 \div 1 = 26{,}000$$

$$26{,}000 \div 10 = 2600$$

$$26{,}000 \div 100 = 260$$

$$26{,}000 \div 1000 = 26$$

Do you see a pattern in these divisions using multiples of 10? These examples suggest the following rule.

> **Dividing by Multiples of 10**
>
> When a whole number has zeros at the far right, divide the number by 10, by 100, or by 1000 by dropping one, two, or three zeros from the whole number.

EXAMPLE 4   Dividing by Multiples of 10

Divide. ─── 0 in divisor

(a) $60 \div 1\underline{0} = \textbf{6}$

─── Drop 0 from dividend to get the quotient.

───── **Continued on Next Page**

❸ Divide.

(a) $24\overline{)3127}$

(b) $52\overline{)10{,}660}$

(c) $39\overline{)15{,}933}$

(d) $78\overline{)23{,}462}$

Answers

3. (a) 130 **R**7   **(b)** 205   **(c)** 408 **R**21
   (d) 300 **R**62

④ Divide by dropping zeros.

(a) $50 \div 10$

(b) $1800 \div 100$

(c) $305,000 \div 1000$

(d) $67,000 \div 100$

⑤ Drop zeros and then divide.

(a) $60\overline{)7200}$

(b) $130\overline{)131,040}$

(c) $2600\overline{)195,000}$

⑥ Check each division. If the quotient is incorrect, find the correct quotient.

(a)
$$18\overline{)774} \quad \begin{array}{r} 43 \\ \hline 72 \\ \hline 54 \\ 54 \\ \hline 0 \end{array}$$

(b)
$$426\overline{)19,170} \quad \begin{array}{r} 42 \textbf{ R}178 \\ \hline 1\,704 \\ \hline 1\,130 \\ 952 \\ \hline 178 \end{array}$$

Answers

4. (a) 5 (b) 18 (c) 305 (d) 670

5. (a) 120 (b) 1008 (c) 75

6. (a) correct (b) incorrect; should be 45

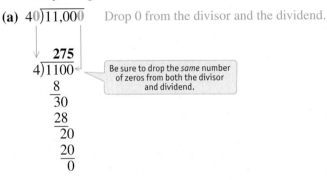

(b) $35\underset{\uparrow}{00} \div 1\underset{\downarrow\downarrow}{00} = \textbf{35}$ ─── 00 in divisor

Drop 00 from dividend to get the quotient.

(c) $91\underset{\uparrow}{5,000} \div 1\underset{\downarrow\downarrow\downarrow}{000} = \textbf{915}$ ─── 000 in divisor

Drop 000 from dividend to get the quotient.

◀ **Work Problem** ④ **at the Side.**

EXAMPLE 5   **Dividing by Multiples of 10**

Divide by using the shortcut of dropping zeros.

(a) $40\overline{)11,000}$ Drop 0 from the divisor and the dividend.

$$4\overline{)1100} \quad \begin{array}{r} 275 \\ \hline 8 \\ \hline 30 \\ 28 \\ \hline 20 \\ 20 \\ \hline 0 \end{array}$$

Be sure to drop the *same* number of zeros from both the divisor and dividend.

(b) $3500\overline{)31,500}$ Drop two zeros from the divisor and the dividend.

$$35\overline{)315} \quad \begin{array}{r} 9 \\ \hline 315 \\ \hline 0 \end{array}$$

◀ **Work Problem** ⑤ **at the Side.**

OBJECTIVE ▶ ③ **Use multiplication to check quotients.** Quotients in long division can be checked just as quotients in short division were checked. Multiply the quotient and divisor, then add any remainder. The result should match the original dividend.

EXAMPLE 6   **Checking Division**

Check the quotient. If incorrect, find the correct quotient.

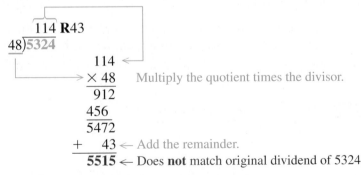

$$48\overline{)5324} \quad 114 \textbf{ R}43$$

$$\begin{array}{r} 114 \\ \times\ 48 \\ \hline 912 \\ 456 \\ \hline 5472 \\ +\quad 43 \\ \hline \textbf{5515} \end{array}$$

Multiply the quotient times the divisor.

← Add the remainder.

← Does **not** match original dividend of 5324

The quotient does *not* check. Rework the original problem to get **110 R44**.

◀ **Work Problem** ⑥ **at the Side.**

R.5 Exercises

FOR EXTRA HELP

Go to MyMathLab *for worked-out, step-by-step solutions to exercises enclosed in a square* and *video solutions to* ▶ *exercises.*

*First, indicate what number to use as the trial divisor. Then use the trial divisor to find the first digit in the quotient and write it in the correct position. Do **not** finish the division.*

1. $24\overline{)768}$

Use _____ as the trial divisor.

2. $35\overline{)805}$

Use _____ as the trial divisor.

3. $18\overline{)4500}$

Use _____ as the trial divisor.

4. $28\overline{)3500}$

Use _____ as the trial divisor.

5. $86\overline{)10,327}$

Use _____ as the trial divisor.

6. $51\overline{)24,026}$

Use _____ as the trial divisor.

7. $52\overline{)38,025}$

Use _____ as the trial divisor.

8. $63\overline{)34,400}$

Use _____ as the trial divisor.

9. $77\overline{)249,826}$

Use _____ as the trial divisor.

10. $92\overline{)247,892}$

Use _____ as the trial divisor.

11. $420\overline{)470,800}$

Use _____ as the trial divisor.

12. $190\overline{)901,050}$

Use _____ as the trial divisor.

*Find each quotient by using long division. **See Examples 1–3.***
*In Exercises 21–24, drop zeros before dividing. **See Examples 4 and 5.***

13. $29\overline{)1859}$

14. $58\overline{)2204}$

15. $47\overline{)11,121}$

16. $83\overline{)39,692}$

17. $26\overline{)62,583}$

18. $28\overline{)84,249}$

19. $63\overline{)78,072}$

20. $55\overline{)43,223}$

21. $150\overline{)499,760}$

22. $720\overline{)52,560}$

23. $400\overline{)340,000}$

24. $900\overline{)153,000}$

Check each division. If a quotient is incorrect, find the correct quotient. See Example 6.

25. $56\overline{)5943}$ — 106 **R**17

26. $87\overline{)3254}$ — 37 **R**37

27. $600\overline{)394,800}$ — 658 **R**9

28. $300\overline{)139,100}$ — 463 **R**200

29. $410\overline{)25,420}$ — 62 **R**3

30. $760\overline{)132,600}$ — 174 **R**360

31. $72\overline{)32,465}$ — 450 **R**65

32. $47\overline{)9570}$ — 23 **R**29

Solve each application problem by using addition, subtraction, multiplication, or division.

33. In 1900, the average workweek was 59 hours. Now, it is 35 hours. How many more hours were worked each year in 1900 than today? Assume 50 weeks of work per year. (Data from statista.com)

34. In the United States, 45,000,000 tons of paper are recycled each year. What is the average amount of paper recycled on each of 250 work days in one year?

35. While planning a vacation, a travel agent found hotel prices ranging from $69 per night to $475 per night. What is the difference in cost between the most expensive and least expensive rooms for a five-night stay?

36. Two divorced parents share their child's education costs, which amount to $3718 per year. If one parent pays $1880 each year, find the amount paid by the other parent over five years.

37. Judy Martinez has a 36-month loan. The total debt, including interest, is $11,088. What is her monthly payment?

38. A consultant charged $41,140 for studying the environmental impact of a new shopping mall. If the consultant worked 220 hours, find the rate charged per hour.

39. Clarence Hanks can assemble 42 circuits in one hour. How many circuits can he assemble in a 5-day workweek of 8 hours per day?

40. Tuition at the community college is $151 per credit. There is also a $10 per credit technology fee, a $2 per credit student health fee, and a one-time application fee of $20. How much will Stephanie, a new student, pay to register for 14 credits?

41. The elementary school PTA brought in $7588 from a fund-raising project. Expenses of $838 were paid first. The balance of the money was divided evenly among 18 classrooms. How much did each classroom receive?

42. Feather Farms Egg Ranch collected 3545 eggs in the morning and 2575 eggs in the afternoon. If the eggs are packed in flats containing 30 eggs each, find the number of flats needed for packing.

Appendix: Inductive and Deductive Reasoning

Inductive and Deductive Reasoning

OBJECTIVES

1 Use inductive reasoning to analyze patterns.
2 Use deductive reasoning to analyze arguments.
3 Use deductive reasoning to solve problems.

OBJECTIVE ▶ **1 Use inductive reasoning to analyze patterns.** In many scientific experiments, conclusions are drawn from specific outcomes. After many repetitions and similar outcomes, the findings are generalized into statements that appear to be true. When general conclusions are drawn from specific observations, we are using a type of reasoning called **inductive reasoning.** The next examples illustrate this type of reasoning.

EXAMPLE 1 Using Inductive Reasoning

Find the next number in the sequence 3, 7, 11, 15, ...

To discover a pattern, calculate the difference between each pair of successive numbers.

$$7 - 3 = 4$$
$$11 - 7 = 4$$
$$15 - 11 = 4$$

Notice that the difference is always 4. Each number is 4 greater than the previous one. Thus, the next number in the pattern is $15 + 4$, or 19.

─── **Work Problem 1 at the Side.** ▶

1 Find the next number in the sequence 2, 8, 14, 20, ... Describe the pattern.

EXAMPLE 2 Using Inductive Reasoning

Find the next number in this sequence.

$$7, 11, 8, 12, 9, 13, \ldots$$

The pattern in this example involves addition and subtraction.

$$7 + 4 = 11$$
$$11 - 3 = 8$$
$$8 + 4 = 12$$
$$12 - 3 = 9$$
$$9 + 4 = 13$$

To get the second number, we add 4 to the first number. To get the third number, we subtract 3 from the second number. To obtain subsequent numbers, we continue the pattern. The next number is $13 - 3 = 10$.

─── **Work Problem 2 at the Side.** ▶

2 Find the next number in the sequence 6, 11, 7, 12, 8, 13, ... Describe the pattern.

Answers
1. 26; add 6 each time.
2. 9; add 5, subtract 4.

3 Find the next number in the sequence 2, 6, 18, 54, ... Describe the pattern.

> **EXAMPLE 3** **Using Inductive Reasoning**
>
> Find the next number in the sequence 1, 2, 4, 8, 16, ...
>
> Each number after the first is obtained by multiplying the previous number by 2. So the next number would be 16 • **2** = 32.

──────────── ◀ **Work Problem ❸ at the Side.**

> **EXAMPLE 4** **Using Inductive Reasoning**
>
> **(a)** Find the next geometric shape in this sequence.
>
>
>
> The figures alternate between a blue circle and a red triangle. Also, the number of dots increases by 1 in each subsequent figure. Thus, the next figure should be a blue circle with five dots inside it.
>
>
>
> **(b)** Find the next geometric shape in this sequence.
>
>
>
> The first two shapes consist of vertical lines with horizontal lines at the bottom extending first *left* and then *right*. The third shape is a vertical line with a horizontal line at the top extending to the *left*. Therefore, the next shape should be a vertical line with a horizontal line at the top extending to the *right*.

4 Find the next shape in this sequence.

──────────── ◀ **Work Problem ❹ at the Side.**

OBJECTIVE ▶ ❷ Use deductive reasoning to analyze arguments. In the previous discussion, specific cases were used to find patterns and predict the next event. There is another type of reasoning called **deductive reasoning,** which moves from general cases to specific conclusions.

> **EXAMPLE 5** **Using Deductive Reasoning**
>
> Does the conclusion follow from the premises in this argument?
>
> All Hondas are automobiles. ← Premise
>
> All automobiles have horns. ← Premise
> ──────────────────────────
> ∴ All Hondas have horns. ← Conclusion
>
> In this example, the first two statements are called *premises* and the third statement (below the line) is called a *conclusion*. The symbol ∴ is a mathematical symbol meaning "**therefore.**" The entire set of statements is called an *argument*.

Answers

3. 162; multiply by 3.

4.

──────────── **Continued on Next Page**

The focus of deductive reasoning is to determine whether the conclusion follows (is valid) from the premises. A set of circles called **Euler circles** is used to analyze the argument.

In **Example 5**, the statement "All Hondas are automobiles" can be represented by two circles, one for Hondas and one for automobiles. Note that the circle representing Hondas is totally inside the circle representing automobiles because the first premise states that *all* Hondas are automobiles.

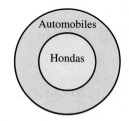

Now, a circle is added to represent the second statement, vehicles with horns. This circle must completely surround the circle representing automobiles because the second premise states that *all* automobiles have horns.

To analyze the conclusion, notice that the circle representing Hondas is *completely* inside the circle representing vehicles with horns. Therefore, it must follow that all Hondas have horns. **The conclusion is valid.**

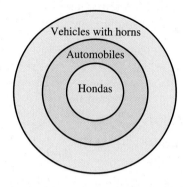

— **Work Problem 5 at the Side.** ▶

EXAMPLE 6 | **Using Deductive Reasoning**

Does the conclusion follow from the premises in this argument?

All tables are round.

All glasses are round.

∴   All glasses are tables.

Use Euler circles. Draw a circle representing tables *inside* a circle representing round objects, because the first premise states that *all* tables are round.

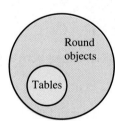

The second statement requires that a circle representing glasses must now be drawn inside the circle representing round objects, but not necessarily inside the circle representing tables. Therefore, the conclusion does ***not*** follow from the premises. This means that ***the conclusion is invalid.***

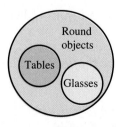

— **Work Problem 6 at the Side.** ▶

---

5 Does the conclusion follow from the premises in the following argument?

All cars have four wheels.
All Fords are cars.
———————————————
∴ All Fords have four wheels.

6 Does each conclusion follow from the premises?

(a) All animals are wild.
All cats are animals.
———————————————
∴ All cats are wild.

(b) All students use math.
All adults use math.
———————————————
∴ All adults are students.

Answers

5. The conclusion follows from the premises; it is valid.

6. (a) The conclusion follows from the premises; it is valid.

(b) The conclusion does *not* follow from the premises; it is invalid.

7 In a college class of 100 students, 35 take both math and history, 50 take history, and 40 take math. How many take neither math nor history? Draw a Venn diagram.

8 A Chevy, BMW, Cadillac, and Ford are parked side by side. The known facts are:

(a) The Ford is on the right end.

(b) The BMW is beside the Cadillac.

(c) The Chevy is next to both the Ford and the Cadillac.

Which car is parked on the left end?

OBJECTIVE ▶ 3 Use deductive reasoning to solve problems. Another type of deductive reasoning problem occurs when a set of facts is given in a problem and a conclusion must be drawn using these facts.

EXAMPLE 7   Using Deductive Reasoning

There were 25 students enrolled in a ceramics class. During the class, 10 of the students made a bowl and 8 students made a birdbath. Three students made both a bowl and a birdbath. How many students did not make either a bowl or a birdbath?

This type of problem is best solved by organizing the data using a drawing called a **Venn diagram.** Two overlapping circles are drawn, with each circle representing one item made by the students, as shown below.

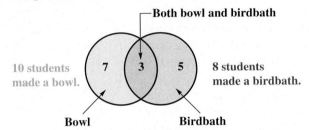

In the region where the circles overlap, write the number of students who made *both* items, namely, 3. In the remaining portion of the birdbath circle, write the number 5, which when added to 3 will give the total number of students who made a birdbath, namely, 8. In a similar manner, write 7 in the remaining portion of the bowl circle, since 7 + 3 = 10, the total number of students who made a bowl. The total of all three numbers written in the circles is 15. Since there were 25 students in the class, this means 25 − 15 or 10 students did not make either a birdbath or a bowl.

◀ **Work Problem 7 at the Side.**

EXAMPLE 8   Using Deductive Reasoning

Four cars in a race finish first, second, third, and fourth. The following facts are known.

(a) Car A beat Car C.

(b) Car D finished between Cars C and B.

(c) Car C beat Car B.

In which order did the cars finish?

To solve this type of problem, it is helpful to use a line diagram.

1. *Write A before C,* because Car A beat Car C (fact **a**).

A   C

2. *Write B after C,* because Car C beat Car B (fact **c**).

A   C   B

3. *Write D between C and B,* because Car D finished between Car C and Car B (fact **b**).

The correct order of finish is shown below.

A   C   D   B

◀ **Work Problem 8 at the Side.**

Answers

7.

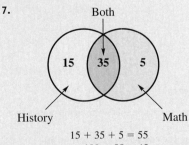

15 + 35 + 5 = 55
100 − 55 = 45

45 students take neither math nor history.

8. BMW

Appendix Exercises

Find the next number in each sequence. Describe the pattern in each sequence.
See Examples 1–3.

1. 2, 9, 16, 23, 30, …

2. 5, 8, 11, 14, 17, …

3. 0, 10, 8, 18, 16, …

4. 3, 9, 7, 13, 11, …

5. 1, 2, 4, 8, …

6. 1, 4, 16, 64, …

7. 1, 3, 9, 27, 81, …

8. 3, 6, 12, 24, 48, …

9. 1, 4, 9, 16, 25, …

10. 6, 7, 9, 12, 16, …

Find the next shape in each sequence. ***See Example 4.***

11.

12.

13.

14.

For each argument, draw Euler circles and then state whether or not the conclusion follows from the premises. **See Examples 5 and 6.**

15. All animals are wild.
 All lions are animals.
 ∴ All lions are wild.

16. All students are hard workers.
 All business majors are students.
 ∴ All business majors are hard workers.

17. All teachers are serious.
 All mathematicians are serious.
 ∴ All mathematicians are teachers.

18. All boys ride bikes.
 All Americans ride bikes.
 ∴ All Americans are boys.

Solve each application problem. **See Examples 7 and 8.**

19. In a given 30-day period, a husband watched television 20 days and his wife watched television 25 days. If they watched television together 18 days, how many days did neither watch television? Draw a Venn diagram.

20. In a class of 40 students, 21 students take both calculus and physics. If 30 students take calculus and 25 students take physics, how many do not take either calculus or physics? Draw a Venn diagram.

21. Tom, Dick, Mary, and Joan all work for the same company. One is a secretary, one is a computer operator, one is a receptionist, and one is a mail clerk.

 (a) Tom and Joan eat dinner with the computer operator.

 (b) Dick and Mary carpool with the secretary.

 (c) Mary works on the same floor as the computer operator and the mail clerk.

 Who is the computer operator?

22. Four cars—a Ford, a Buick, a Mercedes, and an Audi—are parked in a garage in four spaces.

 (a) The Ford is in the last space.

 (b) The Buick and Mercedes are next to each other.

 (c) The Audi is next to the Ford but not next to the Buick.

 Which car is in the first space?

Answers to Selected Exercises

PRETEST Whole Numbers Computation

Adding Whole Numbers
1. 390 **2.** 13,166 **3.** 3053 **4.** 117,510 **5.** 688,226

Subtracting Whole Numbers
1. 350 **2.** 629 **3.** 24,894 **4.** 3909 **5.** 591,387

Multiplying Whole Numbers
1. 0 **2.** 26,887 **3.** 1,560,000 **4.** 1846 **5.** 17,232 **6.** 519,477

Dividing Whole Numbers
1. 23 **2.** undefined **3.** 6259 **4.** 807 **R**6 **5.** 34 **6.** 50
7. 60 **R**20 **8.** 539 **R**62

CHAPTER 1 Introduction to Algebra: Integers

All answers are shown for Concept Checks, Writing exercises, What Went Wrong exercises, Relating Concepts exercises, Summary Exercises, Review Exercises, and Chapter Tests. For other types of exercises, answers are shown only for the odd-numbered exercises.

SECTION 1.1
1. False; we can also use the digit 0. **2.** True **3.** True **4.** False; the left-most 7 has a value of 7 ten-thousands; the right-most 7 has a value of 7 tens. **5.** 15; 0; 83,001 **7.** 7; 362,049 **9.** hundreds **11.** hundred-thousands **13.** ten-millions **15.** hundred-billions **17.** ten-trillions, hundred-billions, millions, hundred-thousands, ones **18.** trillions, ten-billions, hundred millions, ten-thousands, hundreds **19.** eight thousand, four hundred twenty-one **21.** forty-six thousand, two hundred five **23.** three million, sixty-four thousand, eight hundred one **25.** eight hundred forty million, one hundred eleven thousand, three **27.** fifty-one billion, six million, eight hundred eighty-eight thousand, three hundred twenty-one **29.** three trillion, seven hundred twelve million **31.** 46,805 **33.** 5,600,082 **35.** 271,900,000 **37.** 12,417,625,310 **39.** 600,000,071,000,400 **41.** nine thousand, six hundred forty-one **43.** one hundred thirty million, one hundred thousand **45.** seventy-nine billion, two hundred million **47.** 74,059,000 **49.** three million, five thousand **51.** 15,000,000; 5,475,000,000 **53.** largest: 97,651,100; ninety-seven million, six hundred fifty-one thousand, one hundred; smallest: 10,015,679; ten million, fifteen thousand, six hundred seventy-nine **54.** Answers will vary. **55.** sixty-fours; thirty-twos; sixteens; eights **(a)** 101 **(b)** 1010 **(c)** 1111 **56. (a)** Answers will vary but should mention that the location or place in which a digit is written gives it a different value. **(b)** 8 = VIII; 38 = XXXVIII; 275 = CCLXXV; 3322 = MMMCCCXXII **(c)** The Roman system is *not* a place value system because no matter what place it's in, M = 1000, C = 100, etc. One disadvantage is that it takes much more space to write many large numbers; another is that there is no symbol for zero.

SECTION 1.2
1. +29,029 feet or 29,029 feet **2.** −5353 feet **3.** −135.8 degrees
4. +98.6 degrees or 98.6 degrees **5.** −18 yards **6.** +25 yards or 25 yards
7. +$100 or $100 **8.** −$37 **9.** $-6\frac{1}{2}$ pounds
10. $+2\frac{1}{2}$ ounces or $2\frac{1}{2}$ ounces
11.
$$\xleftarrow{\;\;\bullet\;\;+\;\;\bullet\;\;+\;\;+\;\;+\;\;+\;\;\bullet\;\;+\;\;+\;\;\rightarrow}$$
$$-5\;-4\;-3\;-2\;-1\;\;0\;\;1\;\;2\;\;3\;\;4\;\;5$$
13.
$$\xleftarrow{\;\;+\;\;+\;\;+\;\;\bullet\;\;+\;\;+\;\;+\;\;+\;\;\bullet\;\;\bullet\;\;\rightarrow}$$
$$-5\;-4\;-3\;-2\;-1\;\;0\;\;1\;\;2\;\;3\;\;4\;\;5$$
15. (a) Zero is less than five, or, zero is less than positive five. **(b)** Negative ten is greater than negative seventeen. **16. (a)** Three is greater than negative seven, or, positive three is greater than negative seven. **(b)** Twelve is less than twenty-two, or, positive twelve is less than positive twenty-two. **17.** > **19.** < **21.** < **23.** > **25.** < **27.** > **29.** > **31.** < **33.** 15 **35.** 3 **37.** 0 **39.** 200 **41.** 75 **43.** 8042
45.
$$\begin{array}{cccc} A & C & D & B \\ \downarrow & \downarrow & \downarrow & \downarrow \end{array}$$
$$-3\;\;-2\;\;-1\;\;\;0\;\;\;1$$
46. −3, −2, −1, 0 **47.** A: osteoporosis; B: normal; C: at risk for osteoporosis; D: at risk for osteoporosis
48. (a) The patient would think the interpretation was "normal" and wouldn't get treatment. **(b)** Patient B's score of 0; zero is neither positive nor negative.

SECTION 1.3
1. 3
$$-7\;-6\;-5\;-4\;-3\;-2\;-1\;\;0\;\;1\;\;2\;\;3$$
3. −7
$$-7\;-6\;-5\;-4\;-3\;-2\;-1\;\;0\;\;1\;\;2\;\;3$$
5. −1
$$-7\;-6\;-5\;-4\;-3\;-2\;-1\;\;0\;\;1\;\;2\;\;3$$
7. (a) −10 **(b)** 10 **9. (a)** 12 **(b)** −12 **11. (a)** −50 **(b)** 50
13. (a) 158 **(b)** −158 **15.** The absolute values are the same in each pair of answers, so the only difference in the sums is the common sign.
16. Add the absolute values and use the common sign as the sign of the sum. **17. (a)** 2 **(b)** −2 **19. (a)** −7 **(b)** 7 **21. (a)** −5 **(b)** 5
23. (a) 150 **(b)** −150 **25.** Each pair of answers differs only in the sign of the answer. This occurs because the signs of the addends are reversed.
26. Subtract the lesser absolute value from the greater absolute value. Use the sign of the number with the greater absolute value as the sign of the sum.
27. −3 **29.** 7 **31.** −7 **33.** 1 **35.** −8 **37.** −20 **39.** −17

41. −22 **43.** −19 **45.** 6 **47.** −5 **49.** 0 **51.** 5 **53.** −32

55. 13 + (−17) = −4 yards **57.** −62 + 50 = −$12

59. −88 + 35 = −$53 **61.** Jeff: −20 + 75 + (−55) = 0 points; Terry:
42 + (−15) + 20 = 47 points **63.** −2 **65.** 2 **67.** −5 + (−18); −23

69. 15 + (−4); 11 **71.** 6 + (−14 + 14); 6 + 0 = 6

73. (14 + 6) + (−7); 20 + (−7) = 13 **75.** Answers will vary. Some
possibilities are: −6 + 0 = −6; 10 + 0 = 10; 0 + 3 = 3 **76.** Answers
will vary. Some possibilities are: (1 + 2) + 3 = 1 + (2 + 3)
Both sums are 6; (−5 + 0) + 4 = −5 + (0 + 4) Both sums
are −1; (−15 + 6) + 9 = −15 + (6 + 9) Both sums are 0.

77. −4116 **79.** 8686 **81.** −96,077

SECTION 1.4

1. −6; 6 + (−6) = 0 **3.** 13; −13 + 13 = 0 **5.** Valerie did not
change the second number, 6, to its opposite, −6. The correct answer is
−6 + (−6) = −12. **6.** Victor changed the first number, −9, to its opposite, but it should be left as it is. The correct answer is −9 + (−5) = −14.

7. 14 **9.** −2 **11.** −12 **13.** −25 **15.** −23 **17.** −3 + (+8) = 5

19. 20 **21.** 11 **23.** −60 **25.** 0 **27.** 0 **29.** −6 **31. (a)** 8 **(b)** −2

(c) 2 **(d)** −8 **33. (a)** −3 **(b)** 11 **(c)** −11 **(d)** 3 **35.** −6

37. −5 **39.** 3 **41.** −10 **43.** 12 **45.** −5 **47. (a)** 21 °F; 30 − 21 = 9

degrees difference **(b)** 0 °F; 15 − 0 = 15 degrees difference **(c)** −17 °F;
5 − (−17) = 22 degrees difference **(d)** −41 °F; −10 − (−41) = 31

degrees difference **49.** −11 **51.** −5 **53.** −10 **55.** Answers on left:
−8; 8. On right: −1; 1 Subtraction is *not* commutative; the absolute value
of the answer is the same, but the sign changes. **56.** Subtracting 0 from
a number does *not* change the number. For example, −5 − 0 = −5. But
subtracting a number from 0 *does* change the number to its opposite. For
example, 0 − (−5) = 5.

SECTION 1.5

1. (a)

3702 — Next digit is 4 or less, so leave
0 as 0 in the tens place.

Tens place

(b)

908,546 — Next digit is 5 or more, so change
8 to 9 in the thousands place.

Thousands place

2. (a)

65,081 — Next digit is 5 or more, so change 0
to 1 in the hundreds place.

Hundreds place

(b)

723,900 — Next digit is 4 or less, so leave
2 as 2 in the ten-thousands place.

Ten-thousands place

3. 630 **5.** −1080 **7.** 7900 **9.** −86,800 **11.** 42,500 **13.** −6000

15. −78,000 **17.** 6000 **19.** 600,000 **21.** −9,000,000

23. 140,000,000 **25.** 20,000,000,000 **27.** 9,000,000,000 **29.** Mateo
forgot to regroup 1 into the hundred-thousands place. The correct answer is
900,000. **30.** Gabriela forgot to change the digits to the right of the millions
place to zeros. The correct answer is 100,000,000. **31.** 30,000 miles

33. −60 degrees **35.** $10,000 **37.** 2,000,000,000 Internet users

39. 700,000 people in Alaska; 40,000,000 people in California

41. *Estimate:* −40 + 90 = 50; *Exact:* −42 + 89 = 47

43. *Estimate:* 20 + (−100) = −80; *Exact:* 16 + (−97) = −81

45. *Estimate:* −300 + (−400) = −700; *Exact:* −273 + (−399) = −672

47. *Estimate:* 3000 + 7000 = 10,000; *Exact:* 3081 + 6826 = 9907

49. *Estimate:* 20 + (−80) = −60; *Exact:* 23 − 81 = 23 + (−81) = −58

51. *Estimate:* −40 + (−40) = −80; *Exact:* −39 − 39 = −39 + (−39) = −78

53. *Estimate:* −100 + 30 + (+70) = 0; *Exact:* −106 + 34 − (−72) =
−106 + 34 + (+72) = 0 **55.** *Estimate:* 80,000 − 50,000 = $30,000;
Exact: 78,650 − 52,882 = $25,768 **57.** *Estimate:* 2000 − 800 −
400 − 400 − 200 − 200 = $0; *Exact:* 2120 − 845 − 425 − 365 − 182
− 240 = $63 **59.** *Estimate:* −100 + 40 + 50 = −10 degrees; *Exact:*
−102 + 37 + 52 = −13 degrees **61.** *Estimate:* 400 + 100 = 500 doors
and windows; *Exact:* 412 + 147 = 559 doors and windows

SECTION 1.6

1. (a) 63 **(b)** 63 **(c)** −63 **(d)** −63 **2. (a)** −54 **(b)** −54 **(c)** 54

(d) 54 **3. (a)** −56 **(b)** −56 **(c)** 56 **(d)** 56 **4. (a)** 48 **(b)** 48

(c) −48 **(d)** −48 **5.** −35 **7.** −45 **9.** −18 **11.** −50 **13.** −40

15. −56 **17.** 32 **19.** 77 **21.** 0 **23.** 133 **25.** 13 **27.** 0 **29.** 48

31. −56 **33.** −160 **35.** 5 **37.** −3 **39.** −1 **41.** 0 **43.** 5 **45.** −4

47. Commutative property: changing the *order* of the factors does not
change the product. Associative property: changing the *grouping* of the
factors does not change the product. Examples will vary. **48.** Angelo
must work with two factors at a time. (−3)(−3) is 9, and 9(−3) is −27.
The correct answer is −27. **49.** 9(−3 + 5) = 9 • (−3) + 9 • 5
Both results are 18. **51.** 25 • 8 = 8 • 25 Both products are 200.

53. −3 • (2 • 5) = (−3 • 2) • 5 Both products are −30.

55. *Estimate:* 300 • 50 = $15,000; *Exact:* 324 • 52 = $16,848

57. *Estimate:* −10,000 • 10 = −$100,000; *Exact:* −9950 • 12 = −$119,400

59. *Estimate:* 200 • 10 = $2000; *Exact:* 182 • 13 = $2366

61. *Estimate:* 20 • 400 = 8000 hours; *Exact:* 24 • 365 = 8760 hours

63. −512 **65.** 0 **67.** −355,299 **69.** $247 **71.** −22 degrees

73. 772 points **75.** Examples will vary. Some possibilities are:
(a) 6 • (−1) = −6; 2 • (−1) = −2; 15 • (−1) = −15 **(b)** −6 • (−1) = 6;
−2 • (−1) = 2; −15 • (−1) = 15 The result of multiplying any nonzero
number times −1 is the number with the opposite sign.

76. The products are 4, −8, 16, −32. The absolute value doubles each time
and the sign changes. The next three products are 64, −128, and 256.

SECTION 1.7

1. (a) 7 **(b)** 7 **(c)** −7 **(d)** −7 **2. (a)** 6 **(b)** 6 **(c)** −6 **(d)** −6

3. (a) −7 **(b)** 7 **(c)** −7 **(d)** 7 **4. (a)** 9 **(b)** −9 **(c)** 9 **(d)** −9

5. (a) 1 **(b)** 35 **(c)** −13 **(d)** 1 **7. (a)** 0 **(b)** undefined

(c) undefined **(d)** 0 **9.** −4 **11.** −3 **13.** 6 **15.** −11

17. undefined **19.** −14 **21.** 10 **23.** 4 **25.** −1 **27.** 0 **29.** 191

31. −499 **33.** 2 **35.** −4 **37.** 40 **39.** −48 **41.** 5 **43.** 0

45. 2 ÷ 1 = 2 but 1 ÷ 2 = 0.5, so division is not commutative.

46. (12 ÷ 6) ÷ 2 = 2 ÷ 2 = 1; 12 ÷ (6 ÷ 2) = 12 ÷ 3 = 4; different
quotients. Division is not associative. **47.** Similar: If the signs match,
the result is positive. If the signs are different, the result is negative.
Different: Multiplication is commutative, division is not. You can
multiply by 0, but dividing by 0 is undefined. **48.** Examples will vary.
The properties are: Any nonzero number divided by itself is 1.
Any number divided by 1 is the number. Division by 0 is undefined.
Zero divided by any other number (except 0) is 0.

49. Thong forgot that 0 *can be* divided by a number. It is dividing a number by zero that is undefined, such as $12 \div 0$. The correct answer for $\frac{0}{12}$ is 0. **50.** Lien added the numbers instead of dividing. The correct answer is -1. **51.** *Estimate:* $-40{,}000 \div 20 = -2000$ feet; *Exact:* $-35{,}836 \div 17 = -2108$ feet **53.** *Estimate:* $-200 + 500 = \$300$; *Exact:* $-238 + 450 = \$212$ **55.** *Estimate:* $400 - 200 = 200$ days; *Exact:* $365 - 165 = 200$ days **57.** *Estimate:* $-700 \cdot 40 = -28{,}000$ feet; *Exact:* $-730 \cdot 37 = -27{,}010$ feet **59.** *Estimate:* $300 \div 5 = 60$ miles; *Exact:* $315 \div 5 = 63$ miles **61.** Average score of 168 **63.** The back shows $(40)(13)$ or 520 grams, which is 10 grams more than the front. **65.** $-\$15$ **67.** 16 hours, plus another 40 minutes **69.** 33 rooms; one room has only 3 people in it. **71.** -10 **73.** undefined **75.** 31.70979198 rounds to 32 years.

SUMMARY EXERCISES Operations with Integers

1. -6 **2.** 0 **3.** -7 **4.** -7 **5.** 63 **6.** -1 **7.** -56 **8.** -22 **9.** 12 **10.** 8 **11.** -13 **12.** 0 **13.** 0 **14.** -17 **15.** -48 **16.** -10 **17.** -50 **18.** undefined **19.** -14 **20.** -6 **21.** 0 **22.** 16 **23.** -30 **24.** undefined **25.** 48 **26.** -19 **27.** 2 **28.** -20 **29.** 0 **30.** 16 **31.** -3 **32.** -36 **33.** 6 **34.** -7 **35.** -5 **36.** -31 **37.** -2 **38.** -32 **39.** -5 **40.** -4 **41.** -9732 **42.** 100 **43.** 4 **44.** -343 **45.** 5 **46.** -6 **47.** -10 **48.** -5 **49. (a)** The quotient is 0. **(b)** The product is 0. **(c)** The quotient is 1. **50.** In the first expression, Asho forgot that dividing by zero cannot be done. In the second expression, Asho did not work with the numbers two at a time. The first expression is undefined; the second expression equals -1.

SECTION 1.8

1. $4 \cdot 4 \cdot 4$; 4 cubed or 4 to the third power **2.** $10 \cdot 10$; 10 squared or 10 to the second power **3.** 2^7; 128; 2 to the seventh power **4.** 3^5; 243; 3 to the fifth power **5.** 5^4; 625; 5 to the fourth power **6.** 2^6; 64; 2 to the sixth power **7.** 7^2; $7 \cdot 7$; 49 **8.** 6^3; $6 \cdot 6 \cdot 6$; 216 **9.** 10^1; 10; 10 **10.** 4^4; $4 \cdot 4 \cdot 4 \cdot 4$; 256 **11. (a)** 10 **(b)** 100 **(c)** 1000 **(d)** 10,000 **13. (a)** 4 **(b)** 16 **(c)** 64 **(d)** 256 **15.** 9,765,625 **17.** 4096 **19.** 4 **21.** 25 **23.** -64 **25.** 81 **27.** -1000 **29.** 1 **31.** 108 **33.** 200 **35.** -750 **37.** -32 **39. (a)** The answers are 4, -8, 16, -32, 64, -128, 256, -512. When a negative number is raised to an even power, the answer is positive; when raised to an odd power, the answer is negative. **(b)** negative; positive **40.** We need rules so everyone will work problems the same way and get the same answers. We probably want to work from left to right because in many languages we read from left to right. **41.** $2(-3)$; -6 **43.** 0 **45.** -39 **47.** 16 **49.** 23 **51.** -43 **53.** 7 **55.** -3 **57.** 0 **59.** -38 **61.** 41 **63.** -2 **65.** 13 **67.** 126 **69.** 8 **71.** $\frac{27}{-3} = -9$ **73.** $\frac{-48}{-4} = 12$ **75.** $\frac{-60}{-1} = 60$ **77.** -4050 **79.** 7 **81.** $\frac{27}{0}$ is undefined.

Chapter 1 REVIEW EXERCISES

1. 86, 0, 35,600 **2.** eight hundred six **3.** three hundred nineteen thousand, twelve **4.** sixty million, three thousand, two hundred **5.** fifteen trillion, seven hundred forty-nine billion, six **6.** 504,100 **7.** 620,080,000 **8.** 99,007,000,356

9.
$$-5\ -4\ -3\ -2\ -1\ \ 0\ \ 1\ \ 2\ \ 3\ \ 4\ \ 5$$

10. $>$ **11.** $<$ **12.** $>$ **13.** $<$ **14.** 5 **15.** 9 **16.** 0 **17.** 125 **18.** -1 **19.** -13 **20.** -3 **21.** 0 **22.** 1 **23.** -24 **24.** -7 **25.** 3 **26.** 0 **27.** -17 **28.** 5; $-5 + 5 = 0$ **29.** -18; $18 + (-18) = 0$ **30.** -7 **31.** 17 **32.** -16 **33.** 13 **34.** 18 **35.** -22 **36.** 0 **37.** -20 **38.** -1 **39.** -3 **40.** 14 **41.** 15 **42.** -16 **43.** 3 **44.** -8 **45.** 0 **46.** 210 **47.** 59,000 **48.** 85,000,000 **49.** -3000 **50.** $-7{,}060{,}000$ **51.** 400,000 **52.** -200 pounds **53.** -1000 feet **54.** 4,000,000,000 Internet users **55.** 10,000,000,000 people **56.** -54 **57.** 56 **58.** -100 **59.** 0 **60.** 24 **61.** 17 **62.** -48 **63.** 125 **64.** -36 **65.** 50 **66.** -72 **67.** 9 **68.** -7 **69.** undefined **70.** 5 **71.** -18 **72.** 0 **73.** 15 **74.** -1 **75.** -5 **76.** 18 **77.** 0 **78.** 156 days and 2 extra hours **79.** 10,000 **80.** 32 **81.** 27 **82.** 16 **83.** -125 **84.** 8 **85.** 324 **86.** -200 **87.** -25 **88.** -2 **89.** 10 **90.** -28 **91.** $\frac{8}{-8} = -1$ **92.** $\frac{11}{0}$ is undefined.

Chapter 1 MIXED REVIEW EXERCISES

1. associative property of addition **2.** commutative property of multiplication **3.** addition property of 0 **4.** multiplication property of 0 **5.** distributive property **6.** associative property of multiplication **7.** *Estimate:* $\$10{,}000 \cdot 200 = \$2{,}000{,}000$; *Exact:* $\$11{,}900 \cdot 192 = \$2{,}284{,}800$ **8.** *Estimate:* $\$200 + \$400 - \$700 = -\$100$; *Exact:* $\$185 + \$428 - \$706 = -\$93$ **9.** *Estimate:* $900 \div 20 = 45$ miles; *Exact:* $880 \div 22 = 40$ miles **10.** *Estimate:* $(\$40 \cdot 20) + (\$90 \cdot 10) = \$1700$; *Exact:* $(\$39 \cdot 19) + (\$85 \cdot 12) = \$1761$ **11.** $-\$700, -\$100, \$700, \$900, \$0, \$700$ **12.** January; April **13.** \$2100 **14.** $-\$1850$

Chapter 1 TEST

1. twenty million, eight thousand, three hundred seven **2.** 30,000,700,005 **3.**
$$-3\ -2\ -1\ \ 0\ \ 1\ \ 2\ \ 3$$

4. $>$; $<$ **5.** 10; 14 **6.** -6 **7.** -5 **8.** 7 **9.** -40 **10.** 10 **11.** 64 **12.** -50 **13.** undefined **14.** -60 **15.** -5 **16.** -45 **17.** 6 **18.** 25 **19.** 0 **20.** 9 **21.** 128 **22.** -2 **23.** 8 **24.** -36 **25.** An exponent shows how many times to use a factor in repeated multiplication. Examples will vary. Some possibilities are $(2)^4 = 2 \cdot 2 \cdot 2 \cdot 2 = 16$ and $(-3)^2 = (-3)(-3) = 9$. **26.** Commutative property: changing the *order* of addends does not change the sum. Associative property: changing the *grouping* of addends does not change the sum. Examples will vary. **27.** 900 **28.** 36,420,000,000 **29.** 350,000 **30.** *Estimate:* $\$200 + \$300 + (-\$500) = \$0$; *Exact:* $\$184 + \$293 + (-\$506) = -\$29$ **31.** *Estimate:* $-1000 \div 20 = -50$ pounds; *Exact:* $-1144 \div 22 = -52$ pounds **32.** *Estimate:* $30(200 - 100) = 3000$ calories; *Exact:* $31(220 - 110) = 3410$ calories **33.** $-6 - (-79) = -6 + 79 = 73$ degrees difference **34.** 27 cartons because 26 cartons would leave 28 books unpacked

CHAPTER 2 Understanding Variables and Solving Equations

All answers are shown for Concept Checks, Writing exercises, What Went Wrong exercises, Relating Concepts exercises, Summary Exercises, Review Exercises, Chapter Tests, and Cumulative Review exercises. For other types of exercises, answers are shown only for the odd-numbered exercises.

SECTION 2.1

1. constant. **2.** d **3.** m is the variable; -3 is the constant. **4.** n is the variable; -4 is the constant. **5.** h is the variable; 5 is the coefficient. **6.** s is the variable; 3 is the coefficient. **7.** c is the variable; 2 is the coefficient; 10 is the constant. **8.** b is the variable; 6 is the coefficient; 1 is the constant. **9.** x and y are variables. **10.** x and y are variables. **11.** g is the variable; -6 is the coefficient; 9 is the constant. **12.** k is the variable; -10 is the coefficient; 15 is the constant. **13.** **(a)** 664 robes. **(b)** $208 + 10$ is 218 robes. **(c)** $95 + 10$ is 105 robes. **15.** **(a)** $3 \cdot 11$ inches is 33 inches. **(b)** $3 \cdot 3$ feet is 9 feet. **17.** **(a)** $3 \cdot 12 - 5$ is 31 brushes. **(b)** $3 \cdot 16 - 5$ is 43 brushes. **19.** **(a)** $\frac{332}{4}$ is 83 points. **(b)** $\frac{637}{7}$ is 91 points. **21.** $12 + 12 + 12 + 12$ is 48, $4 \cdot 12$ is 48; $0 + 0 + 0 + 0$ is 0, $4 \cdot 0$ is 0; $(-5) + (-5) + (-5) + (-5)$ is -20, $4 \cdot -5$ is -20 **23.** $-2(-4) + 5$ is $8 + 5$, or 13; $-2(-6) + (-2)$ is $12 + (-2)$, or 10; $-2(0) + (-8)$ is $0 + (-8)$, or -8 **25.** A variable is a letter that represents the part of a rule that varies or changes depending on the situation. An expression expresses, or tells, the rule for doing something. For example, $c + 5$ is an expression, and c is the variable. **26.** The number part in a multiplication expression is the coefficient. For example, 4 is the coefficient in $4s$. A constant is a number that is added or subtracted in an expression. It does not vary. For example, 5 is the constant in $c + 5$. **27.** $b \cdot 1 = b$ or $1 \cdot b = b$ **29.** $\frac{b}{0}$ is undefined or $b \div 0$ is undefined. **31.** $c \cdot c \cdot c \cdot c \cdot c \cdot c$ **33.** $x \cdot x \cdot x \cdot x \cdot y \cdot y \cdot y$ **35.** $-3 \cdot a \cdot a \cdot a \cdot b$ **37.** $9 \cdot x \cdot y \cdot y$ **39.** $-2 \cdot c \cdot c \cdot c \cdot c \cdot c \cdot d$ **41.** $a \cdot a \cdot a \cdot b \cdot c \cdot c$ **43.** 16 **45.** -24 **47.** -18 **49.** -128 **51.** $-18,432$ **53.** 311,040 **55.** 56 **57.** $\frac{36}{0}$ is undefined. **59.** **(a)** 3 miles **(b)** 2 miles **(c)** 1 mile **60.** **(a)** $\frac{1}{2}$ mile; take half of the distance for 5 seconds **(b)** $7\frac{1}{2}$ seconds; find the number halfway between 5 seconds and 10 seconds **(c)** $12\frac{1}{2}$ seconds; find the number halfway between 10 seconds and 15 seconds

SECTION 2.2

1. The like terms are $2b^2$ and b^2; the coefficients are 2 and 1. **2.** The like terms are x^3 and $2x^3$; the coefficients are 1 and 2. **3.** The like terms are $-xy$ and $2xy$; the coefficients are -1 and 2. **4.** The like terms are $-a^2b$ and $-3a^2b$; the coefficients are -1 and -3. **5.** The like terms are 7, 3, and -4; they are constants. **6.** The like terms are -5, 1, and 4; they are constants. **7.** $12r$ **9.** $6x^2$ **11.** $-4p$ **13.** $-3a^3$ **15.** 0 **17.** xy **19.** $6t^4$ **21.** $4y^2$ **23.** $-8x$ **25.** $12a + 4b$ **27.** $7rs + 14$ **29.** $a + 2ab^2$ **31.** $-2x + 2y$ **33.** $7b^2$ **35.** cannot be simplified **37.** $-15r + 5s + t$ **39.** $30a$ **41.** $-8x^2$ **43.** $-20y^3$ **45.** $18cd$ **47.** $21a^2bc$ **49.** $12w$ **51.** $6b + 36$ **53.** $7x - 7$ **55.** $21t + 3$

57. $-10r - 6$ **59.** $-9k - 36$ **61.** $50m - 300$ **63.** $10 + 8y + 6$; $8y + 16$ **65.** $6a^2 + 3$ **67.** $9m - 34$ **69.** -25 **71.** $24x$ **73.** $5n + 13$ **75.** $11p - 1$ **77.** A simplified expression usually still has variables, but it is written in a simpler way. When evaluating an expression, the variables are all replaced by specific numbers and the final result is a numerical answer. **78.** The answers are $15x + 10$ and $10 + 15x$. They are equivalent because of the commutative property of addition. **79.** Like terms have matching variable parts, that is, matching letters and exponents. The coefficients do not have to match. Examples will vary. **80.** Add the coefficients of like terms. If no coefficient is shown, it is assumed to be 1. Keep the variable part the same. Examples will vary. **81.** Joaquin changed the variable part from x to x^2 when it should be left as is. The correct answer is $5x + 8$. **82.** In the last step, Paetyn should not change the sign of the first term. The correct answer is $-4a - 5$. **83.** $-2y + 9$ **85.** 0 **87.** $-9x$

SUMMARY EXERCISES Variables and Expressions

1. m is the variable; -1 is the coefficient; -10 is the constant. **2.** c and d are the variables; -8 is the coefficient. **3.** x is the variable; 4 is the coefficient; 6 is the constant. **4.** **(a)** 32 yards **(b)** 120 inches **5.** **(a)** \$13,080 **(b)** \$22,342 **6.** $a \cdot d \cdot d \cdot d \cdot d$ **7.** $b \cdot b \cdot b \cdot c \cdot d$ **8.** $-7 \cdot a \cdot b \cdot b \cdot b \cdot b \cdot c \cdot c$ **9.** 625 **10.** 0 **11.** 0 **12.** 60 **13.** -8 **14.** 120 **15.** -216 **16.** -800 **17.** 120,960 **18.** $24b$ **19.** $7x + 7$ **20.** $-8c - 32$ **21.** 0 **22.** $12c^2d$ **23.** $-6f$ **24.** $6w + 8$ **25.** $-2a - 6b$ **26.** $50x^3y^2$ **27.** $10r^3$ **28.** $7h^2$ **29.** -9 **30.** $32y - 15$ **31.** $36x - 10$ **32.** $19n + 1$ **33.** In the first expression Kalley did not multiply $6 \cdot 2$. The correct answer is $6n + 12$. In the second expression, Kalley has the wrong sign; two negative factors give a *positive* product. The correct answer is $20a$. **34.** Hunter forgot to change the subtraction to adding the opposite, $-8x + x + (-12) + 3$. The correct answer is $-7x - 9$.

SECTION 2.3

1. 58 is the solution. **2.** 25 is the solution. **3.** -16 is the solution. **4.** -17 is the solution. **5.** **(a)** Add 8 to both sides because $-8 + 8$ gives $m + 0$ on the left side. **(b)** Add -5 to both sides because $5 + (-5)$ gives $w + 0$ on the right side. **6.** **(a)** Add -2 to both sides because $2 + (-2)$ gives $n + 0$ on the left side. **(b)** Add 6 to both sides because $-6 + 6$ gives $b + 0$ on the right side.

7. $p = 4$ CHECK $\underbrace{4 + 5}_{9} = 9$
$9 = 9$ ✓

9. $10 = r$ CHECK $8 = \underbrace{10 - 2}_{8}$
$8 = 8$ ✓

11. $n = -8$ CHECK $-5 = n + 3$
$-5 = \underbrace{-8 + 3}$
$-5 = -5$ ✓

13. $k = 18$ CHECK $-4 + k = 14$
$\underbrace{-4 + 18} = 14$
$14 = 14$ ✓

15. $y = 6$ CHECK $y - 6 = 0$
$\underbrace{6 - 6} = 0$
$0 = 0$ ✓

17. $r = -6$ CHECK $7 = r + 13$

$$7 = \underbrace{-6 + 13}$$
$$7 = \quad 7 \checkmark$$

19. $x = 0$ CHECK $x - 12 = -12$

$$\underbrace{0 + (-12)} = -12$$
$$-12 \quad = -12 \checkmark$$

21. $t = -3$ CHECK $-5 = -2 + t$

$$-5 = \underbrace{-2 + (-3)}$$
$$-5 = \quad -5 \checkmark$$

23. $\underbrace{-2 + (-5)} = 3$

Does not balance $-7 \quad \neq 3$

The correct solution is 8. $\underbrace{8 - 5} = 3$

Balances $3 = 3 \checkmark$

25. $7 + x = -11$

$$\underbrace{7 + (-18)} = -11$$

Balances $-11 \quad = -11 \checkmark$

-18 is the correct solution.

27. $-10 = -10 + b$

$$-10 = \underbrace{-10 + 10}$$

Does not balance $-10 \neq 0$

The correct solution is 0. $-10 = \underbrace{-10 + 0}$

Balances $-10 = -10 \checkmark$

29. $c = 6$; CHECK $2 = 2 \checkmark$ **31.** $y = 5$; CHECK $3 = 3 \checkmark$ **33.** $b = -30$; CHECK $-20 = -20 \checkmark$ **35.** $t = 0$; CHECK $-2 = -2 \checkmark$ **37.** $z = -7$; CHECK $-7 = -7 \checkmark$ **39.** $w = 3$; CHECK $5 = 5 \checkmark$ **41.** $x = -10$ **43.** $a = 0$ **45.** $y = -25$ **47.** $x = 15$ **49.** $k = 113$ **51.** $b = 18$ **53.** $r = -5$ **55.** $n = -105$ **57.** $h = -5$ **59.** No, the solution is -14, the number used to replace x in the original equation.   **60.** Does not balance, $-9 \neq -7$. To correct the errors, change $-3 - 6$ to $-3 + (-6)$. Then, add 5 to both sides, not -5. The correct solution is -4.   **61.** $g = 295$ graduates   **63.** $c = 55$ chirps   **65.** $p = \$110$ per month in winter   **67.** $m = -19$ **69.** $x = 2$ **71. (a)** Equations will vary. Some possibilities are $n - 1 = -3$ and $8 = x + 10$.   **(b)** Equations will vary. Some possibilities are $y + 6 = 6$ and $-5 = -5 + b$.   **72. (a)** $x = \frac{1}{2}$ **(b)** $y = \frac{5}{4}$ **(c)** $n = \$0.85$ **(d)** Equations will vary.

SECTION 2.4

1. $6; \dfrac{6z}{6} = \dfrac{12}{6}; z = 2$ CHECK $12 = 12 \checkmark$

2. $8; \dfrac{8k}{8} = \dfrac{24}{8}; k = 3$ CHECK $24 = 24 \checkmark$

3. $r = 4$ CHECK $48 = 12r$

$$48 = \underbrace{12 \cdot 4}$$
$$48 = \quad 48 \checkmark$$

4. $m = 9$ CHECK $99 = 11m$

$$99 = \underbrace{11 \cdot 9}$$
$$99 = \quad 99 \checkmark$$

5. $y = 0$ CHECK $3y = 0$

$$\underbrace{3 \cdot 0} = 0$$
$$0 = 0 \checkmark$$

6. $a = 0$ CHECK $5a = 0$

$$\underbrace{5 \cdot 0} = 0$$
$$0 = 0 \checkmark$$

7. $k = -10$ **9.** $r = 6$ **11.** $b = -5$

13. $r = 3$ CHECK $\underbrace{2 \cdot 3} = 6$

$$6 = 6 \checkmark$$

15. $p = -3$ CHECK $-12 = 5p - p$

$$-12 = \underbrace{5 \cdot (-3)} - (-3)$$
$$-12 = \underbrace{(-15) + (+3)}$$
$$-12 = \quad -12 \checkmark$$

17. $a = -5$ **19.** $x = -10$ **21.** $w = 0$ **23.** $t = 3$ **25.** $t = 0$ **27.** $m = 9$ **29.** $y = -1$ **31.** $z = -5$ **33.** $p = -2$ **35.** $k = 7$ **37.** $b = -3$ **39.** $x = -32$ **41.** $w = 2$ **43.** $n = 50$ **45.** $p = -10$ **47.** Each solution is the opposite of the number in the equation. So the rule is: When you change the sign of the variable from negative to positive, then change the number in the equation to its opposite. In $-x = 5$, the opposite of 5 is -5, so $x = -5$.   **49.** Keng should divide by the coefficient of x, which is 3, *not* the opposite of 3. The correct solution is $x = 5$.   **50.** Maynou forgot the last step of dividing both sides by -1 in order to get a positive coefficient of 1. The correct answer is $x = 8$.   **51.** $s = 15$ feet   **53.** $s = 24$ meters   **55.** $y = 27$ **57.** $x = 1$

SECTION 2.5

1. $p = 1$ CHECK $12 = 12 \checkmark$

2. $k = 2$ CHECK $15 = 15 \checkmark$

3. $1 = y$, or $y = 1$ CHECK $2 = 8y - 6$

$$2 = \underbrace{8(1)} - 6$$
$$2 = \underbrace{8 - 6}$$
$$2 = \quad 2 \checkmark$$

4. $p = 2$ CHECK $10 = 11p - 12$

$$10 = \underbrace{11(2)} - 12$$
$$10 = \underbrace{22 - 12}$$
$$10 = \quad 10 \checkmark$$

5. $a = -2$ CHECK $28 = -9a + 10$

$$28 = \underbrace{-9(-2)} + 10$$
$$28 = \underbrace{18 + 10}$$
$$28 = \quad 28 \checkmark$$

6. $k = 0$ CHECK $-4k + 5 = 5$

$$\underbrace{-4(0)} + 5 = 5$$
$$\underbrace{0 + 5} = 5$$
$$5 = 5 \checkmark$$

7. $m = 0$ **9.** $x = -4$

11. $p = 4$; $4 = p$ CHECK $6(4) - 2 = 4(4) + 6$

$$\underbrace{24} - 2 = \underbrace{16} + 6$$
$$22 = 22 \checkmark$$

13. $k = -2; -2 = k$ CHECK $-2k - 6 = 6k + 10$

$$-2(-2) - 6 = 6(-2) + 10$$
$$4 + (-6) = -12 + 10$$
$$-2 = -2 \checkmark$$

15. $a = 5$ **17.** $t = 0$ **19.** $b = -8$ **21.** $w = 6$ **23.** $y = -9$
25. $t = -5$ **27.** $x = 0$ **29.** $h = 1$ **31.** $y = -2$ **33.** $m = -3$
35. $w = 2$ **37.** $x = 5$ **39.** $a = 3$ **41.** $b = -3$ **43.** $k = 4$ **45.** $c = 0$
47. $y = -5$ **49.** $n = 21$ **51.** $c = 30$ **53.** $p = -2$ **55.** $b = 0$
57. The series of steps may vary. One possibility is:

$-2t - 10 = 3t + 5$ Change subtraction to
adding the opposite.

$-2t + (-10) = 3t + 5$ Add $2t$ to both sides
$\underline{2t \qquad\qquad 2t}$ (addition property).
$0 + (-10) = 5t + 5$ Add -5 to both sides
$\underline{\quad -5 \qquad\quad -5}$ (addition property).

$\dfrac{-15}{5} = \dfrac{5t}{5}$ Divide both sides by 5
(division property).

$-3 = t$

58. Multiplication distributes over both addition and subtraction.
Examples will vary. Some possibilities are $3(2y + 6)$ is $6y + 18$ and
$5(x - 3)$ is $5x - 15$.
59. CHECK $-8 + 4(3) = 2(3) + 2$

$$-8 + 12 = 6 + 2$$
$$4 \neq 8 \quad \text{false}$$

The check does not balance, so 3 is not the correct solution. The student
added $-2a$ to -8 on the left side, instead of adding $-2a$ to $4a$. The correct
solution is 5.
60. CHECK $2(-10 + 4) = -16$

$$2(-6) = -16$$
$$-12 \neq -16 \quad \text{false}$$

The check does not balance, so -10 is not the correct solution. In the first
step on the left side, $2(x + 4)$ is $2x + 8$. The student forgot to multiply
$2(4)$. The correct solution is -12. **61. (a)** $5x + 8$ **(b)** $x = 8$
62. (a) $-6w + 10$ **(b)** $w = 0$ **63. (a)** $9y - 21$ **(b)** $y = -1$
64. Answers will vary. To simplify an expression you write it in a simpler
way by using the distributive property and combining all the like terms.
To solve an equation you find a number that can replace the variable and
make the equation balance.

Chapter 2 REVIEW EXERCISES

1. (a) Variable is k; coefficient is 4; constant is -3. **2.** 106 test tubes
3. (a) $x \cdot x \cdot y \cdot y \cdot y \cdot y$ **(b)** $5 \cdot a \cdot b \cdot b \cdot b$
4. (a) -27 **(b)** -128 **(c)** 720 **5.** $ab^2 + 3ab$ **6.** $-4x + 2y - 7$
7. $16g^3$ **8.** $12r^2t$ **9.** $5k + 10$ **10.** $-6b - 8$ **11.** $6y$ **12.** $20x + 2$
13. $n = -11$ CHECK $16 + n = 5$

$$16 + -11 = 5$$
$$5 = 5 \checkmark$$

14. $a = 4$ CHECK $-4 + 2 = 2a - 6 - a$

$$-4 + 2 = 2(4) - 6 - 4$$
$$-2 = 8 + (-6) + (-4)$$
$$-2 = -2 \checkmark$$

15. $m = -8$ **16.** $k = 10$ **17.** $p = -6$ **18.** $h = -12$ **19.** $w = 4$
20. $c = -2$ **21.** $d = 5$

Chapter 2 MIXED REVIEW EXERCISES

1. (a) 1 **(b)** 10 **(c)** -100 **2.** $-9y + 20$ **3.** $-6b - 18$ **4.** $r = 0$
5. $z = 1$ **6.** $-a^3 + 5$ **7. (a)** $-4 \cdot m \cdot m$ **(b)** $60 \cdot x \cdot y \cdot z \cdot z \cdot z$
8. 180 sandwiches **9.** $a = -5$ **10.** $p = 4$ **11.** $y = 5$ **12.** $m = 3$
13. $x = 9$ **14.** $b = 7$ **15.** $z = -3$ **16.** $n = 4$ **17.** $t = 0$ **18.** $d = 5$
19. $m = 35$ **20.** $b = -2$

Chapter 2 TEST

1. -7 is the coefficient; w is the variable; 6 is the constant.
2. Buy 177 hot dogs. **3.** $x \cdot x \cdot x \cdot x \cdot x \cdot y \cdot y \cdot y$
4. $4 \cdot a \cdot b \cdot b \cdot b \cdot b$ **5.** -200 **6.** $-4w^3$ **7.** 0 **8.** c
9. cannot be simplified **10.** $-40b^2$ **11.** $15k$ **12.** $21t + 28$
13. $-4a - 24$ **14.** $6x - 15$ **15.** $-9b + c + 6$
16. $x = 5$ CHECK $-4 = x - 9$

$$-4 = 5 - 9$$
$$-4 = -4 \checkmark$$

17. $w = -11$ CHECK $-7w = 77$

$$-7(-11) = 77$$
$$77 = 77 \checkmark$$

18. $p = -14$ CHECK $-p = 14$

$$-1(-14) = 14$$
$$14 = 14 \checkmark$$

19. $a = 3$ CHECK $-15 = -3(a + 2)$

$$-15 = -3(3 + 2)$$
$$-15 = -3(5)$$
$$-15 = -15 \checkmark$$

20. $n = -8$ **21.** $m = 15$ **22.** $x = -1$ **23.** $m = 2$ **24.** $b = 54$
25. $c = 0$ **26.** Equations will vary. Two possibilities are $x - 5 = -9$
and $-24 = 6y$.

Chapters 1–2 CUMULATIVE REVIEW EXERCISES

1. three hundred six billion, four thousand, two hundred ten
2. 800,066,000 **3. (a)** $>$ **(b)** $<$ **4. (a)** commutative property of
addition **(b)** multiplication property of 0 **(c)** distributive property
5. (a) 9000 **(b)** 290,000 **6.** -8 **7.** 10 **8.** 30 **9.** 25 **10.** 7
11. 0 **12.** -64 **13.** undefined **14.** -60 **15.** -40 **16.** -9
17. -25 **18.** -28 **19.** $\dfrac{11}{-11} = -1$ **20.** *Estimate:* $600 \div 20 =$
30 miles; *Exact:* $616 \div 22 = 28$ miles **21.** *Estimate:* $-50 + 20 =$
-30 degrees; *Exact:* $-48 + 23 = -25$ degrees **22.** *Estimate:*
$\$2000 - (50 \cdot 8) = \$1600$; *Exact:* $\$2132 - (52 \cdot 8) = \$1716$
23. *Estimate*: $10(800 + 50) = \$8500$; *Exact:* $12(758 + 45) = \$9636$
24. $-4 \cdot a \cdot b \cdot b \cdot b \cdot c \cdot c$ **25.** 120 **26.** h **27.** 0
28. $5n^2 - 4n - 2$ **29.** $-30b^2$ **30.** $28p - 28$ **31.** $-9w^2 - 12$
32. $x = -4$ CHECK $3(-4) = -4 - 8$

$$-12 = -12 \checkmark$$

33. $y = -6$ CHECK $-44 = -2 + 7(-6)$

$$-44 = -2 + (-42)$$
$$-44 = -44 \checkmark$$

34. $k = 7$ CHECK $2(7) - 5(7) = -21$

$$14 - 35 = -21$$
$$-21 = -21 \checkmark$$

35. $m = 4$ CHECK $\underbrace{4 - 6}_{-2} = \underbrace{-2(4) + 6}_{-8 + 6}$

$-2 = -2 \checkmark$

36. $x = -1$ **37.** $r = -18$ **38.** $b = -5$ **39.** $t = 1$ **40.** $y = -12$

CHAPTER 3 Solving Application Problems

All answers are shown for Concept Checks, Writing exercises, What Went Wrong exercises, Relating Concepts exercises, Summary Exercises, Review Exercises, Chapter Tests, and Cumulative Review Exercises. For other types of exercises, answers are shown only for the odd-numbered exercises.

SECTION 3.1

1. A square has four right angles, and all four sides have the same length. Drawings will vary, but should have the same measurement for every side.
2. $P = 4s$; P stands for perimeter; s stands for side (the length of one side).
3. $P = 4 \cdot 9$ cm; $P = 36$ cm   **5.** $P = 100$ in.   **7.** $P = 4$ miles
9. $P = 88$ mm   **11.** 30 ft $= s$ **13.** $s = 23$ yards   **15.** $s = 2$ ft
17. A rectangle is a figure with four sides that meet to form four right angles. Each set of opposite sides is parallel and has the same length. Drawings will vary, but the two longer sides should have the same measurement and the two shorter sides should have the same measurement.
18. $P = 2l + 2w$; P stands for perimeter; l stands for length; w stands for width.   **19.** $P = 16$ yd $+ 12$ yd; $P = 28$ yd   **21.** $P = 70$ cm
23. $P = 72$ ft   **25.** $P = 26$ in.   **27.** $l = 9$ cm   CHECK 9 cm $+$ 9 cm $+$ 6 cm $+$ 6 cm $= 30$ cm   **29.** $w = 1$ mile   CHECK 4 mi $+$ 4 mi $+$ 1 mi $+$ 1 mi $= 10$ mi   **31.** $w = 2$ ft   CHECK 6 ft $+$ 6 ft $+$ 2 ft $+$ 2 ft $= 16$ ft
33. $l = 2$ m   CHECK 2 m $+$ 2 m $+$ 1 m $+$ 1 m $= 6$ m   **35.** $P = 320$ ft
37. $P = 54$ mm   **39.** $P = 48$ ft   **41.** $P = 78$ in.   **43.** $P = 125$ m
45. $? = 40$ cm   **47.** $? = 12$ in.   **49. (a)** Sketches will vary.
(b) Formula for perimeter of an equilateral triangle is $P = 3s$, where s is the length of one side.   **(c)** The formula will *not* work for other kinds of triangles because the sides will have different lengths.   **50. (a)** $P = 5s$
(b) $P = 6s$ **(c)** $P = 10s$ **(d)** $P = ns$ **51. (a)** 140 miles   **(b)** 350 miles
(c) 560 miles   **52. (a)** 70 miles   **(b)** 175 miles   **(c)** 280 miles
(d) The rate is half of 70 miles per hour, so the distance will be half as far; divide each result in **Exercise 51** by 2.   **53. (a)** 50 hours   **(b)** 60 hours
(c) 150 hours   **54. (a)** 61 miles per hour   **(b)** 57 miles per hour
(c) 65 miles per hour

SECTION 3.2

1. $A = s^2$ or $A = s \cdot s$; A stands for area; s stands for side (the length of one side).   **2.** $A = l \cdot w$; A stands for Area; l stands for length; w stands for width.   **3. (a)** 100   **(b)** 20   **(c)** 625   **(d)** 50   **4. (a)** 64   **(b)** 16
(c) 900   **(d)** 60   **5.** $A = 77$ ft^2 **7.** $A = 6$ in. $\cdot$ 6 in.; $A = 36$ in.^{2}
9. $A = 775$ mm^2 **11.** $A = 105$ cm^2 **13.** $A = 72$ ft^2 **15.** $A = 625$ mi^2
17. $l = 6$ ft   CHECK $A = 6$ ft $\cdot$ 3 ft; $A = 18$ ft^2 **19.** $w = 80$ yd
CHECK $A = 90$ yd $\cdot$ 80 yd; $A = 7200$ yd^2 **21.** $l = 14$ in.
CHECK $A = 14$ in. $\cdot$ 11 in.; $A = 154$ in.^{2} **23.** 6; $s = 6$ m   **25.** $s = 2$ ft
27. $h = 20$ cm   CHECK $A = 25$ cm $\cdot$ 20 cm; $A = 500$ cm^2 **29.** $b = 17$ in.
CHECK $A = 17$ in. $\cdot$ 13 in.; $A = 221$ in.^{2} **31.** $h = 1$ m
CHECK $A = 9$ m $\cdot$ 1 m; $A = 9$ m^2 **33.** The first mistake was including height in the perimeter. The second mistake was using

square units, which are for area, not perimeter. The correct answer is
$P = 25$ cm $+$ 25 cm $+$ 25 cm $+$ 25 cm $= 100$ cm   **34.** The first mistake was not applying the exponent correctly. The second mistake was not using square units for area. The correct answer is $A = 7$ ft $\cdot$ 7 ft $= 49$ ft^2.
35. Square; $P = 180$ in.; $A = 2025$ in.^{2}
37. Parallelogram; $P = 60$ cm; $A = 180$ cm^2 **39.** $P = 48$ m; $A = 144$ m^2
41. \$108   **43.** \$725   **45.** 53 yd   **47.** $P = 48$ ft; $A = 144$ ft^2; 18 ft^2 for each camper (rounded)
49.

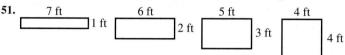

50. (a) 5 ft by 1 ft has area of 5 ft^2; 4 ft by 2 ft has area of 8 ft^2; 3 ft by 3 ft has area of 9 ft^2 **(b)** The square plot 3 ft by 3 ft has the greatest area.
51.

5 ft · 1 ft   6 ft · 2 ft   5 ft · 3 ft   4 ft · 4 ft

52. (a) 7 ft^2, 12 ft^2, 15 ft^2, 16 ft^2 **(b)** Square plots have the greatest area.

SUMMARY EXERCISES   Perimeter and Area

1. Rectangle; $P = 32$ m; $A = 39$ m^2 **2.** Square; $P = 104$ ft; $A = 676$ ft^2
3. Parallelogram; $P = 34$ yd; $A = 56$ yd^2 **4.** Rectangle; $P = 36$ cm; $A = 80$ cm^2 **5.** Square; $P = 36$ in.; $A = 81$ in.^{2} **6.** Parallelogram; $P = 30$ m; $A = 45$ m^2 **7.** Rectangle; $P = 26$ ft; $A = 36$ ft^2
8. Parallelogram; $P = 188$ ft; $A = 2100$ ft^2
9. The first mistake was not using the height when calculating area. The second mistake was forgetting to include square units for area. The correct answer is $A = 16$ cm $\cdot$ 7 cm $= 112$ cm^2
10. The first mistake was forgetting to multiply the length by 2 and the width by 2. The second mistake was using square units, which are for area, not perimeter. The correct answer is $P = 2 \cdot 21$ yd $+ 2 \cdot 55$ yd $= 42$ yd $+ 110$ yd $= 152$ yd
11. $s = 6$ in.

CHECK $A = 6$ in. $\cdot$ 6 in.
$A = 36$ in.^{2}

12. $w = 2$ m

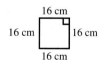

CHECK $A = 42$ m $\cdot$ 2 m
$A = 84$ m^2

13. $s = 16$ cm

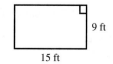

CHECK $P = 16$ cm $+$ 16 cm $+$
16 cm $+$ 16 cm
$P = 64$ cm

14. $l = 15$ ft

CHECK $P = 15$ ft $+$ 15 ft $+$ 9 ft $+$ 9 ft
$P = 48$ ft

15. 504 meters of fencing   **16.** Area of the largest field is 9600 yd^2. Area of the smallest field is 7150 yd^2. Difference is 9600 yd^2 $-$ 7150 yd^2 $= 2450$ yd^2 **17.** 9 inches   **18.** 176 cm of braid
19. $7(15$ ft$^2) = 105$ ft^2 of fabric; $7(16$ ft$) = 112$ ft of binding   **20.** 48 ft

SECTION 3.3

1. (a) false; a number subtracted from 14, or, 14 minus a number.
(b) true   **(c)** false; the quotient of -20 and a number, or, -20 divided by a number.   **2. (a)** true   **(b)** false; the product of 18 and a number, or, 18 times a number   **(c)** false; the quotient of a number and 42, or, a number divided by 42.   **3.** $14 + x$ or $x + 14$ **5.** $-5 + x$ or $x + (-5)$ **7.** $20 - x$
9. $x - 9$ **11.** $-6x$ **13.** $2x$ **15.** $\dfrac{x}{2}$ **17.** $2x + 8$ or $8 + 2x$
19. $7x - 10$ **21.** $2x + x$ or $x + 2x$
23. $n = 7$ **CHECK** $\underline{4 \cdot 7} - 2$ does equal 26
$$\underline{28 \quad -2}$$
$$26$$

25. $2n + n = -15$ or $n + 2n = -15$;
$\quad n = -5$ **CHECK** $\underline{2 \cdot (-5)} + (-5)$ does equal -15
$$\underline{-10 \quad + (-5)}$$
$$-15$$

27. $5n + 12 = 7n;\ n = 6$ **CHECK** $\underline{5 \cdot 6} + 12 = \underline{7 \cdot 6}$
$$\underline{30} + 12 = \underline{42}$$
$$42 = 42$$

29. $30 - 3n = 2 + n;\ n = 7$ **CHECK** $30 - \underline{3 \cdot 7} = \underline{2 + 7}$
$$30 - \underline{21} = \underline{9}$$
$$9 = 9$$

31. $w + 15 - 28 + 5 = 177$. He weighed 185 pounds originally.
33. Let c be the number of cookies the children ate. $18 - c + 36 = 49$. Her children ate 5 cookies.   **35.** Let p be the number of pens in each box. $6p - 32 - 35 = 5$. There were 12 pens in each box.   **37.** Let d be each member's dues. $14d + 340 - 575 = -25$. Each member paid \$15.
39. $4a - 75 = a$. Tamu is 25 years old.   **41.** Let m be the amount Brenda spent. $2m - 3 = 81$. Brenda spent \$42.   **43.** Let p be the number of pieces in each bag. $5p - 3 \cdot 48 = p$. There were 36 pieces of candy in each bag.   **45.** Let d be the daily amount for a young child. $4d + 15 = 75$. A young child should receive 15 mg of vitamin C.

SECTION 3.4

1. (a) known parts; unknown parts   **(b)** least   **2. (a)** label; or, include the units   **(b)** all the facts in the original problem   **3.** $a + a + 9 = 51$. I am 21; my sister is 30.   **5.** m is husband's earnings; $m + 1500$ is Lien's earnings. $m + m + 1500 = 37{,}500$. Husband earned \$18,000; Lien earned \$19,500.   **7.** m is printer's cost; $8m$ is computer's cost. $m + 8m = 396$; printer cost \$44; computer cost \$352.   **9.** Length of shorter piece is x; length of longer piece is $x + 10$. $x + x + 10 = 78$. Shorter piece is 34 cm; longer piece is 44 cm   **11.** Length of longer piece is x; length of shorter piece is $x - 7$. $x + x - 7 = 31$. Longer piece is 19 ft; shorter piece is 12 ft
13. s is number of Senators; $5s - 65$ is number of Representatives. $s + 5s - 65 = 535$. 100 Senators; 435 Representatives   **15.** Length of first part is x; length of second part is x; length of third part is $x + 25$. $x + x + x + 25 = 706$. First part is 227 m; second part is 227 m; third part is 252 m   **17.** length is 19 yd   **19.** length is 12 ft; width is 6 ft
21. length is 13 in.; width is 5 in.   **23.** $P = 52$ in.; $A = 168$ in.^{2}

Chapter 3   REVIEW EXERCISES

1. Square; $P = 112$ cm   **2.** Rectangle; $P = 22$ mi   **3.** Parallelogram; $P = 42$ yd   **4.** $P = 140$ m   **5.** 12 ft $= 4s;\ s = 3$ ft
6. 128 yd $= 2l + 2(31$ yd$);\ l = 33$ yd   **7.** 72 in. $= 2(21$ in.$) + 2w$;

$w = 15$ in.   **8.** $A = 40$ ft^2 **9.** $A = 625$ m^2 **10.** $A = 208$ yd^2
11. 126 ft$^2 = 14$ ft $\cdot w;\ w = 9$ ft   **12.** 88 cm$^2 = 11$ cm $\cdot h;\ h = 8$ cm
13. 100 mi$^2 = s \cdot s;\ s = 10$ mi   **14.** $57 - x$ **15.** $15 + 2x$ or $2x + 15$
16. $-9x$ **17.** $4n + 6 = -30;\ n = -9$ **18.** $10 - 2n = 4 + n;\ n = 2$
19. m is money originally in account. $m - 600 + 750 + 75 = 309$.
\$84 was originally in Grace's account.   **20.** c is number of candles in each box. $4c - 25 = 23$. There were 12 candles in each box.   **21.** p is Reggie's prize money; $p + 300$ is Donald's prize money. $p + p + 300 = 1000$. Reggie gets \$350; Donald gets \$650.   **22.** w is the width; $2w$ is the length. $84 = 2(2w) + 2(w)$. The width is 14 cm; the length is 28 cm

Chapter 3   MIXED REVIEW EXERCISES

1. Parallelogram; $P = 72$ft; $A = 286$ ft^2 **2.** Square; $P = 20$ cm; $A = 25$ cm^2 **3.** Rectangle; $P = 156$ yd; $A = 1080$ yd^2
4. 204 m$^2 = l \cdot 12$m; $l = 17$ m   **5.** 33 in.$^2 = b \cdot 3$ in.; $b = 11$ in.
6. 64 mi $= 4s;\ s = 16$ mi   **7. (a)** 36 ft $= 4s;\ s = 9$ ft   **(b)** $A = 9$ ft $\cdot$ 9 ft; $A = 81$ ft^2 **8.** Rectangles will vary. Two possibilities are: $A = 7$ ft $\cdot$ 3 ft, $A = 21$ ft^2; $A = 6$ ft $\cdot$ 4 ft, $A = 24$ ft^2 **9.** Let f be the fencing for the garden. $36 - f + 20 = 41$. 15 ft of fencing was used on the garden.
10. w is the width; $w + 2$ is the length. $36 = 2(w + 2) + 2 \cdot w$. Width is 8 ft; length is 10 ft

Chapter 3   TEST

1. $P = 262$ m   **2.** $P = 40$ in.   **3.** $P = 12$ miles   **4.** $P = 12$ ft
5. $P = 110$ cm   **6.** $A = 486$ mm^2 **7.** $A = 140$ cm^2 **8.** $A = 3740$ mi^2
9. $A = 36$ m^2 **10.** 12 ft $= 4s;\ s = 3$ ft   **11.** 34 ft $= 2l + 2(6$ ft$)$; $l = 11$ ft   **12.** 65 in.$^2 = 13$ in. $\cdot h;\ h = 5$ in.   **13.** 12 cm$^2 = 4$ cm $\cdot w$; $w = 3$ cm   **14.** 16 ft$^2 = s^2;\ s = 4$ ft   **15.** Answers will vary; one possibility: Linear units like ft are used to measure length, width, height, and perimeter. Area is measured in square units like ft^2 (squares that measure 1 ft on each side).   **16.** $4n + 40 = 0;\ n = -10$ **17.** $7n - 23 = n + 7$; $n = 5$ **18.** Let m represent money spent on groceries. $43 - m + 16 = 44$. Son spent \$15.   **19.** Let d represent daughter's age. $39 = 5d + 4$. Daughter is 7 years old.   **20.** Let p be length of one piece; $p + 4$ be length of second piece. $p + p + 4 = 118$. One piece is 57 cm; second piece is 61 cm   **21.** Let w be width; $4w$ be length. $420 = 2(4w) + 2(w)$. Sketches may vary; length is 168 ft; width is 42 ft   **22.** Let h be Marcella's hours; $h - 3$ be Tim's hours. $h + h - 3 = 19$. Marcella worked 11 hours; Tim worked 8 hours.

Chapters 1–3   CUMULATIVE REVIEW EXERCISES

1. four billion, two hundred six thousand, three hundred   **2.** 70,005,489
3. $<$ $>$ **4. (a)** multiplication property of 1   **(b)** addition property of 0
(c) associative property of multiplication   **5. (a)** 3800   **(b)** 490,000
6. -24 **7.** 27   **8.** -3 **9.** -100 **10.** 5   **11.** 0   **12.** -13 **13.** 25
14. -32 **15.** $\dfrac{27}{0}$ is undefined.   **16.** $10 \cdot w \cdot w \cdot x \cdot y \cdot y \cdot y \cdot y$
17. 240   **18.** $2k$ **19.** $3m^2 + 2m$ **20.** 0   **21.** $-20a$ **22.** $-2x^2 - 3$
23. $-12n + 1$
24. $x = 2$ **CHECK** $\underline{6 - 20} = \underline{2(2)} - \underline{9(2)}$
$$-14 = \underline{4} - \underline{18}$$
$$-14 = -14 \quad \checkmark$$
25. $y = -1$ **CHECK** $\underline{-5(-1)} = \underline{-1 + 6}$
$$5 = 5 \quad \checkmark$$

26. $b = 4$ **27.** $h = -2$ **28.** $x = 0$ **29.** $a = -3$ **30.** $P = 60$ in.; $A = 198$ in.^{2} **31.** $P = 60$ m; $A = 225$ m^2 **32.** $P = 24$ ft; $A = 32$ ft^2 **33.** $5n + (-50) = 0$; $n = 10$ **34.** $10 - 3n = 2n$; $n = 2$ **35.** Let p be number of people originally. $p - 3 + 6 - 2 = 5$. There were 4 people in line originally. **36.** Let m be amount paid by each player. $12m - 2200 = -40$. Each player paid \$180. **37.** Let g be one group; $3g$ the other group. $g + 3g = 192$. There are 48 students in one group; 144 in the other group. **38.** Let w be the width; $w + 14$ be the length. $92 = 2(w + 14) + 2(w)$. Width is 16 ft; length is 30 ft

CHAPTER 4 Rational Numbers: Positive and Negative Fractions

All answers are shown for Concept Checks, Writing exercises, What Went Wrong exercises, Relating Concepts exercises, Summary Exercises, Review Exercises, Chapter Tests, and Cumulative Review Exercises. For other types of exercises, answers are shown only for the odd-numbered exercises.

SECTION 4.1

1. (a) parts of equal size. **(b)** denominator **2. (a)** numerator **(b)** a fraction bar **3.** $\dfrac{\text{shaded parts} \to 5}{\text{number of equal parts} \to 8}$; $\dfrac{3}{8}$ unshaded **5.** $\dfrac{2}{3}$; $\dfrac{1}{3}$ **7.** $\dfrac{3}{2}$; $\dfrac{1}{2}$ **9.** $\dfrac{11}{6}$; $\dfrac{1}{6}$ **11.** $\dfrac{2}{11}$; $\dfrac{1}{11}$; $\dfrac{4}{11}$ **13. (a)** $\dfrac{6}{20}$; **(b)** $\dfrac{19}{20}$ **15.** N: 3; D: 4; 4 equal parts **16.** N: 5; D: 8; 8 equal parts **17.** N: 12; D: 7; 7 equal parts **18.** N: 8; D: 3; 3 equal parts **19.** Proper: $\dfrac{1}{3}, \dfrac{5}{8}, \dfrac{7}{16}$; Improper: $\dfrac{8}{5}, \dfrac{6}{6}, \dfrac{12}{2}$

21. 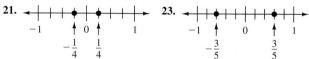 **23.**

25.

27. $-\dfrac{3}{4}$ pound **28.** $\dfrac{1}{3}$ cup

29. $\dfrac{3}{10}$ mile **30.** $-\dfrac{1}{2}$ quart **31.** $\dfrac{2}{5}$ **33.** 0 **35. (a)** 2 times 12; $\dfrac{12}{24}$ **(b)** $\dfrac{8}{24}$ **(c)** $\dfrac{16}{24}$ **(d)** $\dfrac{6}{24}$ **(e)** $\dfrac{18}{24}$ **(f)** $\dfrac{4}{24}$ **(g)** $\dfrac{20}{24}$ **(h)** $\dfrac{3}{24}$ **(i)** $\dfrac{9}{24}$ **(j)** $\dfrac{15}{24}$ **37. (a)** $-\dfrac{1}{3}$ **(b)** $-\dfrac{2}{3}$ **(c)** $-\dfrac{2}{3}$ **(d)** $-\dfrac{1}{3}$ **(e)** $-\dfrac{2}{3}$ **(f)** Some possibilities are: $-\dfrac{4}{12} = -\dfrac{1}{3}$; $-\dfrac{8}{24} = -\dfrac{1}{3}$; $-\dfrac{20}{30} = -\dfrac{2}{3}$; $-\dfrac{24}{36} = -\dfrac{2}{3}$

39. You cannot do it if you want the numerator to be a whole number, because 5 does not divide into 18 evenly. You could use multiples of 5 as the denominator, such as 10, 15, 20, etc. **40.** No, a fraction cannot have a denominator of 0, because division by 0 is not allowed. **41.** 10 **43.** -1 **45.** -6 **47.** 3 **49.** 2 **51.** -9 **53.** 1 **55.** -8 **57.** $\dfrac{2}{5}$ is unshaded. **59.** $\dfrac{5}{8}$ is unshaded.

61. One possibility is shown. ○□□□□□□□△△△

63. One possibility is shown. ⊙!!!,...??? **65. (a)** $\dfrac{1467}{3912}$ **(b)** Divide 3912 by 8 to get 489; multiply 3 by 489 to get 1467

66. (a) $\dfrac{4256}{5472}$ **(b)** Divide 5472 by 9 to get 608; multiply 7 by 608 to get 4256 **67. (a)** $-\dfrac{1}{5}$ **(b)** Divide 3485 by 2, by 3, and by 5 to see that dividing by 5 gives 697. Or divide 3485 by 697 to get 5. **68. (a)** $-\dfrac{1}{6}$ **(b)** Divide 4902 by 4, by 6, and by 8 to see that dividing by 6 gives 817. Or divide 4902 by 817 to get 6.

SECTION 4.2

1. (a) yes **(b)** No, 5 is a common factor. **(c)** yes **(d)** yes **(e)** No, 3 is a common factor. **(f)** No, 7 is a common factor. **2. (a)** No, 2 is a common factor. **(b)** No, 3 is a common factor **(c)** yes **(d)** No, 11 is a common factor. **(e)** yes **(f)** yes **3. (a)** $\dfrac{2}{3}$ **(b)** $\dfrac{2}{3}$ **(c)** $-\dfrac{1}{4}$ **(d)** $-\dfrac{1}{2}$ **(e)** $\dfrac{8}{9}$ **(f)** $-\dfrac{2}{5}$ **5.** comp., prime, comp., neither, prime, prime, comp., comp. **6.** comp., prime, prime, comp., neither, comp., prime, comp. **7.** $6 = 2 \cdot 3$

9. $2 \cdot 2 \cdot 5$ **11.** $5 \cdot 5$ **13.** $2 \cdot 2 \cdot 3 \cdot 3$ **15. (a)** $2 \cdot 2 \cdot 11$ **(b)** $2 \cdot 2 \cdot 2 \cdot 11$ **17. (a)** $3 \cdot 5 \cdot 5$ **(b)** $2 \cdot 2 \cdot 17$ **(c)** $3 \cdot 3 \cdot 3 \cdot 7$ **19.** $\dfrac{\cancel{2} \cdot \cancel{2} \cdot \cancel{2}}{\cancel{2} \cdot \cancel{2} \cdot \cancel{2} \cdot 2} = \dfrac{1}{2}$ **21.** $\dfrac{\cancel{2} \cdot \cancel{2} \cdot \cancel{2} \cdot 2}{\cancel{2} \cdot \cancel{2} \cdot \cancel{2} \cdot 3} = \dfrac{2}{3}$ **23.** $\dfrac{2 \cdot \cancel{7}}{3 \cdot \cancel{7}} = \dfrac{2}{3}$ **25.** $\dfrac{\cancel{2} \cdot 2 \cdot \cancel{3} \cdot 3}{\cancel{2} \cdot \cancel{3} \cdot 7} = \dfrac{6}{7}$ **27.** $\dfrac{2 \cdot 5 \cdot 5}{3 \cdot 3 \cdot 7}$ already in lowest terms **29.** $\dfrac{\cancel{3} \cdot \cancel{3} \cdot 3}{\cancel{3} \cdot \cancel{3} \cdot 5} = \dfrac{3}{5}$ **31.** $\dfrac{\cancel{2} \cdot 2 \cdot \cancel{3}}{\cancel{2} \cdot \cancel{3} \cdot 3} = \dfrac{2}{3}$ **33.** $\dfrac{\cancel{8} \cdot 7}{2 \cdot 2 \cdot 2 \cdot \cancel{8}} = \dfrac{7}{8}$ **35.** $\dfrac{\cancel{2} \cdot 3 \cdot \cancel{3} \cdot \cancel{5}}{\cancel{2} \cdot 2 \cdot \cancel{3} \cdot \cancel{3} \cdot \cancel{5}} = \dfrac{1}{2}$ **37.** $\dfrac{2 \cdot \cancel{3} \cdot \cancel{5} \cdot \cancel{7}}{3 \cdot \cancel{3} \cdot \cancel{5} \cdot \cancel{7}} = \dfrac{2}{3}$ **39.** $\dfrac{\cancel{3} \cdot \cancel{11} \cdot 13}{\cancel{3} \cdot 3 \cdot 5 \cdot \cancel{11}} = \dfrac{13}{15}$ **41. (a)** $\dfrac{1}{4}$ **(b)** $\dfrac{1}{2}$ **(c)** $\dfrac{1}{10}$ **(d)** $\dfrac{60}{60} = 1$ **43. (a)** $\dfrac{1}{3}$ **(b)** $\dfrac{1}{6}$ **(c)** $\dfrac{1}{2}$ **45.** Men, $\dfrac{3}{7}$; Women, $\dfrac{4}{7}$ **47.** When dividing out common factors, 3 divided by 3 is 1, and Marcus did not write 1 above and below all the slashes. The numerator is $1 \cdot 1$, so the correct answer is $\frac{1}{4}$ **48.** Ginny used different numbers to divide the numerator and denominator, but she must divide by the same number. There is no common factor so the fraction is already in lowest terms. The correct answer is $\frac{9}{16}$ **49.** $\dfrac{2c}{5}$ **51.** $\dfrac{4}{7}$ **53.** $\dfrac{6r}{5s}$ **55.** $\dfrac{1}{7n^2}$ **57.** already in lowest terms **59.** $\dfrac{7}{9y}$ **61.** $\dfrac{7k}{2}$ **63.** $\dfrac{1}{3}$ **65.** $\dfrac{c}{d}$ **67.** $\dfrac{6ab}{c}$ **69.** already in lowest terms **71.** $3eg^2$

SECTION 4.3

1. multiply; denominators **2.** positive; negative **3.** $-\dfrac{3}{16}$ **5.** $\dfrac{9}{10}$ **7.** $\dfrac{1}{2}$ **9.** -6 **11.** 36 **13.** $\dfrac{15}{4y}$ **15.** $\dfrac{1}{2}$ **17.** $\dfrac{6}{5}$ **19.** -9 **21.** $-\dfrac{1}{6}$ **23.** $\dfrac{11}{15d}$ **25.** b **27.** $\dfrac{4}{15}$ **29.** $-\dfrac{9}{32}$ **31.** 21 **33.** 15 **35.** undefined **37.** $-\dfrac{55}{12}$ **39.** $8b$ **41.** $\dfrac{3}{5d}$ **43.** $\dfrac{2x^2}{w}$ **45.** $\dfrac{3}{10}$ yd^2 **47.** 80 dispensers **49.** earn \$9300; borrow \$3100 **51.** About 75 infield players **53.** 9 trips

55. 3200 times **57.** $\dfrac{3}{20}$ of 1600 = 240 students **59.** 192 − 64 = 128

students or $\dfrac{2}{25}$ of 1600 = 128 students **61.** 7 million horses

63. 154 million dogs and cats **65.** Evan forgot to write the final answer in lowest terms. The correct answer is $\frac{2}{3}$. **66.** Haddi used the reciprocal in a multiplication problem, but reciprocals are used only for division. The correct answer is $\frac{16}{3}$. **67.** For the first question, Bailey must use the reciprocal of $\frac{4}{1}$ because it is a division problem. The correct answer is $\frac{1}{6}$. For the second question, Bailey forgot that division by zero is undefined. He should have written "undefined" as his answer. **68.** In the first calculation, Jillian forgot to write 1 above and below all the slashes when dividing out the common factors. The correct answer is $\frac{1}{6}$. In the second calculation, she did not divide out the common factor of 2 correctly. The correct answer is $\frac{3}{20}$

SECTION 4.4

1. (a) like fractions **(b)** add them **2. (a)** different denominators

(b) add only the numerators **3.** $\dfrac{7}{8}$ **5.** $-\dfrac{1}{2}$ **7.** $\dfrac{1}{2}$ **9.** $-\dfrac{9}{40}$ **11.** $-\dfrac{13}{24}$

13. $-\dfrac{3}{5}$ **15.** $-\dfrac{7}{18}$ **17.** $\dfrac{8}{7}$ **19.** $-\dfrac{3}{8}$ **21.** $\dfrac{3+5c}{15}$ **23.** $\dfrac{10-m}{2m}$

25. $\dfrac{8}{b^2}$ **27.** $\dfrac{bc+21}{7b}$ **29.** $\dfrac{-4-cd}{c^2}$ **31.** $-\dfrac{44}{105}$ **33.** $\dfrac{a^2b+12}{3a^2}$

35. $\dfrac{-w^3+50}{10w^2}$ **37.** $\dfrac{m^4}{n^2}$ **39.** For the first question, Katelyn added fractions with unlike denominators. She must use the LCD of 20 to get $\frac{15}{20}+\frac{8}{20}$. The correct answer is $\frac{23}{20}$. For the second question, Katelyn did not multiply the numerator and denominator by the same number when rewriting the fractions with the LCD. She should get $\frac{15}{18}-\frac{8}{18}$. The correct answer is $\frac{7}{18}$.
40. For the first question, Sean forgot to write the answer in lowest terms. The correct answer is $\frac{1}{3}$. For the second question, Sean subtracted fractions with unlike denominators. He must use the LCD of 20 to get $\frac{6}{20}-\frac{5}{20}$. The correct answer is $\frac{1}{20}$. **41.** $\dfrac{47}{60}$ in. **43.** $\dfrac{23}{24}$ cup

45. (a) $\dfrac{33}{20}$ miles **(b)** $\dfrac{17}{20}$ mile **47.** $\dfrac{29}{50}$ of U.S. electricity **49.** $\dfrac{1}{4}$ of

U.S. electricity **51.** $\dfrac{5}{16}$ inch **53.** $\dfrac{1}{12}$ mile **55. (a)** $\dfrac{1}{12}; \dfrac{1}{12}$; addition is

commutative. **(b)** $\dfrac{1}{3}; -\dfrac{1}{3}$; subtraction is *not* commutative. **(c)** $-\dfrac{3}{5}; -\dfrac{3}{5}$;

multiplication is commutative. **(d)** $6; \dfrac{1}{6}$; division is *not* commutative.

56. (a) 0; 0; the sum of a number and its opposite is 0. **(b)** 1; 1; when a

nonzero number is divided by itself, the quotient is 1. **(c)** $\dfrac{5}{6}; -\dfrac{17}{20}$;

multiplying by 1 leaves a number unchanged. **(d)** 1; 1; a number times its reciprocal is 1

SECTION 4.5

1.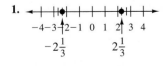

$-2\frac{1}{3}$ $2\frac{1}{3}$

Shade $2\dfrac{1}{3}$ of the circles.

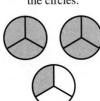

2.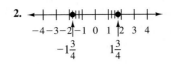

$-1\frac{3}{4}$ $1\frac{3}{4}$

Shade $1\dfrac{3}{4}$ of the rectangles.

3.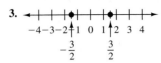

$-\frac{3}{2}$ $\frac{3}{2}$

Shade $\dfrac{3}{2}$ of the squares.

4.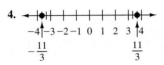

$-\frac{11}{3}$ $\frac{11}{3}$

Shade $\dfrac{11}{3}$ of the circles.

5. $\dfrac{9}{2}$ **7.** $-\dfrac{8}{5}$ **9.** $\dfrac{19}{8}$ **11.** $-\dfrac{57}{10}$ **13.** $\dfrac{161}{15}$ **15.** $4\dfrac{1}{3}$ **17.** $-2\dfrac{1}{2}$

19. $3\dfrac{2}{3}$ **21.** $-5\dfrac{2}{3}$ **23.** $11\dfrac{3}{4}$ **25.** $2 \cdot 4 = 8; \dfrac{63}{8}$ or $7\dfrac{7}{8}$

27. $3 \div 3 = 1; \dfrac{26}{21}$ or $1\dfrac{5}{21}$ **29.** $4 + 2 = 6; \dfrac{11}{2}$ or $5\dfrac{1}{2}$

31. $4 - 1 = 3; \dfrac{11}{3}$ or $3\dfrac{2}{3}$ **33.** $6 \div 6 = 1; \dfrac{17}{18}$

35. $8 - 2 = 6; \dfrac{31}{5}$ or $6\dfrac{1}{5}$ **37.** $P = 7$ in.; $A = 3\dfrac{1}{16}$ in.^{2}

39. $P = 19\dfrac{1}{2}$ yd; $A = 21\dfrac{1}{8}$ yd^2 **41.** $P = 11\dfrac{5}{6}$ ft; $A = 6\dfrac{1}{2}$ ft^2

43. $13 + 9 = 22$ ft; $21\dfrac{1}{6}$ ft **45.** $2 \cdot 6 = 12$ ounces; $9\dfrac{5}{8}$ ounces

47. $4 - 2 = 2$ miles; $2\dfrac{3}{10}$ miles **49.** $4 \cdot 5 = 20$ yd; $18\dfrac{3}{4}$ yd **51.** $\dfrac{5}{8}$ mile;

53. $6\dfrac{3}{8}$ miles **55.** $21\dfrac{3}{8}$ in. **57.** $24 + 35 + 24 + 35 = 118$ in.;

$116\dfrac{1}{2}$ in. **59.** $126 \div 2 = 63$ loads; 84 loads

61. $(4 \cdot 22) + (5 \cdot 23) + (3 \cdot 24) + (12 \cdot 1) = 287$ in.; $280\dfrac{1}{8}$ in.

SUMMARY EXERCISES   Computation with Fractions

1. (a) $\dfrac{3}{8}; \dfrac{5}{8}$ **(b)** $\dfrac{4}{5}; \dfrac{1}{5}$ **2.**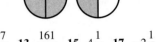

$-1 \uparrow$ 0 $\uparrow$ 1
$-\frac{2}{3}$ $\frac{2}{3}$

3. (a) $-\dfrac{24}{30}$ **(b)** $\dfrac{4}{14}$ **4. (a)** 1 **(b)** -4 **(c)** 9 **5. (a)** $2 \cdot 2 \cdot 2 \cdot 3 \cdot 3$

(b) $3 \cdot 5 \cdot 7$ **6. (a)** $\dfrac{4}{5}$ **(b)** $\dfrac{7}{8}$ **7.** $\dfrac{1}{2}$ **8.** $-\dfrac{5}{24}$ **9.** $\dfrac{17}{16}$ or $1\dfrac{1}{16}$ **10.** $\dfrac{5}{6}$

11. $-\dfrac{2}{15}$ **12.** $-\dfrac{3}{8}$ **13.** 56 **14.** $\dfrac{11}{24}$ **15.** $-\dfrac{7}{6}$ **16.** $-\dfrac{19}{12}$ or $-1\dfrac{7}{12}$

17. $\dfrac{25}{12}$ or $2\dfrac{1}{12}$ **18.** 35 **19.** $5 + 3 = 8; \dfrac{91}{12}$ or $7\dfrac{7}{12}$

20. $2 \cdot 5 = 10; \dfrac{80}{7}$ or $11\dfrac{3}{7}$ **21.** $6 - 3 = 3; \dfrac{33}{10}$ or $3\dfrac{3}{10}$

22. $2 \div 4 = \dfrac{2}{4}$ or $\dfrac{1}{2}; \dfrac{16}{35}$ **23.** $5 \div 1 = 5; 4$ **24.** $3 - 1 = 2; \dfrac{8}{3}$ or $2\dfrac{2}{3}$

25. (a) $2\frac{1}{16}$ in. (b) $\frac{5}{16}$ in. **26.** $P = 3\frac{1}{2}$ in.; $A = \frac{49}{64}$ in.^{2}

27. 12 batches **28.** 6¢ **29.** Not sure, 225 adults; Real, 675 adults;

Imaginary, 600 adults **30.** diameter $= \frac{1}{4}$ in.; $P = 4\frac{1}{2}$ in.

31. 46 bottles **32.** $P = 6$ in.; $A = 2\frac{1}{4}$ in.^{2}

SECTION 4.6

1. (a) $-\frac{3}{8}$ (b) exponent (c) $\left(-\frac{3}{8}\right)\left(-\frac{3}{8}\right)$ **2.** (a) positive

(b) $5\left(-\frac{3}{10}\right)$ and $\left(-\frac{5}{6}\right)\left(\frac{3}{5}\right)$ **3.** $\frac{9}{16}$ **5.** $\frac{8}{125}$ **7.** $-\frac{1}{27}$ **9.** $\frac{1}{32}$

11. $\frac{49}{100}$ **13.** $\frac{36}{25}$ or $1\frac{11}{25}$ **15.** $\frac{12}{25}$ **17.** $\frac{1}{100}$ **19.** $-\frac{3}{2}$ or $-1\frac{1}{2}$

21. $\frac{1}{5} - \frac{42}{10}; \frac{2-42}{10} = -\frac{40}{10} = -4$ **23.** $\frac{5}{16}$ **25.** $\frac{1}{3}$ **27.** $\frac{1}{6}$ **29.** $-\frac{17}{24}$

31. $-\frac{4}{27}$ **33.** $\frac{1}{36}$ **35.** $\frac{9}{64}$ in.^{2} **37.** $1\frac{9}{10}$ miles **39.** 4

41. $-\frac{25}{2}$ or $-12\frac{1}{2}$ **43.** $\frac{1}{14}$ **45.** $\frac{5}{18}$ **47.** -8 **49.** $\frac{9}{100}$

51. (a) The answers are $\frac{1}{4}, -\frac{1}{8}, \frac{1}{16}, -\frac{1}{32}, \frac{1}{64}, -\frac{1}{128}, \frac{1}{256}, -\frac{1}{512}$

(b) When a negative number is raised to an even power, the answer is positive.
When a negative number is raised to an odd power, the answer is negative.

52. (a) Ask yourself, "What number, times itself, is 4?" This is the
numerator. Then ask, "What number, times itself, is 9?" This is the
denominator. The number under the ketchup is either $\frac{2}{3}$ or $-\frac{2}{3}$

(b) The number under the ketchup is $-\frac{1}{3}$ because

$\left(-\frac{1}{3}\right)\left(-\frac{1}{3}\right)\left(-\frac{1}{3}\right) = -\frac{1}{27}$ (c) Either $\frac{1}{2}$ or $-\frac{1}{2}$

SECTION 4.7

1. (a) $\frac{5}{2}$ (b) $\frac{5}{2}$ is the reciprocal of $\frac{2}{5}$. Multiplying $\frac{\cancel{5}^1}{\cancel{2}_1}\left(\frac{\cancel{2}^1}{\cancel{5}_1}c\right)$

gives $1c$, or just c, on the left side of the equal sign. **2.** (a) $\frac{4}{1}$

(b) $\frac{4}{1}$ is the reciprocal of $\frac{1}{4}$. Multiplying $\frac{\cancel{4}^1}{1}\left(\frac{1}{\cancel{4}_1}x\right)$ gives $1x$, or just x,

on the right side of the equal sign.

3. $a = 30$ CHECK $\frac{1}{3}(30) = 10$

$\qquad\qquad\qquad 10 = 10\ \checkmark$

5. $b = -24$ CHECK $-20 = \frac{5}{6}(-24)$

$\qquad\qquad\qquad\quad -20 = -20\ \checkmark$

7. $c = 6$ CHECK $-\frac{7}{2}(6) = -21$

$\qquad\qquad\qquad\qquad -21 = -21\ \checkmark$

9. $d = -\frac{6}{5}$ CHECK $\frac{3}{10} = -\frac{1}{4}\left(-\frac{6}{5}\right)$

$\qquad\qquad\qquad\qquad \frac{3}{10} = \frac{3}{10}\ \checkmark$

11. $n = 12$ CHECK $\frac{1}{6}(12) + 7 = 9$

$\qquad\qquad\qquad\qquad 2 + 7 = 9$

$\qquad\qquad\qquad\qquad\qquad 9 = 9\ \checkmark$

13. $r = -9$ CHECK $-10 = \frac{5}{3}(-9) + 5$

$\qquad\qquad\qquad\qquad -10 = -15 + 5$

$\qquad\qquad\qquad\qquad -10 = -10\ \checkmark$

15. $x = 24$ CHECK $\frac{3}{8}(24) - 9 = 0$

$\qquad\qquad\qquad\qquad 9 - 9 = 0$

$\qquad\qquad\qquad\qquad\quad 0 = 0\ \checkmark$

17. 5; $y = 45$ **19.** $n = -18$ **21.** $x = \frac{1}{12}$ **23.** $b = -\frac{1}{8}$ **25.** $n = 0$

27. (a) $\frac{1}{6}(18) + 1 = -2$

$\qquad\quad 3 + 1 = -2$

$\qquad\qquad\quad 4 \neq -2$

Does *not* balance; correct solution is -18

(b) $-\frac{3}{2} = \frac{9}{4}\left(-\frac{2}{3}\right)$

$\quad -\frac{3}{2} = -\frac{3}{2}\ \checkmark$

Balances, so $-\frac{2}{3}$ is correct solution.

28. (a) $-\frac{3}{4}\left(\frac{5}{6}\right) = -\frac{5}{8}$

$\qquad\qquad -\frac{5}{8} = -\frac{5}{8}\ \checkmark$

Balances, so $\frac{5}{6}$ is correct solution.

(b) $16 = -\frac{7}{3}(6) + 2$

$\quad 16 = -14 + 2$

$\quad 16 \neq -12$

Does not balance; correct solution is -6

29. Let d be the distance traveled.

$\qquad 8 = 2 + \frac{d}{60}$

They traveled 360 miles on the first day.

31. Let d be the distance traveled.

$\qquad 12 = 2 + \frac{d}{50}$

Robb traveled 500 miles on the first day.

33. Let d be the distance traveled.

$\qquad 9 = 2 + \frac{d}{50}$

Robb traveled 350 miles on the third day.

35. Let p be the penny size. $\frac{p}{4} + \frac{1}{2} = 3$. The penny size is 10

37. Let p be the penny size. $\frac{p}{4} + \frac{1}{2} = \frac{5}{2}$. The penny size is 8

39. Some possibilities are: $\frac{1}{2}x = 4; -\frac{1}{4}a = -2; \frac{3}{4}b = 6$

40. Some possibilities are: $\frac{1}{2}y = -6; \frac{2}{3}w = -8; -\frac{1}{12}d = 1$

SECTION 4.8

1. Multiply $\frac{1}{2}$ times 58 m times 66 m; the units in the answer will be m^2.
2. Add 58 m plus 72 m plus 72 m; the units in the answer will be m.

3. $P = 202$ m; $A = 1914$ m^2 **5.** $P = 5$ ft; $A = \frac{27}{32}$ ft^2 **7.** $P = 26\frac{1}{4}$ yd;

$A = 30\frac{3}{4}$ yd^2 **9.** $30\frac{4}{15}$ yd; $A = 38\frac{1}{3}$ yd^2 **11.** $A = 52$ m • 37 m;

$A = 1924$ m^2; $A = 1924$ m$^2 - 208$ m^2; $A = 1716$ m^2 **13.** $A = 7\frac{7}{8}$ ft^2

15. 132 m of curbing; 726 m^2 of grass   **17.** The first mistake was
finding the perimeter instead of the area. The second mistake was
forgetting to include square units for area. The correct answer is
$A = \frac{1}{2} \cdot 15$ ft $\cdot 8$ ft $= 60$ ft^2 **18.** The first mistake was finding
the area instead of the perimeter. The second mistake was using
square units, which are for area, not perimeter. The correct answer
is $P = 10$ cm $+ 13$ cm $+ 13$ cm $= 36$ cm   **19.** Rectangular solid;

$V = 528$ cm^3 **21.** Rectangular solid or cube; $V = 15\frac{5}{8}$ in.^{3} **23.** Pyramid;

$V = 800$ cm^3 **25.** $V = 106\frac{2}{3}$ ft^3 **27.** $V = 18$ in.^{3} **29.** $V = 651,775$ m^3

Chapter 4 REVIEW EXERCISES

1. $\frac{2}{5}$ are squares; $\frac{1}{5}$ are circles   **2.** $\frac{3}{10}$ shaded; $\frac{7}{10}$ unshaded

3.

4. (a) -4 **(b)** 8 **(c)** -1 **5.** $\frac{7}{8}$ **6.** $\frac{3}{5}$ **7.** already in lowest terms

8. $\frac{3x}{8}$ **9.** $\frac{1}{5b}$ **10.** $\frac{4n}{7m^2}$ **11.** $\frac{1}{16}$ **12.** -12 **13.** $\frac{8}{27}$ **14.** $\frac{1}{6x}$

15. $2a^2$ **16.** $\frac{6}{7k}$ **17.** $\frac{5}{24}$ **18.** $-\frac{2}{15}$ **19.** $\frac{19}{6}$ or $3\frac{1}{6}$ **20.** $\frac{3}{2}$ or $1\frac{1}{2}$

21. $\frac{4n+15}{20}$ **22.** $\frac{3y-70}{10y}$ **23.** $2 \div 2 = 1$; $\frac{18}{13}$ or $1\frac{5}{13}$

24. $7 - 5 = 2$; $\frac{5}{2}$ or $2\frac{1}{2}$ **25.** $2 + 2 = 4$; $\frac{81}{20}$ or $4\frac{1}{20}$ **26.** $-\frac{27}{64}$

27. $\frac{1}{36}$ **28.** $-\frac{4}{5}$ **29.** $\frac{7}{6}$ or $1\frac{1}{6}$ **30.** 10   **31.** $-\frac{4}{27}$ **32.** $w = 20$

33. $r = -15$ **34.** $x = \frac{1}{2}$ **35.** $A = 14$ ft^2 **36.** Rectangular solid;

$V = 32\frac{1}{2}$ in.^{3} **37.** Pyramid; $V = 93\frac{1}{3}$ yd^3

Chapter 4 MIXED REVIEW EXERCISES

1. $\frac{8}{15}$ **2.** $\frac{1}{10}$ **3.** $\frac{38}{9}$ or $4\frac{2}{9}$ **4.** $y = -15$ **5.** $d = 6$

6. $x = -\frac{18}{5}$ or $-3\frac{3}{5}$ **7.** $2t^2$ **8.** $\frac{20-3b}{4b}$ **9.** $-\frac{1}{72}$

10. $c = \frac{16}{3}$ or $5\frac{1}{3}$ **11.** $h = 9$ **12.** $x = -\frac{3}{4}$ **13. (a)** $\frac{1}{4}$ pound

(b) $7\frac{1}{2}$ pounds   **(c)** $1\frac{1}{2}$ pounds   **14. (a)** $\frac{5}{6}$ hour   **(b)** $10\frac{11}{12}$ hours

(c) $\frac{11}{12}$ hour   **15.** 12 preschoolers, 40 toddlers, 8 infants

16. $P = 2\frac{1}{10}$ miles; $A = \frac{9}{40}$ mi^2 **17. (a)** $5\frac{5}{6}$ yd   **(b)** $6\frac{1}{2}$ yd   **(c)** Add

your answer $+ 6$ yd $+ 6\frac{1}{2}$ yd. The sum should match the perimeter given in

the problem.

18. (a) $4\frac{1}{2} + 4\frac{1}{2} + 4\frac{1}{2} + 4\frac{1}{2} = \frac{9}{2} + \frac{9}{2} + \frac{9}{2} + \frac{9}{2} = \frac{36}{2} = 18$ in.

(b) $4\left(4\frac{1}{2}\right) = \frac{4}{1} \cdot \frac{9}{2} = \frac{18}{1} = 18$ in.   **(c)** $9\frac{3}{4} \div 4 = \frac{39}{4} \cdot \frac{1}{4} = \frac{39}{16} = 2\frac{7}{16}$ in.

Chapter 4 TEST

1. $\frac{5}{6}$; $\frac{1}{6}$ **2.**

3. $\frac{1}{4}$ **4.** already in lowest terms   **5.** $\frac{2a^2}{3b}$ **6.** $\frac{13}{15}$ **7.** -2 **8.** $-\frac{7}{40}$

9. 14   **10.** $-\frac{2}{27}$ **11.** $\frac{25}{8}$ or $3\frac{1}{8}$ **12.** $\frac{4}{9}$ **13.** $\frac{9}{16}$ **14.** $\frac{4}{7y}$

15. $\frac{24-n}{4n}$ **16.** $\frac{10+3a}{15}$ **17.** $\frac{1}{18b}$ **18.** $-\frac{1}{18}$ **19.** $-\frac{31}{30}$ or $-1\frac{1}{30}$

20. $5 \div 1 = 5$; $\frac{64}{15}$ or $4\frac{4}{15}$ **21.** $3 - 2 = 1$; $\frac{3}{2}$ or $1\frac{1}{2}$ **22.** $d = 35$

23. $t = -\frac{15}{7}$ or $-2\frac{1}{7}$ **24.** $b = 8$ **25.** $x = -15$ **26.** $A = 52$ m^2

27. $A = 58\frac{1}{2}$ yd^2 **28.** Rectangular solid; $V = 6480$ m^3

29. Pyramid; $V = 16$ yd^3 **30.** $14\frac{3}{4}$ hours; $1\frac{5}{6}$ hours

31. $3\frac{1}{2}$ days   **32.** 7392 students work

Chapters 1–4 CUMULATIVE REVIEW EXERCISES

1. five hundred five million, eight thousand, two hundred thirty-eight
2. 35,600,000,916   **3. (a)** 60,700   **(b)** 100,000   **(c)** 3210
4. (a) commutative property of multiplication   **(b)** associative property
of addition   **(c)** distributive property   **5.** -54 **6.** -20 **7.** undefined
8. 81   **9.** 10   **10.** 4   **11.** -16 **12.** $\frac{27}{27} = 1$ **13.** -8 **14.** 35   **15.** $\frac{15a}{2}$

16. $\frac{3}{xy^2}$ **17.** $-\frac{8}{15}$ **18.** $\frac{5}{16}$ **19.** $\frac{14-3b}{21}$ **20.** $\frac{8n+15}{5n}$

21. $\frac{13}{9}$ or $1\frac{4}{9}$ **22.** $\frac{13}{20}$ **23.** -1 **24.** $-\frac{5}{8}$ **25.** $a = 1$ **26.** $y = -5$

27. $k = -36$ **28.** $x = -4$ **29.** $m = 2$ **30.** Square; $P = 18$ in.;

$A = 20\frac{1}{4}$ in.^{2} **31.** Triangle; $P = 20$ yd; $A = 18\frac{3}{4}$ yd^2

32. Parallelogram; $P = 78$ mm; $A = 288$ mm^2 **33.** Let b be the original
weight of each bag. $3b - 16 - 25 = 79$; there were 40 pounds in each bag.
34. Let w be the width; $2w$ is the length. $2(2w) + 2(w) = 102$; width is
17 yd; length is 34 yd.

CHAPTER 5 Rational Numbers: Positive and Negative Decimals

All answers are shown for Concept Checks, Writing exercises, What Went
Wrong exercises, Relating Concepts exercises, Summary Exercises, Review
Exercises, Chapter Tests, and Cumulative Review Exercises. For other
types of exercises, answers are shown only for the odd-numbered exercises.

SECTION 5.1

1. 3   **2.** 4   **3.** 8 tenths   **4.** 3 thousandths   **5.** 7; 0; 4   **7.** 4; 7; 8
9. 1; 8; 9   **11.** 6; 2; 1   **13.** 410.25   **15.** 6.5432   **17.** 5406.045

19. $\dfrac{7}{10}$ **21.** $13\dfrac{2}{5}$ **23.** $\dfrac{7}{20}$ **25.** $\dfrac{33}{50}$ **27.** $10\dfrac{17}{100}$ **29.** $\dfrac{3}{50}$ **31.** $\dfrac{41}{200}$

33. $5\dfrac{1}{500}$ **35.** $\dfrac{343}{500}$ **37.** five tenths **39.** seventy-eight hundredths

41. one hundred five thousandths **43.** twelve and four hundredths

45. one and seventy-five thousandths **47.** 6.7 **49.** 0.32

51. (a) 420.008 **(b)** 0.428 **53. (a)** 0.0703 **(b)** 700.0003

55. 75.030 **57.** Anne should not say "and" because that denotes a decimal point. Anne should have read it as "four thousand three hundred two." **58.** Jerry used "and" twice; Jerry should have read it as "nine and one hundred sixteen-thousandths." Only the first "and" is correct.

59. ten thousandths inch; $\dfrac{10}{1000} = \dfrac{1}{100}$ inch **61. (a)** 12 pounds

(b) 4 pounds **63.** 3-C **64.** 4-C **65.** 4-A **66.** 3-B **67.** one and six hundred two thousandths centimeters **68.** One and twenty-six thousandths centimeters **69. (a)** millionths, ten-millionths, hundred-millionths, billionths **(b)** These names match the words on the left side of the place value chart, with "ths" attached. **70. (a)** First place to the left of the decimal point is ones. **(b)** "Oneths" would mean a fraction with a denominator of 1, which would equal 1 or more. Anything that is 1 or more is to the *left* of the decimal point. **71.** 20,500,000,000

72. 14,600 **73. (a)** 7,820,000 **(b)** 9,600,000,000 **(c)** 42,700

(d) 704.8 trillion **74. (a)** 123,200 **(b)** 503,420,000 **(c)** 36,510,000,000

(d) 90.032 trillion **75.** 81.03 trillion **76** 6.45 millon

77. (a) 3.079 million **(b)** 25.6 thousand **(c)** 590.8 billion

(d) 4.312 trillion **78. (a)** 608.9 billion **(b)** 42.387 million

(c) 85.2 thousand **(d)** 8.74 trillon

SECTION 5.2

1. 0 **2.** 9 **3.** Look only at the 6, which is **5 or more.** Round up by adding 1 thousandth to 5.709 to get 5.710. **4.** Look only at the 2, which is **4 or less.** So the part you are keeping, 10.0, stays the same. **5.** 16.9

7. 0.956 **9.** 0.80 **11.** 3.661 **13.** 794.0 **15.** 0.0980 **17.** 9.09

19. 82.0002 **21.** $0.82 **23.** $1.22 **25.** $0.70 **27.** $48,650

29. $840 **31.** $500 **33.** $1.00 **35.** $1000 **37. (a)** 312 miles per hour **(b)** 149 miles per hour **39. (a)** 186.0 miles per hour

(b) 763.0 miles per hour **41.** Rounds to $0 (zero dollars) because $0.499 is closer to $0 than to $1 **42.** Rounding $0.499 to the nearest cent to get $0.50 is more helpful. Guideline: Round amounts less than $1.00 to the nearest cent instead of the nearest dollar. **43.** Rounds to $0.00 (zero cents) because $0.0015 is closer to $0.00 than to $0.01 **44.** Both round to $0.60. Rounding to the nearest thousandth (tenth of a cent) would allow you to identify $0.597 as less than $0.601

SECTION 5.3

1.
$$\begin{array}{r} 6.420 \\ +10.163 \end{array}$$
 2.
$$\begin{array}{r} 7.000 \\ +9.204 \end{array}$$
 3.
$$\begin{array}{r} 20.0000 \\ +9.1263 \end{array}$$
 4.
$$\begin{array}{r} 137.06 \\ -12.00 \end{array}$$

5. 17.72 **7.** 11.98 **9.** 115.861 **11.** 59.323 **13.** Keneya should have written 6 as 6.00. The correct sum is 46.22 **14.** Caleb should have written 65 as 65.00. The correct sum is 85.36 **15.** 0.3000 written as a fraction in lowest terms is $\frac{3}{10}$, which is equivalent to 0.3

or $0.3000 = \dfrac{3000 \div 1000}{10,000 \div 1000} = \dfrac{3}{10} = 0.3$ **16.** Write 7.0 and 0.7 in fraction

form to see the difference. 7.0 is $7\dfrac{0}{10}$ but 0.7 is $\dfrac{7}{10}$ **17.** 89.7

19. 0.109 **21.** 0.91 **23.** 6.661 **25.** Keagan subtracted in the wrong order; 15.32 should be on top. The correct answer is 7.87

26. The two problems are done in a different order: $8 - 2.9$ is not the same as $2.9 - 8$ because subtraction is not commutative. **27. (a)** 24.75 in.

(b) 3.95 in. **29. (a)** 62.27 in. **(b)** 0.39 in. **31.** 23.013 **33.** -45.75

35. -6.69 **37.** -6.99 **39.** -4.279 **41.** -0.0035 **43.** 5.37

45. *Exact:* 6.275 **46.** *Exact:* 18.63 **47.** *Estimate:* $-7 + 1 = -6$;

Exact: -5.8 **48.** *Estimate:* $-10 + 3 = -7$ *Exact:* -6.58

49. *Estimate:* $-1 + (-9) = -10$; *Exact:* -9.671 **50.** *Estimate:*

$-800 + (-1) = -801$; *Exact:* -803.6 **51.** *Estimate:* $8 + (+50) = 58$;

Exact: 59.23 **52.** *Estimate:* $10 + (+7) = 17$; *Exact:* 21.486

53. *Estimate:* $70 - 50 = 20$ million people; *Exact:* 16.1 million people

55. *Estimate:* $600 + 300 + 200 + 70 + 50 + 40 + 20 = 1280$ million people; *Exact:* 1344.61 million people **57.** *Estimate:* $2 + 2 + 2 = 6$ m;

Exact: 6.14 m, which is less than the rhino's height by 0.26 m

59. *Estimate:* $20 - \$9 = \$11$; *Exact:* $10.88 **61.** *Estimate:*

$7 - 7 = \$0$; *Exact:* $0.50 **63.** *Estimate:* $30 + 3 + 3 + 20 + 4 = \$60$;

Exact: $60.14 **65.** $2249.81 **67.** $71.74 **69.** $b = 1.39$ cm

71. $w = 7.943$ ft

SECTION 5.4

1. 3 **2.** 5 **3.** 2 **4.** 4 **5.** 0.1344 **7.** -159.10 **9.** 15.444

11. $34,500.20 **13.** -43.2 **14.** -43.2 **15.** 0.432 **16.** 0.432

17. 0.0432 **18.** 0.0432 **19.** 0.0000312 **21.** 0.000009

23. *Exact:* 190.08 **25.** *Estimate:* $40 \times 40 = 1600$; *Exact :* 1558.2

27. *Estimate:* $7 \times 5 = 35$; *Exact:* 30.038 **29.** *Estimate:* $3 \times 7 = 21$;

Exact: 19.24165 **31.** unreasonable; $289.00 **32.** unreasonable; 2.5 hours

33. reasonable **34.** unreasonable; $692 **35.** unreasonable; $4.19

36. reasonable **37.** unreasonable; 9.5 pounds **38.** unreasonable; $14.90 or

$149 **39.** *Estimate:* $50 \times 20 = \$1000$; *Exact:* $945.87 (rounded)

41. *Estimate:* $1 \times 4 = \$4$; *Exact:* $2.45 (rounded) **43.** $55.78 for premium unleaded. She would save $6.20 by getting regular unleaded. **45.** $20,265

47. (a) Area before 1929 $\approx$ 23.2 in.^{2} (rounded); Area today $\approx$ 16.0 in.^{2} (rounded) **(b)** 7.2 in.^{2} **49. (a)** 0.43 in. **(b)** 4.3 in.

51. (a) $929.88; $1079.88 **(b)** $1439.28 **53.** $4.09 (rounded)

55. (a) $193.75 **(b)** $216.25 **57. (a)** 5.38; 49.5; 105; 9; 70;

143.4 **(b)** When multiplying by 10, the decimal point moves one place to the right. **(c)** When multiplying by 100, the decimal point moves two places to the right; when multiplying by 1000, it moves three places to the right. **58. (a)** 0.0538; 0.495; 1.05; 0.09; 5.8; 30 **(b)** When multiplying by 0.1, the decimal point moves one place to the left. **(c)** When multiplying by 0.01, the decimal point moves two places to the left; when multiplying by 0.001, the decimal point moves three places to the left.

SECTION 5.5

1. (c) $25.5 \div 5$; $5\overline{)25.5}$ **2. (b)** $42.3 \div 7$; $7\overline{)42.3}$ **3.** $2.2\overline{)8.24}$

4. $5.1\overline{)10.5}$ **5.** -3.9 **7.** 0.47 **9.** 400.2 **11.** 36 **13.** 0.06

14. 0.6 **15.** 6000 **16.** 6 **17.** 60 **18.** 600 **19.** 0.0006 **20.** 0.006

21. 25.3 **23.** 516.67 (rounded) **25.** -26.756 (rounded)

27. 10,082.647 (rounded)

29. unreasonable; *Estimate:* $40 \div 8 = 5$; $8\overline{)37.8}$ with quotient 4.725

30. unreasonable; *Estimate:* $300 \div 3 = 100$; $3\overline{)345.6}$ with quotient 115.2

31. reasonable; *Estimate:* $50 \div 50 = 1$

32. unreasonable; *Estimate:* $2000 \div 5 = 400$; $4.8\overline{)2428.8}$ with quotient 50.6

33. unreasonable; *Estimate:* $300 \div 5 = 60$; $5.1\overline{)307.02}^{\,60.2}$

34. reasonable; *Estimate:* $400 \div 5 = 80$ **35.** $4.00 (rounded)

37. $67.08   **39.** $11.92 per hour   **41.** 28.0 miles per gallon (rounded)

43. $0.03 (rounded)   **45.** 7.37 meters (rounded)   **47.** 0.08 meter

49. 22.49 meters   **51.** The first mistake was multiplying 4.4 by 2 instead of multiplying 4.4 by 4. The second mistake was ignoring the parentheses instead of doing the work inside the parentheses before subtracting. The correct answer is 13.76   **52.** The first mistake was doing the subtraction first instead of doing the work inside the parentheses. The second mistake was writing a plus sign where there is an understood multiplication sign. The correct answer is 44   **53.** 14.25   **55.** 3.8   **57.** -16.155 **59.** 3.714

61. (a) 1,583,333 pieces (rounded)   (b) 26,388 pieces (rounded)
(c) 440 pieces (rounded)   **63.** 200,000 box tops   **65.** 5263 box tops (rounded)   **67.** 0.377; 0.91; 0.0886; 3.019; 40.65; 662.57   (a) Dividing by 10, decimal point moves one place to the left; by 100, two places to the left; by 1000, three places to the left.   (b) The decimal point moved to the *right* when multiplying by 10, by 100, or by 1000; here it moves to the *left* when dividing by 10, by 100, or by 1000.   **68.** 402; 71; 3.39; 157.7; 460; 8730   (a) Dividing by 0.1, the decimal point moves one place to the right; by 0.01, two places to the right; by 0.001, three places to the right.   (b) The decimal point moved to the *left* when multiplying by 0.1, 0.01, or 0.001; here it moves to the *right* when dividing by 0.1, 0.01, or 0.001.

SUMMARY EXERCISES   Computation with Decimal Numbers

1. $\frac{4}{5}$ **2.** $6\frac{1}{250}$ **3.** $\frac{7}{20}$ **4.** ninety-four and five tenths   **5.** two and three ten-thousandths   **6.** seven hundred six thousandths   **7.** 0.05
8. 0.0309   **9.** 10.7   **10.** 6.19   **11.** 1.0   **12.** 0.420   **13.** $0.89
14. $3.00   **15.** $100   **16.** -0.945 **17.** 49.6199   **18.** -50 **19.** 15.03
20. 0.00488   **21.** -2.15 **22.** 9.055   **23.** 18.4009   **24.** -6.995
25. -808.9 **26.** 2.12   **27.** 0.04 (rounded)   **28.** $P = 52.1$ in.
29. $P = 9.735$ meters   **30.** $1.80   **31.** $91.28   **32.** (a) 144 ft^2
(b) $1.18 per square foot (rounded)   **33.** 121 ft^2; $0.82 per square foot (rounded)   **34.** (a) 9 days (rounded)   (b) 15.6 pounds (rounded)
35. Average weight of food eaten by bee is 0.32 ounce.   **36.** 0.112 ounce

SECTION 5.6

1. (a) Not correct; $5\overline{)2}$ (b) $7\overline{)1}$ (c) $8\overline{)3}$ **2.** (a) $9\overline{)4}$
(b) Not correct; $3\overline{)1}$ (c) $6\overline{)5}$ **3.** $4\overline{)3.}^{\,0.}$ To continue dividing, write zeros in the dividend.   **4.** $8\overline{)1.}^{\,0.}$ To continue dividing, write zeros in the dividend.   **5.** (a) less than   (b) greater than   (c) less than   **6.** (a) less than
(b) greater than   (c) greater than   **7.** 0.5   **9.** 0.75   **11.** 0.3
13. 0.9   **15.** 0.6   **17.** 0.875   **19.** 2.25   **21.** 14.7   **23.** 3.625
25. 6.333 (rounded)   **27.** 0.833 (rounded)   **29.** 1.889 (rounded)
31. $\frac{2}{5}$ **33.** $\frac{5}{8}$ **35.** $\frac{7}{20}$ **37.** 0.35   **39.** $\frac{1}{25}$ **41.** 0.2   **43.** $\frac{9}{100}$
45. shorter; 0.72 inch   **47.** more; 0.05 inch   **49.** too much; 0.005 gram
51. 0.9991 cm, 1.0007 cm   **52.** 3.0 ounces; 2.995 ounces; 3.005 ounces
53. (a) $<$ (b) $=$ (c) $>$ (d) $<$ **55.** 0.5399, 0.54, 0.5455
57. 5.0079, 5.79, 5.8, 5.804   **59.** 0.6009, 0.609, 0.628, 0.62812
61. 2.8902, 3.88, 4.876, 5.8751   **63.** 0.006, 0.043, $\frac{1}{20}$, 0.051
65. 0.37, $\frac{3}{8}$, $\frac{2}{5}$, 0.4001   **67.** (a) red box; green box   (b) 0.01 inch

68. (a) green box   (b) no difference; they are equivalent.
69. 1.4 in. (rounded)   **71.** 0.3 in. (rounded)   **73.** 0.4 in. (rounded)

75. (a) A proper fraction like $\frac{5}{9}$ is less than 1, so $\frac{5}{9}$ cannot be equivalent to a decimal number that is greater than 1   (b) $\frac{5}{9}$ means $5 \div 9$ or $9\overline{)5}$ so the correct answer is 0.556 (rounded). This makes sense because both the fraction and decimal are less than 1   **76.** (a) $2.035 = 2\frac{35}{1000} = 2\frac{7}{200}$, not $2\frac{7}{20}$
(b) Adding the whole number part gives $2 + 0.35$, which is 2.35 not 2.035. To check, $2.35 = 2\frac{35}{100} = 2\frac{7}{20}$ but $2.035 = 2\frac{35}{1000} = 2\frac{7}{200}$

77. Just add the whole number part to 0.375. So $1\frac{3}{8} = 1.375$; $3\frac{3}{8} = 3.375$; $295\frac{3}{8} = 295.375$ **78.** It works only when the fraction part has a one-digit numerator and a denominator of 10, or a two-digit numerator and a denominator of 100, and so on.

SECTION 5.7

1. (a) average   (b) the number of values   **2.** (a) the number of times it occurs (frequency)   (b) grade point average (GPA)   **3.** mean $= \frac{558}{8} \approx 69.8$ (rounded)   **5.** $39,622   **7.** 6.1 (rounded)   **9.** 2.60   **11.** (a) 2.8
(b) 2.93 (rounded)   (c) 3.13 (rounded)   **13.** 15 texts   **15.** 516 students
17. 48.5 pounds of shrimp   **19.** 5366 miles (rounded)   **21.** 3865.5 miles
23. 8 samples   **25.** 68 and 74 years (bimodal)   **27.** no mode
29. -10 degrees; 4 degrees; 14 degrees warmer   **31.** Magneilia forgot to write the numbers in order from least to greatest before finding the middle number. The median is 35.   **32.** Leopoldo did not include both of the 86 scores in the numerator. The correct test average is
$\frac{86 + 72 + 90 + 86 + 67}{5} = \frac{401}{5} = 80.2$

SECTION 5.8

1. 4   **3.** 8   **5.** 3.317 (rounded)   **7.** 2.236 (rounded)   **9.** 8.544 (rounded)   **11.** 10.050 (rounded)   **13.** 19   **15.** 31.623 (rounded)
17. 30 is about halfway between 25 and 36, so $\sqrt{30}$ should be about halfway between 5 and 6, or about 5.5. Using a calculator, $\sqrt{30} \approx 5.477$ Similarly, $\sqrt{26}$ should be a little more than $\sqrt{25}$; by calculator $\sqrt{26} \approx 5.099$ And $\sqrt{35}$ should be a little less than $\sqrt{36}$; by calculator $\sqrt{35} \approx 5.916$ **18.** Squaring a number is multiplying the number times itself. Finding the square root "undoes" squaring. It is asking what number was multiplied by itself. Examples will vary.   **19.** $\sqrt{1521} = 39$ ft
21. $\sqrt{289} = 17$ in.   **23.** $\sqrt{144} = 12$ mm   **25.** $\sqrt{73} \approx 8.5$ in. (rounded)
27. $\sqrt{65} \approx 8.1$ yd (rounded)   **29.** $\sqrt{195} \approx 14.0$ cm (rounded)
31. $\sqrt{7.94} \approx 2.8$ m (rounded)   **33.** $\sqrt{65.01} \approx 8.1$ cm (rounded)
35. $\sqrt{292.32} \approx 17.1$ km (rounded)   **37.** Jenoveba's first mistake was using the formula for finding the hypotenuse when the unknown side is a leg. Her second mistake was using square units, which are only for area. The correct answer is $\sqrt{(20)^2 - (13)^2} = \sqrt{231} \approx 15.2$ m (rounded).
38. Ibrahim's first and second mistakes were squaring the numbers incorrectly. 9^2 is $(9)(9)$ or 81, and 7^2 is $(7)(7)$ or 49. His third mistake was rounding the answer to the nearest thousandth instead of the nearest tenth. The correct answer is $\sqrt{81 + 49} = \sqrt{130} \approx 11.4$ in. (rounded)
39. hypotenuse $= \sqrt{65} \approx 8.1$ ft (rounded)

41. leg = $\sqrt{360,000}$ = 600 m **43.** hypotenuse = $\sqrt{48.5}$ ≈ 7.0 ft (rounded) **45.** leg = $\sqrt{135}$ ≈ 11.6 ft (rounded)
47. $\sqrt{16,200}$ ≈ 127.3 ft (rounded)
48. (a)

Second
60 ft
Third — First
60 ft
Home

(b) $\sqrt{7200}$ ≈ 84.9 ft (rounded)

49. The distance from third to first is the same as the distance from home to second because the baseball diamond is a square. **50. (a)** The side length is less than 60 ft. **(b)** 80^2 = 6400; 6400 ÷ 2 = 3200; $\sqrt{3200}$ ≈ 56.6 ft (rounded)

SECTION 5.9

1. Add 6.2 to both sides because −6.2 + 6.2 gives x + 0 on the right side.
2. Divide both sides by −4.5 because $\dfrac{-4.5y}{-4.5}$ gives y on the left side.
3. 0 + n = −1.4, n = −1.4; See MyMathlab for sample *check*.
5. 0.008 = b **7.** 0.29 = a **9.** p = −120 **11.** t = 0.7; See MyMathlab for sample *check*. **13.** x = −0.82 **15.** 0 = z **17.** 110 = h
19. c = 0.45 **21.** w = 28 **23.** p = −40.5 **25.** Let d be the adult dose. 0.3d = 9; The adult dose is 30 milligrams. **27.** Let d be the number of days the saw was rented. 75.95d + 12 = 315.80; The saw was rented for 4 days. **29.** 0.7(220 − a) = 140; The person is 20 years old.
30. 0.7(220 − a) = 126; The person is 40 years old.
31. 0.7(220 − a) = 134; a ≈ 28.57, which rounds to 29. The person is about 29 years old. **32.** 0.7(220 − a) = 117; a ≈ 52.86, which rounds to 53. The person is about 53 years old.

SECTION 5.10

1. perimeter; ft **2.** radius; divide by 2 **3.** d = 18 mm **5.** r = 0.35 km
7. C ≈ 69.1 ft (rounded); A ≈ 379.9 ft² (rounded) **9.** C ≈ 8.2 m (rounded); A ≈ 5.3 m² (rounded) **11.** C ≈ 47.1 cm; A ≈ 176.6 cm² (rounded) **13.** C ≈ 23.6 ft (rounded) A ≈ 44.2 ft² (rounded)
15. A ≈ 76.9 in.² (rounded) **17.** A ≈ 57 cm²
19. A ≈ 7850 yd² **21.** C ≈ 91.4 in. (rounded); Bonus: 693 revolutions/mile (rounded) **23.** A ≈ 70,650 mi² **25.** watch: C ≈ 3.1 in. (rounded) A ≈ 0.8 in.² (rounded); wall clock: C ≈ 18.8 in. (rounded); A ≈ 28.3 in.² (rounded)

Watch Wall Clock

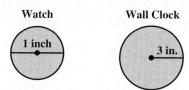

27. A ≈ 3846.5 mi² − 803.8 mi²; Difference ≈ 3042.7 mi² (rounded)

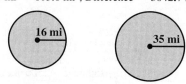

29. (a) d ≈ 45.9 cm (rounded) **(b)** Divide the circumference by π, that is, 144 cm ÷ 3.14 **31.** C ≈ 14.6 ft (rounded) **33.** A ≈ 197.8 cm² (rounded) **35.** V ≈ 471 ft³; S ≈ 345.4 ft² **37.** V = 1617 m³; S = 849.4 m² **39.** V ≈ 763.0 in.³ (rounded); S ≈ 678.2 in.² (rounded)

41. V = 5550 mm³; S = 2150 mm² **43.** S = 163.6 in.²
45. The first mistake was using the diameter of 7 cm instead of the radius of 3.5 cm. The second mistake was labeling the answer as square units, which are used for area, not volume. The correct answer is V ≈ 192.3 cm³ (rounded). **46.** Olivia labeled the answer with the units cm³, which are used for volume, not surface area. The correct answer is S ≈ 186.83 cm²
47. Length, width, and height are all the same. **48.** Multiply the side measurement times itself to get the area of one face of the cube, then multiply by 6 to find the total surface area. $S = 6s^2$, where s is the length of one edge of the cube.

Chapter 5 REVIEW EXERCISES

1. 0; 5 **2.** 0; 6 **3.** 8; 9 **4.** 5; 9 **5.** 7; 6 **6.** $\dfrac{1}{2}$ **7.** $\dfrac{3}{4}$ **8.** $4\dfrac{1}{20}$
9. $\dfrac{7}{8}$ **10.** $\dfrac{27}{1000}$ **11.** $27\dfrac{4}{5}$ **12.** eight tenths **13.** four hundred and twenty-nine hundredths **14.** twelve and seven thousandths
15. three hundred six ten-thousandths **16.** 8.3 **17.** 0.205 **18.** 70.0066
19. 0.30 **20.** 275.6 **21.** 72.79 **22.** 0.160 **23.** 0.091 **24.** 1.0
25. $15.83 **26.** $0.70 **27.** $17,625.79 **28.** $350 **29.** $130
30. $100 **31.** $29 **32.** −5.67 **33.** −0.03 **34.** 5.879 **35.** −6.435
36. *Estimate:* 200 million − 30 million = 170 million people; *Exact:* 120.3 million people **37.** *Estimate:* $400 − $300 − $70 = $30; *Exact:* $15.80
38. *Estimate:* $2 + $5 + $20 = $27; $30 − $27 = $3; *Exact:* $4.14
39. *Estimate:* 2 + 4 + 5 = 11 kilometers; *Exact:* 11.55 kilometers
40. *Estimate:* 6 × 4 = 24; *Exact:* 22.7106 **41.** *Estimate:* 40 × 3 = 120; *Exact:* 141.57 **42.** 0.0112 **43.** −0.000355 **44.** reasonable; *Estimate:* 700 ÷ 10 = 70 **45.** unreasonable; *Estimate:* 30 ÷ 3 = 10; $2.8\overline{)26.6}^{9.5}$
46. 14.467 (rounded) **47.** 1200 **48.** −0.4 **49.** $708 (rounded)
50. $1.99 (rounded) **51.** 133 shares (rounded) **52.** $3.47 (rounded)
53. −4.715 **54.** 10.15 **55.** 3.8 **56.** 0.64 **57.** 1.875 **58.** 0.111 (rounded) **59.** 3.6008, 3.68, 3.806 **60.** 0.209, 0.2102, 0.215, 0.22
61. $\dfrac{1}{8}, \dfrac{3}{20}$, 0.159, 0.17 **62.** mean: 28.5 phones; median: 19.5 phones
63. mean: 44 loans; median: 39 loans **64.** $51.05 **65.** Store J: $107 and $160 (bimodal); Store K: $119 **66.** $\sqrt{289}$ = 17 in.
67. $\sqrt{49}$ = 7 cm **68.** $\sqrt{104}$ ≈ 10.2 cm (rounded) **69.** $\sqrt{52}$ ≈ 7.2 in. (rounded) **70.** $\sqrt{6.53}$ ≈ 2.6 m (rounded) **71.** $\sqrt{71.75}$ ≈ 8.5 km (rounded) **72.** b = 0.25 **73.** x = 0 **74.** n = −13 **75.** a = −1.5
76. y = −16.2 **77.** d = 137.8 m **78.** $r = 1\dfrac{1}{2}$ in. or 1.5 in.
79. C ≈ 6.3 cm (rounded); A ≈ 3.1 cm² (rounded) **80.** C ≈ 109.3 m (rounded); A ≈ 950.7 m² (rounded) **81.** C ≈ 37.7 in. (rounded); A ≈ 113.0 in.² (rounded) **82.** V ≈ 549.5 cm³; S ≈ 376.8 cm²
83. V = 1808.6 m³ (rounded); S ≈ 1205.8 m² (rounded)
84. V ≈ 7.9 ft³ (rounded); S = 25.5 ft²

Chapter 5 MIXED REVIEW EXERCISES

1. 404.865 **2.** −254.8 **3.** 3583.261 (rounded) **4.** 29.0898
5. 0.03066 **6.** 9.4 **7.** −15.065 **8.** −9.04 **9.** −15.74 **10.** 8.19
11. 0.928 **12.** 35 **13.** −41.859 **14.** 0.3 **15.** $2.17 (rounded)
16. $199.71 **17.** $78.50 **18.** −0.7 = y **19.** x = 30 **20.** 15.7 ft long rubber strip; A ≈ 19.6 ft² (rounded) **21.** mean is 67.6; median is 82
22. 20 miles **23.** V ≈ 49.5 in.³ (rounded)

ANSWERS

Chapter 5 TEST

1. $18\frac{2}{5}$ **2.** $\frac{3}{40}$ **3.** sixty and seven thousandths **4.** two hundred eight ten-thousandths **5.** *Estimate:* $8 + 80 + 40 = 128$; *Exact:* 129.2028
6. *Estimate:* $-6(1) = -6$; *Exact:* -6.948 **7.** *Estimate:* $-80 - 4 = -84$; *Exact:* -82.702 **8.** *Estimate:* $-20 \div (-5) = 4$; *Exact:* 4.175
9. 669.004 **10.** 480 **11.** 0.000042 **12.** $\$3.79$ per foot (rounded)
13. Davida, by 0.441 minute **14.** $\$5.35$ (rounded) **15.** $y = -6.15$
16. $x = -0.35$ **17.** $a = 10.9$ **18.** $n = 2.85$ **19.** $0.44, \frac{9}{20}, 0.4506,$
0.451 **20.** 32.09 **21.** 75 books **22.** $103°$ and $104°$ (bimodal)
23. $\$11.25$ **24.** $\$50.55$ **25.** $\sqrt{85} \approx 9.2$ cm (rounded)
26. $\sqrt{279} \approx 16.7$ ft (rounded) **27.** $r = 12.5$ in. **28.** $C \approx 5.7$ km
(rounded) **29.** $A \approx 206.0$ cm^2 (rounded) **30.** $V \approx 5086.8$ ft^3
31. $S \approx 2599.9$ ft^2 (rounded)

Chapters 1–5 CUMULATIVE REVIEW EXERCISES

1. (a) forty-five and two hundred three ten-thousandths (b) thirty billion, six hundred fifty thousand, eight **2.** (a) $160,000,500$ (b) 0.075
3. $46,900$ **4.** 6.20 **5.** 0.661 **6.** $10,000$ **7.** -13 **8.** -0.00006
9. undefined **10.** 44 **11.** 18 **12.** -6 **13.** 56.621 **14.** $\frac{3}{2}$ or $1\frac{1}{2}$
15. $\frac{3}{8b}$ **16.** 11.147 **17.** $-\frac{9}{20}$ **18.** $-\frac{1}{32}$ **19.** 1.3 (rounded)
20. $\frac{5x+6}{10}$ **21.** -8 **22.** -32 **23.** $\frac{18}{18} = 1$ **24.** -5.76
25. $\frac{55}{24}$ or $2\frac{7}{24}$ **26.** 24.2 years (rounded) **27.** 23.5 years
28. 20 and 26 years (bimodal) **29.** $h = -4$ **30.** $x = 5$
31. $r = 10.9$ **32.** $y = -2$ **33.** $n = 4$ **34.** $P = 54$ ft; $A = 140$ ft^2
35. $C \approx 40.8$ m (rounded); $A \approx 132.7$ m^2 (rounded) **36.** $x \approx 12.6$ km
(rounded); $A \approx 37.8$ km^2 **37.** $V \approx 3.4$ in.^{3} (rounded); $S = 13.5$ in.^{2}
38. *Estimate:* $\$60 - \$50 - \$1 = \$9$; *Exact:* $\$11.17$
39. *Estimate:* $2 + 4 = 6$ yards; *Exact:* $6\frac{5}{24}$ yards **40.** *Estimate:*
$\$6 \div 3 = \$2$ per pound; *Exact:* $\$2.29$ per pound (rounded)
41. *Estimate:* $\$80,000 \div 100 = \$800$; *Exact:* $\$729$ (rounded)
42. Let x be first student's money and $3x$ be second student's money. $x + 3x = 30,000$; one student receives $\$7500$; the other receives $\$22,500$.
43. Let w be the width and $w + 4$ be the length. $2(w + 4) + 2w = 48$; The width is 10 in. and the length is 14 in.

CHAPTER 6 Ratio, Proportion, and Line/Angle/Triangle Relationships

All answers are shown for Concept Checks, Writing exercises, What Went Wrong exercises, Relating Concepts exercises, Summary Exercises, Review Exercises, Chapter Tests and Cumulative Review Exercises. For other types of exercises, answers are shown only for the odd-numbered exercises.

SECTION 6.1

1. Answers will vary. One possibility is: A ratio compares two quantities with the same units. Examples will vary. **2.** You can divide out the common units. **3.** 20; $\frac{4}{1}$ **4.** 5; $\frac{4}{15}$ **5.** $\frac{8}{9}$ **7.** $\frac{2}{1}$ **9.** $\frac{1}{3}$ **11.** $\frac{9}{7}$

13. $\frac{5}{6}$ **15.** Because $15 \div 15$ is 1, Chander should write 1 in the numerator. The correct answer is $\frac{1}{2}$ **16.** Reena tried to divide out the fraction parts of the mixed numbers. When a ratio contains a mixed number, the mixed numbers must be rewritten as an improper fraction. $\frac{5\frac{1}{3}}{1\frac{1}{3}} = \frac{\frac{16}{3}}{\frac{4}{3}} =$
$\frac{16}{3} \div \frac{4}{3} = \frac{16}{3} \cdot \frac{3}{4} = \frac{4}{1}$ The correct answer is $\frac{4}{1}$ **17.** 48 in.; $\frac{8}{5}$ **19.** $\frac{1}{12}$
21. $\frac{4}{1}$ **23.** $\frac{1}{7}$ **25.** $\frac{13}{1}$ **27.** $\frac{6}{1}$ **29.** $\frac{38}{17}$ **31.** $\frac{1}{4}$ **33.** $\frac{34}{35}$

SECTION 6.2

1. $\frac{3\text{ feet}}{7\text{ seconds}}$ **3.** $\frac{5\text{ letters}}{1\text{ minute}}$ **5.** $\frac{18\text{ miles}}{1\text{ gallon}}$ **7.** Divide; correct set-up is $\frac{\$5.85}{3\text{ boxes}}$ or $3\overline{)5.85}$ **8.** 5 rooms for 1 nurse **9.** $\$12$ per hour or $\$12/$hour
11. 1.25 pounds/person **13.** $\$103.30/$day **15.** 325.9; 21.0 (rounded)
17. 338.6; 20.9 (rounded) **19.** 4 ounces for $\$3.65$, about
$\$0.913/$ounce **21.** 14 ounces for $\$2.89$, about $\$0.206/$ounce
23. 18 ounces for $\$1.79$, about $\$0.099/$ounce **25.** Answers will vary. For example, you might choose Brand B because you prefer organic or your doctor told you to lower your salt intake, so the additional cost might be worth the added health benefits. **26.** Answers will vary. For example, if you use only half of the larger bag, you really pay $\$0.30$ per pound, so the smaller bag is the best buy. **27.** 1.75 pounds/week **29.** $\$12.26/$hour
31. (a) Uber, $\$15.70$; taxi, $\$17.50$; Lyft, $\$16.40$ (b) Uber, $\$3.14/$mile; taxi, $\$3.50/$mile; Lyft, $\$3.28/$mile; Uber is the best buy. **33.** Uber, $\$30.15$, $\$3.02/$mile; taxi, $\$30.25$, $\$3.03/$mile; Lyft, $\$30.20$, $\$3.02/$mile. They are all about $\$3.00$ per mile. **35.** One battery for $\$1.79$; like getting 3 batteries so $\$1.79 \div 3 \approx \$0.597$ per battery **37.** Brand P with the 50¢ coupon is the best buy. $\$3.39 - \$0.50 = \$2.89$ and $\$2.89 \div 16.5$ ounces $\approx \$0.175$ per ounce **39.** Plan A: $\$0.03/$MB; Plan B: $\$0.06/$MB; Plan C: $\$0.02/$MB (rounded); Plan C is the best buy.
40. (a) Plan A: 50 MB/day; (b) Plan B; 17 MB/day (rounded); (c) Plan C: 83 MB/day (rounded) **41.** 130 MB overage means you need to pay for 100 MB plus another 100 MB to cover the entire overage amount so $2(\$5) = \$10$. (You cannot buy less than 100 MB.) $\$30 + \$10 = \$40$;
$$\frac{\$40}{(500\text{ MB} + 130\text{ MB})} = \frac{\$40}{630\text{ MB}} \approx \$0.06/\text{MB}$$
42. $(90$ extra MB$)(\$1.50) = \$135$; $\$135 + \$60 = \$195$;
$$\frac{\$195}{(2500\text{ MB} + 90\text{ MB})} = \frac{\$195}{2590\text{ MB}} \approx \$0.08/\text{MB}$$
43. Plan A: $\$65$, $\$0.03/$MB (rounded); Plan B: $\$120$, $\$0.05/$MB (rounded) Plan C $\$60$, $\$0.03/$MB (rounded) **44.** Plan A: $\$75$; Plan B: $\$155$; Plan C: $\$810$

SECTION 6.3

1. $\frac{\$9}{12\text{ cans}} = \frac{\$18}{24\text{ cans}}$ **2.** $\frac{28\text{ people}}{7\text{ cars}} = \frac{16\text{ people}}{4\text{ cars}}$
3. $\frac{200\text{ adults}}{450\text{ children}} = \frac{4\text{ adults}}{9\text{ children}}$ **4.** $\frac{150\text{ trees}}{1\text{ acre}} = \frac{1500\text{ trees}}{10\text{ acres}}$
5. $\frac{120}{150} = \frac{8}{10}$ **6.** $\frac{6}{9} = \frac{10}{15}$ **7.** $\frac{3}{5} = \frac{3}{5}$; true **9.** $\frac{5}{8} = \frac{5}{8}$; true

11. $\frac{3}{4} \neq \frac{2}{3}$; false   **13.** Gary did not simplify the second ratio correctly; the numerator and denominator must be divided by the *same* number. Both ratios are already in lowest terms and are not equal, so the proportion is false.   **14.** Margo did not find the cross products. The 5 and 8 need to be multiplied by the numbers in the other ratio. Cross multiply to find that $5 \cdot 40 = 200$ and $8 \cdot 30 = 240$ so this proportion is false.   **15.** $54 = 54$; true   **17.** $336 \neq 320$; false   **19.** $2880 \neq 2970$; false   **21.** $28 = 28$; true   **23.** $42\frac{1}{2} \neq 45$; false   **25.** $66 = 66$; true

27. $\frac{168 \text{ hits}}{600 \text{ at bats}} = \frac{126 \text{ hits}}{450 \text{ at bats}}$ $\begin{array}{l} 600 \cdot 126 = 75{,}600 \\ 168 \cdot 450 = 75{,}600 \end{array}$ Cross products are *equal* so the proportion is *true*; they hit equally well.   **28.** Left-hand ratio compares hours to cartons, but right-hand ratio compares cartons to hours. Correct proportion is: $\frac{3.5 \text{ hours}}{91 \text{ cartons}} = \frac{5.25 \text{ hours}}{126 \text{ cartons}}$ $\begin{array}{l}(91)(5.25) = 477.75 \\ (3.5)(126) = 441\end{array}$ Cross products are *not* equal, so the proportion is *false*; they did not work equally fast.

29. $3 \cdot x = 12$

$\dfrac{\overset{1}{\cancel{3}} \cdot x}{\underset{1}{\cancel{3}}} = \dfrac{12}{3}$

$x = 4$

31. $x = 2$ **33.** $x = 88$ **35.** $x = 91$ **37.** $x = 5$ **39.** $x = 10$
41. $x \approx 24.44$ (rounded)   **43.** $x = 50.4$ **45.** $x \approx 17.64$ (rounded)
47. $x = 1$ **49.** $x = 3\frac{1}{2}$ **51.** $x = 0.2$ or $x = \frac{1}{5}$ **53.** $x = 0.005$ or $x = \frac{1}{200}$ **55.** Find the cross products: $20 \neq 30$, so the proportion is false. $\frac{6\frac{2}{3}}{4} = \frac{5}{3}$ or $\frac{10}{6} = \frac{5}{3}$ or $\frac{10}{4} = \frac{7.5}{3}$ or $\frac{10}{4} = \frac{5}{2}$ **56.** Find the cross products: $192 \neq 180$, so the proportion is false. $\frac{6.4}{8} = \frac{24}{30}$ or $\frac{6}{7.5} = \frac{24}{30}$ or $\frac{6}{8} = \frac{22.5}{30}$ or $\frac{6}{8} = \frac{24}{32}$

SUMMARY EXERCISES   Ratios, Rates, and Proportions

1. $\frac{60 \text{ million}}{20 \text{ million}} = \frac{3}{1}$ **2.** $\frac{20 \text{ million}}{25 \text{ million}} = \frac{4}{5}$ **3.** $\frac{25 \text{ million} + 20 \text{ million}}{60 \text{ million}} = \frac{45 \text{ million}}{60 \text{ million}} = \frac{3}{4}$ **4.** $\frac{505 \text{ million}}{130 \text{ million}} = \frac{101}{26}$ **5.** Comparing the bass to piano, guitar, woodwinds, drums, and strings gives ratios of $\frac{1}{20}, \frac{1}{10}, \frac{1}{8}, \frac{1}{6}$, and $\frac{1}{3}$, respectively.   **6. (a)** woodwinds to bass   **(b)** piano to guitar, or drums to strings   **7.** 2.1 points/min; 0.5 min/point (both rounded)   **8.** 1.7 points/min; 0.6 min/point (both rounded)   **9.** \$16.32/hour; \$24.48/hour for overtime   **10.** \$0.20/channel; \$0.17/channel; \$0.18/channel (all answers rounded)   **11.** 12 ounces for \$7.24, about \$0.60 per ounce (rounded)   **12.** Brand P with the \$2 coupon is the best buy at \$0.57 per pound.

13. $\frac{4}{3} = \frac{4}{3}$ or $924 = 924$; true   **14.** $2.0125 \neq 2.07$; false
15. $68\frac{1}{4} = 68\frac{1}{4}$; true   **16.** $x = 28$ **17.** $x = 3.2$ **18.** $x = 182$
19. $x \approx 3.64$ (rounded)   **20.** $x \approx 0.93$ (rounded)   **21.** $x = 1.56$
22. $x \approx 0.05$ (rounded)   **23.** $x = 1$ **24.** $x = \frac{3}{4}$

SECTION 6.4

1. $\frac{6 \text{ potatoes}}{4 \text{ eggs}} = \frac{12 \text{ potatoes}}{x \text{ eggs}}$ **2.** $\frac{1 \text{ serving}}{15 \text{ chips}} = \frac{x \text{ servings}}{70 \text{ chips}}$
3. $\frac{4 \text{ strips}}{5 \text{ hours}} = \frac{18 \text{ strips}}{x \text{ hours}}$; 22.5 hours   **5.** \$7.20   **7.** 42 pounds
9. \$403.68   **11.** 10 ounces (rounded)   **13.** 5 quarts   **15.** 14 ft; 10 ft
17. 14 ft; 8 ft   **19.** 96 pieces of chicken; 33.6 pounds lasagna; 10.8 pounds deli meats; $5\frac{3}{5}$ pounds cheese; 7.2 dozen (about 86) buns; 14.4 pounds potato salad.   **21.** 44 students is an unreasonable answer because there are only 35 students in the class.   **22.** 2 minutes is an unreasonable answer because it is less than the 30 minutes she gets in just one day.
23. 625 stocks went up   **25.** 4.06 meters (rounded)   **27.** 10.53 meters (rounded)   **29.** 5610 students use cream in their coffee.   **31.** 120 calories and 12 grams of sugar   **33.** You cannot solve this problem using a proportion. Age and weight are not proportional. As Jim's age increases, his weight may decrease, stay the same, or increase.   **34.** You cannot solve this problem using a proportion. The cost per pound for different parts of a chicken is not proportional. Chicken wings are generally much cheaper per pound than chicken breasts.   **35.** 2065 students (reasonable); about 4214 students with incorrect set-up (only 2950 students in the group)   **36.** About 2 people (reasonable); 2000 people with incorrect set-up (only 60 people at the restaurent).   **37.** Tali did not use the correct numbers in the proportion. The problem compares a person's weight to the number of calories burned and does not involve time. The correct proportion is $\frac{150}{210} = \frac{180}{x}$; the answer is 252 calories.   **38.** Amar did not use the correct units in the answer. There are 495 *calories* in 5 cookies.
39. $\frac{2}{3}$ cup water, 1 tablespoon butter/margarine, $\frac{1}{3}$ cup milk, $\frac{3}{4}$ cup flakes
40. 2 cups water, 3 tablespoons butter/margarine, 1 cup milk, $2\frac{1}{4}$ cups flakes

SECTION 6.5

1. Wording may vary slightly. A **line** is a row of points continuing in both directions forever; the drawing should have arrows on both ends. A **line segment** is a piece of a line; the drawing should have an endpoint (dot) on each end. A **ray** has one endpoint and goes on forever in one direction; the drawing should have one endpoint (dot) and an arrowhead on the other end.   **2.** Wording may vary slightly. An **acute angle** measures less than 90°; a **right angle** measures exactly 90° and forms a "square corner"; an **obtuse angle** measures more than 90° but less than 180°; a **straight angle** measures exactly 180° and forms a straight line. Drawings of an acute and obtuse angle may vary.   **3.** line named $\overleftrightarrow{CD}$ or $\overleftrightarrow{DC}$ **5.** line segment named $\overline{GF}$ or $\overline{FG}$ **7.** ray named $\overrightarrow{PQ}$ **9.** perpendicular   **11.** parallel
13. intersecting   **15.** $\angle AOS$ or $\angle SOA$ **17.** $\angle AQC$ or $\angle CQA$
19. right angle (90°)   **21.** acute angle   **23.** straight angle (180°)
25. $\angle EOD$ and $\angle COD$; $\angle AOB$ and $\angle BOC$ **27.** $\angle HNE$ and $\angle ENF$; $\angle ACB$ and $\angle KOL$ **29.** 50°   **31.** 4°   **33.** 50°   **35.** 90°
37. $\angle SON \cong \angle TOM$; $\angle TOS \cong \angle MON$ **39.** $\angle GOH$ measures 63°; $\angle EOF$ measures 37°; $\angle AOC$ and $\angle GOF$ both measure 80°   **41.** True, because $\overleftrightarrow{UQ}$ is perpendicular to $\overleftrightarrow{ST}$.   **42.** True, because they form a 90° angle, as indicated by the small red square   **43.** False; the angles have the same measure (both are 180°).   **44.** False; $\overleftrightarrow{ST}$ and $\overleftrightarrow{PR}$ are parallel.

45. False; $\overleftrightarrow{QU}$ and $\overleftrightarrow{TS}$ are perpendicular. **46.** True, because both angles are formed by perpendicular lines, so they both measure 90°
47. corresponding angles: $\angle 1$ and $\angle 8$, $\angle 2$ and $\angle 5$, $\angle 3$ and $\angle 6$, $\angle 4$ and $\angle 7$; alternate interior angles: $\angle 4$ and $\angle 5$, $\angle 3$ and $\angle 8$ **49.** $\angle 2$, $\angle 4$, $\angle 6$, $\angle 8$ all measure 130°; $\angle 1$, $\angle 3$, $\angle 5$, $\angle 7$ all measure 50° **51.** $\angle 6$, $\angle 1$, $\angle 3$, $\angle 8$ all measure 47°; $\angle 5$, $\angle 2$, $\angle 7$, $\angle 4$ all measure 133°
53. $\angle 6$, $\angle 8$, $\angle 4$, $\angle 2$ all measure 114°; $\angle 7$, $\angle 5$, $\angle 3$, $\angle 1$ all measure 66°
55. $\angle 1 \cong \angle 3$, both are 138°; $\angle 2 \cong \angle ABC$, both are 42°

SECTION 6.6

1. One dictionary definition is "conforming; agreeing." Examples of congruent objects include two matching chairs and two pieces of paper from the same notebook. **2.** One dictionary definition is "resembling, but not identical." Examples of similar objects are sets of different size pots or measuring cups; small and large size cans of beans. **3.** $\angle 1$ and $\angle 4$; $\angle 2$ and $\angle 5$; $\angle 3$ and $\angle 6$; $\overline{AB}$ and $\overline{DE}$; $\overline{BC}$ and $\overline{EF}$; $\overline{AC}$ and $\overline{DF}$.
5. $\angle 1$ and $\angle 6$; $\angle 2$ and $\angle 4$; $\angle 3$ and $\angle 5$; $\overline{ST}$ and $\overline{YW}$; $\overline{TU}$ and $\overline{WX}$; $\overline{SU}$ and $\overline{YX}$ **7.** $\angle 1$ and $\angle 6$; $\angle 2$ and $\angle 5$; $\angle 3$ and $\angle 4$; $\overline{LM}$ and $\overline{TS}$; $\overline{LN}$ and $\overline{TR}$; $\overline{MN}$ and $\overline{SR}$ **9.** SAS **11.** SSS **13.** ASA **15.** use SAS: $BC = CE$, $\angle ABC \cong \angle DCE$, $BA = CD$ **17.** use SAS: $PS = RS$, $m\angle QSP = m\angle QSR = 90°$, $QS = QS$ (common side) **19.** $a = 6$ cm; $\dfrac{7.5}{b} = \dfrac{6}{12}$; $b = 15$ cm **21.** $a = 5$ mm; $b = 3$ mm **23.** $a = 24$ in.; $b = 20$ in. **25.** $x = 24.8$ m; Perimeter $= 72.8$ m; $y = 15$ m; Perimeter $= 54.6$ m **27.** Perimeter $= 8$ cm $+ 8$ cm $+ 8$ cm $= 24$ cm; Area $\approx (0.5)(8$ cm$)(6.9$ cm$) \approx 27.6$ cm² (The area is approximate because the height was rounded to the nearest tenth.) **29.** $h = 24$ ft
31. $x = 50$ m **33.** $n = 110$ m

Chapter 6 REVIEW EXERCISES

1. $\dfrac{3}{4}$ **2.** $\dfrac{4}{1}$ **3.** great white shark to whale shark; whale shark to blue whale
4. $\dfrac{2}{1}$ **5.** $\dfrac{2}{3}$ **6.** $\dfrac{5}{2}$ **7.** $\dfrac{1}{6}$ **8.** $\dfrac{3}{1}$ **9.** $\dfrac{3}{8}$ **10.** $\dfrac{4}{3}$ **11.** $\dfrac{1}{9}$ **12.** $\dfrac{10}{7}$
13. $\dfrac{7}{5}$ **14.** $\dfrac{5}{6}$ **15.** $\dfrac{\$11}{1 \text{ dozen}}$ **16.** $\dfrac{12 \text{ children}}{5 \text{ families}}$ **17.** 0.2 page/minute or $\dfrac{1}{5}$ page/minute; 5 minutes/page **18.** \$20/hour; 0.05 hour/dollar or $\dfrac{1}{20}$ hour/dollar **19.** 8 ounces for \$4.98, about \$0.623/ounce
20. 17.6 pounds for \$18.69 $-$ \$1 coupon, about \$1.005/pound
21. $\dfrac{3}{5} = \dfrac{3}{5}$ or $90 = 90$; true **22.** $\dfrac{1}{8} \neq \dfrac{1}{4}$ or $432 \neq 216$; false
23. $\dfrac{47}{10} \neq \dfrac{49}{10}$ or $980 \neq 940$; false **24.** $\dfrac{5}{8} = \dfrac{5}{8}$ or $4.8 = 4.8$; true
25. $\dfrac{3}{2} = \dfrac{3}{2}$ or $14 = 14$; true **26.** $x = 1575$ **27.** $x = 20$ **28.** $x = 400$
29. $x = 12.5$ **30.** $x \approx 14.67$ (rounded) **31.** $x \approx 8.17$ (rounded)
32. $x = 50.4$ **33.** $x \approx 0.57$ (rounded) **34.** $x \approx 2.47$ (rounded)
35. 27 cats **36.** 46 hits **37.** \$23.74 (rounded) **38.** 3299 students (rounded) **39.** 68 ft **40.** $27\dfrac{1}{2}$ hours or 27.5 hours **41.** 490 calories (rounded) **42.** 14.7 milligrams **43.** line segment named $\overline{AB}$ or $\overline{BA}$
44. line named $\overleftrightarrow{CD}$ or $\overleftrightarrow{DC}$ **45.** ray named $\overrightarrow{OP}$ **46.** parallel
47. perpendicular **48.** intersecting **49.** acute angle **50.** obtuse angle
51. straight angle; 180° **52.** right angle; 90° **53.** (a) 10°
(b) 45° **(c)** 83° **54. (a)** 25° **(b)** 90° **(c)** 147°
55. $\angle 1$ and $\angle 4$ measure 30°; $\angle 3$ and $\angle 6$ measure 90°; $\angle 5$ measures 60°

56. $\angle 8$, $\angle 3$, $\angle 6$, $\angle 1$ all measure 160°; $\angle 4$, $\angle 7$, $\angle 2$, $\angle 5$ all measure 20°
57. SSS **58.** SAS **59.** ASA **60.** $y = 30$ ft; $x = 34$ ft; $P = 104$ ft
61. $y = 7.5$ m; $x = 9$ m; $P = 22.5$ m **62.** $x = 12$ mm; $y = 7.5$ mm; $P = 38$ mm

Chapter 6 MIXED REVIEW EXERCISES

1. $x = 105$ **2.** $x = 0$ **3.** $x = 128$ **4.** $x \approx 23.08$ (rounded)
5. $x = 6.5$ **6.** $x \approx 117.36$ (rounded) **7.** $\dfrac{8}{5}$ **8.** $\dfrac{33}{80}$ **9.** $\dfrac{15}{4}$
10. $\dfrac{4}{1}$ **11.** $\dfrac{4}{5}$ **12.** $\dfrac{37}{7}$ **13.** $\dfrac{3}{8}$ **14.** $\dfrac{1}{12}$ **15.** $\dfrac{45}{13}$ **16.** 24,900 fans (rounded) **17.** $\dfrac{8}{3}$ **18.** 75 ft for \$1.99 $-$ \$0.50 coupon, about \$0.020/ft
19. 21 ft long; 15 ft wide **20. (a)** 1400 milligrams **(b)** 100 milligrams
21. 21 points (rounded) **22.** $\dfrac{1}{2}$ or 0.5 teaspoon **23.** parallel lines
24. line segment **25.** acute angle **26.** intersecting lines
27. right angle; 90° **28.** ray **29.** straight angle; 180°
30. obtuse angle **31.** perpendicular lines **32. (a)** The car turned around in a complete circle. **(b)** The governor took the opposite view, for example, having once opposed taxes but now supporting them.
33. (a) No; because obtuse angles are greater than 90°, their sum would be greater than 180° **(b)** Yes; because acute angles are less than 90°, their sum could equal 90° **34.** $\angle 1$ measures 45°; $\angle 3$ and $\angle 6$ measure 35°; $\angle 4$ measures 55°; $\angle 5$ measures 90° **35.** $\angle 5$, $\angle 2$, $\angle 7$, $\angle 4$ all measure 75°; $\angle 6$, $\angle 1$, $\angle 8$, $\angle 3$ all measure 105°

Chapter 6 TEST

1. $\dfrac{\$1}{5 \text{ minutes}}$ **2.** $\dfrac{9}{2}$ **3.** $\dfrac{15}{4}$ **4.** 16 ounces for \$1.89 $-$ \$0.50 coupon, about \$0.087/ounce **5.** $x = 25$ **6.** $x \approx 2.67$ (rounded) **7.** $x = 325$
8. $x = \dfrac{21}{2} = 10\dfrac{1}{2}$ **9.** 24 orders **10.** 87 students (rounded)
11. 23.8 grams (rounded) **12.** 60 feet **13.** (e) **14.** (a), 90°
15. (d) **16.** (g); 180° **17.** Parallel lines are lines in the same plane that never intersect. Perpendicular lines intersect to form a right angle. Sketches will vary. **18.** 9° **19.** 160° **20.** $m\angle 1 = 50°$, $m\angle 2 = 35°$, $m\angle 3 = 95°$, $m\angle 5 = 35°$ **21.** Measures of $\angle 1$, $\angle 3$, $\angle 5$, and $\angle 7$ are all 65°; measures of $\angle 2$, $\angle 4$, $\angle 6$, and $\angle 8$ are all 115° **22.** ASA (Angle–Side–Angle)
23. SAS (Side–Angle–Side) **24.** $y = 12$ cm; $z = 6$ cm **25.** In the larger triangle, $x = 12$ mm, $P = 46.8$ mm; in the smaller triangle, $y = 14$ mm, $P = 39$ mm

Chapters 1–6 CUMULATIVE REVIEW EXERCISES

1. (a) seventy-seven billion, one million, eight hundred five
(b) two hundredths **2. (a)** 3.040 **(b)** 500,037,000 **3.** 0
4. 14 **5.** $\dfrac{35}{12}$ or $2\dfrac{11}{12}$ **6.** $\dfrac{2h - 3}{10}$ **7.** 99.9905 **8.** $\dfrac{3}{0}$ is undefined.
9. -18 **10.** -14 **11.** $\dfrac{n}{4m}$ **12.** -0.00042 **13.** $\dfrac{1}{2y^2}$ **14.** $\dfrac{27 + 2n}{3n}$
15. -64 **16.** 5.4 **17.** $-\dfrac{1}{18}$ **18.** $y = -1$ **19.** $x = 15$ **20.** $x = 10$
21. Let t be the starting temperature. $t + 15 - 23 + 5 = 71$. The starting temperature was 74 degrees. **22.** Let l be the length; $l - 14$ is the width. $2(l) + 2(l - 14) = 100$. Length is 32 ft; width is 18 ft **23.** $P = 7$ km; $A \approx 2.0$ km² (rounded) **24.** $d = 17$ m; $C \approx 53.4$ m (rounded); $A \approx 226.9$ m² (rounded) **25.** $V \approx 339.1$ ft³ (rounded); $S \approx 282.6$ ft²

26. 250 pounds   **27.** $1\frac{1}{4}$ teaspoons   **28.** $7\frac{3}{20}$ miles   **29.** 22.0 miles per gallon (rounded)   **30.** $0.005, $0.018, $0.02, $0.084, $0.099, $0.10, $0.106   **31.** 202 minutes (rounded)   **32.** 189 minutes (rounded) per $20 card, so buy 2 cards to cover 360 minutes (6 hours).   **33.** $\frac{1000}{200} = \frac{5}{1}$ **34.** 109.5 minutes, but the call would be cut off after 109 minutes.

CHAPTER 7 Percent

All answers are shown for Concept Checks, Writing exercises, What Went Wrong exercises, Relating Concepts exercises, Summary Exercises, Review Exercises, Chapter Tests, and Cumulative Review Exercises. For other types of exercises, answers are shown only for the odd-numbered exercises.

SECTION 7.1

1. (a) 2 places to the left   **(b)** 48.% = 0.48; 120.% = 1.20 or 1.2; 03.5% = 0.035; 06. = 0.06   **2. (a)** 2 places to the right   **(b)** 0.88 = 88%; 2.70 = 270%; 0.105 = 10.5%; 0.30 = 30%   **3.** 0.25   **5.** 0.30 or 0.3   **7.** 0.06   **9.** 1.40 or 1.4   **11.** 0.078   **13.** 1.00 or 1   **15.** 0.005   **17.** 0.0035   **19.** 50%   **21.** 62%   **23.** 3%   **25.** 12.5%   **27.** 200%   **29.** 260%   **31.** 3.12%   **33.** $\frac{1}{5}$ **35.** $\frac{1}{2}$ **37.** $\frac{11}{20}$ **39.** $\frac{3}{8}$ **41.** $\frac{1}{16}$ **43.** $\frac{1}{6}$ **45.** $\frac{13}{10} = 1\frac{3}{10}$ **47.** $\frac{5}{2} = 2\frac{1}{2}$ **49.** 25%   **51.** 30%   **53.** 60%   **55.** 37%   **57.** $37\frac{1}{2}$% or 37.5%   **59.** 5%   **61.** exactly $55\frac{5}{9}$%, or 55.6% (rounded)   **63.** exactly $14\frac{2}{7}$%, or 14.3% (rounded)   **65.** 0.08   **67.** 0.54   **69.** 3.5%   **71.** 200%   **73.** $\frac{95}{100}$ or 95% shaded; $\frac{5}{100}$ or 5% unshaded   **75.** $\frac{3}{4}$ or 75% shaded; $\frac{1}{4}$ or 25% unshaded   **77.** 50%; 0.5; $\frac{1}{2}$ **78.** 0.25; $\frac{1}{4}$; 25%   **79.** 0.01; 1%   **81.** $\frac{1}{5}$; 20%   **83.** $\frac{3}{10}$; 0.3   **85.** 0.5; 50%   **87.** $\frac{9}{10}$; 0.9   **89.** $\frac{3}{2} = 1\frac{1}{2}$; 150%   **91.** 0.08; $\frac{2}{25}$ **93.** 5%; 0.05; $\frac{1}{20}$ **95.** $\frac{1}{4}$; 0.25; 25%   **97.** $\frac{3}{8}$; 0.375; 37.5%   **99.** $\frac{1}{7}$; exactly $14\frac{2}{7}$% or 14.3% (rounded)   **101.** In the first problem Omaria forgot to move the decimal point in 0.35 two places to the right. The correct answer is $\frac{7}{20} = 0.35 = 35\%$. In the second problem, Omaria did the division in the wrong order. She should enter $16 \div 25$ to get 0.64 and then move the decimal point two places to the right. The correct answer is $\frac{16}{25} = 0.64 = 64\%$.
102. In the first problem Ezra moved the decimal point in the wrong direction. The decimal point should move two places to the *right*. The correct answer is 3.2 = 320%. In the second problem Ezra started with the decimal point in the wrong place. The decimal point starts at the far *right* side of 60. The correct answer is 60% = 0.60   **103. (a)** $78   **(b)** $39   **105. (a)** 15 inches   **(b)** $7\frac{1}{2}$ inches   **107. (a)** 2.8 miles   **(b)** 1.4 miles   **109.** 20 children   **111.** 60 credits   **113. (a)** $2509.50   **(b)** 50%   **(c)** $2509.50   **115. (a)** 4100 students   **(b)** 50%   **117. (a)** 35 problems   **(b)** 0 problems   **119.** 50% means 50 out of 100 parts. That's half of the number. A shortcut for finding 50% of a number is to divide the number by 2. Examples will vary.   **120.** Since 100% means 100 parts out of 100 parts, 100% is all of the number. Examples will vary.

SECTION 7.2

1. (a) 10%   **(b)** 3000 runners   **(c)** unknown   **(d)** $\frac{10}{100} = \frac{n}{3000}$

2. (a) 35%   **(b)** 2340 volunteers   **(c)** unknown   **(d)** $\frac{35}{100} = \frac{n}{2340}$

3. (a) unknown   **(b)** 32 pizzas   **(c)** 16 pizzas   **(d)** $\frac{p}{100} = \frac{16}{32}$

4. (a) unknown   **(b)** 140 hours   **(c)** 35 hours   **(d)** $\frac{p}{100} = \frac{35}{140}$

5. (a) 90%   **(b)** unknown   **(c)** 495 students   **(d)** $\frac{90}{100} = \frac{495}{n}$

6. (a) 28%   **(b)** unknown   **(c)** 84 e-mails   **(d)** $\frac{28}{100} = \frac{84}{n}$

7. 4.8 feet   **9.** $\frac{p}{100} = \frac{16}{200}$; 8%   **11.** $28   **13.** $\frac{250}{100} = \frac{n}{7}$; 17.5 hours   **15.** $\frac{p}{100} = \frac{32}{172}$; 18.6% (rounded)   **17.** $\frac{110}{100} = \frac{748}{n}$; 680 e-books   **19.** $\frac{14.7}{100} = \frac{n}{274}$; $40.28 (rounded)   **21.** $\frac{p}{100} = \frac{105}{54}$; 194.4% (rounded)   **23.** $\frac{5}{100} = \frac{0.33}{n}$; $6.60   **25.** His first mistake was not using the correct proportion. His second mistake was forgetting to label his answer with the % symbol. The correct proportion is $\frac{p}{100} = \frac{14}{8}$; the answer is 175%.
26. Her first mistake was not using the correct proportion. Her second mistake was labeling the answer as a percent, when the answer is the number of *children*. The correct proportion is $\frac{30}{100} = \frac{9}{n}$; the answer is 30 children.   **27.** 150% of $30 cannot be *less* than $30; 25% of $16 cannot be *greater than* $16.   **28.** The *percent* is written over 100. It appears with the word *percent* or % after it. The *whole* is the entire quantity. It often appears after the word *of*. The *part* is the number being compared to the whole.

SECTION 7.3

1. 1500 patients   **2.** 96 updates   **3.** $15   **4.** $710   **5.** 4.5 pounds   **6.** 0.7 foot   **7.** $7.00   **8.** $36   **9.** 52 students   **10.** 4.6 miles   **11.** 870 phones   **12.** 7 games   **13.** 4.75 hours   **14.** $0.37   **15. (a)** The shortcut for finding 10% is dividing by 10, or moving the decimal point one place to the left. 10% means $\frac{10}{100}$ or $\frac{1}{10}$. The denominator tells you to divide the whole by 10.   **(b)** Once you find 10% of a number, multiply the result by 2 for 20% and by 3 for 30%.   **16. (a)** The shortcut for finding 1% is dividing by 100, or moving the decimal point two places to the left. 1% means $\frac{1}{100}$ and the denominator of 100 tells you to divide the whole by 100.   **(b)** Once you find 1% of a number, multiply the result by 2 for 2% and by 3 for 3%.   **17.** 231 programs   **19.** 50%   **21.** 680 circuits   **23.** 1080 people   **25.** 845 posts   **27.** $20.80   **29.** 76%   **31.** 125%   **33.** 4700 employees   **35.** 83.2 quarts   **37.** 2%   **39.** 700 tablets   **41.** 325 toddlers   **43.** 1.5%   **45.** 5 gallons   **47.** 1029.2 meters   **49.** In the first test question, he forgot to multiply 0.2 by 100% to change it from a decimal to a percent. The correct answer is 20%. In the second test question, he used the wrong equation. The correct equation is $50 = p \cdot 20$; $p = 250\%$.   **50.** In the first test question, she used the wrong equation. The correct equation is $12 = (0.05)(n)$; $n = 240$ in. In the second test question, she forgot to change 4% to the decimal 0.04 before using the percent equation. The correct equation is $n = (0.04)(30)$; $n = 1.2$ pounds.   **51.** Shortcut for 25%: divide 80 by 4 to get 20. There are 20 boys.

52. 100% is all the children, so 100% − 25% who are boys = 75% who are girls.   **53.** Method 1: 80 children − 20 boys = 60 girls. Method 2: From Exercise 51, 25% of the children is 20 boys. So 75% is 3 times 20 = 60 girls.   **54.** There are 20 boys (from Exercise 51). The shortcut for finding 10% is dividing by 10, or moving the decimal point one place to the left. 10% of 20. is 2 boys who are left handed. There are no left-handed girls, so the total for left-handed children is 2.

SUMMARY EXERCISES   Percent

1. (a) 0.03; 3%   **(b)** $\frac{3}{10}$; 0.3   **(c)** $\frac{3}{8}$; 37.5%   **(d)** $1\frac{3}{5}$; 1.6

(e) 0.0625; 6.25%   **(f)** $\frac{1}{20}$; 0.05   **(g)** 2; 200%   **(h)** 0.8; 80%

(i) $\frac{9}{125}$; 7.2%   **2. (a)** 3.5 ft   **(b)** 19 miles   **(c)** 105 cows   **(d)** $0.08

(e) 500 women   **(f)** $45   **(g)** $87.50   **(h)** 12 pounds   **(i)** 95 students

3. $12\frac{1}{2}$% or 12.5%   **4.** 75 messages   **5.** $0.53 (rounded)   **6.** 175%

7. 5.9 pounds (rounded)   **8.** 600 hours   **9.** 50%   **10.** 500 camp sites

11. 98 golf balls   **12.** 2.7% (rounded)   **13.** 2300 tweets

14. 0.63 ounce   **15.** 145%   **16.** 244 voters (rounded)   **17.** 0.021 inch

18. 400%   **19.** 8%   **20.** $0.68   **21.** Invitations; $1026.80   **22.** $3171

23. $4530   **24.** $1902.60   **25. (a)** $13,288   **(b)** $97.71 (rounded)

SECTION 7.4

1. (a) 10%   **(b)** 5%   **(c)** 1%   **2. (a)** 50%   **(b)** 25%   **(c)** 10%

3. $37.80   **5. (a)** 101.6 pounds (rounded)   **(b)** 10.1 pounds (rounded)

(c) 28.1 pounds (rounded)   **7.** Pay off debt, 59.0%; save or invest, 66.0%; buy something, 22.9% (all rounded).   **9.** 320 million people   **11.** 138%

13. 40 problems   **15.** 1321 shots (rounded)   **17.** 23.7 miles per gallon (rounded)   **19.** 62% (rounded); earns $15,808 more in a year

21. 139% (rounded)   **23.** $535; $535; 64.8% (rounded)   **25.** 9.1% (rounded)   **27.** 40%   **29.** 776% (rounded)   **31.** No. 100% is the entire price, so a decrease of 100% would take the price down to 0. Therefore, 100% is the maximum possible decrease in the price of something.

32. 25% of $80 is $80 ÷ 4 = $20; 75% would be 3 times $20 = $60; 125% would be $80 + $20 = $100   **33.** George ate more than 65 grams, so the percent must be greater than 100%. Use $p \cdot 65 = 78$ to get 120%.

34. The team won more than half the games, so the percent must be greater than 50%. Correct solution is 0.72 = 72%.   **35.** The brain could not weigh 375 pounds, which is more than the person weighs. $2\frac{1}{2}$% = 2.5% = 0.025, so $(0.025)(150) = n$ and n = 3.75 pounds   **36.** If 80% were absent, then only 20% made it to class. 800 − 640 = 160 students, or use $(0.20)(800) = n$.

SECTION 7.5

1. (a) 0.065   **(b)** 0.0125   **(c)** 0.1025   **(d)** 0.0575   **2. (a)** 0.75 year

(b) 11.5 years   **(c)** 4.25 years   **(d)** 8.75 years   **3.** $29.28 (rounded); $395.26   **5.** 3%; $70.04   **7.** $0.12 (rounded); $2.22   **9.** $4\frac{1}{2}$%; $13,167

11. $3 + $1.50 = $4.50; $4.83 (rounded); 2 ($3) = $6; $6.43 (rounded)

13. $8 + $4 = $12; $11.75 (rounded); 2 ($8) = $16; $15.67 (rounded)

15. $1 + $0.50 = $1.50; $1.43 (rounded); 2 ($1) = $2; $1.91

17. $15; $85   **19.** 30%; $126   **21.** $4.38 (rounded); $13.12

23. $3.79 (rounded); $34.09   **25.** $42; $342   **27.** $33.30; $773.30

29. $56.25; $1556.25   **31.** $919.67 (rounded); $18,719.67   **33.** $1170

35. $7978.13 (rounded)   **37.** $26.61 (rounded)   **39.** 8%   **41.** $74.25

43. $25.13   **45.** $395.67; $803.33   **47.** $2016.38 (rounded)

49. $21 (rounded)   **51.** $92.38   **53.** $230.12   **55. (a)** $14,000

(b) $14,859; $859 more with compound interest   **56. (a)** $25,000

(b) $43,219; $18,219 more with compound interest   **57. (a)** $50,000 with simple interest; $469,016 with compound interest; $419,016 more with compound interest   **(b)** 9 times greater (rounded); $\frac{$469,016}{50,000} \approx 9.4$

Chapter 7   REVIEW EXERCISES

1. 0.25   **2.** 1.8   **3.** 0.125   **4.** 0.07   **5.** 265%   **6.** 2%   **7.** 30%

8. 0.2%   **9.** $\frac{3}{25}$ **10.** $\frac{3}{8}$ **11.** $\frac{5}{2} = 2\frac{1}{2}$ **12.** $\frac{1}{20}$ **13.** 75%   **14.** 62.5% or $62\frac{1}{2}$%   **15.** 325%   **16.** 6%   **17.** 0.125   **18.** 12.5%   **19.** $\frac{3}{20}$

20. 15%   **21.** $\frac{9}{5} = 1\frac{4}{5}$ **22.** 1.8   **23.** $46   **24.** $23   **25.** 9 hours

26. $4\frac{1}{2}$ hours   **27.** 242 meters   **28.** 17,000 cases   **29.** 27 smartphones

30. 870 math books   **31.** 9.5% (rounded)   **32.** 225%   **33.** $2.60 (rounded)   **34.** 80 days   **35.** 4%   **36.** $575   **37.** 20 people

38. 175%   **39. (a)** 3000 patients   **(b)** 31.5% decrease (rounded)

40. (a) $364   **(b)** 224%   **41. (a)** 80%   **(b)** 6.3% increase (rounded)

42. 85.7% (rounded)   **43.** $0.11 (rounded); $2.90   **44.** $7\frac{1}{2}$%; $838.50

45. $4 + $2 = $6; $6.41 (rounded); 2 ($4) = $8; $8.55 (rounded)

46. $0.80 + $0.40 = $1.20; $1.21 (rounded); 2 ($0.80) = $1.60; $1.61

47. $3.75; $33.75   **48.** 25%; $189   **49.** $34.13 (rounded); $384.13

50. $183.60; $1713.60

Chapter 7   MIXED REVIEW QUESTIONS

1. $\frac{1}{3}$; $33\frac{1}{3}$% (exact) or 33.3% (rounded)   **2.** Puzzle; 44%; 0.44; $\frac{11}{25}$

3. 277 Americans (rounded)   **4. (a)** puzzle, 122 Americans (rounded); **(b)** arcade, 50 Americans (rounded)   **5.** 85.6% (rounded)   **6.** 10,044 dogs (rounded)   **7.** 71.4% (rounded)   **8.** 9.9% decrease (rounded)

9. 79.1% (rounded)   **10.** 704 animals (rounded)

Chapter 7   TEST

1. 0.75   **2.** 60%   **3.** 180%   **4.** 7.5% or $7\frac{1}{2}$%   **5.** 3.00 or 3   **6.** 0.02

7. $\frac{5}{8}$ **8.** $\frac{12}{5} = 2\frac{2}{5}$ **9.** 5%   **10.** 87.5% or $87\frac{1}{2}$%   **11.** 175%

12. 320 laptops   **13.** 400%   **14.** $19,500   **15.** $8466.75

16. 34% increase (rounded)   **17.** To find 50% of a number, divide the number by 2. To find 25% of a number, divide the number by 4. Examples will vary.   **18.** Round $31.94 to $30. Then 10% of $30. is $3 and 5% of $30 is half of $3 or $1.50, so a 15% tip estimate is $3 + $1.50 = $4.50. A 20% tip estimate is 2 ($3) = $6.   **19.** Exact tip is $4.79 (rounded). Each person pays $12.24 (rounded).   **20.** $3.84; $44.16   **21.** $41.39 (rounded); $188.56   **22.** $815.66 ÷ 6 ≈ $135.94   **23.** $1650; $6650

24. $51.60; $911.60

Chapters 1–7   CUMULATIVE REVIEW EXERCISES

1. (a) ninety and one hundred five thousandths   **(b)** one hundred twenty-five million, six hundred seventy   **2. (a)** 30,005,000,000

(b) 0.0078   **3.** Mean is $871.25; median is $825.

4. 17 ounces for $2.89, which is $0.17/ounce   **5.** 48.901   **6.** 17

7. $\dfrac{9b^2}{8}$ **8.** $\dfrac{1}{2}$ **9.** -7 **10.** $\dfrac{16+5x}{20}$ **11.** -0.001 **12.** 0 **13.** $\dfrac{7m-16}{8m}$

14. -30 **15.** $\dfrac{4}{9n}$ **16.** $\dfrac{7}{2}$ or $3\dfrac{1}{2}$ **17.** 19   **18.** -1 **19.** -24 **20.** 160

21. $h=-6$ **22.** $a=2.2$ **23.** $x=3$ **24.** $n+31=3n+1$; $n=15$

25. Let p be Susanna's pay. Let $2p$ be Neoka's pay. $p+2p=1620$. Susanna receives $540; Neoka receives $1080.   **26.** $P=50$ yd; $A=143.5$ yd^2 **27.** $C\approx31.4$ ft; $A\approx78.5$ ft^2 **28.** $y\approx24.2$ mm (rounded); $A=142.5$ mm^2 **29.** 2100 students   **30.** $P=62$ ft; $A=234$ ft^2 **31.** $\dfrac{1}{6}$ cup more than the amount needed   **32.** $56.70; $132.30   **33.** 93.6 km   **34.** 88.6% (rounded)   **35.** $0.23 (rounded)

CHAPTER 8 Measurement

All answers are shown for Concept Checks, Writing exercises, What Went Wrong exercises, Relating Concepts exercises, Summary Exercises, Review Exercises, Chapter Tests, and Cumulative Review Exercises. For other types of exercises, answers are shown only for the odd-numbered exercises.

SECTION 8.1

1. (a) inch, foot, yard, mile   **(b)** fluid ounce, cup, pint, quart, gallon
2. (a) ounce, pound, ton   **(b)** second, minute, hour, day, week   **3. (a)** 3
(b) 12   **5. (a)** 8   **(b)** 2   **7. (a)** 2000   **(b)** 16   **9. (a)** 60   **(b)** 60
11. (a) 2   **(b)** 240   **13. (a)** $\dfrac{1}{2}$ or 0.5   **(b)** 78   **15.** 14,000 to 16,000 lb
17. $\dfrac{3\text{ ft}}{1\text{ yd}}$; 27 ft   **19.** 112   **21.** 10   **23.** $1\dfrac{1}{2}$ or 1.5   **25.** $\dfrac{1}{4}$ or 0.25
27. $1\dfrac{1}{2}$ or 1.5   **29.** $2\dfrac{1}{2}$ or 2.5   **31.** snowmobile/ATV; person walking
33. 5000   **35.** 17   **37.** 4 to 8 in.   **39.** 216   **41.** 28   **43.** 518,400
45. 48,000   **47. (a)** pound/ounces   **(b)** quarts/pints or pints/cups
(c) minutes/hours or seconds/minutes   **(d)** feet/inches   **(e)** pounds/tons
(f) days/weeks   **48.** Feet and inches both measure length, so you can add them once you've changed 12 inches into 1 ft. But pounds measure weight and cannot be added to a length measurement.   **49.** 174,240   **51.** 800
53. 0.75 or $\dfrac{3}{4}$ **55.** $2.23 per pound (rounded)   **57.** $140   **59. (a)** 660 sec
(b) 11 min   **61. (a)** $12\dfrac{1}{2}$ qt   **(b)** 4 containers, because you can't buy part of a container   **63. (a)** 4560 in.   **(b)** 127 yd (rounded)   **(c)** 9.5 ft
64. (a) 8 yd   **(b)** 75.4 ft (rounded)   **(c)** 25 yd (rounded)
65. (a) 2.5 ft or $2\dfrac{1}{2}$ ft   **(b)** 184 years old (rounded)
66. (a) 120 ft (rounded)   **(b)** 115 ft (rounded)

SECTION 8.2

1. 1000   **2.** 10   **3.** $\dfrac{1}{1000}$ or 0.001; $\dfrac{1}{1000}$ or 0.001   **4.** $\dfrac{1}{10}$ or 0.1; $\dfrac{1}{10}$ or 0.1
5. $\dfrac{1}{100}$ or 0.01; $\dfrac{1}{100}$ or 0.01   **6.** 100; 100   **7. (a)** cm   **(b)** mm   **(c)** cm
9. Answers will vary; about 7 to 12 cm   **11.** Answers will vary; about 15 to 25 mm   **13.** cm   **15.** m   **17.** km   **19.** mm   **21.** cm   **23.** m
25. 700 cm   **27.** 0.040 m or 0.04 m   **29.** 9400 m   **31.** 5.09 m   **33.** 40 cm
35. 910 mm   **37.** less; 18 cm or 0.18 m   **38.** greater; 22 m or 0.022 km

39.   5 mm = 0.5 cm / 1 mm = 0.1 cm
41. 0.061 km   **43.** 163 cm; 1630 mm   **45.** 0.0000056 km

SECTION 8.3

1. (a) milliliters   **(b)** liters   **(c)** milligrams   **(d)** kilograms
2. (a) grams   **(b)** liters   **(c)** milliliters   **(d)** milligrams   **3.** mL
5. L   **7.** kg   **9.** g   **11.** mL   **13.** mg   **15.** L   **17.** kg
19. unreasonable; too much   **20.** reasonable   **21.** unreasonable; too much
22. unreasonable; too much   **23.** reasonable   **24.** reasonable
25. reasonable   **26.** unreasonable; too little   **27.** Marta multiplied by the wrong unit fraction. The correct unit fraction is $\dfrac{1000\text{ g}}{1\text{ kg}}$ so that Marta can divide out kg in the numerator with kg in the denominator. The correct answer is 6500 g   **28.** Maxwell should start with the decimal point on the far right side of 20, not on the left side. The correct answer is 0.02 g
29. 15,000 mL; three places to the right   **31.** 3 L   **33.** 0.925 L
35. 0.008 L   **37.** 4150 mL   **39.** 8 kg   **41.** 5200 g   **43.** 850 mg
45. 30 g   **47.** 0.598 g   **49.** 0.06 L   **51.** 0.003 kg   **53.** 990 mL
55. mm   **57.** mL   **59.** cm   **61.** mg   **63.** 0.3 L   **65.** 1340 g; 9000 g
67. 0.5 L   **69.** 3 kg to 4 kg   **71.** greater; 5 mg or 0.005 g
72. less; 10 mL or 0.01 L   **73.** 200 nickels
75. (a) 1,000,000   **(b)** $\dfrac{3.5\text{ M\kern-0.3em m}}{1}\cdot\dfrac{1{,}000{,}000\text{ m}}{1\text{ M\kern-0.3em m}}=3{,}500{,}000$ m
76. (a) 1,000,000,000   **(b)** $\dfrac{2500\text{ m\kern-0.3em}}{1}\cdot\dfrac{1\text{ Gm}}{1{,}000{,}000{,}000\text{ m\kern-0.3em}}=0.0000025$ Gm
77. (a) 1,000,000,000,000   **(b)** 1000; 1,000,000
78. 1,000,000; 1,000,000,000; 1,000,000,000,000; $2^{20}=1{,}048{,}576$; $2^{30}=1{,}073{,}741{,}824$; $2^{40}=1{,}099{,}511{,}627{,}776$.

SUMMARY EXERCISES   U.S. and Metric Measurement Units

1. Length: inch (in.), foot (ft), yard (yd), mile (mi); Weight: ounce (oz), pound (lb), ton (ton); Capacity: fluid ounce (fl oz), cup (c), pint (pt), quart (qt), gallon (gal)   **2.** Length: millimeter (mm), centimeter (cm), meter (m), kilometer (km); Weight: milligram (mg), gram (g), kilogram (kg); Capacity: milliliter (mL), liter (L)   **3. (a)** ft   **(b)** yd   **(c)** 5280
4. (a) min   **(b)** 60   **(c)** 24   **5. (a)** 8   **(b)** gal   **(c)** 2   **6. (a)** lb
(b) 2000   **(c)** 16   **7.** mL   **8.** m   **9.** kg   **10.** mg   **11.** cm   **12.** mm
13. L   **14.** g   **15.** 0.45 m   **16.** 45 sec   **17.** 600 mL   **18.** 8000 mg
19. 30 cm   **20.** $3\dfrac{3}{4}$ ft or 3.75 ft   **21.** 0.050 L or 0.05 L   **22.** $4\dfrac{1}{2}$ gal or 4.5 gal   **23.** 7280 g   **24.** 36 oz   **25.** 0.009 kg   **26.** 180 in.   **27.** 272 cm; 2720 mm   **28.** 2.64 m; 0.00264 km   **29.** 0.04 m; 4 cm; 40 mm
30. 24 cm; 0.24 m; 240 mm   **31.** $5.19 (rounded)   **32.** 10.3 ft (rounded)

SECTION 8.4

1. (c)   **2. (b)**   **3.** kg; kg; $0.83 (rounded)   **5.** 89.5 kg   **7.** 71 beats (rounded)
9. 180 cm; $0.02/cm (rounded)   **11.** 5.03 m   **13.** 4 bottles   **15.** 1.89 g
17. (a) 1.49 m   **(b)** 298 cm   **19.** 215 g; 4.3 g; 4300 mg
20. 330 g; 0.33 g; 330 mg   **21.** 1550 g; 1500 g; 3 g   **22.** 55 g; 50 g; 0.5 g

SECTION 8.5

1. (a) $\dfrac{1\text{ kg}}{2.20\text{ lb}}$; $\dfrac{2.20\text{ lb}}{1\text{ kg}}$ **(b)** kg; denominator   **2. (a)** $\dfrac{1\text{ qt}}{0.95\text{ L}}$; $\dfrac{0.95\text{ L}}{1\text{ qt}}$
(b) qt; denominator   **3.** 21.8 yd   **5.** 262.4 ft   **7.** 4.8 m   **9.** 5.3 oz
(rounded)   **11.** 111.6 kg   **13.** 30.3 qt (rounded)

15. (a) about 0.2 oz **(b)** probably not **17.** 14.8 L (rounded)
19. 1.3 cm (rounded) **21.** $-8\,°C$ **22.** $80\,°C$ **23.** $40\,°C$ **24.** $25\,°C$
25. $150\,°C$ **26.** $10\,°C$ **27.** $16\,°C$ (rounded) **29.** $-20\,°C$ **31.** $46\,°F$
(rounded) **33.** $23\,°F$ **35.** The first mistake was multiplying 5(42)
before doing the subtraction inside the parentheses. The second mistake
was rounding incorrectly. The correct answer is $6\,°C$, rounded to the nearest
whole degree. **36.** The first mistake was forgetting to multiply 9 by -4
after dividing out the common factor. The second mistake was not including
the degree symbol (small circle) in the answer. The correct answer is $-4\,°F$
37. (a) pleasant weather, above freezing but not hot **(b)** $75\,°F$ to $39\,°F$
(rounded) **(c)** Answers will vary. In Minnesota, it's $0\,°C$ to $-40\,°C$;
in California, $24\,°C$ to $0\,°C$ **39.** $50\,°F$; $104\,°F$ **41.** 1.8 m (rounded)
42. 5.0 kg (rounded) **43.** 3040.2 km (rounded) **44.** 38 hr 23 min
45. 300 m (rounded) **46.** less than 3790 mL **47.** 59.4 mL (rounded)
48. 1.6% (rounded)

Chapter 8 REVIEW EXERCISES

1. 16 **2.** 3 **3.** 2000 **4.** 4 **5.** 60 **6.** 8 **7.** 60 **8.** 5280 **9.** 12
10. 48 **11.** 3 **12.** 4 **13.** $\frac{3}{4}$ or 0.75 **14.** $2\frac{1}{2}$ or 2.5 **15.** 28 **16.** 78
17. 112 **18.** 345,600 **19. (a)** 4666.7 yd (rounded) **(b)** 2.7 mi (rounded)
20. \$2465.20 **21.** mm **22.** cm **23.** km **24.** m **25.** cm **26.** mm
27. 500 cm **28.** 8500 m **29.** 8.5 cm **30.** 3.7 m **31.** 0.07 km
32. 930 mm **33.** mL **34.** L **35.** g **36.** kg **37.** L **38.** mL **39.** mg
40. g **41.** 5 L **42.** 8000 mL **43.** 4580 mg **44.** 700 g **45.** 0.006 g
46. 0.035 L **47.** 31.5 L **48.** 357 g (rounded) **49.** 52.25 kg
50. \$1.42 (rounded) **51.** 6.5 yd (rounded) **52.** 11.7 in. (rounded)
53. 67.0 mi (rounded) **54.** 1288 km **55.** 21.9 L (rounded)
56. 44.0 qt (rounded) **57.** $0\,°C$ **58.** $100\,°C$ **59.** $37\,°C$ **60.** $20\,°C$
61. $25\,°C$ **62.** $-15\,°C$ **63.** $28\,°F$ (rounded) **64.** $120\,°F$ (rounded)

Chapter 8 MIXED REVIEW EXERCISES

1. L **2.** kg **3.** cm **4.** mL **5.** mm **6.** km **7.** g **8.** m **9.** mL
10. g **11.** mg **12.** L **13.** 105 mm **14.** $\frac{3}{4}$ hr or 0.75 hr **15.** $7\frac{1}{2}$ ft or
7.5 ft **16.** 130 cm **17.** $77\,°F$ **18.** 14 qt **19.** 0.7 g **20.** 810 mL
21. 80 oz **22.** 60,000 g **23.** 1800 mL **24.** $30\,°C$ **25.** 36 cm
26. 0.055 L **27.** 1.26 m **28.** 18 tons **29.** 113 g; $177\,°C$ (both rounded)
30. 178.0 lb; 6.0 ft (both rounded) **31.** 13.2 m **32.** 22.0 ft **33.** 3.6 m
34. 39.8 in. **35.** 484,000 lb **36.** 19.0 to 22.7 L **37. (a)** 242 tons
(b) 144 in. **38. (a)** 1.02 m **(b)** 670 cm

Chapter 8 TEST

1. 36 qt **2.** 15 yd **3.** 2.25 hr or $2\frac{1}{4}$ hr **4.** 0.75 ft or $\frac{3}{4}$ ft **5.** 56 oz
6. 7200 min **7.** kg **8.** km **9.** mL **10.** g **11.** cm **12.** mm **13.** L
14. cm **15.** 2.5 m **16.** 4600 m **17.** 0.5 cm **18.** 0.325 g
19. 16,000 mL **20.** 400 g **21.** 1055 cm **22.** 0.095 L **23.** 3.2 ft
(rounded) **24. (a)** 770 mg **(b)** 450 mg or 0.45 g less **25.** $95\,°C$
26. $0\,°C$ **27.** 1.8 m **28.** 56.3 kg (rounded) **29.** 13 gal
30. 5.0 mi (rounded) **31.** $23\,°C$ (rounded) **32.** $10\,°F$ (rounded)
33. 6 m $\approx$ 6.54 yd; \$26.03 (rounded) **34.** Possible answers: Use same
system as rest of the world; easier system for children to learn; less use of
fractional numbers; compete internationally.

Chapters 1–8 CUMULATIVE REVIEW EXERCISES

1. $\frac{3}{4xy}$ **2.** $\frac{18 - 5c}{6c}$ **3.** undefined **4.** $-\frac{5}{12}$ **5.** $\frac{27xy}{4}$ **6.** $\frac{2m + 3n}{3m}$
7. $\frac{17}{6}$ or $2\frac{5}{6}$ **8.** $-\frac{7}{45}$ **9.** 2 **10.** -3.2 **11.** 0 **12.** $n = 2$ **13.** $x = 0.4$
14. $w = 20$ **15.** $y = -5$ **16.** $2n - 8 = n + 7$; $n = 15$ **17.** Let p be
the longer piece; $p - 6$ the shorter piece. $p + p - 6 = 90$ The pieces are
48 ft and 42 ft long. **18.** $P = 9$ ft; $A = 5\frac{1}{16}$ ft^2 or 5.1 ft^2 (rounded)
19. $C \approx 28.3$ mm (rounded); $A \approx 63.6$ mm^2 (rounded) **20.** $y = 12$ yd;
$A = 96$ yd^2 **21.** $V = 96$ cm^3; $S = 128$ cm^2 **22.** 54 in. **23.** 3 days
24. 3700 g **25.** 0.6 m **26.** 58% (rounded) **27.** Brand T at 15.5 oz for
\$2.99 − \$0.30 coupon, about \$0.174/ounce **28. (a)** 93.6 km
(b) 58.0 mi (rounded) **29.** 5.97 kg **30.** $1\frac{1}{12}$ yards
31. \$9.74 (rounded) **32.** \$18.99 (rounded); \$170.95 **33.** 120 rows
34. \$2821.88 (rounded); \$21,571.88 **35.** 23 blankets; \$9 left over
36. \$0.19/application **37.** 37.5% decrease

CHAPTER 9 Graphs

All answers are shown for Concept Checks, Writing exercises, What
Went Wrong exercises, Relating Concepts exercises, Summary Exercises,
Review Exercises, Chapter Tests, and Cumulative Review Exercises.
For other types of exercises, answers are shown only for the odd-numbered
exercises.

SECTION 9.1

1. b **2.** a **3.** b **4.** b **5. (a)** 31,419 points **(b)** Michael Jordan,
Kobe Bryant, Karl Malone, and Kareem Abdul-Jabbar **7. (a)** Kareem
Abdul-Jabbar **(b)** Wilt Chamberlain **9.** 9791 points **11.** For Shaquille:
$23.69 \approx 23.7$ points per game; for Wilt: $31,419 \div 1045 \approx 30.1$ points per
game **13. (a)** 193 calories **(b)** moderate bicycling **15. (a)** moderate
bicycling **(b)** moderate jogging, moderate bicycling, low-impact
aerobics, racquetball, singles tennis **17.** 138 calories + 176 calories =
314 calories **19. (a)** 199 calories (rounded) **(b)** 173 calories
(rounded) **21.** 43 calories (rounded) **23.** 165 million or 165,000,000
arrivals and departures **25.** 5 million or 5,000,000 arrivals and departures
27. 325 million or 325,000,000 arrivals and departures **29.** Her first
mistake was forgetting to include "million" in her answer for the
Atlanta airport. There are $9\frac{1}{2}$ suitcase symbols for Atlanta, so the correct
answer is 95 million or 95,000,000 arrivals and departures. Her second
mistake was multiplying the number of suitcase symbols by 10,000 instead
of 10,000,000 in her answer for the Chicago airport. The correct answer
for Chicago is 70 million or 70,000,000 arrivals and departures. **30.** His
first mistake was not reading the key correctly. Each suitcase represents
10 million, so the number of suitcases for Las Vegas must be multiplied by
10,000,000. The correct answer for Las Vegas is 40 million or 40,000,000
arrivals and departures. His second mistake was not understanding what
half of a suitcase symbol represents. There were $5\frac{1}{2}$ suitcases for New York
(JFK), so he must multiply $5.5 \cdot 10,000,000$. The correct answer for
New York (JFK) is 55 million or 55,000,000 arrivals and departures.

SECTION 9.2

1. b **2.** c **3.** c **4.** a **5.** 500 people **7.** Friday; 180 people **9.** $\dfrac{1}{10}$

11. $\dfrac{180}{120} = \dfrac{3}{2}$ **13.** $\dfrac{300}{20} = \dfrac{15}{1}$ **15.** $(0.35)(400) = 140$ people **17.** raids the buffet table; 16 people **19.** 64 fewer people **21.** 160 people **23.** mustard; 960 people **25.** 64 more people **27.** First, find the percent of the total that is represented by each item. Next, multiply the percent by 360° to find the size of each sector. Finally, use a protractor to draw each sector. **28.** A protractor is used to measure the number of degrees in a sector. First, you draw a line from the center of the circle to the left edge. Next, place the hole of the protractor at the center of the circle, making sure that the 0 on the protractor is on the line. Finally, make a mark at the desired number of degrees. This gives you the size of the sector. **29. (a)** 180° **(b)** 20%; 72° **(c)** 5%; 18° **(d)** 15%; 54° **(e)** 10%; 36° **(f)** See circle graph below.

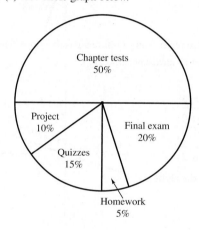

31. (a) \$200,000 **(b)** 6.25%; 20%; 30%; 25%; 18.75% **(c)** 22.5°; 72°; 108°; 90°; 67.5° **(d)** See circle graph below.

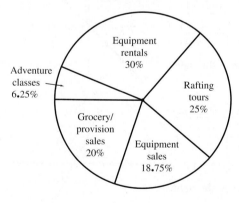

SECTION 9.3

1. c **2.** a **3.** b **4.** b **5.** A few times an hour; 41% **7.** 3160 adults **9.** a few times a day **11.** July; 120,000 unemployed **13.** 10,000 more workers **15.** 120,000 − 80,000 = 40,000 fewer workers; 33% decrease (rounded) **17.** 30% **19.** 2011–12 and 2013–14 **21.** $(0.10)(172) \approx 17$ employers (rounded) **23. (a)** 18 billion dollars, or \$18,000,000,000 **(b)** 5.2 billion dollars, or \$5,200,000,000 **25.** 23 billion dollars, or \$23,000,000,000 **27.** 33.3 billion dollars, or \$33,300,000,000; 185% increase **29. (a)** 3,000,000 DVDs **(b)** 1,500,000 DVDs

31. (a) 2,500,000 DVDs **(b)** 3,000,000 DVDs **33.** Answers will vary. Possibilities include: both stores had decreased sales from 2012 to 2013 and increased sales from 2014 to 2016; Store B had lower sales than Store A in 2012–13 but higher sales than Store A in 2014–2016. **34.** Answers will vary. Some possibilities are that Store B may have started to do more advertising, keep longer store hours, train their staff, employ more help, or give better service than Store A. **35.** 2014; \$25,000 **37.** 2013, 29% (rounded); 2014, 20%; 2015, 17% (rounded); 2016, 38% (rounded) **39.** Answers will vary. Possibilities include: the decrease in sales may have resulted from poor service or greater competition; the increase in sales may have been a result of more advertising or better service. **40.** Most of the time, as sales increase or decrease, so do profits. **41. (a)** Sales have increased every year since 2006. **(b)** Answers will vary. Some possibilities to explain increased sales are: more features and applications are available, the technology has improved, a wider price range makes some models more affordable. **42. (a)** 3% (rounded) **(b)** It is the smallest increase. Answers will vary. Possibilities include: the recession during this period may have caused people to spend less, not enough new features to get people to buy a new phone. **43.** Answers will vary. Possibilities: another financial recession in the United States, or the development of a new device that replaces smartphones. **44.** No. The line graph shows the amount of sales in previous years but cannot accurately predict future sales. We can only make a guess based on the trend in the graph.

SECTION 9.4

1. (a) See dot at center of coordinate grid below. **(b)** See x-axis and y-axis labels on grid below. **2. (a)** I, II, III, IV **(b)** See coordinate grid below.

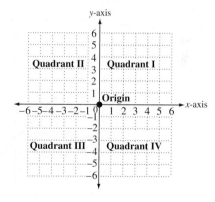

3.

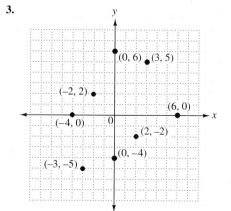

5.

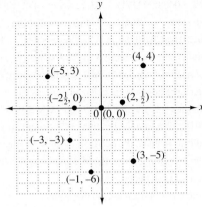

7. A is $(3, 4)$; B is $(5, -5)$; C is $(-4, -2)$; D is approximately $\left(4, \dfrac{1}{2}\right)$;

E is $(0, -6)$; F is $(-5, 5)$; G is $(-2, 0)$; H is $(0, 0)$ **9.** III, none, IV, II
11. (a) any positive number   **(b)** any negative number   **(c)** 0
(d) any negative number   **(e)** any positive number   **13.** Starting at the
origin, move left or right along the x-axis to the number a; then move up if
b is positive or move down if b is negative.   **14.** When a is 0, the point is
on the y-axis. When b is 0, the point is on the x-axis.

SECTION 9.5
1. c   **2.** b   **3.** $(1, 3)$; 2; $(2, 2)$;
All points on the line are solutions.

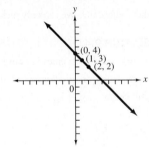

5. -1; $(0, -1)$; -2; $(1, -2)$; -3; $(2, -3)$;
All points on the line are solutions.

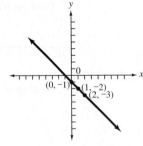

7. 4; -1; -6; 99   **8.** The line in Exercise 4 crosses both the x-axis and
y-axis at -4; the sum of $x + y$ is -4 in the equation. Similarly, in Exercise 6,
the line crosses both the x-axis and y-axis at 1; the sum of $x + y$ is 1 in the
equation.

9.

x	y	(x, y)
1	-1	$(1, -1)$
2	0	$(2, 0)$
3	1	$(3, 1)$

See graph at the right.

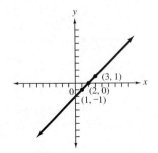

11.

x	y	(x, y)
0	2	$(0, 2)$
-1	1	$(-1, 1)$
-2	0	$(-2, 0)$

See graph at the right.

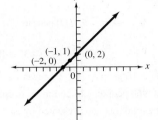

13.

x	y	(x, y)
0	0	$(0, 0)$
1	-3	$(1, -3)$
2	-6	$(2, -6)$

See graph at the right.

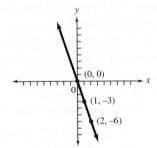

15. The lines in Exercises 9 and 11 have a positive slope. The lines in
Exercises 3, 5, and 13 have a negative slope.

17.

x	y	(x, y)
0	0	$(0, 0)$
3	1	$(3, 1)$
6	2	$(6, 2)$

See graph at the right.

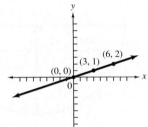

19.

x	y	(x, y)
-1	-1	$(-1, -1)$
-2	-2	$(-2, -2)$
-3	-3	$(-3, -3)$

See graph at the right.

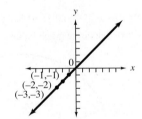

21.

x	y	(x, y)
0	3	$(0, 3)$
1	1	$(1, 1)$
2	-1	$(2, -1)$

See graph at the right.

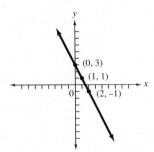

23.

25.

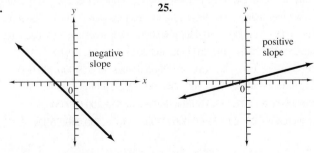

27.

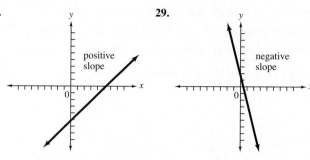

positive
slope

29.

negative
slope

28.

x	y	(x, y)
0	−2	(0, −2)
1	−3	(1, −3)
2	−4	(2, −4)

The graph of $x + y = -2$
has a *negative* slope.

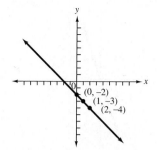

(0, −2)
(1, −3)
(2, −4)

There are many other solutions because all points on the line are solutions. Some possibilities are (−1, −1) and (−2, 0).

Chapter 9 REVIEW EXERCISES

1. (a) gymnastics **(b)** cross country **2. (a)** basketball and cross country
(b) cross country **3. (a)** 4102 men **(b)** 14,409 women **4.** men's, 14.5
(rounded); women's, 15.0 (rounded) **5. (a)** 70 inches **(b)** 15 inches
6. (a) 55 inches **(b)** 65 inches **7. (a)** 25 inches **(b)** 50 inches
8. 85 inches difference between Buffalo and Memphis **9.** lodging; $1400
10. $\dfrac{1400}{1000} = \dfrac{7}{5}$ **11.** $\dfrac{750}{4250} = \dfrac{3}{17}$ **12.** March; 8,000,000 acre-feet
13. 3,000,000 acre-feet; 37.5% decrease **14.** 3,000,000 acre-feet;
60% decrease **15.** $50,000,000 **16.** $20,000,000 **17.** $20,000,000
18. $40,000,000 **19.** Sales decreased for two years and then moved up
slightly. Answers will vary. Perhaps less new construction, remodeling, and
home improvement occurred in the area near Center A. Or better product
selection and service may have reversed the decline in sales. **20.** 36°
21. 35%; 126° **22.** 20%; 72° **23.** 25%; 90° **24.** $2240; 10%
25. See graph below.

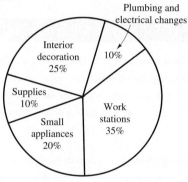

Plumbing and
electrical changes

Interior
decoration
25%

10%

Supplies
10%

Small
appliances
20%

Work
stations
35%

29.

x	y	(x, y)
0	3	(0, 3)
1	4	(1, 4)
2	5	(2, 5)

See graph below.

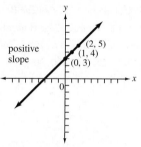

positive
slope

(2, 5)
(1, 4)
(0, 3)

All points on the line are
solutions. Some possibilities
are (−1, 2) and (−2, 1).

30.

x	y	(x, y)
−1	4	(−1, 4)
0	0	(0, 0)
1	−4	(1, −4)

See graph below.

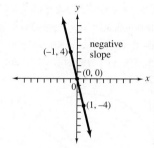

negative
slope

(−1, 4)
(0, 0)
(1, −4)

All points on the line are
solutions. Some possibilities
are $\left(-\dfrac{1}{2}, 2\right)$ and $\left(\dfrac{1}{2}, -2\right)$.

Chapter 9 MIXED REVIEW EXERCISES

1. Painting and wallpapering; 63% **2.** 43%; 147 homeowners (rounded)
3. 184 homeowners (rounded) **4. (a)** Interior decorating and
landscaping/gardening **(b)** Construction work and window treatments
5.

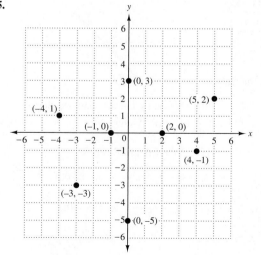

(0, 3)
(5, 2)
(−4, 1)
(−1, 0)
(2, 0)
(4, −1)
(−3, −3)
(0, −5)

(a) (5, 2) **(b)** (−4, 1) **(c)** (−3, −3) **(d)** (4, −1)
(e) (−1, 0); (0, 3); (2, 0); (0, −5)

26.

(1, 7)
(0, 3)
(−7, 1)
$\left(-1\frac{1}{2}, 0\right)$
(−4, −2)
$\left(2, -\frac{1}{2}\right)$
(0, −4)
(5, −5)

27. A is (0, 6); B is approximately $\left(-2, 2\dfrac{1}{2}\right)$; C is (0, 0); D is (−6, −6);

E is (4, 3); F is approximately $\left(3\dfrac{1}{2}, 0\right)$; G is (2, −4).

6.

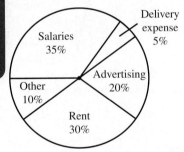

x	y	(x, y)
-1	-3	(-1, -3)
0	0	(0, 0)
1	3	(1, 3)

The graph of $y = 3x$ has <u>positive</u> slope.

Chapter 9 TEST

1. (a) sardines **(b)** cream cheese **2.** 166 calories (rounded)
3. 259 mg (rounded) **4. (a)** fish; 95 fish species **(b)** reptiles;
15 reptile species **5.** 5 species **6.** 325 species **7.** television; $812,000
8. miscellaneous; $84,000 **9.** $308,000 **10.** $644,000 **11.** 2015; $4000
12. $5000; 38% increase (rounded) **13.** 2015; explanations will vary.
Some possibilities are: laid off from work, changed jobs, was ill, cut down
on hours worked. **14.** 5500 students; 3000 students **15.** College B;
1000 students **16.** Explanations will vary. For example, College B may have
added new courses or lowered tuition or added child care. **17.** 35%; 126°
18. 5%; 18° **19.** 20%; 72° **20.** 30%; 108° **21.** $48,000; 10%
22. **23–26.**

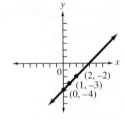

27. (0, 0); no quadrant **28.** (-5, -4); Quadrant III
29. (3, 3); Quadrant I **30.** (-2, 4); Quadrant II

31.

x	y	(x, y)
0	-4	(0, -4)
1	-3	(1, -3)
2	-2	(2, -2)

See graph at the right.

32. (a) Answers will vary; all points on the line are solutions. Some
possibilities are (3, -1) and (4, 0). **(b)** Positive slope

Chapters 1–9 CUMULATIVE REVIEW EXERCISES

1. $\frac{47}{15}$ or $3\frac{2}{15}$ **2.** -0.25 **3.** -39 **4.** -0.00010 or -0.0001 **5.** $\frac{4b}{3a^2}$
6. $-\frac{31}{12}$ or $-2\frac{7}{12}$ **7.** -10.007 **8.** $\frac{2x - 56}{7x}$ **9.** -36 **10.** $\frac{25}{3mn}$
11. $\frac{10}{3}$ or $3\frac{1}{3}$ **12.** undefined **13.** $\frac{59}{8}$ or $7\frac{3}{8}$ **14.** $\frac{18 + wx}{6w}$ **15.** -4
16. 840 **17.** $\frac{0}{-2} = 0$ **18.** -2.06 **19.** $-\frac{7}{20}$ **20.** $y = -6$ **21.** $x = 2$
22. $m = -15$ **23.** $x = 7$ **24.** $h = -9$ **25.** $12 - 5n = n$; $n = 2$
26. Let m be the smaller amount; $m + 500$ the larger amount.
$m + m + 500 = 1800$. The two amounts are $650 and $1150.

27. $P = 36$ ft; $A = 70$ ft² **28.** $C \approx 18.8$ ft rounded;
$A \approx 28.3$ ft² (rounded) **29.** $x = 25$ mi; $P = 56$ mi; $A = 84$ mi²
30. 586 nonsmoking employees (rounded) **31.** $116.78 (rounded)
32. 92.8% (rounded) **33.** $1\frac{5}{12}$ pounds

34.

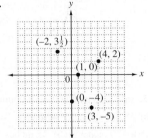

$\left(-2, 3\frac{1}{2}\right)$ is in Quadrant II; no points in Quadrant III

35.

x	y	(x, y)
0	6	(0, 6)
-1	5	(-1, 5)
-2	4	(-2, 4)

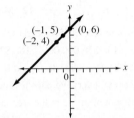

See graph at the right.

Positive slope; all points on the line are solutions, so there are many
possibilities such as (-3, 3) and (-4, 2)

CHAPTER 10 Exponents and Polynomials

All answers are shown for Concept Checks, Writing exercises, What Went
Wrong exercises, Relating Concepts exercises, Summary Exercises, Review
Exercises, Chapter Tests, and Cumulative Review exercises. For other types
of exercises, answers are shown only for the odd-numbered exercises.

SECTION 10.1

1. 1 **2.** three squared; five cubed; seven to the fourth power **3.** false;
$3^3 = 3 \cdot 3 \cdot 3 = 27$ **4.** false; $2^4 + 2^3 = 16 + 8 = 24$ **5.** false;
$(a^2)^3 = a^{2 \cdot 3} = a^6$ **6.** false; $4^5 \cdot 4^3 = 4^8$ **7.** $(-2)^5$ **9.** $\left(\frac{1}{2}\right)^6$
11. $(-8p)^2$ **13.** Betsy did not apply the exponent correctly.
The parentheses show that the base is (-3). The correct answer is
$(-3)^4 = (-3) \cdot (-3) \cdot (-3) \cdot (-3) = 81$ **14.** Devin applied the
exponent only to x, rather than applying it to the entire base of $(5x)$. The
correct answer is $(5x)^3 = (5x) \cdot (5x) \cdot (5x) = 5^3 x^3 = 125x^3$ **15.** Base is 3;
exponent is 5; 243 **17.** Base is -3; exponent is 5; -243 **19.** Base is $-6x$;
exponent is 4 **21.** Base is x; exponent is 4 **23.** The product rule
does not apply to $5^2 + 5^3$ because it is a *sum*, not a product.
$5^2 + 5^3 = 25 + 125 = 150$ **24.** The product rule applies only to
exponential expressions with the same base. $3^2 \cdot 4^3 = 9 \cdot 64 = 576$
25. $5^{2+6} = 5^8$ **27.** 4^{12} **29.** $(-7)^9$ **31.** t^{17} **33.** $-56r^7$
35. $42p^{10}$ **37.** $5x^4 + 9x^4 = 14x^4$; $(5x^4)(9x^4) = 45x^8$
39. $-7a^2 + 2a^2 + 10a^2 = 5a^2$; $(-7a^2)(2a^2)(10a^2) = -140a^6$
41. 4^6 **43.** t^{20} **45.** $7^3 r^3$ **47.** $5^5 x^5 y^5$ **49.** $8q^3 r^3$ **51.** $\frac{1^3}{2^3}$ **53.** $\frac{a^3}{b^3}$
55. $\frac{9^8}{5^8}$ **57.** $(-2)^3 x^6 y^3$ **59.** $3^2 a^6 b^4$ **61.** $A = (4x^3)(3x^2) = 12x^5$

SECTION 10.2

1. negative **2.** positive **3.** negative **4.** negative **5.** positive

6. negative **7.** 1 **9.** -1 **11.** 0 **13.** 0 **15.** 2 **17.** $\dfrac{1}{32}$ **19.** 0

21. $\dfrac{1}{64}$ **23.** $\dfrac{8}{15}$ **25.** $-\dfrac{1}{8}$ **27.** 6^5 **29.** 10^3 **31.** $\dfrac{1}{y^4}$ **33.** c

35. $\dfrac{1}{5^4}$ **37.** $\dfrac{1}{m}$ **39.** 3^4 **41.** a^3 **43.** $\dfrac{1}{2^8}$ **45.** $\dfrac{1}{r^4}$ **47.** 10^{10}

49. $\dfrac{1}{10^5}$ **51.** 10^{11} **53.** $\dfrac{1}{10^8}$ **55.** 10^4 **57.** $\dfrac{1}{a}$ **59.** $\dfrac{1}{2^8}$ **61.** $\dfrac{1}{x^{11}}$

63. 10^2 **65.** $\dfrac{1}{y^2}$ **67.** m^9 **69.** $\dfrac{1}{10^6}$

SUMMARY EXERCISES Using Exponent Rules

1. 9 **2.** 150 **3.** 1 **4.** $-\dfrac{4}{45}$ **5.** 1 **6.** $\dfrac{1}{64}$ **7.** -9 **8.** 72 **9.** -8

10. -1 **11.** $\dfrac{5}{12}$ **12.** 9 **13.** $\dfrac{1}{32}$ **14.** 1 **15.** -8 **16.** 1 **17.** 6^8

18. $\dfrac{1}{x^7}$ **19.** $\dfrac{1}{10^8}$ **20.** 2^9 **21.** $y^0 = 1$ **22.** $2^5 m^{10} n^5$ **23.** $(-5)^{10}$

24. w^3 **25.** m^{24} **26.** $\dfrac{1}{n^2}$ **27.** $\dfrac{1}{10^6}$ **28.** d^9 **29.** $4^2 x^8$ **30.** $\dfrac{1}{c}$

31. 10^8 **32.** 4^5 **33.** $\dfrac{x^6}{y^6}$ **34.** 3^{11} **35.** 7 **36.** h **37.** $-16a^4$

38. $\dfrac{1^4}{2^4}$ **39.** $5^3 x^3 y^3$ **40.** m^5 **41.** $\dfrac{1}{10^7}$ **42.** $\dfrac{1}{b}$ **43.** b^{15} **44.** $\dfrac{1}{5^7}$

45. $\dfrac{1}{10^3}$ **46.** $21y^8$ **47.** 10^{13} **48.** $\dfrac{1}{k^8}$ **49.** $10a^4 b^4$ **50.** $\dfrac{1}{x^9}$

SECTION 10.3

1. yes **2.** yes **3.** no **4.** no **5.** no **6.** no **7.** no **8.** no **9.** yes
10. yes **11.** 5.53×10^{-2}; 3.1783×10^2 **13.** 1.24×10^9; 5.6×10^8
15. 5.876×10^9 **17.** 8.235×10^4 **19.** 7×10^{-6} **21.** 2.03×10^{-3}
23. 750,000 **25.** 5,677,000,000,000 **27.** 6.21 **29.** 0.00078
31. 0.000000005134 **33.** 6×10^{11}; 600,000,000,000 **35.** 1.5×10^7;
15,000,000 **37.** 6.426×10^4; 64,260 **39.** 3×10^{-4}; 0.0003
41. 4×10^1; 40 **43.** 2.6×10^{-3}; 0.0026 **45.** 3.72×10^2 (rounded)
or about 372 people per square mile **47. (a)** 1×10^{-4} or 0.0001 meter
(b) 1×10^{-1} or 0.1 meter **49.** 5×10^2 (rounded) or about 500 items
on average

SECTION 10.4

1. one term; 6; monomial **2.** one term; -9; monomial **3.** one term;
1; monomial **4.** one term; 1; monomial **5.** two terms; -19, -1;
binomial **6.** two terms; 2, -1; binomial **7.** three terms; 1, 8, -8;
trinomial **8.** three terms; 1, -2, 2; trinomial **9.** $-3mn$; mn; $9mn$
10. $4x^2$; $-x^2$ **11.** no like terms **12.** $-2ab^2$; ab^2 **13.** 16; -1
14. no like terms **15.** $2m^5$ **17.** $-r^5$ **19.** cannot be simplified; already
written with descending power **21.** $-5x^5$ **23.** $5p^9 + 4p^7$ **25.** $-2y^2$
27. already simplified and written in descending powers; degree 4;
binomial **29.** already simplified; write in descending powers as
$6m^5 + 5m^4 - 7m^3 - 3m^2$; degree 5; none of these **31.** $x^4 + \dfrac{1}{3}x^2 - 4$;
degree 4; trinomial **33.** $7x$; degree 1; monomial **35. (a)** -1 **(b)** 5
37. (a) 19 **(b)** -2 **39. (a)** 36 **(b)** -12 **41. (a)** -124 **(b)** 5
43. $5m^2 + 3m$ **45.** $-6x^3 + x + 6$ **47.** $3w^3 + 5$ **49.** $4x^4 - 4x^2$
51. $8r^2 + 5r - 12$ **53.** $5m^2 - 14m + 6$ **55.** $4x^3 + 2x^2 + 5x$

57. $12m^3 - 13m^2 + 6m + 11$ **59.** $-11x^2 - 3x - 3$ **61.** $8x^2 + 8x + 6$
63. $8t^2 + 8t + 13$ **65.** $D = 100(1) - 13(1)^2$; $D = 100 - 13$; $D = 87$ ft
66. 0, $(0, 0)$; 47, $(0.5, 47)$; 121, $(1.5, 121)$; 148, $(2, 148)$; 169, $(2.5, 169)$;
183, $(3, 183)$; 191, $(3.5, 191)$

67.

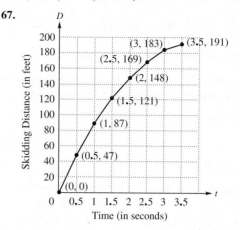

68. Answers will vary. This graph is a curve instead of the straight lines
graphed for linear equations. Also, the rate of change in the distance is not
consistent. From 0.5 sec to 1 sec, the distance increases 40 ft, but from
3 sec to 3.5 sec, the distance only increases 8 ft.

SECTION 10.5

1. (a) 7; 1 **(b)** -1; 4 **2. (a)** two **(b)** three **3. (a)** binomial
(b) monomial **4.** multiplication; multiplication; addition
5. (a) $(9 + 3)b^4 = 12b^4$ **(b)** $(9 \cdot 3)b^{4+4} = 27b^8$
6. (a) $(1 + 5)y^3 = 6y^3$ **(b)** $(1 \cdot 5)y^{3+3} = 5y^6$ **7.** $40y^3 + 15y$
9. $-6m^2 - 4m$ **11.** $24x^4 - 12x^3 + 8x^2$ **13.** $-6k^6 + 9k^5 + 3k^4 - 3k^3$
15. $n^2 + 3n + (-2n) + (-6) = n^2 + n - 6$ **17.** $8r^2 - 10r - 3$
19. $12x^3 + 26x^2 + 10x + 1$ **21.** $20m^4 - m^3 - 8m^2 - 17m - 15$
23. $5x^4 - 13x^3 + 20x^2 + 7x + 5$ **25.** $x^2 + 7x + 12$ **27.** $2x^3 + 7x^2 + 3x$
29. D; B; A; C **31.** Mohamed did not multiply $x(x)$ correctly; $x(x)$ is x^2.
The correct answer is $x^2 - 2x - 3x + 6 = x^2 - 5x + 6$ **32.** Fatima added
unlike terms. Because the exponents are different, $2y^2$ and $10y$ and
$3y$ are not like terms, so they cannot be added. The correct answer is
$2y^2 + 10y + 3y + 15 = 2y^2 + 13y + 15$

Chapter 10 REVIEW EXERCISES

1. $(-5)^{11}$ **2.** $-72x^7$ **3.** $10x^{14}$ **4.** $19^5 x^5$ **5.** $5p^4 t^4$ **6.** $\dfrac{7^6}{5^6}$

7. $3^3 x^6 y^9$ **8.** Does not apply because it is the *sum* of 7^2 and 7^4,
not their *product* **9.** 2 **10.** $\dfrac{1}{32}$ **11.** 10 **12.** $\dfrac{5}{16}$ **13.** $\dfrac{1}{p^{12}}$ **14.** r^4

15. $\dfrac{1}{5}$ **16.** $\dfrac{1}{10^7}$ **17.** $\dfrac{1}{n^3}$ **18.** 10^3 **19.** $\dfrac{1}{2^8}$ **20.** x^6 **21.** 4.8×10^7

22. 8.24×10^{-8} **23.** 24,000 **24.** 0.00000000000995
25. 8×10^2; 800 **26.** 3×10^{-5}; 0.00003 **27.** 1×10^{-2}; 0.01
28. 8.5×10^1 (rounded) or about 85 people per square mile **29.** 1.08×10^8
or 108,000,000 kilometers **30.** $22m^2$; degree 2; monomial
31. $-8y^5 - 7y^4 + 9y$; degree 5; trinomial **32.** $-5a^3 + 4a^2$
33. $11y^2 - 10y + 9$ **34.** $4y^3 - 5y^2 - 4y + 26$ **35.** $7r^4 - 4r^3 - 1$
36. $10x^2 + 70x$ **37.** $-6p^5 + 15p^4$ **38.** $6r^3 + 8r^2 - 17r + 6$
39. $6k^2 - 9k - 6$ **40.** $12p^2 - 48pq + 21q^2$
41. $2m^4 + 5m^3 - 16m^2 - 28m + 9$

Chapter 10 MIXED REVIEW EXERCISES

1. $\dfrac{1}{7^2}$ **2.** $3^4 p^4$ **3.** h **4.** $(-2)^3 c^6 d^{18}$ **5.** $(-3)^7$ **6.** 7^3 **7.** -1

8. $30b^8$ **9.** $\dfrac{1}{m}$ **10.** 10^{12} **11.** $4x^5 y^5$ **12.** $\dfrac{1}{10^5}$ **13.** 0 **14.** $\dfrac{1}{10^5}$

15. $\dfrac{1}{n^{14}}$ **16.** b^8 **17.** $\dfrac{1}{2} + \dfrac{1}{4} = \dfrac{3}{4}$ **18.** 0 **19.** $-8m^7 - 10m^6 - 6m^5$

20. $\dfrac{1}{6} - \dfrac{1}{4} = -\dfrac{1}{12}$ **21.** $-7m^6 + 14m^5 - 42m^3$ **22.** $-6t$

23. $a^3 - 2a^2 - 7a + 2$ **24.** $-2a^3 + 2a^2 - 9$ **25.** 1.67×10^{10} (rounded) or 16,700,000,000 tweets **26.** $\$3.3475 \times 10^{12}$ or $\$3,347,500,000,000$ spent on health care in 2016

Chapter 10 TEST

1. $\dfrac{1}{625}$ **2.** 2 **3.** $\dfrac{7}{12}$ **4.** 6 **5.** $12x^3 y^3$ **6.** $\dfrac{1}{10^4}$ or $\dfrac{1}{10,000}$ **7.** $\dfrac{1}{r^5}$

8. $-21x^7$ **9.** m^3 **10.** $3^2 a^8 b^2$ or $9a^8 b^2$ **11.** Cannot use product rule; $81 + 9 = 90$ **12.** 3.0×10^{11} **13.** 5.57×10^{-6} **14.** $29,000,000$

15. 0.0000000607 **16.** 1.9734×10^{30} kg **17.** $-7x^2 + 8x$; degree 2; binomial **18.** $4n^4 + 13n^3 - 10n^2$; degree 4; trinomial **19.** $4t^4 - 2t^2$

20. $2y^4 - 3y^2 - 6y$ **21.** $-12t^2 + 8$ **22.** $-12x^5 + 20x^4$

23. $3y^2 + 5y - 2$ **24.** $2r^3 + r^2 - 16r + 15$ **25.** $P = 12x + 36$; $A = 9x^2 + 54x + 81$

Chapters 1–10 CUMULATIVE REVIEW EXERCISES

1. -4.171 **2.** -18 **3.** $\dfrac{34}{15}$ or $2\dfrac{4}{15}$ **4.** $\dfrac{0}{-14} = 0$ **5.** $\dfrac{16w}{3x}$ **6.** -0.125

7. $\dfrac{1}{10}$ **8.** $-\dfrac{7}{12}$ **9.** 28 **10.** $-12y^7$ **11.** $\dfrac{25}{3ab}$ **12.** $\dfrac{rt + 48}{8t}$

13. $\dfrac{-24}{0}$ is undefined **14.** 750 **15.** -0.00012 **16.** $2^4 a^{12} b^8$

17. $\dfrac{4m - 54}{9m}$ **18.** $\dfrac{1}{t}$ **19.** $\dfrac{5}{32}$ **20.** $\dfrac{65}{16}$ or $4\dfrac{1}{16}$ **21.** n **22.** -96

23. $1 + \dfrac{1}{2} = 1\dfrac{1}{2}$ **24.** $\dfrac{6}{5}$ or $1\dfrac{1}{5}$ **25.** $h = 5$ **26.** $x = -7.9$ **27.** $a = 6$

28. 145% **29.** $\$4.87$ (rounded); $\$69.82$ **30.** $P = 23.8$ m; $A = 30.6$ m^2

31. $C \approx 28.3$ cm (rounded); $A \approx 63.6$ cm^2 (rounded)

CHAPTER R Whole Numbers Review

SECTION R.1

1. (a) 23 **(b)** 19 **3. (a)** 8928 **(b)** 59,224 **5.** 150 **7.** 1651

9. 1004 **11.** 9253 **13.** 11,624 **15.** 16,658 **17.** 5009 **19.** 3506

21. 15,954 **23.** 10,648 **25.** incorrect; should be 769 **27.** correct

29. 33 miles **31.** 38 miles **33.** $212 **35.** 699 people

37. 65,759,458 people **39.** 294 inches **41.** 708 feet

SECTION R.2

1. incorrect; should be 62 **3.** incorrect; should be 121 **5.** 8 **7.** 25

9. 16 **11.** 61 **13.** 519 **15.** 9177 **17.** 7589 **19.** 8859 **21.** 3

23. 23 **25.** 1942 **27.** 5687 **29.** 19,038 **31.** 65,556 **33.** 19,984

35. incorrect; should be 2494 **37.** correct **39.** 70 calories

41. 180 passengers **43. (a)** Greatest is accountant; lowest is auto mechanic. **(b)** $17,365 **45.** 364 feet

SECTION R.3

1. 9 **3.** 63 **5.** 0 **7.** 24 **9.** 36 **11.** 245 **13.** 168 **15.** 19,092

17. 258,447 **19.** 12,500 **21.** 44,550 **23.** 270,000 **25.** 86,000,000

27. 1496 **29.** 3735 **31.** 15,200 **33.** 32,805 **35.** 183,996

37. 3,321,934 **39.** 18,980,480 **41.** 252 inches; 540 inches

43. 576 plants **45.** 528 miles **47.** 2440 miles; 16,170 miles

49. 7406 calories

SECTION R.4

1. 1; $12\overline{)12}$; $12 \div 12$ **3.** undefined; $\dfrac{24}{0}$; $0\overline{)24}$ **5.** 0; $4\overline{)0}$; $0 \div 4$

7. 0; $\dfrac{0}{12}$; $12\overline{)0}$ **9.** undefined; $\dfrac{21}{0}$; $21 \div 0$ **11.** 21; dividend: 84; divisor: 4; quotient: 21 **13.** 36; dividend: 324; divisor: 9; quotient: 36

15. 1520 **R**5 **17.** 309 **19.** 1006 **21.** 5006 **R**1 **23.** 6671

25. 12,458 **R**3 **27.** 10,253 **R**5 **29.** 18,377 **R**6

31. incorrect; should be 670 **R**2 **33.** incorrect; should be 3568 **R**1

35. correct **37.** correct **39.** $23; $14; $19 **41.** $20,720

43. 378 for $5 tickets; 270 for $7 tickets; 210 for $9 tickets

45. $7; $210 **47. (a)** 358; 190 **(b)** 10,404; 78 **(c)** 13,740; 985

49. 2, 3, 5, 10 **51.** 2 **53.** 5 **55.** 3 **57.** 2, 3 **59.** none of the above

SECTION R.5

1. $24\overline{)768}$ $\overset{3}{}$ Use 2 as the trial divisor. **3.** $18\overline{)4500}$ $\overset{2}{}$ Use 2 as the trial divisor.

5. $86\overline{)10,327}$ $\overset{1}{}$ Use 9 as the trial divisor. **7.** $52\overline{)38,025}$ $\overset{7}{}$ Use 5 as the trial divisor.

9. $77\overline{)249,826}$ $\overset{3}{}$ Use 8 as the trial divisor. **11.** $420\overline{)470,800}$ $\overset{1}{}$ Use 4 as the trial divisor. **13.** 64 **R**3

15. 236 **R**29 **17.** 2407 **R**1 **19.** 1239 **R**15 **21.** 3331 **R**110 **23.** 850

25. incorrect; should be 106 **R**7 **27.** incorrect; should be 658

29. incorrect; should be 62 **31.** correct **33.** 1200 hours **35.** $2030

37. $308 **39.** 1680 circuits **41.** $375

APPENDIX

1. 37; add 7 **3.** 26; add 10, subtract 2 **5.** 16; multiply by 2

7. 243; multiply by 3 **9.** 36; add 3, add 5, add 7, etc.; or $1^2, 2^2, 3^2$, etc.

11. 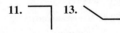 **13.**

15. Conclusion follows. **17.** Conclusion does not follow.

19.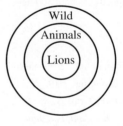

Neither watched TV on 3 days.

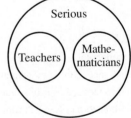

$2 + 18 + 7 = 27$
$30 - 27 = 3$

21. Dick is the computer operator.

Photo Credits

CHAPTER 1: **1** (top) Bikeriderlondon/Shutterstock, (bottom) Dainis Derics/Shutterstock; **4** Vladitto/Shutterstock; **9** (top) Peter Bernik/123RF, (bottom) Labrador Photo Video/Shutterstock; **13** (top) Ivan Cholakov/Shutterstock, (bottom) Sergey Dubrov/Shutterstock; **17** Aleksandr Sadkov/Shutterstock; **18** Burger/Phanie/SuperStock; **28** Daniel Padavona/Shutterstock; **29** Dave & Les Jacobs/Getty Images; **33** Elenathewise/Fotolia; **43** (left) Orhan Cam/Shutterstock, (right) Kzenon/Shutterstock; **52** (left) Golden Pixels LLC/Shutterstock, (right) Gutaper/Fotolia; **53** (left) March Cattle/Shutterstock, (right) NOAA Undersea Research Center; **61** IIIadam36III/Shutterstock; **62** Bob Denelzen/Shutterstock; **86** Iofoto/Shutterstock; **87** Beth Anderson/Pearson Education, Inc.

CHAPTER 2: **93** Egd/Shutterstock; **101** (left) Bikeriderlondon/Shutterstock, (right) Alexsvirid/Shutterstock; **102** (left) Zadorozhnyi Viktor/Shutterstock, (right) Joe Belanger/Shutterstock; **104** Anna Omelchenko/Shutterstock; **132** (left) Bikeriderlondon/Shutterstock, (right) Alexsvirid/Shutterstock; **163** Andrea Izzotti/Fotolia

CHAPTER 3: **165** Isitsharp/iStock/Getty Images; **174** Sergey Tryapitsyn/123RF; **183** (left) Flairimages/Fotolia, (right) Shock/Fotolia; **184** Oliveromg/Shutterstock; **186** (left) Beboy/Shutterstock, (right) Beth Anderson/Pearson Education, Inc.; **195** (left) Monkey Business/Fotolia, (right) Photobar/Shutterstock; **201** (left) Richard A. McMillan/Shutterstock, (right) Sashkin/Fotolia

CHAPTER 4: **217** Syda Productions/Fotolia; **225** Nicholas Sutcliffe/Shutterstock; **250** Patryk Kosmider/Fotolia; **251** (top) Aliaj/Fotolia, (bottom left) Sorbis/Shutterstock, (bottom right) Hxdbzxy/Shutterstock; **252** (top) Alena Ozerova/Shutterstock, (bottom) RSnapshotPhotos/Shutterstock; **262** Earl Orf; **277** (top left) Blend Images/Shutterstock, (top right) Wavebreakmedia/Shutterstock, (bottom) Earl Orf; **280** Steve Mann/Fotolia; **286** (left) Key05/Fotolia, (right) Beth Anderson/Pearson Education, Inc.; **295** Frontpage/Shutterstock; **296** Kelly Boreson/Shutterstock; **304** Rahmo/Shutterstock

CHAPTER 5: **323** Leung Cho Pan/123RF; **332** Pressmaster/Fotolia; **333** Wavebreakmedia/Shutterstock; **340** (both) Beth Anderson/Pearson Education, Inc.; **341** (top) John Foxx Collection/Imagestate, (bottom) Tan Kian Khoon/Fotolia; **351** Goodluz/Fotolia; **368** Ostill/123RF; **369** (left) Feng Yu/Fotolia, (right) Stephen Coburn/Fotolia; **371** (top) Sadik Gulec/Shutterstock, (bottom left) Very Olive/Fotolia, (bottom center) Frank Greenaway/Dorling Kindersley Ltd., (bottom right) Steve Byland/Shutterstock; **377** (left) Damon Yancy/Shutterstock, (right) Monkey Business Images/Shutterstock; **378** NASA; **385** Dell/Fotolia; **386** Chris Gloster/Fotolia; **393** (both) Earl Orf; **400** Fototheobald/Fotolia; **410** (left) Beth Anderson/Pearson Education, Inc., (right) Jacque Descloitres/NASA; **411** (left) Robert Crum/Shutterstock, (right) SeanPavonePhoto/Shutterstock; **422** (left) Oneblink/123RF, (right) Beth Anderson/Pearson Education, Inc.

CHAPTER 6: **431** Tyler Olson/Fotolia; **438** (top) Earl Orf, (bottom) Giovanni Cardillo/Fotolia; **445** Stephen Coburn/Shutterstock; **446** Edbockstock/Fotolia; **455** (left) Actionpics/Fotolia, (right) Monkey Business Images/123RF; **458** AP Images; **464** Blend Images/Shutterstock; **466** Ella1977/Shutterstock; **501** (left) Arena Creative/Shutterstock, (right) Dmitry Kalinovsky/123RF

Index

Study Skills

Are you feeling a little uncertain of your ability to learn and remember mathematics? You'll find the help you need in the special study skills activities in this text. Before you start, read "Your Brain *Can* Learn Mathematics" to find out how your brain learns and remembers best. Then, throughout the other chapters, watch for the study skills activities to learn brain-friendly ways to use your text, take lecture notes, do homework, make study cards, prepare for and take math tests, and much more. It's your ticket to success in mathematics! (See the Contents pages for the titles and locations of the activities.)

Formulas

(*See next page for geometry formulas.*)

Percent proportion: $\dfrac{\text{percent}}{100} = \dfrac{\text{part}}{\text{whole}}$

Percent equation: $\text{percent} \cdot \text{whole} = \text{part}$

Simple interest: $\text{Interest} = \text{principal} \cdot \text{rate} \cdot \text{time}$

Converting from Celsius to Fahrenheit: $F = \dfrac{9C}{5} + 32$

Converting from Fahrenheit to Celsius: $C = \dfrac{5(F-32)}{9}$

$\text{Mean} = \dfrac{\text{sum of all values}}{\text{number of values}}$

Order of Operations

1. Do all operations inside *parentheses* or *other grouping symbols*.
2. Simplify any expressions with *exponents* and find any *square roots*.
3. Do the remaining *multiplications and divisions* as they occur **from left to right**.
4. Do the remaining *additions and subtractions* as they occur **from left to right**.

Example 1

$6^2 - 8 \div 2 + (7 + 3)$ Work inside parentheses.

$6^2 - 8 \div 2 + 10$ Simplify exponents.

$36 - 8 \div 2 + 10$ Divide.

$36 - 4 + 10$ Add and subtract from left to right.

$32 + 10$

42

Example 2

$4 + 10 \div 2(6) - 3(8 \div 4)$ Work inside parentheses.

$4 + 10 \div 2(6) - 3(2)$ Multiply and divide from left to right.

$4 + 5(6) - 3(2)$

$4 + 30 - 3(2)$

$4 + 30 - 6$ Add and subtract from left to right.

$34 - 6$

28

A phrase to help you remember the order of operations:

Please	**Excuse**	**My**	**Dear**	**Aunt**	**Sally**
Parentheses;	Exponents;	Multiply **and** Divide;	Add **and** Subtract		

Fraction-Decimal-Percent Equivalents

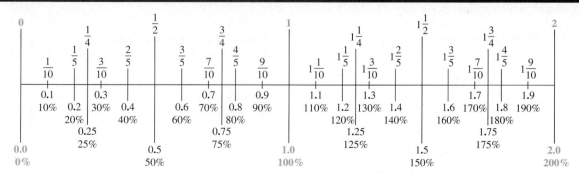

Geometry Formulas

Rectangle
Perimeter: $P = 2l + 2w$

Area: $A = lw$

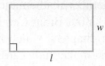

Right Triangle (one 90° angle)

hypotenuse $= \sqrt{(\text{leg})^2 + (\text{leg})^2}$

leg $= \sqrt{(\text{hypotenuse})^2 - (\text{leg})^2}$

Square
Perimeter: $P = 4s$

Area: $A = s^2$, which means $s \cdot s$

Rectangular Solid
Volume: $V = lwh$

Surface area:
$S = 2lw + 2lh + 2wh$

Parallelogram
Perimeter: Add the lengths
of the sides.

Area: $A = bh$

Right Circular Cylinder
Volume: $V = \pi r^2 h$

Surface area: $S = 2\pi rh + 2\pi r^2$

Use 3.14 as the approximate
value of π.

Triangle
Perimeter: Add the lengths
of the sides.

Area: $A = \dfrac{1}{2}bh$

or $A = 0.5bh$

Pyramid
Volume: $V = \dfrac{1}{3}Bh$

or $V = \dfrac{Bh}{3}$

B is area of the base.

Circle
diameter: $d = 2r$

radius: $r = \dfrac{d}{2}$

Circumference: $C = \pi d$
or $C = 2\pi r$

Area: $A = \pi r^2$, which means $\pi \cdot r \cdot r$

Use 3.14 as the approximate value of π.

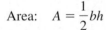

Classifying Angles: Four Types of Angles
Acute angles measure less than 90°.

Right angles measure *exactly* 90°.

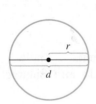

Obtuse angles measure more
than 90° but less than 180°.

Straight angles measure
exactly 180°.